A Textbook of VERTEBRATE ZOOLOGY

A Textbook of VERTEBRATE ZOOLOGY

REVISED FOURTEENTH EDITION

S N PRASAD
Ex-Professor of Zoology
Allahabad University
Allahabad, Uttar Pradesh

V KASHYAP
Assistant Professor
Government Postgraduate College
Rishikesh, Uttarakhand

NEW AGE INTERNATIONAL (P) LIMITED, PUBLISHERS
LONDON • NEW DELHI • NAIROBI
Bangalore • Chennai • Cochin • Guwahati • Hyderabad • Kolkata • Lucknow • Mumbai
Visit us at **www.newagepublishers.com**

Published by New Age International (P) Ltd., Publishers
First Edition: 1991
Revised Fourteenth Edition: 2011
Reprint: 2020

GLOBAL OFFICES

- **New Delhi** **NEW AGE INTERNATIONAL (P) LIMITED, PUBLISHERS**
7/30 A, Daryaganj, New Delhi-110002, INDIA
Tel.: (011) 23253771, 23253472, **Telefax:** 23267437, 43551305
E-mail: contactus@newagepublishers.com • Visit us at www.newagepublishers.com
- **London** **NEW AGE INTERNATIONAL (UK) LTD.**
27 Old Gloucester Street, London, WC1N 3AX, UK
E-mail: info@newacademicscience.co.uk • Visit us at www.newacademicscience.co.uk
- **Nairobi** **NEW AGE GOLDEN (EAST AFRICA) LTD.**
Ground Floor, Westlands Arcade, Chiromo Road (Next to Naivas Supermarket), Westlands, Nairobi, KENYA
Tel.: 00-254-713848772, 00-254-725700286
E-mail: kenya@newagepublishers.com

BRANCHES

- **Bangalore** 37/10, 8th Cross (Near Hanuman Temple), Azad Nagar, Chamarajpet, Bangalore- 560 018
Tel.: (080) 26756823, **Telefax:** 26756820, **E-mail: bangalore@newagepublishers.com**
- **Chennai** 26, Damodaran Street, T. Nagar, Chennai-600 017, **Tel.:** (044) 24353401, **Telefax:** 24351463
E-mail: chennai@newagepublishers.com
- **Cochin** CC-39/1016, Carrier Station Road, Ernakulam South, Cochin-682 016, **Tel.:** (0484) 2377303, **Telefax:** 4051304
E-mail: cochin@newagepublishers.com
- **Guwahati** Hemsen Complex, Mohd. Shah Road, Paltan Bazar, Near Starline Hotel, Guwahati-781 008,
Tel.: (0361) 2513881, **Telefax:** 2543669, **E-mail: guwahati@newagepublishers.com**
- **Hyderabad** 105, 1st Floor, Madhiray Kaveri Tower, 3-2-19, Azam Jahi Road, Near Kumar Theater, Nimboliadda Kachiguda, Hyderabad-500 027, **Tel.:** (040) 24652456, **Telefax:** 24652457
E-mail: hyderabad@newagepublishers.com
- **Kolkata** RDB Chambers (Formerly Lotus Cinema) 106A, 1st Floor, S N Banerjee Road, Kolkata-700 014
Tel.: (033) 22273773, **Telefax:** 22275247, **E-mail: kolkata@newagepublishers.com**
- **Lucknow** 16-A, Jopling Road, Lucknow-226 001, **Tel.:** (0522) 2209578, 4045297, **Telefax:** 2204098
E-mail: lucknow@newagepublishers.com
- **Mumbai** 142C, Victor House, Ground Floor, N.M. Joshi Marg, Lower Parel, Mumbai-400 013
Tel.: (022) 24927869, **Telefax:** 24915415, **E-mail: mumbai@newagepublishers.com**
- **New Delhi** 22, Golden House, Daryaganj, New Delhi-110 002, **Tel.:** (011) 23262368, 23262370, **Telefax:** 43551305
E-mail: sales@newagepublishers.com

ISBN: 978-81-224-2672-4
C-19-05-12081

Printed in India at Print-O-Pack, Delhi.
Typeset at Kalyani Computer Services, Delhi.

NEW AGE INTERNATIONAL (P) LIMITED, PUBLISHERS
7/30 A, Daryaganj, New Delhi-110002
Visit us at **www.newagepublishers.com**
(CIN: U74899DL1966PTC004618)

PREFACE TO THE FOURTEENTH EDITION

It is a great pleasure to present the revised fourteenth edition of the book **A Textbook of Vertebrate Zoology** after a lapse of over fifteen years. The text has been completely revised and updated covering the syllabus of almost all universities of India. An up-to-date account of all primitive and higher chordates has been given in the book as simply and as clearly as possible. In compiling this book outstanding work of various authors, to whom I am greatly indebted, has been consulted. The figures have been drawn from various authoritative sources. It is a complete book for students of U.G. as well as P.G. classes.

I would like express my sincere thanks to my family for their cooperation in the completion of this book. My heartfelt gratitude is due to my collegues, particularly Smt. Usha Joshi, Principal, Govt. College, Laksar, for motivating me to revise this book and for her valuable suggestions. Last but not the least, I would like to thank the publishers and the staff as without their invaluable assistance the present form of the book could not have been attained.

Mistakes, if any, are highly regretted. Constructive suggestions from readers, especially students shall be highly appreciated and incorporated in the next edition.

S. N. Prasad
V. Kashyap

PREFACE TO THE FIRST EDITION

PREFACE TO THE FIRST EDITION

Like its predecessor, **A Textbook of Vertebrate Zoology,** this book has also been written to be a complete account by itself. It, therefore, provides a more or less thorough description of those vertebrate types that from the basis of the study. The plan of the book, in essence, is to tell the story of vertebrates and their evolution as simply and as clearly as possible.

The book has been divided into three parts. *Part One* includes the description of primitive chordates such as **Balanoglossus,** an ascidian and **Amphioxus**. Apart from the morphology and anatomy, a complete account of the development and phylogeny of each type has been provided. *Part Two* describes the structure of the vertebrates in general, giving full information that throws light on important problems of Comparative Anatomy. *Part Three* furnishes as complete a description of vertebrate types as possible. The types have been drawn from all the classes and include **Petromyzon** (Cyclostomata), **Scoliodon** (Chondrichthyes); **Labeo** (Osteichthyes); **Rana** (Amphibia), **Uromastix** and **Kachuga** (Reptile), **Columba** (Aves) and **Lepus** (Mammalia). The description of each type includes complete account of Embryology and Systematics.

In order to make the subject as complete and up-to-date as possible had to seek help of outstanding works of various authors to whom I am greatly indebted. It is not possible to list all such literature here. The most notable among these are: *The Development of Saccoglossus* by A.L. Colwin and L.H. Colwin; *The Frog, Reproduction and Development* by Roberts Rugh; *Development of the Chick* by Viktor Hamburger and Hamilton; *Development of the Chick* by Lillie; *Origin of Vertebrates* by Berril, the *Tunicata* by Berril; *Evolution Emerging* by Gregory, *Skull of **Varanus*** by Bahl; *Circulatory System of **Uromastix*** by Bhatia; ***Herdmania*** by Das; *Vertebrate Body* by Romer; *Biology of the Vertebrates* by Walter and Sayles, *The Life of Vertebrates* by Young; *An Introduction to Ornithology* by Wallace; various volumes of the Fauna of British India, and many more.

An acute awareness of the importance of illustrations in presenting any phase of zoology led me to spend much time in the planning and execution of each drawing prepared from authoritative sources, a book or a journal article. All the illustrations are simple and self-explanatory. In response to numerous suggestions from my friends and students the use of abbreviations or numbers has been avoided in most of the illustrations and full labels have been provided.

The most pleasant thing about working on this book has been the interest and friendly helpfulness of my students and colleagues. Among my colleagues Mr. B.B. Srivastava has been an unfailing source of help towards the last phase of the preparation of the manuscript. He also revised the classification of the Teleostomi. I am also indebted to Dr. P.N. Srivastava, who went through the manuscript of the chapter on **Labeo**, and Sri C.B.L. Srivastava, who checked up the portions dealing with origin of Amphibians, Reptiles, Birds and Mammals.

My indebtedness to my numerous students is beyond expression. I wish I could acknowledge all my debts to them individually, but they are far too numerous. Their critical evaluation of approaches and their constructive suggestions have been of special value because they were made from their (students') point of view, difficult for a teacher to appreciate.

Lastly my thanks are due to Sri Surem Chandra and Sri Suresh Kumar, who toiled for several days to prepare the Index and to S. Hardeo Singh who helped me in preparing the illustrations. The execution of the task and the finished form presented here could not have been attained but for the invaluable assistance given by the publishers and their staff.

S. N. Prasad

CONTENTS

1

CHAPTER

INTRODUCTION:

THE PHYLUM CHORDATA

DEFINITION

The **Chordata** are animals characterised by the presence of a notochord, a dorsal gelatinous stiffening rod, present during some stage in development; tubular nerve cord, situated dorsal to the notochord and pharyngeal gill slits.

THE CHORDATA

The name Chordata was first coined by Balfour in 1880. It is derived from two Greek words, *'chorde'* meaning a string or a cord and *'ata'* meaning bearing. The cord refers to the gelatinous stiffening rod present in all members of this phylum during some stage in embryonic development.

The Chordata is the last and most important major phylum of the animal kingdom. It includes a vast variety of living and extinct animals including man. Ecologically the chordates are very successful because they are able to occupy all types of habitats. In fact, they play a very important role in their ecosystem.

The Chordata can be grouped into two major subdivisions or subphyla: the Vertebrata or Craniata and Protochordata or Acraniata. The Vertebrata include animals like fishes, amphibians, reptiles, birds and mammals. They are the best known and most numerous representatives of this phylum, exceeded in number of species only by Arthropoda and Mollusca, and widely distributed in all types of habitats. The other chordates are the Acraniata which include tunicates, lancelets and acorn worms. They are less well known and limited to marine environment only.

FUNDAMENTAL CHORDATE CHARACTERS

All chordates show three fundamental characters (Fig. 1.1) during early embryonic life. They rarely persist in the adult stage except in *Branchiostoma*. In higher chordates they are modified or lost in the adults.

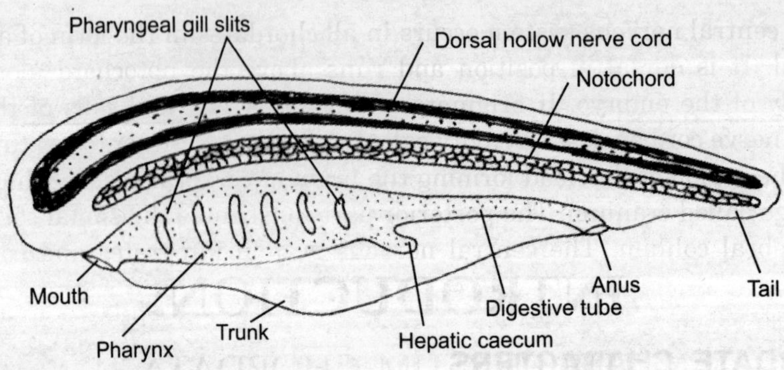

Fig. 1.1 Diagram showing three fundamental chordate characters (side view)

The primary characters are:

1. The pharynx and pharyngeal gill clefts.
2. The stiffening rod or notochord.
3. The dorsal, tubular nerve cord.

1. The pharynx

The pharynx is the anterior portion of the digestive canal and in all chordates it acquires a somewhat complicated structure. The complication is due to a series of lateral symmetrically arranged paired outgrowths from the pharynx called gill pouches. These gill pouches open to the outside by paired lateral openings called gill clefts or branchial clefts or gill-slits. The gill clefts appear during the development of all chordates. The gill slits are functional in the adults in protochordates and lower aquatic vertebrates which respire with the help of gills. The primary function of the gill slits is to allow the passage of water through the pharynx, thus bathing the gills for respiration. In the protochordates the gill slits serve a secondary function, the incoming water current brings food particles which are retained in the pharynx for feeding. In higher vertebrates respiration in adults is pulmonary, so gill slits disappear.

2. The notochord

The notochord is the most important feature of the phylum Chordata. It is a rod of large vacuolated cells containing a gelatinous matrix and surrounded by outer fibrous and inner elastic notochordal sheath. It is derived from the endodermal cells present in the roof of the archenteron of the embryo. It lies in the median plane extending nearly the entire length of the embryo, a little below the nerve cord and just above the gut. In protochordates it is the only structural axis ever produced and in *Branchiostoma* it persists throughout life. In adult vertebrates, it is partially or completely replaced by the vertebral column. In these animals the notochord is found to run through the vertebrae but it gradually diminishes and in adult mammals it persists only as a series of disconnected thickenings, the nuclei pulposi of the intervertebral discs.

3. The nerve cord

The tubular central nervous system occurs in all chordates in the form of a longitudinal hollow nerve cord. It is dorsal in position and runs above the notochord along the whole length of the body of the embryo. It originates from the ectodermal cells of the embryonic neural plate. The nerve cord encloses a cavity called neural canal or neurocoel. In vertebrates it is enlarged in the region of the head forming the brain or cerebral vesicle which is enclosed in a protective cage called cranium. The posterior part constitutes the spinal cord which runs through the vertebral column. The central nervous system integrates and coordinates all body activities.

OTHER CHORDATE CHARACTERS

The higher chordates possess some characters which are also of great importance. They are:

The heart: The heart is **ventral** in position and the flow of blood in the dorsal vessel is from the anterior to the posterior end. In higher vertebrates the heart is contained within a modified part of the coelom called the **pericardial cavity**. The bounding wall of the pericardial cavity is called the **pericardium**. In non-chordates, the heart, if present, is a modification of the dorsal vessel, hence dorsal in position.

The mouth: The position of the mouth differs fundamentally from that of the invertebrates in which the brain is more or less above the mouth and the initial position of the gut is surrounded by a circumoral ring of nervous material. In the vertebrates such a ring is lacking and the entire nervous system is on one side of the body and dorsal to the mouth.

The limbs: There are **two pairs** of limbs which are **lateral extensions** of the surface of the body and aquires an **internal skeleton** for support and to provide attachment to muscles. Among the invertebrates there are no structures homologous with the vertebrate limbs.

The tail: The posterior part of the vertebrate body takes the form of a post-anal metamerically segmented tail which differs markedly from the unsegmented condition of the terminal part of the invertebrate body. The tail is extremely **flexible** and **muscular** and in aquatic forms constitutes the main **propulsive organ**. It is modified in various ways in the terrestrial forms and may be vestigeal in the adult.

The gut: The division of the digestive tract into a **stomach** with **gastric glands** internally and **intestine,** and the **mesentry** suspending these to the dorsal wall of the abdomen is another important feature. The position of the **liver**, its development as a special **diverticulum** of the gut, and its associated veins are equally distinctive.

The hypophysis: The hypophysis is the embryological name applied to the structure which in the adult is known as the **anterior lobe** of the pituitary body. It develops as an outgrowth from the cavity of the mouth of the embryo. Comparatively early in development it becomes entirely separated from the cavity of the embryo and forms a closed vesicle. It exists in all known vertebrate animal, has been much studied but still remains an organ, the full significance of which is not clearly understood.

Characters shared with higher nonchordates

There are a few characters which the chordates share with the higher nonchordates:

Symmetry. The vertebrates are **bilaterally symmetrical**, that is, assuming the animal to be standing on a horizontal substratum, a median vertical longitudinal section divides the animal into right and left equivalent parts, as if one was the mirror image of the other. Some higher nonchordates like Annelida and Arthropoda also exhibit bilateral symmetry.

Surfaces and directions of the body: With the appearance of bilateral symmetry the animal usually develops the habit of keeping one particular side of the body either in contact with the substratum or facing downwards. This undersurface is the **ventral surface** of the animal while the upper surface is the **dorsal.** The forward end is **anterior** and the opposite end **posterior** and the sides are **lateral.** The point of attachment to a structure is its **proximal** end and the free end is **distal**. Sometimes the term **cephalic** or **cephalad** is used to denote the direction towards the head and **caudad** towards the tail (Fig. 1.2).

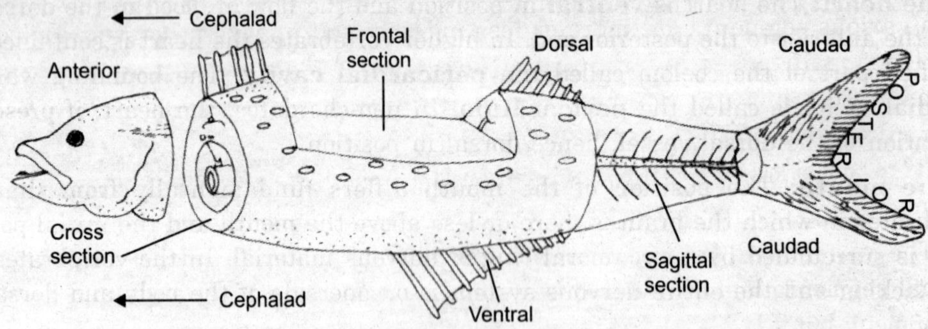

Fig. 1.2 Planes and surfaces of a vertebrate body

Mechanical axis: The longitudinal axis of the body running from the head to the tail is called the **antero-posterior axis.** In the vertebrates the mechanical axis of the body is the vertebral column. In fishes and tailed amphibians, this axis lies nearly in a straight line. In land vertebrates it is more or less curved in the vertical plane, especially in the region of the head and neck. In **quadruped** this long axis is ordinarily in **horizontal position** and in a **biped**, which assumes an erect position, it is **vertical**. The quadruped walks with its head forward and the biped is erect, as such the surface which is forward during locomotion in a biped corresponds with the surface which is beneath in the quadruped.

The axiate body organisation in the two groups however, is not strictly homologous, for eg., the head. During embryonic development the blastoporal end forms the mouth (**protostomia**) in the nonchordates whereas in the chordates it forms the anus (**deuterostomia**).

Planes: There are three planes used as landmarks in the description of animals. If a bilaterally symmetrical animal is cut exactly in the mid-longitudinal axis it is a **sagittal section.** A **longitudinal section** is any cut in the vertical plane parallel with the longitudinal axis. Section in the vertical plane at right angles to the longitudinal axis is a **transverse** or **cross section.** Section in the horizontal plane is **horizontal** or **frontal** section.

Metamerism: In the phyla Chordata, Arthropoda and Annelida a structural condition occurs in which all or most of the paired parts or structures are repeated at regular intervals along the antero-posterior axis of the body. The body of such animals is made up of a longitudinal series of elements, in each of which all or most of the systems of the body are represented, either by entire paired structures or by a portion of the median unpaired structures. Each such division of the body is called a metamere, somite or segment. In some cases such as the Annelida and Arthropoda the anterior and posterior boundaries of each segment may be marked externally by a constriction of the body wall, therefore, the animal is said to exhibit both **external** and **internal** metamerism. In others, however, there is no external metamerism, only internal segmentation is present. Such animals in which various segments are nearly alike are said to display homonomous segmentation, but in majority of segmented animals various segments differ from each other in many respects, such animals are said to possess heteronomous segmentation. Homonomous segmentation is a primitive condition in which various segments are more or less independent of each other and each is capable of performing all the necessary functions of life. In heteronomous condition the segments become unlike and there exists a division of labour among them, some segments performing some functions and others other functions. It follows therefore that in the evolution of segmented animals there has been a continuous progression from homonomous to the extreme heteronomous condition.

The vertebrates display heteronomous segmentation, that too internal. There is no external segmentation. The vertebrates show segmentation, most clearly in the embryonic development, but it is confined to the muscular, skeletal and nervous systems. In the embryonic development of vertebrates the change from a somewhat homonomous condition to an extreme heteronomy can be clearly followed.

Cephalization: The segments of the anterior end of the body tend to become more and more distinctly separated from the rest of the body forming the **head.** This differentiation consists chiefly of the localization, within the head, of the main parts of the **nervous system** – the brain and sense organs. Thus the head becomes the dominant part of the body. This localization of the nervous system in the head along with the dominance of head is called cephalization. In the vertebrates cephalization is well marked.

Cephalization is correlated with the advance of heteronomous condition. As it progresses more and more anterior segments are appropriated by the head, and in general it may be said that higher the degree of cephalization, the greater is the number of segments composing the head. In advanced cephalization, however, it is very difficult to find out the number and boundaries of the segments actually forming the head.

Coelom or Body cavity: A true coelom is a **cavity** present between the **body wall** and the **alimentary canal** and is lined by **mesoderm.** It is present in higher non chordates and chordates. The mode of origin of coelom varies in different groups of animals. In Annelida, Arthropoda, Mollusca and higher Chordata, the mode of origin of coelom is schizocoelous, i.e., it is formed by the splitting of mesodermal bands originating from the blastoporal region. In

Echinodermata and Protochordata the mode of origin is called **enterocoelous**, i.e., it is formed by fusion and expansion of a number of pouches splitting off from the roof of archenteron.

Three germ layers: All the animals above the level of Coelenterata are called **triploblastic** because they have **three germ layers**—the **ectoderm, endoderm** and **mesoderm.** The mesoderm is the third germ layer which in schizocoelous animals arises as a solid band from the junction of the ectoderm and endoderm during gastrulation and in enterocoelous animals arises as lateral outpushings from the archenteron during the gastrula stage.

Organ grade of organisation: All animals from phylum **Nemertina** (a minor invertebrate phyla) onwards exhibit organ grade of organisation. They have definite organ systems for performing various life functions like digestion, circulation, respiration, excretion, etc. Each organ system is made up of several organs which together perform the same function. In vertebrates, however, the organ system is more well developed than in the invertebrates.

From the structural similarities discussed above it is probable that the chordates and the non-chordates may have a common ancestry but enough evidences are not available. However, efforts are being made to study the lineage and few evidences are available indicating the evolution of chordates from the invertebrates.

CHORDATE ANCESTRY

During the course of organic evolution innumerable varieties of animals ranging from the simplest *Amoeba* to the most complicated forms, such as man, have arisen, but it is not clearly understood how the vertebrates have originated. Some 550 million years back there was apparently a revolution in the general organisation of the animal body with the consequent emergence of the vertebrate body.

One of the earliest attempts to explain the origin of chordates was made by Eienne Geoffroy St. Hilaire (1818). According to him a vertebrate represents an insect lying on its back. If an insect turns on its back and swims naturally the ventral nerve cord becomes dorsal and the heart ventral. St. Hilaire believed that a new mouth was formed on the ventral side after the old had degenerated. His colleague Cuvier showed that there were many notable objections to this theory. This idea of reversal of an invertebrate body was applied by many other workers like Dohrn (1875), Semper (1875–76), Minot (1897), Patten (1891–1912), Owen (1883), Gaskell (1896, 1898–1906), etc. but with different details. After more than a century of debate, the methods remain fundamentally the same and the amount of details has become enormous. The present position is different and most of the greater phyla of typical invertebrates have been eliminated from the list of possible ancestors of vertebrates. It is, however, necessary to sum up all these theories only because of their historic standpoint before the present solutions are discussed.

Coelenterate theory

The three layred (triploblastic) vertebrates are supposed to have evolved from two-layered (diploblastic) coelenterates. Masterman (1897) assumed that the bilaterality and triploid subdivision of an early embryonic stage of *Amphioxus* has been derived directly from the secondary bilaterality of a circular coelenterate, resulting from contact with the substratum and subsequent efforts to creep forward.

Nemertean theory

Hubrecht (1887) took the long proboscis and its sheath of the nemerteans as the primary source of the vertebrate notochord and shifted the horizontal lateral blood vessels to form the main dorsal and ventral vessels. This theory is not supported by any evidence and intermediate stage, hence it is given up.

Annelid theory

Semper (1875–76), Dohrn (1875) and Minot (1897) with a few others suggested the possibility of annelids for being the ancestors of vertebrates. The annelids are segmented, the vertebrates are also segmented, at least the backbone, muscles and nerves, and both are bilaterally symmetrical. The annelids have a nerve cord and well developed blood vascular system. Both these are markedly different from the vertebrates as they are located in opposite positions. If, however, the annelid gets up side down, these differences can be corrected (Fig. 1.3). Though reversal corrects some differences, it raises some problems, for instance the position of the mouth. In both the mouth lies on the underside, hence it was contended that the old mouth closed after reversal and the new one appeared. Then there are structures like the notochord and the gill slits that are absent in the annelids. Hence the theory was ultimately given up.

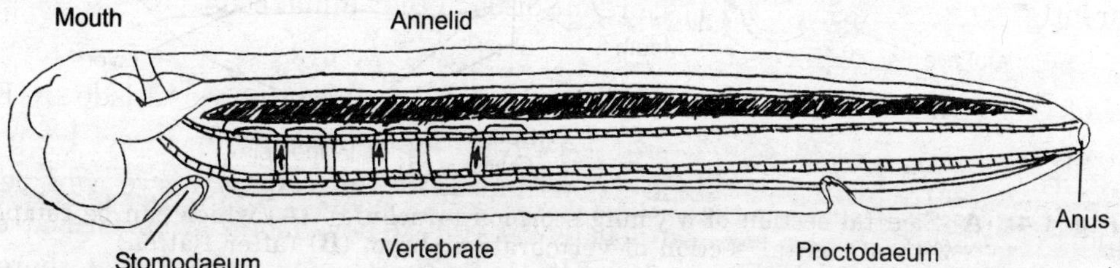

Fig. 1.3 Reversible diagram illustrating the annelid theory. If the annelid is reversed, the vertebrate plan will be apparent (after Wilder)

Arachnid theory

Some authors such as Gaskell and Patten sought the vertebrate ancestors among the arthropods. The insects, crustaceans and centipedes and their relatives, however, have not been given serious consideration. The arachnids which include the scorpions, horse-shoe crabs (*Limulus*) and its palaeozoic relative the eurypterids or water scorpions have been considered as the probable vertebrate ancestors.

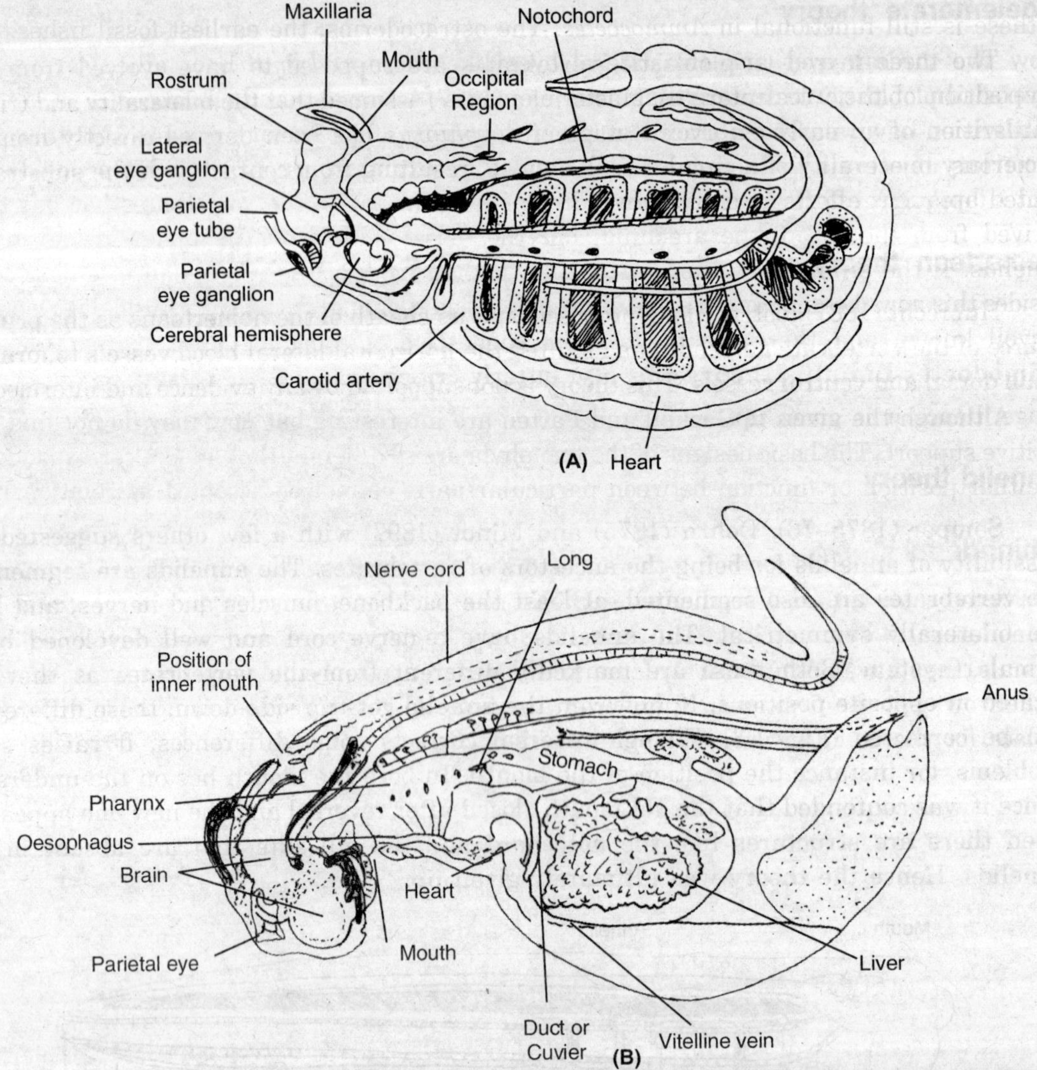

Fig. 1.4 **(A)** Sagittal section of a young scorpion (arachnid). **(A)** which can be compared with a sagittal section of vertebrate embryo; **(B)** (after Patten)

The eurypterids, like other arthropods, were armoured types, with an external rather than an internal skeleton. Some of the earlier vertebrates (ostracoderms) were also armoured and also possessed some deeper-lying skeletal structures. The fact that the eurypterid armour resembled the armour of the ostracoderms cannot be denied. The degree of likeness is often so great that it becomes a difficult task for the palaentologist to adjudge whether a particular fossil is of a fish or an arthropod. Patten made ingenious use of the available palaeontologic evidence in favour of his arachnid theory. This theory emphasises that *Limulus* is the nearest living survivor of the invertebrate ancestors of the vertebrates.

The resemblances between the larval lamprey (*Ammocoetes*) and *Limulus* were described by Gaskell in 1908 and Patten in 1912. Remarkable is the presence of the pineal gland which is all that remains of a pair of median eyes of eurypterid ancestor of the vertebrates. One of these is still functional in *Ammocoetes*. The ostracoderms, the earliest fossil fishes, also show two median eyes and two lateral eyes like those of *Limulus* or allied forms. The composition of the internal cartilaginous elements of *Limulus* and *Ammocoetes* is similar. Similarities have also been traced in the respiratory and olfactory organs. But there are numerous objections to these views. It cannot be explained as to what happened to the jointed appendages of the arthropods. Further it is the top of the vertebrate that has been derived from the top of the arachnid, but the nerve cord is on the underside as in the annelids. If the animal is turned up-side-down then most of the resemblance is destroyed. Besides this now the anatomy of some of the oldest and seemingly most scorpion-like vertebrates is well known and there are no features suggestive of arachnid relationship. Gaskell's **Ammocoetes-Limulus theory** was also bitterly criticized by his co-workers.

Although the views of Gaskell and Patten are interesting but now they do not find any positive support. The basic designs of the two phyla are so different that a mere correspondence in either position or function between particular parts cannot be accepted as relationship.

Echinoderm theory

Some zoologists believe that the place of the ancestors of the vertebrates is among the echinoderms. No typical chordate characters are found among these radially symmetrical animals, they have no internal skeleton and have too many peculiar and complicated organs of their own. At the first sight, therefore, this group seems to be most unpromising of all as a place for the potential ancestor of the vertebrates.

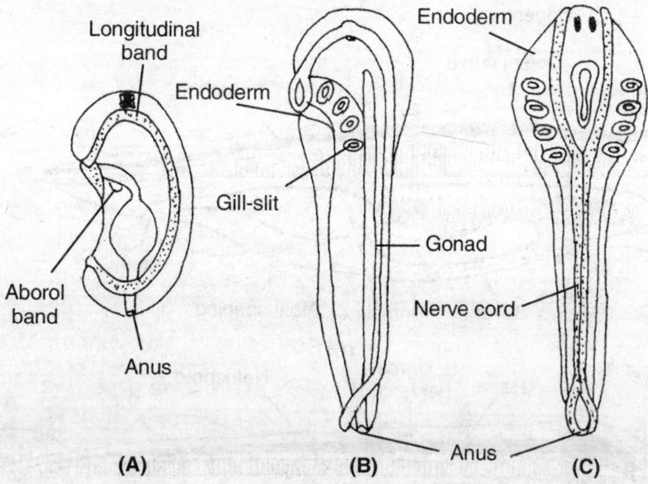

Fig. 1.5 Figure showing how a protochordata might have been derived from an echinoderm larva, auricularia (**A**); (**B**) Protochordata, side view; (**C**) Same in dorsal view (after Garstang)

Inspite of the extreme dissimilarity between the echinoderms and the chordates, both are enterocoelic type of animals. The early embryo of echinoderm is a tiny free-swimming creature and floats freely in water. Unlike the adult it is bilaterally symmetrical, suggesting that the radial symmetry of the adult echinoderm is a secondary affair. This larva resembles the **tornaria larva** of *Balanoglossus*. The type of development of some of the body cavities is identical with that found in the embryos of some primitive vertebrates. What seems to be the most primitive known echinoderms, the Cambrian and Ordovician Carpoidea, might be conceived tornaria-like forms which had begun to settle down to sedentary life, acquiring an exoskeleton of many plates, together with a water pore system developed out of ciliated grooves.

Here are some positive facts. According to Willey, echinoderms are descended from bilaterally symmetrical ancestors represented in their development by such larval forms as the **bipinnaria** and **auricularia**. The general likeness between the auricularia of echinoderm and tornaria of *Balanoglossus* is so great that it can be accounted for on the ground of genetic affinity. *Balanoglossus* is a protochordate supplied with gill slits and a dorsal tubular nervous system and requires only the notochord and segmental myotomes to approximate the amphioxid type.

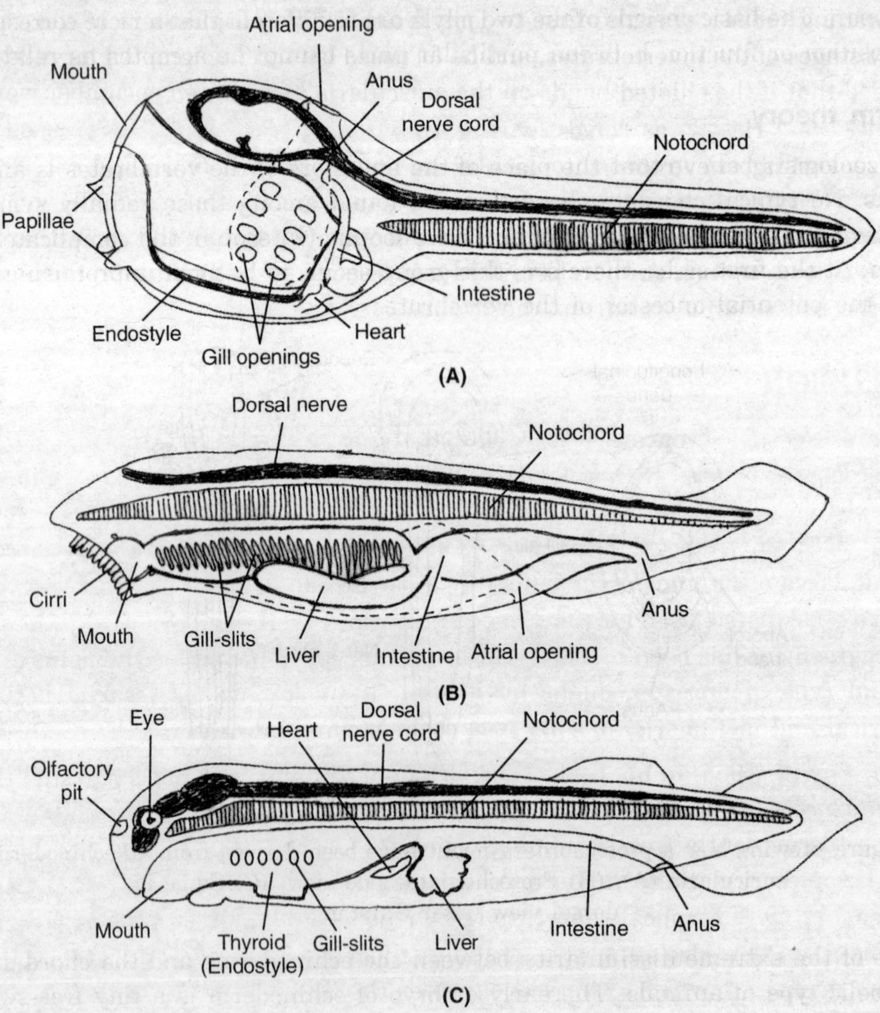

Fig. 1.6 (**A**) Larval urochordate; (**B**) Cephalochordate; (**C**) Vertebrate (cyclostome). The figure shows transformation of an ascidian, the tadpolelike animal, into a vertebrate

The evidence from the larvae strongly suggests that in the dawn of the world there existed some type of small, bilaterally symmetrical animal of very simple structure possessing many of the features of the larval echinoderm or acorn worm but lacking the specialization of either vertebrate group or echinoderm. By the assumption of radial symmetry and sessile mode these forms gave rise to echinoderms. But from those very forms arose types that retained their original bilateral symmetry and developed specialized breathing organs in the form of gill slits, better possibility of mobility with the development of powerful musculature a notochord for its support and better nervous control of activity through the development of a dorsal nerve cord. From this line, it is believed, the chordates appeared and these finally gave rise to the true vertebrates.

Neotenous larva theory

Garstang (1894) was the first to look for the trace of ancestors of the vertebrates in the early stages of the invertebrates, and he focussed his attention on the larva of echinoderm. He showed that if the ciliated bands on the auricularia larva of sea cucumber were to become accentuated and rise up as ridges leaving a groove between them and if these ridges were to fuse converting the groove into a tube, a structure will be produced which has all the relations to the vertebrate nervous system. de Beer points out that this view about the origin of vertebrate nervous system has several advantages. Garstang's theory further asserts that if the larval form of such animals persisted and became sexually mature, they would provide exactly the necessary material for the evolution of the chordates.

CONCLUSION

There are some workers who regard the similarity between the larval forms due to similar ecological factors. H.B. Fell (1948) for instance, concludes that the similarity between the free swimming larvae of the echinoderms and *Balanoglossus* provides no trustworthy evidence of common ancestry. Gregory (1951) is of the view that bag-like, ciliated free-swimming larva of *Balanoglossus* is merely a parallel adaptation among different groups for securing a wide distribution for the sessile or slow moving adult. From this it is apparent that no final agreement has been reached. However, ther is no dispute on the point that whatever the actual type of ancestor might have been, it must have necessarily been bilaterally symmetrical and metameric, moving in a cephalo-caudal direction.

N.J. Berrill (1955) in his book, "The Origin of Vertebrates", has suggested the following larval sequence — echinoderm-auricularia → hemichordate-tornaria → protochordate–ascidian tadpole → permanently free swimming chordate. On the basis of this theory the ascidian tadpole becomes essentially an interpolation in the life cycle of the ascidian; and by the suppression of metamorphosis and through further evolution it gives rise to the vertebrates as a whole. This view also supports Garstang's theory. Berrill, however, agrees with this view not in the manner in which Garstang supposes the changes to have taken place. He places the ascidian in the main line of the chordate origin, at least as larvae, although not necessarily as the basic stock.

GENERAL CHARACTERS OF PHYLUM CHORDATA

1. Aquatic, terrestrial or aerial.
2. All free living, no fully parasitic forms known.
3. Small to large in size.
4. Bilaterally symmetrical.
5. Metamerically segmented.
6. Body triploblastic.
7. Presence of a true coelom.
8. Exoskeleton well developed in most vertebrates.
9. Notochord present at some stage of development ; persists in adult *Amphioxus* and replaced by vertebral column in the vertebrates.
10. Endoskeleton cartilaginous in some and bony in most of the vertebrates.
11. Pharyngeal gill slits present at some stage of development, may persist in the adult stage in some chordates but disappear in vertebrates with pulmonary respiration.
12. Digestive system, a complete tube with digestive glands.
13. Circulatory system closed with a ventral heart and a well developed hepatic portal system.
14. The central nervous system dorsal and hollow. The anterior end enlarged to form a brain.
15. Excretory system consists of a proto-meso or metanephric kidneys.
16. A postanal tail present at some stage in development which may or may not persist in the adult.
17. Sexes generally separate, with rare exceptions.

CLASSIFICATION OF PHYLUM CHORDATA

Phylum Chordata includes a wide variety of animals differing from one another in many respects. Different taxonomists have proposed different schemes of classification. In the present book an attempt has been made to present the classification in a simple way by combining the best parts of various classification. It is summarised in the following Table 1.1:

Table 1.1. Various Classifications of Phylum Chordata

	Group	Subphylum	Division	Superclass	Class	
Phylum Chordata	Acraniata or Protochordata (Lower chordates)	1. Hemichordata			1. Enteropneusta	
					2. Pterobranchia	
					3. Planctosphaeroidea	
					4. Graptolita	
		2. Urochordata			1. Ascidiacea	
					2. Thaliacea	
					3. Larvacea	
		3. Cephalochordata			1. Leptocardii	
	Craniata or Euchordata (Higher chordates)	Vertebrata	1. Agnatha		1. Ostracodermi	Anamniota (Lower vertebrates)
					2. Cyclostomata	
			2. Gnathostomata	1. Pisces	1. Acanthodii	
					2. Placodermi	
					3. Chondsichthyes	
					4. Osteichthyes	
				2. Tetrapoda	1. Amphibia	Amniota (Higher vertebrates)
					2. Reptilia	
					3. Aves	
					4. Mammalia	

Group ACRANIATA (Gr. *a*, absent ; *kranion*, head). Exclusively marine, small sized, primitive or lower chordates, without head, cranium, brain and vertebral column. It is divided into three subphylum on the basis of position of notochord.

Subphylum HEMICHORDATA (Gr. *hemi,* half ; *chorde*, cord). Notochord short, doubtful, not similar to other chordates, nerve tissue in epidermis.

Class 1. Enteropneusta: Tongue worms or acorn worms. Large worm-like hemichordates with many gill slits and straight intestine. *Balanoglossus*, *Saccoglossus*.

Class 2. Pterobranchia: Minute hemichordates with two or no gill slits, intestine U-shaped. *Cephalodiscus, Rhabdopleura*.

Class 3. Planctosphaeroidea: Round, transparent pelagic larva, like tornaria with profusely branching ciliary bands, adults not known. *Planctosphaera*.

Class 4. Graptolita: Extinct colonial hemichordates abundant in the Ordovician and Silurian period. *Dendrograptus*.

Subphylum UROCHORDATA (Gr. *oura*, tail ; *chorde*, cord). Adults contained in secreted tunic, notochord and nerve cord only in larva.

Class 1. Ascidiacea: Sessile, solitary, colonial or compound tunicates with many gill slits and tunic with scattered muscles. Ascidians or sea squirts. *Herdmania, Ciona, Molgula*.

Class 2. Thaliacea: Free swimming or pelagic tunicates in which tunic is provided with circular muscles. Chain tunicates. *Salpa, Doliolium, Pyrosoma*.

Class 3. Larvacea: Tadpole-like, minute, transparent, free-floating tunicate with two gill slits and temporary tunic. *Oikopleura*.

Subphylum CEPHALOCHORDATA (Gr. *kephale*, head ; *chorde*, cord). Notochord and nerve cord extend along entire body and persistent gill slits persistent.

Class 1. Leptocardii: Lancelets. slender, fish-like segmented, free-swimming and burrowing cephalochordate with many gill slits. *Branchiostoma (= Amphioxus), Asymmetron*.

Group CRANIATA Small to large-sized chordates with distinct head, brain protected in cranium and vertebral column. It includes only one subphylum, the Vertebrata.

Subphylum VERTEBRATA (L. *vertebratus,* backbone). Chordates with visceral arches and vertebral column, body divisible into head, neck, trunk and tail. It is divided into two divisions.

Division 1. AGNATHA (Gr. *a*, not ; *gnathos*, jaw) Primitive fish-like vertebrates without true jaws and paired appendages.

Class 1. Ostracodermi (Gr. *ostrakon*, shell ; *derma*, skin) Ancient extinct armoured fishes with large scales often fused as cephalothoracic shield, dominant in the Palaezoic period. *Cephalaspis, Drepanaspis*.

Class 2. Cyclostomata (Gr. *cyklos,* circular ; *stoma*, mouth) Skin without scales, mouth round and suctorial, gill slits 5 to 16 pairs. *Petromyzon, Myxine, Bdellostoma*.

Division 2. GNATHOSTOMATA (Gr. *gnathos,* jaw ; *stoma*, mouth) Vertebrates with jaws and usually with paired appendages. It is further divided into two superclasses.

Superclass 1. PISCES (Fishes)

Paired and median fins, gills, skin with scales.

Class 1. Acanthodii: Acanthodians. Small shark-like fishes covered by diamond-shaped scales, *Climatius*.

Class 2. Placodermi: Placoderms. Ancient extinct Palaeozoic fishes with primitive jaws. *Dinichthys, Pterichthyodes*.

Class 3. Chondrichthyes (Gr. *chondros*, cartilage ; *ichthys*, fish) Skin with placoid scales, cartilaginous endoskeleton, gill slits 5 to 7 pairs, not covered by operculum, male with claspers. Sharks and rays. *Scoliodon, Chimaera*.

Class 4. Osteichthyes (Gr. *Osteon*, bone ; *ichthys,* fish) Skin with cycloid or ctenoid scales, endoskeleton bony, gill slits four pairs, covered by operculum, male without claspers. *Labeo, Hippocampus, Protopterus*.

Superclass 2. TETRAPODA (Gr. *tetra,* four ; *podos,* foot). Amphibious and terrestrial vertebrates, with paired limbs, cornified skin, lungs and bony endoskeleton.

Class 1. Amphibia (Gr. *amphi,* both ; *bios,* life). Skin moist, soft with no external scales. Adults terrestrial and larval stage usually aquatic. Frog, toad, salamander.

Class 2. Reptilia (L. *reptilis,* creeping). Skin dry, covered with scales or scutes. Terrestrial tetrapods, cold blooded. Lizard, snake, turtle, crocodile.

Class 3. Aves (**L.** *avis,* bird). Flying vertebrates, forelimbs modified into wings, skin covered with feathers, warm blooded. Parrot, Pigeon, Ostrich.

Class 4. Mammalia (**L.** *mamma*, breast) Skin covered with hairs. Female with mammary glands which secrete milk for suckling the young one. Warm blooded. Spiny ant eater, kangaroo, bat, dog, man.

SUBPHYLUM PROTOCHORDATA

The **Protochordata** (Gr. *protos*, first; *chorde*, cord) are simple form of chordate animals in which the notochord is wholly or partially formed and the vertebral column and skull are lacking, hence **Acrania**. This division includes a heterogenous group of animals of the phylum Chordata, most of which are fairly related to the vertebrates but none possesses the full complement of vertebrate characters.

The Protochordata is divided into three classes:

(i) The **Hemichordata** which comprises worm-like forms whose vertebrate affinities are much more remote (status doubtful).

(ii) The **Tunicata** or **Urochordata** or the **ascidians** which in the adult stage are entirely different from the vertebrates. The ascidians, through their larval forms constitute the actual stock from which the chordates have emerged.

(iii) The **Cephalochordata** is a group of small animals, *Branchiostoma (Amphioxus),* which looks like fish but has no fish-like skeleton. *Amphioxus* is rather a degenerate form and is not a satisfactory vertebrate prototype.

Significance of protochordates

Although protochordates are not of much economic importance, but they are of great phylogenetic significance because they possess the three basic chordate features which may persist throughout life (*Branchiostoma*). They are the most primitive chordates but they share affinities with the vertebrates. They are considered close to the ancestors which finally gave size to the vertebrates.

2

CHAPTER

HEMICHORDATA

Hemichordata has recently been removed from the Chordata and placed as an independent phylum of invertebrates close to Echinodermata. Bateson (1885) included the Hemichordata in the phylum Chordata, a classification that has largely been followed in Zoology textbooks. But many French and German authors have separated it from the chordates and Hyman (1959) has placed it with the invertebrates. In the present system of classification Hemichordata occupies a controversial position and exhibit no more than a convergent resemblance to the chordate organisation. In the present book Hemichordata is being described as a subphylum of the Chordata for convenience of comparative study of the protochordates.

There has been controversy regarding the name of this group. Many other names have been suggested by various authors. In 1948 Dawydoff proposed a new name **Stomochordata** and another name **Adelochordata** (*aldelo*, hidden) was given because the notochord is not distinct, but both these names have been given up because there is no use in changing any appropriate name that has been in use for such a long time. Hyman (1959) retained the name Hemichordata as suitable and just name of the group because it does not imply the possession of a notochord by hemichordates; it simply says that hemichordates are "part-chordates", that is, related to chordates.

The Hemichordata are exclusively marine, solitary or colonial, soft bodied, worm-like, mostly tubicolous animals with the body divided into proboscis, collar and trunk. The coelom is **enterocoelous,** divisible into three regions corresponding with the three external divisions of the body. They have many paired gill slits. The nervous system consists of a network of nerve fibres and nerve cells embedded in the epidermis. They have a **buccal diverticulum** in the **preoral** region which is often regarded as the **notochord**. Sexes are separate. Development is **indirect** and the larva is called **tornaria**. In some direct development has been reported.

TYPE BALANOGLOSSUS

Balanoglossus (**Gr.** *balano*, acorn; *glossus*, tongue) are peculiar worm-like animals living an obscure life under stones and burrows in the sand down the abyssal region of the sea. The name refers to the tongue-shaped muscular proboscis ; hence commonly called **tongue worm**. It is also called **acorn worm** because the proboscis resembls acorn, the fruit of oak. It was discovered by J.F. Eschscholtz (1825) in Marshal islands and was described as a worm-like **holothurian** (Echinodermata). Several years later Kowalevsky (1866) discovered numerous pairs of gill slits piercing the body wall. On this account Gagenbaur created a special class **Enteropneusta**. Bateson (1883–1886) showed by embryological studies that the Enteropneusta exhibit chordate affinities in respect of the coelomic, skeletal, nervous and respiratory system and that the gill slits are formed on a similar plan to that of the gill slits of *Amphioxus*.

DISTRIBUTION

Some 20 species of *Balanoglossus* are known and are distributed all over the world, particularly in the tropical and subtropical seas. Some common species are *B. clavigerus, B. capensis, B. gigas, B. jamaicensis,* etc.

HABIT AND HABITAT

Balanoglossus is a burrowing and exclusively marine animal found in shallow coastal waters, a few species occur in deeper waters. It lives in self excavated **U-shaped burrows** (Fig. 2.1) in sandy bottom. One of the openings of the burrow is concealed with spiral coils of faecal matter and is called the **posterior opening**. It leads into the **posterior limb** of the burrow. The other end is the **anterior opening**, it is funnel-shaped and lies exposed. It leads into the **anterior limb** of the burrow which gives off a number of branches each opening by independent funnel shaped opening. The walls of the tube are lined by **mucous** secreted by the mucous glands of the animal. The mucous binds the sand particles together and thus prevents the tube from collapsing. Males and females live in separate burrows.

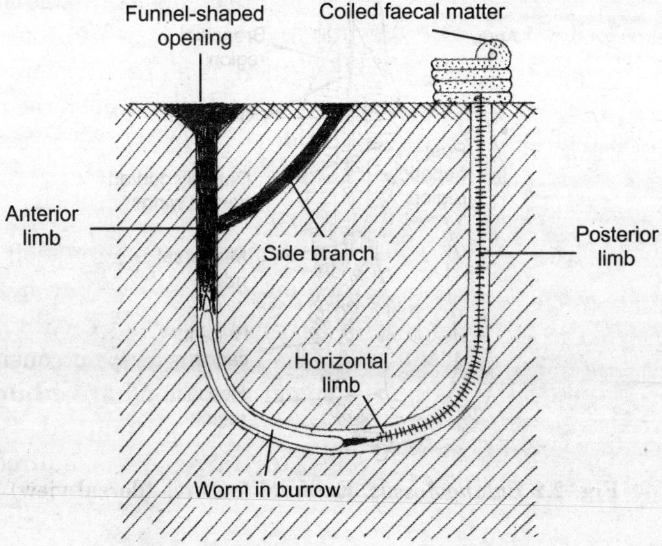

Fig. 2.1 *Balanoglossus.* U-shaped burrow

The worm is sluggish and not much affected by external stimuli. It lies in a twisted condition but the anterior and posterior ends are straight. The animal moves in the burrow with the help of cilia present all over the body (Knight Jones, 1952). The muscular proboscis is the most active part of the body. It protrudes it anterior end to survey the surrounding and posterior end to cast out faecal matter. It feeds on microorganism and organic matter ingested along with sand or mud.

EXTERNAL MORPHOLOGY

Shape, size, colour and odour

The body of *Balanoglossus* is elongated, cylindrical and bilaterally symmetrical. The size range from 10–25 cm in length, however, some species may be as small as 2 cm and still others as long as 2.5 metres (*B. gigas*). Natural **colouration** of *Balanoglossus* is a brilliant mixture of several colours. The proboscis is yellowish and the collar is orange red with a posterior white ring. The trunk has a variety of colours, the branchio-genital region is yellowish orange, the hepatic region is greenish and the caudal region is semi-transparent. The **odour** of *Balanoglossus* is very offensive and irritative, smelling somewhat like iodoform. It is probably protective in nature. The body is covered with **mucous**. Some species are luminous because of mucous.

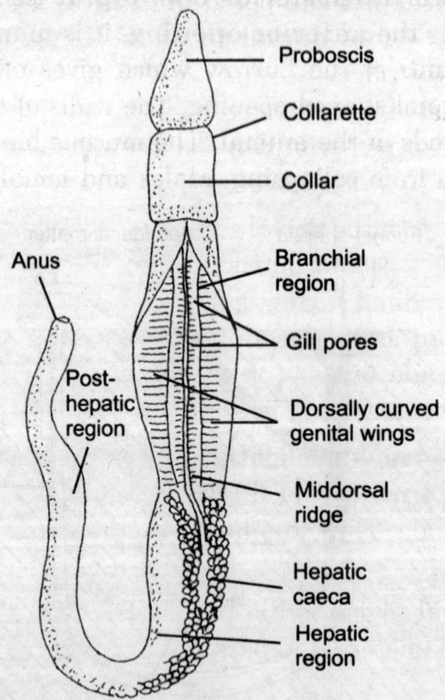

Fig. 2.2 *Balanoglossus.* External features (dorsal view)

Division of body

The body is divided into three regions: **proboscis** or **protosome**; **collar** or **mesosome** and **trunk** or **metasome** (Fig. 2.2).

Proboscis: The proboscis or protosome is the anteriormost region of the body. It is a short, rounded or conical and highly muscular structure. It is generally circular in cross-section, but in some species bears a mid-dorsal groove. The wall of proboscis is thick and muscular. The cavity of the proboscis is called **proboscis coelom** which communicates with the exterior through **proboscis pore** situated mid-dorsally at the base of the proboscis. Posteriorly the proboscis narrows forming a slender **proboscis stalk** which is surrounded by the anterior part of the collar. Below the stalk the base of proboscis bears a U-shaped **preoral ciliary organ** which tests the quality of water and food entering the mouth. The proboscis is a good digging organ.

Collar: The collar or mesosome is the middle region of the body, just behind the proboscis. It is a short, cylindrical, muscular structure about as wide as long and mostly shorter than the proboscis. The anterior part of the collar is funnel-like and is called **collarette**. It surrounds the proboscis stalk and sometimes posterior part of the proboscis. On the ventral side the collarette encloses a wide opening, the **mouth**, which remains permanently open. It is prevented from collapsing by a pair of **skeletal cornua**. Posteriorly the collar is sharply demarcated from the trunk by a circular constriction. The collar encloses a cavity called collar coelom which opens to the outside by a pair of collar pores.

Trunk: The trunk or metasome is the posterior and the longest portion of the body following the collar. This region is also muscular but these muscles do not help much in locomotion. They serve mainly to promote peristaltic contractions of the body by which the food is carried through the gut. A **mid-dorsal** and a **mid-ventral longitudinal ridge** runs along its whole length. The two ridges accommodate their corresponding nerves and blood vessels. The trunk is differentiated into three regions : an anterior **branchio-genital region**, a middle **hepatic region** and a posterior **abdominal** or **post-hepatic** or **caudal region** (Fig. 2.2).

The branchiogenital region, on the dorsal surface, bears a double row of minute apertures called **branchial** or **gill pores**, one row on each side of the mid-dorsal ridge. The pores are situated on a prominent longitudinal furrow. The number of pores increases with the age of the animal. There may be as many as 35 (sometimes more) pairs of gill pores. The branchiogenital region also has two rows of **gonads**, one row on each lateral side. The lateral regions containing the gonads are thin and flat forming wing-like lateral folds called **genital wings**.

The hepatic region of the trunk is slightly smaller than the branchiogenital region. It is marked by the presence of a number of finger-like processes known as **hepatic caecae** or **diverticula** on the dorsal side. They are darkly coloured and appear as folds of the body wall from outside.

The post-hepatic or caudal region is the posteriormost and longest part of the body. It is more or less uniform in diameter. It tapers posteriorly and terminates into the **anus**. There is no true tail.

BODY WALL

The body wall of *Balanoglossus* consists of an outer **epidermis** followed by a **nervous layer**, a **basement membrane** and the **muscle layer** (Fig. 2.3).

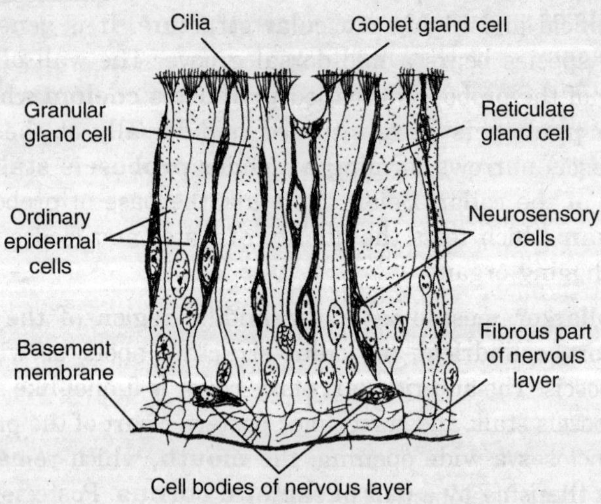

Fig. 2.3 *Balanoglossus*. Vertical section of the body wall

Epidermis: The epidermis is the outermost layer. It consists of a single layer of tall, slender, columnar, **epithelial cells**. The free ends of the cells bear numerous cilia. Numerous **gland cells** and **neurosecretory cells** are scattered in the epidermis. The gland cells are of three kinds :

(i) The **globlet cells** are flask shaped and secrete mucous.

(ii) The **reticulate cells** are long cells with vacuolated cytoplasm. They also secrete mucous. The secretion of these cells is very copious with the result the animal is always covered with slime. The slime may also give obnoxious smell.

(iii) The **mulberry** or **granular gland cells** are long cells containing granular cytoplasm. They secrete amylase. The epidermis of the collar region is very glandular. It is divided into three to five histologically different transverse zones like heavily, moderately or less glandular zones.

The **neurosecretory cells** are present in the epidermis of the proboscis and anterior part of the collar.

Nervous layer: At the base of the epidermis lies a thick layer of nervous tissue. It consists of nerve cells and fibres which form a network in close contact with the epidermis. It is transversed by the filamentous bases of the epidermal cells which are connected with the basement membrane.

Basement membrane: Internally the nervous layer is bounded by a strong basement membrane which serves for the attachment of the underlying muscles and also supports the epidermis.

Muscle layer: The basement membrane is followed by the muscle layer which is not very well developed. The muscle fibres are smooth. The muscle layer consists of an outer layer of **circular muscle fibres** and an inner layer of **longitudinal muscle fibres. Diagonal**

muscle fibres are also present. The proboscis consists of a thin layer of circular muscle and a thick layer of longitudinal muscle fibres. In the collar the musculature is confined to the collarette and consists of an inconspicuous layer of circular muscle fibres and thick bands of longitudinal and diagonal muscle fibres. In the trunk region only longitudinal muscle fibres are present. The muscle fibres are interrupted by dorsal and ventral mesenteries.

MOVEMENT

The animal moves by pushing the proboscis and collar forwards in the sand first and then pulling the body behind. As the circular muscles are weak the protrusion of the proboscis cannot be vigorous. It is however brought about by ciliary action, distension of coelom and contraction of circular muscles. The cilia probably play the main part in locomotion and burrowing.

COELOM

Balanoglossus is truely coelomate. The coelom is enterocoelous in origin. In the embryonic stage the coelom is a spacious cavity with a definite **coelomic epithelium.** In the adult the coelomic epithelium gives rise to connective tissue and muscle fibres which fills up the coelomic cavity and so it becomes very much reduced. In correspondence with the external triregional differentiation of the body the coelom is divided into three portions completely separated from one another. These are: (*i*) the **proboscis coelom** (*ii*) the **collar coelom** and (*iii*) the **trunk coelom** (Fig. 2.4).

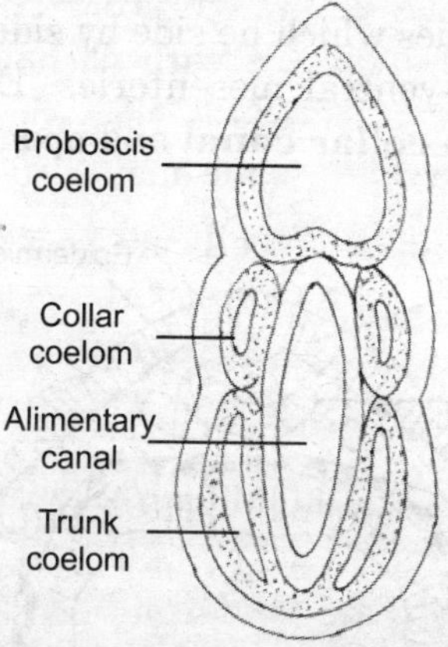

Fig. 2.4 *Balanoglossus.* Coelomic cavities (diagrammatic)

Proboscis coelom or protocoel

It is a single cavity largely occupied by muscles and connective tissue and the **proboscis complex** which projects into it. The proboscis complex consists of three structures—**buccal diverticulum**, **glomerulus** and **central sinus** (Fig. 2.5).

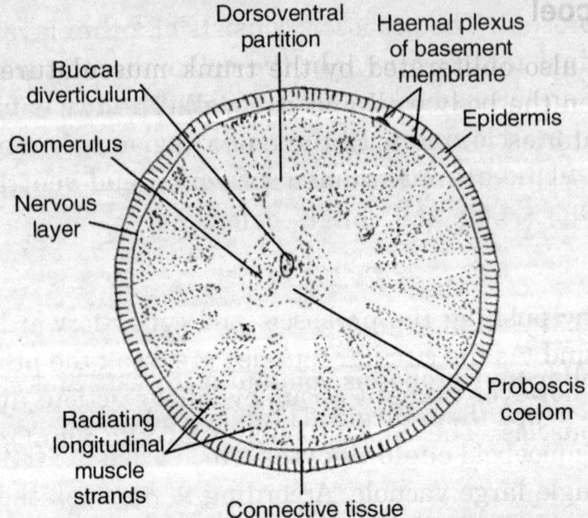

Fig. 2.5 *Balanoglossus*. T.S. through proboscis

Dorsally the proboscis coelom, towards its posterior side, is divided by a **dorsal mesentery** into right and left dorso-lateral compartments which extend into the proboscis stalk, the left compartment is larger than the right. Ventrally it is divided by a **ventral mesentery** into right and left ventrolateral compartments which become continuous behind the mesentery. The proboscis coelom opens to the exterior by a single **proboscis pore** situated mid-dorsally at the base of the proboscis stalk.

Collar coelom or mesocoel

The collar coelom is also very much reduced by the collar musculature and connective tissue. It is represented by two cavities which lie side by side in the collar. They are separated by incomplete mid-dorsal and mid-ventral mesenteries. Each collar coelom communicates with the first gill sac of its side by a **collar canal** and opens to the outside by a **collar pore** (Fig. 2.6).

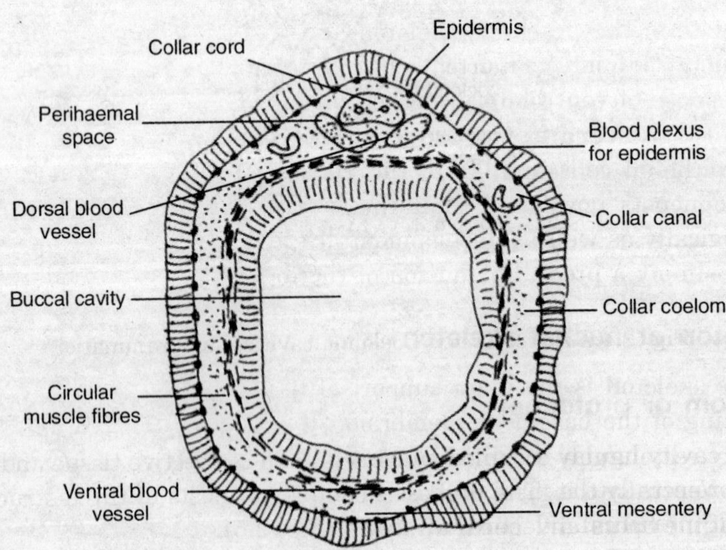

Fig. 2.6 *Balanoglossus*. T.S. through collar

Trunk coelom or metacoel

The trunk coelom is also obliterated by the trunk musculature. It is represented by two closed cavities between the body wall and gut wall. The two cavities are separated by dorsal and ventral mesenteries. They do not open to the exterior. In the branchiogenital region each cavity is further divided by **lateral septum** into a **dorso-lateral** and a **ventro-lateral compartment**. The trunk and collar coelom are partitioned by a **collar-trunk septum**.

Coelomic fluid

The cavities of the collar and proboscis communicate with the outside and are largely filled with sea water which keeps them turgid. The two trunk coelom are filled with a watery **coelomic fluid** containing amoeboid **coelomocytes** which originate from coelomic epithelium. Each coelomocyte has a single large vacuole. According to Spengel, the coelomocytes behave like leucocytes because they secrete a membrane around any foreign material that enters the animals body.

ENDOSKELETON

A definite endoskeleton of bone or cartilage is absent in *Balanoglossus* but some hard structures supportive in nature, are present. These are **buccal diverticulum, proboscis skeleton, branchial skeleton** and **pygochord.**

Buccal diverticulum

From the roof of the buccal cavity a short, stiff, hollow, thick-walled dorsal extension named **buccal diverticulum** arises. It opens into the buccal cavity. This structure was named **notochord** by Bateson (1885) and **stomochord** by Willey (1899b). Since words formed from *'chord'* seem to imply solid construction, Hyman (1959) has used the name buccal diverticulum for it. The diverticulum begins as a narrow neck that passes anteriorly just above the main plate of the proboscis skeleton and continues in the posterior part of the proboscis ending blindly in front. It often expands considerably and usually presents sacculations because of the presence of ventrolateral pockets inside. The enlarged portion of the buccal diverticulum is called the **vermiform process**. The wall of buccal diverticulum consists of tall, vacuolated epithelial cells similar to the epithelium of the buccal cavity. All recent workers on enteropneusts now agree that the buccal diverticulum does not represent a notochord. Histologically as well as developmentally, it is very different from the notochord of chordates. It is simply a preoral extension of the digestive tract (Silen 1950).

Proboscis skeleton or nuchal skeleton

The proboscis skeleton is the main support of the proboscis stalk. It is a non-cellular, laminated thickening of the basement membrane. It is **Y-shaped** structure consisting of a broad, flat, roughly rectangular **median plate** from which arise two diverging arms called **horns** or **cornua**. Ventrally the median plate is produced into a **keel**. The median plate lies beneath the base of the buccal diverticulum and the horns extend backwards along the outer side of the buccal cavity (Fig. 2.8).

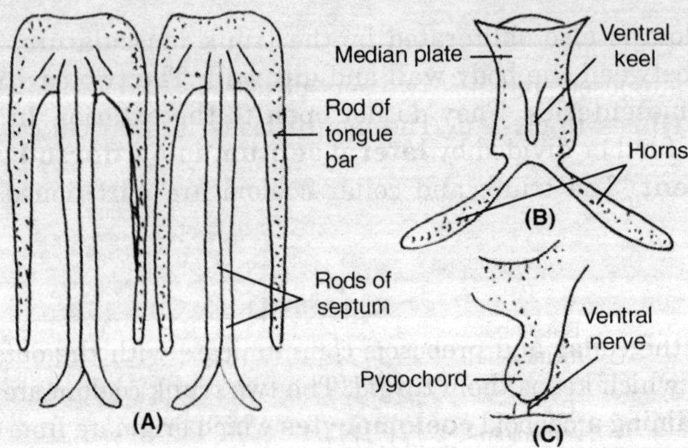

Fig. 2.7 *Balanoglossus*. Skeleton. (A) Branchial skeleton; (B) Proboscis sketeton; (C) Pygochord

Branchial skeleton

The U-shaped gill slits present in the branchial region of the pharynx are supported by **M-shaped** skeletal rods formed by the thickening of the basement membrane (Fig. 2.7B). The middle arm of the **'M'** is called the **primary gill rod**. It is thicker than the other two outer arms and is forked at the ventral free end. This indicates that it is formed by the fusion of two arms of the adjacent U-shaped rods that fuse to form the M-shaped rods. The two outer arms are called **secondary gill rods**, they are thin and not forked.

Pygochord

The pygochord (Fig. 2.7C) is a longitudinal hard rod-like structure found in the caudal region of the trunk. It is a median derivative of the alimentary canal and lies mid-ventrally between the body wall and intestine. It consists of vacuolated cells. It supports the soft caudal region. It probably performs some other function yet not known.

DIGESTIVE SYSTEM

Alimentary canal

The alimentary canal is a straight tube between the **mouth** and the **anus.** It is supported along its whole length by dorsal and ventral mesenteries. The wall of the alimentary canal is made up of ciliated epitheluim and basement membrane, the muscle layers are absent. The alimentary canal is divisible into **mouth, buccal cavity, pharynx, oesophagus, intestine** and **anus.**

Mouth. The mouth is the anterior opening of the alimentary canal (Fig. 2.8). It is a wide circular opening situated in a groove between the collarette and the proboscis stalk. The mouth remains premanently open, therefore, the food is brought to it by ciliary current of the proboscis. Knight Jones (1952) is of the view that the mouth does not remain permanently open as previously supposed, but has two sets of muscle fibres—the **radial fibres** to open it and **concentric fibres** to close it. The mouth leads into buccal cavity.

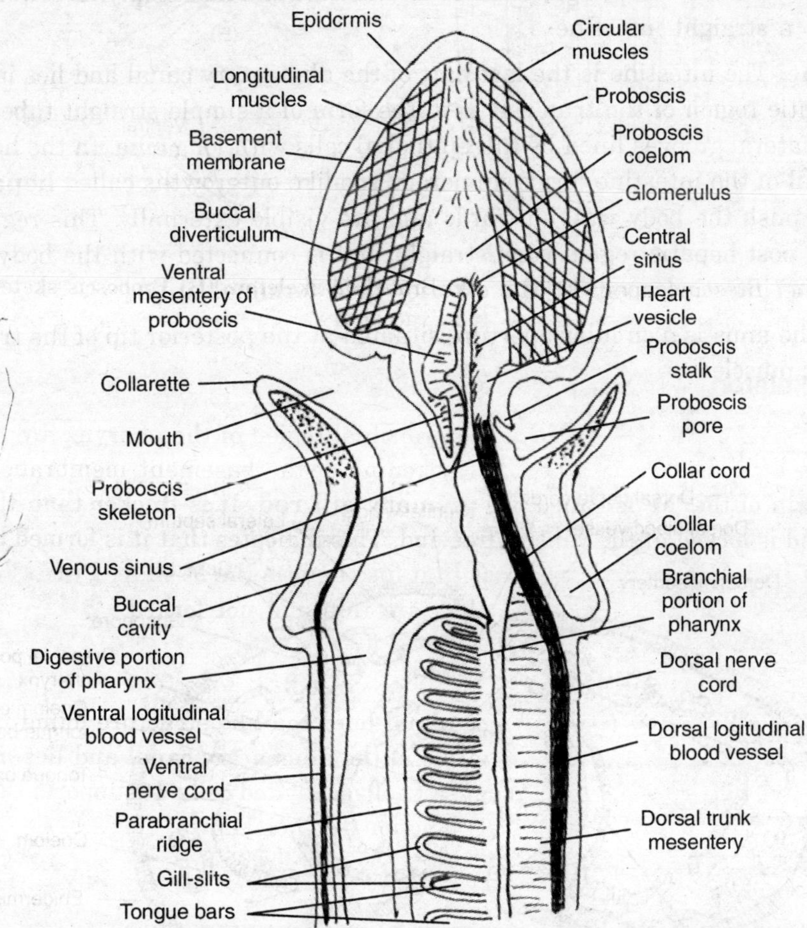

Fig. 2.8 *Balanoglossus*. V.L.S. Anterior region showing mouth, buccal cavity and pharynx

Buccal cavity: The buccal cavity lies in the collar region. It is lined by ciliated glandular epithelial cells. From the roof of the buccal cavity arises a stiff and hollow **buccal diverticulum** (Fig. 2.8) that projects into the proboscis coelom and ends blindly in front. Posteriorly it extends upto the collar-trunk septum and opens into the pharynx.

Pharynx: The buccal cavity leads into a wide pharynx lying in the branchial region of the trunk. The pharyngeal wall bears a longitudinal constriction on each lateral side. These constriction grow into the lumen of the pharynx as **parabranchial ridges** and divide it into a **dorsal branchial region** and a **ventral digestive region** (Fig. 2.8). The branchial region consists of a series of U-shaped gill slits and is concerned with respiration. The digestive region has ciliated glandular lining and is concerned with digestion and absorption of food.

Oesophagus: The pharynx, behind the last pair of gill slits, passes into a short oesophagus. The dorsal and the ventral region of the pharynx continue for some distance into the oesophagus. At the posterior end the oesophagus reduces in diameter and has deeply furrowed epithelium. It passes into a straight intestine.

Intestine: The intestine is the last part of the alimentary canal and lies in the hepatic and post-hepatic region of the trunk. It is in the form of a simple straight tube and bears a pair of dorso-lateral grooves lined by tall epithelial cells with long cilia. In the hepatic region the dorsal wall of the intestine forms numerous sac-like outgrowths called **hepatic caecae**. These caecae push the body wall outwards and are visible externally. This region is highly vascular. The post hepatic region runs straight and is connected with the body wall by the pygochord. The intestine opens to the exterior by a terminal **anus**.

Anus: The anus is a circular aperture situated at the posterior tip of the trunk. It often has sphincter muscles.

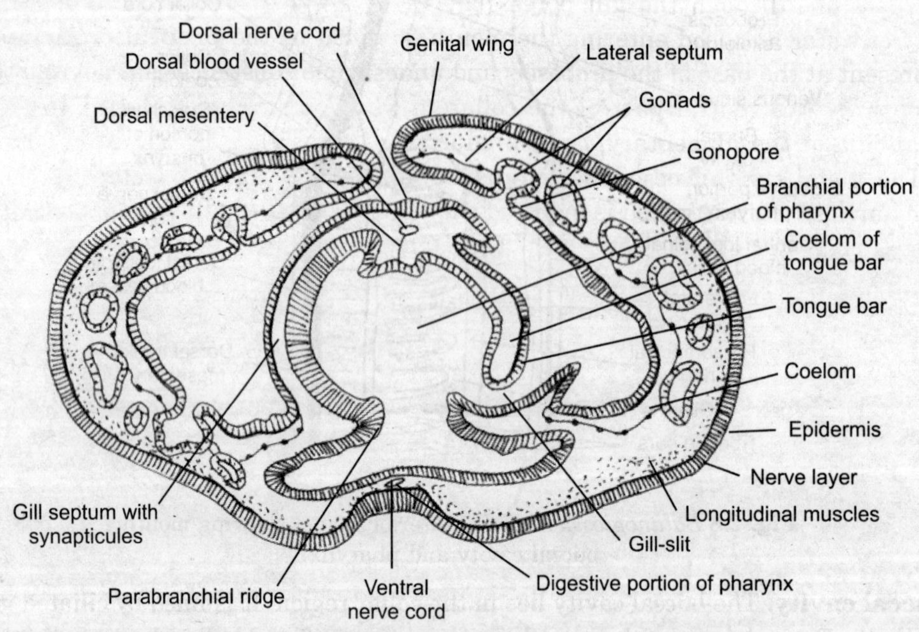

Fig. 2.9 *Balanoglossus*. T.S. through branchiogenital region

Food and feeding mechanism

Balanoglossus is a ciliary feeder. The large number of microscopic marine organisms living in the sand constitutes its food. As it makes its burrow the sand is passed through the body and the nutrient matter that adheres to it is extracted during its passage through the intestine. The exhausted sand is finally ejected through the anus at the posterior opening of the burrow in the form of worm casting. It is in response to this type of feeding that the mouth remains permanently open. *Balanoglossus,* thus literally feeds its way through the sand. A large amount of water also enters along with sand. This is brought about by the ciliary currents produced by cilia of gill slits (Fig. 2.10). The water that enters through the

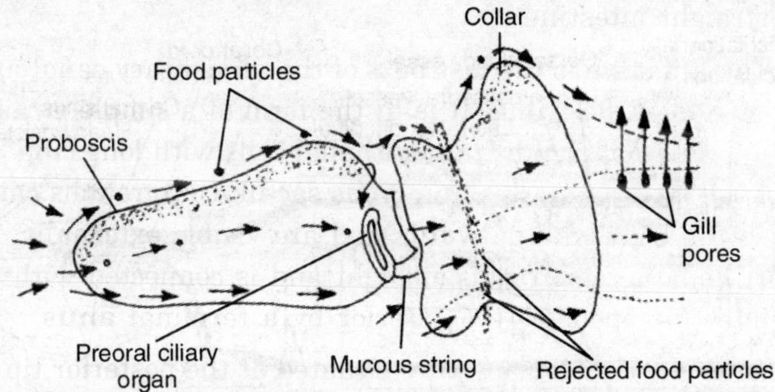

Fig. 2.10 *Balanoglossus.* Anterior region showing direction of feeding current

mouth takes its course through the buccal tube, pharynx, gill slits, branchial sac and finally pass out through the gill pores. This is the **respiratory-cum-food current**. The qualilty of water and food entering the mouth is tested by the **preoral ciliary organ** (Fig. 2.11) present at the base of the proboscis and undesirable substances are not allowed to enter the mouth by the ventral part of the collarete which does so by covering the mouth. The internal cilia of the alimentary canal produce currents directed backwards for moving the food. The gland cells of oesophagus produce enzymes. Mucous secreted by the proboscis contain amylase enzyme which is ingested with food. The hepatic caecae also secrete amylase, maltase, lipase and protease enzymes. These enzymes digest the organic matter present in mud.

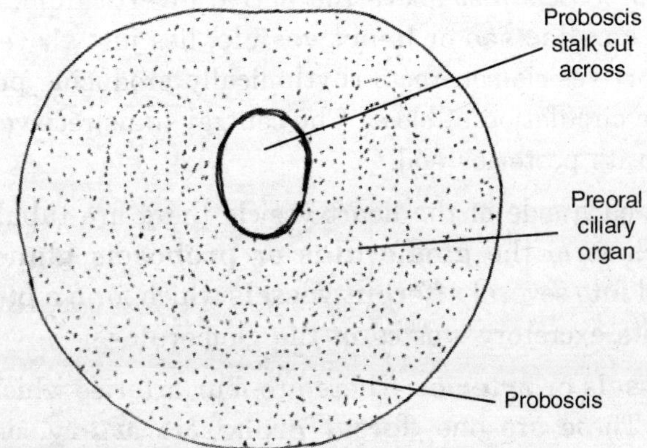

Fig. 2.11 *Balanoglossus.* Posterior view of proboscis showing preoral ciliary organ

Formation of definite **mucous strands** for feeding has been reported in *Glossobalanus.* Food particles are caught in the mucous on the surface of the proboscis. The cilia of the proboscis direct the mucous strands towards the preoral ciliary organ at the base of the proboscis. Strong ciliary currents carry the mucous string to the collar and cilia at the anterior end of the collar either reject the particles or drive them ventrally into the mouth.

CIRCULATORY SYSTEM

The circulatory system of *Balanoglossus* is of **closed type**. It is quite peculiar consisting of **sinuses** or **lacunae** and **blood vessels.**

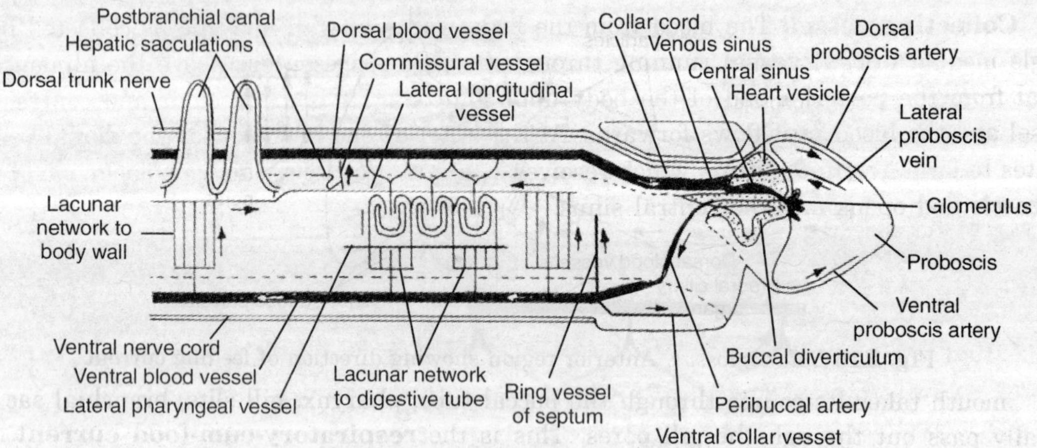

Fig. 2.12 *Balanoglossus.* Blood vascular system

Blood: The blood is a colourless fluid that circulates in the system. It may contain a few colourless amoeboid corpuscles. There is no respiratory pigment.

The main blood channels of the circulatory system are (*i*) a **central sinus** and **heart vesicle**, (*ii*) **glomerulus**, (*iii*) **distributing vessel** and (*iv*) **collecting vessel.**

Central sinus and **heart vesicle:** The central sinus is a small non-contractile elongated space situated in the proboscis just above the buccal diverticulum. A closed muscular and contractile vesicle, the **cardiac sac** or **heart vesicle**, lies just above the central sinus. The ventral wall of the heart vesicle contracts rhythmically producing pulsations in the central sinus which help in the circulation of blood. The central sinus receives blood from the dorsal vessel which open into its posterior end.

Glomerulus: On each side of the heart vesicle there are tubular projections into the proboscis coelom which form the **glomerulus** or **proboscis gland**. The blood from the central sinus is pumped into several **afferent vessels** which form a **plexus** in the glomerulus. The blood gets rid of its excretory wastes in the glomerulus.

Distributing vessels or **arteries:** There are four arteries which carry the blood away from the glomerulus. These are one **dorsal proboscis artery**, one **ventral proboscis artery** and two **efferent glomerular arteries**.

The dorsal and ventral proboscis arteries supply the proboscis. The two efferent glomerular arteries run backwards along the sides of the buccal diverticulum, encircle the buccal tube as **peribuccal vessels** which form a blood plexus and unite to form a single **ventral vessel** which runs backwards up to the posterior end of the body through the ventral mesentery below the alimentary canal. The ventral vessel has muscular contractile walls and the blood in it flows backwards. All along its length it supplies blood through nerve plexus to the body wall and the wall of the alimentary canal. It also supplies the collar by a **ventral collar vessel**, to the collar trunk septum by a **ring vessel** and to each gill by **afferent branchial artery** (Fig. 2.12).

Collecting vessel: The blood from the body wall, gut wall and gills is collected by a single median **dorsal vessel** running through the dorsal mesentery above the alimentary canal from the posterior end of the body upto the collar. It is a muscular and contractile vessel and the blood in it flows forwards. At the anterior end of the collar the dorsal vessel dilates to form a **venous sinus** which receives a lateral proboscis vein from each side of the proboscis and opens into the central sinus.

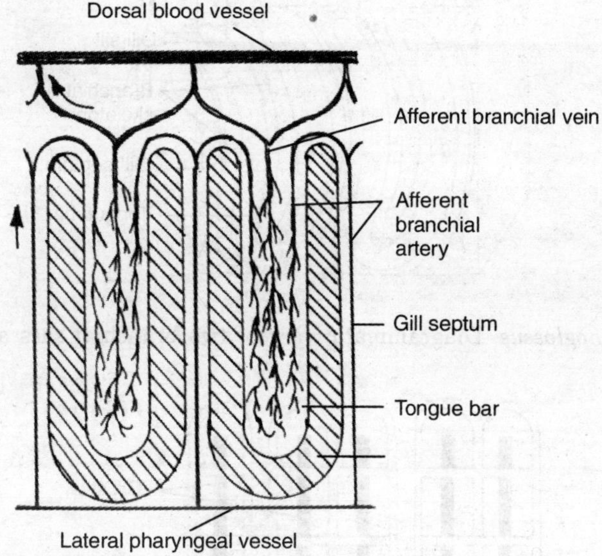

Fig. 2.13 *Balanoglossus.* Blood supply of branchial wall

Aeration of blood probably takes place in the branchial region although there is no respiratory pigment (Fig. 2.13).

RESPIRATORY SYSTEM

The respiratory system of *Balanoglossus* is similar to that of other protochordates. It lies in the branchial part of the pharynx and consists of a series of **gill slits** on each dorso-lateral side. In the early stages the gill slits are broad and oval but later become **U-shaped** by the development of hollow downgrowth of the pharyngeal wall from its dorsal side. The downgrowth is called **tongue bar**. It separates the two limbs of the 'U'. The portion of the pharyngeal wall between two adjacent U-shaped gill slits is called **gill septa**. The tongue bars and gill septa are supported by **M-shaped skeletal rods.** The middle arm of M-shaped rod is called **primary gill rod** and is forked at the free end. It lies in the gill septum. It is regarded that there were two rods but later they fused except at the ventral free end. The outer arms of M are called **secondary gill rods** and they lie in the tongue bars (Fig. 2.15).

Thus each tongue bar contains two arms, one from the two adjacent skeletal rods. The gill rods are merely thickenings of the basement membrane which forms the gill septa and tongue bars. The gill slits do not open directly to the outside. Each gill slit opens into a **branchial sac** or **gill pouch** which lies between the pharynx and the body wall. Each branchial sac opens independently to the outside by a **gill pore** or **branchial pore** situated dorsolaterally in the body wall (Fig. 2.14).

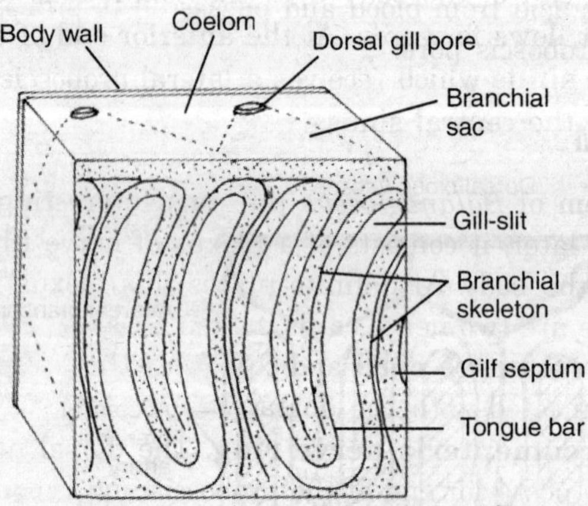

Fig. 2.14 *Balanoglossus.* Diagrammatic view of two branchial sacs and two gill slits

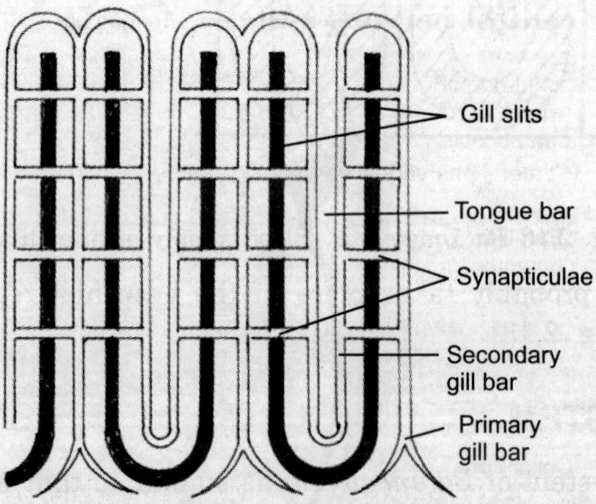

Fig. 2.15 *Balanoglossus.* Branchial wall showing branchial skeleton

Mechanism of respiration: Each gill slit is richly lined by cilia called **lateral cilia.** The beating of these cilia set a current of water that enters the mouth, goes into the pharynx and through the gill slits into the gill pouches and finally pass out through the gill pores. The pharyngeal wall is richly supplied with blood vessels, gaseous exchange takes place between the blood and current of water. Oxygen dissolved in water is taken up and carbon dioxide is returned to water.

EXCRETORY SYSTEM

Some authors (Young, 1955) maintain that there is no specialized structure for excretion in any part of the body. The **proboscis gland** or **glomerulus**, lying on one side of the buccal diverticulum and projecting into the proboscis coelom has been described as the excretory organ. It is made up of numerous blind tubular projections formed by the peritoneum covering the buccal diverticulum, central sinus and heart vesicle. It has been reported that this gland separates urea and uric acid from blood and passes on to proboscis coelom from where it is excreted through the proboscis pore.

NERVOUS SYSTEM

The nervous system of *Balanoglossus* presents interesting resemblances with that of echinoderms. As in the later, it consists of a plexus of nerve fibres and cells lying beneath the epidermis all over the body. At definite places this plexus is thickened to form **nerve cords** (Fig. 2.16). There are two nerve cords extending along the whole length of the trunk, one is the **median dorsal nerve cord** and the other is the **median ventral nerve cord**. The ventral cord is connected with the dorsal nerve cord in the trunk-collar septum by a circular tract called **circumenteric nerve ring.** The dorsal nerve cord extends up to the proboscis stalk, and the posterior end of the proboscis and is connected with a circular tract called **anterior nerve ring.** In the region of the collar the dorsal nerve cord leaves the epidermis, runs through the collar coelom above the buccal cavity and is invaginated to form a tubular **collar cord** or **neural tube** open at both ends. This part lying above the alimentary canal is known as the **central nervous system.** From this tube **dorsal roots** proceed to

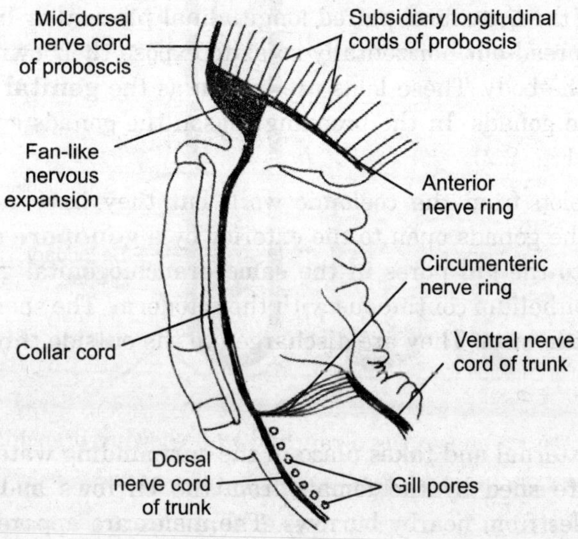

Fig. 2.16 *Balanoglossus.* Nerve cords in the anterior region of the body

the skin of the back of the collar. Thus, it is evident that the nervous system of *Balanoglossus* presents unmistakable resemblances with the centralized subepithelial plexus of echinoderms on the one hand and with the tubular dorsal nerve cord of the vertebrates on the other.

SENSE ORGANS

Organs of special sense are lacking in *Balanoglossus.* In the epidermis of the proboscis and the anterior part of the collar neurosensory cells are present and they are connected with nerve net. Some people claim that a few neurosensory cells are sensitive to light (**photoreceptors**). The U-shaped **preoral ciliary organ** present ventrally at the base of the proboscis is made up of ciliated cells and is a **chemoreceptor**.

BEHAVIOUR

The behaviour of *Balanoglossus* suggests the presence of only one clearcut reflex, that is contraction of the longitudinal muscles in response to tactile stimulation. If separate pieces of the body are subjected to light or to tactile stimuli, they react by moving away, suggesting, thereby, the presence of uncentralized nature of nervous system as that of echinoderms. There are no special conducting pathways. This is proved by the stimulation of flaps of body partly cut off from the rest that produce a generalized contraction. Thus, it is evident that conduction can occur in all directions. The dorsal and ventral nerve cords act only as quick conduction pathways.

REPRODUCTIVE SYSTEM

The sexes are separate. When mature the colour of the genital region may be slightly different. Both male and female **gonads** consist of more or less **lobulated sacs** connected with the epidermis by short ducts. They may be either **uniserial, biserial** or **multiserial** in disposition. They occur in the branchial region and also extend to a variable distance behind it. The sides of the body form paired longitudinal **pleural** or **lateral folds** which are movable and can be spread out horizontally so as to expose their own upper side as well as the dorsal surface of the body. These folds are known as the **genital pleurae** because they contain the bulk of the gonads. In the breeding season the gonads swell and bulge forming a soft **genital wing** (Fig. 2.2).

The gonads develop from the coelomic wall, but they have no connection with the coelom in the adult. The gonads open to the exterior by a **gonopore** or **genital pore** which lies usually external to the gill pores in the same branchiogenital groove. The gonads are lined with germinal epithelium continuous with the ectoderm. The sperms and ova proliferate from this germinal epitheluim. They are discharged to the outside through the genital pores.

Fertilization

Fertilization is external and takes place in the surrounding water. First masses of eggs covered by mucous are shed by the female from the burrows and then the sperms are discharged by the males from nearby burrows. The males are apparently stimulated by the released eggs. Soon the mucous covering is broken by the tidal current and the eggs are scattered. The eggs are opague and irregularly ovoid when shed. The number of eggs released at a shedding ranges from a few dozen to more than a thousand, usually there are about one to three hundred, the greater number coming from larger females. Available evidence indicate that maturation begins some hours before ovulation and that the egg is generally in the metaphase of the first meiotic division when shed. It is at this stage the egg is fertilizable. Fertilization of the egg within 6-7 hours after shedding yields a high percentage of normal development. The spermatozoa is able to enter at any point over the entire surface.

Development

Development in *Balanoglossus* is **indirect,** with a free swimming larval stage called **tornaria.** After fertilization the egg undergoes an orderly series of successive changes in shape till cleavage sets in.

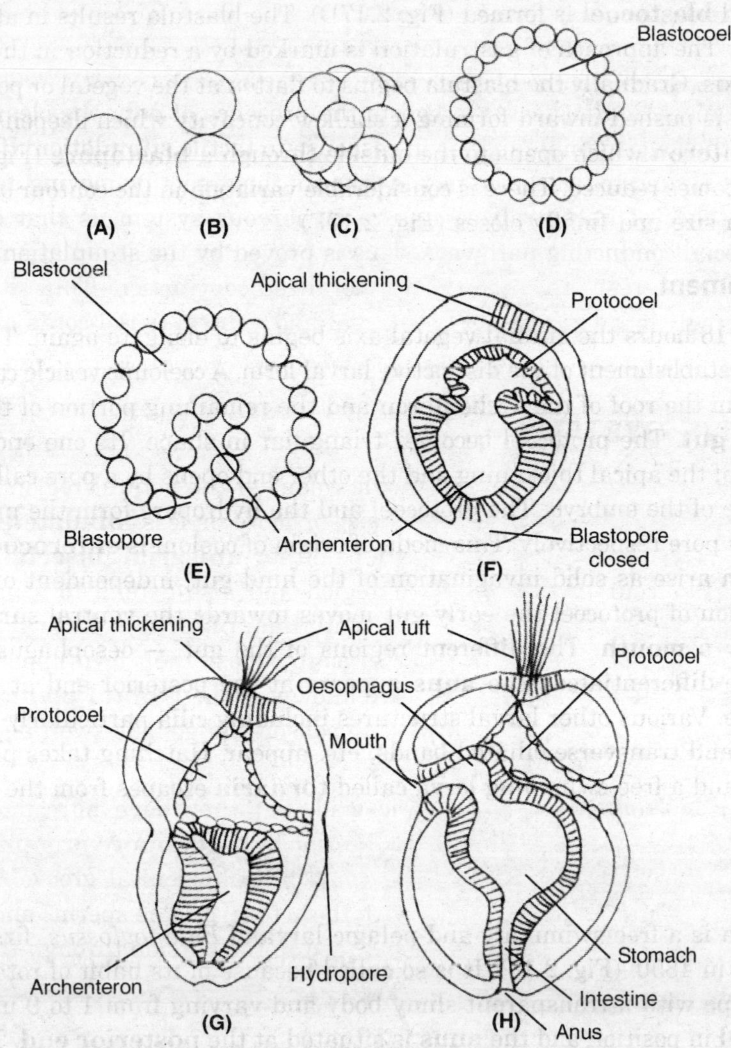

Fig. 2.17 *Balanoglossus.* Stages in early development. **(A–C)** Cleavage; **(C)** Morula; **(D)** Blastula; **(E)** Gastsulation begins; **(F)** Gastrula; **(G) (H).** Development of tornaria larva

Cleavage

The fertilized egg or **zygote** undergoes cleavage which is **complete (holoblastic),** almost **equal** and of **radial type.** The first cleavage is vertical. It begins about two hours after fertilization and produces two, generally but not invariably, equal cells, the **blastomeres**. The second cleavage is also like the first and produces four usually approximately equal cells. As a result of third cleavage two tiers of cells are produced, a tier of four smaller **animal cells** and a tier of four larger **vegetal cells.** The cleavage continues till a multicellular spherical **morula** (Fig. 2.17C) is formed. The morula undergoes reorganization of **blastomeres** and forms a spherical, single-layered, hollow **blastula** or **coeloblastula**. In the interior a fluid-filled cavity called **blastocoel** is formed (Fig. 2.17D). The blastula results in about 6-15 hours after fertilization. The approach of gastrulation is marked by a reduction in the volume in the animal-vegetal axis. Gradually the blastula begins to flatten at the vegetal or posterior end and the flattened end is pushed inward forming a shallow concavity which deepens gradually and forms the **archenteron** which opens to the outside through a **blastopore** (Fig. 2.17E) . Soon the blastopore becomes reduced. There is considerable variation in the contour of the blastopore as it decreases in size and finally closes (Fig. 2.17F).

Larval development

After about 18 hours the animal-vegetal axis begins to elongate again. This heralds the beginning of the establishment of the distinctive larval form. A coelomic vesicle called **protocoel** is pinched off from the roof of the archenteron and the remaining portion of the archenteron forms the future **gut**. The protocoel becomes triangular in shape. Its one end gets attached to the underside of the apical thickening and the other end opens by a pore called **hydropore** on the dorsal side of the embryo. The protocoel and the hydropore form the proboscis coelom and the proboscis pore respectively. This mode of origin of coelom is **enterocoelic.** The collar and trunk coelom arise as solid invagination of the hind gut, independent of the protocoel. After the formation of protocoel the early gut moves towards the ventral surface and opens to the outside by a **mouth**. The different regions of the gut — oesophagus, stomach and intestine, become differentiated. The **anus** appears at the posterior end at the side of the former blastopore. Various other larval structures including cilia particularly the apical tuft, the longitudinal and transverse ciliated bands, etc. appear. Hatching takes place usually on the seventh day and a free swimming larva called **tornaria** escapes from the egg membrane (Fig. 2.17 G,H).

Tornaria larva

The tornaria is a free-swimming and pelagic larva of *Balanoglossus,* first described by Johannes Müller in 1850. (Fig. 2.18). It is so called because of its habit of rotating in circles. It is ovoid in shape with a transparent slimy body and varying from 1 to 9 mm in size. The **mouth** is **ventral** in position and the **anus** is situated at the **posterior end**. The alimentary canal is simple having an oesophagus, stomach and intestine. The cilia are arranged in the form of ciliated band on the body surface. There is a **preoral ciliary band** anterior to the mouth, a **postoral ciliary band** behind the mouth and a **circumanal ring** or **posterior ciliary band** or **telotroch** in front of the anus. The cilia of the preoral and postoral ciliated bands are short and serve to collect food by directing the water current towards the mouth. The cilia of the telotroch are long and powerful, hence it is the chief locomotory organ of tornaria. A ciliary wave passing along the telotroch causes the larva to rotate in circles constantly while swimming. At the anterior end the larva has a thickened epidermal plate called the **apical plate** which bears a pair of pigmented **eye spots** and a tuft of sensory cilia called **apical tuft** or **ciliary organ.** A thin-walled sac called **protocoel** and its opening called **hydropore** is present in the body of the larva. A pulsating **heart vesicle** develops in the later stages to the right of the hydropore.

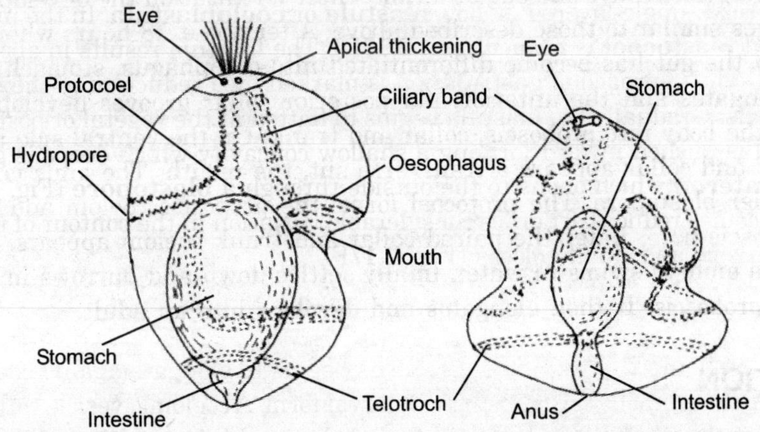

Fig. 2.18 *Balanoglossus.* Tornaria larva. A. Early; E. Fully developed

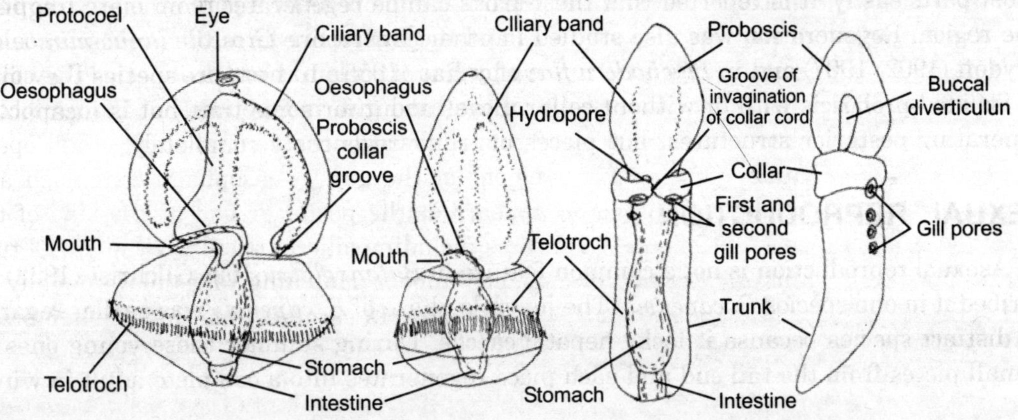

Fig. 2.19 *Balanoglossus.* Stages in metamorphosis

Metamorphosis

After leading a free-swimming life for some time the larva sinks down to the bottom and metamorphoses into an adult (Fig. 2.19) During metamorphosis the transparency of the larva, the ciliary bands and eyes are lost. There is a reduction in size probably due to loss of water. The body becomes differentiated into proboscis, collar and trunk and at the same time the buccal diverticulum, gill slits and coelomic sacs are also formed. The buccal diverticulum and gill slits are formed as outgrowth of the alimentary canal. The protocoel persists as the proboscis coelom and the hydropore as the proboscis pore. The collar and trunk coelom form later. The reproductive organs develop probably from the mesoderm. The trunk region elongates and the larva gradually changes into adult worm.

DIRECT DEVELOPMENT

Direct development has been observed in *Saccoglossus kowalevskii*. There is no tornaria larva stage. *Saccoglossus* lays one egg at a time. After fertilization the development proceeds by gradual stages similar to those described above. After about 18 hours when the protocoel has formed and the gut has become differentiated into oesophagus, stomach and intestine, the embryo elongates and the anterior and posterior collar grooves develop. These collar grooves divide the body into proboscis, collar and trunk. On the ventral side at the junction of the proboscis and collar appears a transverse slit, the mouth. The anus is formed at the position of earlier blastopore. The protocoel forms the proboscis coelom and the hydropore forms the proboscis pore. Later the paired collar and trunk coelom appears. Now hatching takes place. The embryo swims in water, finally settles down and burrows in the sand with the help of its proboscis. It then elongates and develops into an adult.

REGENERATION

Balanoglossus possesses great power of regenerating lost parts. If cut, the posterior part of the body is readily formed. The proboscis, collar and branchial parts can also regenerate the lost parts easily. It is reported that these parts can be regenerated from mere fragments of the region. Regeneration was also studied in other genera like *Glossobalanus minutus* by Dawydoff (1902, 1907) and in *Ptychodera flava* by Rao (1955). In both the species they found that isolated proboscis with or without collar moves about for some time but is incapable of regenerating posterior structures, but pieces of trunk regenerate completely.

ASEXUAL REPRODUCTION

Asexual reproduction is not a common feature in *Balanoglossus* but Gilchrist (1923) has described it in one species *B. capensis*. The juvenile phase of *B. capensis* was earlier regarded as a distinct species because it lacks hepatic caecae. During summer these young ones cut off small pieces from the tail end and each piece regenerates into a complete adult in winter.

AFFINITIES

The class Enteropneusta, to which *Balanoglossus* belongs, was established in 1870 by Gagenbaur. The name Hemichordata was given by Bateson in 1885 and he included it with the Chordata. Hyman in 1959, however gave independent phylum status to Hemichordata in the invertebrate. In establishing its phylogenetic position, *Balanoglossus* has been linked to several diverse groups, the chief ones are Annelida, Echinodermata and Chordata. It was also associated with a number of minor phyla like Nemertinea, Pogonophora and Phoronida but due to lack of evidence these relationships were rejected.

AFFINITIES WITH ANNELIDA

The relationship between Annelida and Hemichordata was first established by Spengel (1893).

Similarities

1. The habits of *Balanoglossus* like burrowing, ingesting mud, extraction of nutrients from ingested mud and leaving the waste in the form of castings, and the general body form suggests it to be an earthworm.
2. The proboscis of *Balanoglossus* and prostomium of *Nereis* are similar and preoral.
3. The collar of acorn worm and the clitellum of earthworm are similar.
4. The blood vascular system of both the groups is similar–the heart is dorsal in position and the blood flows forward in the dorsal vessel and backward in the ventral vessel.
5. The tornaria larva of *Balanoglossus* and the trochophore larva of polychaete annelids are similar in being pelagic and ciliated and in having an apical plate, eye spots and well developed alimentary canal.

Dissimilarities

The two groups differ in the following characters:

1. Absence of gill slits and buccal diverticulum in Annelida.
2. Organisation of the nervous system — the nerve cord is tubular and dorsal in *Balanoglossus* and in Annelida the nerve cord is double, ventral and solid.
3. Nephridia of annelids absent in *Balanoglossus*.
4. The tornaria larva of *Balanoglossus* differs from the trochophore larva of polychaete worms in the following characters:
   (i) Absence of nephridia in tornaria.
   (ii) Absence of preoral coelom in trochophore.
   (iii) The blastopore becomes the mouth in trochophore and anus in tornaria.

The differences between the two groups are greater as compared to the resemblances which are superficial and insignificant. It probably indicates a convergent evolution resulting from similar habits and habitat. In the light of these differences, therefore, the phylogenetic relationship between the two groups has been given up.

AFFINITIES WITH ECHINODERMATA

There is no similarity in the appearance of the adults of both hemichordates and echinoderms so it is difficult to suspect any relationship between them. The only anatomical similarity between them lies in the organisation of the nervous system which is poorly developed and forms epidermal plexus. They have common habitat and power of regeneration. The mode of origin of coelom is enterocoelic in both.

The similarity between the two groups, however, is based on embryological evidence and the similarity between larval stages. The resemblance between young tornaria larva of *Balanoglossus* and young auricularia and bipinnaria larvae of echinoderms is so great that it misled even Johannes Müller who regarded tornaria as the larva of starfish. Metschnikoff (1865) proved it to be the larva of Enteropneusta and thus established its relationship with the Echinodermata.

Larval resemblances

1. The larvae are small, oval transparent and pelagic.
2. The course of ciliated bands is identical.
3. The shape and division of the digestive tract is similar.
4. The cleavage pattern and formation of gastrula are similar.
5. The formation and arrangement of coelomic cavities is similar. The origin of coelom is enterocoelous and is divided into three regions which in hemichordates are called **protocoel, mesocoel** and **metacoel** and in echinoderms the **exocoel, hydrocoel** and **somatocoel.** The proboscis pore of *Balanoglossus* compares well with the water pore of the echinoderms.
6. The heart vesicle of hemichordates is regarded homologous to the madreporic vesicle of the echinoderm larvae. In the adult the heart vesicle is closely connected with the glomerulus in the hemichordates and the madreporic vesicle with axial gland in echinoderms, both these structures combine excretory and vascular function.
7. The blastopore becomes the larval anus.

Differences

1. Absence of apical plate, eye spot and telotroch in bipinnaria.
2. The protocoel is echinoderm larva is paired but in tornaria it is single.

The great similarity in embryological development cannot be accidental or due to convergent evolution as believed by Fell (1963). It is likely that they have evolved from a common ancestor. The echinoderms have, however, deviated greatly from the common ancestral type while the hemichordates are closer to it.

AFFINITIES WITH CHORDATA

The branchial apparatus of *Balanoglossus* shows a rough analogy with that of *Amphioxus*. In fact, it is this that stimulated a search of additional chordate characters in *Balanoglossus*. Bateson (1885) was the first to identify the structure that he regarded as notochord, a characteristic feature of all chordates and included Hemichordata in phylum Chordata in 1887.

Resemblances

The hemichordates appear to resemble the chordates in possessing the three fundamental chordate characters:

1. **Notochord:** The buccal diverticulum of hemichordates has been regarded as the notochord by Bateson.
2. **Central nervous system:** The central nervous system of hemichordates and chordates shows a marked resemblance in structure and development. The dorsal position of the collar cord, its mode of construction and occasional presence of a neuropore in it is comparable with the dorsal tubular nerve cord of the chordates.
3. **Pharyngeal gill slits:** The most striking link between the hemichordates and the chordates is the presence of pharyngeal gill slits which have the same general construction with tongue bars, synapticulae and M-shaped skeletal supports as those of *Amphioxus.* Hyman (1959) says that this relationship does not, however, justify the inclusion of the hemichordates in the phylum Chordata.

Differences

These resemblances are out weighed by important differences in the features mentioned above:

1. From the account of the buccal diverticulum, it is evident that the name notochord was erroneously given to this structure by Bateson and his followers. The following differences between the two structures support this view

Buccal diverticulum	Notochord
1. Formed as a forward evagination of the buccal wall.	1. Formed as a solid rod-like elevation from the roof of archenteron.
2. Made up of ordinary epithelial cells.	2. Consists of large vacuolated cells.
3. Lack sheath	3. Sheath present.
4. Not related to blastopore	4. Related to blastopore.
5. Not supporting in function.	5. Supporting in function.
6. Ventral to the mid-dorsal blood channel.	6. Dorsal to the main dorsal blood vessel.
7. Some sort of preoral gut confined to the proboscis.	7. Axial stiffening rod, extends far backwards.

It can thus be concluded that there is no notochord in hemichordates.

2. The nervous system of hemichordates presents typical invertebrate pattern. It is intraepidermal in position, the main nerve cord is ventral and circumenteric connectives are present (not found in chordates).

The other significant differences are:

1. The coelomic regions characterstic of Hemichordata are not found in Chordata.
2. The chordates are metamerically segmented whereas in hemichordates there is no trace of segmentation.

3. There is no post-anal tail in hemichordates.
4. The blood vascular system is of invertebrate type.

CONCLUSION

From the above it can be concluded that the hemichordates are more like invertebrates in their general organisation and morphology. It is likely that the three groups – Echinodermata, Hemichordata and Chordata, evolved from a common ancestor which is probably a **dipleura larva.** The echinoderms have deviated from the main ancestral stock while the hemichordates are closer to it. Thus the main ancestor gave rise to the Hemichordata and Chordata along the main line of evolution and the Echinodermata as a blind off shoot.

3

CHAPTER

HEMICHORDATA:
GENERAL CHARACTERS AND CLASSIFICATION

GENERAL CHARACTERS

1. The hemichordates are exclusively marine, solitary or colonial, soft-bodied, worm-like, mostly tubicolous animals.
2. Body bilaterally symmetrical and triploblastic
3. Body is not metamerically segmented but divisible into three distinct regions: **proboscis, collar** and **trunk**.
4. Body wall has a single layered epidermis.
5. A **buccal diverticulum** which arises form the roof of the buccal cavity projects into the proboscis coelom. It was earlier regarded as the notochord.
6. The coelom is divided into three distinct regions corresponding with the external divisions of the body
7. There are many paired permanent **gill slits**.
8. The alimentary canal is a complete straight or U-shaped tube.
9. The animals are **ciliary filter feeder**.
10. Nervous system primitive, nerve tissue occur both dorsally and ventrally embedded in the epidermis.
11. Blood vascular system simple and the flow of blood is reverse to that in true vertebrates.
12. Excretory organ is a single **glomerulus** situated in the proboscis.
13. Sexes separate, sometimes hermaphrodite, gonads one to several pairs.
14. Reproduction sexual.
15. Development generally **indirect** with a free-swimming **tornaria larva**. In some development is direct.

CLASSIFICATION

The Hemichordata includes about 80 known species and a very few fossil species. The adult form of one class is yet not known. It is divided into **four** classes.

CLASS 1: ENTEROPNEUSTA

1. Solitary worm-like burrowing animals commonly known as **acorn** or **tongue worm**.
2. Body elongated and vary from moderate to great length.
3. The epidermis is ciliated and glandular.
4. The digestive tract is straight with a terminal anus.
5. There are numerous paired U-shaped gill slits.
6. Sexes separate and gonads numerous, sac-like.
7. Development generally includes a free swimming tornaria larva

Examples: *Balanoglossus, Saccoglossus (=Dolichoglossus), Ptychodera.*

Saccoglossus

Saccoglossus is almost world wide in distribution. The smallest known species of Enteropneusta is *Saccoglossus pygmaeus* which measures about 2–3 cm in length. *Saccoglossus* (Fig. 3.1) is similar to *Balanoglossus* is habits, habitat and structure. It lives in spirally twisted burrows. The proboscis of this animal is exceptionally long and pointed. The collar slightly overhangs the anterior part of the trunk covering first 3-4 pairs of gill slits. The hepatic caecae and genital wings are absent. The gonads are marked externally by dorso-lateral genital folds. Sexes are separate, fertilization is external and development is direct.

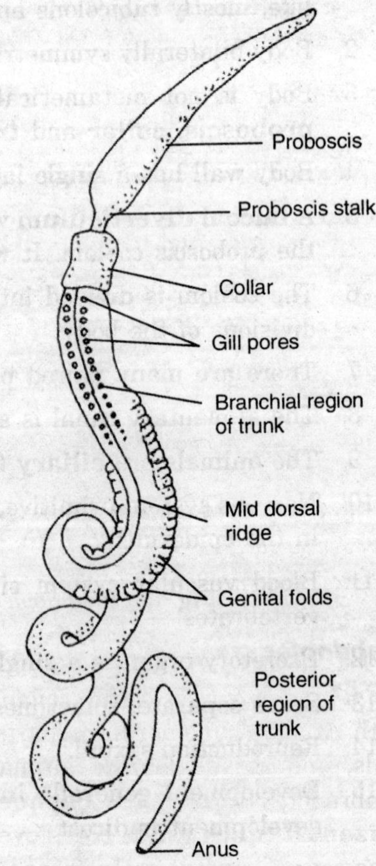

Fig. 3.1 *Saccoglossus*

CLASS 2: PTEROBRANCHIA

1. They are sedentary, solitary or colonial, tubicolous animals living in secreted chitinous tubes.
2. Body is short, compact and with a stalk for attachment.
3. Proboscis is shield-shaped.
4. Collar bears hollow ciliated arms with tentacles called **lophophores**. They are meant to produce ciliary feeding currents of water.

5. Gill slits two or absent.
6. Alimentary canal U-shaped. Mouth and anus situated close to each other.
7. Sexes separate or hermaphrodite, gonads one or two.
8. Development direct or indirect.
9. Asexual reproduction by budding.

Order 1: Rhabdopleurida

1. Form true colonies, the individuals of a colony are connected by a stolon.
2. Collar bears a pair of tentaculated arms.
3. Gill slits are absent.
4. Gonad single.

Example: *Rhabdopleura* (Fig. 3.2)

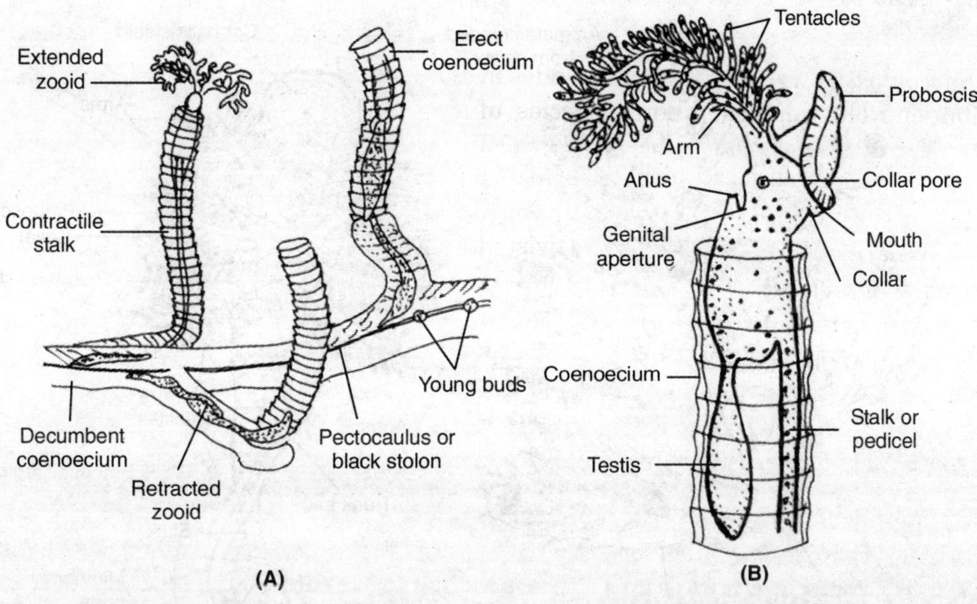

Fig. 3.2 *Rhabdopleura.* **(A)** Part of colony; **(B)** Individual zooid

Rhabdopleura

Rhabdopleura is a true colonial form occurring in various parts of the world including North Atlantic and northern parts of the North Sea. The colonies are attached to stones, corals, and sessile marine animals. The colony lies enclosed in a system of branching membranous tubes called **coenoecium**. The tubes are ringed and of two types: (*i*) The **horizontal tubes** which creep over the substratum and through them the living substance of the colony runs as a black cord, the **pectocaulus** or **black stolon**, and (*ii*) **erect tubes** which arise at intervals from horizontal tubes and each contains a **zooid**. Each zooid is attached to the black stolon by a long contractile stalk with the help of which they withdraw into the coenoecuim. They protrude out by the creeping action of the proboscis. The zooids have usual structures - proboscis, collar and trunk, ciliated arms, coelomic spaces with pores but no gill slits. The alimentary canal is U-shaped and the mouth and anus lie close to each other. The glomerulus is absent. Sexes are separate and gonads single. The eggs are large and yolky. The original individual is produced sexually. The trunk has a peduncle which bears buds which do not become free but form zooids. Each zooid secretes its own vertical tube.

Order 2: Cephalodiscida

1. Solitary or aggregation of several zooids not connected with each other but housed in a common gelatinous encasement.
2. Collar bears several tentaculated arms.
3. Gill slits and gonads single pair.

Example: *Cephalodiscus.* (Fig. 3.3)

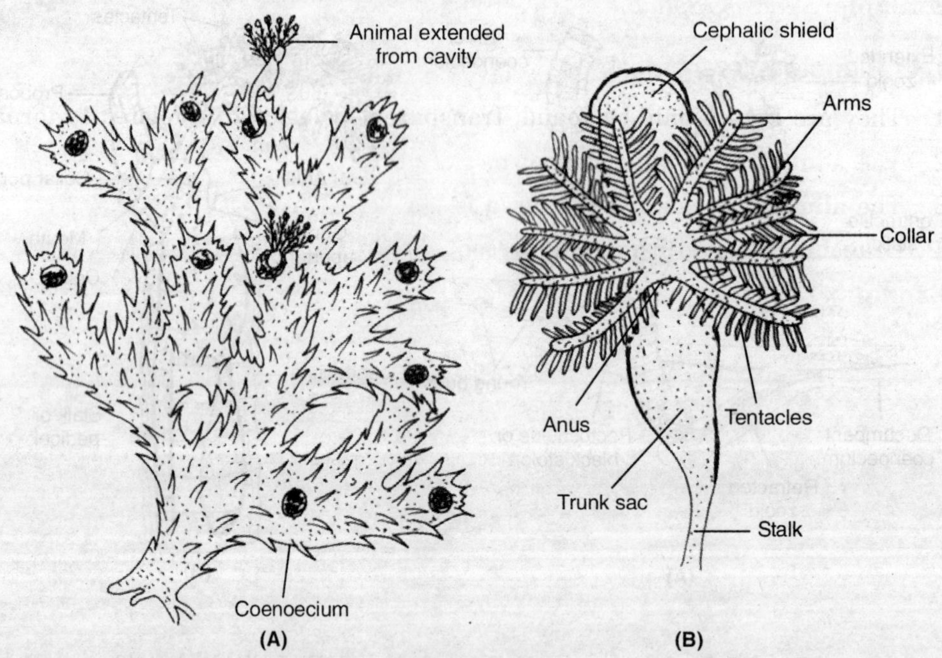

Fig. 3.3 *Cephalodiscus.* **(A)** Part of colony; **(B)** Individual zooid

Cephalodiscus

Cephalodiscus is found on the sea bottom at various depths, mainly in the southern hemisphere. Many individuals live together in many chambered gelatinous house called **coenoecium,** therefore they have been called colonial. Each individual or **zooid** is 2-3 mm long and has a proboscis, collar and trunk but these structures are visible only under careful examination. Ordinarily *Cephalodiscus* does not look like *Balanoglossus.* The proboscis is shield shaped and overhangs the mouth. It is richly ciliated. The proboscis coelom opens out by two proboscis pores. The collar is produced into a number (8-16) of ciliated arms which bear numerous fine tentacles by means of which the animal feeds. The arms contain continuations of the collar coelom. There are two collar coelom, each opening out by separate collar pores. The trunk is short and plump. A narrow elongated stalk arising from the aboral side of the trunk attaches the zooid with the wall of the cavity in which it lives. The alimentary canal is U-shaped with a ventral mouth and dorsal anus situated at the distal end. The pharynx has only one pair of gill slits through which water of the feeding current drawn in by the cilia of the tentacles passes out. The blood vascular system is similar to the acorn worm. Sexes are separate, gonads single pair and development direct. They reproduce asexually by budding but the zooids do not remain connected with each other.

Class 3: Graptolita

1. Includes extinct colonial hemichordates.
2. Known from the fossil structure of their tubes.
3. They were found in the Ordovician and Silurian period.

Example: *Dendrograptus.*

Class 4: Planctosphaeroidea

1. They are known only by round, transparent pelagic larva related to tornaria.
2. The adult form is yet not known.
3. The alimentary canal of the larva is U-shaped.
4. Branching ciliated bands are present on the aboral surface of the larva.

Example: *Planctosphaera.*

4
CHAPTER

TUNICATA

The **Tunicata** or **Urochordata** are hermaphrodite marine chordate animals contained in secreted tunic in adult stage. They which are usually sessile and degenerate and may be solitary or colonial, fixed or free. The chordate characters in these animals are represented in their larval stage in which the notochord is restricted to the posterior part of the body (tail). The larva is free swimming, non-feeding, highly motile and short-lived. Metamorphosis is retrogressive during which the advanced chordate characters like the notochord, nerve cord and the sense organs are lost. The adult is a ciliary feeder. There is no metameric segmentation. Coelom is absent and represented by the atrial cavity. They are found practically in all parts of the sea and at all depths.

TYPE HERDMANIA

The most common species of the subphylum Urochordata inhabiting the Indian seas is *Herdmania* (= *Rhabdocynthia*). It belongs to class **Ascidiacea**, hence commonly known as **ascidian**. It is also known as **sea-squirt** because when disturbed it forcefully ejects jets of water from the two siphons by contracting the body suddenly. The name *Rhabdocynthia* was given by Herdman in 1891 and the name *Herdmania* was given by Lahille in 1888. In 1910 Hartmeyer changed the name to *Herdmania* according to the law of priority.

Herdmania was first described by S.M. Das as one of the India animal types in 1936. His work is mainly on the species *H. pallida*. The description in the following pages is based on his work.

DISTRIBUTION

Herdmania is cosmopolitan in distribution. It is represented by 12 species of which only 4 species are found in the Indian ocean. Two of these species, *H.pallida* and *H. ceylonica* occur in coastal water and *H.mauritiana* and *H. ennurensis* are found in deep waters.

HABIT AND HABITAT

Herdmania is the most common ascidian occuring in coastal waters of the Indian seas. It is solitary and sedentary attached to the substratum by means of a **foot**. Sometimes it attaches to a mollusc shell. The mollusc carries it from place to place, thus helps in the distribution of the sedentary animal, *Herdmania* in return protects the mollusc from its enemies being unpalatable (**commensalism**). The animal is a **ciliary feeder**, feeding on microorganisms like algae, diatoms, etc. The water current entering through the branchial aperture carries oxygen and food particles while the water leaving through the atrial aperture carries carbon dioxide, undigested food, excretory wastes and sex cells. *Herdmania* is **hermaphrodite**. Fertilisation takes place in sea water. Development is **indirect** with a free-swimming **tadpole larva** showing basic chordate characters. After a short life it undergoes **retrogressive metamorphosis** and becomes degenerate sedentary adult.

EXTERNAL MORPHOLOGY

Shape, size, colour and orientation

Herdmania is about 9–13 cm long, 7 cm broad and 4 cm thick. The **size** increases with age. So far as the **shape** is concerned no two animals are absolutely alike. The body is

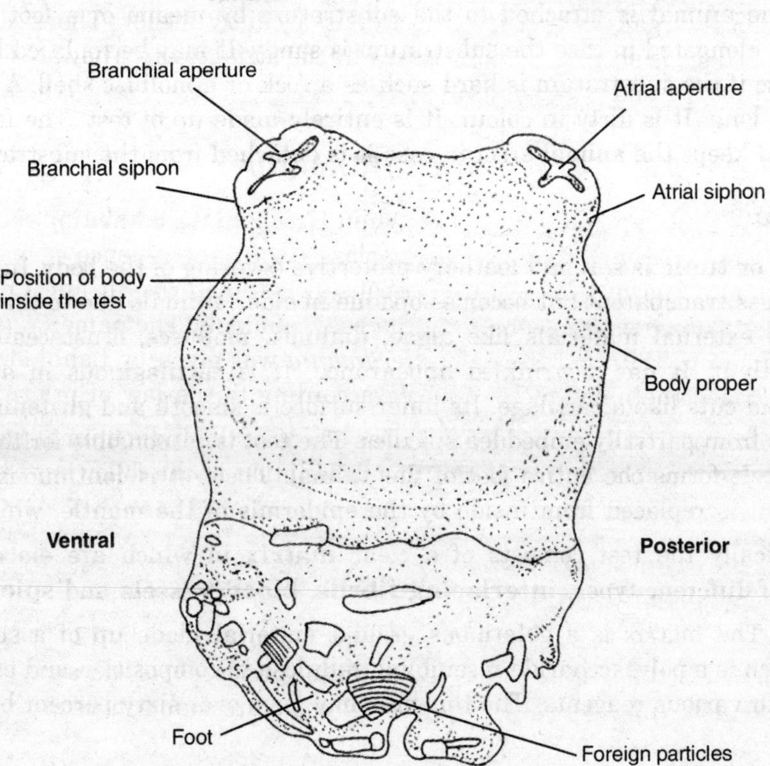

Fig. 4.1 *Herdmania.* External features

laterally compressed and somewhat oblong in shape, with the foot slightly narrower than the free end of the body. The animal looks like a purse or a bag. The **colour** of live animals is pink whereas preserved animals are of yellowish brown colour. The body of *Herdmania* exhibits a peculiar **orientation** due to rotatory changes, which occur during metamorphosis. The two flat surfaces represent the lateral sides. The branchial aperture marks the anterior side whereas the atrial aperture marks the dorsal side. The opposite sides are posterior and ventral respectively (Fig. 4.1).

Division of body

The body is divisible into two parts: (*i*) **body proper** and (*ii*) **foot.**

Body proper: The distal free end of a the animal is called body proper. It bears two openings – **branchial** or **oral** or **incurrent aperture** or **mouth** and **atrial** or **excurrent aperture** at the distal end. The branchial aperture opens on the morphologically **anterior end** and the atrial aperture on the **dorsal end** of the body. Both the apertures are situated on small protuberances of the body called **branchial** and **atrial siphons** respectively. The branchial siphon is about 1cm long and is directed outwards whereas the atrial siphon is about 1.5 cm long and is directed upwards. The branchial aperture is broader than the atrial aperture to allow better inflow of water. Both the apertures are guarded by four lobes or lips formed by the test. When disturbed the animal can eject jets of water with considerable force from the two apertures.

Foot: The animal is attached to the substratum by means of a foot which may be expanded and elongated in case the substratum is sandy. It may be replaced by a broad, flat or concave disc if the substratum is hard such as a rock or a mollusc shell. A fully expanded foot is 3-4 cm long. It is dirty in colour. It is entirely made up of test. The foot also acts as a balancer and keeps the animal erect in case it is detached from the substratum accidently.

Test or tunic

The **test** or **tunic** is soft and leathery protective covering of the body. In young animals it is more or less transparent but becomes opaque in older animals due to thickness and also because some external materials like algae, diatoms, molluscs, crustaceans, etc. become associated with it. It has a wrinkled appearance. It is cartilaginous in appearance and consistency and cuts like a cartilage. Its inner surface is smooth and glistening, interrupted by projections from partially embedded spicules. The test is responsible for the definite form of the animal. It forms the entire foot of the animal. The test is continuously worn out at the surface but is replaced from inside by the epidermis of the mantle, which secretes it.

Histologically the test consists of a clear **matrix** in which are embedded **cells** or **corpuscles** of different types, **interlacing fibrils**, **blood vessels** and **spicules** (Fig. 4.2).

Matrix: The matrix is a gelatinous ground material made up of a substance called **tunicine** which is a polysaccharide resembling cellulose in composition and behaviour under treatment with various reagents. The tunicine may form over sixty percent by weight of the entire test.

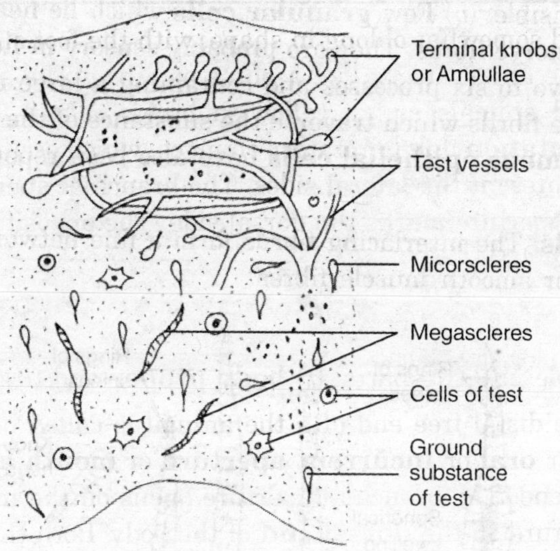

Fig. 4.2 *Herdmania*. V.S. of test

Cells or Corpuscles: The cells present in the matrix have, most probably, migrated into it from the body. These cells are mesodermal in origin. Six different types (Fig. 4.3) of cells have been reported: (*i*) Large spherical or ovoid cells staining easily in eosin, hence called **eosinophilous cells**. They are few in number. (*ii*) Few small **amoeboid cells** without any definite shape (*iii*) Numerous small **eosinophilous cells** with eccentric nucleus scattered throughout. (*iv*) Small spherical **vacuolated cells** filled with small or large number of

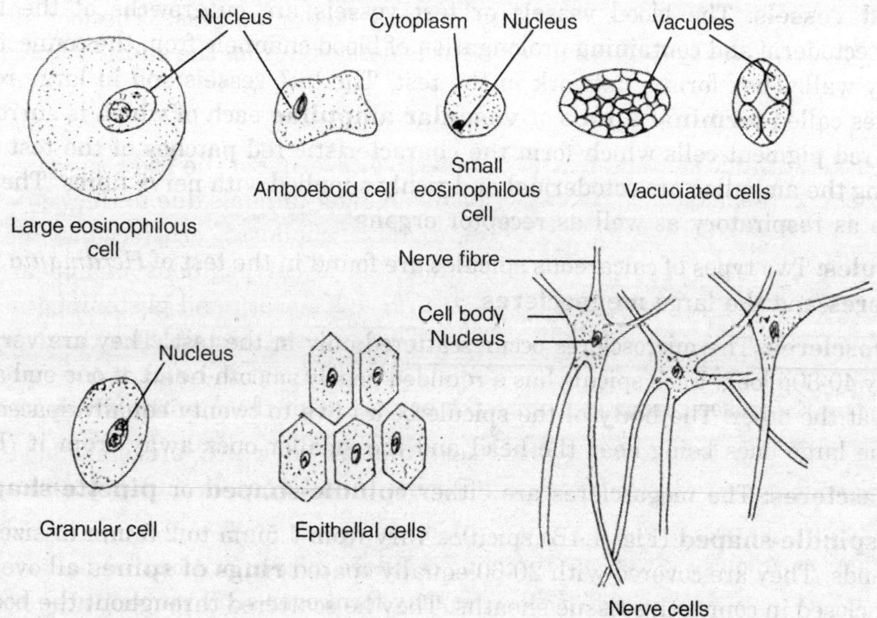

Fig. 4.3 *Herdmania*. Different types of cells found in the test

vacuoles, nucleus is invisible. (*v*) Few **granular cells** which lie near the surface of the test and are surrounded by nerve fibres. They are probably sensory in function. (*vi*) Small **nerve cells** each giving out two to six processes and containing a large nucleus. These processes are continued into nerve fibrils which traverse the substance of the test and connect various nerve cells. Some **squamous epithelial cells** have also been reported to be present in the matrix.

Interlacing fibrils. The interlacing fibrils form a fine network in the test. They may resemble nerve fibres or smooth muscle fibres.

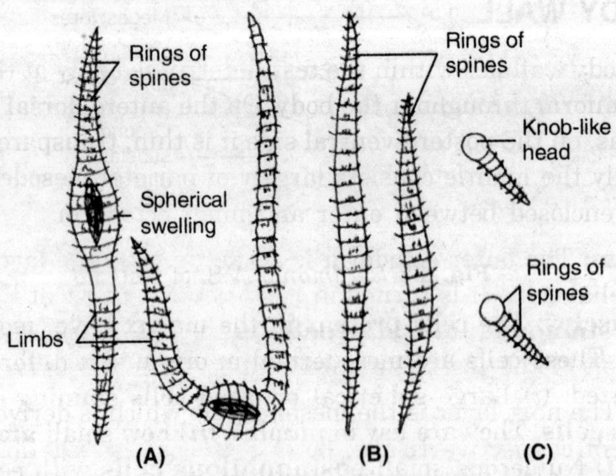

Fig. 4.4 *Herdmania*. Different types of spicules. **(A)** Pipette-shaped megascleres; **(B)** Spindle-shaped megascleres; **(C)** Microscleres

Blood vessels: The blood vessels or test vessels are outgrowths of the body wall covered by ectoderm and containing prolongation of blood channels from the connective tissue of the body wall. They form a network in the test. The test vessels end in large rounded or oval vesicles called **terminal knobs** or **vascular ampullae** each of which is surrounded by a layer of red pigment cells which form the characteristic red patches of the test. The cells surrounding the ampullae are ectodermal and richly supplied with nerve fibres. The ampullae thus serve as respiratory as well as receptor organs.

Spicules: Two types of calcareous spicules are found in the test of *Herdmania*, the small **microscleres** and the large **megascleres**.

Microscleres: The microscleres occur scattered only in the test. They are very small in size, nearly 40-80μ long. Each spicule has a rounded nearly smooth **head** at one end and tapers to a point at the other. The **body** of the spicule bears five to twenty equally spaced **rings of spines**, the large ones being near the head and the smaller ones away from it (Fig. 4.4C).

Megascleres: The megascleres are either **spindle-shaped** or **pipette-shaped**.

The **spindle-shaped** (Fig. 4.4B) spicules vary from 1.5mm to 2.5 mm in size and have tapering ends. They are covered with 20-60 equally spaced **rings of spines** all over the body and are enclosed in connective tissue sheaths. They lie scattered throughout the body in most of the tissues. In the posterior half of the test they are regularly arranged and form a sheath round the vessels traversing the test.

The **pipette-shaped** (Fig. 4.4A) spicules are larger, upto 3.5mm long, having a large spherical swelling in the middle giving it the shape of a pipette. Their ends are pointed. Often the spicules are U– or V–shaped, not straight. They also have equidistant **rings of spines** and are enclosed in connective tissue sheath.

The spicules serve as a rigid internal skeleton. They keep the test firmly attached to the mantle. Small spicules that surround the blood vessels stiffen the walls of the vessel, so that they may not collapse. They also help in driving away the predators.

MANTLE OR BODY WALL

The mantle or body wall lies within the test, interrupted only at the branchial and atrial apertures. It is not uniform throughout the body. On the antero-dorsal side it is thick, opaque and muscular whereas, on the postero-ventral side it is thin, transparent and almost without muscles. Histologically the mantle consists largely of parietal mesoderm with blood sinuses and muscle bundles enclosed between outer and inner ectoderm.

Outer ectoderm: The outer ectoderm is made up of single layer of flattened more or less hexagonal epithelial cells. It is turned in for a short distance at the branchial and atrial apertures as a short **stomodaeum** and **proctodaeum** respectively, lined in each case with a delicate prolongation of the test. The outer ectoderm secretes the test.

Mesenchyme: The next layer is the mesenchyme which is derived from the mesoderm. It is largely formed of connective tissue, both homogenous and fibrous, with cells, blood sinuses, muscle fibres and nerve fibres. The muscles are formed of long, fusiform, non-striped fibres and are restricted to the antero-dorsal half of the body, the postero-ventral part being without muscles. In correlation to the sessile habits the muscles are not well developed except at the branchial and atrial apertures to help squirting of sea-water. Three groups of muscles have been distinguished: (*i*) the circularly disposed **annular muscles** surrounding each siphon, (*ii*) a set of **longitudinal muscles** starting from the atrial and branchial siphon and running upto the middle of the body on each side and (*iii*) **branchio-atrial muscles** running between the two siphons. The circular muscles bring about the contraction of the siphons while the contraction of the body is brought about by the longitudinal muscles. The connective tissue cells are chiefly amoeboid and vacuolated. They migrate into the test through the ectoderm and secrete matrix.

Inner ectoderm: The inner ectoderm or epidermis is made up single layer of more or less flattened polygonal cells. It forms the lining of the atrial cavity.

CAVITIES OF THE BODY

Atrium

The mantle encloses a large cavity, the **atrial** or **peribranchial cavity**, which opens to the exterior dorsally by the **atrial aperture**. Before opening it forms a wide **cloaca**, along the dorsal edge of the body, which leads into the atrial siphon. At the base of the **atrial** siphon there is a strong sphincter and a ring of serrated **atrial tentacles**. At the anterior and ventral region, the wall of the branchial sac is fused with mantle with the result that the atrial cavity is separated into right and left halves which, however, are continuous dorsally and lead into the cloaca. Thus the atrial cavity surrounds the branchial sac with which it communicates by the **secondary gill slits** or **stigmata**. The atrial cavity is lined by the inner ectoderm of the mantle.

COELOM

There is no extensive body cavity lodging the viscera as in the vertebrates. The true coelom formed in the mesoderm is very much suppressed, as a result of over development of the atrium. It is represented in the adult by pericardium and by small cavities in the renal and reproductive organs and ducts. This great reduction of the coelom has been regarded to be a result of degeneration.

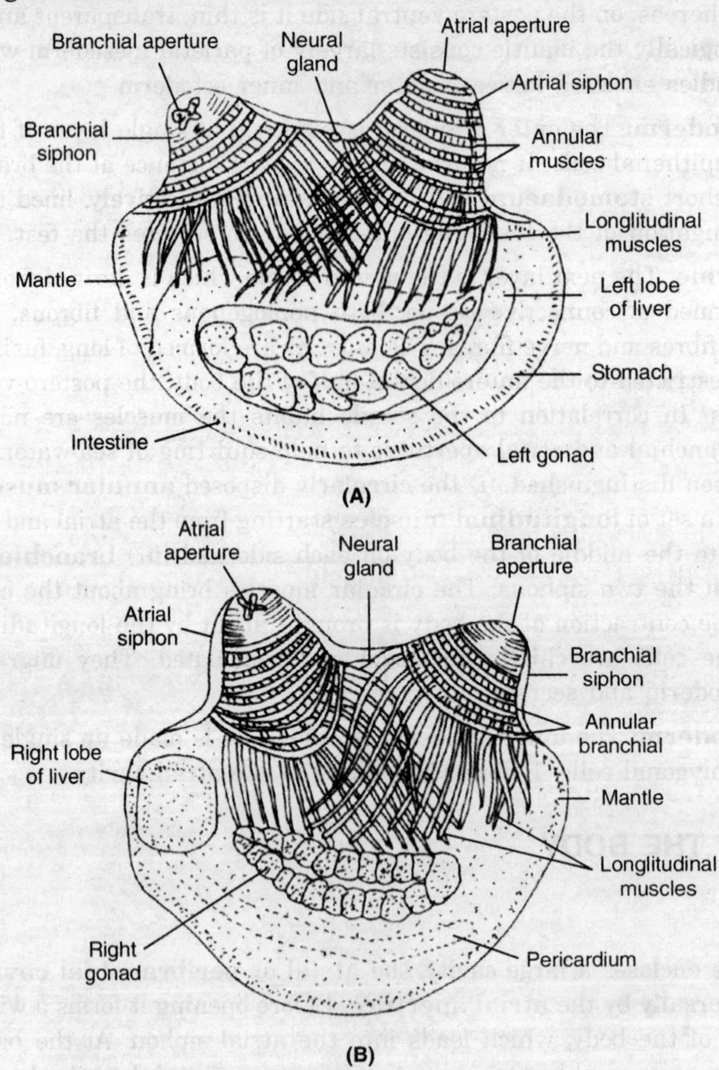

Fig. 4.5 *Herdmania*. Test removed to show the viscera. **(A)** From left side; **(B)** From right side

DIGESTIVE SYSTEM

The digestive system includes the **alimentary canal** and the **digestive glands**.

Alimentary canal

The alimentary canal is a coiled and complete tube consisting of a **mouth**, tubular **buccal cavity**, a large modified **pharynx**, a short **oesophagus**, **stomach**, **intestine**, **rectum** and **anus** (Fig. 4.6).

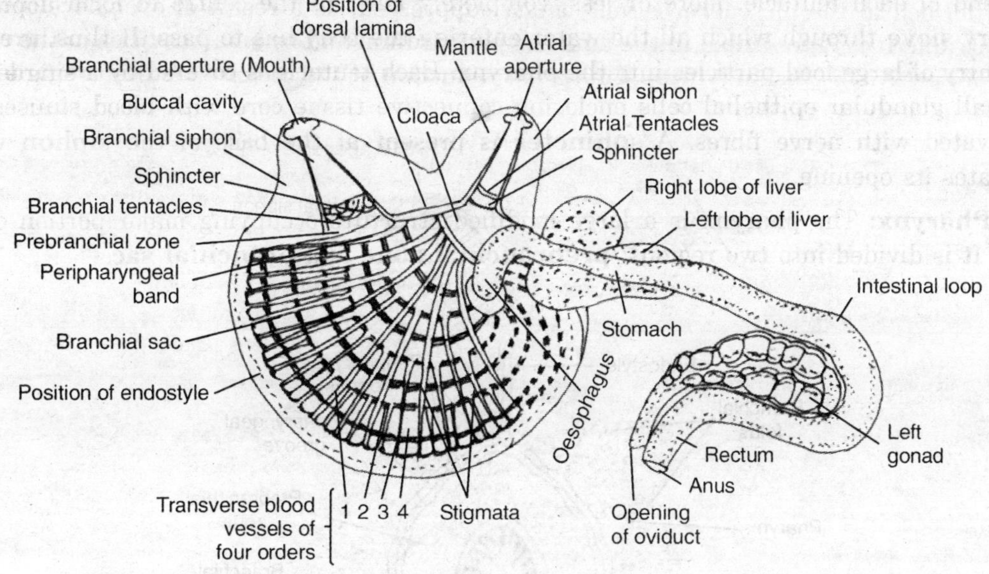

Fig. 4.6 *Herdmania.* Alimentary canal

Mouth: The mouth or the branchial aperture is situated on top of the branchial siphon. It is guarded by four lobes or lips formed by the test.

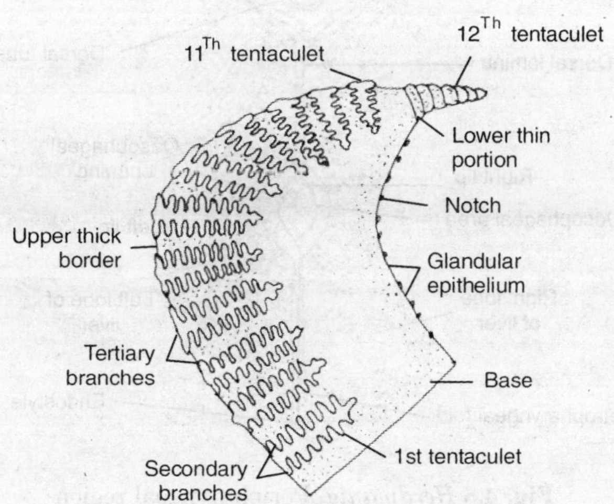

Fig. 4.7 *Herdmania.* Side view of a branchial tentacle

Buccal cavity: The mouth opens into the cavity of the branchial siphon called the **buccal cavity** or **stomodaeum**. It is lined by epidermis. At the base of the siphon is placed a circle of delicate, hair-like branched **tentacles**. There are about **64** tentacles of four sizes: **8 large**, about 5mm in length; **8 medium** about half this size (2.5 mm) alternating with them; **16 small**, about 1.5 mm in length placed alternately between large and medium- sized tentacles and **32 very small**, about 0.5 mm in length.

Each tentacle (Fig. 4.7) is a curved **sickle-shaped** structure bearing a large number of paired lateral branches called **tentaculets**. It is attached to the wall by a broad base. The free end of each tentacle, more or less, completely meet in the centre to form a delicate sensory **sieve** through which all the water entering the body has to pass. It thus, prevents the entry of large food particles into the pharynx. Each tentacle is covered by a single layer of small glandular epithelial cells enclosing connective tissue core with blood sinuses and innervated with nerve fibres. A **sphincter** is present at the base of the **siphon** which regulates its opening.

Pharynx: The pharynx is a large modified structure occupying major portion of the body. It is divided into two regions: **prebranchial zone and branchial sac**.

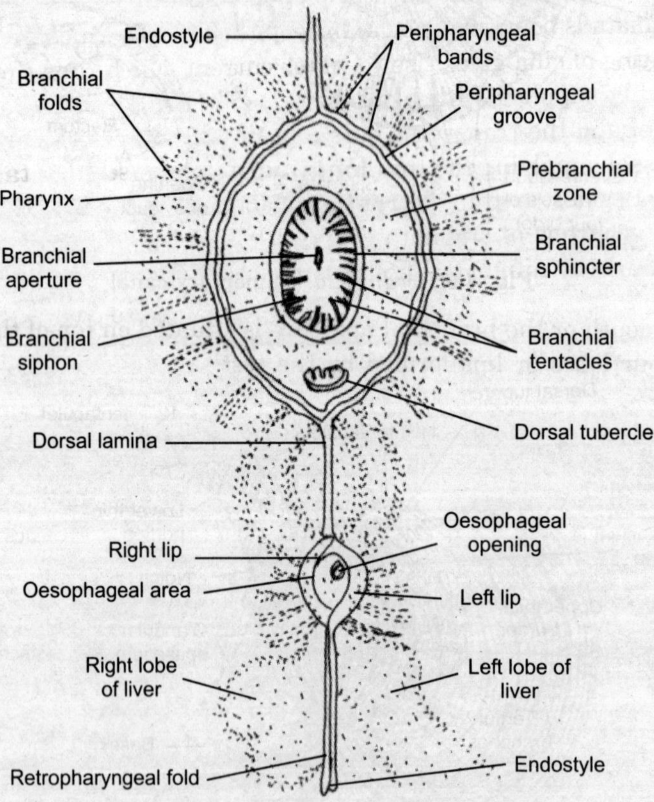

Fig. 4.8 *Herdmania.* Peripharyngeal region

Prebranchial zone: The smaller anterior region of the pharynx is the **prebranchial zone**. It has smooth walls, without folds, cilia and stigmata. It is bounded by a pair of thin parallel closely placed ciliated ridges-the **anterior** and **posterior peripharyngeal bands** separated by a narrow ciliated **peripharyngeal groove** (Fig. 4.8). The cilia of the bands are larger than that of the groove. The anterior band forms a complete ring. In front of it in the mid-dorsal position lies a swollen **dorsal tubercle** made up of two spiral coils. The posterior band is interrupted mid-dorsally by the **dorsal lamina** and mid-ventrally by the **endostyle**.

Branchial sac: The larger posterior region of the pharynx is the **branchial sac** or the **branchial basket** (Fig. 4.9). It is the largest part of the body extending almost up to its posterior end while the rest of the alimentary canal lies on its left side. The branchial sac communicates with the atrial cavity by a large number of apertures called **stigmata or gill slits**. The inner wall of the branchial sac is raised into a number of broad **longitudinal folds**, thus increasing the surface area. The number of folds is generally ten. The wall of the branchial sac consists of a number of **longitudinal** and **transverse bars** at regular intervals, which cross one another at right angles and enclose more or less rectangular areas called **stigmatic areas**: There may be more than 10,000 stigmatic areas on each side and each area may contain upto 20 stigmata. Thus each side of the branchial sac may have upto 2,00,000 stigmata. Each stigmata is bounded by columnar epithelial cells which bear long **lateral cilia** (Fig. 4.11). The beating of cilia gives rise to water current which runs through the body. The branchial sac is richly supplied with blood vessels. The longitudinal bars contain internal **longitudinal vessel** and the transverse bars contain external **transverse vessel**. There is **interstigmatic vessel** supplying the thin **longitudinal intersegmental bars** between two adjacent stigmata. Besides, a thin **intrastigmatic bar** and **intrastigmatic vessel** run transversely across each row of stigmata (Fig. 4.10). The outer wall of the branchial sac is connected to the mantle by number of hollow strands called **trabecullae** each containing a blood vessel.

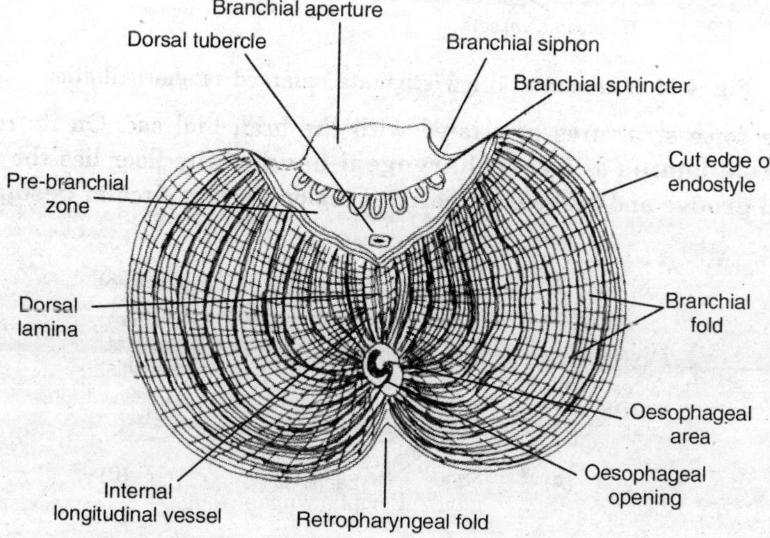

Fig. 4.9 *Herdmania.* L.S. of branchial sac along mid-ventral line showing internal structure

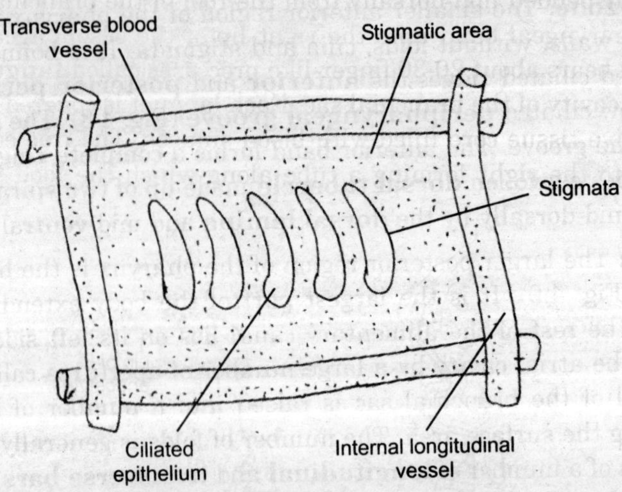

Fig. 4.10 *Herdmania.* **A stigmatic area showing stigmata**

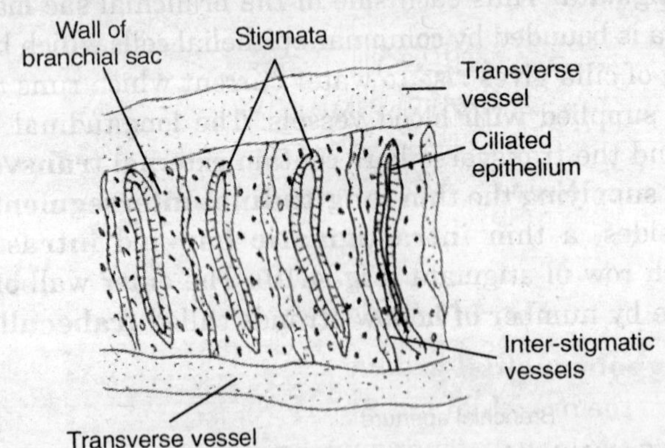

Fig. 4.11 *Herdmania.* **A few stigmata enlarged to show ciliation**

There are some structures associated with the branchial sac. On its roof there is a ridge called **dorsal lamina** or **hyperpharyngeal band,** on its floor lies the **endostyle** or **hypobranchial groove** and in the posteriormost region there is circular **oesophageal area**.

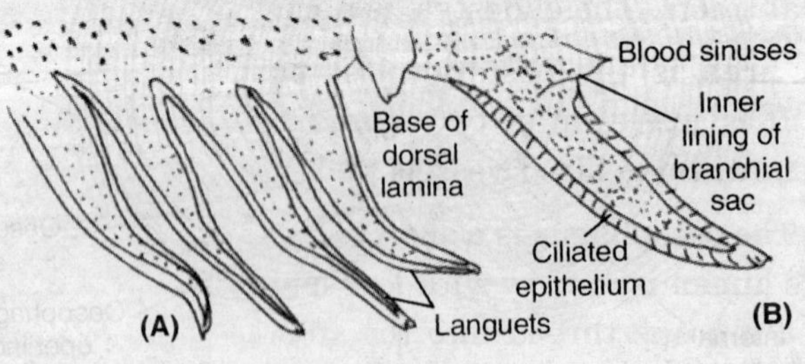

Fig. 4.12 *Herdmania.* **(A) Part of dorsal lamina; (B) One languet in T.S.**

Dorsal lamina or hyperpharyngeal band is a small (about 1.5 cm long), thin, more or less ridged fold suspended mid-dorsally from the roof of the branchial sac. It extends from the posterior peripharyngeal band upto the right lip of the oesophageal area. The free edge of the dorsal lamina bears about 20-30 finger-like processes called **languets** (Fig. 4.12), which hang down into the cavity of the branchial sac. Each languet is covered by ciliated epithelium enclosing a connective tissue core filled with blood sinuses. The free edge of the lamina and languets is curved to the right forming a tube along which the food passes.

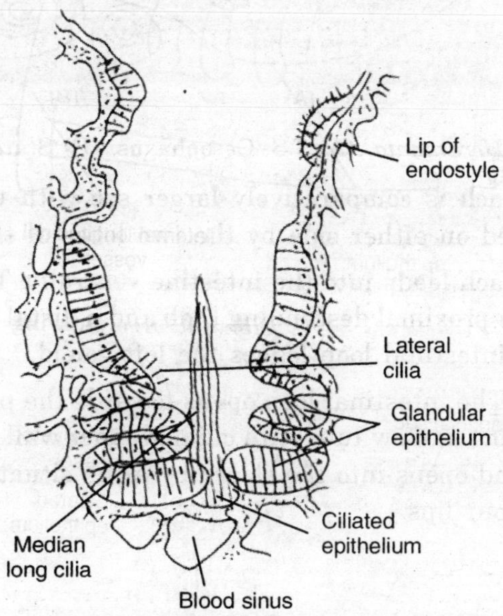

Fig. 4.13 *Herdmania*. T.S. Endostyle

Endostyle or hypobranchial groove is a long, glandular, ciliated canal situated mid-ventrally on the floor of the branchial sac. Anteriorly the endostylar groove is continuous with the peripharyngeal groove and its marginal folds merge with the peripharyngeal bands. Posteriorly the groove terminates before the oesophageal opening and its marginal folds reach the oesophageal opening as **retropharygeal folds** (Fig. 4.8). The sides and the floor of the endostyle are richly ciliated. Structurally, the endostyle consists of **five** glandular ciliated **longitudinal tracts, one median** and **four lateral**. The cilia of the median tract are larger than that of the lateral tract. There are **four mucous secreting tracts** alternating with the longitudinal tracts. The endostyle is homologous with the vertebrate thyroid gland.

Oesophageal area is the posteriormost part of the branchial sac from which the oesophagus arises. It is made up of two **semicircular lips** which guard the oesophageal opening. This area is without blood vessels and stigmata.

Oesophagus: The oesophagus is a short curved, thick-walled tube, which leads ventrally into the stomach. Its lumen is narrow with four longitudinal ciliated grooves (Fig. 4.14) which direct the food-laden mucous thread into the stomach.

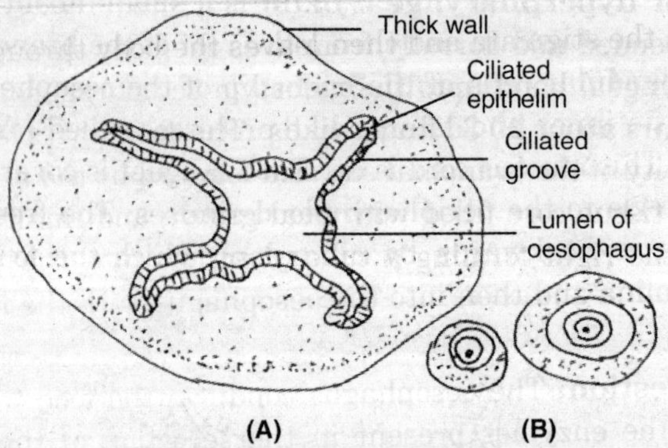

Fig. 4.14 *Herdmania.* **(A)** T.S. Oesophagus; **(B)** Starch granules

Stomach: The stomach is comparatively larger sac with thin wall, which is smooth internally. It is surrounded on either side by the two lobes of the liver.

Intestine: The stomach leads into the intestine ventrally. The intestine is thin-walled U-shaped loop formed by a proximal descending limb and a distal ascending limb both united at the anterior end. The intestinal loop lodges the left gonad.

Rectum and anus: The intestinal loop opens towards the posterior side and leads into the **rectum**, which is a short narrow tube with ciliated inner wall. The rectum curves slightly towards the dorsal side and opens into the cloaca by **anus** situated near the atrial opening. The anus is guarded by four lips.

Digestive glands

Liver: The liver is a bilobed, brown coloured digestive gland lying against the stomach. The left lobe is larger than the right lobe. The liver appears papillated in general appearance because it is made up of a large number of branched tubules called **hepatic tubules** or **hepatic caecae** embedded in a connective tissue matrix containing blood sinuses. The hepatic tubules unite to form about a dozen **hepatic ducts** which open separately into the stomach. The secretion of liver contains protein, carbohydrate and fat digesting enzymes.

Pyloric gland: The pyloric gland is a system of delicate much branched tubules which ramify into the walls of the stomach and intestine. These tubules lead into a number of ducts, which join together at the beginning of the intestine and open into it by a single opening in the middle of the proximal limb of the intestinal loop. The pyloric gland probably functions as the pancreas and also as an excretory organ.

Food and feeding

The **food** of *Herdmania* includes microscopic organisms like algae, diatoms, protozoans etc. *Herdmania* is a **ciliary** or **filter feeder**. The beating of the lateral cilia of the stigmata set up a water current which enters the pharynx through the mouth and buccal cavity. The branchial tentacles act as a sieve and allow only minute food particles to enter the pharynx along with the water current. The larger particles are thrown out of the mouth by a **reverse current**. The food particles are retained in the pharynx while the water current passes into the atrial cavity through the stigmata and then leaves the body through the cloaca and atrial aperture. The ciliated glandular cells of the endostyle secrete mucous which is released in the branchial sac along its inner and lateral walls. The food particles get entangled in the mucous and form a cord. This mucous cord is carried forward by the action of the cilia present on the floor of the endostyle to the peripharyngeal groove and finally to the dorsal lamina. The food cord then passes backwards by ciliary action into the groove formed by the languets of the dorsal lamina and then into the oesophagus with the help of the oesophageal lips.

Physiology of digestion: The oesophagus conducts the food into the stomach where the food is digested by the enzymes present in the secretion of the liver. It contains the enzyme **protease**, which digests proteins, **amylase**, which digests carbohydrates and **lipase**, which digests fats. The pyloric gland serves as an accessory organ of digestion and is pancreatic in nature. Digestion is completed in the intestine. Absorption of digested food also takes place in the intestine. The carbohydrate is stored in the form of starch granules in the liver and to a small extent in the oesophagus, stomach and intestine. The undigested food goes into the rectum where it forms a coiled **faecal cord** with mucous. The faecal cord then goes into the cloaca through the anus. It is then thrown out forcefully through the atrial aperture by sudden contraction of the body.

RESPIRATORY SYSTEM

The branchial sac is the main respiratory organ in *Herdmania*. It has a rich supply of blood vessels where exchange of gases takes place. The water entering the branchial sac by beating of cilia of the stigmata contains dissolved oxygen which diffuses into the blood and carbon dioxide in the blood diffuses into the water. The general external surface of the test also acts as accessory respiratory organ. Exchange of gases takes place in the blood vessels and vascular ampullae present in the test.

CIRCULATORY SYSTEM

The circulatory system of *Herdmania* consists of a tubular **heart, blood vessels** and the circulating fluid, the **blood**.

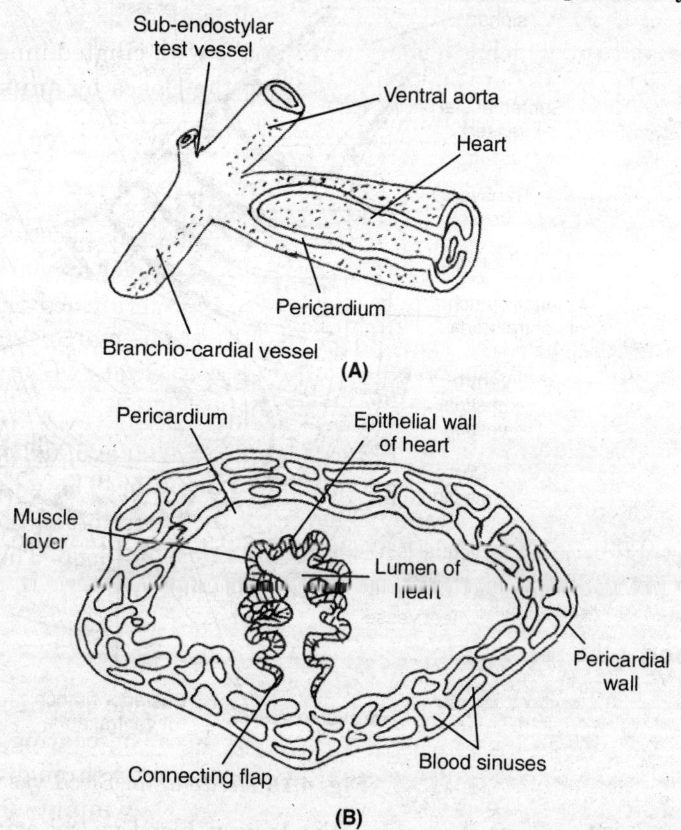

Fig. 4.15 *Herdmania.* **(A)** Wall of pericardium cut open to show the attachment of heart; **(B)** T.S. of pericardium and heart

Heart and pericardium: The heart is a cylindrical contractile tube enclosed within the **pericardium** (Fig. 4.15A). The pericardium is a closed transparent, non-contractile tube about 7cm long and 3mm wide. It is filled with a colourless fluid with corpuscles. The pericardial wall is thick and convoluted made up of connective tissue containing blood sinuses, and lined internally by a single layer of squamous epithelial cells surrounded by a thin layer of striated muscle fibres. The heart is formed by the infolding of the pericardial wall. It is open at both the ends. It is attached to the pericardial wall by a thin mesentery-like flap of connective tissue (Fig. 4.15B). The wall of the heart is thin and made up of a layer of epithelial cells and striated muscle fibres. There are no valves in the heart. A **pear-shaped body** present in the pericardium regulates the flow of blood.

Blood vessels: There are many well-developed blood vessels. The major ones are: (*i*) **ventral aorta**, (*ii*) ***dorsa aorta,*** (*iii*) **brachio-visceral vessel** and (*iv*) **cardio-visceral vessel**.

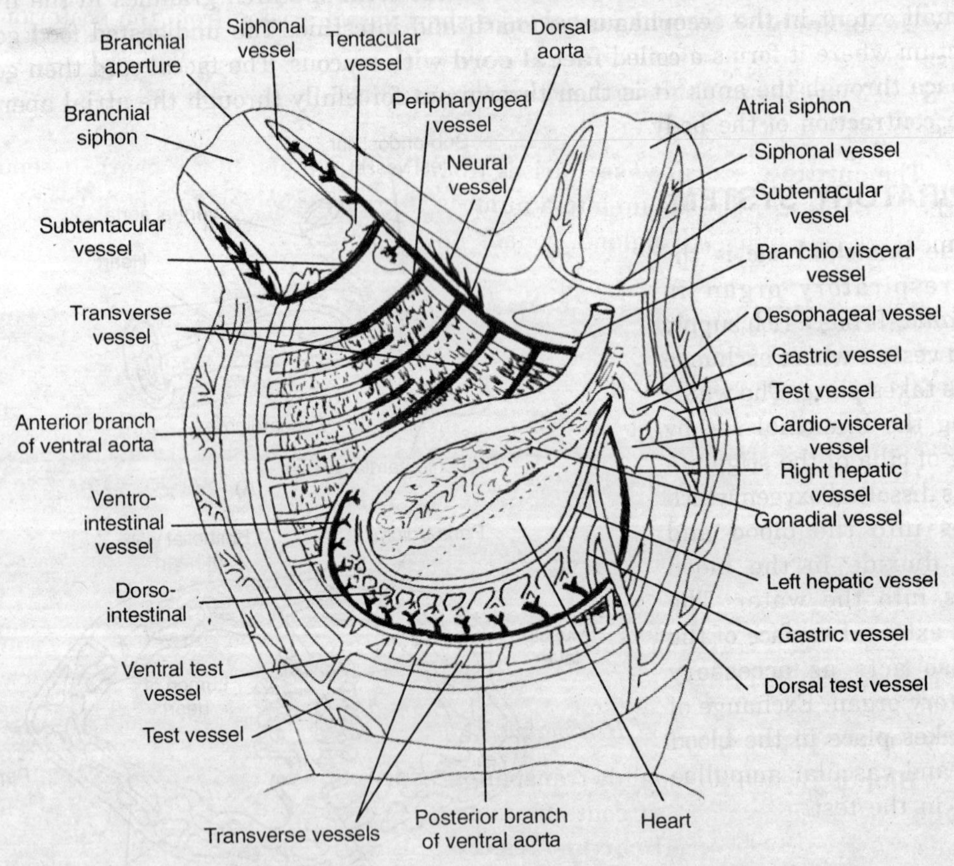

Fig. 4.16 *Herdmania*. Blood vascular system

The **ventral aorta** is the largest blood vessel originating from the ventral end of the heart. At the point of origin it gives out a stout **ventral test vessel** or **subendostylar vessel** which supplies the test. It continues forward and then divides into two branches—the

anterior and **posterior hypobranchial vessels** which run below the endostyle and send fine branches to the endostyle and the mantle. The anterior hypobranchial vessel continues forward as the **peripharyngeal vessel** which runs below the peripharyngeal groove and a **subtentacular vessel** that sends a **tentacular vessel** to each branchial tentacle and 6-8 **siphonal vessels** to the branchial siphon. A fine branch from the posterior hypobranchial vessel supplies the oesophageal area. Throughout its length the ventral aorta gives out paired **transverse** vessels which run between the rows of stigmata on each side of the pharyngeal wall.

The **dorsal aorta** is a stout vessel which runs along the dorsal edge of the branchial sac and lies just above the dorsal lamina. It does not originate from the heart but is connected to it through the ventral aorta. It communicates with the dorsal end of all the transverse vessels of the branchial sac. From its middle portion it sends a branch (**neural vessel**) to the neural gland and from its anterior end it sends a fine branch to the dorsal tubercle and 6-8 siphonal to the branchial siphon.

The **branchio-visceral vessel** is a short-vessel arising from the posterior end of the dorsal aorta. It soon divides into two branches—a short **right oesophageal vessel** supplying the right liver lobe and the right side of the oesophagus and a long stout **ventro-intestinal vessel** supplying the left liver lobe, the left side of the alimentary canal and the left gonad.

The **cardio-visceral vessel** arises from the dorsal end of the heart. It sends a branch to the test and then breaks up into a number of branches which supply different parts of the alimentary canal, digestive gland, gonads and the atrial siphon and tentacles.

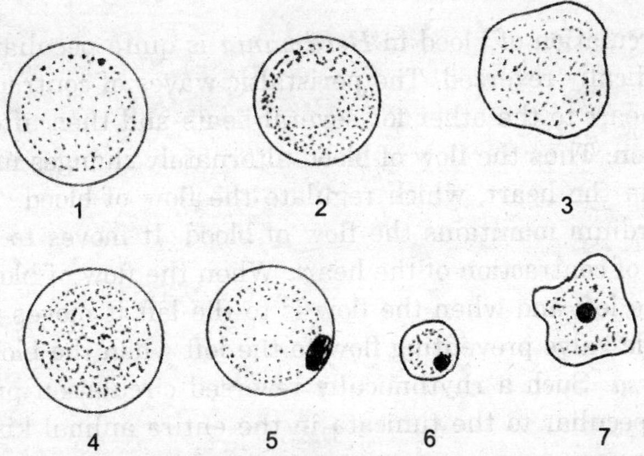

Fig. 4.17 *Herdmania*. Different types of blood corpuscles

Blood: The blood is mainly transparent but usually contain certain pigmented corpuscles giving it a reddish tinge. It contains many colourless **amoeboid cells** or **leucocytes** and about **7** types of **coloured corpuscles** (Fig. 4.17). They include: (*i*) nearly spherical, large, non-nucleated corpuscles containing uniformly granulated orange-coloured protoplasm; (*ii*) non-nucleated spherical corpuscles containing a single large vacuole; (*iii*) nearly spherical non-nucleated yellowish green corpuscles containing 4 or 5 large and small vacuole ; (*iv*) very large non-nucleated corpuscles containing many vacuoles and light yellow pigment granules; (*v*) small eosinophilous corpuscles containing uniformly granulated cytoplasm with brown pigment and a large eccentric nucleus; (*vi*) very small corpuscle with a large central nucleus and cytoplasm with uniformly distributed light brown pigment; (*vii*) a large colourless nucleated amoeboid corpuscle capable of amoeboid movement. There is no clear demarcation between blood and body fluid due to intermixing.

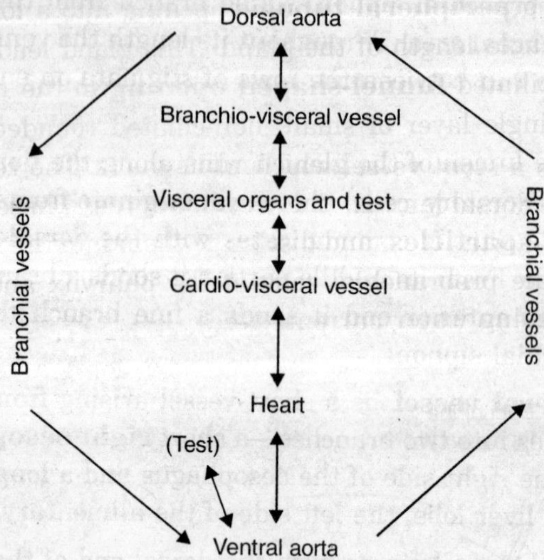

Fig. 4.18 *Herdmania.* Course of blood circulation

Course of circulation

The course of circulation of blood in *Herdmania* is quite peculiar. It is not constant in direction but is periodically reversed. The peristaltic waves of contraction of the heart pass from one end of the heart to the other for several beats and then after a short pause, pass in the opposite direction. Thus the flow of blood alternately changes in direction. This is due to absence of valves in the heart, which regulate the flow of blood. The pear-shaped body present in the pericardium maintains the flow of blood. It moves to and fro in a direction opposite to the waves of contraction of the heart. When the flow of blood is to the right side, this body moves to the left and when the flow is to the left it moves to the right. Probably, it functions as a mobile valve preventing flow to the left when the blood is being pumped to the right and *vice versa*. Such a rhythmically reversed circulation presents an example of **double circulation** peculiar to the tunicata in the entire animal kingdom.

The blood, after being oxygenated in the branchial sac and the test is returned to the heart by the ventral aorta. When the heart beats **ventro-dorsally**, the oxygenated blood is pumped into the cardio-visceral vessel and distributed to various organs, gonads, mantle and test. The deoxygenated blood is collected by the branchio-visceral vessel and sent to dorsal aorta from where it goes to the branchial sac, through transverse vessels, for oxygenation.

When the heart beats **dorso-ventrally** the flow of flood is reversed. The oxygenated blood is collected by the dorsal aorta from the branchial sac and distributed to the visceral organs, gonads, mantle and test by the branchio-visceral vessel. The deoxygenated blood is collected by the cardio-visceral vessel and sent to the ventral aorta from where it goes to the branchial sac and the test for oxygenation.

EXCRETORY SYSTEM

In *Herdmania pallida* the **neural gland** is the only excretory organ (Das 1936). It is oval or elliptical mass (Fig. 4.19) of light brown colour, about 4 mm long and 1 mm wide. It lies embedded in the mantle mid-dorsally, just above the nerve ganglion. The gland consists of a large number of branching **peripheral tubules** opening into a long non-ciliated **central canal** running along the whole length of the gland. This gland leads into a duct that runs anteriorly and opens by a ciliated **funnel-shaped opening** in the middle of the basal part of the dorsal tubercle. A single layer of small, non-ciliated rounded cells containing large nuclei form the lining of the lumen of the gland and its ducts. The rest of the gland consists of large number of small excretory cells, the **nephrocytes**. These cells collect excretory wastes **(xanthine** and **urate particles)** and discharge them into the lumen of the gland from where they pass out into the prebranchial zone of the pharynx and finally carried to the outside along with the outgoing water current.

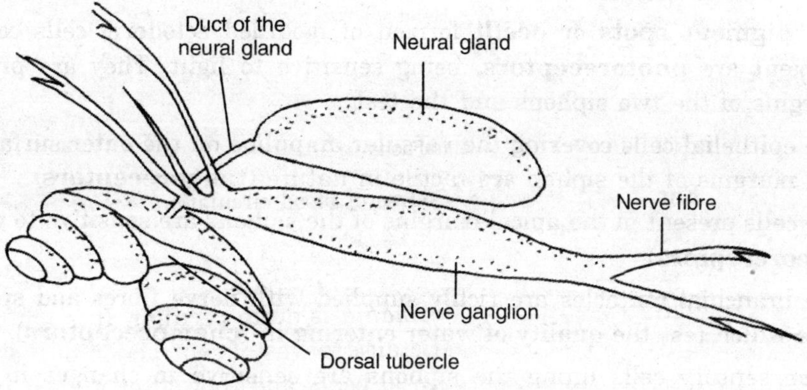

Fig. 4.19 *Herdmania.* Neural complex

Some scientists like Julin (1881), Metcalf (1901), B. Florkin and Das (1956), are of the view that the neural gland is **homologous** to the **vertebrate pituitary gland** because it secretes a hormone, which helps in egg laying, development and metamorphosis.

NERVOUS SYSTEM

The nervous system of the adult tunicate is very poory developed. There is an elongated, about 4 mm long, solid **nerve ganglion** or the **brain** or **cerebral ganglion** (Fig. 4.19). It lies embedded in the mantle in the mid-dorsal line between the two siphons just below the neural gland. It is made up of rounded nerve cells with interlacing nerve fibres in the centre. Nerves arise from both the ends of the ganglion. In *Herdmania* there are five nerves, three at the anterior end, which innervate the branchial siphon and two at the posterior end running around the atrial siphon (Fig. 4.20).

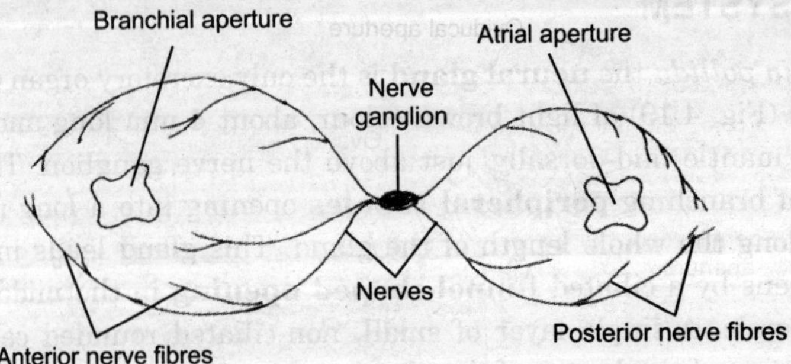

Fig. 4.20 *Herdmania*. Nervous system

SENSE ORGANS

There are no well defined sense organs in the adult tunicate, but sensory cells of various kinds occur in various organs.

(i) The **pigment spots** or **ocelli** formed of modified ectoderm cells containing red pigment are **photoreceptors,** being sensitive to light. They are present on the margins of the two siphons and the test.

(ii) The epithelial cells covering the vascular ampullae on the outer surface of the test and margins of the siphon are tactile in nature (**tangoreceptors**).

(iii) The cells present in the apical margins of the siphons are sensitive to water current (**rheoreceptors**).

(iv) The branchial tentacles are richly supplied with nerve fibres and special sensory cells which test the quality of water entering in (**chemoreceptors**).

(v) Some sensory cells lining the siphons are sensitive to changes in temperature (**thermoreceptors**).

(vi) The **dorsal tubercle** (Fig. 4.19) is also believed to be sensory in nature. It is a small horse-shoe-shaped structure situated in the dorsal part of the prebranchial zone just in front of the dorsal lamina. It consists of two conical spirally coiled lobes resting on a broad base. The cells on the surface of the lobe are ciliated. Each lobe has a narrow open channel running from the base to its tip. Recent view is that the dorsal tubercle is both olfactory and gustatory in function. It tests the smell and taste of water entering the pharynx.

The neural gland, the nerve ganglion and the dorsal tubercle are collectively called the **neural complex** (Fig. 4.19).

REPRODUCTIVE SYSTEM

Herdmania is **hermaphrodite.** The **gonads** are present on both sides of the body embedded in the mantle and bulges out into the prebranchial cavity. The **left gonad** lies in the loop of the intestine and the **right** one occupies a space dorsal to the heart.

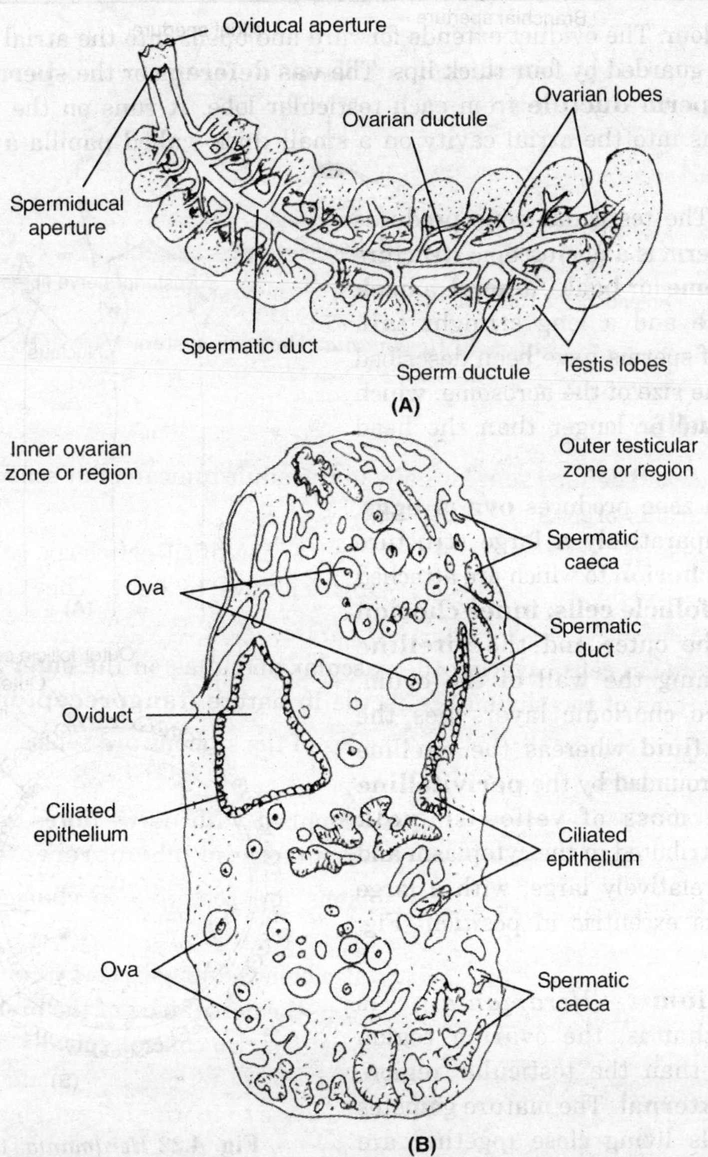

Fig. 4.21 *Herdmania.* **(A)** Gonad; **(B)** T.S. of gonad

Each gonad (Fig. 4.21) is about 3–4 cm long and 1 cm wide. It is a lobulated structure consisting of about **20 lobes** arranged in pairs along a central axis. The posteriormost lobe is unpaired and largest. Each lobe of the gonad is hermaphrodite. It consists of two **zones** – the outer or peripheral zone containing spermatic caeca is the **testicular zone** and the inner zone containing ova in all stages of development is the **ovarian zone**. The two zones can be distinguished by appearance and colouration. The testicular region is large, papillated in appearance and brick red in colour. The ovarian region is small, pink in colour and because of the presence of ova it shows large number of rounded areas. Each gonad has two **gonoducts**. From all ovarian lobes arise short **ovarian ductules**, which pass into a wide **oviduct**, which is also pink in colour. The oviduct extends forward and opens into the atrial cavity by an **oviducal aperture** guarded by four thick lips. The **vas deferens** or the **sperm duct** is narrow and receives a **sperm ductule** from each testicular lobe. It runs on the inner side of the oviduct and opens into the atrial cavity on a small thick-walled papilla a little behind the oviducal aperture.

Sex cells: The testicular zone produces **sperms**. Each sperm is a microscopic structure having an **acrosome** (or **beak**), a **head**, a **neck** or **middle piece** and a long straight **tail**. Different types of sperms have been described on the basis of the size of the acrosome, which may shorter, equal or longer than the head (Fig. 4.22A).

The ovarian zone produces **ova** or **eggs**. An ovum is comparatively a large structure having an **outer chorion** to which are attached vacuolated outer **follicle cells; inner chorion** lying beneath the outer and the **vitelline membrane** forming the wall of the ovum. Between the two chorionic layers lies the **interchorional fluid** whereas the vitelline membrane is surrounded by the **perivitelline fluid**. A dense mass of yellowish **yolk substance** is distributed in the cytoplasm and the **nucleus** is relatively large, with a large **nucleolus** and is excentric in position (Fig. 4.22B).

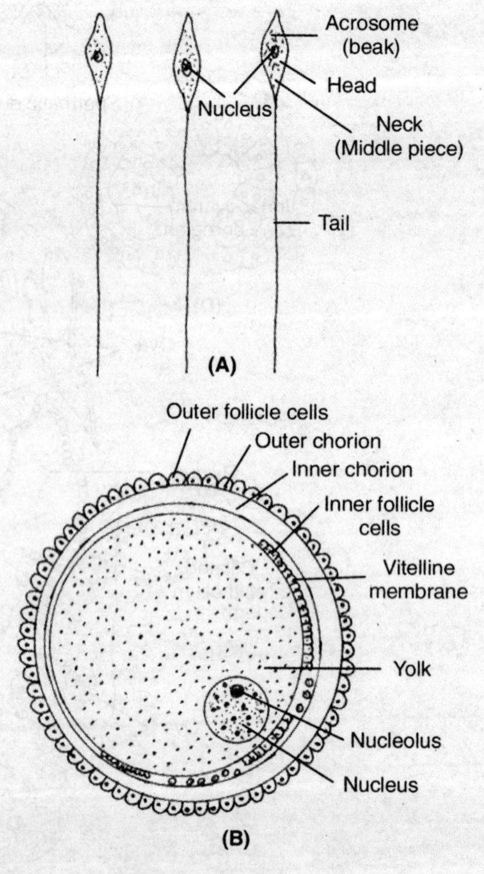

Fig. 4.22 *Herdmania.* **(A)** Sperms; **(B)** Ovum

Fertilization: *Herdmania* is **protogynous**, that is, the ovarian region matures earlier than the testicular region. Fertilization is **external**. The mature gametes from the animals living close together are released in the surrounding water where **cross-fertilization** takes place. Self-fertilization is generally rendered impossible because ova and spermatozoa of an individual mature at different times. The fertilized egg or the zygote shows presumptive areas.

Development

Development in *Herdmania* is **indirect**. The embryo hatches into a free swimming **tadpole larva** which is more advanced than the adult and shows all basic chordate characters. It provides a clue to their inclusion in phylum Chordata. The larva undergoes retrogressive metamorphosis and changes into a degenerate and sedentary adult. This phenomenon is peculiar to the tunicates and is not exhibited by any other chordate.

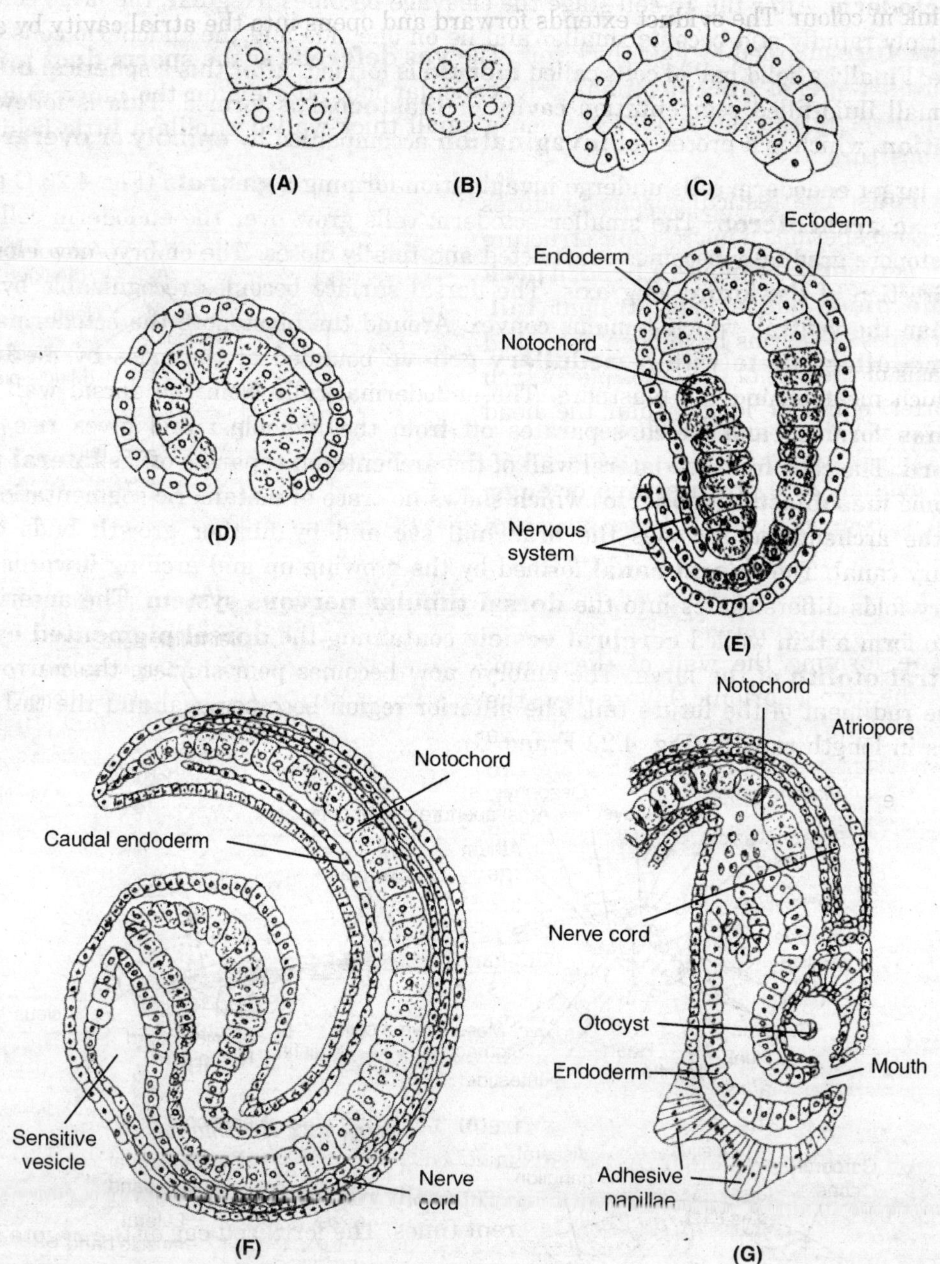

Fig. 4.23 *Herdmania*. Stages of development. **(A–B)** Cleavage; (C) Gastrulation begins; **(D)** Gastrula; **(E)** Advanced embryo; **(F)** and **(G)** Development of tadpole larva; **(F)** Early stage; **(G)** Later stage

Embryology: Cleavage or **segmentation** is **complete (holoblastic)** and **determinate** (*i.e.*, fate of blastomeres is predetermined). It begins about half an hour after fertilization. The first two cleavage are **vertical** followed by a third **horizontal** division (Fig. 4.23 and B). At the eight cell stage four smaller cells are the **endoderm (hypoblast)** and four larger cells are the **ectoderm**. After the 16-cell stage the cleavage becomes irregular, the large ectoderm cells multiply rapidly and become smaller and lie on one side and the endoderm cells on the other side. Finally a solid ball of cells called **morula** is formed. After this a spherical **blastula** with a small fluid-filled segmentation cavity or **blastocoel** is formed. This is followed by **gastrulation,** which is a process of **invagination** accompanied by **epiboly** or **overgrowth**.

The larger endoderm cells undergo invagination forming a **gastrula** (Fig. 4.23 C and D) enclosing an **archenteron**. The smaller ectoderm cells grow over the endoderm cells. The wide blastopore gradually becomes constricted and finally closes. The embryo now elongates in the direction of the future long axis. The dorsal surface becomes recognizable by being flatter than the ventral, which remains convex. Around the blastopore the ectodermal cells form a **medullary plate** with a **medullary groove** bounded at the sides by **medullary folds** which meet behind the blastpore. The endodermal cells from the dorsal wall of the archenteron form a band which separates off from the hypoblast and gives rise to the **notochord**. The cells from the lateral wall of the archenteron separate off as **lateral plates** and become **mesoblast** (**mesoderm**) which shows no trace of metameric segmentation. The rest of the archenteron becomes the branchial sac and by further growth buds off the alimentary canal. The **neural canal** formed by the growing up and arching inwards of the medullary folds differentiates into the **dorsal tubular nervous system**. The anterior end dilates to form a thin walled **cerebral vesicle** containing the **dorsal pigmented eye** and the **ventral otolith** of the larva. The embryo now becomes pear-shaped, the narrow part being the rudiment of the future tail. The anterior region becomes oval and the tail region increases in length rapidly (Fig. 4.23 F and G).

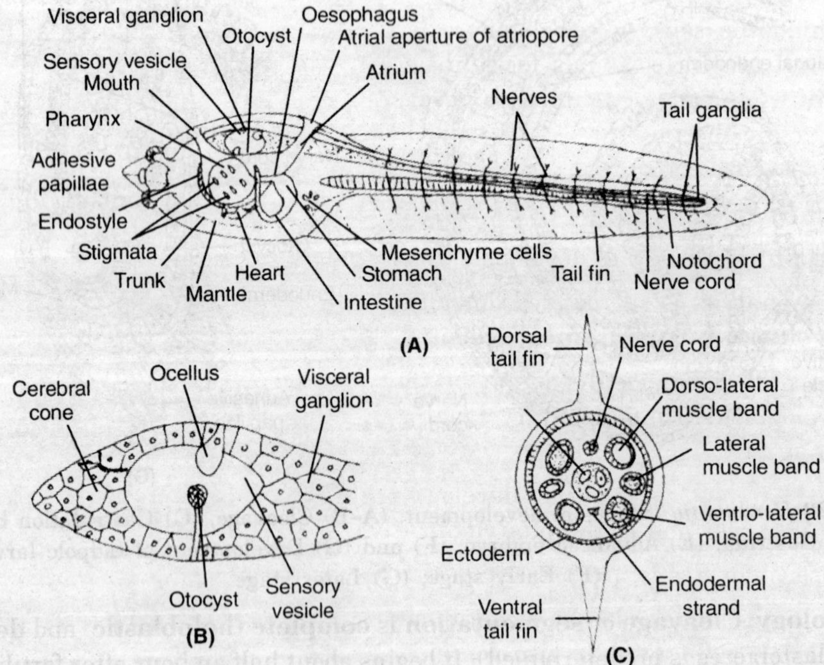

Fig. 4.24 *Herdmania.* **(A)** Tadpole larva (lateral view); **(B)** Sensory vesicle showing otolith and ocellus; **(C)** T.S. of tail of tadpole

The larva

The embryo hatches as a larva called **ascidian tadpole** (because of its resemblance to the tadpole of a frog), which leads a free swimming life for a short time (3 to 4 hours). It is a transient non-feeding stage. The larva is very small in size, about 1.2 mm long and 0.3 mm wide. The body is transparent and is covered by a thin tunic or test secreted by the ectoderm. It is divided into two distinct regions—an oval **head** or **trunk** and a slender tapering **tail** (Fig. 4.24A). This larval stage is the highest in its chordate organization. It is unquestionably a chordate more advanced than *Amphioxus* is some respects.

The notable characters of the larva are:

1. The anterior end of the trunk bears three **adhesive papillae**, one **ventro-median** and two **dorso-lateral**. They help the larva to attach to the substratum during metamorphosis.
2. A hollow oval **sensory vesicle** or **brain** (Fig. 4.24B) is present in the dorsal part of the trunk.
3. The sensory vesicle opens anteriorly into the pharynx near the mouth at a place called **neuropore** probably homologous with that of *Amphioxus.*
4. The sensory vesicle has a single **median eye** or **photoreceptor** derived from the internal wall of the hollow vesicle and possess **retina, pigment layer** and a **lens** or **cornea** (Fig. 4.24B).
5. An organ of equilibrium, the **otocyst** or **otolith** (Fig. 4.24C) is present on the ventral floor of the sensory vesicle.
6. Behind the sensory vesicle there is a solid mass of nerve cells called **visceral ganglion** which extends forward and runs up to the tip of the tail as a **dorsal hollow nerve cord**. It bears several ganglia from which fine nerves arise and innervate the tail.
7. **Notochord** runs below the nerve cord but is restricted to the tail region. It may slightly extend into the trunk region.
8. The **pharynx** is well developed with an **endostyle** and three pairs of **stigmata**.
9. The pharynx is surrounded dorsally and laterally by the **atrium**, which opens to the outside by **atrial aperture** situated mid-dorsally.
10. The post pharyngeal portion of the **alimentary canal** is in early stages of development. The alimentary canal is not functional because the mouth is covered by the test.
11. The heart and pericardium develop as a diverticulum from the stomach just below the posterior end of the pharynx.
12. Mesenchyme cells are present in the posterior part of the trunk.
13. The tail is surrounded by a continuous **caudal fin**, which is marked by **striations** resembling fin rays. It also contains **muscle bands**, which resemble vertebrate myotomes (Fig. 4.24C).

Metamorphosis

The free swimming life of the larva is very short as it is without any arrangement for feeding. It swims for a few hours or so, then becomes sluggish, sinks to the bottom and attaches itself by the anterior adhesive papillae. For sometime it stands erect head down and then enters into a phase of metamorphic changes, in which some already well differentiated tissues are lost and some undergo differential growth. Such a metamorphosis is called **retrogressive** or **retrograde metamorphosis**.

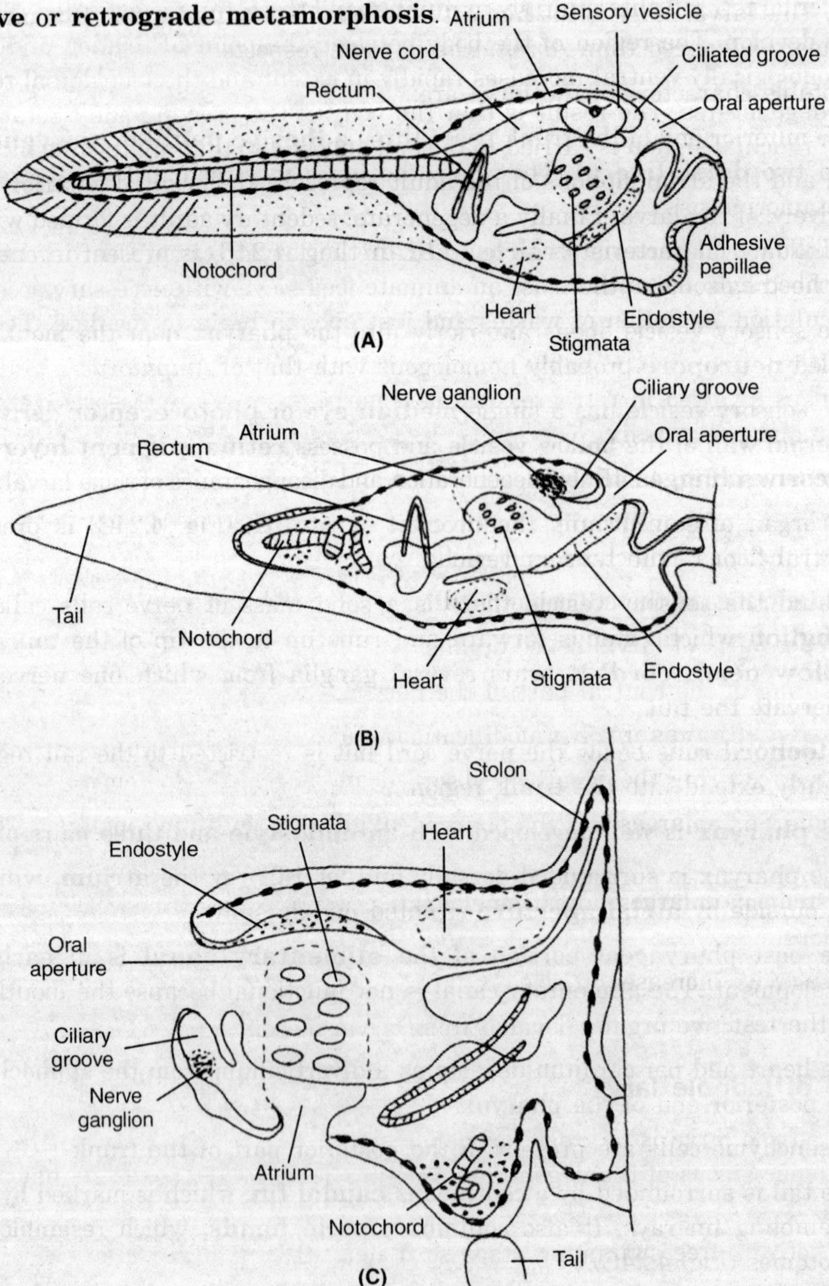

Fig. 4.25 *Herdmania.* Free-swimming tadpole metamorphoses into sessile adult

During metamorphosis very rapid increase of test takes place (Fig. 4.25). The entire tail with its notochord is partly withdrawn and absorbed by phagocytosis and partly it is cast off in shreds. The posterior part of the nerve tube disappears with the tail, and the anterior part, the brain degenerates. The remaining part is reduced to form a small neural ganglion above which the neural gland is formed. With the degeneration of the locomotory, nervous and sensory organs there is a progressive growth of the digestive and reproductive organs. The branchial sac enlarges, gill slits multiply in number and the atrium is extended. The stomach and intestine develop. The region of the body between the point of fixation and the mouth which is morphologically ventral increases rapidly in extent and the true dorsal region stops growing and degenerates. The result is that the body of the growing adult rotates through 180° and the branchial siphon is carried to the end opposite to that of attachment. Therefore, the branchial and the atrial siphons of the adult represent the original anterior and dorsal sides, respectively, of the larva. Finally a degenerate, sedentary adult is formed with little or no trace of chordate characteristics so marked in the larval stage. This degenerate adult has been described by some authors as "an animate food-sieve with necessary accessories for digestion, circulation, excretion of wastes and just enough brain to regulate the activities" (Newman).

The changes involved in retrogressive metamorphosis can be categorised into 2 types: **retrogressive** and **progressive**.

Retrogressive changes involve degeneration and disappearance of some larval structures.

(i) Diminution and eventually complete disappearance of the tail with the contained caudal muscles, notochord and caudal part of the nerve cord.

(ii) Disappearance of the adhesive papillae.

(iii) Disappearance of larval sense organs.

(iv) Dwindling of the central part of the nervous system to a single nerve ganglion.

Progressive changes involve modification of certain larval structures in the adult.

(i) The test covering the mouth is absorbed and the mouth becomes functional.

(ii) The pharynx enlarges and the pharyngeal gill slits multiply rapidly and increase in number.

(iii) The stomach enlarges, intestine elongates and gets curved and the digestive gland develops.

(iv) Atrial cavity increases in size.

(v) The reproductive organs develop from larval mesenchyme cells.

Significance of tadpole larva

1. The tadpole larva is of great taxonomic significance because without the larva the taxonomic status of the degenerate sedentary adult ascidian would have remained doubtful.
2. It is actively free swimming stage so it helps in the dispersal of these sedentary animals.

PHYLOGENY

A typical tunicate is one of the most inanimate being among chordates rivalling only a sponge in inertness. For this reason they were believed to be related to either **sponges** or **Alcyonaria** for a long period. Aristotle was the first to describe a simple ascidian under the name of *Tethyum* whereas a compound ascidian *Botryllus* was described by Schlosser and Ellis in 1756, but for some time their affinities remained obscure. In 1816 Lamarck, chiefly on the basis of anatomical discoveries of Savigny (on compound ascidians in 1816) and Cuvier (on simple ascidians in 1815), instituted the class **Tunicata** but placed it between the Radiata and Vermes.

Affinities with non-chordates

Urochordata shares some characters with a few non-chordates.

1. Its sessile nature with Porifera and some Coelenterata.
2. Its filter feeding mechanism and respiration through water current are similar to that of sponges and molluscs.
3. Larval sensory organs the median eye and otocyst, are found in many non-chordates.
4. Many non-chordates exhibit colonial mode of life like the urochordates.

These features, however, do not exhibit its relationship with the nonchordates. It is only because of similar mode of life.

Affinities with chordata

Kowalevsky's (1866) working on the development of a simple ascidian placed the tunicates in the right place. The tailed larva was discovered previously but without any clear understanding with regard to its position. Kowalevsky showed the striking similarity in structure and development between the larval ascidian and the vertebrate embryo. This discovery clearly indicates that the Tunicata is closely allied to *Amphioxus* and the Vertebrata and the tailed larva represents the primitive or ancestral form from which the adult ascidian has evolved by degeneration (retrogression).

Affinities with cephalochordata

The ascidian tadpole possesses the following features, which are also shared by larval *Amphioxus:*

(i) Notochord present.

(ii) Pharynx well defined with endostyle and three pairs of stigmata.

(iii) Dorsal tubular nerve cord with the anterior end enlarged to form the sensory vesicle or the brain.

(iv) Larval sensory organs, median eye and otocyst present.

(v) Atrium and post pharyngeal digestive tract are in developing stage.

(vi) Tail with dorsal and ventral fins.

Not only the larva but the adults also possess characters that establish relationship between the tunicates and cephalochordates.

(i) Similar food concentrating mechanism and respiratory mechanism.

(ii) Large pharynx with endostyle and other accessories.

(iii) Similar atrial complex.

(iv) Branchial tentacles similar to velar tentacles.

Evidently these characters show a close homology and furnish convincing evidences of common ancestry.

Affinities with vertebrata

The organization of the tadpole larva can be compared with larval fish in having

(i) Dorsal tubular nerve cord.

(ii) Axial stiffening rod- the notochord.

(iii) Pharynx and pharyngeal gill slits.

(iv) Tail covered with caudal fin.

(v) Embryology- the type of cleavage and gastrulation.

The neural gland and endostyle of the adult tunicate is homologous with the vertebrate pituitary and thyroid glands, respectively.

CONCLUSION

During recent years most zoologists have become convinced that the tunicates are primitive and ancestral to the chordates as a whole. Accordingly, the ascidian tadpole larva is regarded as a relic of a past free swimming chordate ancestor and the adult ascidian is regarded as secondarily sessile and degenerate only to the extent that they have become adapted to a fixed habit of life (Wiley 1894). It means that they are not altogether degenerate forms produced by the degeneration of a complex chordate, but simple primitive forms that have lost certain parts, that were not needed as they took to fixed habit of life.

5

CHAPTER

TUNICATA:
GENERAL CHARACTERS AND CLASSIFICATION

GENERAL CHARACTERS

1. The Urochordata are marine animals found practically in all parts of the sea and at all depths.
2. They are solitary or colonial, fixed or free-swimming or pelagic.
3. Size vary from 0.25 to 250 mm; shape and colour also variable.
4. Adults are degenerate with a sac-like unsegmented body.
5. Body is contained in a secreted tunic made up of a cellulose-like substance **tunicine** (hence the name **Tunicata**).
6. Chordate characters present in the larval stage in which the notochord is restricted to the tail, hence the name **Urochordata**.
7. Coelom absent, represented by the atrial cavity which opens to the outside by atrial aperture.
8. Alimentary canal complete. Pharynx or branchial sac large with several pairs of gill slits and endostyle.
9. The urochordates are ciliary feeder.
10. Respiration through branchial sac and test.
11. Circulatory system open, consists of a tubular heart, blood vessels and blood.
12. Nervous system in adults is poorly developed.
13. Generally hermaphrodite. Fertilization external.
14. Development indirect. Larva is called **tadpole** and it shows the basic chordate characters.
15. Metamorphosis is retrogressive.
16. Asexual reproduction by budding.

CLASSIFICATION

Subphylum Urochordata includes about 2,000 sessile species and about 100 pelagic forms. It is divided into 3 classes:

CLASS 1: ASCIDIACEA

1. Sedentary animals usually fixed to a substratum.
2. Solitary, colonial or compound animals.
3. Test well developed and permanent
4. Pharynx large with many persistent gill slits.
5. Hermaphrodite. Reproduction both asexual and sexual.
6. Metamorphosis complete and retrogressive.

Order 1: Enterogona

1. Body may be divided into thorax and abdomen.
2. Neural gland usually ventral to neural ganglion.
3. Gonad single, lying in the intestinal loop, sometimes behind the loop.
4. Asexual reproduction by budding.

Examples: *Ascidia, Ciona, Clavelina.*

Ascidia

Ascidia is commonly called **sea squirt**. It is quite similar to *Herdmania*. It is a solitary, marine sedentary animal commonly found in cold temperate regions, usually inhabiting shallow water, some may occur in deeper waters. The body is 10-15 cm long, sac-like or cylindrical, with a broad base attached to a rocky substratum (Fig. 5.1). The test is thick, translucent, wrinkled and brownish in colour. The branchial and the atrial siphon do not lie at the same level. The branchial aperture is eight-lobed and terminal, while the atrial aperture is six-lobed and situated half way down the right lateral side. The pharynx is large with numerous gill slits but there are no internal longitudinal folds. The intestine has a typhlosole. The liver is absent. Dorsal lamina is a longitudinal ridge without languets. Dorsal tubercle is horse-shoe-shaped. Brain is dorsal to the neural gland. Excretory vesicles lie on

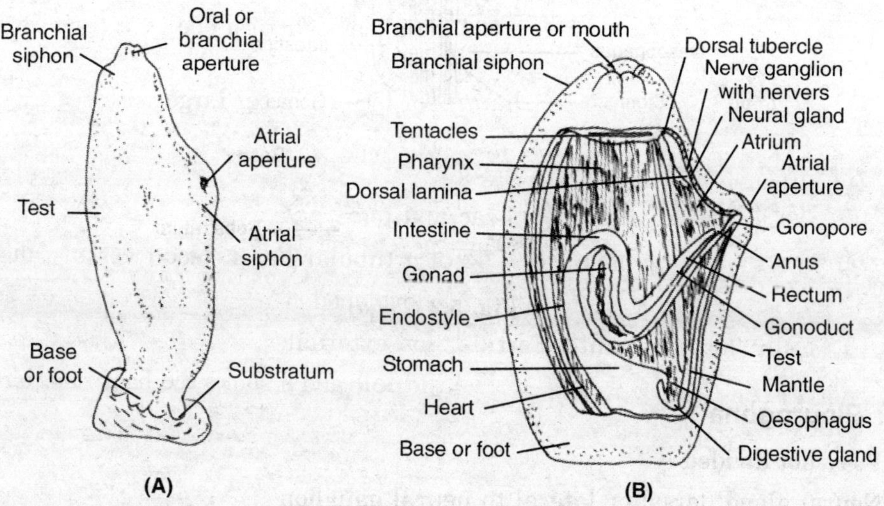

Fig. 5.1 *Ascidica.* **(A)** Entire animal; **(B)** Animal dissected to show viscera

the stomach and intestine. There is a single hermaphrodite gonad in the intestinal loop. Larva is a tadpole and metamorphosis is retrogressive.

Ciona

Ciona, comonly known as **tube sea squirt,** is a solitary, sedentary animal found in cold temperate seas. It occurs in shallow waters attached to rocks by a broad base. The body is cylindrical, about 10-12 cm long. The test is thin, elastic and transparent (Fig. 5.2). The thick longitudinal bands of muscles can be seen through the test. The branchial and atrial siphons are cylindrical, and the former is higher up in position than the latter. The branchial aperture is eight-lobed whereas the atrial aperture is six-lobed. The pharynx is well-developed with numerous gill slits and without internal folds. The dorsal lamina has languets. Dorsal tubercle is horse-shoe-shaped. The brain is dorsal to the neural gland. There is a single hermaphrodite gonad lying in the intestinal loop.

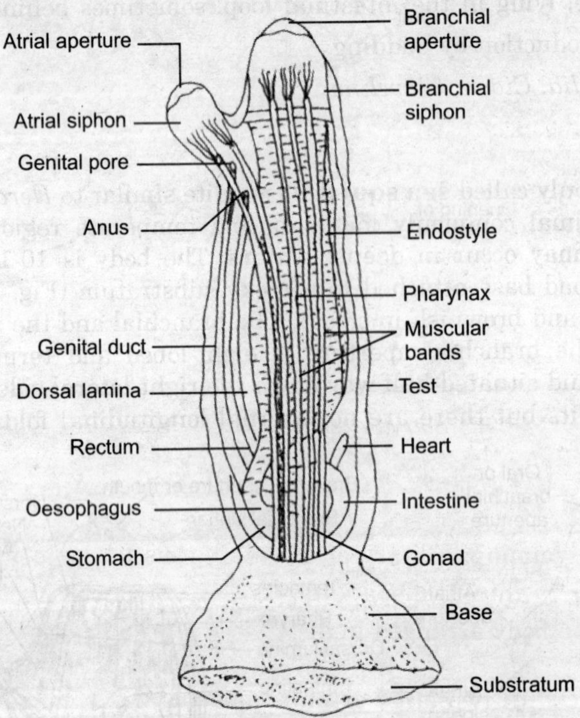

Fig. 5.2 *Ciona*

Order 2: Pleurogona

1. Body not divided
2. Neural gland dorsal or lateral to neural ganglion.
3. Gonads two or more lying embedded in the mantle wall.

Examples: *Herdmania, Botryllus, Molgula.*

Botryllus

Botryllus is a colonial compound ascidian (Fig. 5.3) widely distributed in the Atlantic and the Mediterranean seas. It is found in shallow waters fixed to wharfs, piers and ships. The common test grows as an encrusting gelatinous mass over the surface of rocks, shells or sea weeds. The individuals or zooids are grouped together and so arranged as to form a star or rosette. The branchial apertures or mouths of the zooids are separate and lie at the tip of the rays of the stars whereas, the atrial apertures open into a common cloacal chamber which opens by a single opening at the centre of the star. The structure of each zooid is like that of a simple ascidian. Dorsal lamina is without languets, the gill slits are numerous and rectangular. A pair of hermaphrodite gonad is present. Asexual reproduction is by budding and the new individuals thus produced remain permanently in the colony.

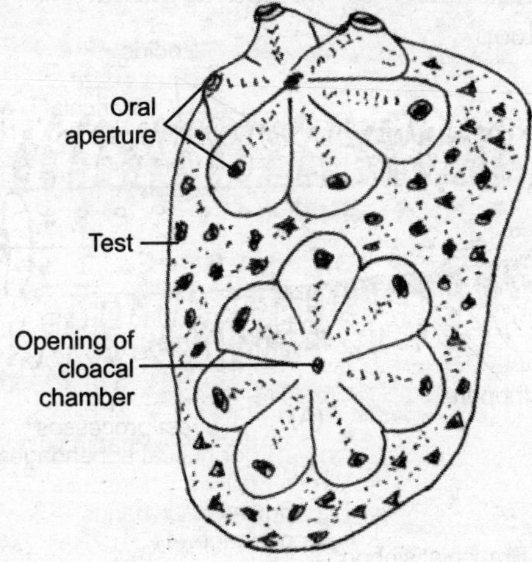

Fig. 5.3 *Botryllus* showing two groups of zooids

CLASS 2: THALIACEA

1. Large free swimming, solitary or colonial forms found in warm temperate seas.
2. Body covered by thin, transparent, gelatinous test.
3. Muscles of the body arranged in distinct circular bands, encircling the body wholly or partially.
4. Branchial sac with two large or numerous small gill slits.
5. Adults without tail, notochord and nerve cord.
6. Hermaphrodite, with or without larval stage.
7. Two distinct phases in life history (alternation of generation). The two phases differ from one another structurally (polymorphism), besides one reproduces sexually and the other asexually.

Order 1: Pyrosomida

1. Free swimming, pelagic, colonial forms with transparent gelatinous test.
2. Colony of zooids compact, tubular, closed at one end, enclosed in a common test such that branchial aperture opens on the outer surface and the atrial aperture on the inner surface.

3. Muscle bands confined to the ends of the body.
4. Colony is phosphorescent due to the presence of symbiotic luminous bacteria.
5. Reproduction by budding.
6. Larval stage absent.

Example: *Pyrosoma.*

Pyrosoma

Pyrosoma presents an interesting example of pelagic colonial form. It inhabits warm parts of Atlantic, Indian and Pacific oceans. The colony is cylindrical or tubular about 100 cm long, closed at one end and open at the other. The colony consists of several individuals called **blastozooids** enclosed in a common gelatinous test (Fig. 5.4). The branchial aperture of each

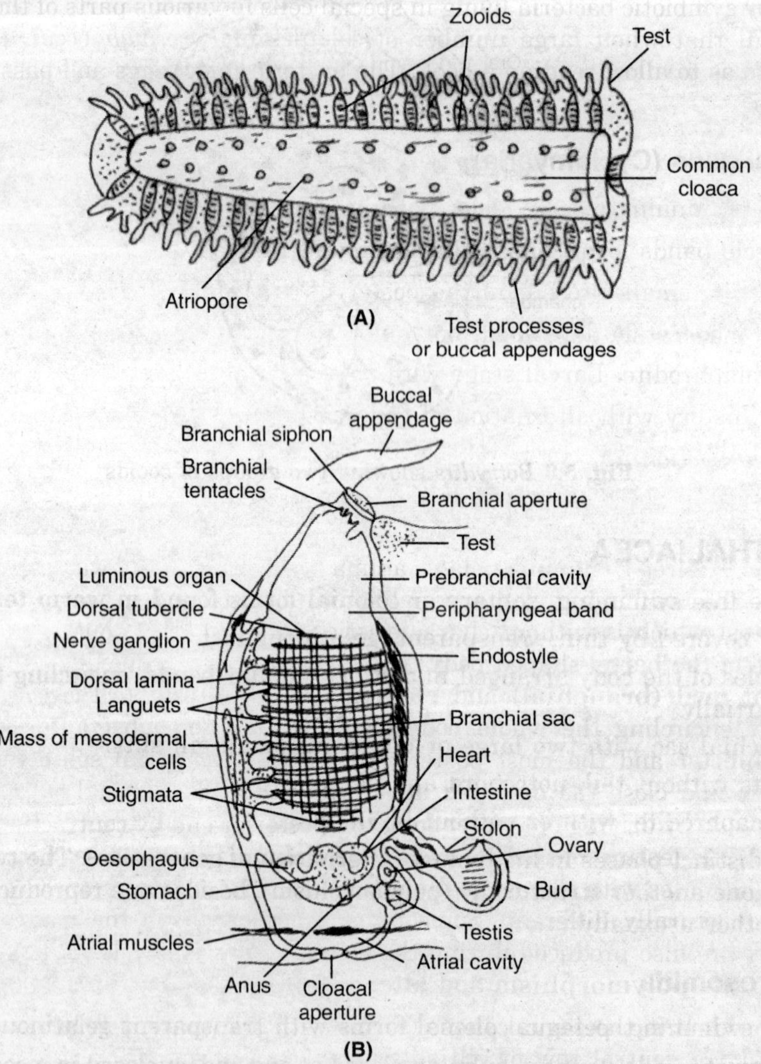

Fig. 5.4 *Pyrosoma.* **(A)** Colony in L.S; **(B)** Single zoid

zooid opens to the outside and is guarded by a tongue-shaped **buccal appendage.** The atrial aperture of each zooid opens into the common cloacal chamber of the colony which opens to the outside by a single cloacal aperture guarded by a **velum** or a **diaphragm**. The velum acts as a regulator of water flow. It also acts as an organ of locomotion which depends on the flow of water through it. All the zooids in the colony are oriented in the same way, with anterior end outward, posterior end inward, dorsal side towards the open end and the ventral side towards the closed end of the cylinder. The general structure of the zooid is like a simple ascidian. The pharynx has numerous gill slits and an endostyle. The dorsal lamina has languets. At the inner end of the pharynx there is an atrium. The heart lies behind the pharynx. There is a **stolon** near the endostyle which gives rise to new blastozooid by **budding.** There is a single hermaphrodite gonad made up of a single yolky egg and a lobed testis. Development is direct and takes place within the body of the parent. There is no larval stage. The outstanding characterstic of *Pyrosoma* is the powerful light which they emit. This is produced by symbiotic bacteria living in special cells in various parts of the body. This light is so powerful that when large number of colonies of *Pyrosoma* occur, the whole sea is illuminated so as to allow reading a book. The bacteria infect eggs and pass from generation to generation.

Order 2: Doliolida (Cyclomyaria)

1. Free swimming pelagic forms with barrel-shaped body.
2. Muscle bands form complete ring around the body.
3. Gill slits small, present in the posterior part of the body.
4. Test moderately developed, not much thickened.
5. Hermaphrodite. Larval stage with a tail and notochord.
6. Life history with alternation of generation.

Example: *Doliolum.*

Doliolum

Doliolum is a free swimming pelagic animal inhabiting warm water in tropical and sub-tropical seas. The body is barrel-shaped, about 1 to 2 cm long, covered by a thin, tough and transparent test which is without characteristic test cells. The mouth and atriopore are at opposite ends of the barrel-shaped body. The open ends are fringed with lobes or tentacles and represent **oral (branchial)** and **cloacal (atrial) funnels.** There are eight rings of muscle bands encircling the whole body (cyclomyaria), the most anterior one forms the branchial sphincter and the most posterior one forms the atrial sphincter. The sphincter serves to open and close the oral and atrial apertures. The animal moves forward by jet propulsion caused by driving out water from the posterior end by contraction of muscles. The most characterstic feature of the life cycle of *Doliolum* is that the sexually mature solitary **gonozooid** alternates with the asexually reproducing gregaria phase, the **oozooid.** The two phases are structurally different. Some other stages such as the **phorozooids** and the **gastrozooids** are also produced during the course of the life cycle. Thus the life history of *Doliolum* exhibits **polymorphism** and **alternation of generation.**

Gonozooid: It is the sexual phase (Fig. 5.5A). The pharynx is large and is perforated by a dorsal and a ventral row of gill slits. There are about 60 or more gill slits and are generally regarded as primary gill slits. A distinct endostyle is present on the ventral edge of the branchial sac. The dorsal tubercle is present but the dorsal lamina is absent. The dorsal nerve ganglion and neural gland are present behind the third muscle band. *Doliolum* is hermaphrodite. Both the ovary and testis lie to the left of the middle line alongside the alimentary canal and their ducts open in the atrial cavity. Fertilization is external. The zygote develops into a tailed larva enclosed in a perivitelline envelope which acts as a float during development. The tail contains chordal cells. The tail has no connection with the anterior region where neural tissue develops. It possesses no sense organs. Metamorphosis is represented by absorption of the tail. Soon after metamorphosis the oozoid is formed.

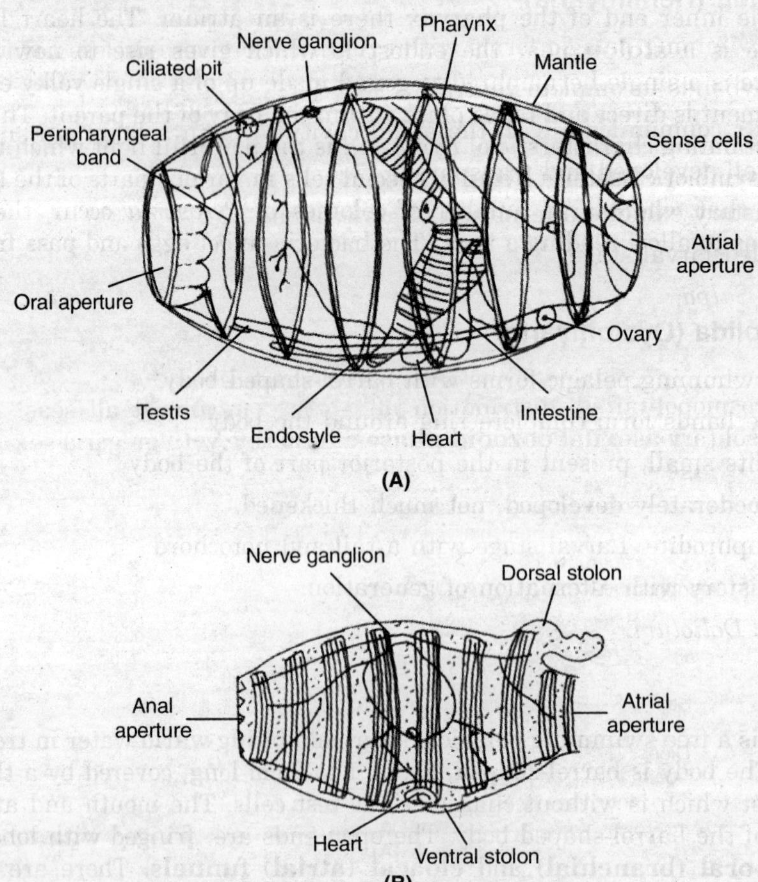

Fig. 5.5 *Doliolum.* **(A)** Gonozooid; **(B)** Oozooid

Oozooid: It represent the asexual phase (Fig. 5.5B). It differs in some respect from the gonozooid. The muscle bands are 9 in number. The gill slits are fewer and confined to the posterior part of the branchial sac. The most characterstic feature of the oozooid is the presence of a ventrally placed **stolon (rosette organ)** from which buds arise by constriction and a **dorsal appendix (cadophore)** near the posterior end of the body.

Bud formation takes place by successive constrictions from the distal end of the stolon. The buds or **probuds** migrate along the surface of the parent and arrange in three lines along the dorsal appendix. The migration of buds is possible by certain amoeboid test cells which become attached to them. Being polymorphic the buds develop into three types of individuals of which two types do not develop reproductive organs and only one becomes sexually mature forming gonozooids. The lateral buds form nutritive individuals called **trophozooids** or **gastrozooids** which remain attached to the parent and help in respiration and nutrition. The buds in the median row develop into **phorozooids.** They are sterile, have doliolum-like appearance and detach from the parent colony. They carry very young buds on their ventral outgrowth which are destined to grow into **gonozooids**. They have been called foster forms because they carry these young gonozooids which are released when fully mature.

Order 3: Salpida (Hemimyaria)

1. Free swimming pelagic forms with cylindrical or more or less fusiform body.
2. Muscle bands incomplete ventrally.
3. Pharynx communicate with the atrial cavity through a large gill slit.
4. Test well developed and transparent.
5. Life history exhibits alternation of generation.
6. No tailed larval stage.

Example: *Salpa*.

Salpa

Salpa is cosmopolitan in distribution and occurs in almost all seas. In contrast to *Doliolum* it has solitary **asexual oozooid phase** alternating with gregaria **sexual gonozooid** or **blastozooid phase**.

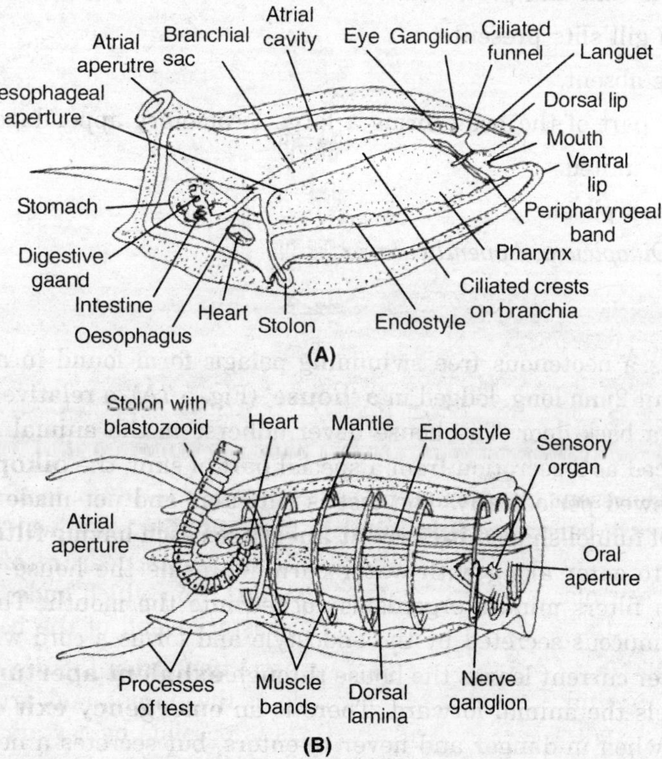

Fig. 5.6 *Salpa*. (A) Sexual blastozooid; (B) Oozooid

Oozooid: It is the solitary asexual phase (Fig. 5.6B) quadrangular in shape with branchial (mouth) and atrial (anus) apertures at opposite ends. It is bilaterally symmetrical. The test is thin and transparent. There are nine muscle bands of which the anterior most and the posteriormost bands are complete and form sphincters and the remaining seven are incomplete ventrally. The mouth is guarded by dorsal and ventral lips. It leads into the pharynx which communicates with the atrium by a single large gill slit. The endostyle lies on the ventral side of the pharynx. The dorsal lamina forms a ciliated gill bar. The oesophagus, stomach, intestine and pyloric gland form a dark compact mass called **nucleus.** A single ocellus is present above the nerve ganglion. The heart lies mid-ventrally behind the endostyle. A process called **stolon** arises between the heart and the endostyle. It elongates, projects near the atriopore and forms a chain of buds which break off in groups to form sexual blastozooids or gonozooids.

Blastozooids: It is the sexual gregaria phase (Fig. 5.6A). It differs from the asexual phase in being smaller and asymmetrical, having fewer muscle bands, in having gonads and there is no stolon. There is one testis and an ovary with a single ovum. *Salpa* is **viviparous.** Fertilization and development are internal. There is no larval stage. Embryo is nourished by a placenta and finally grows into asexual oozooid with stolon which escapes into the surrounding water.

CLASS 3: LARVACEA (APPENDICULARIA)

1. Free swimming pelagic forms retaining larval characters in the adult stage.
2. The test is thin and periodically cast off (not permanent).
3. Only two gill slits present.
4. Atrium is absent.
5. Posterior part of the body forms a large locomotory appendage, the tail.
6. Sexes are united.
7. No metamorphosis.

Examples: *Oikopleura, Appendicularia.*

Oikopleura

Oikopleura is a neotenous free swimming pelagic form found in almost all seas. The body is small, about 2mm long, lodged in a **'house'** (Fig. 5.7A), a relatively jelly-like dwelling with a front and a back door. The house never adheres to the animal. The house building material is produced as a secretion from a special part of skin, the **oikoplastic epitheluim**. The house is renewed periodically. The test is cuticular and not made up of tunicine. The house has a pair of funnel-shaped **incurrent apertures** each having **filters** which allow only minute particles to enter along with water current. Inside the house there is a **filtering apparatus** which filters minute organisms sucked into the mouth. The food particles are entangled in the mucous secreted by the endostyle and forms a cord which passes into the stomach. The water current leaves the house through **exhalent aperture** with a valve. This jet of water propels the animal forward. There is an **emergency exit door** through which the animal exits when in danger and never re-enters, but secretes a new house. The house is thus protective, hydrostatic and a filtering apparatus.

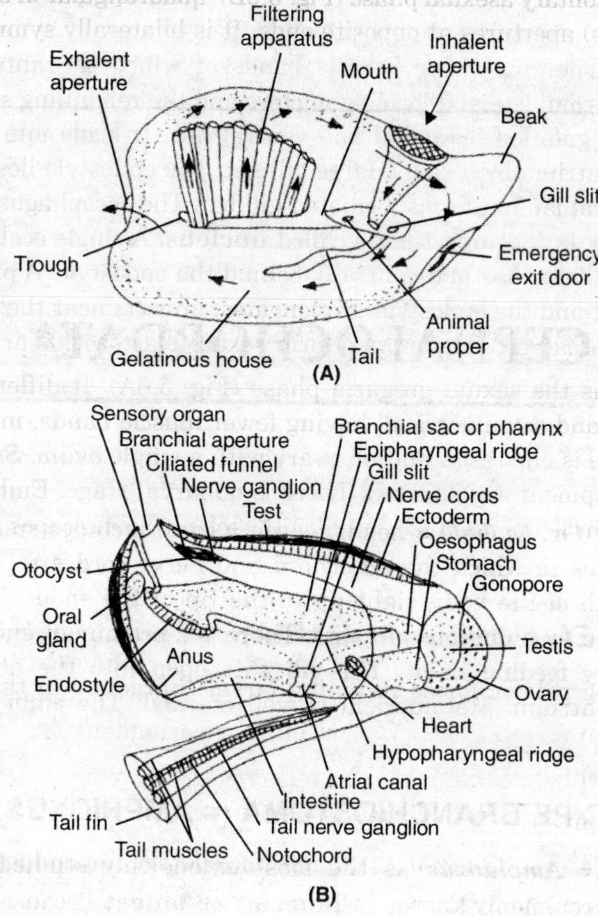

Fig. 5.7 *Oikopleura*. **(A)** Inside test; **(B)** Detailed structure

The **body proper** consists of a stout rounded **trunk** and a laterally compressed **tail**. The mouth is anterior and it leads into a large pharynx with a short endostyle, two peripharyngeal bands and a pair of ciliated gill slits which open directly to the outside as there is no atrium. The alimentary canal is V-shaped and the anus opens antero-ventrally. A cerebral ganglion and an otocyst lie near the mouth, neural gland is absent, heart is present below the stomach and a hermaphrodite gonad lies at the posterior end of the trunk. The animal is protandrous. A nerve cord arises from the cerebral ganglion and runs upto the tip of the tail. The tail is narrow laterally compressed, vibratile and surrounded by a continuous tail fin. Besides nerve cord the tail has two rows of muscles. It is the chief locomotory organ. In the adult the tail is bent underneath the trunk. The larva does not undergo metamorphosis but becomes sexually mature by developing gonads (neotenous).

6

CHAPTER

CEPHALOCHORDATA

Cephalochordata (**Gr**. *kephale* = head; *chorde* = cord) includec small, marine, and fish-like acraniate chordates in which the notochord and nerve cord are persistent and extend along the entire length of the body, right up to the tip of the snout. The pharynx is large and sac-like, perforated by numerous gill slits. There is a prominent endostyle and a definite ciliary tract for ciliary feeding habit. The gill slits open into the atruim. The coelom is reduced due to large atruim. Metamerism is well marked. The embryonic development is quite complicated.

TYPE BRANCHIOSTOMA (= AMPHIOXUS)

Branchiostoma (= *Amphioxus*) is the most extensively studied type of subphylum Cephalochordata. It is commonly known as '**lancelet**' or '**lancet**' because of the sharp, pointed, lance-like ends of the body. It was first discovered by P.S. Pallas in 1778 who took it to be a slug (Mollusca) and naturally gave the name *Limax lancoelatus*. It was in 1834 that O. G. Costa recognized its true position in the animal kingdom and named it *Branchiostoma lanceolatus*. Yarrel in 1836 named it *Amphioxus lanceolatus*. But according to the law of priority the generic name *Branchiostoma* is retained and the name *Amphioxus* which is more famous and familiar to zoologists is retained as the common name for the cephalochordates in general.

Amphioxus (Fig. 6.1A) is of great theoretical interest. In the constitution of its digestive, respiratory, vascular, excretory, skeletal, nervous and muscular systems, and in its manner of development from the egg, it exhibits what appears to be a primordial condition of vertebrate organization, a condition which is in fact, partly recapitulated in the course of the embryonic development of craniate vertebrates. The absence of paired organs of special sense, olfactory, optic and auditory, which are characteristic of higher vertebrates suggest that the animal is degenerate rather than primitive. This is a puzzle that surrounds *Amphioxus*. It is true, however, that a definite gap separates this animal from the lowest fishes in many ways.

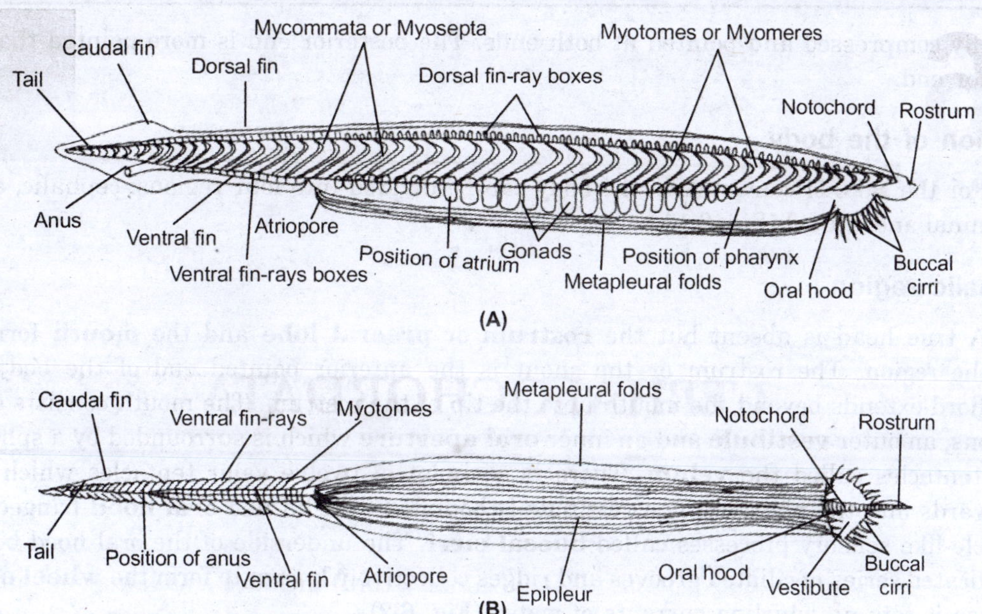

Fig. 6.1 *Branchiostoma.* Entire animal. **(A)** Lateral view; **(B)** Ventral view

DISTRIBUTION

Branchiostoma is cosmopolitian in distribution but more common in warm seas like the Mediterranean. It is abundant near the coast of China and Japan where it serves as food. About 9 species are known of which *B. indicum, B. pelagieum* and *B. ceribbaeum* are found in Indian sea.

HABIT AND HABITAT

Branchiostoma is found in the shore water and on the sandy beaches of the temperate subtropical and tropical portions of the world. It favours littoral habitat and rarely descends below 50 fathom line. *Branchiostoma* is essentially a burrowing animal. It burrows rapidly, head first in the sand by means of vibratory action of the entire body, but comes to rest with anterior end exposed to the water. It rests with head protruding and mouth wide agape. It swims in water vertically by means of lateral strokes of the posterior portion of the body. When disturbed it jumps out of its burrow, swims for a short distance, dives into the sand with head downwards, takes a U-turn so that the anterior end protrudes above the sand. *Branchiostoma* is a ciliary feeder. A respiratory-cum-food current enters the mouth and passes out through the atriopore. The food consists of planktonic microorganisms. The animal leaves the burrow particularly at night and during breeding season. Sexes are separate. Fertilization is external. Development is indirect.

EXTERNAL MORPHOLOGY

Shape, size and colour

Branchiostoma lanceolatus is about 2-3 inches long, fish-like creature, semi-transparent in appearance and showing iridescent play of colour. The body is narrow, stream-lined, laterally compressed and pointed at both ends. The posterior end is more pointed than the anterior end.

Division of the body

For the sake of convenience the body can be divided into four regions: cephalic, atrial, abdominal and caudal (Fig. 6.1A).

Cephalic region

A true head is absent but the **rostrum** or **preoral lobe** and the **mouth** form the cephalic region. The rostrum or the snout is the anterior pointed end of the body. The notochord extends beyond the mouth upto the tip of the rostrum. The mouth consists of two portions, an outer **vestibule** and an inner **oral aperture** which is surrounded by a sphincter with tentacles called the **velum**. There is a circlet of twelve velar tentacles which hang backwards into the pharynx. The vestibule is bounded by a lid-like **oral hood** fringed with tentacle-like sensory processes called **buccal cirri**. The underside of the oral hood bears a complicated series of ciliated grooves and ridges collectively known to form the **wheel organ**, because it sets up whirling currents of water (Fig. 6.2).

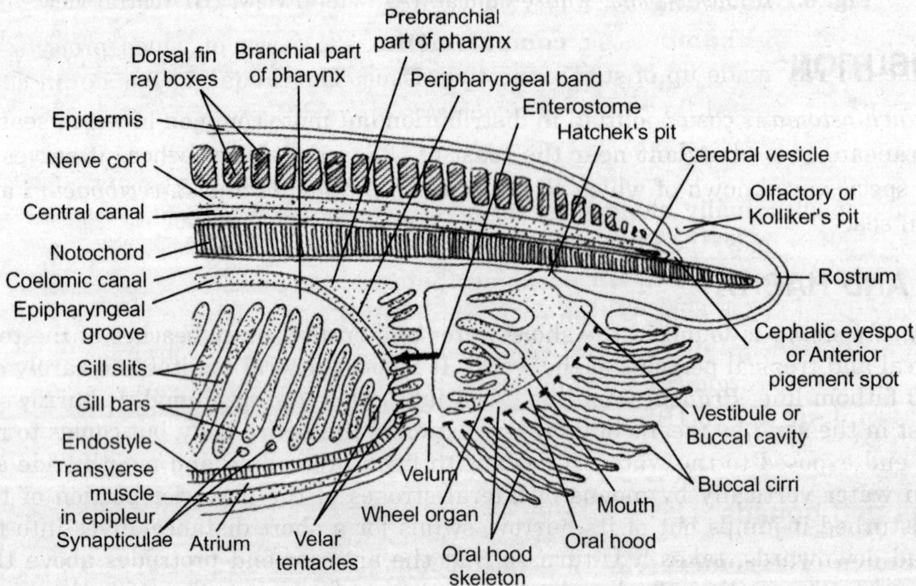

Fig. 6.2 *Branchiostoma.* Anterior region L.S.

Atrial region

The atrial region extends from the mouth over about two-third of the length of the body, terminating at a large median aperture, the **atriopore** (Fig. 6.1A). It is an aperture for the exit of the respiratory current of water and also serves for the release of reproductive products. For this reason the atrial region is also called the **branchio-genital region** but the external features do not indicate its reproductive nature. The ventral side of the body in the atrial region is broadened and convex so that the body presents a somewhat triangular appearance.

Abdominal region

It is a small region of the body between the atriopore and the anus and is characterized by the presence of a **ventral fin.** The atrial and abdominal region together constitute the **trunk.**

Caudal region

It is the posterior most, postanal region of the body surrounded by a **caudal fin.** The anal opening lies to the left side of the fin.

Fins

Branchiostoma has three unpaired fins – the **dorsal, ventral** and **caudal** (Fig. 6.1). The **dorsal fin** runs mid-dorsally along the entire length of the body and posteriorly it is continuous with the much broader **caudal fin** surrounding the tail. The **ventral fin** runs mid-ventrally between the atriopore and anus. Its lower keel-like portion is continued behind the anus as the expanded caudal fin. The cavity of the dorsal and the ventral fins is divided into a number of **fin chambers** or **compartments**, into each of which projects a stout column–like **fin ray** made up of stiff connective tissue. The caudal fin has no fin chambers and fin rays. The fins and fin rays of *Branchiostoma* are different from that of fishes.

Folds

Running longitudinally along the ventro-lateral edge of the atrial region of the body there are two hollow folds, the **metapleural folds** (Fig. 6.1B) which terminate symmetrically shortly behind the atriopore. Really speaking the oral hood is continuous posteriorly with metapleural folds on each side. Within each metapleural fold lies a continuous longitudinal lymph-space, the **metapleural canal**. The metapleural folds probably help in burrowing rapidly in sand due to turgidity caused by flow of lymph in the metapleural canal. A horizontal fold called **epipleur** (Fig. 6.1B) connects the two metapleural fold and form the floor of the atrial cavity.

Myotomes and gonads

Myotomes or **myomeres** are a series of V-shaped muscle bands which are visible through the transparent body wall on each lateral side of the body. Similarly a series of **gonads** are visible between the mouth and the atriopore on either side below the myotomes.

BODY WALL

The different layers of the body wall (Fig. 6.3) are **epidermis, musculature** and **peritoneum.**

Epidermis: The body is covered by a thin layer of epidermis which differs from that of the higher chordates in being made up of a single layer of **cuboidal epithelum.** These cells are ciliated in the young, but in the adult the outer border becomes slightly cuticularized.

The **cuticle layer** is thin, non-pigmented and perforated at places. The epidermis rests on a thin **basement** membrane. Below the epidermis lies a tough connective tissue layer distinguished into **cutis** and **subcutis**. The cutis is thin and compact and made up of fibrous connective tissue. The subcutis is thick and spongy and made up of a gelatinous matrix containing nerve fibres, blood cells and blood vessels. There are a few scattered nomadic cells which secrete the cutis and the subcutis. There are no gland cells and chromatophores in the epidermis, which is more or less transparent and the underlying musculature can be seen through it.

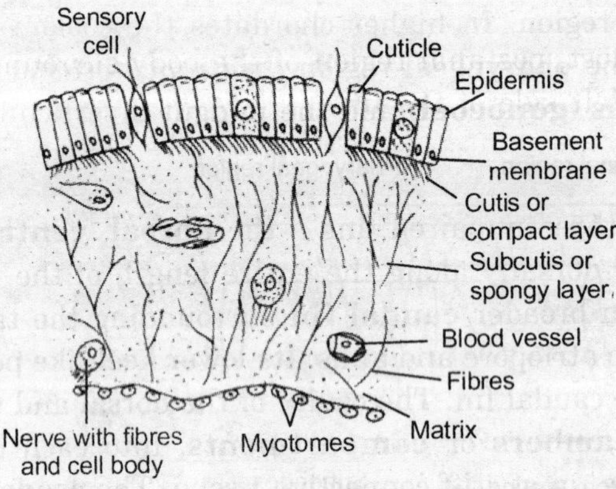

Fig. 6.3 *Branchiostoma.* V.S. of body wall

Musculature: The musculature forms the greater part of the body wall. It consists of sixty segmentally arranged muscle segments or blocks called **myotomes** arranged on either side of the body along its whole length. The myotomes of the two sides alternate with each other. Each muscle block is **V**-shaped, the vertices of 'V' being directed anteriorly (Fig. 6.1). Each myotome is enclosed in a complete box of connective tissue called **mycommata** or **myosepta** which separates adjacent myotomes. In the myotomes the muscles are **striped** and arranged **longitudinally**. They help in lateral undulating movements of the body. There are transverse muscles running across the ventral surface of the body between the metapleural folds. They help in driving water out of the atrial cavity.

Peritoneum: The musculature is followed by a thin parietal layer of peritoneum.

LOCOMOTION

Branchiostoma is a sluggish animal but when disturbed it swims in water by the contraction of the longitudinal muscle fibres of the myotomes. The myotomes contract rhythmically in antero-posterior direction. The notochord prevents shortening of the body during contraction but its elasticity makes contraction more efficient. The animal moves in the forward direction by the propulsive movement of the myotomes. Since the fins are not very well developed, they do not help in locomotion. The caudal fin may sometimes accelerate transverse movements.

COELOM

The body cavity of *Branchiostoma* is a true coelom lined by mesoderm and filled with coelomic fluid. It is enterocoelic in origin. It is a fairly spacious cavity in the posterior part of the body where it lodges the mid- and hindgut regions of the alimentary canal which are suspended by a **mesentery**, as in other vertebrates. In the pharyngeal region, however, the coelom in adult condition is restricted to only a pair of **dorsal longitudinal canals** running on either side above the pharynx enclosing the brown funnels, a **mid-ventral longitudinal** or **sub-endostylar coelomic canal** running below the endostyle, and the **vertical coelomic or spaces** which run through the primary gill bars and connect the dorsal and sub-endostylar coelomic canals (Fig. 6.4). In the larval forms the coelom in the pharyngeal region is uniform and surrounds the pharynx. As development proceeds it becomes restricted because the gill slits arise as outgrowth of the pharynx which meet and fuse with corresponding inpushings of the body wall of that region. In higher chordates the coelom is further reduced and is completely absent from the pharyngeal region. In *Branchiostoma* small coelomic spaces are present around the gonads (**gonocoel**) and the midgut diverticulum.

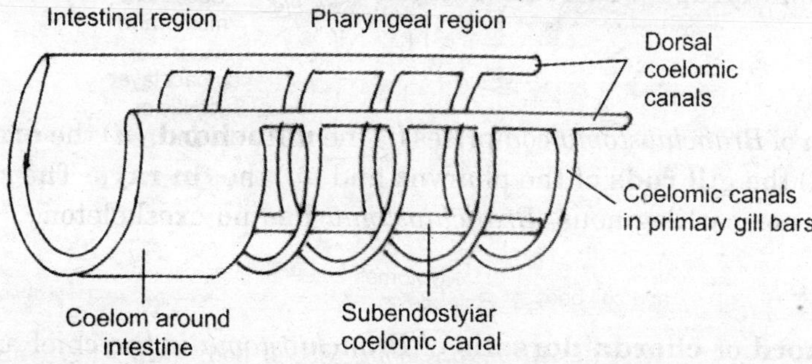

Fig. 6.4 *Branchiostoma.* Coelom diagrammatic representation

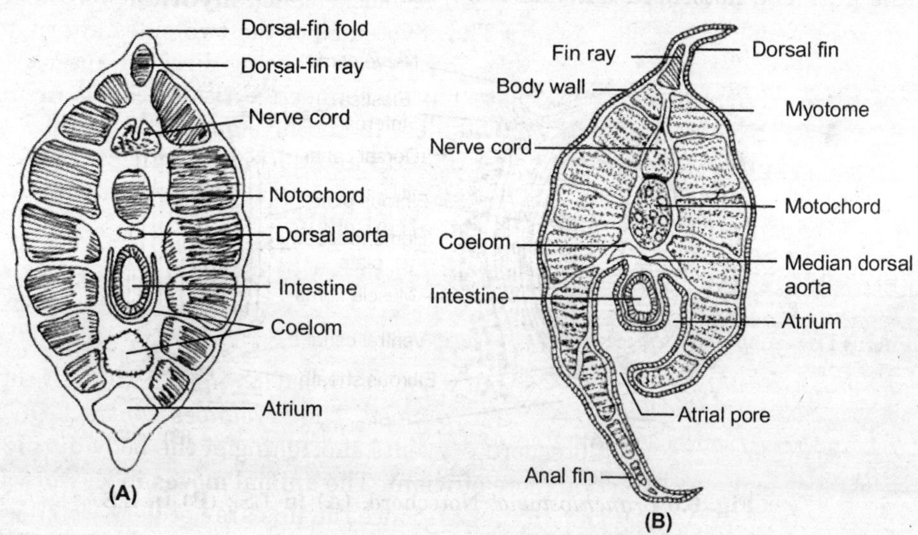

Fig. 6.5 *Branchiostoma.* **(A)** T.S. through atrium and coelom; **(B)** T.S. through atiopore

ATRIUM

The entire branchial apparatus of the adult *Branchiostoma* is surrounded by a large water-filled chamber called the **atrium**. The atrium is a peripharyngeal cavity of secondary origin and opens to the exterior by a small aperture called the **atriopore** situated on the ventral side of the body just in front of the ventral fin. The atrium is lined by ectoderm. It arises as long ectodermal folds of the body wall called the **metapleural folds** which are united ventrally by a transverse shelf, the **epipleur** (Fig. 6.18). The atrial cavity surrounds the branchial apparatus and the anterior part of the intestine. The gill slits open into the atrium and not directly to the outside as in other chordates. The water that comes through the pharyngeal gill clefts passes out through the atriopore. At the level of the posterior end of the pharynx the atrium gives rise to two forwardly directed pockets, one on each side, pushing into the dorsal coelomic canal with which it is reported to communicate. These pockets are called the **brown funnels** and are of unknown function. The atrium is protective in function, it protects the delicate pharynx while the animal is in its sandy burrow and helps to maintain an uninterrupted current of water.

SKELETON

The skeleton of *Branchiostoma* comprises (*i*) the **notochord**, (*ii*) the **oral ring** supporting the oral hood, (*iii*) the **gill rods** of the pharynx and (*iv*) the **fin rays.** The skeletal structures are neither bony nor cartilaginous. *Branchiostoma* has no exoskeleton.

Notochord

The **notochord** or **chorda dorsalis** of *Branchiostoma* is the chief axial endoskeleton. It is an elongated, narrow, cylindrical structure lying mid-dorsally below the nerve cord and above the gut. The notochord extends from the tip of the snout up to the end of the tail,

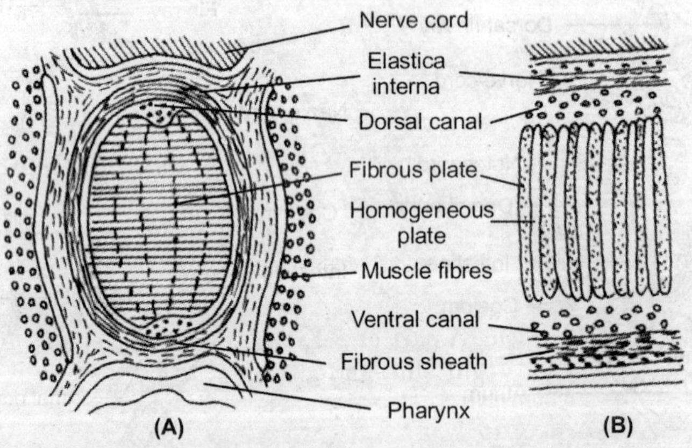

Fig. 6.6 *Branchiostoma.* Notochord. **(A)** In T.S.; **(B)** In L.S.

projecting beyond the level of myotomes. This is probably related with the burrowing habit of the animal. The notochord is complex in structure. It is surrounded by a thick fibrous connective tissue sheath the **notochordal sheath** which is covered by a thin elastic membrane. In early stages of development the chord proper consists of large **vacuolated cells** filled with a fluid. In adults it consists of a series of **flattened plates** arranged with their flat surfaces in the transverse plane of the body. The plates are of two types—**fibrous** and **homogenous** which alternate with each other (Fig. 6.6). Each plate develops from a single vacuolated cell, the nuclei being pushed aside to the dorsal and ventral edges. The notochord does not support the myotomes and other visceral structures like the vertebral column. Its elasticity allows body movement and its rigidity prevents shortening of the body during locomotion.

Oral ring

The oral hood is supported by a thin oral ring made up of separate rod-like pieces. From each piece, rod-like prolongations extend into each oral cirrus to support it (Fig. 6.2). The oral skeleton is made up of gelatinous fibrous connective tissue resembling soft cartilage.

Gill rods

The gill rods are skeletal rods made up of gelatinous substance. They support the gill bars of the gill slits. The gill rods are of two types : the **primary gill rods** with forked ends and the **secondary gill rods** with unforked ends (Fig. 6.7). Dorsally the two rods are united. Besides, the forked gill rods are connected by **transverse skeletal rods** supporting the synapticulae or transverse bars. The primary gill rod encloses a blood vessel and a coelomic canal throughout its length. The secondary gill rod encloses only a blood vessel, the coelomic canal is absent. A blood vessel is also present in the synapticula.

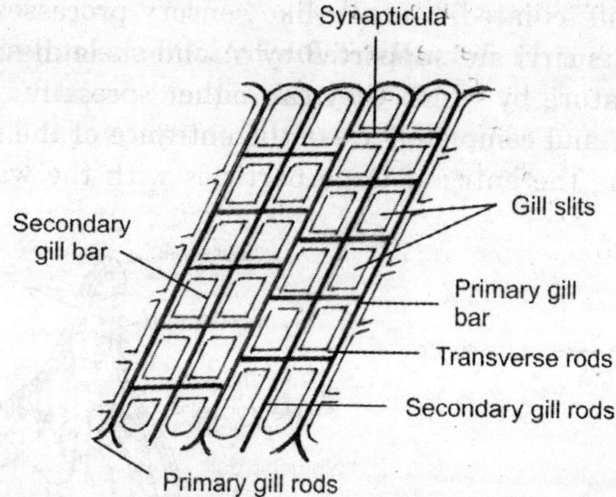

Fig. 6.7 *Branchiostoma.* A part of pharyngeal wall showing gill rods, gill slits and gill bars

Fin ray boxes

The fin ray boxes support the fins. They are made up of connective tissue containing a gelatinous substance. The dorsal fin has a single row of such boxes, while the ventral fin has double row of fin ray boxes.

DIGESTIVE SYSTEM

The digestive system includes the **alimentary canal** and the **digestive glands**.

Alimentary canal

The alimentary canal is a straight tube lined throughout by ciliated epithelium. It begins at the mouth and terminates at the anus (Fig. 6.8).

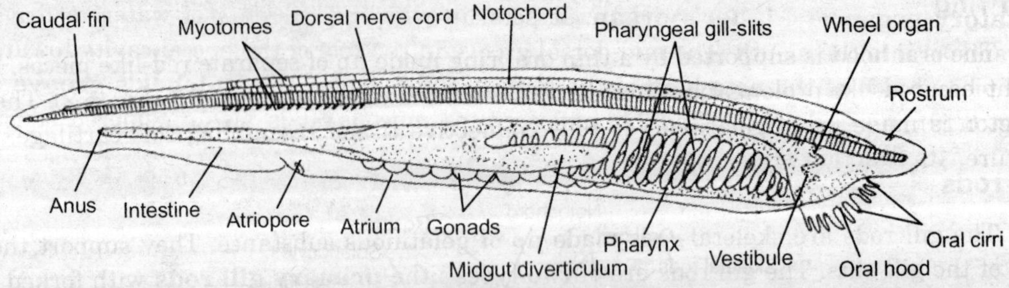

Fig. 6.8 *Branchiostoma.* Alimentary canal

Mouth: The mouth is a wide oval aperture situated antero-ventrally below the rostrum. It is guarded by a frilly membrane, the **oral hood**.

Oral hood: The oral hood (Fig. 6.9) is a lid-like outgrowth around the mouth. It is formed by the dorsal and lateral projections of the anterior end of the trunk. It is fringed with 10–11 pairs of slender, stiff, ciliated, tentacle-like sensory processes called **buccal** or **oral cirri.** The oral hood and its cirri are supported by a solid skeletal axis. They have a special nerve supply and musculature by which they can either spreadout and bend inwards. The cirri may cross each other and completely close the entrance of the mouth, thus they act as a sieve or filter preventing the entry of large particles with the water current.

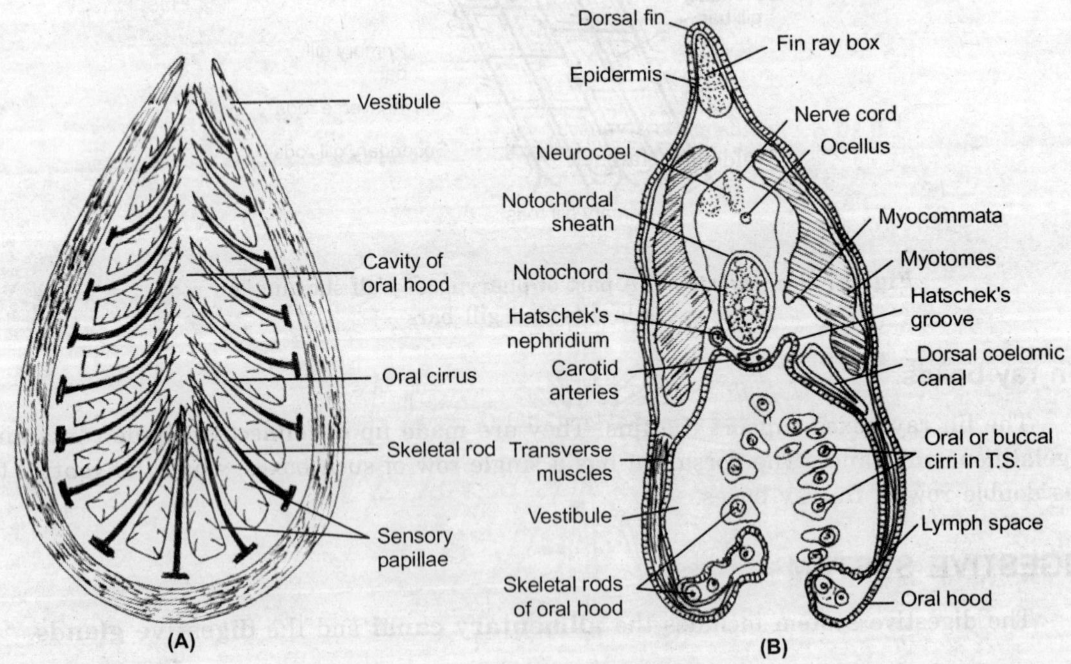

Fig. 6.9 *Branchiostoma.* **(A)** Oral hood; **(B)** T.S. through oral hood

Vestibule or buccal cavity: The mouth leads into the buccal cavity or vestibule. It is funnel-shaped and enclosed by the oral hood. It is lined by ectoderm.

Wheel organ: The underside of the oral hood bears 6–8 pairs of finger-like processes which form a series of ciliated grooves and ridges collectively known as the **wheel organ** or **rotatory organ** or **Muller's organ**. It sets up whirling currents of water to drive food organisms into the mouth. The mid-dorsal groove in the roof of the buccal cavity towards the right hand side of the notochord is the largest and is called the **Hatchek's groove** or **pit** (Fig. 6.10). It is ciliated, glandular and secretes mucous. It is also considered sensory in nature. Its function is not known.

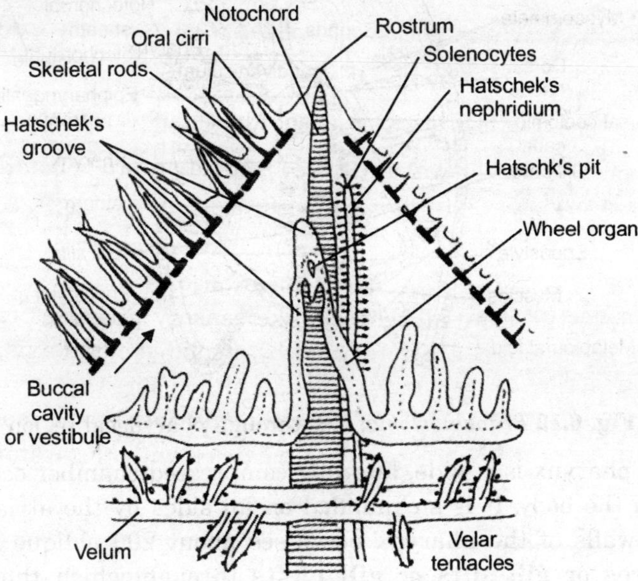

Fig. 6.10 *Branchiostoma.* A ventral view of anterior end showing wheel organ, veldeum, Hatchek's groove, pit and nephridium

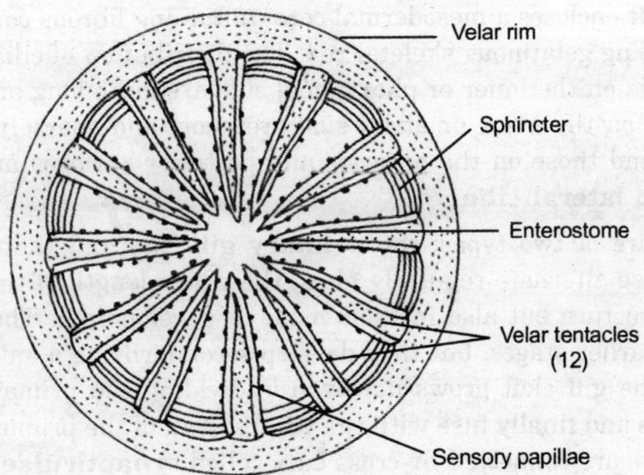

Fig. 6.11 *Branchiostoma.* Velum

Velum: The vestibule is closed posteriorly by a circular vertical membrane called **velum** (Fig. 6.11). It bears an aperture in the centre, called **enterostome** which opens into the pharynx. The velum has a **sphincter** to open and close the enterostome. It is also provided with a circlet of 10–12 slender, ciliated and sensory **velar tentacles** which hang backwards into the pharynx. It acts as a sieve and strains the water entering the pharynx.

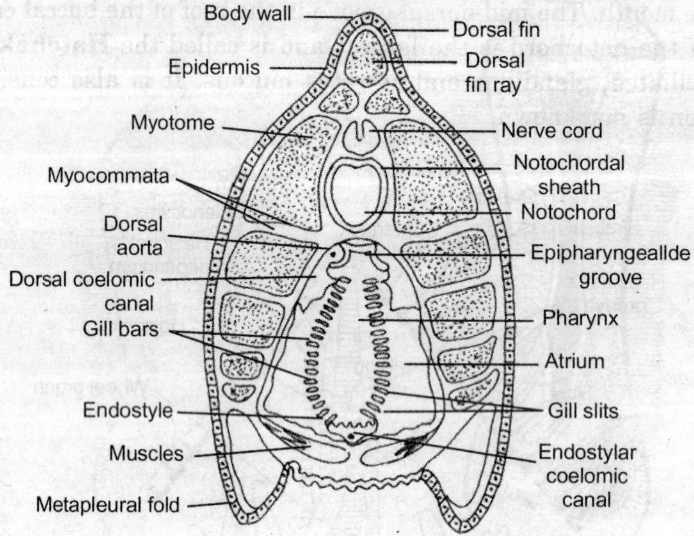

Fig. 6.12 *Branchiostoma.* T.S. through pharyngeal region

Pharynx: The pharynx is a wide, laterally compressed chamber occupying nearly one-half anterior part of the body. It is surrounded on all sides by the atrial cavity, except the dorsal. The lateral walls of the pharynx possesses about 200 oblique vertical slits called **branchial apertures** or **gill slits** or **gill clefts** through which the pharyngeal cavity communicates with the atrial cavity. The number of gill slits increases with age, the new ones being added constantly at the posterior end of the pharynx. The pharyngeal wall between two adjacent gill cleft is called **branchial lamella** or **gill bar**. The gill bar is covered by ciliated epithelium. It encloses a mesodermal core containing fibrous connective tissue, blood vessel and a supporting gelatinous skeletal rod. The distribution of cilia over the gill bars is not uniform. The cilia on the inner or pharyngeal side are quite long and few and are called **frontal cilia**, those on the outer or atrial side are short and sparsely distributed and are called **atrial cilia** and those on the anterior and posterior surfaces are long powerful and dense and are called **lateral cilia**.

The gill bars are of two types: the **primary gill bar** and the **secondary gill bar** (Fig. 6.13). Both these alternate regularly along the whole length of the pharynx and differ not only in their structure but also in their mode of development. The secondary gill bars do not exist in the earlier stages, but they develop secondarily as a small **tongue bar** from the upper wall of the gill cleft grows downwards dividing the primary gill cleft into two **secondary gill slits** and finally fuse with the ventral wall of the primary gill slit (Fig. 6.14). The primary gill bars are connected by cross bars called **synapticulae** or **transverse bars** that run right across the gill openings giving the pharynx a basket-like appearance like the branchial sac of the tunicates. The primary, secondary and transverse bars are supported by primary, secondary and transverse skeletal rods respectively.

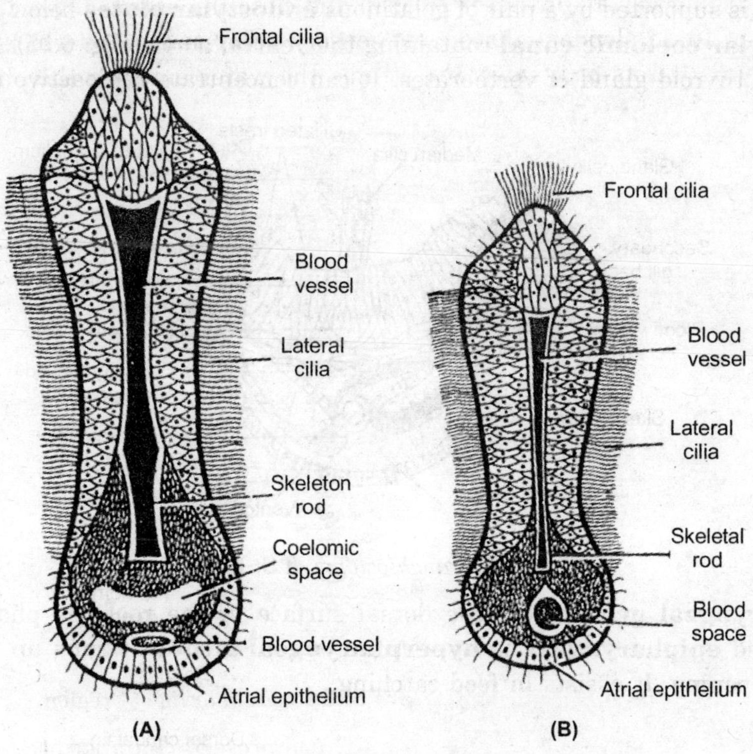

Fig. 6.13 *Branchiostoma.* **(A)** T.S. of primary gill bar; **(B)** T.S. of secondary gill bar

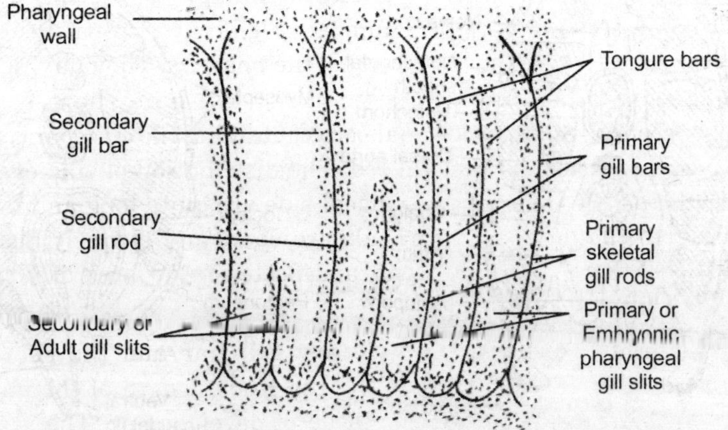

Fig. 6.14 *Branchiostoma.* Development of secondary gill bars and gill slits

Endostyle: The endostyle of *Branchiostoma* is similar to that of *Herdmania.* It is present mid-ventrally along the floor of the pharynx. It consists of five **ciliated longitudinal**

grooves alternating with four **longitudinal tracts** of **mucus secreting gland cells**. The cilia of the median longitudinal groove are longer than that of lateral longitudinal grooves. The endostyle is supported by a pair of gelatinous **endostylar plates** below which is present a **subendostylar coelomic canal** containing the ventral aorta (Fig. 6.15). The endostyle is similar to the thyroid gland of vertebrates. It can concentrate radioactive iodine.

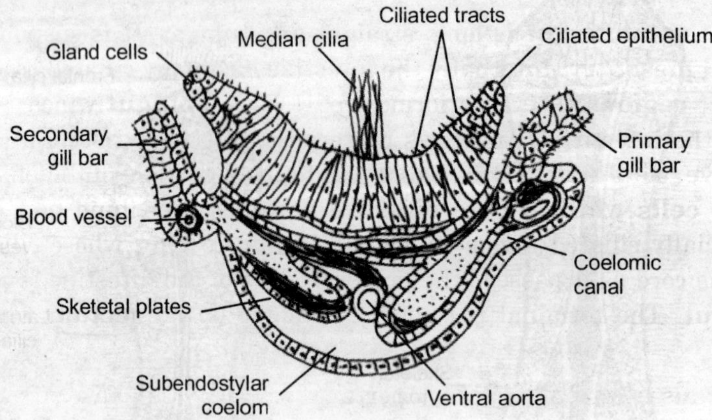

Fig. 6.15 *Branchiostoma.* T.S. Endostyle

Epipharyngeal groove: On the dorsal surface in the roof the pharynx there is a median ciliated **epipharyngeal** or **hyperpharyngeal** groove. It runs up to the tip of the oesophageal opening. It assists in food catching.

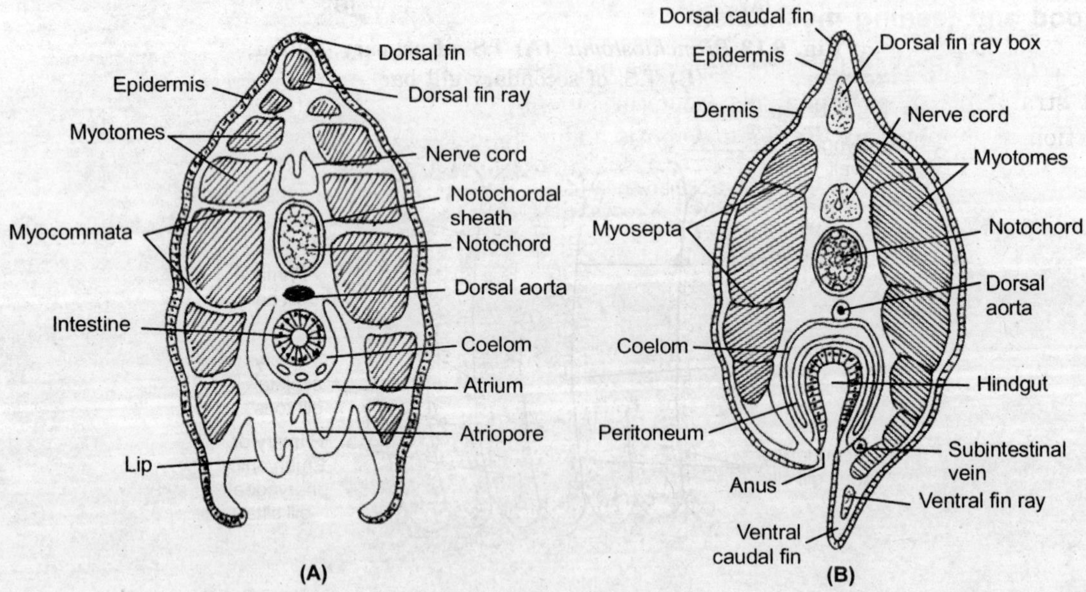

Fig. 6.16 *Branchiostoma.* **(A)** T.S. through intestinal region; **(B)** T.S. through anal region

Peripharyngeal bands: The peripharyngeal bands are a pair of narrow ciliated tracts arising from the anterior end of the endostyle running upwards along the lateral sides of the pharynx and uniting with the epipharyngeal groove dorsally. It separates a small antero-dorsal **prebranchial region** of the pharynx without gill slits. It diverts the food particles captured by the sticky threads of the endostyle to the epipharyngeal groove from where they are conducted backwards to the midgut.

Oesophagus: The pharynx opens into a short, narrow, tubular and ciliated **oesophagus** which leads into the **gut** or **intestine.**

Gut: The **intestine** or **gut** is long, straight and tubular. It is suspended in the atrial cavity by a **dorsal mesentry**. Internally the intestine has many ciliated tracts. The intestine has **three** distinct **regions.** The anteriormost part is the **midgut** which is wider than rest of the portion of the intestine. It narrows posteriorly and gives rise to a large blind pouch called **midgut diverticulum** or **liver**, which lies to the right side of the pharynx. It contains numerous **gland cells** which secrete digestive enzymes. The hind end of the midgut is marked by a specially ciliated region called the **ilio-colic ring** whose cilia rotate the food containing mucous cord as it passes. The posterior part of the intestine is a narrow straight tube, the **hind-gut**. The terminal part of the hindgut is the **rectum** which opens to the outside by **anus**.

Anus: The anus is a small circular aperture present at the base of the caudal fin. It is controlled by a sphincter.

Digestive gland

The **midgut diverticulum** or **liver** is the only digestive gland in *Branchiostoma*. It has already been described.

Food and feeding mechanism

Branchiostoma feeds upon pelagic microscopic organisms like diatoms, protozoans, etc. It strains off these organisms from the surrounding sea water by the automatic filtering action of the pharynx. *Branchiostoma* is a sluggish animal. It does not move about in search

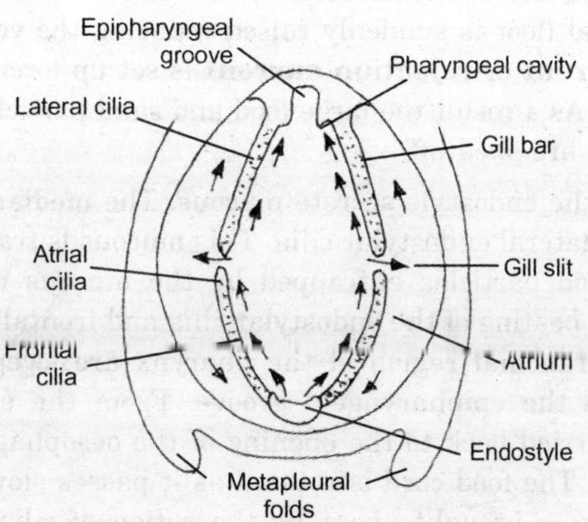

Fig. 6.17 *Branchiostoma.* Diagrammatic T.S. showing direction of feeding current

of food. The water is constantly entering the mouth and passing into the atrium through the gill slits, it is expelled out at the atriopore. The food bringing water current is produced and maintained by the action of cilia, that is why the animal is described as a **ciliary feeder** (Fig. 6.17). The food particles are filtered from the current of water.

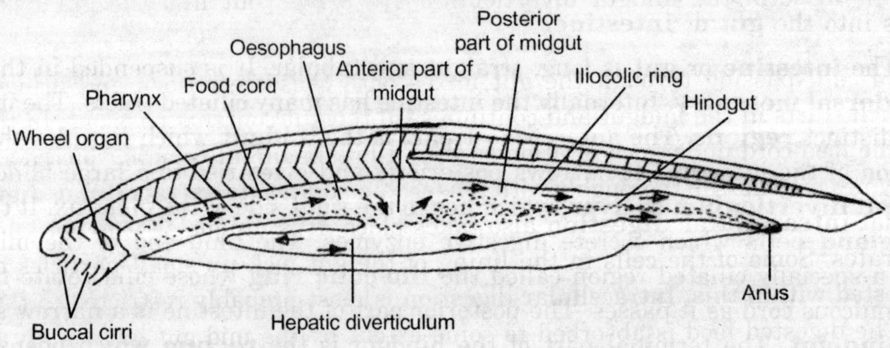

Fig. 6.18 *Branchiostoma.* Movement of food cord in the alimentary canal

The mode of feeding is complicated but quite interesting. The rotatory movement of the cilia of wheel organ causes a whorling current of water into the mouth. The buccal cirri fold over one another forming a sieve for straining the larger particles during feeding. When water current passes through the enterostome the velar tentacles further filter water, so that only fine food particles enter the pharynx. The chemoreceptors present on the buccal cirri and velar tentacles test the quality of food and water. The food particles are concentrated by mucous secreted by the Hatschek's groove and are pushed through the enterostome into the pharynx. The lashing movements of the lateral and atrial cilia of the gill bars drive water out of the pharynx through the gill slits into the atrium and finally out through the atriopore. Periodically the atriopore closes, the muscles on the floor of the pharynx undergo sudden contraction and the atrial floor is suddenly raised reducing the volume of the atrial cavity. A powerful **reverse current** or **rejection current** is set up forcing out water through the enterostome and mouth. As a result the large food and sand particles accumulated on buccal cirri and velar tentacles are shed off.

The gland cells of the endostyle secrete mucous. The median tract of endostylar cilia drive the mucous to the lateral endostylar cilia. This mucous is transferred to the side walls of the pharynx. The food particles entrapped by the mucous move dorsalwards to the epipharyngeal groove by beating of the endostylar cilia and frontal cilia of gill bars. The food particles from the prebranchial region of the pharynx are swept up by the cilia of the peripharyngeal band to the epipharyngeal groove. From the epipharyngeal groove the mucous laden food is carried back to the opening of the oesophagus in the form of a **food cord** or **food cylinder**. The food cord is rotated as it passes slowly backwards in the gut (Fig. 6.18). The rotation is brought about by the action of ciliated cells in the ilio-colic region and behind this region. Any portions which breaks away from the cord fail to rotate behind this region.

Physiology of digestion

The digestive enzymes are secreted by the epithelial lining of gut and the midgut diverticulum. According to Barrington, the digestive enzymes in *Branchiostoma* are **amylase, lipase** and **protease**. All the regions of the gut behind the oesophagus produce their own secretions and those of the midgut diverticulum are driven out into the gut by ciliary action.

The food cord passing through the gut mixes with these secretions and bring about digestion which starts in the midgut and continues till the gut contents pass out. In the ilio-colic region the food cord undergoes rotation so that it breaks up into smaller pieces and the digestive enzymes may mix up thoroughly with the food. In addition to **extracellular digestion** it is likely that **intracellular digestion** also takes place. It is unique as it is not known in other vertebrates. Some of the cells in the lining of the gut pick up small particles of food that are digested within this. Intracellular digestion is most probably restricted to the hind gut region. The digested food is absorbed to some extent by the mid gut but mainly by the hind gut.

CIRCULATORY SYSTEM

The circulatory system of *Branchiostoma* (Fig. 6.19) appears peculiar in that it has no heart and the blood contains neither haemoglobin nor other respiratory pigment. It is closed type and presents the fundamental plan on which the circulation of other chordates is based. The circulation is brought about by slow waves of contraction that pass along the major arterial vessels in such a way that the blood is driven forwards in the ventral vessel and backwards in the dorsal vessel. The blood vessels are muscular and contractile in nature. There is no structural difference between the arteries and veins but they are named so due to their similarities with blood vessels of higher chordates. The main components of the circulatory system are:

Sinus venosus: It is a large, thin-walled sac lying just below the posterior end of the pharynx. It receives blood from all parts of the body.

Ventral aorta: From the sinus venosus arises a large branchial artery, the **ventral aorta** or **subendostylar aorta** or **truncus arteriosis**, which runs forwards in the ventral wall of the pharynx below the endostyle in the subendostylar coelom. It is contractile and drives blood forwards. From the ventral aorta arise several fine vessels, the **afferent branchial arteries**, on each side. Each branchial artery has a small contractile dilation at the base called **bulbillus or bulbule**, presumably acting as a **branchial heart**. The afferent vessels of the primary gill lamella communicate with similar vessels in the secondary gill lamella by **transverse vessels** running through the synapticulae. While traversing the vessels the blood is exposed to the aerating influence of the respiratory current. The afferent vessels leave the branchial lamellae dorsally as **efferent branchial arteries** which open on each side into longitudinal vessels, the right and left **lateral dorsal aortae**. Before opening into the dorsal aorta each efferent vessel splits up into a capillary network called **nephric glomerular sinus** or **glomus** in the nephridium.

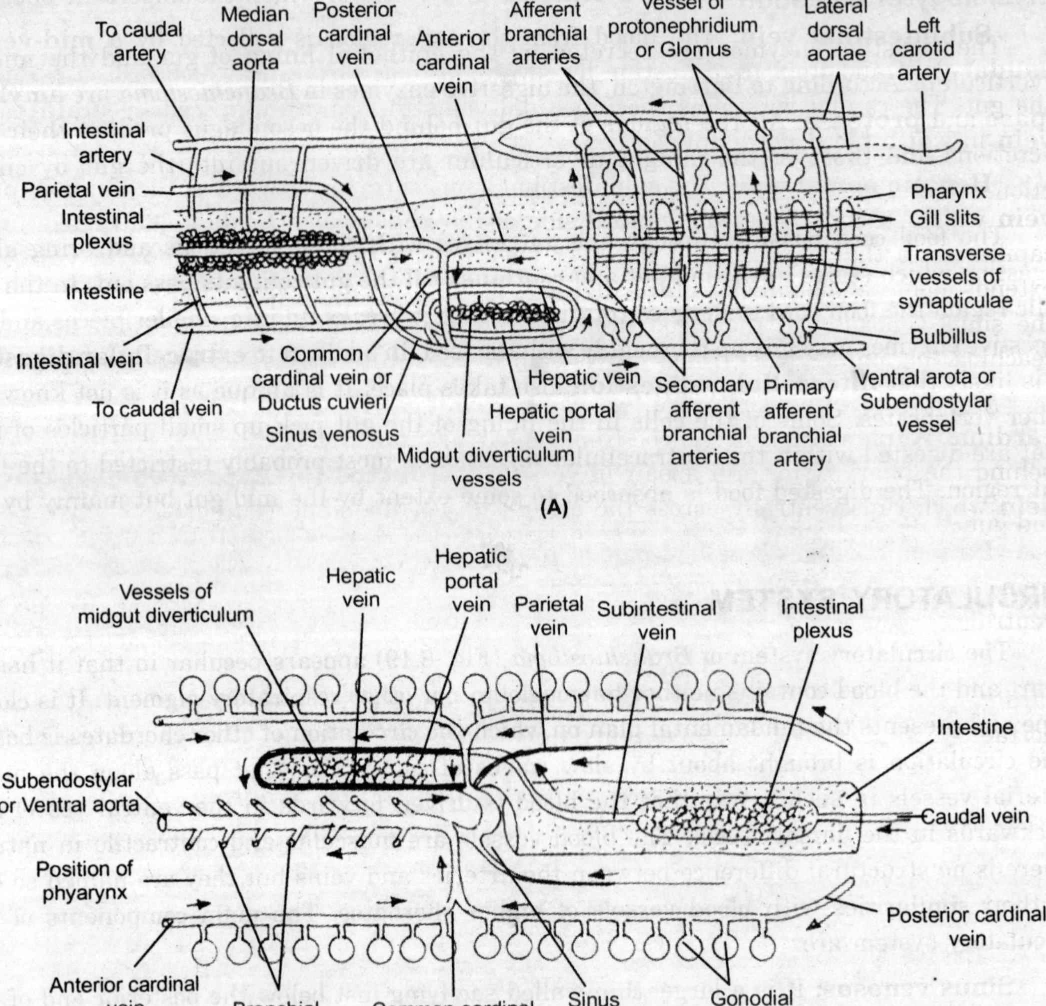

Fig. 6.19 *Branchiostoma.* Blood vascular system.
(A) Lateral view; **(B)** Dorsal view

Dorsal aorta: The dorsal aorta is a pair of longitudinal vessels, the **right** and **left lateral dorsal aortae** lying one on either dorso-lateral side of the pharynx. Anteriorly both the aortae are continued forward as **internal carotid arteries** to supply the snout region. The right artery is relatively delicate. Posteriorly both the aortae unite behind the pharynx to form a single **median dorsal aorta** which runs between the notochord and intestine and continues in tail region as the **caudal artery**. The blood flows backwards in the dorsal aorta. The median dorsal aorta send a number of **intestinal arteries** to the intestine which form a capillary network or **plexus** in the intestinal wall. From the lateral and median dorsal aortae blood is carried by a system of lacunae to supply myotomes, body wall, lymph space, etc. There are no true capillaries.

From the lacunae the blood is collected into **veins** of which the important ones are:

Subintestinal vein: The blood from the tail region is collected by a **mid-ventral caudal vein** and from the intestine by several **intestinal veins** which form a **plexus** on the gut. The caudal vein joins the intestinal veins to form a large median **subintestinal vein** in which blood flows anteriorly.

Hepatic portal vein: The subintestinal vein continues forward as the **hepatic portal vein** which extends along the ventral side of the midgut diverticulum and breaks up into capillaries in that organ. From the liver blood makes its way into a **hepatic vein** which extends along the dorsal side of the digestive gland, runs downwards and forwards and joins the sinus venosus. All the blood from the intestine has to pass through the liver before reaching the sinus venosus, hence it is said to constitute the hepatic portal system.

Cardinal veins: There are two pairs of cardinal veins, the **anterior** and **posterior cardinal veins** which collect blood from ventro-lateral region of the body. The two unite just behind the pharynx to form a pair of vessels, the **ductus Cuvieri** or **common cardinal vein** which runs ventrally across the atrium to join the sinus venosus.

Parietal veins: There is a pair of **parietal veins** one on either side, collecting blood from the dorsal body wall. For some distance they run above the intestine and then turn ventrally to join the sinus venosus.

Course of circulation: Anteriorly, the blood flows in parietal, subintestinal and posterior cardinal veins and in ventral aorta. Posteriorly the blood flows in lateral and median dorsal aortae and the anterior cardinal vein. The course of circulation is as follows:

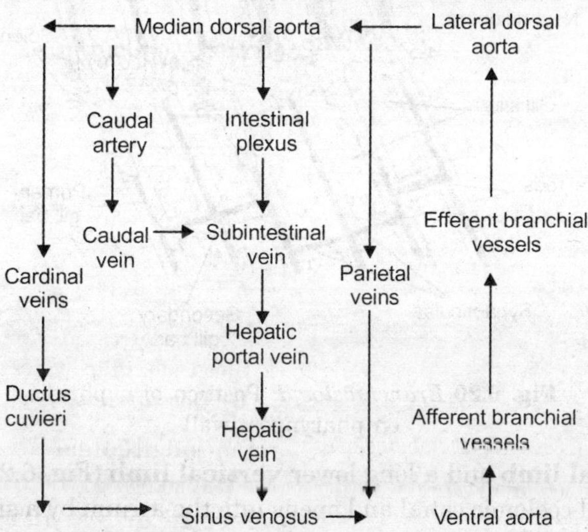

Lymphatic system

In *Branchiostoma* lymph is absent but **lymph sinuses** or **spaces** are present inside the fins, metapleural folds and atrial wall. These are filled with blood without leucocytes. Maximum gaseous exchange takes place in these sinuses.

RESPIRATORY SYSTEM

There are no special respiratory organs in *Branchiostoma*. The pharynx which is mainly a feeding organ, has richly vascular walls. It is, therefore, assumed that some gaseous exchange between blood and sea water takes place in the pharynx. No experimental evidence, however supports this view. Further it has been suggested that the exchange of gases takes place through the whole surface of the body, particularly the wall of the atrium, fins and metapleural folds containing lymph spaces.

EXCRETORY SYSTEM

The excretory system of *Branchiostoma* does not agree with the basic chordate plan. It consists of **protonephridia** which is similar to the protonephridia of many invertebrates such as the platyhelminths, annelids, branchiopods, etc. It is ectodermal in origin. In the vertebrates the excretory organs are kidneys which are mesodermal in origin.

Protonephridia

There are 90–100 pairs of protonephridia in *Branchiostoma*. They are present segmentally in the dorso-lateral wall of the pharynx (Fig. 6.20), one above each gill slit. Each nephridium is small, simple, ciliated, closed, thin-walled, bent tube of ectodermal origin. It consists of a

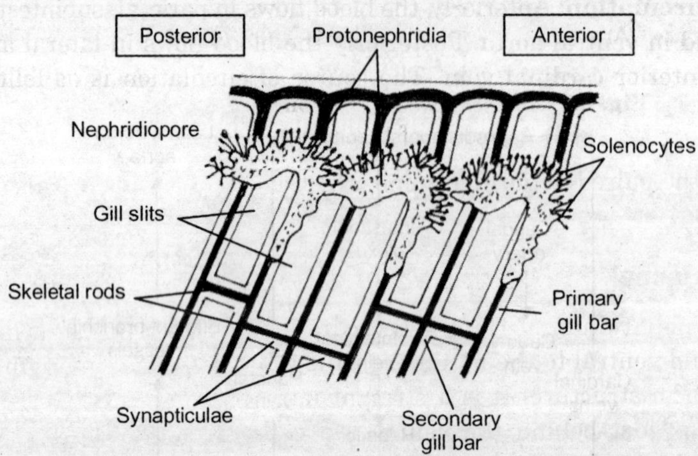

Fig. 6.20 *Branchiostoma.* Position of nephridia on pharyngeal wall

small upper **horizontal limb** and a long lower **vertical limb** (Fig. 6.21). The horizontal limb is situated in the dorsal coelomic canal and opens into the atrium by a small ciliated **excretory pore** or **nephridiopore** situated on the dorsal edge of the gill slit. The vertical limb also lies in the coelomic canal of the primary gill lamella and terminates blindly. The anterior and dorsal surfaces of each nephridium bear numerous small branches, each studded with numerous elongated **flame cells** or **solenocytes**. A simple nephridium may bear upto 500 solenocytes. Each solenocyte is an elongated cell, about 50μ long, consisting of a tiny circular, nucleated **cell body** and a long **stalk** or **tubule** that opens into the nephridial lumen by a separate aperture. A long vibratile flagellum, arising from a basal body, runs through the tubule and projects freely into the lumen of the nepridium (Fig. 6.21). It is constantly moving like a flickering flame. The nephridium has a rich supply of blood vessels. The tuft of solenocytes project into the dorsal coelomic canal and become directly bathed in the coelomic fluid, this facilitates extraction of nitrogenous wastes. The nitrogenous wastes extracted by the solenocytes is discharged into the atrium by the nephridiopore and passed out through the atriopore with the outgoing water.

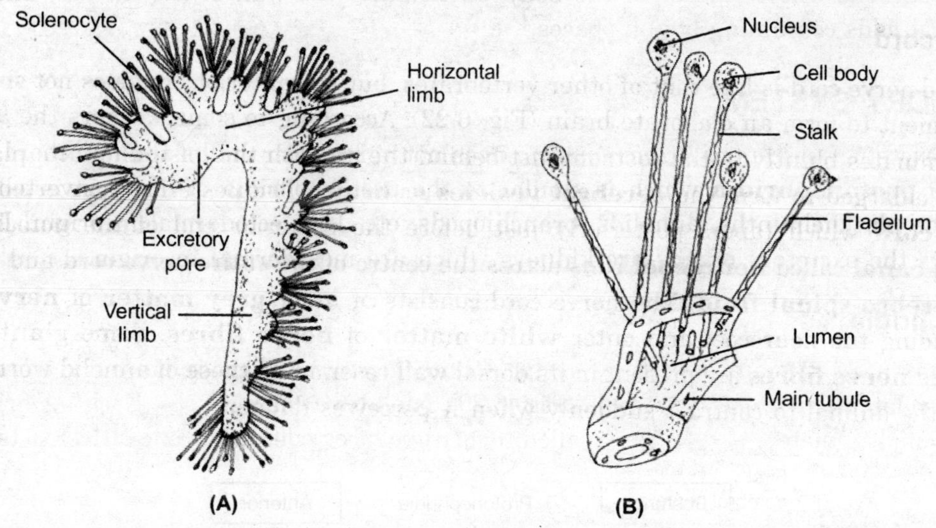

Fig. 6.21 *Branchiostoma.* **(A)** Single nephridium; **(B)** A portion of the same magnified

In addition to the paired nephridia, some other cells and organs are also regarded to be excretory in function.

Nephridium of hatschek

A single large nephridium of Hatschek occurs in the roof of the oral hood (Fig. 6.10), slightly to the left and ventral to the notochord. It is also ectodermal in origin and resembles the paired nephridia in structure. It is a straight narrow tube which opens at its posterior end into the pharynx just behind the velum and ends blindly anteriorly just in front of Hatschek pit. Its entire surface is covered by several solenocytes which are associated with a network of capillaries. The solenocytes extract nitrogenous wastes.

Brown funnels

A pair of sac-like brown funnels are present dorsally near the posterior end of the pharynx. The narrow anterior end of the funnel opens into the dorsal coelomic canal of its side, while the broad posterior end opens into the atrium. According to some workers they are excretory in function, while others regard them as receptor organs.

Renal papillae

Groups of cells called renal papillae are present in the floor of the atrium and are probably excretory in function.

NERVOUS SYSTEM

The nervous system of *Amphioxus* is simple, consisting of a **central nervous system** which includes the hollow, tubular **nerve cord** situated on the dorsal side of the notochord, the **peripheral nervous system** which includes the nerves arising from the nerve cord and the **autonomic nervous system** which includes the nerve plexes.

Nerve cord

The nerve cord is like that of other vertebrates, but the anterior end does not show any enlargement to form an elaborate brain (Fig. 6.22). According to some workers the anterior end terminates bluntly in the rostrum just behind the anterior end of the notochord, and is slightly enlarged to form the **cerebral vesicles** or **brain**. The posterior part is called the **spinal cord** which tapers posteriorly just before the posterior end of the notochord. A **central canal** called **neurocoel** runs across the centre of the entire nerve cord and is filled with **cerebro-spinal fluid**. The nerve cord consists of inner **grey matter** of **nerve cells** surrounding the neurocoel and outer **white matter** of **nerve fibres**. Some **giant nerve cells** and **nerve fibres** are present in its dorsal wall resembling those of annelid worm. They enable the animal to contract suddenly when it perceives danger.

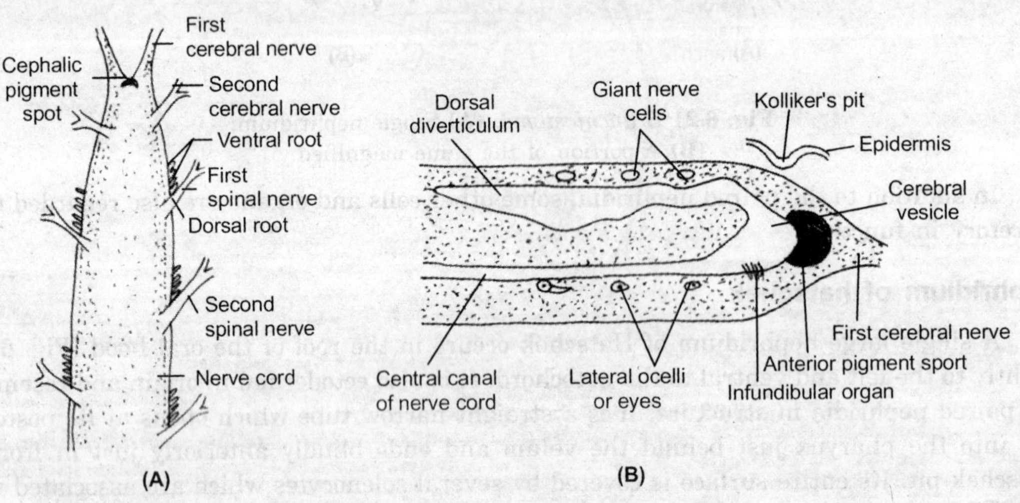

Fig. 6.22 *Branchiostoma.* **(A)** Anterior part of nerve cord in dorsal view; **(B)** V.L.S. of the same

Nerves

The peripheral nervous system consists of two simple sets of **nerves** which arise from the nerve cord. The first two pair of nerves which arise from the cephalic vesicle are called **cephalic nerves** (Fig. 6.22). They innervate the snout, oral hood and buccal cirri. They are purely sensory in nature. They do not have ventral roots.

The paired nerves which arise from the nerve cord behind the cephalic vesicle are called **spinal nerves** (Fig. 6.22). One pair arises in each segment. Each spinal nerve has a **dorsal**

root and a **ventral root**, but the roots do not join to form a single mixed spinal nerve as in vertebrates. The dorsal root is a **mixed fibre**. It runs out between the myotomes and carries all the **sensory** or **afferent** fibres of the segment and **motor** or **efferent** fibres of the muscles of the ventral part of the body. The ventral roots lie opposite to the myotomes, to which they carry **motor** fibres and these end in the muscle fibres with end plates. This pattern of nerve arrangement conforms with that of the vertebrates. The dorsal root is single but the ventral root is made of several branches. The two roots do not originate at the same level, the dorsal root of one side lies opposite to the ventral root of the other side (Fig. 6.23). The nerve fibres in *Amphioxus* differ from those of vertebrates in that they have no thick myelin sheath. The sensory fibres are more like neuro-secretory cells of the invertebrates. The cell bodies are not collected into spinal ganglia, but lie either close below the epidermis or scattered along the course.

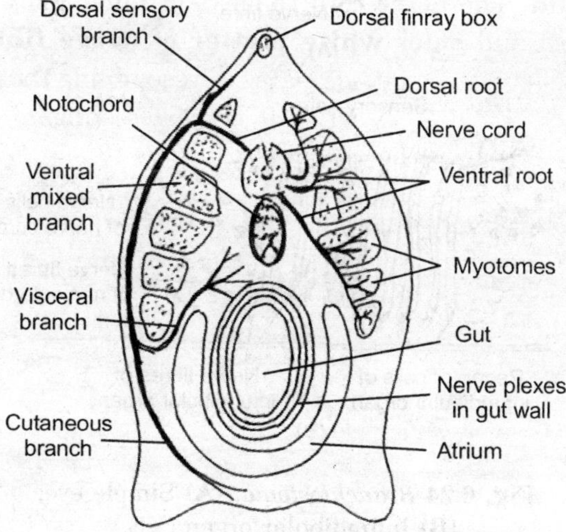

Fig. 6.23 *Branchiostoma.* Diagrammatic T.S. showing distribution of nerve roots

Autonomic nervous system

It includes two nerve plexes present in the gut wall. The plexes are connected to the nerve cord by visceral nerves arising from dorsal root in each segment. This system controls the involuntary muscles of the gut wall.

SENSE ORGANS

There are various types of organs of special sense in *Amphioxus*.

Eye spots or simple eyes or ocelli

These are **photoreceptors** or **light sensitive cells** (Fig. 6.24A) present in definite tracts along the ventrolateral sides of the nerve cord. They are pigmented and resemble those of the planarians. Each eyespot consists of a single large **photosensitive cell** bearing a **cup-shaped cap** of blackish brown **pigment** on one side of the cell. The pigment cap has a nucleus. The cytoplasm between the photosensitive cells and the pigment cup is differentiated to form a **striated layer** which probably acts like a **lens**. The photosensitive cell sends a single nerve fibre to the nerve cord from the side opposite to that bearing the pigment cup.

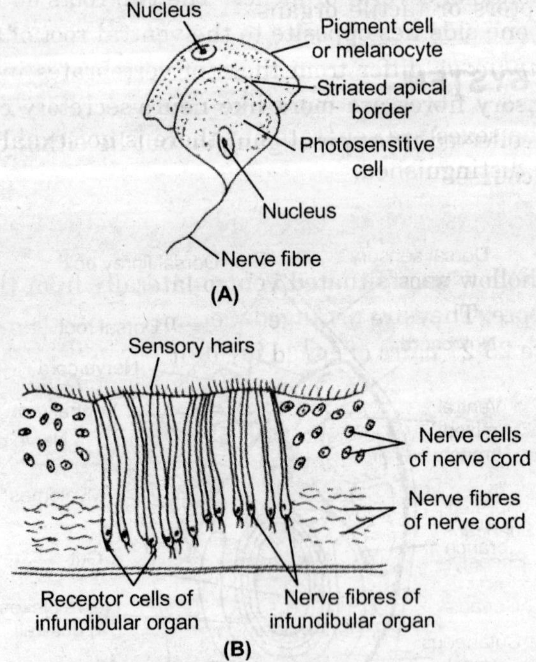

Fig. 6.24 *Branchiostoma.* **(A)** Simple eye; **(B)** Infundibular organ (V.S.)

Cephalic pigment spot

It is often referred to as cerebral eye, but a lens is absent and it is not photosensitive. It is simply a large pigment spot on the anterior wall of cerebral vesicle. It probably acts as a thermoreceptor.

Infundibular organ

It is situated in the wall of the cerebral vesicle and is so called because it is homologous to the infundibulum of vertebrate brain. It is composed of tall, columnar epithelial cells with long cilia which beat in opposite direction to those of the rest of the vesicle (Fig. 6.24B). Each cell gives out a fine **Reissner's fibre** on the inner side, which runs posteriorly inside the neurocoel. It probably acts as a gravity or pressure receptor.

Kolliker's pit or olfactory pit

It is presumably a chemoreceptor, situated at the anterior end of the cerebral vesicle.

In young stages cerebral vesicle opens by an anterior neuropore which closes later. At this place develops a pit lined by ciliated epithelial cells supplied with nerve endings. This is the **Kolliker's pit**.

Sensory cells and papillae

The oral cirri and velar tentacles are provided with groups of sensory cells and papillae which are chemoreceptors or tactile organs.

REPRODUCTIVE SYSTEM

In *Amphioxus*, the sexes are separate but there is no sexual dimorphism, i.e., the male and female cannot be distinguished.

Gonads

The gonads are hollow sacs situated ventro-laterally from the middle of the pharygeal region up to the atriopore. They are arranged metamerically, one pair in each of the segments 25 to 51, thus there are 26-27 pairs of gonad in adult *Amphioxus* (Fig. 6.25A). They are clearly

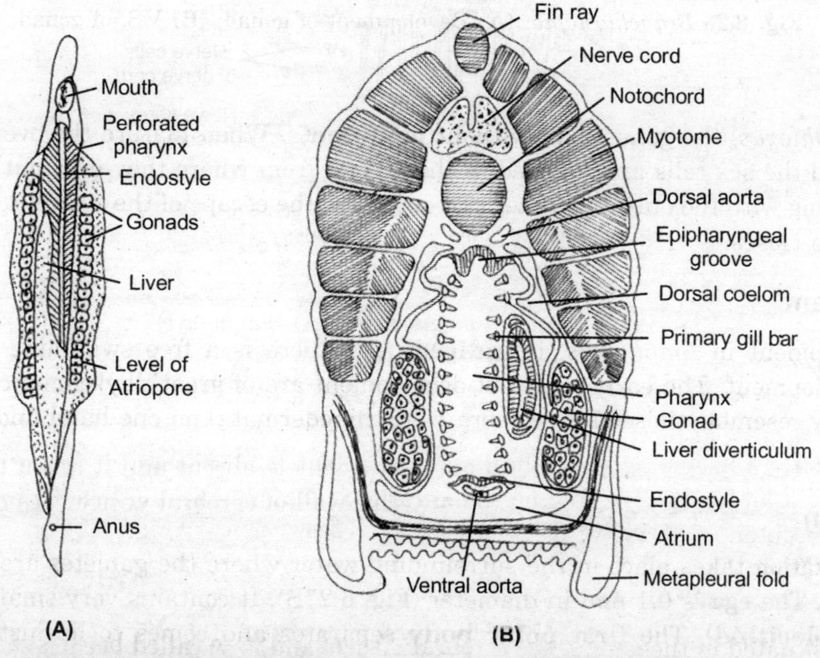

Fig. 6.25 *Branchiostoma.* **(A)** Animal dissected to show the position of gonads; **(B)** T.S. through posterior part of the pharynx showing gonad

visible through the transparent body wall. They are mesodermal in origin. They appear as series of bulges inwards towards the atrium. They are covered on the outer side by the body wall and on the inner side by the atrial epithelium. Each gonad has an inner **primary gonadial cavity** which is completely surrounded by an outer **secondary** or **perigonadial cavity** (Fig. 6.26). The germ cells are budded off from the wall of the primary gonadial cavity. The sex cells or gametes are the sperms in male and ova in female.

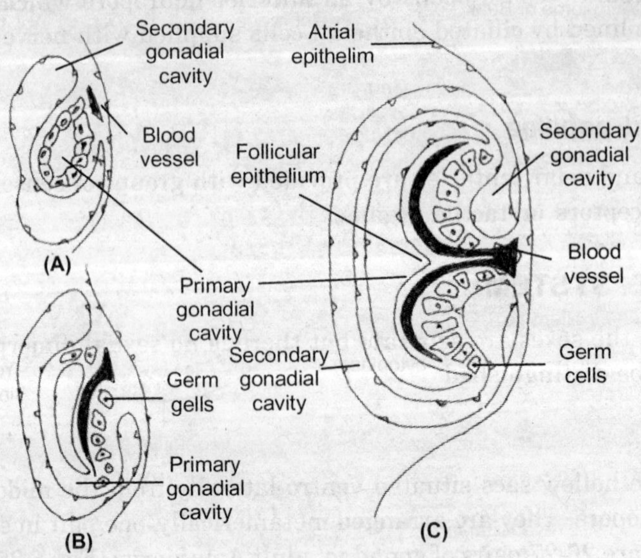

Fig. 6.26 *Branchiostoma.* **(A)** Development of gonad; **(B)** V.S. of gonad

Gonoduct

In *Amphioxus,* the gonads are without **gonoducts**. When mature the overlying tissue ruptures and the sex cells are released in the atrium from where they pass out through the atriopore along with the outgoing water current. After the escape of the sex cells the aperture in the tissue closes.

Development

Development in *Amphioxus*, is **indirect**, *i.e.*, there is a free swimming larval stage during development. The early stages of development are of great phylogenetic significance because they resemble those of invertebrates (Echinodermata) on one hand and vertebrates on the other.

Fertilization

Fertilization takes place in the surrounding water where the gametes are shed, i.e., it is **external**. The egg is 0.1 mm in diameter (Fig. 6.27B). It contains very small quantity of yolk (**microlecithal**). The **first polar body** separates and comes to lie just outside the **vitelline membrane** by the time the egg reaches the sea. The cytoplasm of the egg is arranged in two layers : a **peripheral layer**, free from yolk, lying just below the vitelline membrane ; and a mass of **yolky cytoplasm** which lies mainly towards the vegetative pole of the egg. A major part of the animal pole is occupied by the **egg nucleus (germinal vesicle)**.

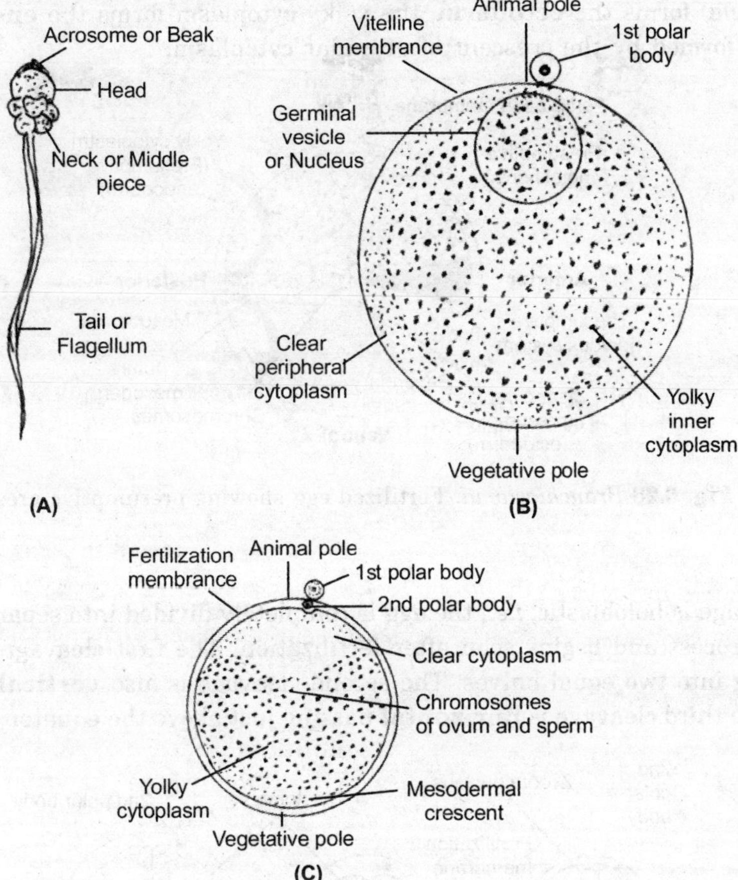

Fig. 6.27 *Branchiostoma.* **(A)** Sperm; **(B)** Unfertilized egg; **(C)** Fertilized egg

The sperm (Fig. 6.27A) of *Amphioxus* is the smallest among the chordates. It is about 18 μ long. It enters the egg near the vegetal pole and this stimulates maturation of the egg. The second polar body is given off and comes to lie near the animal pole just within the vitelline membrane. The nucleus in the **fertilized egg** or **zygote** (Fig. 6.27C) lies near the centre of the egg.

Presumptive area

Following fertilization the cytoplasmic contents are completely rearranged. As a result of rupture of the germinal vesicle an area of clear cytoplasm is left almost in contact with the peripheral layer. Later this flows towards the vegetative pole and condenses to form a crescent - shaped structure called **grey crescent** at the future posterior end. The arms of the crescent reach round the sides of the egg and partially embrace the yolky area of the egg. The various cytoplasmic areas give rise to definite structures in the embryo and if any area is removed or damaged its correlated structure may not be formed. The different cytoplasmic areas are called **organ-specific areas** or **presumptive areas** (Fig. 6.28). The zygote is said to be of mosaic type due to this regional differentiation. The clear cytoplasm at the anterior end (animal pole) forms the **ectoderm**, the yolky cytoplasm forms the **endoderm** and the **mesoderm** is formed by the crescent of granular cytoplasm.

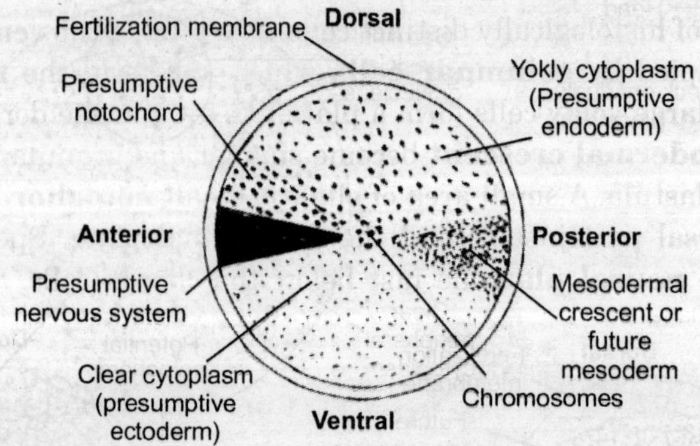

Fig. 6.28 *Branchiostoma.* Fertilized egg showing presumptive areas

Cleavage

The cleavage is holoblastic, *i.e.*, the egg is completely divided into separate blastomeres. It is a rapid process and begins soon after fertilization. The first cleavage is **vertical** and divides the egg into two equal halves. The second cleavage is also **vertical**, at right angles to the first. The third cleavage is **horizontal** passing just above the equator dividing the four

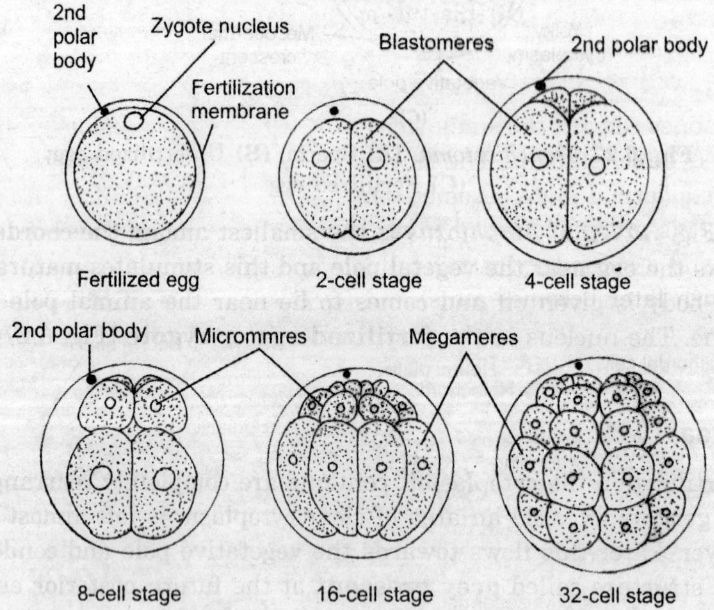

Fig. 6.29 *Branchiostoma.* Early cleavage

blastomeres into an upper tier of four small cells, the **micromeres** and a lower tier of four large cells, the **macromeres** (Fig. 6.29). At the sixth cleavage a ball of 64 cells called **morula** is formed. After this the cleavage becomes irregular and the cells tend to flatten out against each other. The cleavage continues till somewhat spherical **blastula** is formed. It is somewhat pear-shaped, the pointed end being posterior. It has a fluid-filled cavity called **blastocoel** (Fig. 6.30). In the next stage the blastula flattens on one side and becomes hemispherical. Its walls show areas of histologically distinct cells. The bulk of the ventral half of the blastula (flat surface) is occupied by **columnar cells** which will form the **future ectoderm.** The **endoderm** forming large yolky cells form a plate-like area in the dorsal part of the blastula. The cells of the **mesodermal crescent** become smaller and granular and lie on the postero-lateral sides of the blastula. A small area of slightly small **notochord** forming **chorda cells** lie in the antero-dorsal part just behind the endodermal plate. Specialised ectoderm cells destined to form the **neural plate** lie just below the chorda cells.

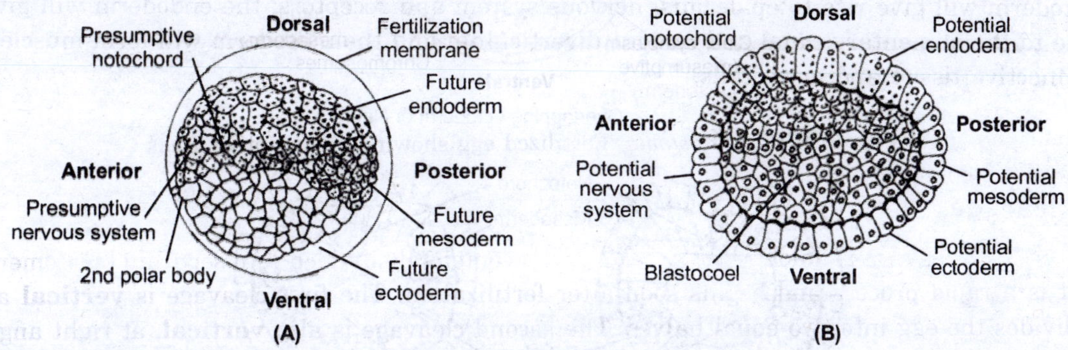

Fig. 6.30 *Branchiostoma.* **(A)** Blastula showing presumptive areas; **(B)** Blastula in sagittal section

Gastrulation

Further changes convert the single layered blastula into double layered gastrula and the process is termed **gastrulation**. The flattening becomes more and more pronounced and ultimately the endodermal plate becomes concave and sinks inwards as if pushed in by some invisible finger. This process is called **invagination** or **emboly**. The blastocoel is gradually reduced in size and a secondary cavity called **archenteron** is formed, its opening is called **blastopore** (which later becomes anus) and its **rim** the **lips of blastopore**. At this stage

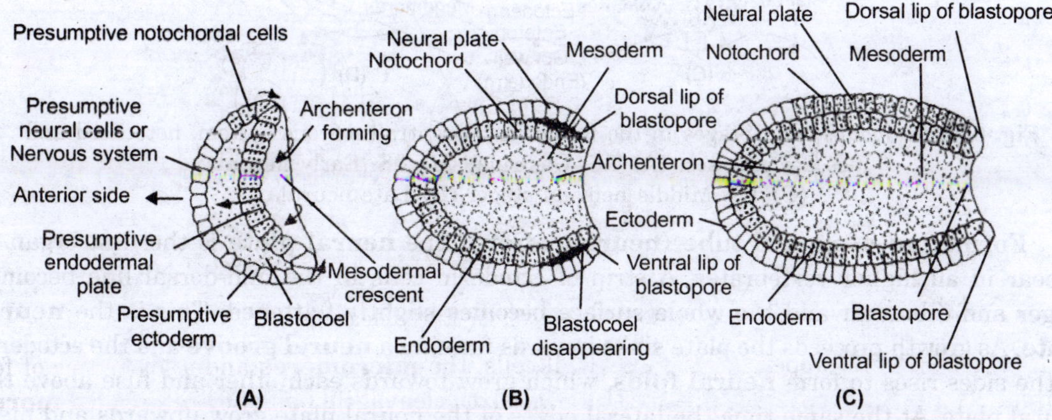

Fig. 6.31 *Branchiostoma.* **(A)** Easily gastrula; **(B)** Middle stage; **(C)** Complete gastrula

the embryo is a cup-shaped, double-layered **gastrula** (Fig. 6.31). The presumptive notochordal cells roll inwards along the dorsal lip of the blastopore and cells of the mesodermal crescent roll inwards along the lateral and ventral lips of the blastopore. This process is called **involution**. At about this stage the gastrula becomes covered with cilia by which it rotates in the egg case. The embryo now elongates and its dorsal side flattens. The dorsal lip grows backward as a result of which the blastopore shifts from a nearly dorsal to posterior position, becomes reduced in size and persists as the anus.

Organogenesis

Formation of first definite organ begin now. This is known as **organogenesis**. The ectoderm will give rise to epidermis, nervous system and receptors; the endoderm will give rise to the alimentary canal and midgut diverticulum and the mesoderm will form muscles, connective tissue and germ cells.

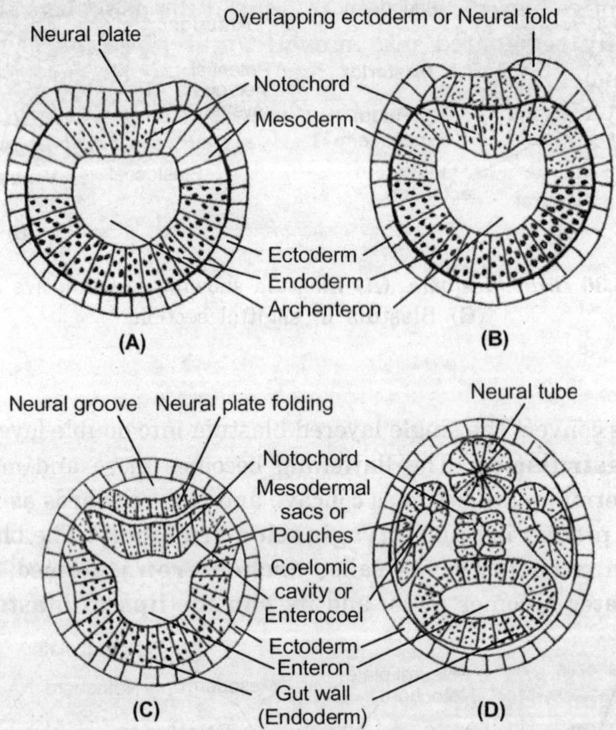

Fig. 6.32 *Branchiostoma.* Stages in the formation of central nervous system, notochord and mesodermal sacs. **(A)** T.S. Gastrula; **(B)** T.S. Early neurula; **(C)** T.S. Middle neurula stage; **(D)** Late neurula.

Formation of neural tube (neuralization). The **neural plate** is the first organ to appear in all higher vertebrates. A strip of ectoderm cells in the mid-dorsal line becomes larger and columnar and the whole surface becomes slightly flattened. This is the **neural plate**. As growth proceeds the plate sinks inwards forming a **neural groove** and the ectoderm at the sides rises to form **neural folds**, which grow towards each other and fuse above the neural plate. At the same time the lateral edges of the neural plate grow upwards and meet to form a hollow longitudinal **neural tube**. The cavity of neural tube is called **neurocoel** and it opens anteriorly by a minute aperture, the **neuropore** which closes in the adult forming the **olfactory** or **Kolliker's pit**. The neural tube communicates with the hinder end of the archenteron by a small **neurenteric canal** which closes after some time (Fig. 6.32). Unlike vertebrates the neural tube in *Branchiostoma* is formed entirely by the neural plate and neural folds do not take part in its formation.

Formation of notochord (notogenesis). As soon as the blastopore is closed, the **chorda cells** (Fig. 6.32) situated mid-dorsally in the roof of the archenteron rearrange themselves to form the **notochordal plate**. It bulges up towards the neural plate and finally pinched off from the archenteron to form the **notochord**. It elongates with the embryo. In early stages it is made up of a single row of disc-shaped cells. Later on these cells become vacuolated and a **notochordal sheath** surrounds them.

Formation of mesoderm, myotomes and coelom. As the development of notochord proceeds the presumptive mesodermal cells present in the dorso-lateral roof of archenteron, become metamerically segmented into **mesodermal pouches** or **coelomic pouches** (Fig. 6.33) by development of tranverse partitions. These make their appearance anteriorly and proceed with growth to the posterior end of the body. The formation of mesodermal

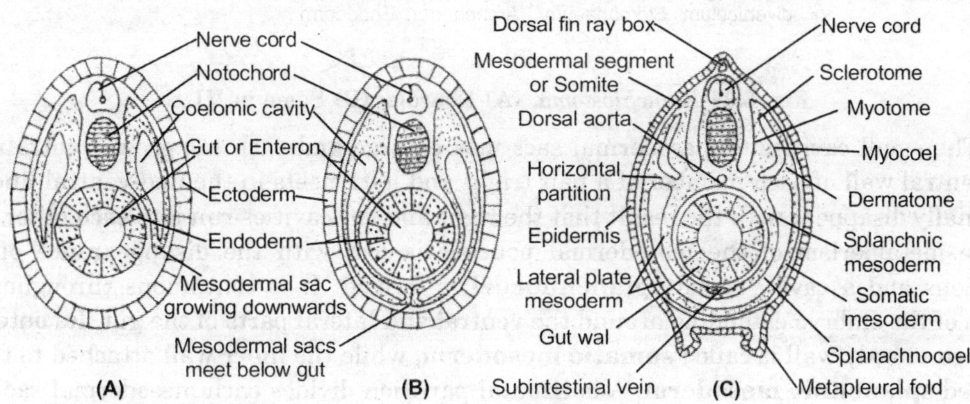

Fig. 6.33 *Branchiostoma.* Stages of development of mesodermal sacs and coelom in T.S.

pouches marks the beginning of metameric segmentation. As growth proceeds the mesodermal pouches become cut off from the wall of archenteron and finally their openings in the archenteric cavity close up, forming a series of closed **mesodermal** or **coelomic sacs**, each enclosing a small cavity. Thus, the origin of coelom is **enterocoelic**. The archenteron is reduced to **enteron** and develops the midgut diverticulum. Soon after the formation of mesodermal sacs, the embryo grows in length rapidly and is called *neurula* (Fig. 6.34). It has an elongated fish-like body. The neurula hatches by rupturing the vitelline membrane into a minute free swimming larva which is incapable of feeding as the mouth and anus are not yet formed.

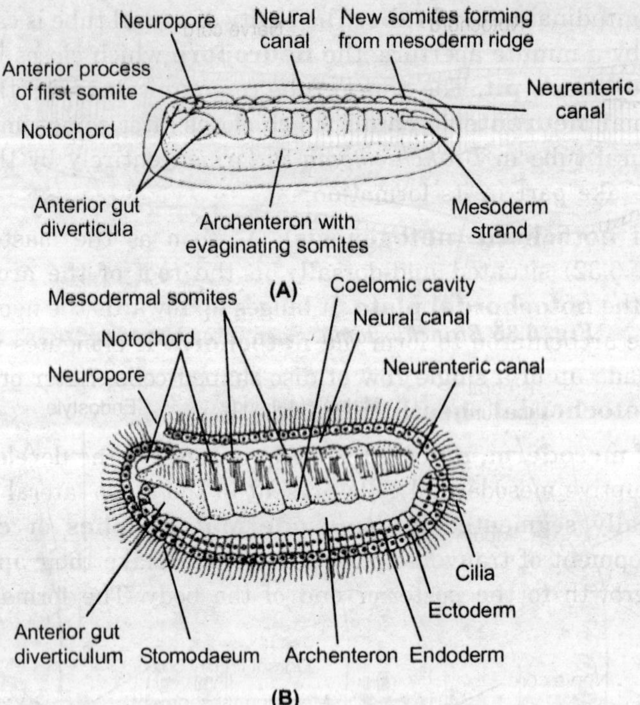

Fig. 6.34 *Branchiostoma.* **(A)** Neurula; **(B)** Same in V.L.S.

The small cavities of mesodermal sacs now enlarge and each sac grows ventralwards. The ventral wall of each member of a pair (right and left) meets in the mid-ventral line, fuse and finally disappear with the result that the right and left cavities run into each other. Later on the metamerism of the mesodermal pouches is lost with the disappearance of cross partitions and a cavity called **splanchnocoel** is formed. It is continuous throughout the length of the embryo extending around the ventral and lateral parts of the gut. Its outer wall attached to body wall is called **somatic mesoderm,** while the inner wall attached to the gut is called **splanchnic mesoderm**. A horizontal partition divides each mesodermal sac into a dorsal and a ventral portion. The dorsal part is distinguished as the **somite** which form the **myotomes**. The cavity of myotomes is called **myocoel**. Thus the dorsal part of the mesoderm retains metamerism. The ventral part is called the **lateral plate mesoderm** and its cavity forms a part of the coelom.

Larval development

The larva keeps on swimming without feeding for sometime and elongates rapidly. Then mouth develops. A patch of ectoderm fuses with the endoderm of the gut on the left side of the ventral surface of the head and then perforation occurs at that point. This is the mouth. The anus is formed in a similar manner on the left side of the mid-ventral line near the posterior end of the body. The development now becomes asymmetrical due to spiral movement and method of feeding (Fig. 6.35). The first two coelomic pouches also grow asymmetrically, the right one grows large and becomes the preoral body cavity or the cavity of the head and the left becomes the preoral pit or Hatchek's pit.

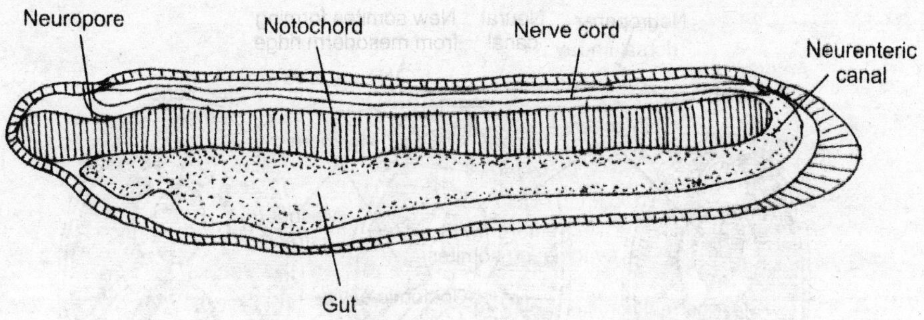

Fig. 6.35 *Branchiostoma.* Soon after hatching

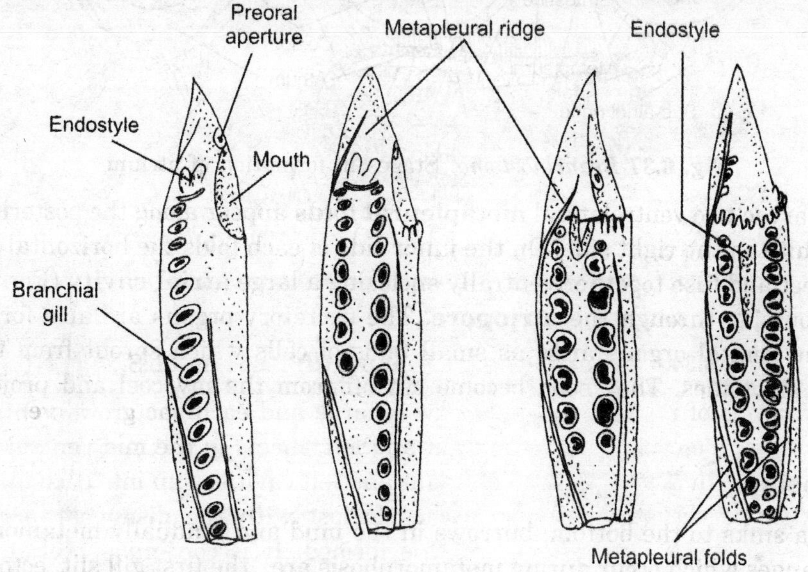

Fig. 6.36 *Branchiostoma.* Larva showing changes in the position of gill openings

Simultaneously with these changes the pharynx becomes enlarged and gill clefts also appear but not in a regular sequence from front to rear. The first cleft arises before mouth opens, as a diverticulum from the anterior end of the pharynx on the right side. Ultimately it develops into a club-shaped gland which opens to the exterior just ventral to the mouth. Other gill slits appear in the mid-ventral line. They are the primary gill slits, fourteen in all, and later on form the gill slits of the left side. The first series of slits of the right side are eight in number and are formed just above the fourteen previously formed slits (Fig. 6.36). Thus two linear series of gill slits are formed on the same side of the body, the right. With further growth the first fourteen gill slits move to the left side across the mid-ventral line and take up their definite position. Now six slits close thus forming eight slits on each side. The gill slits of both side then become subdivided by tongue bars and further slits are added to the hinder end of both series. The wall of the pharyngeal floor become modified to form a V-shaped ridge of ciliated cells, this is the **endostyle** and its cells become glandular later on.

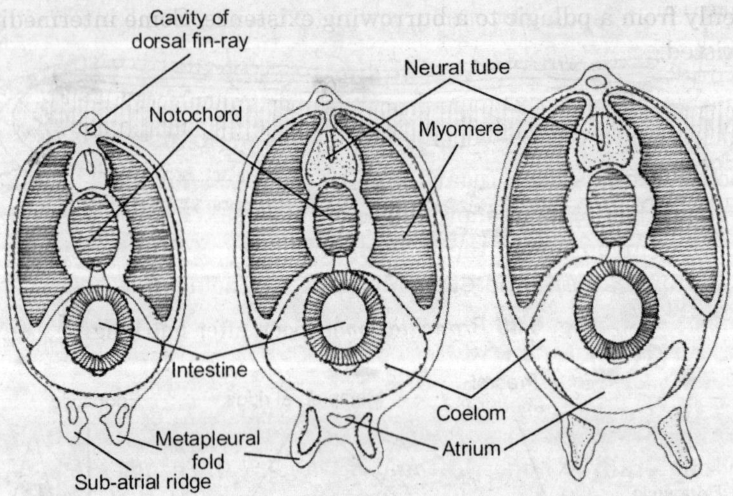

Fig. 6.37 *Branchiostoma.* Stages in formation of atrium

In older larvae two ventrolateral **metapleural folds** appear along the posterior extension of the first somite on the right side. On the inner side of each folds the horizontal **epipleures** grow. They meet and fuse together ventrally enclosing a large **atrial cavity** (Fig. 6.37) which opens to the outside through the **atriopore**. The excretory organs are also formed during larval life. The genital organs arise as small buds of cells which sprout from the anterior angles of the myotomes. They soon become cut off from the myocoel and project into the gonocoel.

Metamorphosis

The larva sinks to the bottom, burrows in the mud and gradually metamorphoses into adult. The changes which occur during metamorphosis are : the first gill slit, ectodermal cilia and the club-shaped gland disappear, and the mouth becomes anterior. Oral hood and cirri, wheel organ and velum develop, the gill slits increase in number because the primary gill slits are subdivided by tongue bars. The liver begins to develop as a little pouch of the gut just behind the pharynx. Nephridia develop. The tail region elongates and the notochord extends into the rostrum.

From the above description it is evident that some features of the larva are of special interest, especially its asymmetry. The strange sequence of gill formation, the large left-sided larval mouth and the club-shaped gland are some asymmetrical features of the larva. It is difficult to explain how this extraordinary asymmetrical larva of *Amphioxus* becomes metamorphosed into an adult which is nearly bilaterally symmetrical. It must however, be admitted that the larva of *Branchriostoma* is so extraordinary as to present a serious problem for those who believe that the larval stage represents, in modified form, an ancestral condition of affairs. That a bilaterally symmetrical embryo should pass through an asymmetrical plagic larval stage in order to attain the form of a practically symmetrical burrowing adult seems to be completely contrary to what should be expected. A very plausible solution has been suggested by Stafford. He points out that the ancestor of *Branchiostoma* could not have jumped suddenly from a pdlagic to a burrowing existence. Some intermediate condition of life must have existed.

PHYLOGENY

Cephalochordata has been regarded to be phylogenetically related to chordates as well as many non-chordates. Since it shows more chordate-like features its relationship with non-chordates has been given up. In the following paragraph its affinities with some important non-chordate groups like Annelida and Mollusca are being described.

AFFINITIES WITH ANNELIDA

Similarities

1. Body bilaterally symmetrical and metamerically segmented.
2. Coelom well developed.
3. Blood vascular system closed type.
4. Filter feeders as some polychaete worms.
5. Excretory organs protonephridia as in Polychaeta.

Differences

1. Metameric segmentation occurs only in myotomes and gonads in Cephalochordata.
2. In Cephalochordata development of coelom is enterocoelic whereas in Annelida it is schizocoelic.
3. The flow of blood in the main blood vessels in both the groups is in opposite direction.
4. The three basic chordate characters present in Cephalochordata are lacking in Annelida.

AFFINITIES WITH MOLLUSCA

Branchiostoma was first discovered and described by Pallas in 1778 who regarded it as a kind of slug (Mollusca) and named it *Limax lanceolatus.* The two groups, Cephalochordata and Mollusca, show similar feeding and respiratory mechanism, but this is due to similar mode of life. Their anatomy, otherwise, is completely different.

AFFINITIES WITH CHORDATA

Cephalochordata possesses three basic chordata features. It also shows phylogenetic relationship with the major groups of Chordata.

AFFINITIES WITH HEMICHORDATA

Similarities

1. Pharyngeal apparatus, respiratory and feeding mechanism similar.

2. Origin of coelom is enterocoelic.
3. Gonads numerous and without gonoducts.

Differences

1. The muscles of Hemichordata are not segmented.
2. Gill slits dorsal in Hemichordata but lateral in Cephalochordata.
3. Postanal tail absent in Hemichordata.

The position of Hemichordata in the Chordata is doubtful because of the nature of notochord. It is regarded as more primitive than Cephalochordata.

AFFINITIES WITH UROCHORDATA

Recent studies link Cephalochordata with Urochordata.

Similarities

1. Similar mode of development of the nervous system in ascidian larva and *Amphioxus* larva.
2. The notochord appears similarly in both and presents similar structure.
3. The pharynx and the associated complicated structures like the endostyle, epipharyngeal groove, dorsal lamina, etc., are similar in both.
4. Atrium and atriopore in the larval forms are similar.

Differences

The adult Urochordata is different from adult Cephalochordata in several features :

1. Unsegmented body covered by test.
2. Notochord and nerve cord absent.
3. Digestive gland is liver.
4. Nephridia absent.
5. Adult hermaphrodite.

The natural conclusion is that the two groups, Urochordata and Cephalochordata, show impressive similarity in their structure and embryonic development but Cephalochordata is better evolved than Urochordata. The similarity between the two groups are most probably due to common ancestry.

AFFINITIES WITH VERTEBRATES

Similarities

1. Metameric segmentation (myotomes).
2. Presence of true coelom lined by mesoderm.
3. The digestive gland, midgut diverticulum can be compared with the liver.

4. Hepatic portal system well developed.
5. The flow of blood is forward in ventral blood vessel and backward in the dorsal blood vessel.

Differences

1. Skull or cranium absent in cephalochordata.
2. No muscular heart and red blood cells.
3. Brain and specialized sense organs absent.
4. Atrium present in Cephalochordata.
5. Excretory organ in Cephalochordata is nephridia.
6. In *Branchiostoma* gonads are numerous and without gonoducts.

Systematic position

Although *Branchiostoma* possesses all the basic chordate features but still its systematic position remains controversial. It was first discovered by Pallas in 1778 who named it *Limax lanceolatus* (a slug). Later, an Italian naturalist, Costa (1834) identified it correctly as a lower vertebrate and called it *Branchiostoma lanceolatum.* A couple of years later Yarrel rediscovered it and gave it the most popular name *Amphioxus,* although the law of priority claims the use of the name *Branchiostoma.* Yarrel included it with fishes in the order Marsipobranchii, but any animal without gills, heart, brain, red blood corpuscles, endoskeleton, and paired eyes and other sense organs cannot be grouped under Pisces. Therefore a new group Cephalochordata was established for *Amphioxus* and the same has been raised to the status of a subphylum now.

CONCLUSION

From the study of morphology and anatomy of *Amphioxus* it is evident that it possesses practically the full complement of chordate characters and differs from the craniates in the relative simplicity of its bodily organization. The opinions, however, are divided on its position. Some regard it as primitive whereas, others consider it to be modified or degenerate.

The **primitive features** which *Amphioxus* seems to share with the prechordate ancestor include

1. Naked notochord from end to end.
2. Complete myotomic segmentation from end to end.
3. Absence of jaws for biting.
4. Absence of paired fins.
5. Simple one-layered epidermis.
6. Simple straight tubular intestine with a simple liver diverticulum and the ciliary method of feeding.
7. Simple circulatory system without a specialized heart.

8. Segmental nephridia which are not coelomoduct.
9. Segmental gonads with no ducts.
10. Separate dorsal and ventral roots of spinal nerves.
11. Small almost yolkless egg, hollow spherical blastula, embolic gastrulation and formation of anterior coelomic pouches.

Amphioxus also possesses certain **specialized features**.

1. The peculiar **pharynx**, with very large number of gill slits is one such important feature. The ciliary feeder relies on it for the maintenance of water current from which food is sifted. The food sifting is made possible by the presence of elaborate velum and oral hood equipment.
2. The atrium and atriopore is another such specialized feature.

The **degenerate** characters are:

1. The reduced brain and sense organs probably in response to its sedentary life.
2. Then extension of notochord up to the anteriormost tip of the body.

Amphioxus has been regarded as a generalized chordate. This has been very clearly expressed by the statement "That if *Amphioxus* had not been discovered it would have to have been discovered." Its structure throws much light on the form of ancestral chordate, but unfortunately gives no hint of the way in which it might have evolved from any known type of non-chordate animal.

7

CHAPTER

CEPHALOCHORDATA:
GENERAL CHARACTERS AND CLASSIFICATION

GENERAL CHARACTERS

1. Exclusively marine, found in shore water and on sandy beaches.
2. Burrowing animal leading a sedentary life.
3. Body about 5 to 8 cm long, fish-like, slender, semi-transparent, with metameric segmentation, head absent.
4. Paired fins absent, only median fins present.
5. Muscles in the form of segmental myotomes.
6. Coelom enterocoelous in origin. In the region of pharynx it is very much reduced due to development of atrium.
7. Notochord persistent and extend along the entire body, right up to the tip of the snout.
8. Alimentary canal is a perfectly straight tube lined by ciliated epithelium.
9. Pharynx with large number of gill slits which open into the atrium.
10. Ciliary feeders.
11. No special respiratory organ. Exchange of gases takes place through the whole surface of the body.
12. Excretion by protonephridia.
13. Circulatory system closed type, without heart and red blood corpuscles. Hepatic portal system well devoloped.
14. Nerve cord hollow, tubular situated on the dorsal side of the notochord. The dorsal and ventral nerve roots do not join.
15. Sexes separate. Gonads numerous and without gonoducts.
16. Fertilization takes place in sea water.
17. Development indirect.

CLASSIFICATION

The subphylum Cephalochordata has only one class.

Class: Leptocardi: Slender fish-like Chordata in which epidermis is single layered, many gill slits occur and there are no scales.

Example: *Branchiostoma.*

SUBPHYLUM VERTEBRATA

Vertebrata is a subphylum of the phylum chordata. It includes animals called **vertebrates** characterised by the presence of **vertebral column** or **backbone** made up of serially arranged **vertebrae**, hence the name. Another important character of all vertebrates is the presence of **cranium** or **skull** or **brain base** enclosing the brain and various sense organs, hence the name **Craniata**.

Often the terms Chordates and vertebrates are considered synonymous but actually they are not because phylum chordata includes all animals having a notochord at some stage in their life while Vertebrata is a subphylum of the phylum Chordata and includes animals having vertebral column and cranium in the adult stage. It shows the basic chordate features (notochord, dorsal tubular nerve cord and pharyngeal gill slits) in the embryonic stage.

Vertebrata may defined as a special group of chordates having cartilagineous or bony endoskeleton, consisting of a cranium enclosing the brain and a vertebral column, through which runs the spinal cord. Although the vertebrates include majority of the chordate animals consisting of about 40,000 living species but it is many times smaller than some groups of invertebrates. The animals belonging to this group are highly diverse and show many interesting features.

Basic characters of vertebrates

The three fundamental chordate features, i.e., notochord, dorsal tubular nerve cord and pharyngeal gill slits and the vertebral column and cranium form the basic characters of all vertebrates.

General characters of vertebrates

1. Vertebrates inhabit all types of habitats. Lower vertebrates (cyclostomes, fishes and amphibians) are aquatic and higher vertebrates (reptiles, birds and mammals) are terrestrial.
2. Size medium to large.
3. Body bilaterally symmetrical and metamerically segmented.
4. Body divisible into head, trunk and tail; in terrestrial forms a neck is present between the head and trunk.
5. Two pairs of lateral appendages are present on the trunk. They help in locomotion or some special function.
6. Skin made up of outer epidermis and inner dermis, glandular in aquatic forms, glandular in reptiles and birds and with secondary glands (sweat and sebaceous) in mammals.

7. Skin with exoskeleton in the form of scales, feathers, hairs, nails, horns, etc. (amphibians naked).
8. Coelom large, schizoeoclic and occupied by viscera.
9. Endoskeleton axial (skull and vertebral column) and appendicular (limbs and girdles), mostly bony; in some cartilaginous. The notochord replaced by the vertebral column.
10. Muscular system well developed.
11. Alimentary canal long and more or less convoluted.
12. Respiratory organs are paired gills in aquatic forms and a pair of lungs in terrestrial forms.
13. Blood vascular system is of closed type. Heart is ventral, muscular and contractile. Blood contains respiratory pigment haemoglobin. Both white and red blood corpuscles present.
14. Excretory organs are a pair of kidneys.
15. Anterior part of the nerve cord forms the brain enclosed in cranium and the posterior part forms the spinal cord which runs through the vertebral column.
16. Sense organs well developed.
17. A system of ductless glands (endocrine glands) is present which regulates various physiological processes of the body.
18. Sexes separate.
19. Development direct or indirect.

Classification of vertebrata

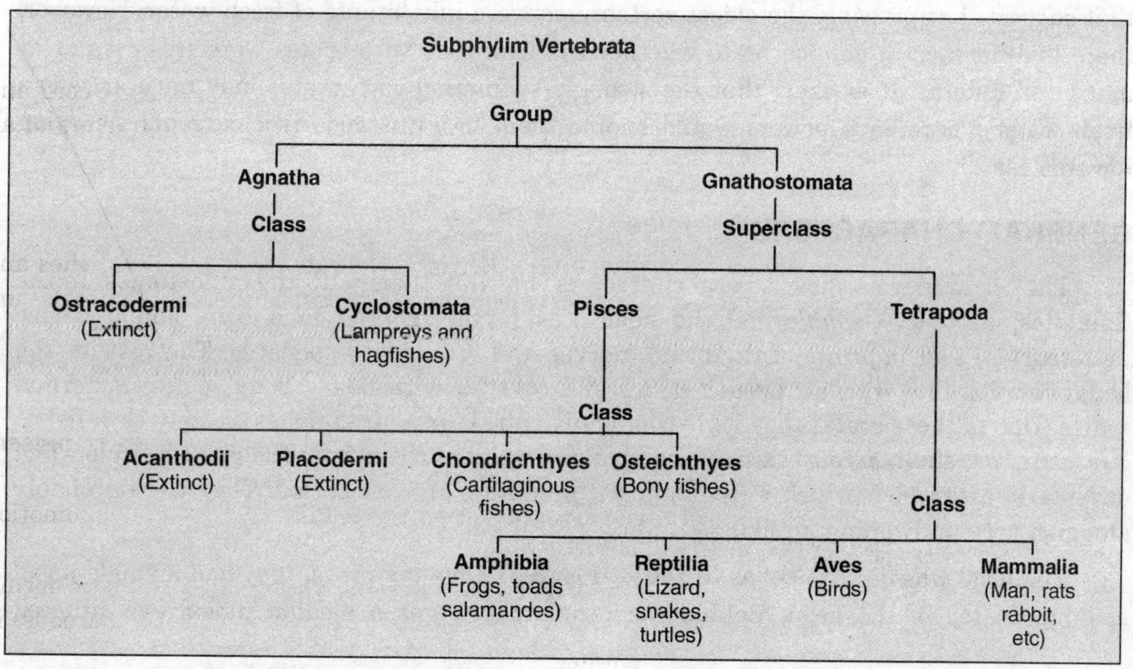

8
CHAPTER

CLASS OSTRACODERMI

The member of the class **Ostracodermi** are commonly known as **ostracoderms**. They are probably the earliest known jawless, fish-like vertebrates belonging to the group Agnatha. They closely resemble the present day cyclostomes.

OCCURRENCE

The ostracoderms present the first opportunity of the study of fossil record. In the second period of the Palaeozoic era, the Ordovician, dating perhaps 500 million years back in the earth's history, the first faint traces of the oldest known vertebrates are available. It is in the form of small flakes of bones in rock deposits apparently formed in shallow waters that suggest that probably the oldest vertebrates were inhabitants of fresh water. However, there is difference in opinion as to whether the first true vertebrates were fresh water or marine in habitat. It is likely that the strongly swimming vertebrates may have arisen in fresh water where such powers would enable them to withstand river currents sweeping towards sea.

GENERAL CHARACTERS

The ostracoderms show several characters that link them with the cyclostomes. In the Caphalaspids (e.g., *Cephalaspis*) the head was large, encased in a bony shield whose undersurface had imprints of the brain, nerves and branchial apparatus. The rest of the body was fish-like with an upturned tail and covered with heavy bony scales or dermal scales. Due to these scales they were commonly called armoured fishes or bony skin fishes (**Gr.** *ostrakon*, shell; *derma*, skin). These scales protected them from the giant scorpion-like eurypterid arthropods which were dominant predators of that period. They were probably sluggish bottom dwelling animals.

The head was quite peculiar in shape (Fig. 8.1). Like lampreys, they had a single nasal opening on top of the head. Behind the nostril there was a median pineal eye situated between a pair of large lateral eyes. The head had sensory fields which were, probably, a part of the lateral line system characteristic of fishes. A small mouth without jaws or teeth was present on the underside. They were probably filter feeders. The internal ear had two semicircular canals.

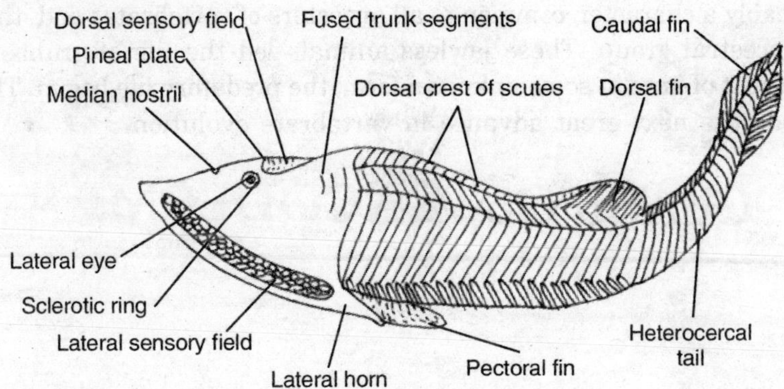

Fig. 8.1 Ostracoderm. *Cephalaspis*

Another ostracoderms dominant in this period was *Pteraspis* (order Pteraspida). The body (Fig. 8.2) was large with the snout prolonged into a rostrum. The head was encased in a shield. The tail was heterocrcal, laterally compressed and covered with small scales. The trunk was covered with small rectangular scales. A dorsal spine projected backwards and upwards from the posterior end of the cephalic shield. Lateral line was present but no other sensory organs were present. It is likely that they gave rise to hagfishes (*Myxine*).

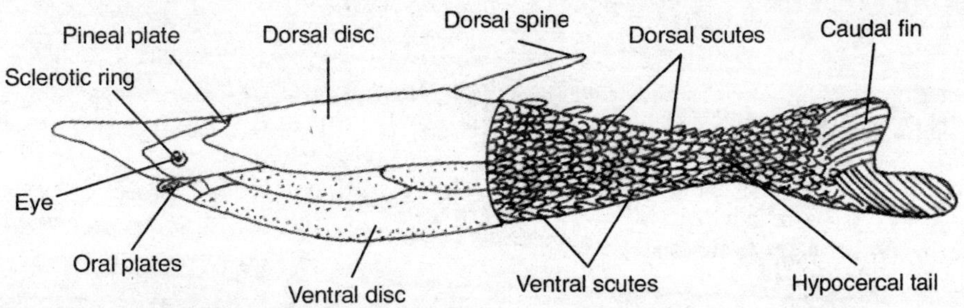

Fig. 8.2 Ostracoderm. *Pteraspis*

BIOLOGICAL SIGNIFICANCE

It seems likely that the jawless and limbless condition is one to be expected in the oldest vertebrates. Ostracoderms present the most primitive stage in vertebrate evolution, but it cannot be said that they were ancestors of lampreys. Recent work, however, has tended strongly to show that some of the ostracoderms probably are exceedingly close to the line of descent of both cyclostomes and higher fishes. Bone, in the form of both the surface armour and internal skeletal parts, seems to be an extremely old vertebrate character. It is probable that many vertebrates now lacking such structures are degenerate rather than primitive. We know for example, sturgeon (a fish), with hardly a bone in its body, has descended from forms with a well ossified skeleton and it is highly probable that other modern forms which have no bones such as lampreys and sharks, are in reality degenerate rather than primitive types. Armour was probably a character, common to all ancestors of vertebrates and the ostracoderms are really an ancestral group. These jawless animals led the life of grubbers in the mud. With the development of jaws in some unknown form, the predatory life begun. The development of jaws thus made the next great advance in vertebrate evolution.

9

CHAPTER

CYCLOSTOMATA

The **Cyclostomata (Gr.** *cyklos,* circular; *stoma,* mouth) are the most primitive living vertebrates belonging to the group **Agnatha (without jaws)**. They include lampreys, hagfishes and slime eels. The cyclostomes are commonly known as round-mouthed eels because they superficially resemble the eels, but they are very different from the eels that belong to the class Osteichthyes ***(bony fishes)***. Many of the customary fish-like characters such as paired fins, scales, swim bladder, true mesodermal teeth, vertebral centra, ribs, etc. are lacking in cyclostomes.

GENERAL CHARACTERS

1. Body elongated and eel-like.
2. Skin smooth, soft, scaleless, with unicellular mucous glands.
3. The muscles of trunk and tail are regular segmental myotomes separated by mycommata.
4. Mouth is ventral, circular, suctorial and devoid of jaws.
5. Tongue is muscular, piston-like, with horny epidermal teeth.
6. Paired fins absent, median fins with cartilaginous fin rays, tail fin diphycercal.
7. External nasal opening single, median in position.
8. They breathe by means of 6-15 pairs of internal gills situated in lateral sac-like gill pouches present in the pharynx.
9. Heart is four-chambered, with one auricle, one ventricle, a sinus venosus and a conus arteriosus anteriorly.
10. Kidneys two, mesonephric, with ducts to the urinogenital papilla.
11. Brain is differentiated with 8 or 10 pairs of cranial nerves . Each auditory organ has one or two semicircular canals.

12. The skeleton consists of the notochord which remains well developed throughout life as an axial stiffening rod and is enclosed in a thick fibrous sheath continuous with a layer of connective tissue which also encloses the spinal cord. Certain cartilaginous thickenings **(arcualia)** over the sheath represent rudimentary vertebrae. In the tail region these pieces fuse to form a continuous plate.
13. Sexes separate or united. Gonad single, large and without duct.
14. Fertilization external, development direct (hagfishes, slime eels) or indirect (lampreys). The larval stage is called **ammocoete**.
15. Parasitic in habit, sometimes may lead to the death of the unfortunate host.
16. Usually marine, although move to fresh water to breed. Some species permanent inhabitants of fresh water.

TYPE LAMPREY

Lampreys are found both in fresh and marine waters. The fresh water forms are generally smaller than the marine species. The common fresh water species is *Lampetra fluviatilis* and the marine species is *Petromyzon marinus.* The lampreys are worldwide in distribution and occur in the temperate zone of both the hemispheres.

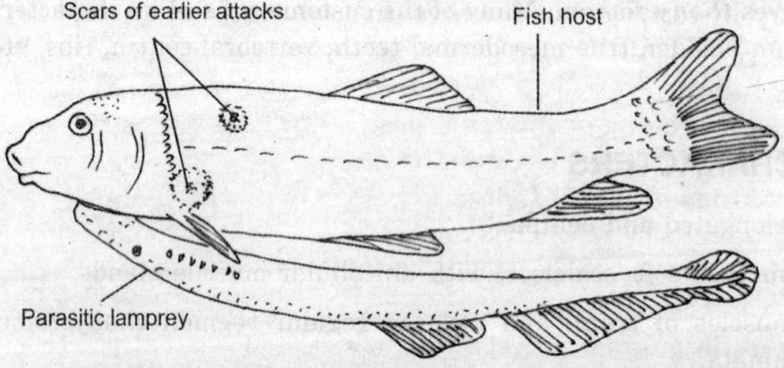

Fig. 9.1 *Petromyzon*. The animal attached to a bony fish

HABITS

The marine lamprey *Petromyzon marinus* is a parasitic species. It feeds on fishes, turtles, etc. during the adult stage. They are often very harmful resulting in the death of the unfortunate host. They attach themselves to the side of fishes (Fig. 9.1) or other objects by their suctorial mouth and then cut the tissue by the file-like horny teeth that cover their tongue. *Petromyzon* is not a strong swimmer. It swims near the bottom of the sea by undulations of its body. During the breeding season it leaves the sea and goes to fresh water streams to spawn. Such animals are called **anadromous**. The adults die soon after spawning and their body disintegrates rapidly as the animal is without much skeletal material. The larval stage is called **ammocoete**. It is a harmless, fresh water, filter feeder, sedentary stage with a life period of 3 to 7 years depending upon the species. The young lampreys, after attaining maturity migrate to the sea.

EXTERNAL FEATURES

The lampreys have an elongated, cylindrical, eel-like body, (Fig. 9.2) which is differentiated into **three**, not clearly demarcated, regions: **head, trunk** and **tail**. The tail is laterally compressed. They are **greenish brown** in colour with the upper surface **dark** and lower surface **light**. They are of varying size. *Petromyzon marinus* is about one metre long, *Lampetra fluviatilis* is 90 cm long and *L. planeri* is about 45 cm long.

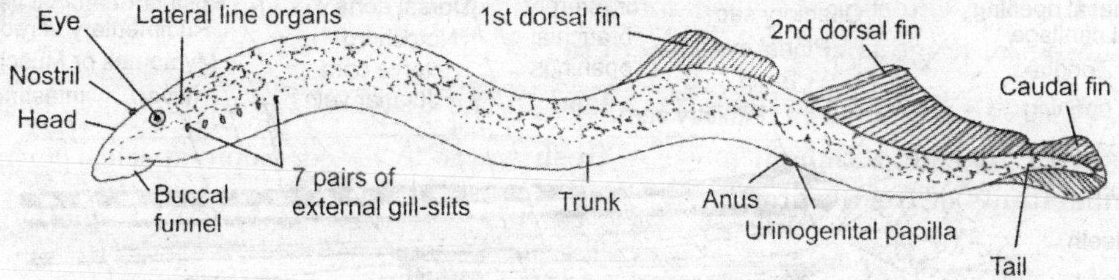

Fig. 9.2 *Petromyzon*. External features

Paired fins are not present but two unequal **median dorsal fins** and a **caudal fin** are present. The caudal fin surrounds the tail. Both the dorsal fins are situated near the posterior end of the body. The second dorsal fin is bigger than the first and is continuous with the upper lobe of the caudal fin. The dorsal fin of mature animal may show sexual dimorphism. The fins are supported by cartilaginous **fin rays**. In some lampreys an **anal fin** may be present in the female but in male it is reduced to a **copulatory papilla**. A pair of laterally placed **eyes** covered by transparent skin is present on the head. On the dorsal side of the head a single median **nasal opening *or* nostril** is present. Behind the nostril there is a transparent area of skin which indicates the position of **pineal eye**. There are seven pairs of small rounded **gill openings** or **gill slits** in a longitudinal row on each lateral side, behind the eye. Along each lateral side of the body occurs a row of small lateral sensory pores of the **lateral line system**. On the ventral side, just at the junction of the trunk and tail, the **anal opening** is present. Close to the anus a stit-like depression, the cloaca is present. A small **urinogenital papilla** protrudes through the cloaca. On the ventral side of the head there is a large cup-like **buccal funnel** or **sucker** with the help of which the lampreys attach themselves to their host. The mouth lies in the centre of the buccal funnel.

INTERNAL ANATOMY

Integument

The **body surface** or skin is soft, smooth with no scales and is many layered. The **epidermis** is the outermost layer, followed by **cutis, subcutis** and finally the **musculature**. The outermost layer of cells of the epidermis have a striated cuticular border. The inner layers of epidermis are protoplasmic and become progressively cornified as they pass outwards. The cornification takes place due to production of a compound called **keratin**. Such a protective layer occurs in all vertebrates. Many gland cells occur in the epidermis as in other aquatic vertebrates. These cells produce slime, thus the skin of lampreys is slimy. The **cutis** or **dermis** is a layer of bundles of collagen and elastic fibers running mostly in a circular direction. Below this lies a distinctly marked-off layer, the **subcutis**, consisting of connective tissue, fat and blood vessels. Star-shaped pigment cells occur in the cutis and also form a thick pigment layer at the junction of cutis and subcutis. The pigment in these cells possess the ability to migrate, making the animal pale or dark. In the larval lamprey this change is especially well marked.

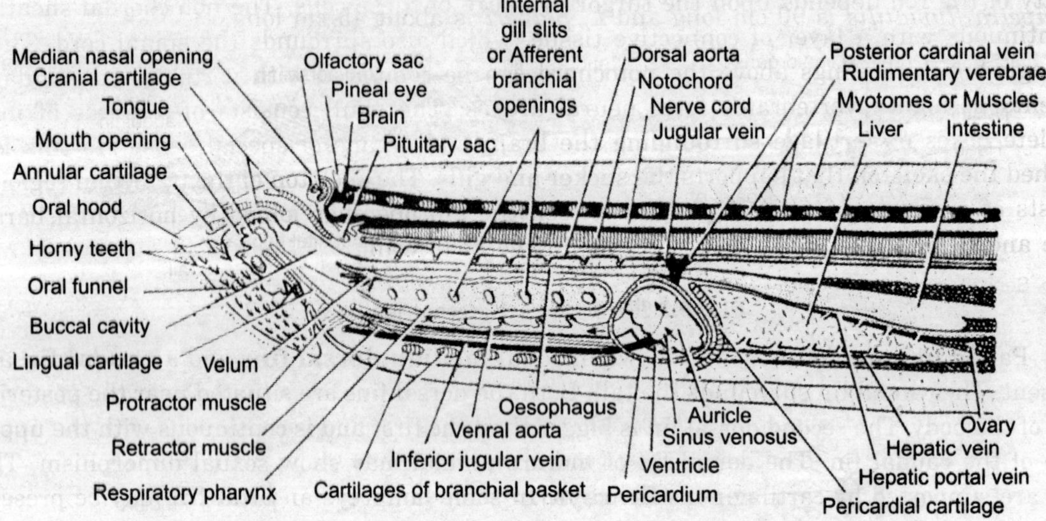

Fig. 9.3 *Petromyzon*. M.L.S. of anterior region showing internal structures

MUSCULAR SYSTEM

The segmental musculature of cyclostomes presents the same generalized plan as in *Amphioxus* but the myotomes on the back of the head are W-shaped instead of V-shaped as in the latter. Regular segmental **myotomes** or **myomerees** (Fig. 9.3) run uniformly from head to tail. They are separated by tough connective tissue partitions called **myocommata.** No horizontal septum divides the myotome into dorsal and ventral halves, as in true fishes because the lateral line is absent. The musculature of the tongue is an intricate system of muscle bands consisting of small **protractor** and large **retractor muscles**. They help in the rasping movement of the tongue. The buccal funnel consists of **radial muscle fibres**.

MOVEMENT

The movement of lamprey is mechanically inefficient ; they show great activity but move slowly. They swim in an eel-like manner, using myotomes serially. Contraction of myotomes on one side results in bending the entire body in that direction. Often it comes to rest by attaching itself to its prey or some stone. In this position, water enters through the gill openings. When swimming the backward jet of water may help in locomotion.

SKELETON

The skeleton of lamprey includes (*i*) the **notochord** with cartilaginous **arcualia,** dorsal to the notochord, two pairs in most body segments; (*ii*) the complex **skull**; (*iii*) the elaborate paired **visceral arches** or **branchial basket**; (*iv*) the stout **lingual cartilage** of the tongue; (*v*) a ring of **annular cartilage** surrounding the buccal funnel; and (*vi*) the **cartilaginous structure** around the heart and the pericardium.

The notochord remains well developed throughout life as axial skeleton. It is a slender gelatinous rod made up of large vacuolated cells enclosed in a thick fibrous sheath. The rigidity of the rod depends upon the turgor pressure of these cells. The notochordal sheath is continuous with a layer of connective tissue, which also surrounds the spinal cord. The cartilaginous thickenings above the notochord can be compared with vertebrae as similar elements in higher vertebrates form neural arches. The skull consists of a series of incomplete boxes of cartilage surrounding the brain and organs of special sense. To this is attached the skeleton that supports the sucker and gills. The skeleton of the branchial region consists of a system of vertical plates between the gill openings joined by horizontal bars above and below. This whole structure lies in the body wall.

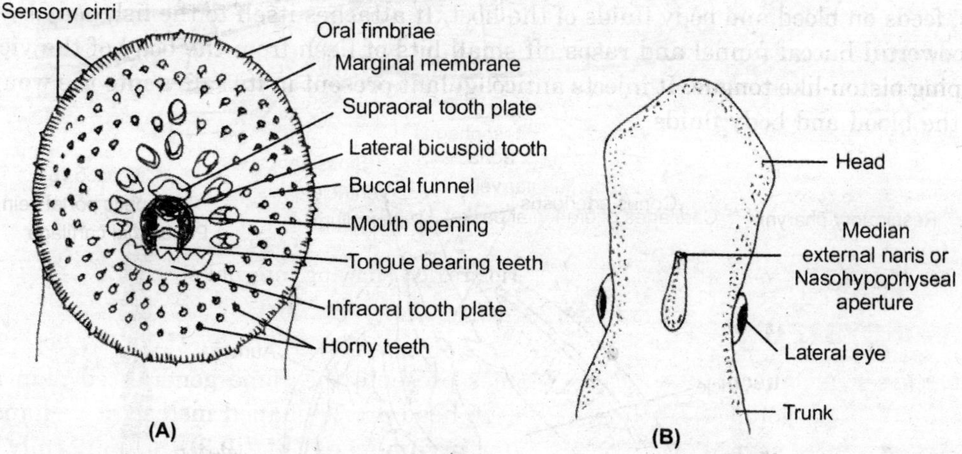

Fig. 9.4 *Petromyzon*. **(A)** Head, Buccal funnel, ventral view; **(B)** Head, doreal view

DIGESTIVE SYSTEM

On the ventral side of the head a large cup-like **buccal funnel** or **sucker** (Fig. 9.4 A) is present. Its margin bears numerous small soft projections called **oral fimbriae** or **papillae**. In between the papillae long sensory processes called **cirri** are present. The inside of the buccal funnel is lined by rows of yellow conical horny **epidermal teeth**, which show definite arrangement. The **mouth** lies in the centre of the buccal funnel. It is surrounded by concentric rows of bicuspid teeth, the arrangement of which varies in different species. The mouth is closed or opened by forward and backward movement of the **piston-like tongue**, which also bears large sharp horny teeth. The tongue is used as a rasping organ. The mouth leads into a large **buccal cavity** (also called **pharynx** by some). At the hinder end it divides into a dorsal food passage, the **oesophagus** and a ventral **respiratory tube**. The latter is a blind sac bearing 7 internal gill slits on each side. Its entrance is guarded by a **velum** bearing a series of sensory **velar tentacles**, which prevent the passage of food into the respiratory tube. The oesophagus opens directly into the straight intestine as there is no true stomach in lamprey. The anterior part of the intestine is dilated, representing the stomach. The surface area of the intestine is increased by a **typhlosole** or **spiral valve**. The intestine ends at the small anus (Fig. 9.3).

A pair of pigmented sacs, the **salivary glands** are embedded in the hypobranchial muscles. Each gland has a folded wall and opens by a duct below the tongue. The secretion of the salivary glands contains an anticoagulant which prevents coagulation of blood of the fish on which the lamprey feeds. The structures are of special interest, as salivary glands do not occur in other vertebrates except mammals. A large bilobed **liver** surrounds the anterior part of intestine. A gall bladder and a bile duct are absent in the adult (present in the larval stage). The presence of a pancreas is uncertain. The epithelium of anterior part of intestine contains two types of secretary cells: **zymogen cells** which secrete pancreatic enzymes and are tryptic in nature and **cells of the follicles of Langerhans** which are endocrine in nature. They maintain the level of blood glucose in the animal.

Lampreys are ectoparasite feeding mainly, on bony fishes. It is **sanguivorous** in habit, that is, feeds on blood and body fluids of the host. It attaches itself to the fish body by means of its powerful buccal funnel and rasps off small bits of flesh from the body of the victim by its rasping piston-like tongue. It injects anticoagulant present in its saliva into the wound and sucks the blood and body fluids.

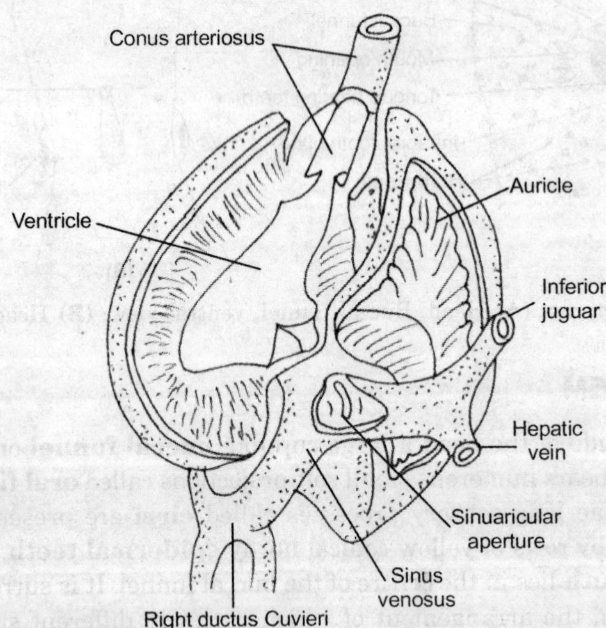

Fig. 9.5 *Petromyzon*. Diagrammatic longitudinal section of heart to show the internal structure

CIRCULATORY SYSTEM

The circulatory system of lamprey is similar to that of *Amphioxus,* but a definite **heart** has developed. This lies just behind the gills within the cup-shaped posterior end on the branchial basket. It is enclosed in a thick-walled **pericardium** (Fig. 9.3) which communicates with the coelom. The pericardium is supported by a cartilaginous plate. The heart is **S-shaped**. In the larva it is a **straight tube** but due to lack of space it twists and becomes S-shaped. The heart is divided into four chambers (Fig. 9.5) the posteriormost thin-walled chamber is the **sinus venosus**, it opens into a thin walled **auricle** lying just above. The auricle opens into a thick-walled **ventricle** below, followed by the **conus arteriosus**. The contractions proceed forwards from behind.

The sinus venosus receives blood from various parts of the body by veins and conducts it first to the auricle and then into the ventricle, which pumps it into the ventral aorta. It runs forwards and gives off seven pairs of **afferent branchial arteries** to the gills where aeration takes place. From the gills the aerated blood is collected by seven pairs of **efferent branchial arteries** into a **median dorsal aorta** which carries it to all parts of the body and to the myotomes through a series of **segmental arteries**. Most of the major blood vessels are provided with valves at their beginning to reduce the pressure of blood in the arteries.

The venous system presents simple vertebrate plan. A single **caudal vein** collecting blood from the tail divides into two **posterior cardinals** on entering the abdomen. These receive blood from the kidneys, gonads, etc. and open into the heart by a single **ductus cuvieri** on the right hand side. **Anterior cardinals** collect blood from anterior side of the body. A **hepatic portal system** collecting blood from the alimentary canal is present but there is no **renal portal system**. **Lymphatic vessels** are also present.

The **blood** contains the respiratory pigment, haemoglobin, within nucleated red blood corpuscles. White blood corpuscles are also present. They are similar to lymphocytes and polymorphs of higher vertebrates. It is not known how the blood corpuscles are produced as the spleen is absent. It is, however, believed that they are produced in lymphoid tissue in the kidneys, spiral valve and spinal cord.

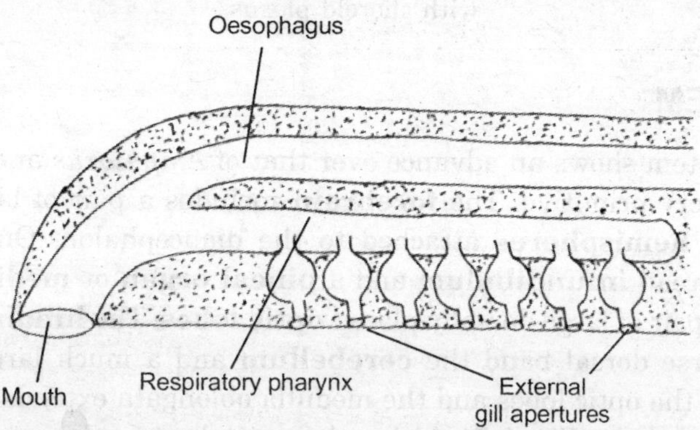

Fig. 9.6 *Petromyzon*. Diagrammatic representation of the gill system

RESPIRATORY SYSTEM

There are seven **gills** placed in **gill pouches** on either side of the body between the respiratory tube and the body wall. The gills are separated from each other by **inter-branchial septa**. The inner wall of each gill pouch contains many **gill filaments** or **lamellae** with fine capillaries in which aeration takes place. The outer wall is highly muscular. Each gill pouch communicates with the exterior by **external gill slit** and with the respiratory tube by **internal gill slit** (Fig. 9.6). The water enters and leaves the gill pouches through the same opening. It is not so in true fishes. The reason for this is that the lamprey attaches to its prey or other objects by its buccal funnel, as such water cannot enter through the mouth. In the larval stages the normal course exists, that is, water enters through the mouth and comes out of the gill openings as in fishes.

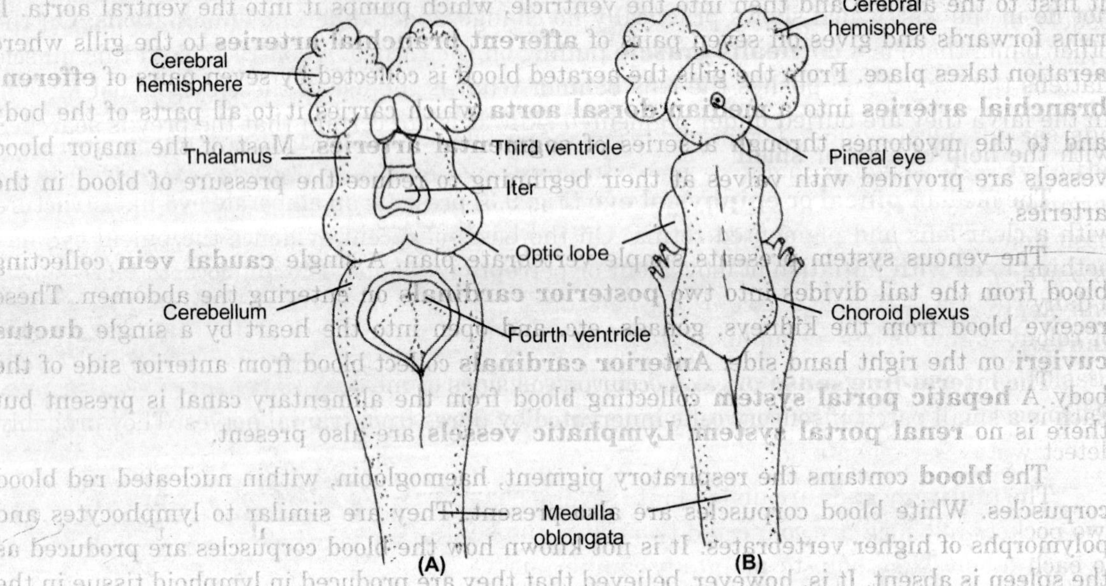

Fig. 9.7 *Petromyzon*. Brain, **(A)** Dorsal view with choroid plexus removed; **(B)** Dorsal view with choroid plexus

NERVOUS SYSTEM

The nervous system shows an advance over that of *Amphioxus* and presents vertebrate plan. The **brain** is very primitive. The **forebrain** includes a pair of large **olfactory lobes** and small **cerebral hemispheres** attached to the diencephalon. On the ventral side of diencephalon lies a broad **infundibulum** and a **pineal organ** or **median eye** lies dorsally. The **midbrain** has a pair of large dorsally placed **optic lobes**. The **hindbrain** is differentiated into a small transverse dorsal band the **cerebellum** and a much larger ventral **medulla oblongata**. Between the optic lobes and the medulla oblongata extends the **choroid plexus** containing rich blood supply (Fig. 9.7). As in other vertebrates, there are four **ventricles** in the brain. There are **ten** pairs of **cranial nerves**.

The **spinal cord** is dorso-ventrally flattened and band-like structure of transparent grey colour. The **spinal nerves** are similar to those of *Amphioxus* in that the dorsal and ventral roots do not join. There are no blood vessels in the spinal cord. The sympathetic nervous system is poorly developed. The nerve fibers in lamprey are non-myelinated, so the rate of conduction of nerve impulse is very slow.

SENSE ORGANS

The single nasal aperture leads into a small **olfactory sac** (Fig. 9.3), the organ of smell. It is innervated by two olfactory nerves, one from each olfactory lobe of the brain. The olfactory sac gives off a long posteriorly directed sac, the **naso-hypophysial sac**, of unknown function.

The paired **eyes** present the same structure as found in normal vertebrate eye. The eyes are formed as a result of evagination of the wall of the diencephalon. There are **six extrinsic muscles** controlling the movement of each eye. The power of accomodation does not lie in the iris, which shows practically no change under different illuminations. On the other hand, there is a **cornealis muscle** pulling on the cornea. A contraction on this muscle flattens the cornea and pushes the lens behind. What is the use of these eyes is not known. In the larva they are buried under pigmented skin. It is not certain that the prey is searched with the help of eyes or smell.

The median **pineal** or **epiphysial eye** (Fig. 9.3) presents an elaborate eye-like structure with a clear lens and pigmented retina. On the basis of recent evidence the pineal eye has nothing to do with initiation of movement or perception of light. The ammocoete larva shows a daily rhythm of change of colour (becomes dark during day and pale at night). This change of colour is due to nerve impulses from the pineal eye.

The **lateral-line sense organs** occur on the sides of the body under the surface of skin. Each is a small patch of sensory cells innervated by fibres from cranial nerves. They probably detect water movements.

The **auditory sacs** are divided into several chambers. The **sacculus** and **utriculus** are two pockets to which are connected two tubes, the **semicircular canals**, set at right angles to each other. They help to maintain equilibrium.

Taste buds occur in the pharynx.

Some light sensitive cells are found in the skin (**integumentary photoreceptors**). They are numerous in the tail and if the tail is exposed to light, the animal moves away quickly.

ENDOCRINE GLAND

The **pituitary gland** lies beneath the infundibulum. Its major portion is a mass of secreting cells in which two parts can be recognized—**anterior** and **intermediate**. The secretion of pituitary gland is presumably under the control of pineal eye. Its functions are similar to those of other vertebrates. The gland is associated with the nasohypophysial sac.

EXCRETORY SYSTEM

The excretory system includes a pair of **mesonephric kidneys**, which lie in the body cavity suspended mid-dorsally by mesenteries. From each kidney arises a tubular **ureter** which runs, along its free ventral edge, to the **urinogenital sinus** which opens to the exterior on the **urinogenital papilla**. The nitrogenous wastes are extracted by uriniferous tubules and poured into the ureters to be discharged outside.

REPRODUCTIVE SYSTEM

The differentiation of gonad takes place quite late in life. In the ammocoete larva the gonad contains developing oocytes and spermatocytes *i.e.*, it is **hermaphroditic**. But later on the gonad becomes either testis or ovary, *i.e.*, sexes in the adult are separate (dioecious). At sexual maturity a single large gonad fills most of the abdominal cavity. The ripe **ovary** consists of **ova** each surrounded by a single layered **follicle**. The **testis** consists of a number of follicles containing **sperms**. The gonads have no ducts. Ova and sperms are released in the coelom by the rupture of gonads. The sex cells escape through **genital pores**, developed just before mating, into the urinogenital sinus from where they are discharged into water through the urinogenital papilla.

During the breeding season, *i.e.*, spring or early summer, lampreys leave the sea and go to fresh water streams to spawn. Not being a strong swimmer it makes its way to the breeding ground with some difficulty. It moves by making short spurts, fastening its suctorial funnel to a stone or other object, resting a bit and then repeating the process. A small nest, 6 to 24 inches in diameter is hollowed out in the sandy bottom with the help of their buccal funnel. The female lays the eggs in the nest. The male winds its tail around the female and discharges the sperm. A single female lays more than 1,00,000 eggs. **Fertilization is external**. Soon after spawning the adults die. The eggs are yolky (**telolecithal**). Cleavage is holoblastic and equal. After about two or three weeks the eggs hatch into tiny transparent larvae called **ammocoete.**

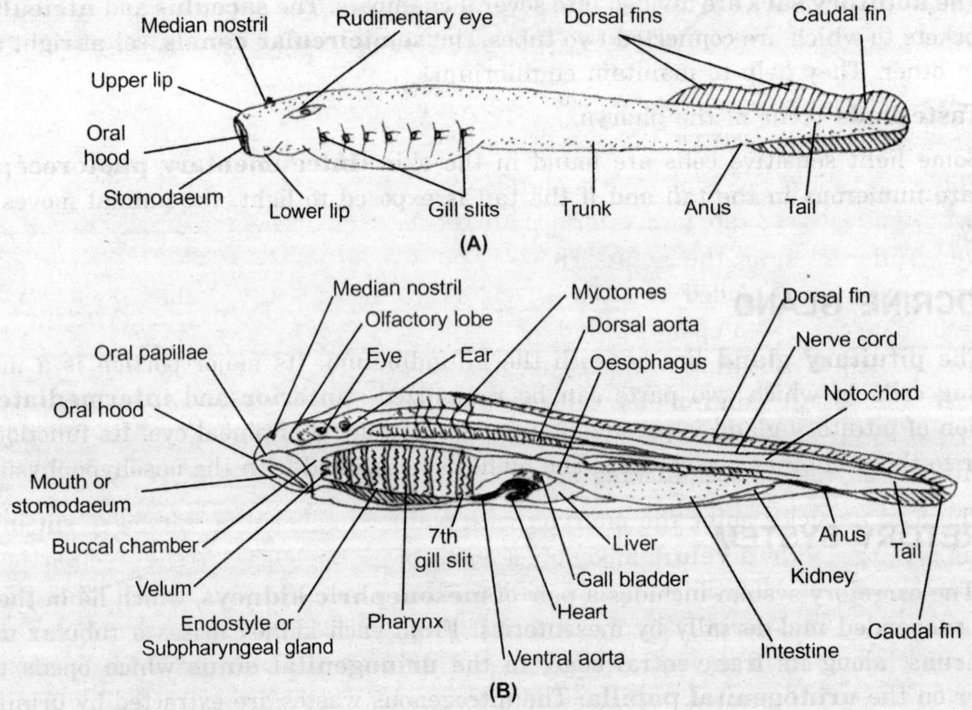

Fig. 9.8 *Petromyzon*. Ammocoete larva, **(A)** External features. **(B)** Showing internal structure

Ammocoete larva. On hatching the larva is about 7 to 8 mm in length (Fig. 9.8). It leaves the nest when it becomes 15 mm long. It swims for a shortwhile and soon burrows in the soft mud of river bottom or banks. The burrows are U- or V-shaped. The larval period is relatively long, about three to seven years during which the larvae grow into an opaque eel-like creature about 170 mm long. When undisturbed the larvae are fully burried in mud. Food is drawn in by the water current produced by the rhythmic movement of the gills. The burried larvae emerge occasionally to change their feeding ground, when they find the mud less nutritious.

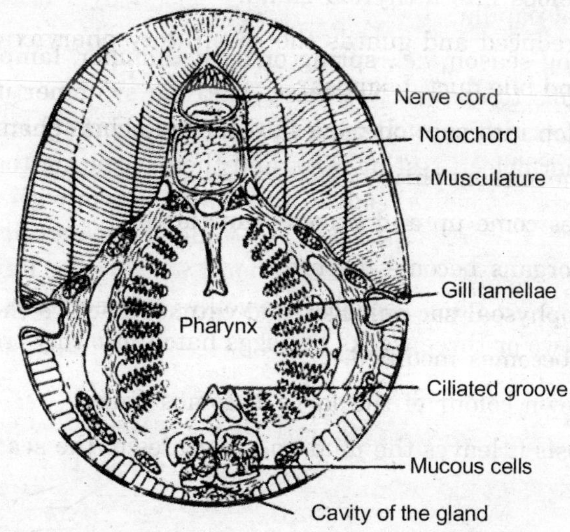

Fig. 9.9 *Petromyzon*. Cross-section of larva through the pharyngeal region

The ammocoete larva is of great evolutionary significance because it probably represents the most primitive vertebrate form intermediate between the cephalochordates and vertebrates. It is very different from the adult. These is no sucker, the mouth is lodged in a small depression and is surrounded by an oral hood as in *Branchiostoma*. The developing lateral eyes become covered with opaque skin and the larva is blind. The pineal eye remains light sensitive but is not concerned with any form of vision. It is concerned with diurnal change in pigmentation of the skin from dark during the day and pale at night. Active light response comes from photoreceptors in the tail which informs a burrowing larva whether the rear end is within the burrow or still sticking out of it.

The food particles enter the pharynx with water. The entry of water current into the pharynx is regulated by a **velum** made up of a pair of muscular flaps. The food particles are caught by a column of flowing mucous cord produced by a short complex **endostyle**. The mucous food cord rotates in the current and passes back to the **oesophagus**. No enzymes have been discovered in the endostyle. It is not an open groove as in *Branchiostoma* but presents similar arrangements of the secretary columns. In ammocoete water current is set up by muscular activity and not by ciliary action as in *Branchiostoma*. This enables the larva to engulf larger particles; the ciliary action imposes limitation to size of food particles. The use of muscles to move the gills is regarded as an important step in chordate evolution. The endostyle, on the other hand, shows the survival of primitive feeding method. Liver, bile duct and gall bladder are present. A protonephric kidney is present. The heart is enclosed in a pericardial cavity.

Metamorphosis: The larva metamorphoses into parasitic adult after a period of 3 to 7 years. It undergoes following structural changes.

A suctorial buccal funnel with strong and sharp teeth, rounded mouth, tongue and complex musculature appears and replaces the oral hood.

The endostyle develops into a thyroid gland.

- The velum is reduced and guards the respiratory pharynx only.
- Gall bladder and bile duct disappear.
- The gills develop into sacs opening into the branchial chambers.
- Kidneys become mesonephric.
- The paired eyes come up and become functional.
- The olfactory organs become mature.
- The naso–hypophyseal sac enlarges and extends back to the gills.
- The intestine becomes modified.
- The yellow brown colour of the larva becomes dark.

After metamorphosis it leaves the mud and migrates to the sea as adult and begins its parasitic life.

CLASSIFICATION

Class Cyclostomata includes eel-shaped vertebrates with no jaws, paired fins and with a single median dorsal nostril. The group is divided into two orders-Petromyzontiformes and Myxiniformes, both showing marked morphological differences.

Order 1: Petromyzontiformes

The members of this order are commonly called lampreys or lamprey eels.

1. The mouth is ventral, surrounded by a large suctorial buccal funner bearing many horny teeth.
2. Nostril dorsal in position. The nasohypophyseal sac is not connected to the pharynx.
3. Gill pouches seven pairs, opening separately into the respiratory pharynx.
4. Branchial basket completely developed.
5. Ear with two semicircular canals.
6. Eggs small and numerous.
7. Development indirect, with a long larval period and metamorphosis.
8. Occurs along sea-coasts and in streams and lakes all over the world.

Examples: More than 30 species are known *Petromyzon marinus* (Fig. 9.2) is the most common species. Other examples include *Entospherus tridentus* (Pacific lamprey), *Ichthyomyzon concolor* (silvery lamprey), *Lampetra*.

Order 2: Myxiniformes

The members of this order are commonly known as hagfishes or slime eels.

1. The mouth is terminal, with four pairs of tentacles present around the margin and few teeth, there is no buccal funnel.
2. Nostril terminal in position. Nasohypophyseal sac connected to the pharynx by a duct.
3. The eyes are degenerate and lie beneath the skin.
4. There are 6 to 15 pairs of gill pouches that open into a common chamber before opening by a single gill slit.
5. Branchial basket is poorly developed.
6. Ear with single semicircular canal.
7. Eggs large and few.
8. Development is direct.
9. The members of this order are exclusively marine.
10. They eat flesh of fishes.

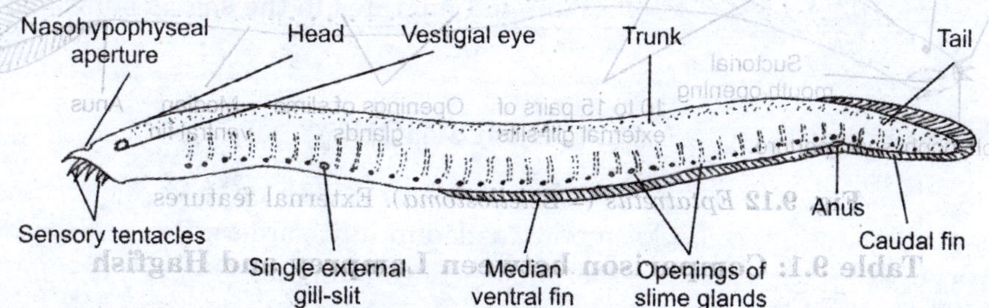

Fig. 9.10 *Myxine*. External features

Examples: *Myxine glutinosa* is the most common species. Other examples are *Myxine limosa* (Atlantic coast), *Paramyxine*, *Pelistorema*, the borer, *Eptatretus* (*=Bdellostoma)* (Fig. 9.12).

Myxine. *Myxine* is the most common species of the order Myxiniformes. It is found on the coasts of America, Africa, Japan and Europe. It is a colonial species and lives in mud burrows on the bottom of the sea. They feed on polychaete worms. They burrow into the body of injured or dead fish and eat their flesh, hence also called borer. They damage valuable fishes, hence pose a serious threat to the fish industry in some regions. The body is eel-like, elongated, about 50-60 cm long and differentiated into head, trunk and tail (Fig. 9.10). The skin is soft scaleless and pinkish in colour. The mouth is terminal and surrounded by soft wrinkled lips. Eyes are degenerated and covered by skin. The tonque is protrusible and bordered by two horny plates bearing numerous teeth. It serves as a powerful rasping organ. The nostril is single, median and terminal. There are 6 pairs of gill pouches which open by a single gill slit (Fig. 9.11). This is an adaptation for burrowing. The openings of the mucous glands are present along the sides of the body. They produce slime, hence commonly known as slime eel. The dorsal fin is absent, the ventral and caudal fins are poorly developed. They do not migrate to fresh water bodies for spaning. Eggs are large and laid in marine water. Development is direct. The eggs hatch into miniature forms which soon mature into adults.

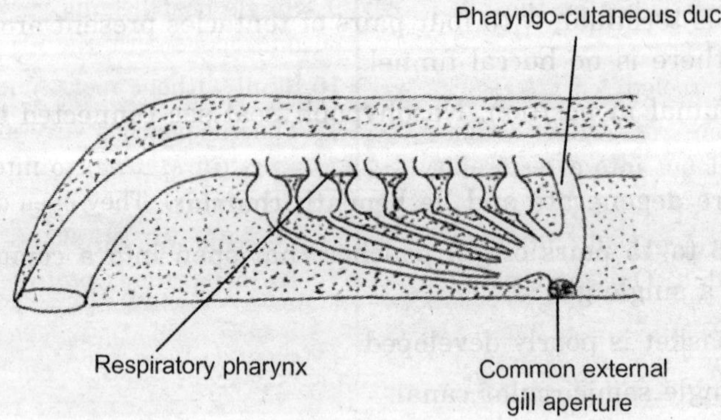

Fig. 9.11 *Myxine*. Diagrammatic representation of the gill system

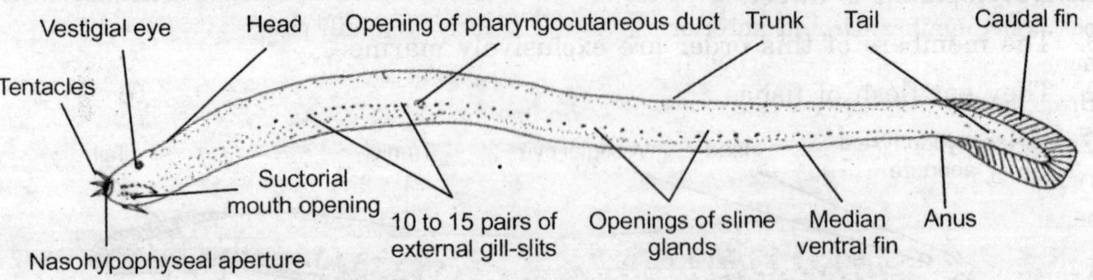

Fig. 9.12 *Eptatretus* (= *Belellostoma*). External features

Table 9.1: Comparison between Lamprey and Hagfish

Lamprey *(Petromyzon)*	Hagfish *(Myxine)*
1. Usually found in sea but migrate to rivers for breeding.	1. Exclusively marine, live in burrows in the sand.
2. The adults are usually ectoparasites, some species nonparasitic, the larval stage nonparasitic.	2. Usually scavengers.
3. Feed by rasping away flesh and sucking blood of the host fish.	3. Burrow inside dying or dead fish and consume flesh.
4. Body stout and cylindrical.	4. Body feeble, slender and eel-like.
5. Dorsal fin notched, caudal fin well developed.	5. Fins poorly developed.
6. Mouth subterminal due to over development of upper region to	6. Mouth terminal, buccal funnel absent.

form a buccal funnel. Horny teeth present in the buccal funnel.	
7. No tentacles around the mouth.	7. Three or four pairs of tentacles present around the mouth.
8. Tongue less developed with large teeth. Salivary glands paired.	8. Tongue strongly developed with small teeth. Salivary gland absent.
9. Single median nostril lies on the dorsal side of head.	9. Single nostril lying very close to the mouth and opens terminally.
10. Skull imperfectly roofed.	10. Skull without roof.
11. Gill pouches seven pairs which do not open into the gut but into a respiratory tube.	11. The gill pouches vary in number from six pairs (in *Myxine*) to fifteen pairs (in *Eptatretus*). They open directly into the gut.
12. Seven pairs of external gill slits are present in the adult. One more pair may be present in the larva.	12. External gill slits one pair in *Myxine* and 13-15 pairs in *Eptatretus*.
13. Pharynx ends blindly as a respiratory tube.	13. Pharynx continuous with the oesophagus.
14. Branchial region well developed.	14. Branchial region poorly developed.
15. Intestine with spiral fold or typhlosole.	15. Intestine without longitudinal folds.
16. Each aortic arch divides and supplies posterior hemibranch and anterior hemibranch of adjacent gill pouches.	16. Each aortic arch supplies hemibranchs of the same gill pouch.
17. Brain better developed.	17. Brain poorly developed.
18. Cranial nerves 10 pairs.	18. Cranial nerves 8 pairs.
19. Dorsal and ventral spinal nerve root separate.	19. Spinal nerve roots united.
20. Eyes functional and well developed.	20. Eyes rudimentary and covered by skin.
21. Pineal eye present.	21. Pineal eye absent.
22. Nasohypophyseal duct ends blindly.	22. Nasohypophyseial duct opens into the pharynx.
23. Internal ear with 2 semicircular canals.	23. Internal ear with single semicircular canal.
24. Kidneys advanced mesonephros. The kidney tubules complicated and without segmental disposition. Larval kidneys pronephros.	24. Pronephros retained in the adult but functional kidney is mesonephric. The kidney tubules show segmental disposition.
25. Urinogenital sinus present.	25. Urinogenital sinus absent.
26. Sexes separate.	26. Sexes united, hermaphrodite, anterior part of the gonad acts like ovary and posterior part as testis.
27. Eggs small, without shell.	27. Eggs large, enclosed in shell.
28. Cleavage holoblastic.	28. Cleavage meroblastic.
29. Development indirects, larva is called ammocoete.	29. Development direct, no larval stage.

10

CHAPTER

PLACODERMI

APPEARANCE OF JAW

The next evolutionary step is the appearance of well jawed predators from the jawless mud grubbers. It is certain that the **Placodermi,** now extinct, were the earliest jawed vetebrates. Their oral arch was more or less differentiated to form jaws. The jaws of these ancient fishes were in the experimental stage and were sometimes only poorly developed and sometimes quite aberrant in nature. The fishes bearing them were armoured without exception. Most characteristic of these fishes were the spiny sharks (**Acanthodians**). The placoderms and acanthodians were earlier grouped togethed but now the taxonomists place them in two separate classes **Placodermi** and **Acanthodii**. The acanthodians, e.g. *Climatius* (Fig. 10.1) were fresh water, superficially shark- like organisms with elongated fusiform body which in the final stages of evolution became almost eel-like. It is probable that later forms were sea dwellers. They were well armoured, covered with scales and plates of bone-like material. A peculiar side branch of these primitive forms was the jointed-neck fishes (**Arthrodires**) of the Devonian. *Dinichthys* (Fig. 10.2) was best example of this group. It was a giant fish about 26 feet long, with massive jaw-like structure which were operated in a peculiar fashion. From such forms have appeared the modern sharks with true jaws. Once the sharks reached the sea

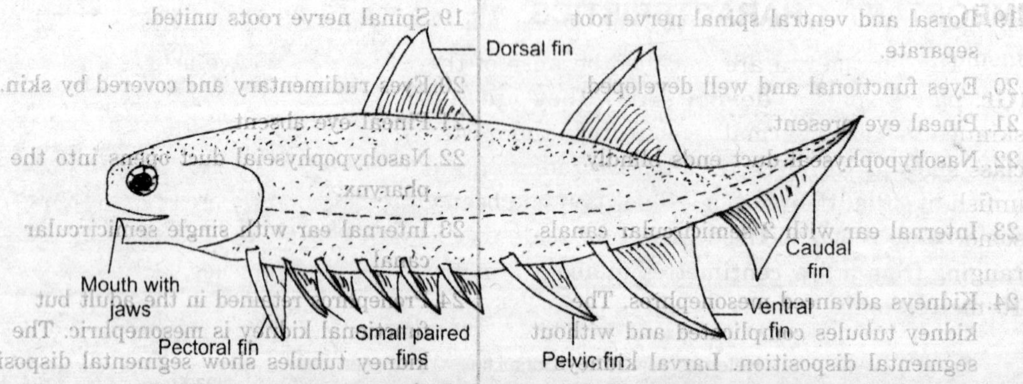

Fig. 10.1 Acanthodii. *Climatius*, earliest known gnathostome

abundant food supply forced them to turn into predators. Along with the development of jaws and teeth, they left the bottom and became predacious preying on large things. This mode of life required faster locomotion, therefore the body became stream-lined and paired locomotory fins also developed.

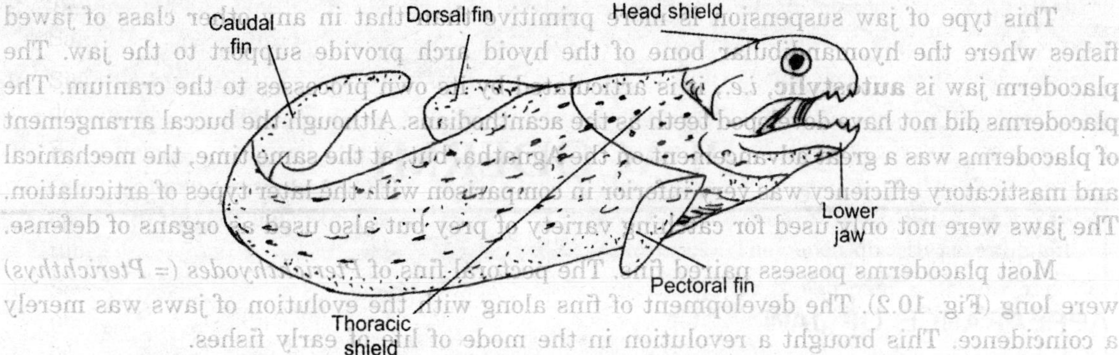

Fig. 10.2 Placodermi. *Dinichlthyes*

OCCURRENCE

Placoderms were the earliest jawed vertebrates. They appeased in the Silurian, flourished in the Devonian and Carboniferous and became extinct in the Permian. They probably occurred both in fresh and marine water.

IMPORTANT CHARACTERISTICS

The Placodermi are so called because of their heavy defensive armour of **bony plates** (**Gr**. *plakos*: plates; **derma**: skin) They are commonly known as armoured fish or plate-skinned fish. The dermal armour links them with the jawless ostracoderms. Members of this class show great degree of adaptive radiation and in some instances the appearance was so unfish-like that their zoological position had been uncertain. Some were thought to be Agnatha, some being allied to the Dipnoi and some to the elasmobranchs. They vary greatly in size ranging from a few centimetres to about 3 metres. The giant species, *Dunkleosteus* is about 10 metres long.

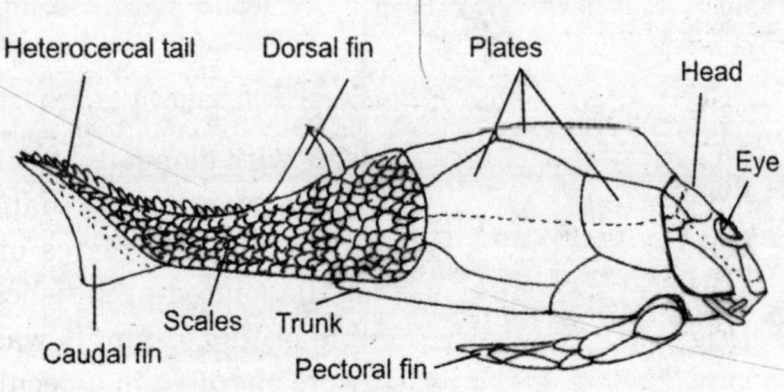

Fig. 10.3 Placodermi. Pterichthyodes (= *Pterichthys*)

All placoderms possess **jaws**. It seems probable that jaws have been derived from the first pair of gill bars. The hyoid gill slits peristed and so the spiracles failed to arise and further, the hyoid arch remained unmodified and did not support the jaws—the **apheto-hyoidean condition** was retained.

This type of jaw suspension is more primitive than that in any other class of jawed fishes where the hyomandibular bone of the hyoid arch provide support to the jaw. The placoderm jaw is **autostylic**, *i.e.*, it is articulated by its own processes to the cranium. The placoderms did not have developed teeth as the acanthodians. Although the buccal arrangement of placoderms was a great advancement on the Agnatha, but, at the same time, the mechanical and masticatory efficiency was very inferior in comparison with the later types of articulation. The jaws were not only used for catching variety of prey but also used as organs of defense.

Most placoderms possess paired fins. The pectoral fins of *Pterichthyodes (= Pterichthys)* were long (Fig. 10.2). The development of fins along with the evolution of jaws was merely a coincidence. This brought a revolution in the mode of life of early fishes.

BIOLOGICAL SIGNIFICANCE

The placoderms are the earliest known jawed vertebrates. They lived for a very short period and hence are often considered an **unsuccessful experimental stage** in the evolution of gnathostomes. It was once believed that the cartilaginous fishes represented the most primitive type and gave rise to the bony fishes. But now it appears evident that this group of bony animals were the forerunners of some of the modern bony fishes and even the elasmobranchs. Both the bony fishes (Osteichthyes) and cartilaginous fishes (chondrichthyes) arose in early Devonian from some primitive placoderms. In fact, bone and not cartilage, is a primitive character. The appearance of bone early provided a basis for the later development of animals of extraordinary complexity and adaptive efficiency.

11

CHAPTER

CHONDRICHTHYES

The **Chondrichthyes** (**Gr.** *chondra*, cartilage; *ichthyes*, fish) is a class of vertebrate including cartilaginous fishes. They are also known as **elasmobranchs** (plate-gilled). They include sharks, rays, skates, chimaeras and some fossil groups. In shape these fishes vary from graceful stream-lined specimens such as sharks to much flattened rays and skates. The cartilaginous fishes are very numerous at present but were far more numerous in ancient times as is confirmed by their fossil records. In Chondrichthyes true bone is absent although deposition of calcium takes place in the cartilage. These are almost exclusively marine and unlike other fish classes, have been so throughout their known history, but their ancestors were probably fresh water forms. The cartilaginous endoskeleton was formerly regarded as a primitive feature, but it is not so. Bone is quite old in evolutionary history. The absence of bone is, therefore, a secondary feature. They have given up their defensive armour because of the development of other means of protection. These include swift running, good sense organs and brain and powerful jaws. Nearly all cartilaginous fishes are carnivorous or scavengers.

TYPE SCOLIODON

Scoliodon, commonly known as dogfish, is one of the most abundant cartilaginous fish of the Indian seas. It is a suitable study material because it is easily available, easy to handle because of its appropriate size and easy to dissect because its endoskeleton is cartilaginous and not bony.

DISTRIBUTION

Scoliodon has a worldwide distribution. It has been reported from the Indian ocean, Bay of Bengal, East Indies and Philippine islands, Eastern Pacific, West Indies and Eastern coasts of South America. Nine species of *Scoliodon* have been reported of which four species are found in Indian waters. They are *S. sorrakowah*, *S. palasorrah*, *S. dumerilii* and *S. walbeehmi*. The most common species is *S. sorrakowah* and it is commonly known as **"black shark"** because of its colour.

HABIT AND HABITAT

Scoliodon is a marine fish. It is an inhabitant of the open sea and for the capture of its prey and protection it depends on swift movement. The body is perfectly streamlined, hence it is a fast swimmer. It is a carnivorous fish, feeding on other fishes, worms and arthropods. Sexes are separate. It is a viviparous fish, that is, it gives birth to young ones that develop in the uterus of females.

EXTERNAL FEATURES

Shape, size and colour

The body is elongated, laterally compressed, streamlined, spindle-shaped, tapering at both ends and broadest in the region of the pectoral fins. They are about 30 to 60 cm long while the monster sharks may be about 30 feet long. The body colour is dark grey above and pale white below.

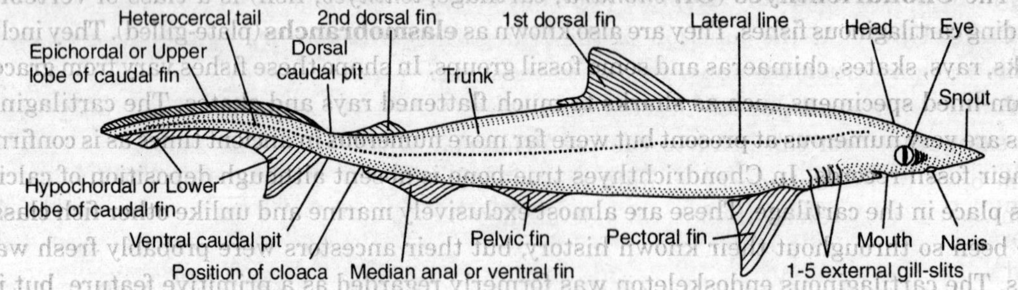

Fig. 11.1 Lateral view showing external features

Divisions of the body

The body is divisible into three regions: the **head**, **trunk**, and **tail**, although there is no distinct demarcations between these regions. (Fig. 11.1)

The **head** is dorsoventrally compressed and is produced into a pointed snout anteriorly. It bears two large circular **eyes** at its sides, the **eyelids** are immovable and can hardly be distinguished from the surrounding integument. Each eye is provided with a special eyelid, the **nictitating membrane**, which lies along the anteroventral border and can be drawn over the eye in time of danger. The **pupil** is narrow and vertical. The head merges into the **trunk** which is almost oval in transverse section. The middle region of the body is the thickest part of the trunk. The trunk passes into the **tail** which is laterally compressed and slightly bent upwards. The tail is **heterocercal**.

Body apertures

Mouth: The mouth is a wide crescentic opening situated on the ventral side of the snout. It is bounded by the upper and lower jaws, each bearing one or two rows of sharp pointed backwardly directed teeth adapted for holding and tearing the prey.

Nostrils: The nostrils or nares are a pair of obliquely placed crescentic apertures present on the under surface of the head in front of the mouth. A small fold of skin from the anterior edge covers each nostril partially. The nostrils are only olfactory in function and are not respiratory as they do not open internally into the mouth cavity.

External gill clefts: For breathing there are a series of five vertical slits, the **gill clefts** on each side, in front of and slightly above the pectoral fin. They open obliquely backwards into gill pouches and the pharynx.

Cloacal aperture: On the ventral surface of the body between the pelvic fins there is an elongated opening, the **cloacal aperture** which leads into a small chamber, the **cloaca**. The urinogenital pore in male and the urinary pore in female, situated on a conical papilla, lie just posterior to the anus.

Abdominal pores: They are a pair of openings situated on elevated papillae, the **cloacal papillae** on either side of the cloaca. The coelom communicates with the exterior through these minute pores. Their exact function is not known.

Ampullary pores: On careful examination the head shows several groups of small openings on the sides and on the snout. There are the apertures of ampullae of Lorenzini. If the head is squeezed tiny droplets pour out of these openings.

Caudal pits: At the junction of the trunk and tail there are two shallow pits—one dorsally and the other ventrally, these are the caudal pits.

Lateral line and pores: A careful look reveals the presence of a faint line, the **lateral line**, extending from the head to the posterior end of the body. The sensory lateral line system open to the outside by minute pores which are present at intervals along the lateral side of the animal.

Fins

The body supports several **fins** that help the animal in locomotion and in maintaining balance. The fins are flattened expansions of the skin, usually triangular in shape and supported by cartilaginous rods and horny fin rays.

Median fins: Among the median fins there are two **dorsal**, one **ventral** and one **caudal** fin.

The **first** or **anterior dorsal fin** is large, triangular and situated a little in front of the middle of the body. The **second** or **posterior dorsal fin** is situated a little behind the first dorsal fin. It is similar in shape to the first but smaller in size.

The **caudal fin** forms a vertical fringe around the posterior portion of the tail. It is divided into two unequal lobes. The **upper** or **dorsal lobe** (**epichordal**) is reduced, it forms a low ridge along the greater part of the upper surface of the tail in the median line. The lower or **ventral lobe** (**hypochordal**) is well developed and is divided into a larger anterior and a smaller posterior portions.

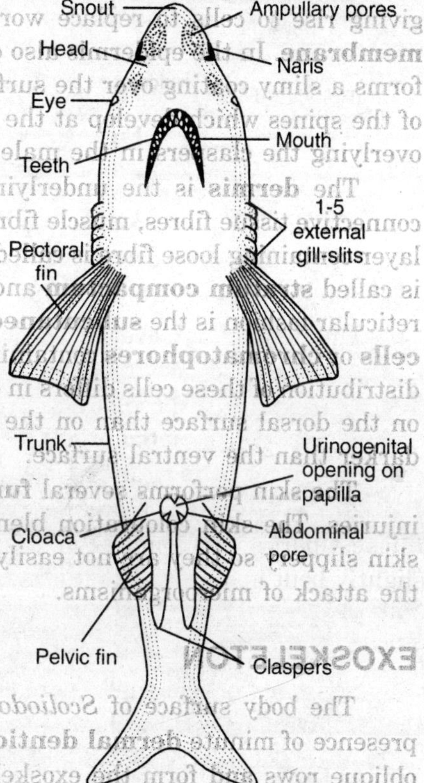

Fig. 11.2 *Scoliolon.* Male in ventral view

The **ventral median fin** or **anal fin** is situated in the mid-ventral line slightly in front of the caudal fin. This is almost opposite to the second dorsal fin.

Paired fins: Comparable with the limbs of higher vertebrate there are two sets of **paired** or **lateral fins** : the **pectoral** and **pelvic fins**. Originating from the ventro-lateral margins of the body immediately behind the gill clefts are the **pectoral fins**, triangular in shape and ending over a long space. These are the largest fins. The **pelvic fins** are smaller in size but similar in shape. They arise close together from the ventral surface of the body a little in front of the middle of its length. Their inner borders touch each other and in male are fused together behind. They enclose the cloacal chamber. In the male a part of the fin, on each side, forms a copulatory organ, the **claspers**. Each clasper is a stiff rod-like appendage grooved dorsally. The groove leads into a cavity, the **siphon**, beginning at the base of the clasper. (Fig. 11.2)

SKIN

The integument or skin of dogfish comprises of two layers: an ectodermal **epidermis** and an underlying mesodermal **dermis.**

The **epidermis** is several layers in thickness. It is made up of ciliated epithelial cells but the cilia are lost in the adult. The **outer cells** may form a thin **squamous layer** covering the external surface. The **basal** or **inner layer** is the **germinative layer**, the **stratum Malpighi** or **germinativum** and retains its germinative capacity throughout life giving rise to cells to replace worn out cells of the outer layer. It rests on a **basement membrane**. In the epidermis also occur certain **mucous secreting glands**. Their secretion forms a slimy coating over the surface of the skin. Multicellular glands appear at the bases of the spines which develop at the anterior margins of the dorsal fins and in the epidermis overlying the claspers in the male.

The **dermis** is the underlying layer composed of scattered cells intermingled with connective tissue fibres, muscle fibres and nerve fibres. It consists of **three** layers—the outer layer containing loose fibre is called **stratum laxum**, middle layer containing compact fibres is called **stratum compactum** and the innermost layer containing few fibres arranged in a reticular fashion is the **subcutaneous layer,** it is of variable thickness. Numerous **pigment cells** or **chromatophores** containing black pigment occur beneath the Malpighian layer. The distribution of these cells differs in different parts of the body. They are much more numerous on the dorsal surface than on the ventral, that is why the dorsal surface of the animal is darker than the ventral surface.

The skin performs several **functions**. It protects the internal organs from mechanical injuries. The skin colouration blends with the dark surroundings. The mucous makes the skin slippery so they are not easily caught by the predators. The mucous also protects from the attack of microorganisms.

EXOSKELETON

The body surface of *Scoliodon* is rough like a sand paper. The roughness is due to presence of minute **dermal denticles** called **placoid scales** which are arranged in regular oblique rows and form the exoskeleton.

The scales are derived partly from the epidermis and partly from the dermis. Each scale consists of two parts: a **basal plate** and a trident **spine** (Fig. 11.3). The basal plate is diamond-shaped and lies embedded in the skin. It is derived from the dermis. The basal plate is formed of calcified tissue called **cement** closely allied to the cement of our teeth. A small opening is present in the centre of the plate. It leads into a cavity, the **pulp cavity** which traverses the spine and is filled with vascular connective tissue called **pulp** containing blood vessels, nerve endings, lymph vessels and numerous dentine forming **odontoblast cells**. The spine projects out from the basal plate and is backwardly directed. It is a trident structure, *i.e.*, formed by three spines—one **median** and two **lateral**. The spine is made up of hard calcareous substance, the **dentine** which is externally covered by another harder and shiny enamel-like substance called **vitrodentine**. The dentine is derived from the dermis while vitrodentine is derived from the epidermis. The dentine is relatively spongy and numerous fine branches or **canaliculi** ramify through it.

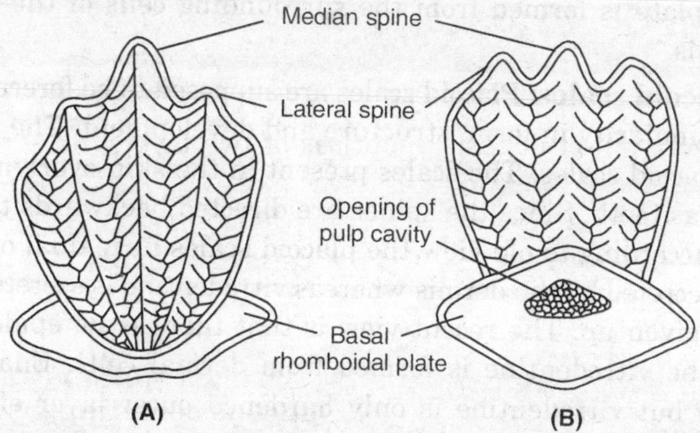

Fig. 11.3 *Scoliodon*. Structure of placoid scale. **(A)** Dorsal view; **(B)** Ventral view

The form of the spine differs in different parts of the body. On the ventral surface the spine has one point and on the dorsal three points. The scales in the skin which cover the mouth and jaws and which continue in the buccal cavity are larger and stronger. They have a trilobed base and a five pointed spine and function as **teeth**.

The scales are being constantly worn away and equally constantly replaced, so that in vertical section of the skin they appear in various stages of development. Normally, the scales

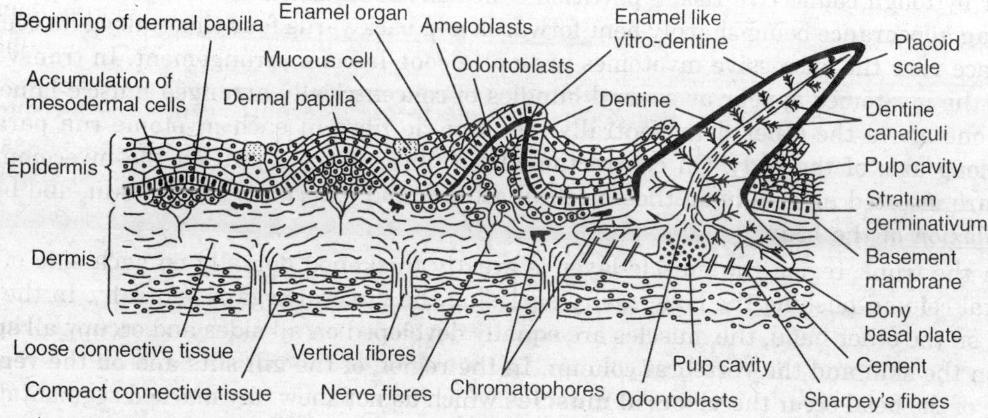

Fig. 11.4 *Scoliodon*. Vertical section of skin showing development of placoid scale

are set closely together in the skin. The enamel-covered spines are so strong that the dried skin of some dogfish, known as **shagreen**, is used for making the surface smooth.

Development of placoid scales: The first indication of the developing scales is the accumulation of mesenchyme cells in the dermis to form a **dermal papilla** (Fig. 11.4). As this increases in size it projects upwards into the dermis, pushing the Malpighian layer outwards. The cells of the outer margin of the dermal papilla form a layer of **odontoblast cells** which secretes a hard bone-like substance resembling dentine of the teeth of higher vertebrates. The cells of the basal epidermal layer, overlying the dentine-like substance, form an **enamel organ** composed of columnar **ameloblast cells** which produce a hard enamel-like coating over the outer portion of the conical mass of dentine. As this tooth-like structure grows in size it pushes the epidermis aside and emerges above the surface as a **placoid scale** (Fig. 11.4). The basal plate is formed from the surrounding cells of the dermis. It remains embedded in the dermis.

Homology of placoid scales: Placoid scales are supposed to be forerunners of vertebrate teeth as both present similarity in form, structure and development. The teeth, in *Scoliodon*, are merely modified placoid scales. The scales present in the skin overlying the jaws become enlarged and function as teeth. Since the spines are directed backwards they help in holding and tearing the prey. According to old view the placoid scales have dual origin, *i.e.*, the basal plate and dentine are secreted by the dermis whereas vitrodentine is secreted by the epidermis. But, now this view is given up. The recent view is that there is no epidermal enamel-layer in placoid scale and the vitrodentine is formed from dermal cells. Enamel is the hardest substance in the body but vitrodentine is only hardened outer layer of dentine. Thus the entire placoid scale has a mesodermal origin just like the scales of bony fishes. A tooth has a layer of enamel which is derived from the ectoderm.

MUSCULATURE

The greater part of the body-bulk consists of muscles packed between the skin and the skeleton. These muscles are known as **parietal muscles**. The muscles are highly developed in the region of the trunk and tail. They present a perfectly segmental arrangement comprising of paired series of muscles segments called **myotomes** or **myomeres** separated from one another by tough connective tissue partition called **myocommata**. Each myotome presents a **zig-zag** appearance being sharply bent forwards and backwards four times along its course. In surface view the successive myotomes present a cone-in-cone arrangement. In transverse section the myotomes appear as several bundles of concentrically arranged muscle lamellae placed one above the other dorsoventrally. The muscle fibre in each myotome run parallel to the long axis of the body. On contraction these fibres exert a pull on the myocommata which are inserted internally on the vertebral column and externally on the skin, and bring about flexion of the body.

In the trunk region the muscle-layer is greatly thickened dorsally on each side of the vertebral column pushing the body cavity enclosing the visceral organs ventrally. In the tail region, on the other hand, the muscles are equally developed on all sides and occupy all space between the skin and the vertebral column. In the region of the gill slits and on the ventral surface of the head occur the **visceral muscles** which do not show any marked segmentation. They serve to move the gill arches, eyes and the jaws. These muscles are of many types but here only the following may be mentioned : the **trapezius**, the long muscles above the gill slits and below the myotomes; the **dorsal constrictors** between the dorsal portions of the gill arches, with their fibres directed obliquely forward; the **ventral constrictor** between the ventral portions of the gill arches, extending forward to the mouth with fibres directed backwards; and the **adductor mandibularis**, the thick muscles at the angle of the jaws, used to close the lower jaw.

The muscles undergo special modification in association with the fins. Those responsible for the movement of fins bud off as muscle slips from the myotomes in the neighbourhood of the appendage. The **elevators** of the pectoral fin arise from the posterior side of the scapula and run on the dorsal surface of the fin which they elevate (draw dorsally) and pull it towards the body. The **depressors** arise from the lateral side of coracoid, extend on the ventral surface of the fin which they depress (draw ventrally). The small **adductors** arise from the middle part of the girdle and extend to the anterior border of the fin. They pull the fin away from the body.

Thus it is apparent that the primitive segmental arrangement of myotomes is lost in the head region and greatly modified in the region of gills and fins. The muscles of dogfish persist in higher vertebrates, although the gills have been lost. They remain associated with structures derived from the gill arches, namely, the jaws, the hyoid and the laryngeal cartilages.

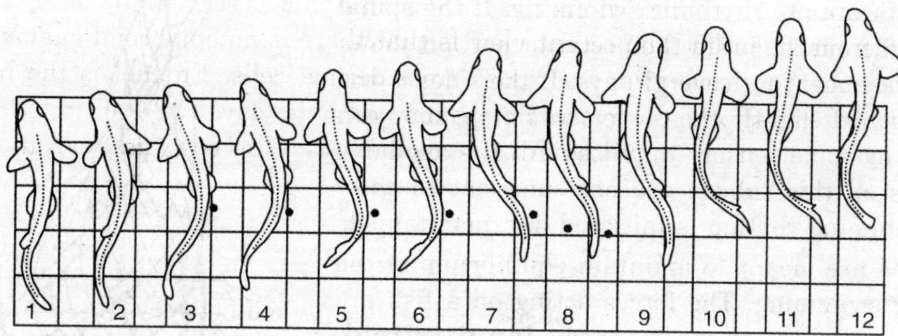

Fig. 11.5 *Scoliodon*. Illustrations of locomotion showing curvatures passing antero-posteriosly over the body (After Gray)

LOCOMOTION

The fins are not the organs of locomotion as was believed for a long time. They simply help the animal in keeping balance in water, in keeping constant course and also in changing the direction. The actual propulsive forces that move a fish through the water are produced by the longitudinal muscles fibres of the myotomes. As the muscle fibres in any myotome contract, the body is bent. The waves of curvature pass down alternately on each side (Fig. 11.5) of the body from head to tail because the myotomes contract one after the other from front to behind. The bending is more evident behind the head region, especially in the tail, because there is a progressive increase in the transverse amplitude of the waves as they pass backwards. As one wave of contraction passes backward on one side, a similar second wave of contraction follows on the other side of the body. Thus, it is evident that waves of curvature of the body are produced by a succession of longitudinal waves of contraction on alternate sides of the fish. This contraction of myotomes causing waves of curvature is called

metachronal contraction. The muscle contractions produce movements of the parts in a transverse direction to its path of swimming in such a way that it is bent at an angle or obliquely backwards, so that this part transmits a backward momentum to push the water backwards. Thus each portion of the side of fish can be considered as moving like the blade of an oar of a boat. Gray (1933) has shown this by superimposed drawings of an eel (Fig. 11.6) made from successive photographs. The region marked xy is moving from right to left and x_1y_1 from left to right. There is, therefore, a definite angle between the surface of the fish and its path of motion. Each point of the body evidently travels along a figure of '**8**' curve as is indicated by dotted lines at the tail point of the eel. Alternate pushes in quick succession give a forward thrust to the body of the fish. According to Gray the magnitude of this forward thrust depends on:

(a) the angle which the surface of the fish makes with its own path of motion,

(b) the angle between the surface of the fish and the axis of forward movement of the whole fish,

(c) the velocity of transverse movement of the body. The efficiency of movement depends on the shape of the body and the action of its muscles.

Nervous system also plays important part in the production of swimming waves. In dogfish isolated spinal cord is able to initiate rhythmic swimming. If the spinal cord is cut off from the brain the posterior portion shows swimming movement for several days. If the sides of the body are touched slightly the movements stop, they can be initiated by some sensory impulses. All the available informations on this subject are not sufficient to show how the swimming rhythm is initiated and maintained.

The fins are meant to maintain equilibrium of the body during swimming. The forces acting on a fish are measured along three main axes—**longitudinal**, **horizontal** and **vertical**. Deviation from the line of motion about the vertical axis is known as **yawing**, about the longitudinal axis is **rolling** and about the horizontal axis is **pitching**. The forces along these three axes are known as **drag**, **lateral force** and **lift**. Recently, Harris studied the function of fins by making use of models of fishes. He established that the median dorsal fins of a dogfish act as stabilising keels and prevent rolling. They thus maintain equilibrium in horizontal plane. The paired and caudal fins prevent pitching, that is, maintain equilibrium in the vertical plane. The strong lateral movements of the heterocercal tail and tail fin during swimming, besides giving a forward thrust to the fish, produce a negative pitching reaction making the fish dive head downwards; this negative pitching reaction is counter balanced by a positive pitching reaction (lift force) produced by the paired fins. These fins, in fact, are essentially **elevating planes**

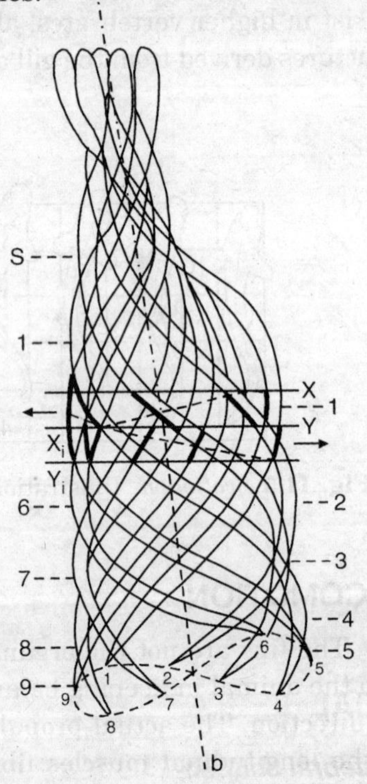

Fig. 11.6 Enlarged drawings of young eel *(Anguilla)* illustrating the movements of short segments of the body during the passage of a complete wave past the segments (after Gray). The segement xy is moving from right to left and x_1y_1 from left to right. The tip of the tail is moving in a figure of '8'

in the dogfish. In other words, while the movements of tail and tail fin tend to force the fish downwards, the trailing movement of the pectoral and pelvic fins force the fish upwards nearly with the same amount of force. The tail and the paired fins, therefore, neutralize each others opposed pitching effects and thus lead to the normal equilibrium of swimming dogfish.

ENDOSKELETON

The endoskeleton is well developed in dogfish but consists entirely of **cartilage** calcified at places. The primary function of endoskeleton is to provide a system of rigid levers to which muscles can be attached, but evolutionary history shows that it may take on a protective function by developing around the heart, lungs, etc. The endoskeleton of dogfish can be divided into: (i) **axial skeleton** comprising the **skull** and **vertebral column** and (ii) **appendicular skeleton** consisting of the **pectoral** and **pelvic girdles** and the **skeleton of fins**.

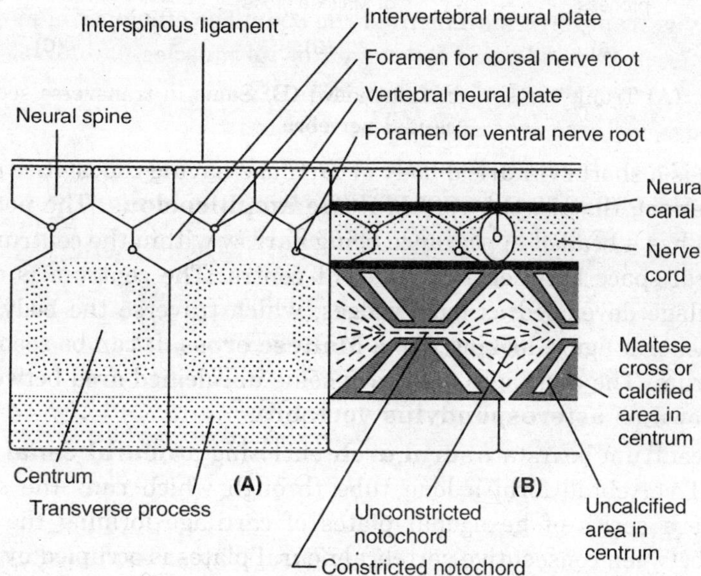

Fig. 11.7 *Scoliodon*. A portion of vertebral column of trunk region. **(A)** Two vertebrae in lateral view; **(B)** Two in sagittal section

AXIAL SKELETON

Vertebral column

The vertebral column consists of a chain of **cartilaginous vertebrae** which develop around the persistent notochord. It appears like an intricate series of cartilaginous plates bound together by fibrous connective tissue. In adult dogfish there are about 130 vertebrae. The vertebrae vary only slightly in shape. There are only two regions in the vertebral column—**trunk region** and **caudal regions**.

Trunk vertebra: A typical trunk vertebra consists of a ventral **centrum** around the notochord, a dorsal **neural arch** around the spinal cord and a pair of **transverse processes** projecting ventrolaterally from the centrum (Fig. 11.8A and B).

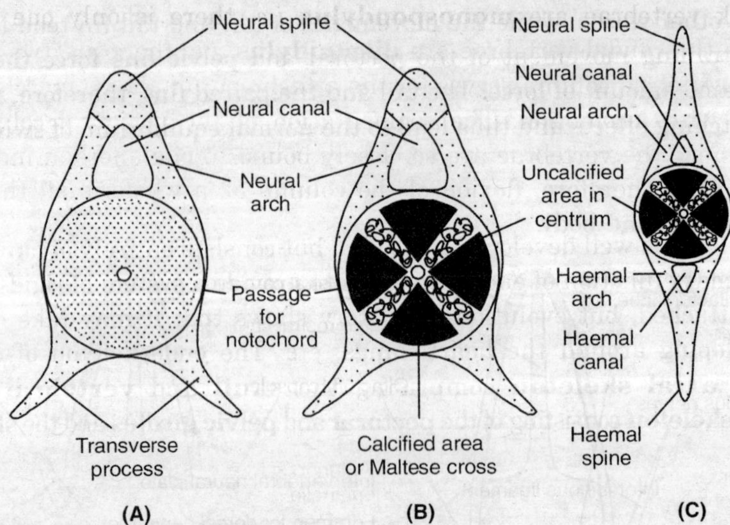

Fig. 11.8 *Scoliodon.* **(A)** Trunk vertebra in front view; **(B)** Same in transverse section; **(C)** T.S. of caudal vertebra

The **centrum** is a short cylindrical body of cartilage having concavities on both surfaces (anterior and posterior), that is, the vertebrae are **amphicoelous**. The notochord running within the centrum has a beaded appearance being narrow within the centrum and expanded in the intervertebral space between two adjacent centra. The centrum is strengthened by calcified fibro-cartilage developed as four wedges, which traverse the body of the centrum giving rise to a cruciform figure looking like a **Maltese cross**. It can be seen in a transverse section of the centrum. The cartilages also have some uncalcified area between them, hence the vertebra is known as **asterospondylus vertebra**.

Dorsally the centrum bears a **neural arch** enclosing a **neural canal** (Fig. 11.7). The neural canal of all vertebrae form a long tube through which runs the spinal cord. The **neural plates** are a series of hexagonal plates of cartilage forming the sides of neural arches. The space between consecutive vertebral neural plates is occupied by **intervertebral neural plate**. The neural arch is formed by two kinds of structures:

(i) the **neural process** (**neurapophysis**) projecting upward from the sides of each centrum and

(ii) the **neural spine**, a median piece of cartilage completing the neural arch above.

The **transverse processes** are a pair of slender horizontal projections attached to the ventrolateral border of the centrum. They are short and extend outwards. At their anterior ends they bear movably articulated cartilaginous ribs about half an inch in length.

Caudal vertebra: The caudal vertebra is slightly modified than the trunk vertebra. The neural arch is short and the neural spine is large and pointed. The transverse processes bend downwards beneath the centrum and fuse in the mid-ventral line to form an arch, the **haemal arch** enclosing a **haemal canal** containing caudal artery and vein (Fig. 11.8C). The ribs are absent. The haemal arch is produced into a backwardly directed and flattened **haemal spine** which provides support to the ventral lobe of the caudal fin. The dorsal lobe of the caudal fin is supported by the prolongations of the neural spines. The posterior part of the vertebral column is bent dorsally to form an asymmetrical **heterocercal tail**.

The trunk vertebrae are **monospondylus** *i.e.* there is only one vertebra in each myotome while the caudal vertebrae are **dispondylus**, *i.e.* there are two vertebrae in each myotome. The vertebral column serves for the attachment of all the muscles of the body and tail concerned with locomotion. The segmentation of the vertebral column increases its flexibility but since the vertebrae are so closely bound to one another, individual vertebrae do not move freely. Therefore, flexion of the column as a whole is all that is required for lateral movement of the body.

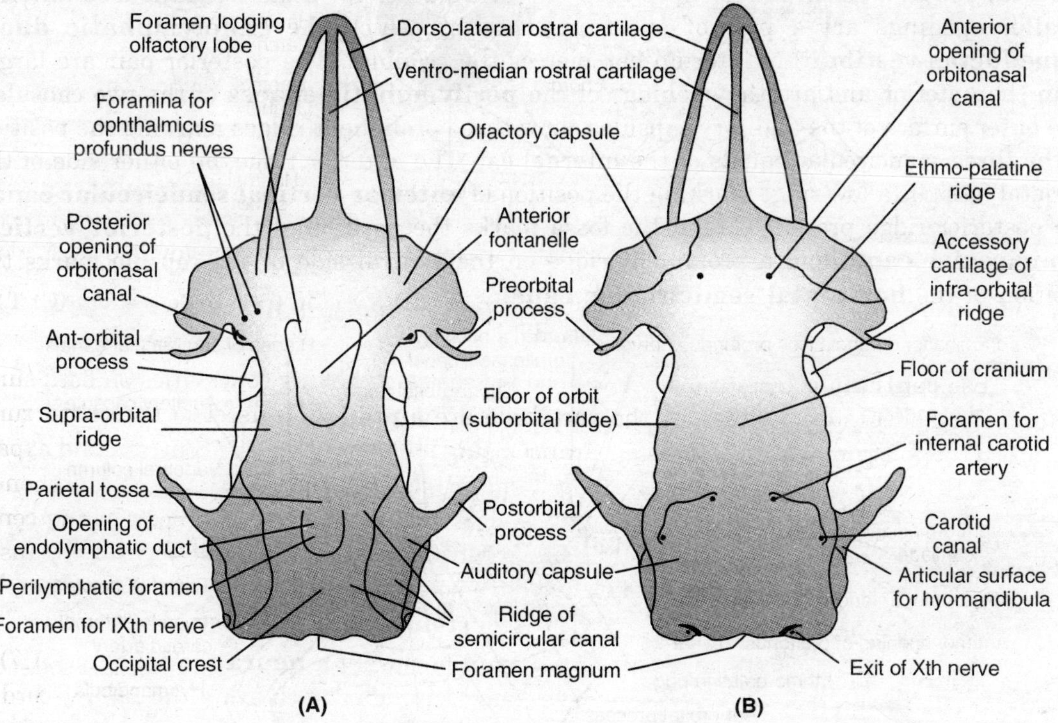

Fig. 11.9 *Scoliodon.* **(A)** Skull in dorsal view; **(B)** Same in ventral view

Skull

The skull (Fig. 11.9) of dogfish is called **chondrocranium** because cartilaginous condition is retained in adult. It consists of (*i*) **cranium** or **brain box** and **sense capsules** and (*ii*) **visceral skeleton** forming the jaws and supporting the pharynx and gills.

Cranium

The cranium encloses and protects the brain. It is a simple cartilaginous cylinder open in front and behind. The roof or dorsal side is arched and remains incomplete leaving two spaces or **fontanelles**. The floor is flat. Anteriorly it gives off three cartilaginous bars which are prolonged forward to form the **rostrum**. The cranium can be roughly divided into **four** regions: (*i*) **occipital**, (*ii*) **auditory**, (*iii*) **orbital** and (*iv*) **ethmoid**.

Occipital region: It forms the posterior part of the cranium enclosing a large median opening, the **foramen magnum**, through which the brain is continued into the spinal cord. On either side of the foramen magnum there is a small rounded **occipital condyle** which articulates with the first vertebra. Above the foramen magnum on the roof of occipital region there is a prominent median ridge, the **occipital crest**. External to the occipital condyles, on either side, lies a large **foramen** for the exit of the tenth nerve, the **vagus**.

Auditory region: It lies in front of the occipital region and consists of the **auditory** or **otic capsules** and the part of cranium they are fused with. In the embryo the auditory capsules remain separate from the cranium but in the adult they fuse with it firmly. The large laterally projected auditory capsules lodge the **auditory organs**. On the roof of the cranium lying between the two auditory capsules there is a marked oval depression, the **parietal fossa**, with a pair of apertures on each side of its posterior part. The anterior smaller openings are a pair of apertures through which the **endolymphatic ducts** (**aqueductus vestibuli**) of internal ear pierces the cranium. The posterior pair are larger than the anterior and are the openings of the **perilymphatic spaces** of the two capsules. The outer surface of the auditory capsules show three prominent ridges marking the position of the three semicircular canals of the internal ear. The anterior ridge, on either side of the parietal fossa, is a low ridge marking the position of **anterior vertical semicircular canal**, the posterior ridge present behind the fossa marks the position of the **posterior vertical semicircular canal** and a prominent ridge on the ventral side of the capsule marks the position of the **horizontal semicircular canal**.

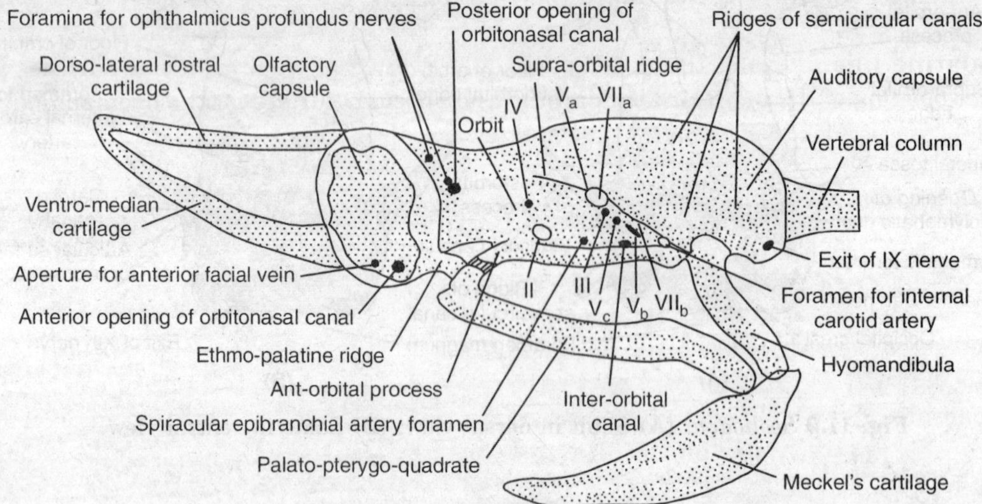

Fig. 11.10 *Scoliodon.* Skull and jaws in lateral view

Orbital region: The orbital region lies almost in the middle part of the cranium anterior to the auditory region. On each lateral side it bears large hollow **orbit** which lodges the eye, its muscles, nerves and blood sinuses. The dorsal boundary of the orbit is formed by **supraorbital ridge** which is a prominent ridge of cartilage running along the sides of the skull from the olfactory to the auditory capsules. The ventral boundary is formed by the **infraorbital** or **suborbital ridge** which arises from the base of the cranium as a lateral outgrowth. Anteriorly this ridge is produced into **antorbital process** to which a small cartilage is attached for the insertion of ligaments of the upper jaw. The anterior boundary of the orbit is bordered by a short slender **preorbital process** arising from the roof of each orbit. The posterior boundary is formed by a long and stout **postorbital process** which arises from auditory capsule and curves forward along the upper border of the orbit. The orbit has many foramina for penetration of nerves.

Ethmoid region: The ethmoid region is the anterior part of the cranium. It consists of two **nasal** or **olfactory capsules** and the **rostrum**. The olfactory capsules are two large oval cartilaginous cups situated laterally at the anterior end of the skull, firmly united to the cranium in the adult condition. A thin median cartilage, the **internasal septum** or **mesethmoid**, separates the two olfactory capsules from each other. A short and stout **ethmopalatine ridge** is present on the postero-ventral side of each capsule. The cranial roof in this region is incomplete, there being a large anterior **fontanelle** covered over by a sheet of connective tissue. Within each olfactory capsule there is a large opening leading into the cranial cavity for the entrance of the olfactory nerve.

The **rostrum** is formed by three cartilaginous bars. In front of the anterior fontanelle arises a pair of **dorso-lateral cartilages** one from the roof of each olfactory capsule. These run forward to converge and meet in front with a **median ventral cartilage** which projects from the base of the cranium.

The floor of cranium is broad and flat and bears towards its hinder end two obliquely transverse grooves, the **carotid canals**. The anterior half of each carotid canal bears two apertures, one leading into the cranium and transmitting the **internal carotid artery** and the other leading into the orbit transmitting the **stapedial artery**. Immediately behind the olfactory capsules lie two large and prominent articular surfaces for the attachment of the **ethmo-palatine ligaments** of the upper jaw, while in front of these articular surfaces lie the anterior opening of the **orbito-nasal canals**. The articular surface for the hyomandibular cartilage is a concave depression on the side of the hinder end of the skull, below the auditory capsule.

Foramina of the skull: The cranium has many **foramina** or **apertures** for various nerves and blood vessels. The **optic foramen** is a large circular aperture for the optic nerve at about the middle of the orbit. The foramen for the **third nerve** is a small hole in the inner wall of the orbit behind and above the optic foramen. The foramen for the **fourth nerve** is also a small hole vertically above the optic foramen often pushed a little posteriorly. The foramen for the **fifth** and **seventh nerves** is a large hole immediately in front of the auditory capsule. The **sixth nerve** also enters the orbit through the same opening. Each olfactory capsule has a foramen for the **olfactory nerves**. The aperture for the **interorbital canal** lies in front of the opening for the fifth and seventh nerves. The interorbital canal traverses the base of the skull and places the orbital blood sinuses of the two sides in communication. The foramen for the **spiracular epibranchial artery** into the cranium lies slightly anterior to the aperture of the interorbital canal while the foramen for the **hyoidean artery** lies a little below the same. The foramina for the **ophthalmic branches** of the fifth and seventh nerves are two separate apertures near the posterior end of the orbit just in front of the auditory capsule and above the foramen for the main branches of the fifth and seventh nerves.

Visceral skeleton

It is made up of a series of seven, **U-shaped visceral arches** which develop in the wall of the pharynx between the visceral clefts. It forms the jaws and pharyngeal skeleton. Typically each visceral arch is an incomplete ring of cartilage consisting of a basal piece in the mid-ventral line with which articulates a series of pieces on each side extending upwards almost to the mid-dorsal line (Fig. 11.11).

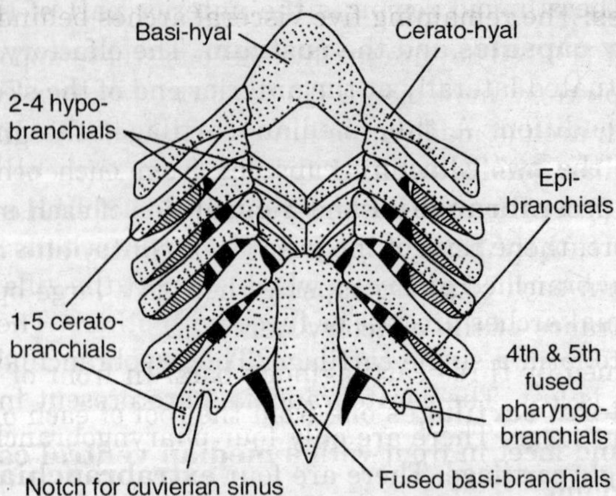

Fig. 11.11 *Scoliodon*. Ventral view of visceral skeleton

Mandibular arch: It is the first pair of visceral arch bordering the mouth and modified to form the lower and upper jaws. It has lost its original arch form and is a complete ring consisting of four pieces of cartilage, two on each side of the mouth and all joined by ligaments. The upper segment called **palato-** or **pterygo-quadrate**, forms the upper jaw and the lower segment called **Meckel's cartilage**, forms the lower jaw. The two rami of the upper and lower jaws meet anteriorly in the middle. The posterior end of each Meckel's cartilage has an articular surface for a movable joint with the upper jaw. The jaws have no direct connection with the chondrocranium.

Hyoid arch: It is a loop of cartilage made up of 5 pieces: a median **basihyal**, two **ceratohyals** and two **hyomandibulars** (Fig. 11.12). The **basihyal** is a ventro-median piece of cartilage lying between the rami of the lower jaw. It is rounded at the anterior end. It supports the floor of the buccal cavity and the tongue. The **ceratohyals** are two long and slender cartilages which run forwards and inwards in the floor of the buccal cavity behind the Meckel's cartilage. It articulates with the basihyals at its lower end. The **hyomandibulars** are two cartilages dorsal to ceratohyals. Each hyomandibular is a short, stout rod fitting at its upper end into depressions in the auditory capsules of the chondrocranium to which they are attached by ligaments. These cartilages are the only direct connection between the jaws and cranium forming, on each side, the **suspensorium** of the upper and lower jaws. Such a skull in which the hyomandibulars form the suspensorium between the cranium and jaws is called **hyostylic skull**.

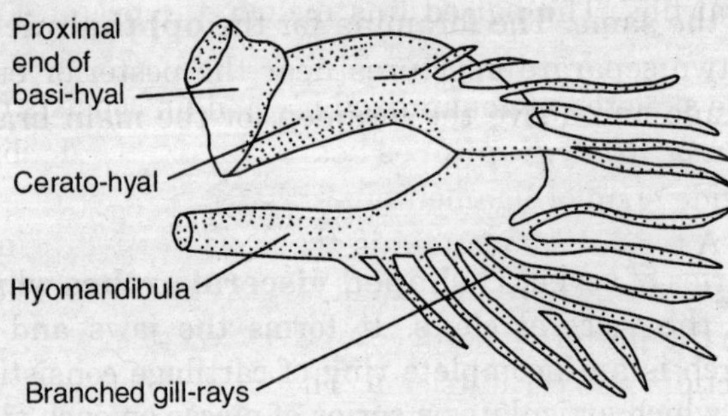

Fig. 11.12 *Scoliodon*. Skeleton of hyoid arch in ventral view

Branchial arches: The remaining five visceral arches behind the hyoid arch are called **branchial arches** (Fig. 11.13). They are so called because they support the gills and pharyngeal walls and help in respiratory movement. The branchial arches gradually diminish in size antero-posteriorly. Each arch is made up of nine cartilaginous segments connected by joints. There is a single median **basibranchial** to which are attached, on each side, a dorsal **pharyngobranchial**, an **epibranchial** and a **ceratobranchial** supporting the sides and a **hypobranchial** supporting the floor of the jaw. The epibranchials and ceratobranchials bear slender cartilaginous unbranched **gill rays** which support the gills. In *Scoliodon* some of the segments of the branchial arches are lost or fused with others. The basibranchials of all five branchial arches fuse to form a single cartilage. The hypobranchial of first branchial arch is very small and of fifth is lost. The ceratobranchials are present in all five branchial arches and the fifth one is very large. There are only four pharyngobranchials, the fourth and fifth are fused to form a single cartilage. There are four **extrabranchial arches** present at right angles to the first four branchial arches.

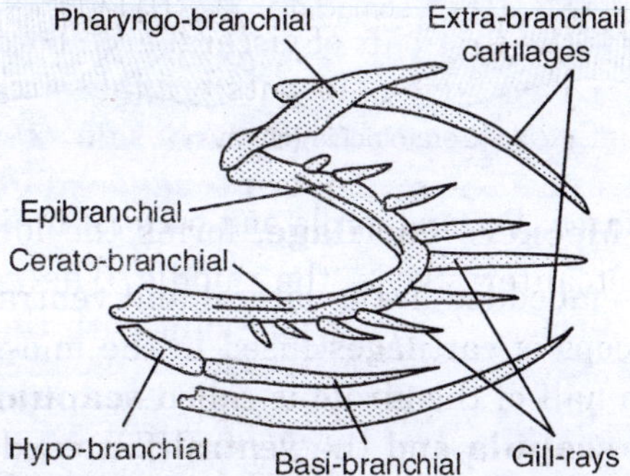

Fig. 11.13 *Scoliodon*. Skeleton of a typical branchial arch

The skull or cranium forms a protective covering of the brain and sense organs. It forms the skeleton of the jaws, gills and pharynx. The presence of jaws in fishes is a great step in the evolution of vertebrates. It forms the skeleton of the mouth, help in biting and feeding and also in holding objects.

Appendicular skeleton

The appendicular skeleton includes the **median fins**, the **paired fins** and their **girdles, pectoral** and **pelvic**. Of the median fins there are two dorsal fins (anterior and posterior dorsal) and two ventral fins. The paired fins include the pectoral and pelvic fins and their girdles.

Median fins: The skeleton of generalized median fin consists of a series of cartilaginous rods called **somactidia** or **pterygiophorcs** bearing distally a double series of horny fin rays or **ceratotrichia**. A wide strip of ligamentous tissue connects the somactidia of the fin with the vertebral column. A typical somactidia is three-segmented being made up of **proximal, mesial** and **distal** segments. There are about twenty-one somactidia, the proximal ends of which are fused along the posterior border forming an oblique axis of the fin. In the caudal fins there are no somactidia and ceratotrichia. The lobes of the fins are supported by elongated and flattened neural and haemal spines of the vertebral column.

Pectoral region: It includes the pectoral girdle and pectoral fin.

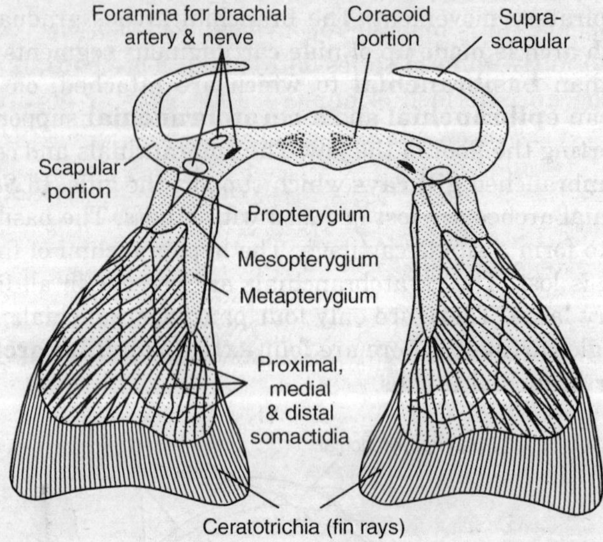

Fig. 11.14 *Scoliodon*. Pectoral girdle and pectoral fin in dorsal view

Pectoral girdle: It lies embedded in the lateral and ventral body wall posterior to the gills. It consists of two half loops of cartilages fused in the mid-ventral line and free at the dorsal ends (Fig. 11.14). Each half of the girdle is called **scapulocoracoid**. The dorsal, rod-like tapering region is called **scapula** and the ventral flattened region is called **coracoid**. Between the two regions there is a **glenoid region** which has three **foramina** for the branchial artery and nerve. In the mid-ventral portion of the girdle, at the junction of the two coracoids, there is a well-defined depression, the **pericardial depression**, which lodges the pericardium and the heart. A **suprascapular cartilage** is present at the dorsal end. The pectoral girdle is not attached to the vertebral column. It provides attachment to the pectoral fins and its muscles, and the muscles responsible for swallowing and breathing. It also protects the heart.

Pectoral fin: The pectoral fin articulates with the glenoid region of the pectoral girdle (Fig. 11.14) by three **basal cartilages—propterygium**, **mesopterygium** and **metapterygium**. Extending from the basal cartilage there are numerous **radial cartilages** or **somactidia**, each made up of three segments (poximal, mesial and distal), thus forming three rows. The second and third rows are not uniform because the pieces forming them take the form of interfitting polygonal plates. The periphery of the fin is formed by upper and lower series of thread, stiff and horny **fin-rays** or **ceratotrichia**. They support the peripheral membranous part of the fin blade.

Pelvic region: It includes the pelvic girdle, pelvic fin and claspers (copulatory organ of male *Scoliodon*).

Pelvic girdle: It is very simple in structure. (Fig. 11.15). It consists of a simple rod of cartilage called **ischiopubic bar** embedded transversely in the ventral body muscles in front of cloaca. On each side of the bar there is a small blunt **iliac process** and below it there is an **acetabular facet** for the articulation of the basal cartilage of the pelvic fin. The pelvic girdle is more primitive than the pectoral girdle. Like pectoral girdle, it also has no connection with the vertebral column.

Pelvic fin: Each pelvic fin is made up of a single basal cartilage, the **basipterygium** (Fig. 11.15) supporting a fairly uniform series of fifteen or more **radials** or **somactidia**. The peripheral portion of the fin is formed of numerous **fin rays** or **ceratotrichia**.

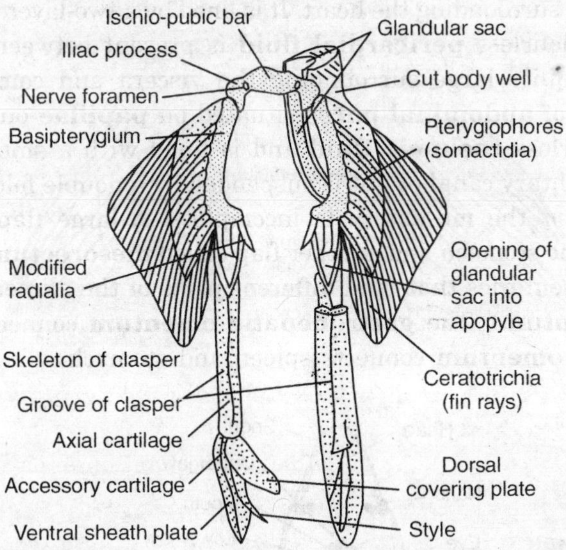

Fig. 11.15 *Scoliodon*. Pelvic girdle and Pelvic fin of a male in dorsal view

Claspers: In male dogfish the two pelvic fins become modified to form the **copulatory organ** or **claspers** (Fig. 11.15), posteriorly. It is attached to the distal end of the basipterygium. Each clasper is a tubular, backwardly directed, cartilaginous prolongation, grooved dorsally. The groove is open at both the ends. The anterior opening is called **apopyle**. It communicates with the cloaca and a blind, muscular **glandular sac** or **siphon** present at the base of the clasper. When filled with water the siphon is used to force out spermatic fluid. The posterior opening is called **hypopyle**. It opens on a sharp pointed style enclosed by a dorsal and ventral **covering** or **sheathing plates**. At the upper end of the style lies a small **accessory plate** with a serrated margin.

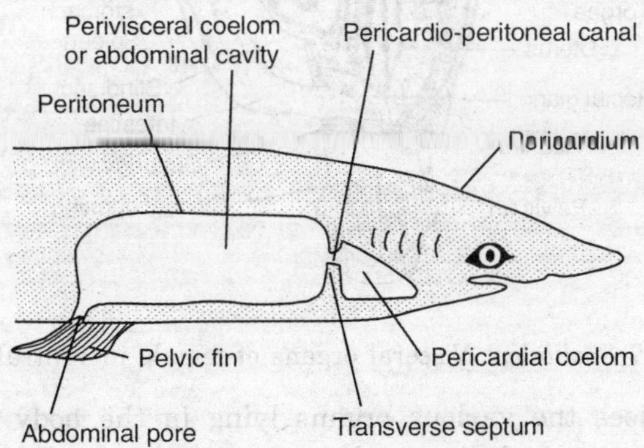

Fig. 11.16 *Scoliodon*. Coelon in lateral view

COELOM AND VISCERA

The spacious **coelom** or **body cavity** (Fig. 11.16) of dogfish is divided into two unequal cavities, anterior **pericardial cavity** and posterior **abdominal cavity**, separated from each other by a membranous partition of tough connective tissue, the **septum transversum**. The septum is pierced by a **pericardio-peritoneal canal** through which the two cavities communicate with each other. The **pericardial cavity** is a small triangular space lying beneath the pharynx and surrounding the heart. It is lined by a two-layered coelomic epithelium, the **pericardium**. A colourless **pericardial fluid** is present between the two layers. The **abdominal cavity** is quite large surrounding the viscera and communicating with the exterior through a pair of **abdominal pores** situated on **papillae** on either side of cloaca. It is filled with a colourless **coelomic fluid** and is lined with a smooth membrane called **peritoneum**. The alimentary canal or gut is suspended by a double fold of peritoneum called **mesentery**. In *Scoliodon* the mesentery is incomplete. A large flap of mesentery called **mesogaster** suspends the stomach and another flap called **mesorectum** suspends the hinder part of the gut. The mesenteries that bind different parts of the alimentary canal are called **omenta** (singular **omentum**). The **gastrohepatic omentum** connects liver and stomach and the **gastrosplenic omentum** connects spleen and stomach.

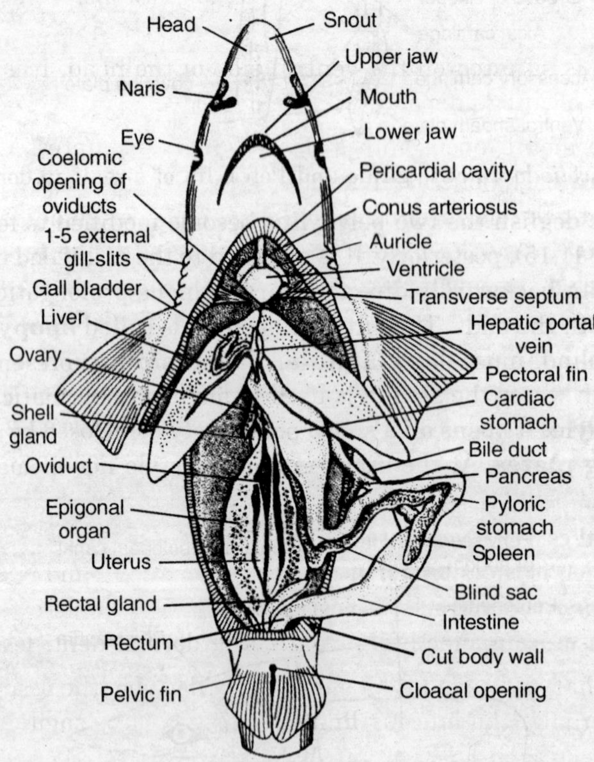

Fig. 11.17 *Scoliodon*. Visceral organs of female in ventral view

The **viscera** comprises the various organs lying in the body cavity. The important organs are heart, alimentary canal, liver, spleen, pancreas, gonads and kidneys.

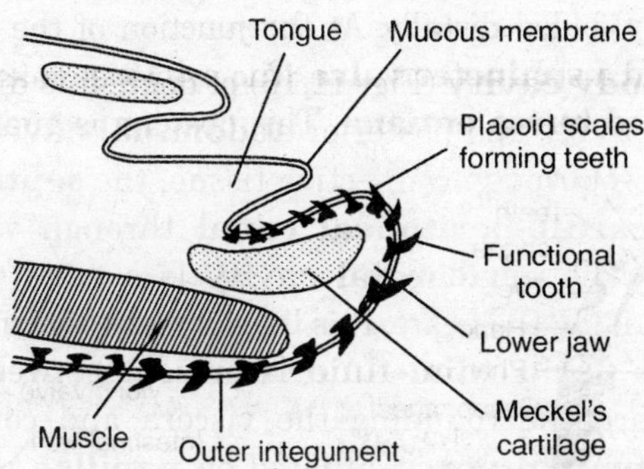

Fig. 11.18 *Scoliodon*. Lower jaw in section showing formation and replacement of teeth

DIGESTIVE SYSTEM

The digestive system includes the alimentary canal or gut and the digestive gland.

Alimentary canal

It comprises the mouth, buccal cavity, pharynx, oesophagus, stomach, intestine and rectum (Fig. 11.17).

Mouth: The **mouth** is situated on the ventral side of the head. It is a wide crescentic opening bounded by folds of integument.

Buccal cavity: The mouth opens into a dorsoventrally compressed spacious **buccal cavity** bordered by the **jaws**. The buccal cavity is lined with thick mucous membrane which is raised ventrally into a thick fold forming a non-muscular, non-glandular, non-protrusible **tongue** which is supported internally by flat basihyal cartilage. The **teeth** (Fig. 11.18) are oblique, backwardly directed, having sharp, more or less compressed cusps, the edges of which are smooth and non-serrated. They are not attached to the jaws but are simply embedded in the skin like placoid scales. All teeth are alike in shape, *i.e.*, **homodont** and are borne in several parallel rows (**lyodont**) on the upper and lower jaws. The teeth of dogfish are **polyphyodont**, *i.e.*, if lost or destroyed they are replaced a number of times during a life time. The teeth are not used for biting the prey but simply meant to hold the prey and prevent it from escaping.

Pharynx: The buccal cavity opens into the **pharynx**. On either side of the pharynx lie inconspicuous oval pit which is the internal opening of the spiracle and five internal gill slits of gill pouches. In *Scoliodon* the spiracle is vestigeal, *i.e.*, not visible externally. Internally the pharynx is lined with mucous membrane containing dermal denticles.

Oesophagus: The pharynx leads into a short but wide tube, the oesophagus which has thick muscular wall. Internally the mucous lining is thrown into longitudinal folds.

Stomach: The stomach is horse-shoe shaped large muscular tube. The long and wide proximal limb is called **cardiac stomach** while the short and narrow distal limb is called **pyloric stomach** (Fig. 11.19). The opening of the oesphagus into the cardiac stomach is guarded by an **oesophageal valve**. The inner mucous lining of cardiac stomach is also thrown into longitudinal folds like that of oesophagus. The lining of pyloric stomach is quite smooth proximally but slightly folded distally. At the junction of the two limbs there is a blind outgrowth called **blind sac** and a **sphincter valve**. The pyloric stomach opens into a small, thick-walled muscular chamber called **bursa entiana**. This opening is guarded by a **pyloric valve.**

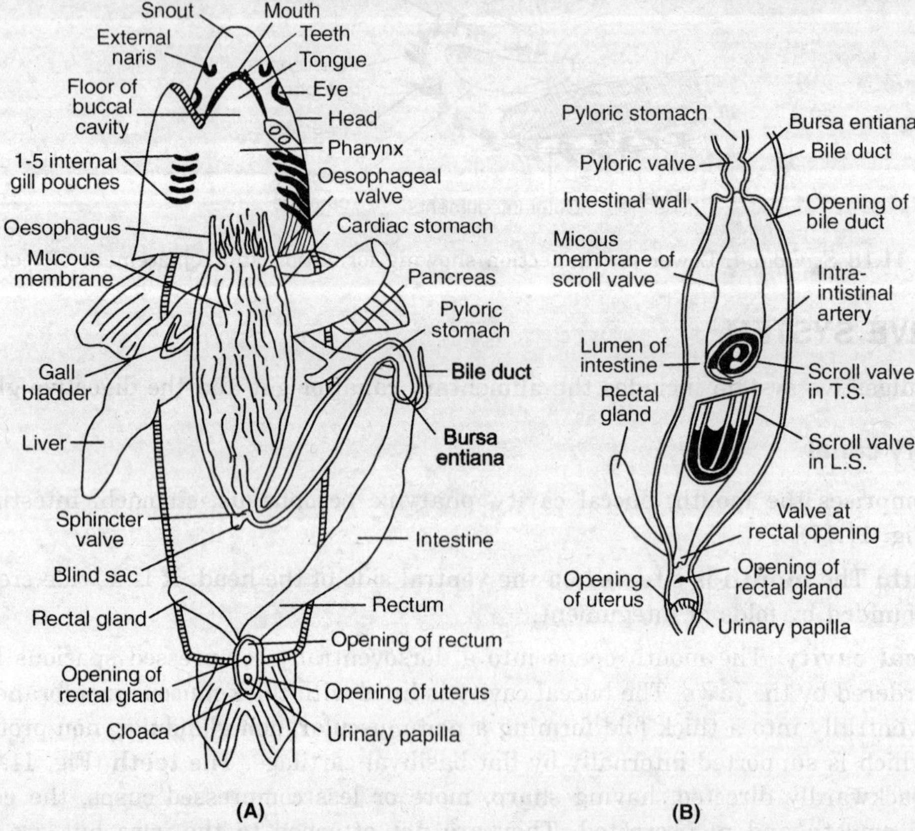

Fig. 11.19 *Scoliodon*. **(A)** Animal cut open to show the digestive system; **(B)** Intestine cut open to show scroll valve

Intestine

The bursa entiana continues into the intestine which is a wide tube running straight backwards into the abdominal cavity. The internal surface of the intestine is increased by a characteristic fold of mucous membrane called the **scroll valve** or **spiral valve** (Fig. 11.19B). Its one end is attached to the inner wall of the intestine and the other end is rolled up longitudinally on itself into a scroll making an anticlockwise spiral of about two and a half turns. In transverse section it looks like a watch spring. In some genera of dogfish, for example, *Scylliorhinus* and *Squalus*, the spiral valve is made up of 14-15 turns. The scroll valve delays the passage of food through the intestine and increase surface area for absorption of food.

Rectum: The intestine opens posteriorly into a short and narrow tube, the **rectum** which is the last part of the gut. The **caecal** or **rectal gland** opens into the rectum dorsally. The rectum opens behind through an **anus** into a ventral chamber, the **cloaca**. The urinogenital duct also opens into the cloaca.

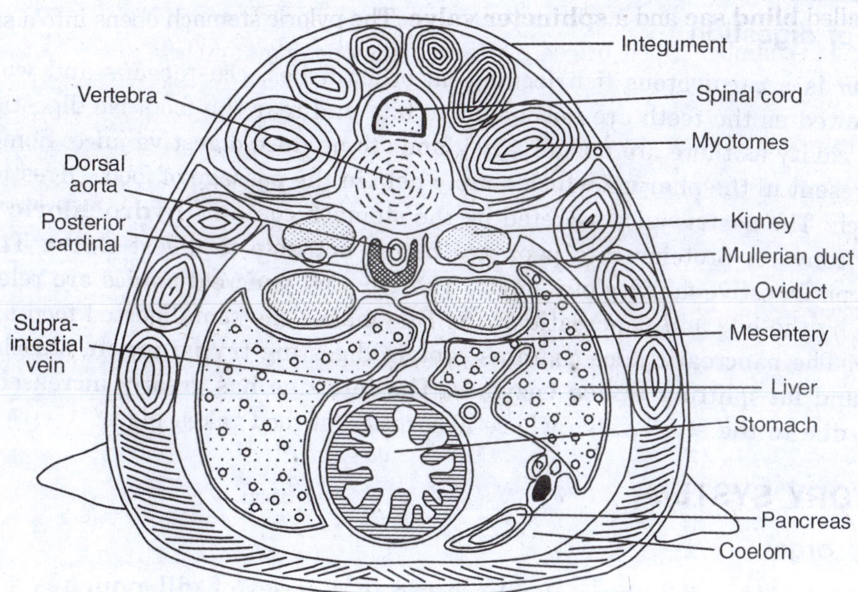

Fig. 11.20 *Scoliodon*. T.S. passing through the stomach region

Glands of digestive system

Liver: It is a large, compact, elongated, yellowish gland made up of two lobes united anteriorly. They extend backwards freely occupying a greater part of the abdominal cavity (Fig. 11.19). A median **suspensory ligament** connects the liver lobes with anterior wall of the abdominal cavity. A thin-walled **V**-shaped sac, the **gall bladder**, lies embedded in the right lobe of the liver. The gall bladder stores bile secreted by the liver. The **bile duct**, about one inch long, leaves the gall bladder between the two lobes of the liver, almost in the median plane and opens into the anterior end of the intestine. It also receives fine ducts from the liver lobes. The liver secretes bile and stores fat and glycogen. The **Küpffer cells** present in the liver destroy worn out red blood cells.

Pancreas: It is a whitish, laterally compressed, compact gland consisting of two lobes (Fig. 11.19A). The **dorsal lobe** is longer and runs parallel to the posterior part of cardiac stomach while the **ventral lobe** is smaller and lies closely applied to the pyloric stomach. The **pancreatic duct** runs forward through the pancreas close to its ventral border and opens into the intestine just opposite to the opening of the bile duct.

Rectal or caecal gland: It is a short, thick-walled tube lying in the abdominal cavity dorsal to the rectum into which it opens. It discharges a fluid of unknown function, into the lumen of the intestine. It is highly vascular and composed of lymphoid tissue.

Spleen: It is a large gland found attached to the stomach but is not associated with the digestive system. It consists of two lobes, one extending along the outer curvature of the cardiac stomach while the other lying parallel to the pyloric stomach. It is a lymphoid organ and produces lymphocytes.

Physiology of digestion

Scoliodon is a carnivorous fish feeding on other fishes, crustaceans and worms. The prey is swallowed as the teeth are not meant to bite and tear the food. No digestion occurs in the buccal cavity as there are no salivary glands to produce digestive juice. Some mucous glands are present in the pharynx which simply help in the passage of food. Digestion begins in the stomach. The gastric juice secreted by the stomach contains **hydrochloric acid** and **pepsin** that converts proteins into peptones but is not able to digest chitin. The pyloric stomach has no digestive activity of its own. The bile and pancreatic juice are released into the intestine by the bile and pancreatic ducts. Bile makes the semidigested food alkaline in the intestine. The pancreatic juice contains protein digesting **trypsinogen**, starch-splitting **amylopsin** and fat-splitting **lipase** enzymes. The intestine has greatly increased internal surface area due to the scroll valve where absorption of food takes place.

RESPIRATORY SYSTEM

Respiratory organ

Scoliodon breathes by means of **gills** borne in a series of **gill pouches** situated on either side, in the lateral wall of the pharynx behind the hyoid arch. There are **five** pairs of gill pouches. Each gill pouch is compressed antero-posteriorly. It communicates with the pharynx by a large **internal branchial aperture** and with the exterior by a narrow **external branchial aperture** or **gill slit**. Successive gill pouches are separated from each other by a stout fibromuscular partition called **interbranchial septum** or **gill septum** which is covered by an epithelium and contains blood vessels and nerve fibres. The inner border of each gill septum is supported by a visceral arch and its branchial or gill rays. The gill septum extends beyond the branchial lamellae and forms flaps which protect the gills and external gill slits. The mucous membrane of the gill septum is raised into a number of highly vascular horizontal folds called **branchial** or **gill lamellae** or **gill filaments** which form the **gill proper**. Each gill septum has two sets of gill lamellae, one on its anterior and the other on

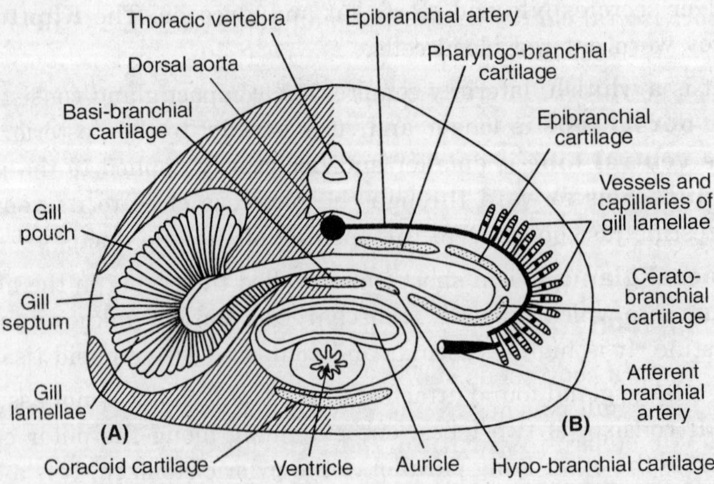

Fig. 11.21 *Scoliodon*. T.S. of body showing respiratory system (diagrammatic); **(A)** Gill pouch opened to show the gill in surface view; **(B)** Blood supply of gill lamella

its posterior wall. Each set of gill lamellae is called a **half gill** or **hemibranch** or **demibranch** thus each gill pouch has two hemibranchs. The posterior hemibranch is larger than the anterior one. Two sets of branchial lamellae attached to a visceral arch and its interbranchial septum forms a **complete gill** or **holobranch** (Fig. 11.22). In a complete gill the visceral arch supports the posterior branchial lamellae or hemibranch of one gill pouch and the anterior hemibranch of the gill pouch behind. The visceral arch also gives out some rigid comb-like processes called **gill rakers** which project inwards and protect the internal gill slits from food.

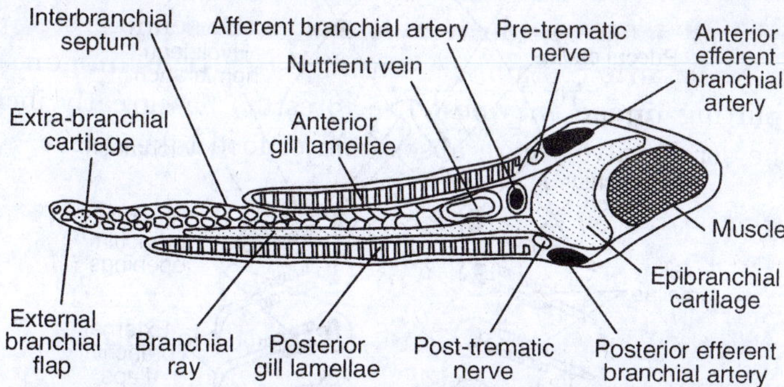

Fig. 11.22 *Scoliodon*. Horizontal section of a holobranch

In *Scoliodon* the hyoid arch bears a demibranch on its posterior border. The first four branchial pouches bear holobranchs and the fifth branchial pouch has no gills (**abranch**). Thus there are nine hemibranchs on each side. The form of gill in *Scoliodon* is **lamelliform**. It is regarded as more primitive than the **filiform** or **pectinate gills** of bony fishes.

There is a dense network of blood vessels within each gill lamella (Fig. 11.21B). These receive venous blood from the **afferent branchial vessels** (running on the outer side of the arch) and pass it to the **efferent branchial loop**. Oxygenation takes place while the blood passes through the network of blood vessels.

The first pair of visceral cleft (between the mandibular and hyoid arches) gives rise to the spiracle which is vestigeal in *Scoliodon*. It is in the form of oval pits present on either lateral side of the pharynx. It is without gill lamella and external opening. In other elasmobranchs minute gill lamellae are present forming a false gill or **pseudobranch** which opens to the outside by an aperture behind the eye. It is called false because its blood supply is not derived directly from an afferent branchial artery but from an efferent branchial artery of the following gill arch which brings oxygenated blood. The spiracles are used to take water in for respiration.

Mechanism of respiration

In order to maintain a continuous supply of oxygen for respiration, a current of water is maintained through the gill pouches by the movement of pharyngeal wall. The breathing mechanism involves two steps : **inspiration** and **expiration** (Fig. 11.23).

Water enters the buccopharyngeal cavity through the mouth (perhaps also through the spiracles, if they are functional). During **inspiration** the floor of the pharynx is lowered with the help of hypobranchial muscles and the mouth is opened. The increased volume of the bucco-pharyngeal cavity draws water in. **During expiration** the mouth is closed by the action of adductor muscles, the floor of the pharynx is raised by the action of constrictor and interbranchial muscles and the volume of the buccopharyngeal cavity is reduced. As a result the contained water is forced through the internal openings of the branchial clefts into the gill pouches (the entrance of the oesophagus is closed by contraction of its muscles). The water bathes the gill lamellae and then passes out through the gill slits.

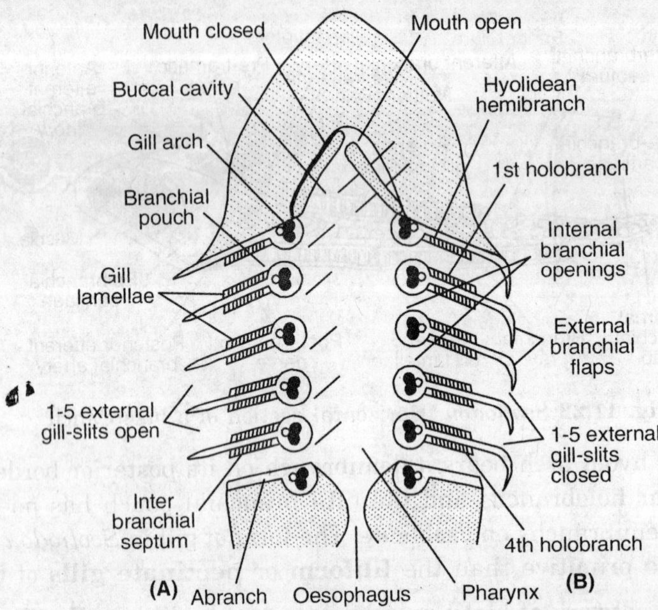

Fig. 11.23 *Scoliodon*. Breathing mechanism; **(A)** Expiration; **(B)** Inspiration

The buccopharyngeal cavity acts as a force pump that maintains water current only in one direction, *i.e.*, in through the mouth and out through the gill slits. Along the anterior border of each gill slit there are flaps of skin which prevent the entry of water through the slits when the mouth is open and the pharynx enlarging. Likewise when the pharynx contracts the water is prevented from rushing out through the mouth which is closed and sealed by the lips. By alternate contraction and relaxation of their intrinsic musculature, the gill pouches also contract and relax, thus helping in maintenance of water current.

Physiology of respiration

The water that enters the gill pouches contains a large amount of dissolved oxygen. The wall of blood capillaries is very thin and permeable. The oxygen dissolved in water passes in through the thin wall of the blood vessels by **endosmosis** and carbon dioxide of the venous blood passes out by **exosmosis** into the water. The carbon dioxide is got rid off by the outgoing respiratory current of water and oxygen picked up by blood is transported to different parts of the body. This is known as **external respiration**. The oxygen taken by blood is used to oxidise food and release carbon dioxide and energy within the tissue. The carbon dioxide is released in the blood which becomes venous. This is called **internal respiration** or **tissue respiration**.

BLOOD VASCULAR SYSTEM

The **blood vascular system** or the **circulatory system** comprises the **heart**, the **blood vessels** (**arteries** and **veins**) and the **blood.**

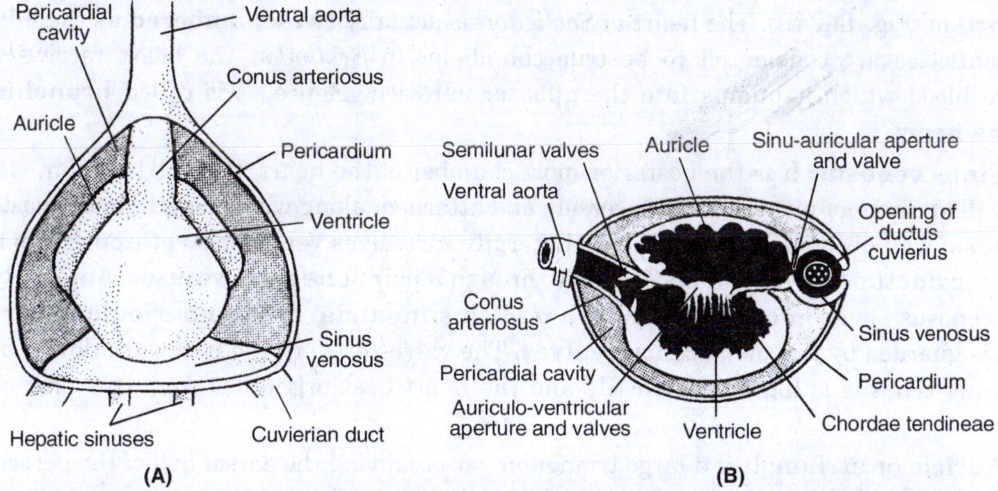

Fig. 11.24 *Scoliodon*. Heart and pericardm; **(A)** Ventral view; **(B)** Sagittal section

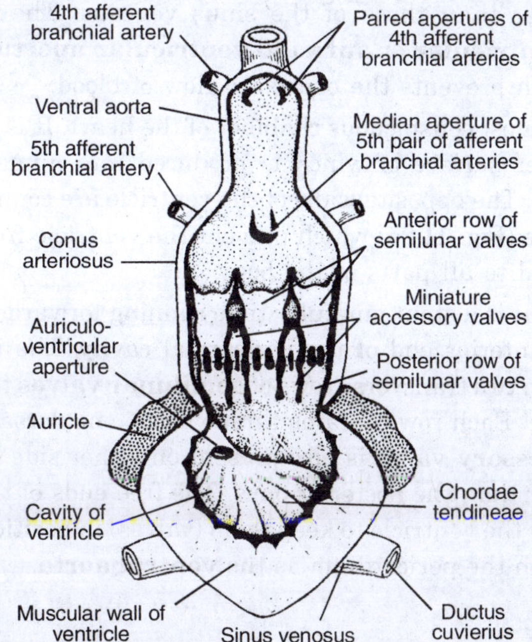

Fig. 11.25 *Scoliodon*. Internal structure of heart

Heart

The heart (Fig. 11.24A) lies in the **pericardial cavity** which is a median triangular cavity lying ventrally between the gills beneath the pharynx, with the apex directed forward.

It is supported from below by the coracoid cartilage of the pectoral girdle. The heart almost completely occupies the pericardial cavity. It is surrounded by a double layered membranous **pericardium**. It is a muscular **S**-shaped tube bent on itself forming **four chambers** : **sinus venosus**, **auricle**, **ventricle** and **conus arteriosus** separated from each other by transverse constriction (Fig. 11.24B). The heart of *Scoliodon* is actually **two-chambered** as the auricles and ventricles are considered to be true chambers. In *Scoliodon* the heart receives only impure blood which it pumps into the gills for aeration. Hence, it is called **branchial** or **venous heart**.

Sinus venosus: It is the posterior-most chamber of the heart. It is a triangular, elastic, thin-walled chamber elongated transversely and attached, along whole length, to the posterior and dorsal wall of the pericardial cavity. Laterally it receives venous blood through two large veins, the **ductus Cuvieri** and posteriorly through a pair of **hepatic sinuses**. Anteriorly the sinus venosus opens into the atrium by a median **sinu-atrial** or **sinu-auricular aperture** which is guarded by two membranous **valves**. The valves prevent the backward flow of blood. The sinus venosus is highly contractile and the heart beat originates from this part of the heart.

Auricle or atrium: It is a large triangular sac occupying the dorsal half of the pericardial cavity. It lies in front of the sinus venosus and dorsal to the ventricle. Its posterior angles are produced into processes which project laterally at the sides of the ventricle. The walls of the atrium are thicker than those of the sinus venosus. The atrium opens into the ventricle by the **atrio-ventricular** or **auriculo-ventricular aperture** (Fig. 11.25) guarded by a bilabiate valve which prevents the backward flow of blood.

Ventricle: It is the most conspicuous chamber of the heart. It is pear-shaped, with thick muscular walls, the inner surface of which is produced into numerous muscular strands giving it a spongy texture. The opposite walls of the ventricle are connected by tendinous cord called **chordae tendinae** (Fig. 11.25) which prevent the ventricle from over expanding. The ventricle pumps the blood to all parts of the body.

Conus arteriosus: It is a stout muscular tube running forward from the anterior end of the ventricle up to the anterior end of the pericardial cavity. The inner wall of the conus arteriosus is provided with two transverse rows of **semilunar valves** to prevent backward flow of blood into the ventricle. Each row contains three valves, one dorsal and two ventrolateral. In addition, a small **accessory valve** is also present on either side of the dorsal valve. The anterior valves are larger than the posterior ones. The free ends of the valves are connected to the chordae tendinae of the ventricle to keep them (valves) in position. The conus arteriosus continues forward through the pericardium as the **ventral aorta**.

Arterial system

It includes a set of blood vessels taking blood away from the heart. In dogfish it consists of two sets of blood vessels : the **afferent branchial blood vessels** which carry impure blood from the heart to the gills for purification, and the **efferent branchial blood vessels** which collect purified blood from the gills and conduct it to different parts of the body (Fig. 11.26).

Afferent branchial arteries

The ventral aorta is a stout artery which arises as an anterior prolongation of the conus arteriosus. Passing through the pericardium it runs along the ventral surface of the pharynx right up to the posterior border of the hyoid arch where it bifurcates into two short branches, the **innominate arteries**, each of which again divide into two forming **first** and **second afferent branchial artery**. The first afferent branchial artery runs along the posterior

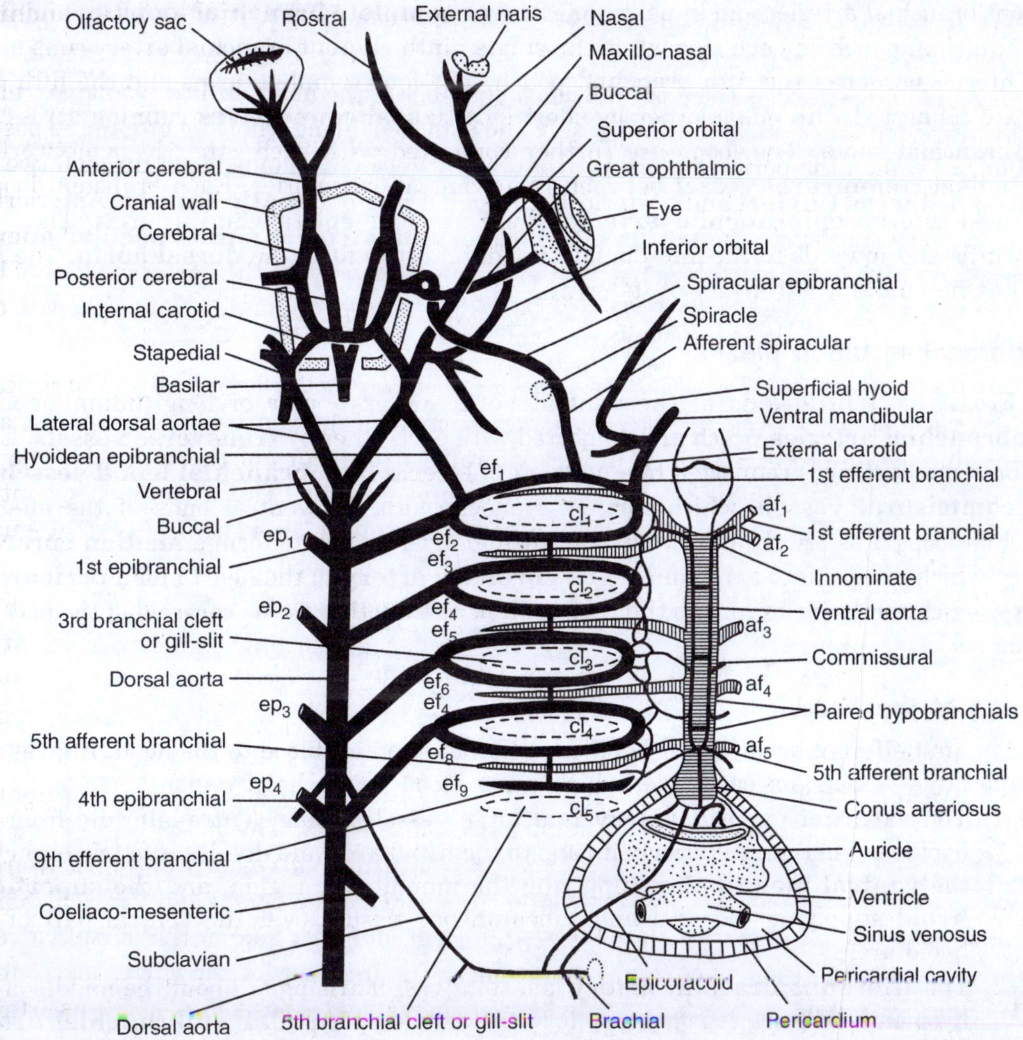

Fig. 11.26 *Scoliodon*. Arterial system. Afferent arteries (dotted lines) and efferent arteries (black lines)

border of the hyoid arch and supplies arterial branches, to the gill lamellae of hyoidean demibranch. The second afferent branchial artery supplies arterial branches to the anterior and posterior gill lamellae of the first branchial arch. The **third, fourth** and **fifth afferent branchial arteries** arise independently from the ventral aorta almost equidistant from each other and supply arterial branches to the anterior and posterior gill lamellae of the second, third and fourth branchial arches, respectively. These arteries break up into fine capillaries in the gill lamellae.

Efferent branchial arteries

The aerated blood from the gills is collected by **nine efferent branchial arteries** which run along the anterior and posterior borders of the branchial pouches. The first-eight efferent branchial arteries join in pairs to form four **complete branchial loops** around first-four branchial pouch on each side while the single ninth efferent branchial artery runs along the anterior border of the fifth branchial pouch. The four complete loops and the fifth half loop are connected with one another by short longitudinal **connectives** running across the inter-branchial septa. The loops are further connected with each other by a network of longitudinal **commisural vessel** between their ventral extremities. Each branchial loop is continued into an **epibranchial artery**. There are four epibranchial arteries. They run backwards and inwards to the mid-dorsal line and meet to form the **dorsal aorta**. The half loop has no epibranchial artery of its own.

Hypobranchial blood plexus

From the ventral side of the ventral aorta arises a pair of longitudinal median **hypobranchial arteries** which are connected with each other by **transverse vessels**. Each hypobranchial artery is connected to a network of **lateral hypobranchial blood vessels** by four **commisural vessels** which further connects with the ventral ends of the efferent branchial loops. Posteriorly, the two hypobranchial arteries join to form a **median coracoid artery** which divides into two branches : a **coronary artery** to the heart and a **pericardial artery** which sends one branch to the pericardium and another to the coracoid of the pectoral girdle.

Arteries of the head

The first efferent branchial or hyoidean efferent supplies blood to the head. It gives rise to three large vessels, on each side, which carry blood to the head region.

1. The **external carotid artery** is a large vessel arising ventro-laterally from the hyoidean efferent. Anteriorly it runs to the hyoid arch and divides into two branches: the **ventral mandibular** supplying the mandibular region, and the **superficial hyoid** supplying the skin and subcutaneous tissues over the ventral part of the hyoid arch.
2. The **afferent spiracular artery** is a stout vessel arising at about the middle of the hyoidean efferent. It continues forward as the **spiracular epibranchial artery** which runs on the outer side of the hyomandibular and epibranchial artery across the orbit and enters the cranial cavity through a small foramen. Before entering the cranium it gives a branch, the great **ophthalmic artery**, to the eyeball. Within the cranium it unites with a branch of the internal carotid to form a short **cerebral artery** which soon bifurcates to form **anterior** and **posterior cerebral arteries** supplying blood to the brain.
3. The **hyoidean epibranchial** arises from the dorsal end of the hyoidean efferent. It runs forwards and inwards to join a lateral branch of the dorsal aorta and immediately divides into two branches: the **stapedial** and the **internal carotid**.

The stapedial artery runs forwards and enters the orbit where it gives two branches: (*i*) the **inferior orbital** which supply the eye muscles and the superficial tissue in the region above the auditory capsules, and (*ii*) the **superior orbital** which supply the anterior boundary of the orbit. It sends a **buccal artery** to the lower jaw muscles, a **nasal artery** to the olfactory sac, fine **maxillary arteries** to the upper jaw muscles and a **rostral artery** to the rostrum.

The internal carotid artery passes inwards to the buccal cavity and enters the cranium where it divides into two branches, one of which unites with its fellow of the opposite side forming a loop while the other unites with the spiracular epibranchial to form the **cerebral artery**.

Dorsal aorta

The dorsal aorta is formed by the union of four pairs of epibranchial arteries. It is a large median vessel running along the whole length of trunk and tail, lying beneath the vertebral column. In the tail region it continues into the haemal canal of the tail vertebrae as **caudal artery.** The principal branches of dorsal aorta are :

1. Anteriorly the dorsal aorta bifurcates into two branches, the **lateral aortae** or **radices** (singular radix). Each lateral aorta joins the hyoidean epibranchial artery of its side.
2. The **subclavian arteries** are a pair of vessels arising close to the union of four epibranchial arteries. They run outwards and backwards to supply the pectoral girdle and pectoral fin.
3. **Buccal arteries** are a pair of vessels supplying the roof of the buccal cavity.
4. The **coeliaco-mesenteric** is a large median artery arising from the dorsal aorta behind the junction of the fourth pair of epibranchial arteries. It divides into two unequal branches :
   (*i*) the **coeliac** that supplies the stomach and liver,
   (*ii*) the **mesenteric** supplying the pancreas, intestine and rectum.
5. The **lienogastric artery** arises a short distance behind the origin of the coeliaco-mesenteric. It is a median vessel supplying the genital organs, stomach, posterior part of intestine and spleen.
6. The **posterior mesenteric** is a small median vessel arising a little behind the lienogastric. It runs backwards supplying the mesentery and rectal gland.
7. The **parietals** are a series of paired vessels arising at intervals along the whole length of the dorsal aorta and supplying the body wall.
8. The **renal arteries** are small paired vessels arising from the parietal arteries and supplying the kidneys, which they enter from the dorsal surface.
9. The **iliacs** are a pair of vessels extending into the pelvic fins.
10. The **vertebrals** are a pair of vessels supplying the vertebral column.

Venous system

Venous blood is returned to the heart by narrow tubular **veins** which have thinner walls than the arteries. In dogfish many of the veins are in the form of large irregular spaces called **sinuses** which do not have definite walls. The venous system can be described under the following heads.

Anterior cardinal system (Fig. 11.27)

It collects blood from the part of the body anterior to the heart by two sinuses.

1. Inferior jugular sinus: It is a small median ventral sinus collecting blood from the floor of buccopharyngeal region, gill pouches and pericardial region and finally opens into the ductus Cuvieri.

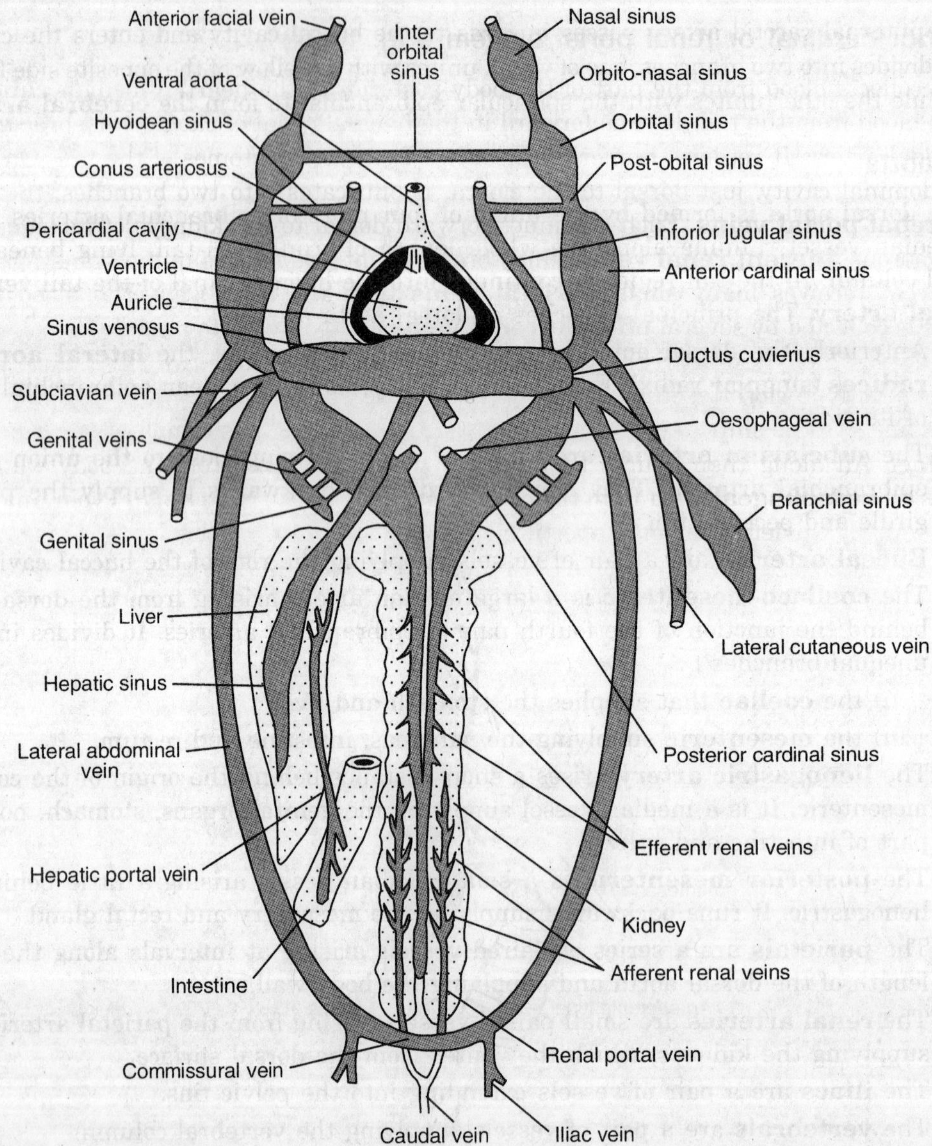

Fig. 11.27 *Scoliodon*. Dogrammatic representation of the venous system

2. Internal jugular sinus: It is a large dorso-lateral sinus collecting blood from the dorsal region of head and gill pouches. It comprises a number of veins or sinuses. It begins as an **anterior facial vein** from the rostrum which joins an **olfactory** or **nasal sinus** behind the olfactory sac. A large **orbital sinus** lies in the orbit and is connected with its fellow of the opposite side by an **interorbital vein** and is joined by an **orbito-nasal vein** to the nasal sinus. The orbital sinus opens into a narrow **post-orbital sinus** and finally into a large **anterior cardinal sinus** which lies between the dorsal ends of the gill pouches and the muscles of the body wall. It opens into the ductus Cuvieri behind. A **hyoidean sinus** lying just in front of the first gill connects the anterior cardinal and inferior jugular sinus. It also receives five **nutrient branchial sinuses** from the gills.

Posterior cardinal or renal portal system (Fig. 11.27)

It collects blood from the part of the body posterior to the heart. A median **caudal vein** collects blood from the tail. It runs forward in the haemal canal of caudal vertebrae ventral to the caudal artery. It receives numerous branches from the myotomes of the tail. On entering the abdominal cavity, just dorsal to the cloaca, it bifurcates into two branches, the **left** and **right renal portal veins**. Each continues forward dorsal to the kidneys of its side and give out numerous **afferent renal veins** which break up into capillaries in the kidney. Each renal portal vein receives many small **parietal veins** from the body wall. The blood from the kidneys is collected by several **efferent renal veins** and poured into a single vessel, the **inter renal vein** which lies between them. This vein bifurcates into two **posterior cardinal sinuses** which lie close together along the roof of the abdominal cavity. In the region of the oesophagus both the sinuses enlarge to form wide thin walled sacs communicating freely with each other. All along their course they receive **parietal veins** from the body wall. It also receives an **oesophageal vein** from the oesophagus. Each posterior cardinal vein finally opens into the ductus Cuvieri by a small aperture opposite the opening of the anterior cardinal sinus.

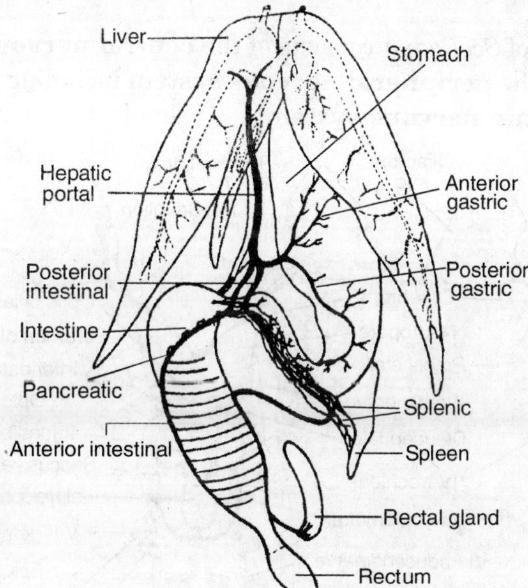

Fig. 11.28 *Scoliodon*. Hepatic portal system

Hepatic portal system

The venous blood from the alimentary canal and the digestive glands is collected by several veins, namely, **anterior** and **posterior intestinal vein**, **lienogastric** and **anterior** and **posterior gastric veins**, which unite to form a single long vein, the **hepatic portal vein** (Fig. 11.28). It runs forwards, divides into two branches and enters the right and left lobes of the liver where it breaks up into capillaries. The blood from the liver lobes is collected by two large thin walled **hepatic sinuses** which open anteriorly into the sinus venosus by two small apertures.

Lateral abdominal system

It consists of a pair of **lateral abdominal vein** which collects blood from the inner lateral part of the body wall, a **cloacal vein** from the cloacal region and the **femoral vein** from the pelvic fins. In the posterior part of the abdominal cavity the two lateral abdominal veins are connected with each other by a **commisural vein**. Anteriorly each joins a **branchial vein** from the pectoral fin. The **subclavian vein** is formed by the union of the branchial vein and the lateral abdominal vein. It enters the ductus Cuvieri laterally.

Cutaneous system

It includes a **mid-dorsal**, a **mid-ventral** and two **lateral cutaneous veins**. The mid-dorsal cutaneous veins runs along the mid-dorsal line beneath the skin. The mid-ventral cutaneous vein runs along the mid-ventral line beneath the skin and joins the lateral abdominal vein in front of the cloacal vein. The two lateral cutaneous veins run along each side beneath the lateral line canal and open into the branchial vein in front. The venous blood from different parts of the body is collected in ductus Cuveiri and finally emptied into the sinus venosus.

NERVOUS SYSTEM

The nervous system of *Scoliodon* consists of the **central nervous system** including the **brain** and **spinal cord**, the **peripheral nervous system** including the **cranial** and **spinal nerves** and the **autonomic nervous system**.

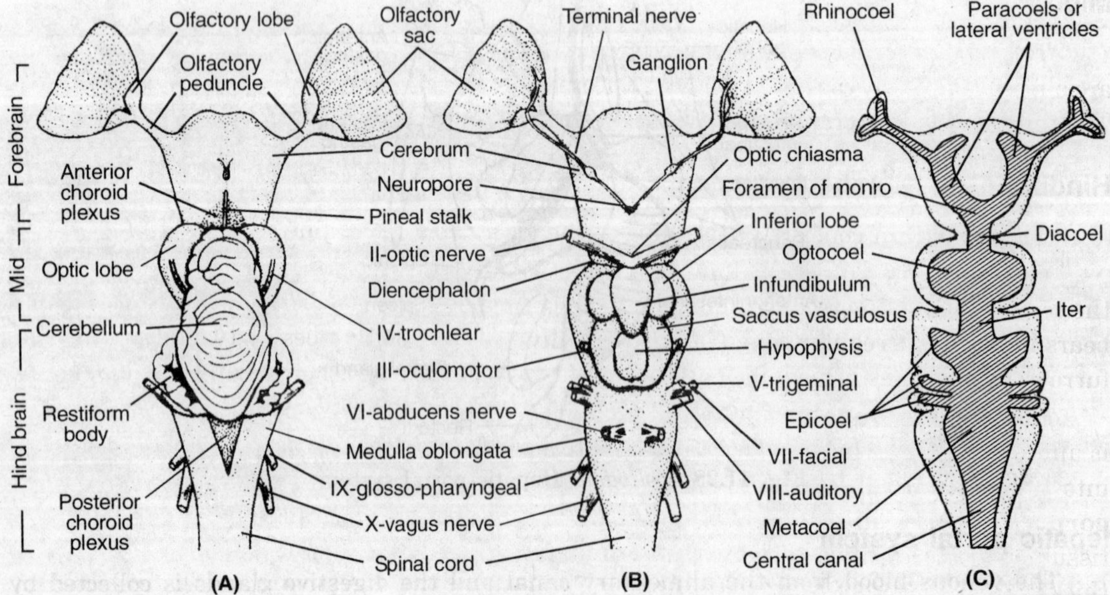

Fig. 11.29 *Scoliodon*. Structure of brain; **(A)** Dorsal view; **(B)**Ventral view; **(C)** Sagittal section showing ventricles

CENTRAL NERVOUS SYSTEM

Brain

The brain of *Scoliodon* consists of three parts: **forebrain, midbrain** and **hindbrain.**

Forebrain or prosencephalon

The forebrain consistes of two regions—**cerebrum** and **diencephalon.**

Cerebrum: The cerebrum is the large undivided anterior part of the forebrain. There is no median groove to divide it into right and left cerebral hemispheres. From its antero-lateral ends arise a short-stout **olfactory peduncle** or **tract** which ends into a large bilobed **olfactory lobe** situated behind the olfactory sac. It opens to the outside through the **nostrils** or **external nares**. The dorsal surface of the cerebrum is smooth but its mid-ventral surface has a small aperture called **neuropore** through which two delicate **terminal** or **preolfactory nerves** come out and innervate the mucous membrane of the olfactory sacs.

Diencephalon: The diencephalon lies posterior to the cerebrum. It is short and narrow and completely covered by the forward prolongation of the cerebellum. The roof of diencephalon is thin, membranous and highly vascular and is called the **anterior choroid plexus**. From the postero-dorsal end of the diencephalon extends upwards a slender **pineal stalk** ending in a small rounded **pineal body** or **epiphysis**. Its function is not known. On the floor of ventral surface of the diencephalon a hollow outgrowth called **infundibulum** is present. It terminates in a rounded sac-like body called **hypophysis.** They together form the **pituitary body**. On either side of the infundibulum lies a thick walled sac called **inferior lobe,** the distal end of which it continued into a glandular sac, the **saccus vasculosus**. In front of the infundibulum near the anterior margin lies the **optic chiasma** formed by the crossing of two optic nerves.

Midbrain or mesencephalon

The midbrain consists of a pair of large rounded **optic lobes** or **corpora bigemina**. They are almost completely concealed dorsally by cerebellum and ventrally by infundibular outgrowths. The third and fourth cranial nerves arise from the floor and roof of the midbrain, respectively.

Hindbrain or rhombencephalon

The hindbrain consists of two parts—the **cerebellum** and **medulla oblongata.**

Cerebellum: The cerebellum is a large rhomboid structure extending forwards up to the cerebrum and covering part of medulla oblongata behind. The dorsal surface of cerebellum bears numerous irregular convolutions. It is divided into three lobes by two deep transverse furrows. Sometimes a median longitudinal furrow divides it into right and left halves.

Medulla oblongata: The medulla oblongata is the posterior-most part of the brain. It is almost triangular in shape with broad anterior end and narrow posterior end which passes into the spinal cord. Its antero-dorsal end, on either side, bears a hollow outgrowth called **corpora restiformia**. It is overlapped by the cerebellum. A transverse band of nervous tissue connects the two corpora restiformia. The floor and sides of medulla oblongata are thick walled. The roof bears a thin non-vascular membranous **posterior choroid plexus**. The fifth to tenth cranial nerves arise from the medulla oblongata.

Ventricles or cavities of the Brain (Fig. 11.29C)

Sections of the brain reveal a number of cavities called **ventricles**. They communicate with each other and are continuous with the central canal of the spinal cord. They are filled with **cerebrospinal fluid**. The cavity of each olfactory lobe is known as **rhinocoel**. It communicates with the **lateral ventricles** or **paracoels**, the cavities of cerebrum, also called **ventricles I** and **II**. They communicate with the large third ventricle behind, each by the **foramen of Monro**. The third ventricle is also called **diacoel**, the cavity of diencephalon. The third ventricle extends into the infundibulum of the pituitary body and also into the base of pineal stalk. The cavity of optic lobes is called **optocoel**. The fourth ventricle or **metacoel** is the cavity of the medulla oblongata into which opens the cavity of the cerebellum (**epicoel**). A common space called **iter** or **aqueduct of Sylvius**, connects the third and fourth ventricle, the optocoels also open into it. Iter is the cavity of the midbrain.

Spinal cord

The spinal cord is of the typical vertebrate type. It begins from the medulla oblongata and extends up to the end of the tail. It runs through the neural canal of the vertebrae. There is a shallow **dorsal fissure** and a well marked **ventral fissure** externally and internally there is the **central canal** surrounded by **grey matter** consisting chiefly of **neurons** and outside the grey matter lies the **white matter** comprising a non-nervous material made up of spider-like **neuroglia** cells and **medullated nerve fibres**. The grey matter is arranged into **dorsal** and **ventral horns**. The dorsal horns are united to form a single broad band giving the grey matter an **inverted 'T' Shape**. The spinal cord is surrounded by **pia matter** only.

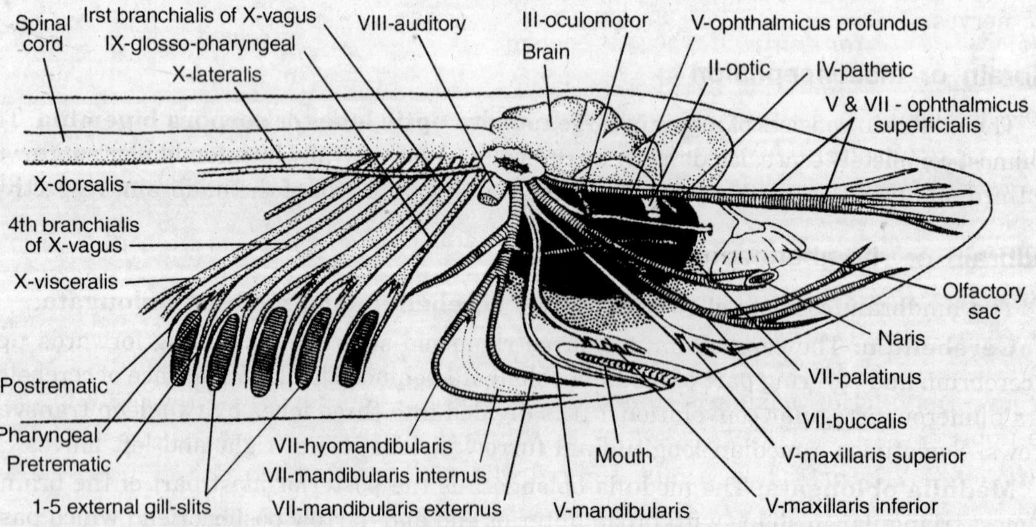

Fig. 11.30 *Scoliodon*. Cranial nerves in lateral view

PERIPHERAL NERVOUS SYSTEM

Cranial nerves

There are **ten pairs** of cranial nerves in *Scoliodon* and they have been numbered in Roman numbers **I** to **X**. Besides, an additional pair, the **anterior terminal** or **preolfactory nerve**, which was discovered much later, is also present and has been numbered zero (0).

Preolfactory nerve (0): Anterior terminal or preolfactory nerve arises from the ventral surface of the cerebrum through the neuropore. It runs along the olfactory peduncle and innervates the olfactory or nasal septum and external nostrils. It is a **sensory** nerve concerned with **sense of smell**.

Olfactory nerve (I): Olfactory nerve consists of a group of nerve fibres which arise from each olfactory lobe and innervate the olfactory sac of its side. It is also **sensory** and concerned with **sense of smell**.

Optic nerve (II): Optic is a thick short nerve which arises from the optic thalamus and innervates the retina of eye. Before entering the orbit the nerve of each side cross each other to form the **optic chiasma**. It is **sensory** in nature and is concerned with **sense of sight**.

Occulomotor nerve (III): Occulomotor is a slender **motor** nerve arising from the ventral surface of the midbrain. It divides into four branches which innervate the anterior, superior and inferior recti muscles and inferior oblique muscles of the eyeball.

Trochlear or pathetic nerve (IV): Trochlear is a slender **motor** nerve which arises dorsolaterally from the midbrain and innervates the superior oblique muscles of the eyeball.

Trigeminal nerve (V): Trigeminal is a large nerve which arises from the anetero-lateral side of the medulla oblongata along with the VII and VIII nerves. It is a **mixed** nerve having both sensory and motor functions. It divides into following branches:

***(i) Ophthalmicus superficialis*:** It is a small branch which enters the orbit along with the ophthalmicus superficialis of the seventh nerve. It is **sensory** in nature and supplies the skin of the snout.

***(ii) Ophthalmicus profundus*:** It arises from the anterior border of the roof of the main nerve, runs forward for a short distance within the cranial cavity, enters the orbit, passes forward to the olfactory capsule and emerges again through the roof of the olfactory capsule and innervates the skin of the dorsal surface of snout. It is **sensory** in function.

***(iii) Maxillaris*:** It occurs in two groups—**superior** and **inferior**. The superior maxillaris forms a flat ribbon-like band which runs along the floor of the orbit ventral to the eyeball and eye muscles, passes over the lower margin of the orbit and innervates the skin of the upper jaw. The inferior maxillaris runs directly outwards and innervates the posterior part of the upper jaw.

***(iv) Mandibularis*:** It is a mixed nerve which innervates the skin of the lower jaw, tongue and gill region.

Abducens nerve (VI): Abducens is a slender **motor** nerve which arises midventrally from the medulla oblongata and innervates the posterior rectus muscles of the eyeball.

Facial nerve (VII): Facial is a large mixed nerve which arises from the medulla oblongata immediately behind the fifth nerve and divides into four main branches.

(i) *Ophthalmicus superficialis*: It penetrates the orbit close and dorsal to the ophthalmic branch of the fifth nerve and innervates the lateral line receptors and ampulla of Lorenzini on the dorsal side of the snout.

(ii) *Buccalis*: It runs forward on the floor of the orbit along with the maxillaris superior branch of the fifth nerve and they together join to form a large **infraorbital nerve** which penetrates the infraorbital canal of the snout and associated ampullae of Lorenzini.

(iii) *Palatinus*: It is a slender nerve which runs across the floor of the orbit and innervates the roof of the buccal and pharyngeal cavity.

(iv) *Hyomandibularis*: It lies in the posterior wall of the orbit and divides into three branches :

(a) **Hyoidean** which innervates the muscles of the throat.

(b) **Mandibularis externus** which innervates the mandibular canal.

(c) **Mandibularis internus** which innervates the mucous membrane of the floor of buccal cavity.

Auditory nerve (VIII): Auditory is a small **sensory** nerves which arises from the medulla oblongata just behind the seventh nerve and passes into the auditory capsule. It is concerned with the sense of hearing. It divides into two branches: **vestibular** which innervates the membranous labyrinth and **saccular** which innervates the cochlea.

Glossopharyngeal nerve (IX): Glossopharyngeal is a mixed nerve which arises from the ventrolateral side of the medulla oblongata just behind the sixth nerve. It divides into three branches : a **pre-trematic** branch passing over the anterior border of the first gill cleft, a **post-trematic** running along the posterior border of the first gill cleft and a **pharyngeal** to the mucous membrane of the pharynx.

Vagus or Pneumogastric nerve (X): Vagus is a mixed nerve, perhaps the largest of all. It arises from the postero-lateral side of the medulla oblongata behind the ninth by a series of rootlets. It has three main branches :

(i) Lateralis: It is a long and prominent nerve which runs posteriorly along the whole length of the body and innervates the lateral line receptor organs of the trunk.

(ii) Visceralis: It enters into the body cavity and sends branches to the heart, digestive tract and other visceral organs.

(iii) Branchialis: It branches into four subsidiary nerves which innervate second to fifth gill pouches. Each nerve divides into **anterior pre-trematic** and **posterior post-trematic** around the gill pouch and a **pharyngeal** branch to the mucous membrane of the pharynx.

Spinal nerves

From the spinal cord arise paired metameric spinal nerves at regular intervals along its entire length. Each spinal nerve arises by two **roots**—a **dorsal sensory root** and a **ventral motor root**. The dorsal root bears a slight enlargement called **ganglion** and always arises in front of the ventral root. Both run backward for a short distance within the neural canal and pass out separately through a foramen in the neural arch of the vertebrae. They unite outside to form a mixed spinal nerve. The dorsal root has somatic and visceral nerve fibres while the ventral root has somatic and visceral motor fibres. There are a large number of spinal nerves, nearly equal to the number of vertebrae.

Each spinal nerve divides into three branches:

(i) Ramus dorsalis supplying the skin and muscles of dorsal body wall.

(ii) Ramus ventralis supplying the skin and muscles of ventral body wall.

(iii) Ramus communicans which joins the autonomic nervous system.

In the region of the pectoral fin, third to sixth spinal nerves run posteriorly and ventrally and join each other by cross connections forming the **cerricobranchial plexus** which sends branches to the pectoral fin and muscles of the body. The pelvic fin is supported by nine or ten spinal nerves, the posterior ones of which form the **lumbo-sacral plexus**. A branch of the occipital nerve which arises behind the vagus nerve unites with the first-two spinal nerves to form a **hypobranchial nerve** which supply muscles in the floor of the buccal cavity. It is likely that this nerve is homologous with the twelfth cranial nerve **(hypoglossus)** of higher vertebrates.

AUTONOMIC NERVOUS SYSTEM

Autonomic nervous system is poorly developed in *Scoliodon*. It consists of a series of paired and irregularly arranged ganglia situated anteriorly in the dorsal wall of the posterior cardinal sinus and posteriorly in the dorsal part of the kidney, on each side of the mid-dorsal line in close relation to the dorsal aorta. There is atleast one ganglion in each segment but sometimes there are two or three in a segment. Sometimes successive ganglion are joined by longitudinal connectives but there is no well defined continuous chain. The **gastric ganglion** is the largest as it is formed by the fusion of several ganglia. It is situated behind the post-branchial plexus. It gives off branches to the viscera and is connected to numerous fibres from the spinal nerves. The succeeding ganglia are small. The posterior ones send small branches to the kidney and genital organs.

SENSE ORGANS

The dogfish is an active free-swimming predatory organism so it possesses an elaborate system for guiding and controlling its movements. All the three primary sense organs—**eyes, ears** and **nose** are well developed. In addition to these there are the **lateral line organs** and the **ampullae of Lorenzini**, the functions of which seem to be associated with the aquatic mode of life.

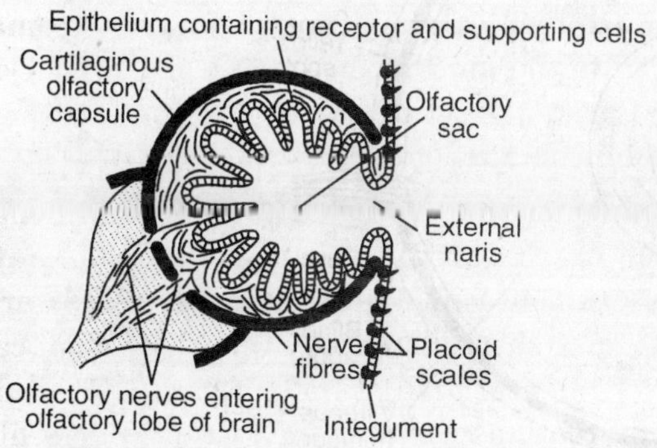

Fig. 11.31 *Scoliodon*. Olfactory sac in T.S.

Fig. 11.32 *Scoliodon*. Narsal sac dissected to show its relation to the nasal flaps

Olfactory organs

The olfactory organs are a pair of blind **olfactory** or **nasal sacs** situated ventrally in the snout one on each side. Each olfactory sac is formed as an ectodermal invagination, but does not communicate with the buccal cavity. It is a large oval sac, the mucous membrane of which is thrown into a large number of closely set **folds or lamellae** called **Schneidarian folds** which are held in place by a median band of connective tissue called **median raphe**. The folds are lined by olfactory epithelium containing **olfactory receptors** mixed with stiff supporting cells. The olfactory receptors receive fibres from the olfactory nerve. Sensory hairs project from the free ends of the receptor cells. Externally a thin membrane covers each sac. They are lodged in the cartilaginous olfactory capsule on each side of skull. The olfactory sacs open to the outside by **external nares** situated ventro-laterally on the snout. Each external naris is more or less completely divided into two, a **lateral incurrent siphon** and a **median excurrent siphon** by three **nasal flaps**. Of these flaps, two project inward fitting closely one above the other while the third hangs down from the dorsal side slightly overlapping the other two.

The olfactory sense is very highly developed in all cartilaginous fishes and is correlated with a highly developed sense of smell that enables them to detect food. If the external narial openings are plugged the dogfish will swim over food material without detecting it and with one naris open it will detect even concealed food.

Eye

The eyeball is almost spherical in shape. Only its outer surface is flat. It is held in position by six **eye muscles** and a cartilaginous stalk the **optic peduncle**. The eyeball consists of three concentric layers : **sclerotic**, **choroid** and **retina**.

The sclerotic forms the outer protective layer. It is opaque and cartilaginous in nature. It becomes transparent in front forming the **cornea** which is flat and fused with an outer **conjunctiva** continuous with the lining of the immovable eyelids.

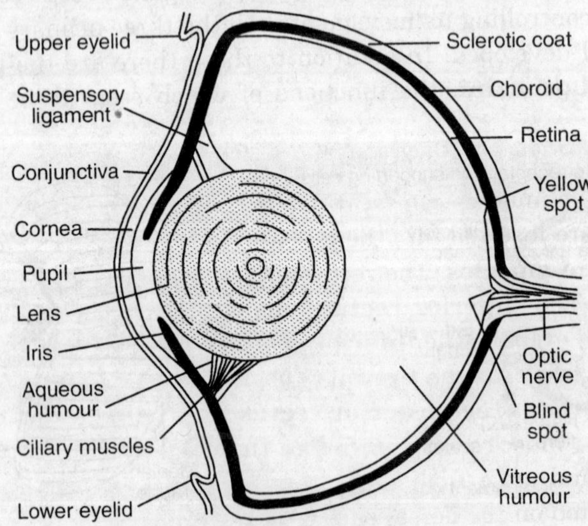

Fig. 11.33 *Scoliodon*. V.S. of the eye

The choroid lies next to the sclerotic. It is a connective tissue coat, pigmented and richly supplied with blood vessel. It is continued in front as a strongly pigmented **iris** bearing a median vertical slit, the **pupil** which is not sphinctered, *i.e.*, cannot be dilated or contracted. In cartilaginous fishes the choroid has some cells containing **guanin crystals**. It forms a light reflecting silvery surface called **tapetum lucidum**. In dim light it reflects light back to retina to increase their stimulation. In bright light it is covered by pigment so that light is not reflected back.

The retina is the innermost layer and forms the sensitive part of the eye. It is made up of special nerve or receptor cells called **rods** which are connected with the fibres of the optic nerves. The cones are absent so dogfish is colour blind.

A large spherical crystalline **lens** closely attached to the posterior surface of the iris lies immediately behind the pupil and is held by suspensory ligaments with ciliary muscles. It divides the inner space of the eye into two unequal chambers each filled with a semifluid substance. **Aqueous humour** is a transparent saline fluid which fills the small anterior chamber in front of the lens and a jelly-like substance called **vitreous humour** fills the large posterior chamber behind the lens.

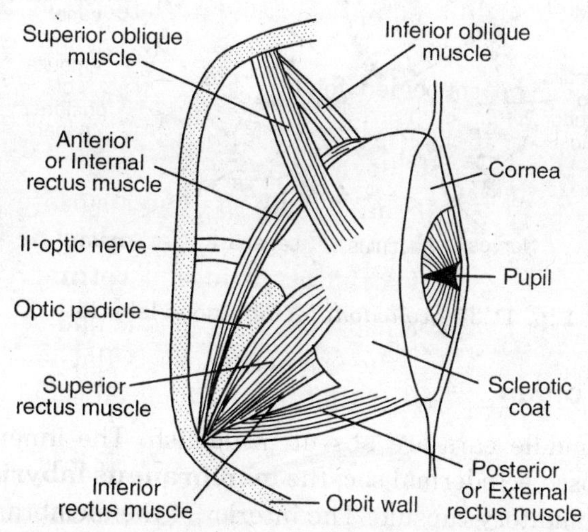

Fig. 11.34 *Scoliodon*. Diagrammatic sketch showing eye muscles

Eye muscles: The eye muscles consist of six narrow muscular bands which arise from the wall of the skull and are inserted into the eyeball. They are arranged in two groups : the **recti muscles** and **oblique muscles**. The recti muscles are four in number arising from the posterior end of the orbit. The **superior rectus** runs outwards and upwards and is inserted on the dorsal surface of the eyeball. The **inferior rectus** runs outwards and downwards to be inserted on the ventral surface of the eyeball. The **anterior rectus** runs forward and is inserted on the anterior surface. The **posterior rectus** runs backward and is inserted on the posterior surface. The recti muscles serve to rotate the eyeball in the horizontal plane. The oblique muscles are two and arise from the anterior angle of the orbit close together. The **superior oblique** is inserted on the dorsal surface of the orbit in front of the superior rectus. The **inferior oblique** is inserted on the ventral surface of the eyeball.

Working of eye: The eye of dogfish, like the human eye, is normally focussed for distant objects and it has to be accommodated for near vision. The power of accommodation is poor because the pupil is non-contractile. The accommodation, however, is brought about by moving the lens forwards, *i.e.*, away from the retina, to catch more light. The retina is largely composed of rods which means that the fish is adapted for life in relatively low light intensities. Another adaptation to vision in dim light is provided by the presence of tapetum lucidum which reflects extra light on the retina. In strong light the tapetum is covered by a black pigment layer of contractile cells so that light is not reflected. In strong light the iris also closes to an oblique slit.

The two eyes are set wide apart as such they cannot be focussed together on any one point. Their field of vision are separate. This arrangement is called **monocular vision.**

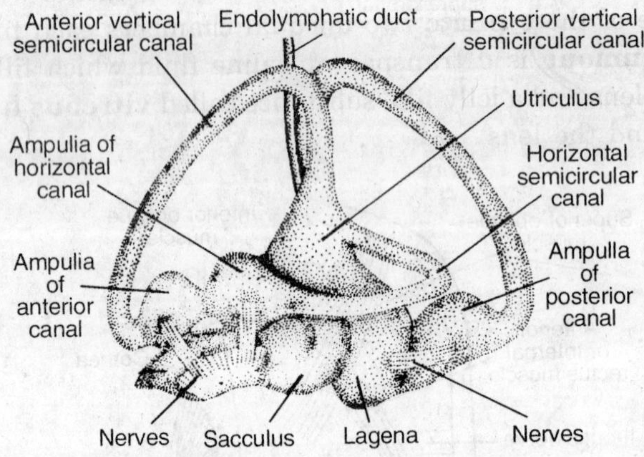

Fig. 11.35 *Scoliodon*. Membranous labyrinth

Ear or stato-acoustic organ

The external and middle ears are absent in dogfish. The inner ear or stato-acoustic organ of each side is a closed ectodermal sac, the **membranous labyrinth**, which is enclosed within the cartilaginous auditory capsule. The interior of the membranous labyrinth is filled with a fluid known as **endolymph** containing numerous calcareous bodies called otoliths. The space between the inner ear and the auditory capsule is filled with a fluid called **perilymph**. The membranous labyrinth consists of a central portion, the **vestibule**, and three **semicircular canals**. The vestibule is a laterally compressed sac differentiated by a constriction into a dorsal chamber, the **utriculus** and a lower chamber, the **sacculus**, the posterior end of which forms an outgrowth, the **lagena**. From the sacculus arises a narrow tube the **ductus endolymphaticus** which runs upwards and pierces the cranium. The semicircular canals are three tubular loops opening into the utriculus. These canals are mutually at right angles, one being horizontal, the **horizontal canal**, and the other two vertical, the **anterior vertical canal** and the **posterior vertical canal**. The three canals are arranged in the three planes of space. At one end where it joins the utriculus each semicircular canal is swollen out to form an ampulla. The inner ear is innervated by the auditory nerves.

Functions: The ear of dogfish is not only concerned with hearing but also performs other functions. It is primarily a **static organ**, *i.e.*, helps in maintaing balance or equilibrium in water. This is achieved by the movements of otoliths and endolymph which stimulate the sensory nerve endings of ampulla and vestibule. The sacculus and lagena serve acoustic function, *i.e.*, they are concerned with hearing. The animal can also adjust according to changes in speed, direction and orientation.

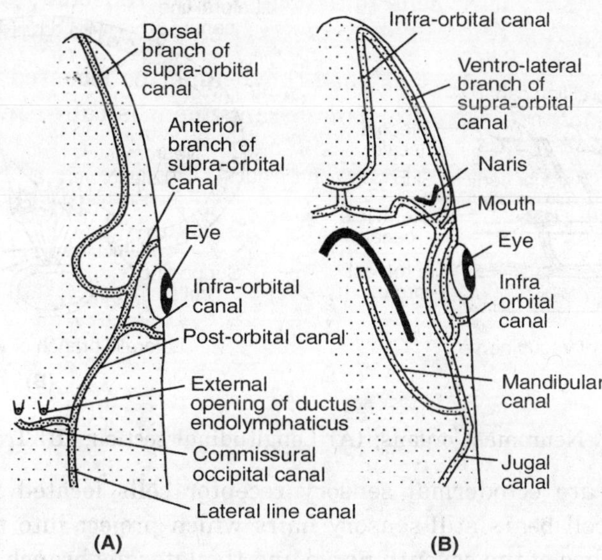

Fig. 11.36 *Scoliodon*. Lateral view of head showing canals of the lateral line system; **(A)** Dorsal view; **(B)** Ventral view

Lateral line system

The lateral line system is an extremely sensitive and accurate method of locating objects in the neighbourhood. It includes **lateral lines, sensory receptors** called **neuromasts** and **pit organs**.

The **lateral line proper** extends from head to tail and is visible as a faint line on either lateral side of trunk and tail. It lies over the lateral septum which separates the epaxial and hypaxial muscles of the body wall. It consists of number of slender mucus-filled **ectodermal canals** (Fig. 11.36) lying embedded in the dermis. The canals open on the surface by numerous **pores**. On reaching the head each canal bends dorsally and is connected with its fellow of the opposite side by a transverse commissure the **occipital** or **supratemporal canal** lying immediately behind the openings of the endolymphatic ducts. In front of this commissure each lateral line canal (**cephalic canal**) bends a little outward and forms the **post-orbital canal**. On reaching the posterior margin of the orbit it divides into a **supra-orbital** and an **infra-orbital canal**. The supra-orbital canal runs forward dorsal to the eye and the infra-orbital lies between the eye and the spiracle and runs forward below the eye. The supra-orbital canal divides into two branches : **(i)** the **dorsal branch** running up to the tip of the snout and bending sharply backwards as the **ventro-lateral** branch running along the ventro-lateral edge of the snout and **(ii)** the **anterior branch** that bends downwards just in front of the eye and joins the infra orbital canal. At the junction of the infra-orbital canal and the ventro-lateral branch of the supra-orbital canal arises the **jugal canal** which runs parallel to and beneath the infra-orbital canal. It gives rise to a **mandibular canal** which runs along the lower jaw right up to the lip.

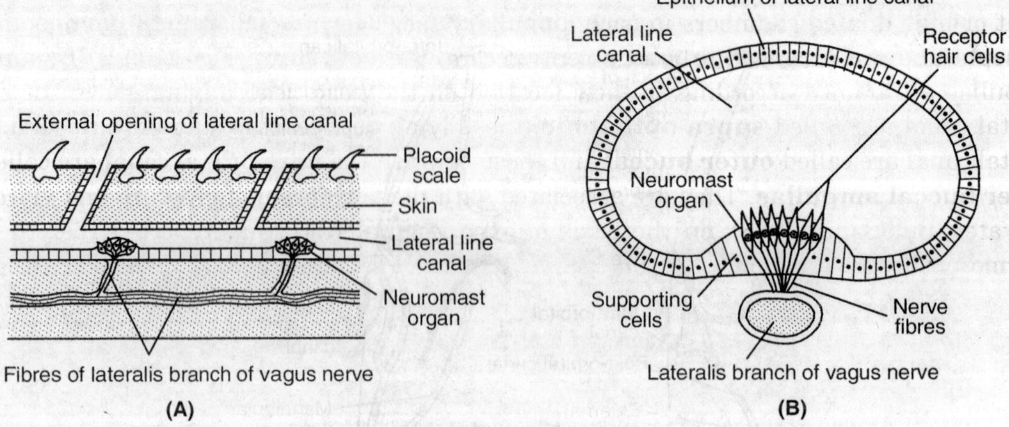

Fig. 11.37 *Scoliodon*. Neuromast organs; **(A)** Longitudinal section; **(B)** Transverse section

The **neuromasts** are ectodermal sensory receptor cells located in the canals (Fig. 11.37B). Each sensory cell bears stiff-sensory hairs which project into the canal. They are innervated by the branches of the seventh nerve and the lateralis branch of the vagus. These receptor cells (Fig. 11.37A) are stimulated by slow vibrations of the watery environments and control the orientation of the body in relation to waves and currents (**rheoreceptors**). The lateral line system thus enables the animal to detect even the slightest disturbance in the surrounding medium even when the fish is in complete darkness of the ocean.

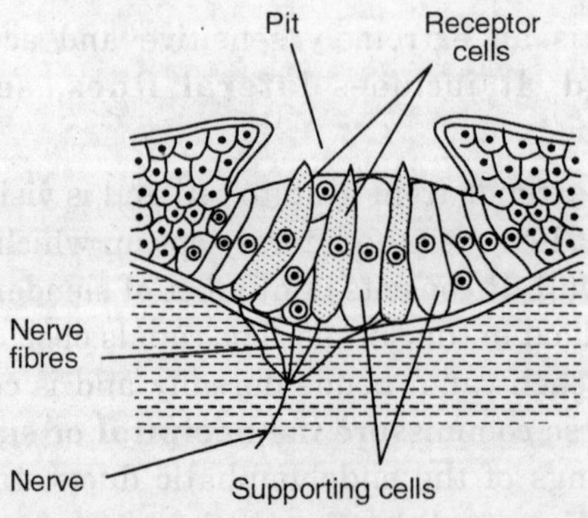

Fig. 11.38 *Scoliodon*. Vertical section of pit organ

The **pit organs** (Fig. 11.38) consist of small ectodermal receptors located in pits on the dorsal and lateral surfaces of the head. They are innervated by branches from seventh cranial nerve. They are found in large numbers in rays among elasmobranchs.

Ampullae of lorenzini

Over the surface of the head numerous small **pores** are scattered in definite groups. These represent the openings of the **ampullae of Lorenzini**. Each pore is connected with a mucous filled canal called the **ampullary tubule**. The terminal portion of the tubule is enlarged to form the **ampulla** or **ampullary sac** composed of **dilated chambers** supplied by delicate nerve fibres coming from branches of facial nerve (Fig. 11.39A). There are about eight or nine dilated chambers in each ampullary sac. They are separated from one another by septa and are arranged around a central core, the **centrum**. (Fig. 11.39B) Usually the ampullae are named according to their location on the head. Those lying above the supra-orbital canal are called **supra-opthalmic**, those lying between the supra-orbital and infra-orbital canal are called **outer buccal** and those beneath the infra-orbital canal are called the **inner buccal ampullae**. They are associated with the perception of pressure and vibrations in water. But Sand (1938), on the basis of experimental work, has reported that they are thermosensory in function.

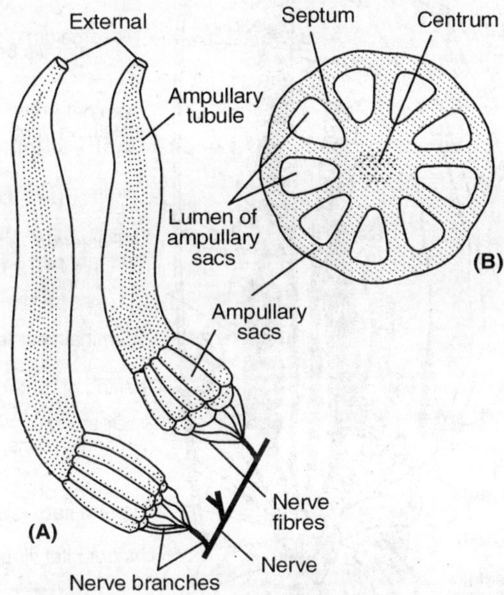

Fig. 11.39 *Scoliodon*. Ampulla of Lorenzini; **(A)** Entire; **(B)** Cross section

URINOGENITAL SYSTEM

Male

Excretory organs: A pair of **kidneys** are the excretory organs of dogfish. They are **mesonephric**. Each kidney is a long, strap-shaped glandular structure lying against the dorsal body wall, dorsal to the peritoneum, on either side of the median line. The kidneys extend through the entire length of the body cavity—from the root of the liver in front to the side of the cloaca behind. Each kidney is made up of an anterior slender portion, the

genital kidney or **cranial mesonephros** and a posterior thickened portion, the **renal kidney** or **caudal mesonephros**. The anterior posterior is reduced, without excretory tubules and genital in function (conveys genital products) and therefore is called **epididymis.** The posterior portion is greatly developed and is the functional excretory organ of adult dogfish. Its substance is made up of coiled glandular **uriniferous tubules** comprising a **Bowman's capsule** enclosing the **glomerulus** and a coiled **renal tubule**, many of which open into a common **collecting tubule**. The collecting tubules of each kidney open independently into a thin walled common duct the **ureter**. Both the ureters open posteriorly into a median wide chamber the **urinogenital sinus** which finally opens into the cloaca by a small aperture lying on a small **urinogenital papilla**.

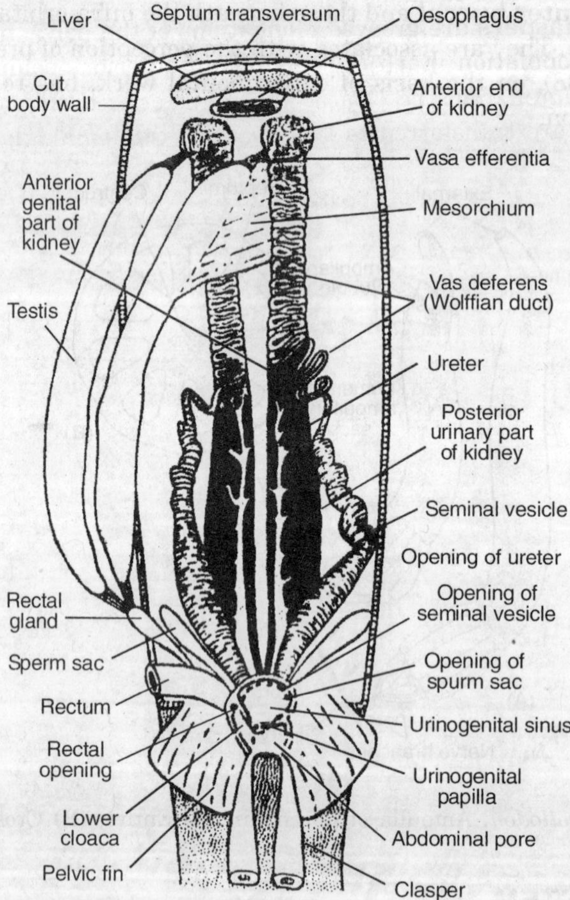

Fig. 11.40 *Scoliodon*. Male urinogenital system

Reproductive organs: The male reproductive organs include a pair of **testes**, a pair of **vas deferens** and a pair of **seminal vesicles**.

The testes are large elongated structures extending from the base of the liver in front to the caecal or rectal gland behind. They are attached to the mid-dorsal body wall by a fold of peritoneum called **mesorchium** and to the caecal gland by non-glandular tissue. The

Wolffian or **mesonephric duct**, now known as **vas deferens** is well developed and is thrown into a series of fine coils overlying the ventral surface of genital portion of the kidney. The **vasa efferentia** are minute tubules which open into the anterior end of the vas deferens. The vasa deferentia become enlarged posteriorly forming the **seminal vesicles** on each side. The seminal vesicles open separately into a large triangular chamber, the **urinogenital sinus** which finally open into the cloaca on an elevated **urinogenital papilla**. The **sperm sacs** are two blind elongated outgrowths of the urinogenital sinus, whose function seems doubtful. The **spermatozoa** are produced in the seminiferous tubules of the testis, carried to the vas deferens, nourished by a fluid in the genital part of the kidney and stored in the seminal vesicles. During copulation the sperms from the cloaca pass into the claspers which are inserted into the cloaca of the female. The **claspers** and **siphons** are accessory male reproductive organs. The **claspers** are grooved elongations of the pelvic fins which serve as intromitent organ during copulation. Each clasper has an anterior opening called **apopyle** though which the sperms enter from the cloaca, and a posterior opening called **hypopyle**, through which the sperms are transferred to the cloaca of the female during capulation.

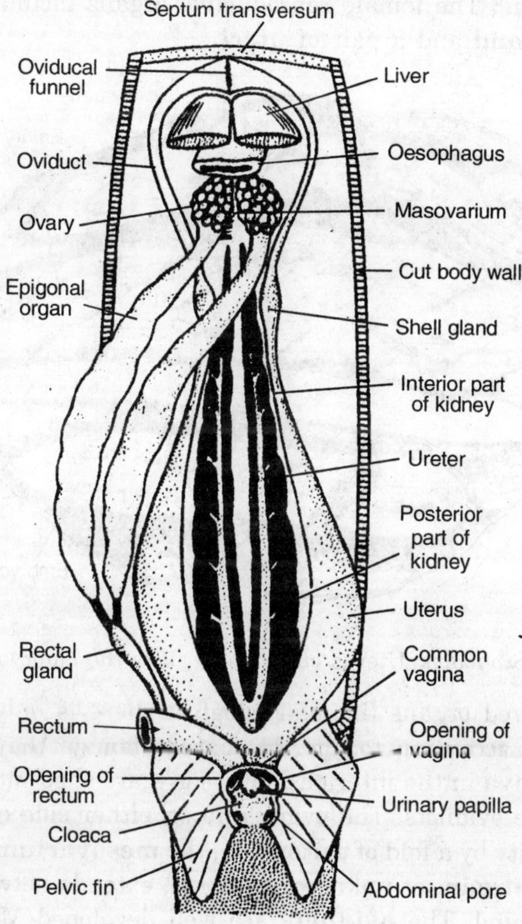

Fig. 11.41 *Scoliodon.* Female urinogenital system

There are certain muscular sacs called **siphons** lying beneath the skin of the ventral surface in the pelvic region. Posteriorly these sacs extend as **siphon tubes**, each of which open into the apopyle of the clasper of its side. The use of siphon seems to be obscure, though it is believed to flush spermatozoa accumulated in the claspers into the female cloaca by squirting out the contained sea water.

Female

Excretory organs: The **kidneys** are paired, each showing differentiation into anterior and posterior portions. The **anterior** portion is reduced to a narrow strand and has no connection with the kidney or genital organ. The **posterior** functional portion of the kidney contains numerous coiled **uriniferous tubules**. From the inner margin of each kidney arises a **mesonephric** or **Wolffian duct** or the **ureter**. Each ureter commences in the anterior part, continues backwards and in the posterior part unite to form a common duct opening into a large median triangular chamber, the **urinary sinus**. The sinus opens into the cloaca at the tip of a small **urinary papilla**.

Reproductive organs: The female reproductive organs include a pair of **ovaries**, a pair of **oviducts**, **shell gland** and a pair of **uteri**.

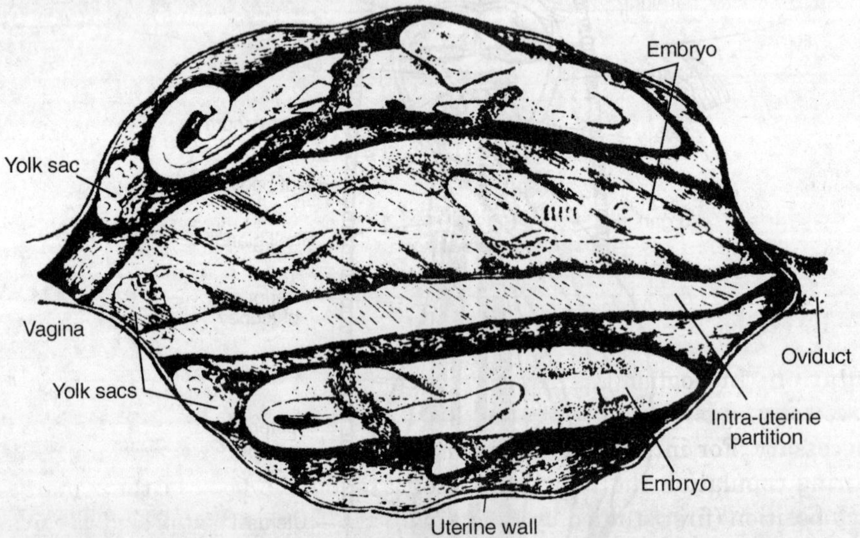

Fig. 11.42 *Scoliodon*. Uterus cut open to show the embryos inside

The **ovaries** are paired organs (in many dogfish there is only a single large ovary) varying in size and form according to age. In mature female they are lobulated due to pressure of large **eggs** or **ova** in the interior. Mature ova are shed into the abdominal cavity from where they enter the oviducts. The ovaries lie on either side of the vertebral column suspended in the body cavity by a fold of perioneum, the **mesovarium**. A long tubular strand of tissue called **epigonal organ**, of unknown function, extends between the ovary in front and the caecal glands behind. The **oviducts** are well developed Müllerian ducts, each of which is a stout, thick-walled, muscular tube. The anterior coelomic opening of oviducts fuse to form a single wide longitudinal aperture, the **oviducal funnel** or **ostium**, just below the oesophagus near septum transversum. From the common aperture each oviduct narrows slightly and passes backwards at the sides of the coelom to the cloaca. At about one-third of the way down its anterior end each oviduct swells out to form the **oviducal** or **shell gland** which is responsible for the secretion of the egg case over the descending eggs which are fertilized between the oviducal funnel and shell gland. Behind the gland the oviduct again becomes narrow but towards the posterior end each dilates to form a wide chamber called **uterus** in which the development of embryo takes place (Fig. 11.42). It contains a variable number of embryos in different species (7 in *Scoliodon sorrakowah*) and fills the greater part of the abdominal cavity. The mucous lining of the uterus becomes divided into as many compartments as there are embryos. The two uteri unite to form a median chamber, the **vagina**, which opens by a large aperture into the cloaca. A fold of mucous membrane between the vagina and cloaca acts as a **valve** and closes the **vagino-cloacal aperture**. The cloaca is a shallow depression between the pelvic fins. In female there are three openings into it, the rectum in front, the genital aperture in the middle and the urinary sinus behind.

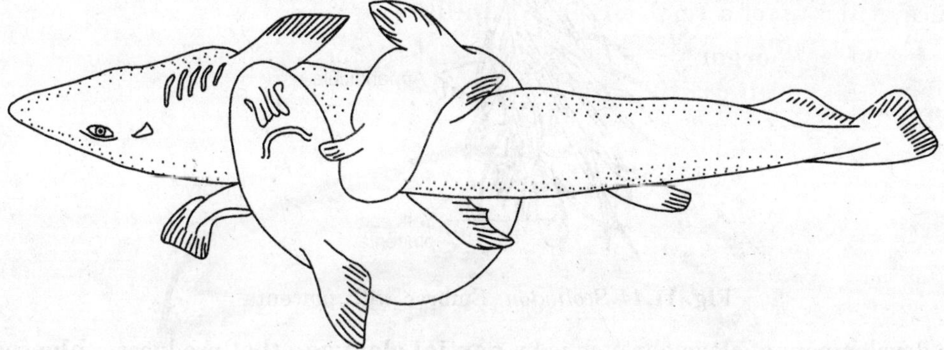

Fig. 11.43 *Scoliodon*. Copulation

REPRODUCTION

Copulation: In dogfish reproduction occurs throughout the year. All elasmobranchs produce large yolky eggs, generally protected by a horny egg case, therefore internal fertilization becomes necessary. For internal fertilization the claspers are inserted into the cloaca of the female. During copulation the male coils itself around the body of the female in a curious back to back position (first studied by Bolan in 1881). This method provides a firm grip in the rough-skinned forms. The sperms are transferred through the grooved claspers.

Fertilization: Fertilization in **internal**. It takes place in the anterior part of the oviduct just in front of the shell glands.

Development: *Scoliodon* is **ovoviviparous** because the eggs develop inside the uteri and they give birth to young ones. In *S. sorrakowah* one uterus may contain as many as seven embryos but in *S. palasorrah* three embryos have been reported from each uterus. In *Squalus acanthias*, which is also ovoviviparous, a beautiful clear amber-coloured egg-case is formed around several eggs and the early development takes place within this egg case.

The ripe eggs are released into the body cavity by the rupture of the mature follicles. They are carried by the coelomic fluid to the anterior part of the coelom where they are picked up by the ostium. Passage down the right or left oviduct seems to be a matter of chance. Spermatozoa in the upper part of the oviduct fertilize the eggs as they pass down. Development of the embryo takes place in the uterus.

The fertilized eggs develop very slowly. A 48-day old dogfish embryo is at almost the same stage as a 48-hour chick embryo. This difference is because of the different temperature of development. If the temperature of development of dogfish is raised, the rate of development is accelerated. Normally, the time taken for development in dogfish is about 16 to 25 months.

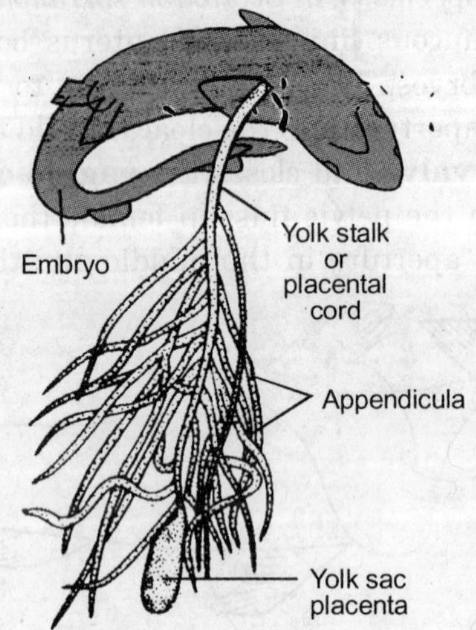

Fig. 11.44 *Scoliodon*. Embryo with placenta

The development of elasmobranch is by **partial cleavage** that produces a **blastoderm**, perched on the top of a large mass of **yolk**. As soon as this grows further it develops a tubular **yolk stalk** connected at one end with the intestine of the embryo and at the other with a **yolk sac**, filled with yolk for the nourishment of the developing embryo (Fig. 11.44). In the later stages when the yolk is more or less used up the yolk sac becomes folded and get embedded in the wall of the uterus forming a **yolk sac placenta**. Thus the embryo starts drawing its nourishment from maternal tissue. As the placenta is formed blood vessels develop in the yolk stalk which now becomes **placental cord** attached to the embryo in the mid-ventral line at a point in level with the anterior edge of the pectoral fins. The embryo is attached by the placental cord to the uterine wall. Each placental cord is provided with numerous delicate tubular processes, the **appendicula**, each of which is made up of several layers of epithelial cells surrounding a central core of loose connective tissue. The function of appendicula is probably absorption of nutritive material secreted by the uterine wall of the mother. This conclusion is based on the fact that appendicula are better developed in species that have vestigeal placenta.

The young dogfish which is produced when the egg hatches is called a **fingerling**. It is just a miniature of the adult. Later life history of dogfish is not known. Attempts to catch recently hatched fingerlings have not succeeded. It is not known where they spend the first year or so of their life. Since all dogfish caught are of surprisingly uniform size and sexually mature, it seems as though the early life may be spent off-shore, the mature fish returning to shallow water.

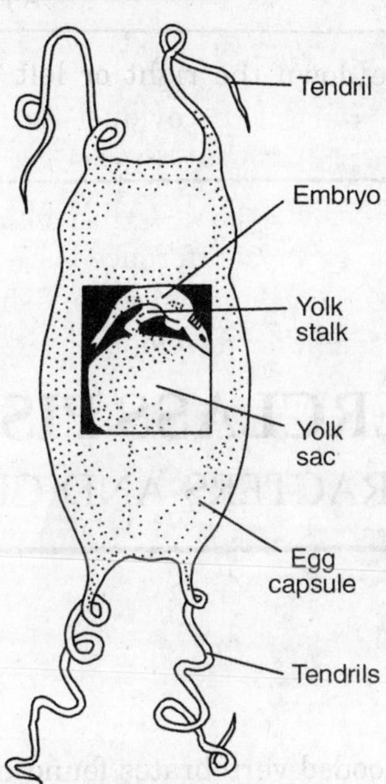

Fig. 11.45 Egg-case or mermaids purse of the dogfish *Scyllium*

Some sharks and rays lay fertilized eggs, i.e., they are **oviparous**. The eggs are encased in protective, horny egg cases or capsules secreted by the shell gland. The egg case of *Scyllium* is a rectangular box-like structure called **mermaids purse** (Fig. 11.45). The four corners of the egg case are produced into long, coiled thread-like structures called **tendrils** with the help of which they attach to the sea weeds. Usually one embryo develops in one egg case. The fully formed young one hatches out by rupturing the egg case in a period of 6 to 9 months.

12 CHAPTER

SUPERCLASS PISCES:
GENERAL CHARACTERS AND CLASSIFICATION

GENERAL CHARACTERS

1. Pisces (fishes) are cold blooded vertebrates found in almost every place where there is water.
2. Body usually stream-lined, laterally compressed, spindle-shaped, but some are elongated, snake-like and a few are dorso-ventrally compressed.
3. They differ widely in shape, size, habits and habitats.
4. Some are a few inches in length while others may attain a length of about 18 metres.
5. The skin is covered with scales, dermal denticles or bony plates. Some species are naked secondarily.
6. They have paired and unpaired fins supported by soft or spiny fin rays. Fins help in locomotion.
7. Dorsal, caudal and anal fins are unpaired, the pectoral and pelvic fins are paired.
8. Gills are the organs of respiration borne by gill arches and open to the outside by gill slits never more than seven pairs, usually five pairs present. Gill slits naked or covered by an operculum.
9. Some fishes possess accessory respiratory organ called air bladder or swim bladder.
10. Lateral line system well developed.
11. Some species possess barbels which are excellent organs of touch.
12. Heart two-chambered (one auricle and one ventricle). Renal and hepatic portal systems present. Red blood cells with a nucleus.
13. Endoskeleton cartilaginous or bony. Notochord more or less replaced by cartilaginous or bony vertebrae.
14. Cranial nerves ten pairs.

15. There is no middle ear, only inner ear present.
16. Kidneys mesonephric. Excretion ureotelic.
17. Sexes separate. Gonads paired with gonoducts opening into a cloaca or independently.
18. Fertilization internal or external. Females oviparous or ovoviviparous. Eggs small or large and yolky. Development usually direct and without metamorphosis.

CLASS: CHONDRICHTHYES (Gr. *chondros* = cartilage; *ichthyes* = fish)

General characters

1. Exclusively marine, with a very few exceptions.
2. Body streamlined and spindle-shaped.
3. Skin tough, covered with dermal placoid scales and contains mucous glands.
4. Both paired and unpaired fins supported by fin rays present. Caudal fin heterocercal. Pelvic fin provided with claspers in male.
5. Mouth large and crescentic, situated on the ventral surface. Jaws well developed, with teeth which are modified placoid scales.
6. Endoskeleton cartilaginous, notochord persistent, vertebrae separate and girdles present.
7. Stomach J-shaped. Intestine short, with a scroll valve in the lumen.
8. Respiration by lamelliform gills which open to the outside by 5–7 pairs of lateral gill slits, not covered by operculum. Air bladder absent. Spiracles. (small openings just behind the eyes) present.
9. Heart two-chambered, with sinus venosus and conus arteriosus. Venous system with large sinuses. Hepatic and renal portal systems present.
10. Kidneys mesonephric and excretion ureotelic.
11. Cloaca present. Cloacal aperture serves as a common outlet for rectum, renal and reproductive ducts, all open separately into the cloaca.
12. Brain primitive with large olfactory lobes and cerebrum. Inner ear with three semicircular canals. Lateral line system present. Cranial nerves ten pairs. Ampullae of Lorenzini present.
13. Sexes separate, gonads paired. Oviparous or ovoviviparous. Eggs large, yolky, encased in a horny case. Fertilization internal.
14. Development direct. Young one born is just a miniature of the adult.

CLASS: OSTEICHTHYES (Gr. *osteon* = bone; *ichthyes* = fish)

General characters

1. Found in all types of waters—fresh water, marine, brackish water, etc.

2. Body streamlined and spindle-shaped.
3. Exoskeleton of dermal scales which may be gonoid, cycloid or ctenoid. Some species are without scales.
4. Skin contains numerous mucous glands.
5. Both paired and unpaired fins supported by cartilaginous or bony fin rays present. Caudal fin homocercal.
6. Mouth terminal or subterminal, variable in shape and size. Jaws usually with teeth.
7. Endoskeleton bony. Notochord replaced by vertebrae. Pectoral girdle well developed but pelvic girdle usually small or absent.
8. Stomach variable in shape, absent in some forms. Intestine long, without scroll valve.
9. Respiration by 4 pairs of filiform gills. Gill slits 5 pairs, covered by an operculum.
10. Accessory respiratory organ, air or swim bladder present. In Dipnoi air bladder functions as lungs. Spiracles absent.
11. Heart two-chambered with sinus venosus and bulbus arteriosus, conus arteriosus is absent. Red blood cells with a nucleus.
12. Kidneys mesonephric and excretion ureotelic.
13. Cloaca absent. Anus present. Urinary and genital ducts open independently to the exterior.
14. Brain advanced with small olfactory lobe and cerebrum but large optic lobes and cerebellum. Cranial nerves ten pairs. Lateral line system well developed. Ampullae of Lorenzini absent. Inner ear with three semicircular canals.
15. Sexes separate, with paired gonads Fertilization external. Male without claspers Eggs minute to 1.2 cm in size. Development direct, no metamorphosis.

CLASSIFICATION

There is no agreement among various authors regarding classification of fishes. Different authors have given different classification of fishes. Among the recent authors Goodrich (1930), Berg (1940), Grasse (1958) and Romer (1959) have given a detailed classification of fishes. They have tried to include the recent and fossil fishes. Modern classification schemes are generally based on evolutionary relationship among fishes. In this book classification of fishes is being presented in a simple manner. All the important living classes and orders are being included.

CLASS: CHONDRICHTHYES

It includes cartilaginous fishes like sharks, skates, rays and chimaeras. About 600 living species are known.

Subclass 1. SELACHII (**Gr.** *selachos* = a shark)

1. Paired fins well developed, without a median axis.
2. Base of pectoral fin constricted. Pectoral fin has cartilaginous rods (propterygium, mesopterygium and metapterygium) supporting it.

3. Multiple gill slits present on lateral sides, protected by individual gill flaps.
4. A spiracle is present behind each eye.
5. Caudal fin heterocercal.
6. Cloaca present.

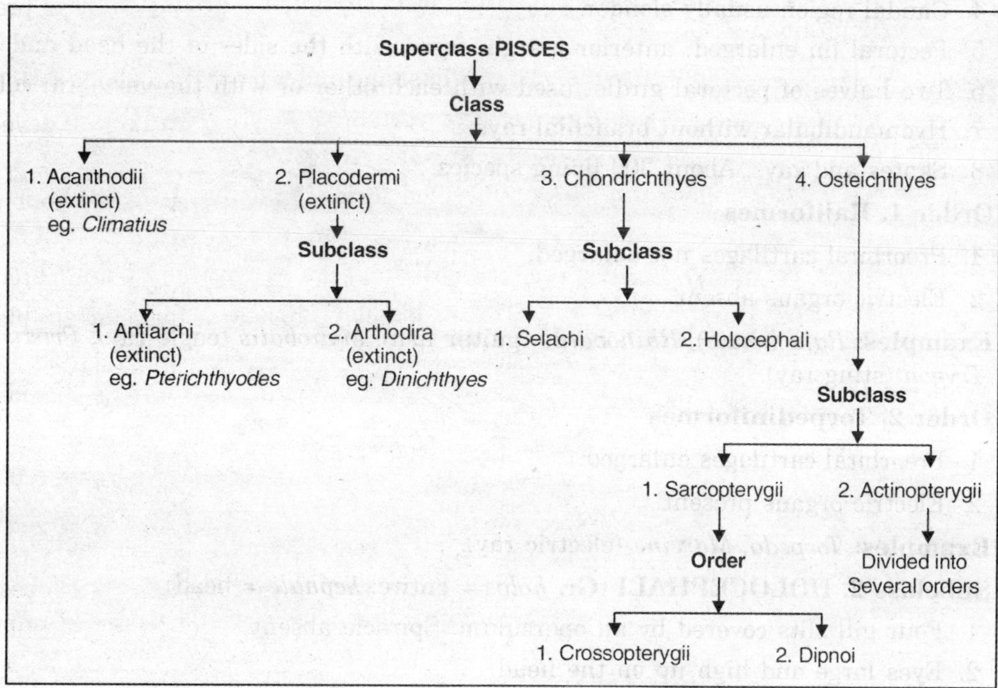

Superorder 1. PLEUROTREMATA (Gr. *pleuro* = side; *trema* = opening)

1. Body spindle-shaped.
2. Gill slits on lateral sides, 5–7 pairs. Spiracles small, non-functional.
3. Pectoral fin constricted at the base, anterior margin free.
4. Caudal fin heterocercal.
5. Hyomandibular bearing branchial rays.
6. Two halves of pectoral girdle separate.

Order 1. Lamniformes

1. Two dorsal fins without spines.
2. Gill arches 5 in number.
3. Anal fin present.
4. Palato-quadrate loosely articulated or not articulated to the skull.
5. True sharks. About 250 living species.

Examples: *Scoliodon*, *Chiloscyllium* (dogfish), *Sphyrna* or *Zygaena* (hammer-headed shark), *Stegostoma* (tiger or zebra shark), *Heptanchus* (seven-gilled shark), *Rhineodon* (whale shark).

Superorder 2. HYPOTREMATA (Gr. *hypo* = below; *trema* = opening)

1. Body flattened dorsoventrally.
2. Gill slits on ventral side, 5 pairs.
3. Spiracles large and functional.
4. Caudal region usually slender.
5. Pectoral fin enlarged, anterior margin fused with the sides of the head and body.
6. Two halves of pectoral girdle fused with each other or with the vertebral column.
7. Hyomandibular without branchial rays.
8. Skates and rays. About 300 living species.

Order 1. Rajiformes

1. Preorbital cartilages not enlarged.
2. Electric organs absent.

Examples: *Raja* (skate), *Rhinobatus* (guitar fish), *Myliobatis* (eagle ray), *Pristis* (saw fish), *Trygon* (sting ray).

Order 2. Torpediniformes

1. Preorbital cartilages enlarged.
2. Electric organs present.

Examples: *Torpedo*, *Marcine* (electric ray).

Subclass 2. HOLOCEPHALI (Gr. *holos* = entire; *kephale* = head)

1. Four gill slits covered by an operculum. Spiracle absent.
2. Eyes large and high up on the head.
3. Mouth small and ventral and bounded by lip-like folds.
4. Teeth in the form of plates.
5. Skin without scales.
6. Cloaca absent.
7. Male with claspers.
8. About 25 species known

Example: *Hydrolagus* (= *Chimaera*).

CLASS: OSTEICHTHYES

It is a large class including true bony fishes. About 40,000 living species are known.

Subclass 1. SARCOPTERYGII (Gr. *sarcos* = fleshy; *pterygium* = fin)

1. Paired fins lobate, fleshy and covered with scales, and have a bony central axis bearing radials.
2. Caudal fin diphycercal, some have a median lobe.
3. Internal nares called **choanae** connect the olfactory sac with the mouth (hence formerly called subclass **Choanichthyes**)
4. Commonly known as air breathing fishes.

Order 1. Crossopterygii

1. Paired fins with scale covered lobe. Skeleton of pectoral fin with a jointed central axis bearing radials on each side.
2. Air bladder vestigeal.
3. Internal nares present or absent.
4. Spiracles present.
5. Caudal fin with 3 lobes–epichordal, hypochordal and median.
6. Many species extinct.

Example: *Latimaria* (only living species).

Order 2. Dipnoi

1. Commonly known as lung fishes.
2. Paired fins acutely lobate.
3. Skeleton largely cartilaginous.
4. No vertebrae, notochord persistent as an unsegmented rod.
5. Air bladder lung-like, single in *Neoceratodus* and paired in others, function as efficient respiratory organ.
6. Internal nares present, spiracles absent.

Examples: Only three living genera known, *Neoceratodus* (Australia), *Protopterus* (Africa) and *Lepidosiren* (South America).

Subclass 2. ACTINOPTERYGII (Gr. *actis* = ray; *pterygium* = fin)

1. Paired fins thin, ray-like, without fleshy lobe and central axis.
2. Internal nostrils absent.
3. Scales ganoid, in some species scales completely absent.
4. Tail fin homocercal, in some primitive forms heterocercal or semi-heterocercal.
5. Spiracles absent, gill slits covered by operculum.
6. Air bladder present or secondarily lost.
7. Commonly known as ray-finned fishes.

Superorder 1. Chondrostei

1. Includes primitive ray-finned fishes with large mouth.
2. Skeleton largely cartilaginous or bony.
3. Scales ganoid.
4. Caudal fin heterocercal.

Order 1. Polypteriformes

1. Body covered with rhombic ganoid scales.
2. Pectoral fin with a small scale covered fleshy lobe at the base.
3. Caudal fin symmetrical externally.

4. Dorsal fin with many peculiar finlets.
5. Spiracles present. Air bladder bilobed, help in respiration.
6. Internal nostrils absent.
7. Skeleton bony.

Example: *Polypterus* (bichir of African rivers).

Order 2. Acipenseriformes

1. Body elongated and covered with five rows of bony (ganoid) scutes or naked.
2. Snout elongated, with simple or branched preoral barbels ventrally.
3. Endocranium cartilaginous.
4. Jaws without teeth.
5. Caudal fin heterocercal.

Examples: *Acipenser* (sturgeon), *Polyodon* (paddle fish).

Superorder 2. Holostei

1. Endoskeleton bony.
2. Body covered with ganoid or cycloid scales.
3. Mouth opening small.
4. Caudal fin heterocercal.
5. Occupy intermediate position between chondrostei and teleostei.

Order 1. Amiiformes

1. Body with cycloid scales on trunk and ganoid on head.
2. Caudal fin abbreviated heterocercal.
3. Dorsal fin long.
4. Snout rounded.

Example: *Amia* (bowfin).

Order 2. Lepidosteiformes

1. Body elongated, covered with rhmboid ganoid scales.
2. Snout elongated.

Example: *Lepidosteus* (garpike).

Superorder 3. Teleostei

1. Mouth small and terminal.
2. Endoskeleton exclusively bony.
3. Scales cycloid or ctenoid or may be absent.
4. Caudal fin perfectly homocercal.
5. Air bladder usually present. Spiracles absent.
6. Conus arteriosus replaced by bulbus arteriosus.
7. Includes modern ray-finned fishes.

Order 1. Clupeiformes

1. Scales usually cycloid.
2. Caudal fin homocercal.
3. Dorsal fin single, ventral fin present or absent.
4. Dorsal and anal fins without spines.
5. Weberian apparatus absent.

Examples: *Salmo* (american trout), *Clupea* (herring), *Esox* (pike), *Notopterus* (chital fish), *Hilsa sardinops* (pacific sardine).

Order 2. Scopeliformes

1. Deep sea fishes with photophores.
2. Body nearly cylindrical and silvery white.
3. Mouth large, with numerous small teeth.
4. Dorsal and anal fins with no spines.
5. Important food fish.

Example: *Harpodon* (bombay duck), only species found in Indian waters.

Order 3. Cypriniformes

1. A peculiar Webarian apparatus connects inner ear with the air bladder.
2. Air bladder connected to the pharynx by a duct.
3. Dorsal, anal and pectoral fins with a spine each or spines may be absent.

Division 1. Cyprini

1. Body covered with scales or naked, bony plates never present.
2. Third and fourth vertebrae not fused with each other.

Examples: *Cyprinus* (carp), *Labeo* (rohu), *Catla*, *Carassius* (goldfish).

Division 2. Siluri

1. Body not covered by scales.
2. Maxillary bone reduced to support the barbels.
3. Second, third, fourth and fifth vertebrae fused.

Examples: *Clarias* (magur), *Heteropneustes* or *Saccobranchus* (singhi), *Mystus* (tengra), *Wallago* (lachi), *Electrophorus* (electric eel).

Order 4. Anguilliformes

1. Body elongated, slender and eel-like.
2. Scales absent, if present minute and embedded in the skin.
3. Dorsal and anal fins long narrow and confluent. Pelvics, if present, abdominal.
4. Fins without spines.
5. Air bladder connected with the intestine by a duct.

Examples: *Anguilla* (fresh water eel), *Muraena* (moray), *Ophichthyes.*

Order 5. Beloniformes

1. Physoclistous fishes with elongated body covered with cycloid scales.
2. Fins without spines.
3. Ventral fins abdominal.
4. Pectoral fin inserted high up in the body.

Examples: *Belone* (garfish), *Hemirhamphus* (half beak), *Exocoetus* and *Cypselurus* (flying fishes).

Order 6. Syngnathiformes

1. Physoclistous fishes with a small mouth situated at the tip of an elongated tubiform snout.
2. Body covered with protective scales or bony rings.
3. Dorsal fin, if present, is spinous.
4. Male with a brood pouch for the development of the young.

Examples: *Hippocampus* (sea horse), *Syngnathus* (pipe fish), *Fistularia* (flute fish).

Order 7. Symbranchiformes

1. Eel-like body.
2. Gill slits join to form a single transverse ventral slit. Gills usually reduced, respiration chiefly buccopharyngeal and intestinal.
3. Air bladder, paired fins and fin rays absent.
4. Dorsal, caudal and anal fins continuous.

Examples: *Amphipnous*, *Symbranchus* (eels).

Order 8. Ophiocephaliformes or Channiformes

1. Commonly called snake headed fish.
2. Head depressed, covered with large scales.
3. Air bladder prolonged into the caudal region.
4. Fins without spines.
5. Accessory respiratory organ present.

Example: *Ophiocephalus* (= *Channa*).

Order 9. Perciformes

1. Vast order of marine fishes.
2. Fins usually with spines.
3. Dorsal fin usually two.
4. Air bladder without duct (physoclistous)
5. Weberian apparatus absent.

Examples: *Anabas* (climbing perch), *Perca* (yellow perch), *Fierasfer*.

Order 10. Mastacembeliformes

1. Eel-shaped, freshwater, physoclistous fish.
2. Dorsal, caudal and anal fins confluent. Sometimes a small caudal fin is separate. Ventral fin absent. Pectoral fins present.
3. Some free spines present before dorsal fin. Anal fin with three spines.
4. Gill opening small and ventral.

Examples: *Mastacembelus*, *Macrognathus*.

Order 11. Pleuronectiformes

1. Body flat, adapted for living at the bottom.
2. Skull asymmetrical.
3. Both eyes situated on the upper side.
4. Fins usually without spines.
5. Adults without air bladder.

Examples: *Pleuronectus* (flatfish), *Synaptura, Solea* (sole fish), *Pseudorhombus*.

Order 12. Echeneiformes

1. First spinuous dorsal fin transformed into an adhesive disc placed on the head.
2. No spines in second dorsal and anal fins.
3. Air bladder absent.
4. Scales cycloid.

Examples: *Echeneis* or *Remora* (sucker fish).

Order 13. Tetradontiformes

1. Body usually globular, covered with bony plates or scutes.
2. Scales modified into spines.
3. Jaws modified into a strong beak-like structure.
4. Gill opening restricted, in the form of a small hole in front of pectoral fin.

Examples: *Tetradon* (globe fish), *Diodon* (porcupine fish), *Mola* (giant sun fish), *Ostracion* (trunk fish).

Order 14. Scorpaeniformes

1. Bottom dwelling, predatory marine fishes with large head, mouth and eyes.
2. Scales reduced or absent.
3. Pectoral fins enlarged.
4. Numerous stout spines projecting from the fins, head and operculum.
5. Possess deadly poison which is neurotoxic.

Example: *Pterois* (scorpion fish).

Order 15. Lophiformes

1. Dorsal fin two, first one usually modified into a tentacle.

2. Scales absent, skin has flaps.
3. Lumuniscent organs present.
4. Mouth large with long pointed teeth.

Examples: *Lophius, Antennarius* (angler fish).

Order 16. Dactylopteriformes

1. Pectoral fins very large, divided into two portions.
2. Fins usually with spines.
3. Enlarged keel-like scales on lower part of body.

Example: *Dactylopterus* (flying gurnard).

Order 17. Batrachoidiformes

1. Body elongated and naked.
2. Head large flat and toad-like, with a large mouth.
3. Eyes protruding, situated dorsally on the head.

Examples: *Batrachus*, *Opsanus* (Toad fishes).

DIPNOI

Dipnoi is an order of subclass Sarcopterygii of class Osteichthyes. It includes specialized group of bony fishes commonly known as **lung fishes**. They respire through gills as well as lungs. They show affinities with the teleost on the one hand and amphibians on the other. They appeared in mid-Devonian as an early offshoot of Crossopterygian stem and flourished during Permian and Triassic. The earliest known fossil lungfish is *Dipterus* which dominated in the Devonian. The other fossil lungfish is *Ceratodus* which had a worldwide distribution in the Triassic. Dipnoans have not changed much since they have appeared.

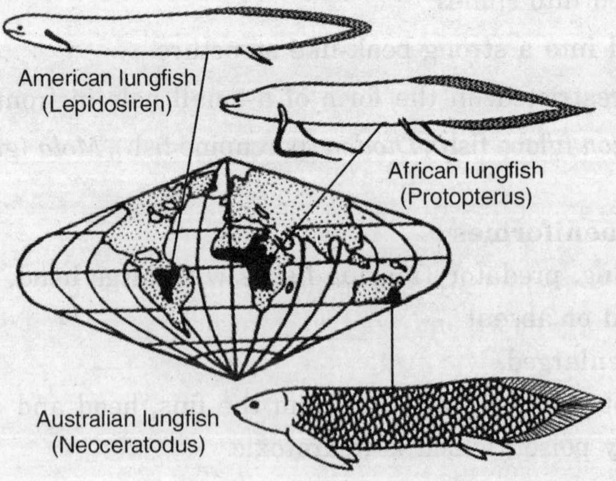

Fig. 12.1 Discontinuous distribution in Dipnoi

Distribution

At present Dipnoi is represented by three living genera which are discontinuously distributed in the southern hemisphere. The modern lungfish *Neoceratodus* is found in the rivers and pools of Australia. It is a direct descendent of fossil lungfish *Ceratodus*. It retains many primitive features so it is described as a living fossil. The other living lung fish is *Protopterus* which is found in the lakes and rivers of Tropical Africa. The third living species is *Lepidosiren* which is found in swampy marshes and rivers of South America.

Habit and habitat

All the lung fishes are inhabitants of fresh waters and swamps. They are large, sluggish, bottom dwelling carnivorous fishes, feeding on worms, crustaceans, molluscs, small fishes and even members of their own species (*Protopterus*). They swim by undulating movements of the body and even walk on the bottom using their paired fins as limbs. They can survive drought periods by undergoing **aestivation** or **summer sleep** in burrows (*Neoceratodus*) or mucous cocoons (*Lepidosiren* and *Polypterus*). During aestivation they use their air bladder as lungs. The dipnoans show parental care. They male guards the eggs and larvae and even aerate the water surrounding them.

External characters

The lung fishes are large in size, about one to two metre long, with slender body. They are characterised by the possession of narrow lobate or filamentous paired fins and overlapping cycloid scales. The scales are thin and embedded in connective tissue of the dermis. The paired fins are supported by dermal fin rays which are jointed, branched bony structures. The tail is diphycercal and the dorsal and anal fins are continuous with it. Eyes are small. Mouth is situated on the ventral surface. The nostrils lie on the ventral surface of the snout and open into the mouth cavity by a pair of internal nares. Spiracles are absent. The gills are covered by operculum and there is a single branchial aperture. Cloaca present. The lateral line system is well developed.

Internal characters

Endoskeleton is mainly cartilaginous. The notochord is persistent as an unsegmented rod. Vertebrae are absent. The vertebral column consists of cartilaginous vertebral arches but no centra. Anterior vertebrae fused with the skull. The skull is cartilaginous. Maxillae and premaxillae are absent. Jaws are short and autostylic. Teeth are in the form of crushing plates. The hyoid and branchial arches are cartilaginous. The pectoral girdle is cartilaginous and consists of dorsal scapular and ventral coracoid region and two investing bones the cleithrum and intraclavicle. Pelvic girdle is well developed and cartilaginous.

Intestine has a scroll valve. The digestive glands, liver and pancreas, present. Anus opens into the cloaca.

Respiration is both aquatic and aerial. Gills may be well developed or reduced. In *Protopterus* there are six branchial arches with five gill slits but only fourth and fifth arches bear gill filaments. It depends more on aerial respiration. It comes to the surface to inhale air. In *Neoceratodus* four pairs of gills are developed and each branchial arch bears a double row of gill filaments. The air bladder is vascular and alveolar and functions as lungs for aerial respiration. In *Protopterus* and *Lepidosiren* there are two lungs but in *Neoceratodus* there is only one lung. The lungs of Dipnoi resemble those of amphibians.

The blood vascular system resembles that of amphibians. The atrium is completely divided into two, the sinus venosus is imperfectly divided, the ventricle is undivided and leads into a contractile conus arteriosus which is divided into two by means of valves. The auriculo-ventricular aperture is closed by means of a fibrous plug or open. Pulmonary arteries arise from paired dorsal aortae and carry blood to the lungs. Pulmonary veins open directly into the left auricle. Two precavals and only one (left) posterior cardinal vein present.

The brain has well developed cerebral hemispheres, small cerebellum and optic lobes fused to form a single oval body. Pineal body present.

The excretory system is similar to dogfish. The kidneys are mesonephric. The ureters open into the cloaca. In female a pair of ovaries is present close to the kidneys. A pair of oviducts open anteriorly by a funnel-shaped aperture into the abdominal cavity and posteriorly into the cloaca. In male a pair of testis lies close to the kidneys. The two vasa deferentia join to open into the cloaca. Fertilization external, development indirect, with metamorphosis. The larvae have suckers and external gills.

CLASSIFICATION

Dipnoi is divided into two **suborders**

1. Monopneumona. One lung present and lateral rays of paired fins well developed.

Examples. *Neoceratodus*, *Ceratodus* (extinct).

2. Dipneumona. Lungs paired and lateral rays of paired fins vestigeal or absent.

Examples: *Lepidosiren* and *Protopterus.*

AFFINITIES

Dipnoi possesses many special features of its own but at the same time shows resemblance with different groups of fishes and amphibians.

Affinities with elasmobranchs

Similarities

1. Presence of scroll valve in the intestine.
2. Kidneys mesonephric and without nephrostomes.
3. Female reproductive system similar in both.
4. Each gill arch with two efferent arteries.
5. Conus arteriosus similar in both.

Differences

Dipnoi shows following differences from elasmobranchs:

1. Presence of lungs.
2. Operculum present.

3. Tail diphycercal.
4. Claspers absent.
5. Fertilization external.

Affinities with holocephali

Similarities

Jarvick (1964, 1967) gave the following similarities between the two groups:

1. Operculum present.
2. Intestine with a scroll valve.
3. Teeth in the form of crushing plates.
4. External nostrils open into the mouth cavity.
5. Jaw suspension is autostylic.
6. Urinogenital system similar.
7. Two efferent arteries in each gill arch.

Differences

Dipnoi differ from holocephali in the following characters :

1. Presence of lungs.
2. Claspers absent.
3. Fertilization external.

Affinities with crossopterygii

Similarities

Like Dipnoi, Crossopterygii is also an order of subclass sarcopterygii. In fact, Dipnoi is regarded as an offshoot of the crossopterygian stem. The two groups show the following similarities :

1. Scales covered with a layer of cosmine.
2. Similar powerful and lobate paired fins.
3. Tail fin diphycercal, the vertebral column extends upto the tip of the tail fin.
4. Internal nares in the roof of the mouth cavity in some cases.
5. Air bladder functions as lungs and has good pulmonary circulation.
6. Bones of skull are similar in both.
7. Intestine with a spiral valve.
8. In some cases the larval form has external gills for respiration.

There is great similarity between the fossil forms of the two groups. *Osteolepis* is a fossil crossopterygian of the Devonian period and *Dipterus* is the fossil dipnoi of the same period.

1. Shape and size of the body similar.
2. Tail heterocercal, dorsal fin two, anal fin single, all separate. Fins supported by dermal bony rays.
3. Paired fins lobate with a fleshy central axis.

4. Skin with exoskeleton of cycloid scales.
5. Bones of the skull and pectoral girdle similar, lower jaw similar, vertebrae without centra.
6. Lateral line system present.

On the basis of these similarities some scientists regarded Dipnoi as degenerate descendent of the crossopterygians. Jarvick (1968) and others discarded this view because according to them Dipnoi had more specialized characters than crossopterygians.

Differences

Dipnoi differ from crossopterygians in several characters of which some important ones are:

1. Fin skeleton and visceral skeleton.
2. Teeth modified into crushing plates.
3. Division of heart.
4. Blood supply of the air bladder.

Affinities with amphibians

Similarities

1. Dipnoi is capable of living in marshy habitat.
2. Presence of lungs for aerial respiration.
3. Heart similar in having two auricles and conus divided by valves.
4. Pulmonary artery and vein present.
5. Inferior vena cava and anterior abdominal vein present.
6. External nostrils open into the roof of the mouth cavity by a pair internal nares.
7. Spiracles absent.
8. Jaw suspension autostylic.
9. Structure of cerebral hemispheres and cerebellum of the brain similar.
10. Structure and development of eggs similar.
11. Larvae with external gills and suckers, similar to the tadpole of frog.

According to Dollo, these resemblances are due to convergent evolution, an adaptation to similar environment. However, Dipnoi possesses certain specialized characters not found in amphibians.

Differences

1. Presence of lobate fins.
2. Fin skeleton different from primitive tetrapod.
3. Presence of scales.
4. Skull mainly cartilaginous, a few anterior vertebrae fused with the skull.
5. Teeth in the form of crushing plates.
6. Lungs situated dorsal to the gut.

Primitive characters of dipnoi

Dipnoi possesses a few characters which may be considered as primitive.

1. Presence of cloaca.
2. Spiral valve in the intestine.
3. Diphycercal tail.
4. A contractile conus.

Specialized characters of dipnoi

1. Skull cartilaginous with no ossification.
2. Maxillae and premaxillae bone absent in the skull.
3. A few anterior vertebrae fused with the skull.
4. Teeth in the form of crushing plates.
5. Dorsal position of lungs.

Phylogenetic significance of dipnoi

Different ichthyologists have presented different views regarding the relationship of Dipnoi with different groups of animals. Some authors are of the view that Dipnoi, although fish-like in many characters, possesses certain features which are not present in fishes but are found in amphibians. However it cannot be included in Amphibia. According to some authors the Dipnoi and Crossopterygii have evolved from a common stock, but this view is also not widely accepted. Modern Dipnoi has not changed much during the period of its existence. It is true that Dipnoi is not ancestral to Amphibia but it is probable that the two groups have been derived from some common stock. Dipnoi is therefore of great significance in tracing the evolution of Tetropoda.

REPRESENTATIVE TYPES

Cartilaginous fishes

(A) *Sphyrna*: (= *Zygaena*): ***Heptanchus***. It is commonly known as **seven-gilled shark**. It is one of the most primitive shark found in warm waters of Atlantic ocean and Mediterranean sea. It has seven pairs of gill slits and a pair of spiracle. Another primitive shark is *Hexanchus* which has six pairs of gill slits. The body is elongated with a subterminal mouth and single soft dorsal fin. The tail is heterocercal, vertebral column is cartilaginous and notochord is persistent. *Heptanchus* is viviparous but there is no placenta. It is commonly known as **hammer-headed shark** (Fig. 12.2A) because its head looks like a hammer. It is a ferocious shark attacking the prey with its head. The body of *Sphyrna* is elongated and may grow upto 15 feet in length. The head is flattened in front and is extended laterally into two prominent **lateral lobes** which are supported by cartilaginous outgrowths from the post-orbital or nasal regions of the skull. The eyes are present at the extremity of the head lobes. They have three eyelids and a nictitating membrane. The mouth is present on the ventral surface. The nostrils are also situated ventrally at the base of lateral lobes. It has five pairs of gill slits situated laterally. There are two dorsal fins, one in front of the pectoral fin and the other opposite to the anal fin. Spiracles are lacking. *Sphyrna* is viviparous, *i.e.*, it gives birth to young ones. The liver oil of this fish is useful, but the yield is not very high. The fish inhabits tropical and subtropical seas. Two species commonly found along the Indian coast are *S. blochii* and *S. zygaena*.

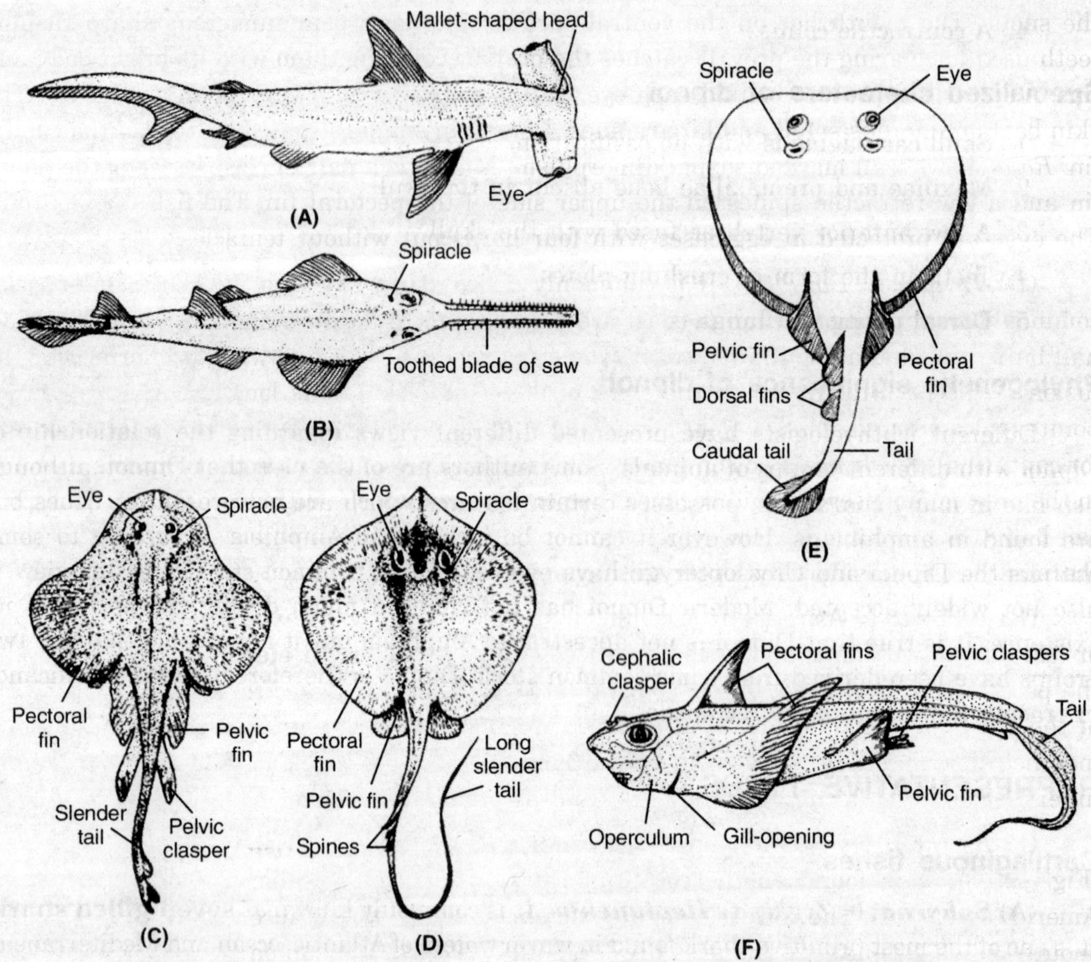

Fig. 12.2 Some cartilaginous fishes; **(A)** *Sphyrna*; **(B)** *Pristis*; **(C)** *Raja*; **(D)** *Trygon*; **(E)** *Torpedo*; **(F)** *Chimaera*

(B) ***Pristis***: It is commonly known **saw-fish** (Fig. 12.2B) because its snout is produced into a flattened saw-like **rostral process** or **rostrum**, the lateral margins of which are provided with a series of strong, teeth-like **denticles**, giving it the appearance of a saw. The body is elongated shark-like with a thick-heterocercal tail. It may attain a length of about 10-20 feet. The jaw teeth are minute. The fish is capable of tearing the flesh of the prey by a lateral stroke of the saw. There are two dorsal fins, one opposite to the pelvic fin and the second behind the first. A spiracle is present behind each eye. *Pristis* is viviparous and the saw is formed when the young is in the uterus of the mother. It in valued for liver oil which is rich in vitamin A. It is found in Mediterranean sea and Atlantic ocean. *P. cuspidatus* and *P. microdon* are common along the coast of Mumbai.

(C) *Raja*: It is commonly called **skate** (Fig. 12.2C). It is a sluggish fish living at the bottom of the sea. It inhabits seas in the temperate regions but is more abundant in the northern hemisphere. The body is dorso-ventrally flattened. A broad rhomboidal disc formed by the pectoral fins extends along the margins of the head and trunk from the pelvic fin to the snout. The mouth lies on the ventral surface. The jaws bear numerous sharp rasping teeth used for tearing the prey. It catches the prey by covering them with its broad body and fins. A spiracle is present behind each eye. Tail is narrow with a very small caudal fin. The skin bears numerous small and large spines. Dorsal fin single, situated in front of the caudal fin. *Raja* shows well marked sexual dimorphism. Male has a pair of claspers near the pelvic fin and a few retractile spines on the upper side of the pectoral fin. The fish is oviparous. The eggs are protected in egg cases with four horns but without tentacles.

(D) *Trygon* (= *Dasyatis*): It is commonly called **sting ray** (Fig. 12.2D). Some species commonly found along the Indian coast are *T. imbricata*, *T. blackeri* and *T. sephen*. They live half buried in sand along the sea coast. The sting ray has a flat and wide disc formed by the fusion of the pectoral fins with the sides of the head. The tail is long and slender with a pointed and serrated spine is place of dorsal fin. The spine is used as an organ of defense and for capturing prey. It has a poison gland which secretes poisonous mucous. It inflicts a wound in the prey causing severe pain because the poisonous mucous is introduced into it. The spine can be replaced by another one which develops behind it. *Trygon* shows sexual dimorphism, the male being considerably larger than the female.

(E) *Torpedo* (= *Astrape*): It is commonly known as **electric ray** (Fig. 12.2 E). It is more or less circular in shape and is characterized by having powerful **electric organs** between the pectoral fins and head. There are two large and two small electric organs on each side of the head, the smaller one lying enclosed within the larger one. The electric organs are modified dorsal muscle capable of giving powerful electric shock to their prey. They are innervated by cranial nerves. They are used as organs of defense.

(F) *Hydrolagus* (= *Chimaera*): It is commonly known as **rat fish** or **king of Herrings** (Fig. 12.2F). It is found on the coasts of Europe, Japan, Australia, New Zealand, North America and Africa. The body is elongate and shark-like in appearance with a bluntly conical snout. The skin is naked. The eyes are large and situated high up on the head. The mouth is ventral and bounded by lip-like folds. The nostril is confluent with the mouth. The jaws are covered by hard flat plate with the help of which they crush and eat their prey which are generally molluscs. The spiracle is absent. Lateral line is in the form of an open groove. It has four pairs of gill slits which are covered by operculum. The first dorsal fin has a strong spine on its front border. Pectoral fins are large and ventrally placed. Tail is long and filamentous and appears to be diphycercal. There is a club-shaped frontal clasper on the head of the male in addition to the ordinary claspers present behind the pelvic fin. The notochord is unsegmented and with numerous calcified rings in place of vertebrae. The upper jaw is immovably fused with the cranium and not suspended by ligaments as in other elasmobranchs. *Hydrolagus* is oviparous and lays one egg at a time.

Bony fishes

1. *Latimaria*: Crossopterygians were dominant in the Devonian period, some 350 million years ago. They were found both in fresh water and sea. Several of their fossils have been discovered from the Mesozoic era but no fossils were found in the post Mesozoic era. In 1938, a live specimen of Coelocanthi, a side branch of lobed fin fishes, was found near Comoro islands between Africa and Madagascar, and was named *Latimaria* (Fig. 12.3). It is the most primitive vertebrate and is called a living fossil because it has remained unchanged for over 100 million years.

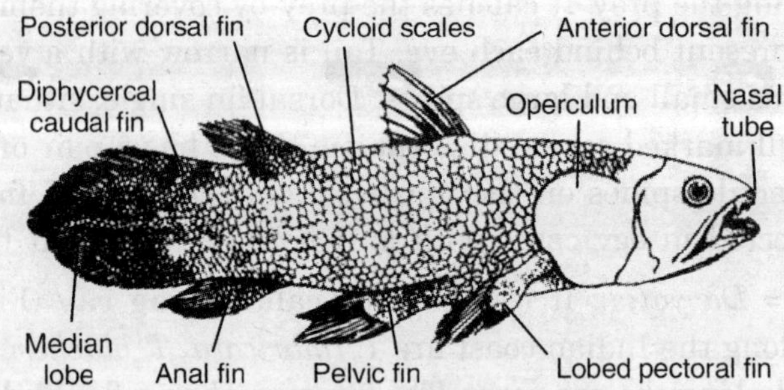

Fig. 12.3 *Latimaria,* a living coelocanth

Latimaria is steel blue in colour with a few scattered white spots and golden eyes. It is about 5 feet long. The scales are cycloid and massive with hollow tubercles. They overlap each other forming a thick protective covering. The paired fins are pendunculate each having a short scale covered lobe with fin rays attached around it like a fan. The fins are not suitable for active swimming but adapted for locomotion on the bottom of the sea. The endoskeleton of the fin consists of a single basal plate attached to the girdle and a few atrial rods round which the fin rays are arranged. The anterior dorsal fin is supported by a bony plate and the posterior one has endoskeleton like that of pectoral fin. The caudal fin is diphycercal with a small central projecting lobe. The vertebral column is a hollow cartilaginous tube with ossified neural and haemal spines. Notochord is unconstricted. The spines on the back are hollow (hence called Coelocanthi). There are five gill arches. There is no internal nostril. The swim bladder is extremely variable and even vestigeal. It has no pulmonary function. Heart is extremely primitive with the atrium lying posterior to the ventricle. The oesophagus is a thick muscular tube followed by a spacious double-limbed stomach. The intestine is long and has a spiral valve. The rectum opens into the cloaca. Liver is bilobed. A large gall bladder is present and pancreas is well developed. Each kidney has a ureter which expand to form urinary bladder. Thus there are two urinary bladder in *Latimaria*. The urinogenital aperture is situated between the pelvic fins. Gonads are paired. Males without copulatory organ. Fertilization is internal. The flesh of *Latimaria* is very oily.

2. *Dipnoi*: Dipnoi or **lung fishes** are the most interesting group of bony fishes. They show close affinity with the teleost on the one hand and with the amphibians on the other. They appeared in the Devonian period, flourished during Permian and Triassic and at present are represented by three living genera existing in widely separated regions of the world in fresh water habitats. These are *Neoceratodus* of Queensland (Australia), *Lepidosiren* of South America and *Protopterus* of Africa.

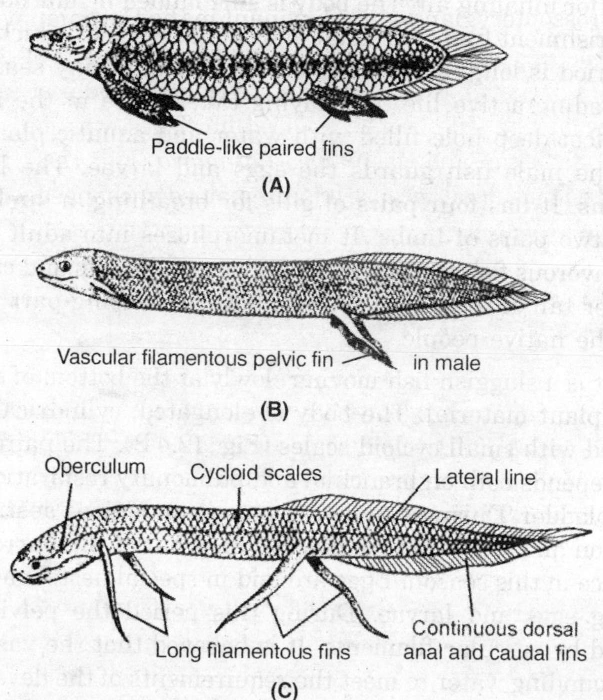

Fig. 12.4 *Dipnoi.* **(A)** *Neoceratodus;* **(B)** *Lepidosiren;* **(C)** *Protopterus*

***Neoceratodus*:** It is a large fish, about 5 to 6 feet long (Fig. 12.4 A). Body is covered with thin non-ganoid scales. Paired fins are leaf-like or paddle-like. The dorsal and ventral fins are continuous and tail is diphycercal. The nostrils are present on the ventral surface of the snout and open into the mouth by a pair of internal nares. The dental plates are oval or triangular, terminating in a smooth or slightly denticulated biting margins. The lower jaw has a small toothless dentary on each side. There are four pairs of gills. Each branchial arch bears a double row of gill filaments. The fifth arch is without gills. There is only one air bladder which functions as lungs when oxygen concentration declines. When water is well oxygenated it makes use of branchial respiration. During summer when the water level falls, the fish lives in isolated pools where oxygen percentage is low due to decomposition of animal and vegetable matter, and depends on pulmonary respiration. It comes to the surface to inhale air and the mouth remains closed. It makes a peculiar grunting sound when the air is expelled out from the mouth. The lateral line system is degenerated. *Neoceratodus* is a sluggish fish lying motionless at the bottom of the river. It feeds on molluscs, worms and crustaceans. The breeding season is September-October. Egg laying generally occurs in morning hours. The eggs are laid among aquatic plants but later they sink to the bottom.

***Protopterus*:** It is a large fish with elongated, cylindrical, eel-like body, reaching a length of 6 to 7 feet (Fig. 12.4 C). It lives in marshy places near rivers. The body is covered by small cycloid scales completely embedded in the skin. The paired fins are long and filamentous. They do not help in locomotion. Five gill arches are present but the anterior ones are without gill filaments. Air bladder double. It respires both by gills and lungs. During summer when the marshes dry up it undergoes **aestivation** or **summer sleep** (metabolism is lowered). During this period the fish moves down into deeper mud and coils up in a **cocoon** made up of special clay mixed with mucous secreted by the skin. The mouth of the cocoon has a small aperture for inhaling air. The body is surrounded by mucous which keeps the skin moist. It derives nourishment from stored fat and caudal muscles which undergo degeneration if the aestivation period is long (about six months). During rainy season the fish leaves the cocoon and starts leading active life. Egg laying takes place in the rainy season, in small **nests** which is one foot deep hole filled with water and aquatic plants. The eggs hatch in about eight days. The male fish guards the eggs and larvae. The larva is similar to the tadpole of amphibians. It has four pairs of gills for breathing, a **sucker** or **cement organ** for attachment and two pairs of limbs. It metamorphoses into adult in about seven weeks. *Protopterus* is a carnivorous fish feeding on frogs, worms, insects and crustaceans. Sometimes they bite the limbs or tail of their own species, but the missing part is soon regenerated. It is used as food by the native people.

***Lepidosiren*:** It is a sluggish fish moving slowly at the bottom of swamps feeding mainly on snails and some plant material. The body is elongated, cylindrical, eel-like, about 4 feet in length and covered with small cycloid scales (Fig. 12.4 B). The paired and median fins are like *Protopterus*. It depends both on branchial and pulmonary respiration. It has five branchial arches and two air bladder. During summer season it undergoes aestivation and depends on pulmonary respiration and stored fat for life. It comes out of the burrow in the rainy season. Egg laying takes place in this season. Eggs are laid in special nests made by males. He guards the nests containing eggs and larvae. During this period the pelvic fin of male becomes enlarged and covered by vascular filaments. It is believed that the vascular filaments release oxygen into the surrounding water to meet the requirements of the developing eggs and larvae. The larvae are similar to that of *Protopterus*. It is also relished as food by the native people.

3. *Polypterus*: It has survived from ancient times so it is regarded as one of the most primitive actinopterygian. It is found in the rivers of Africa. It is a small fish, measuring less than one metre in length (Fig. 12.5 A) and is commonly known as **bichir**. The body is covered with thick ganoid scales. The dorsal fin is divided into a number of separate **finlets**, hence the name. Each finlet has a single bony ray supporting it. The pectoral fin has a fleshy scale covered lobe at the base. The tail has become secondarily diphycercal. It has double air bladder, situated ventrally, which serves as lungs. Nostrils are dorso-laterally situated. Spiracles are present. The intestine has a spiral valve. The vertebrae are bony and the notochord is constricted. The larva resembles amphibian tadpole.

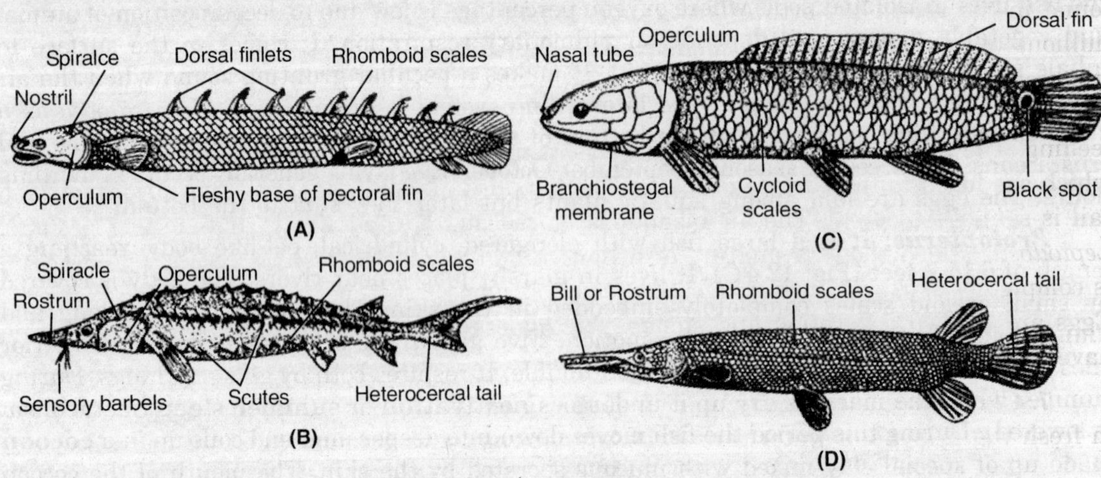

Fig. 12.5 *Bony fishes.* **(A)** *Polypterus;* **(B)** *Acipenser;* **(C)** *Amia;* **(D)** *Lepidosteus*

4. *Acipenser*: It is commonly known as **sturgeon**. It is a huge fish measuring 1 to 6 metre in length. The snout is elongated into a pointed **rostrum** (Fig. 12.5 B) which bears sensory barbels on the undersurface. The body is not covered with scales but has five rows of bony dermal scutes which are grooved and bear spines. It has five pairs of gill slits, a pair of spiracle and a single air bladder connected with the gut dorsally. The jaws are reduced and without teeth. They feed on small invertebrates, small fishes and plants. It shows some primitive features like incomplete vertebrae, unconstricted notochord, and heterocercal tail. It also shows some secondarily modified characters like reduction of jaws, loss of ganoine of scales and reduction of scales. Sturgeons are found in the rivers of Europe, North America and Asia. About 23 species are known. One or two species are exclusively fresh water but most of them are migratory fishes. They live in sea and move to rivers for breeding. They are of great economic importance as they are used as food.

5. *Amia*: It is commonly known as bowfin. It is a powerful predatory fish represented by only one species found in rivers of North America. The body is fusiform, 0.5 to 1 metre long (Fig. 12.5 C)and covered uniformly with thin cycloid scales which overlap each other. A few scales on the head are ganoid. The mouth is wide with sharp teeth. Two peculiar comb-like structures are present on the throat. It has a long continuous dorsal fin, hence the name bow-fin. The tail is almost homocercal. Air bladder is single and opens dorsally into the pharynx. It functions as lung when the water is deficient in oxygen. It shows primitive as well as advanced characters. Primitive features are presence of dorsal cellular air bladder, spiral valve in intestine, and a small valvular conus arteriosus in the heart. Some advanced features are thin cycloid scales, homocercal tail, absence of spiracles and amphicoelous vertebrae.

Amia is a voracious feeder. Its food consists of other fishes, crustaceans, insects and worms. It shows sexual dimorphism. The male is smaller than the female and has a shiny black spot at the base of caudal fin. During breeding season they move to one end of the lake where there is abundance of aquatic plants. The male builds a crude saucer-shaped nest. Egg-laying and nest building generally occurs at night. The male guards the nest. Hatching takes place in about ten days. The male takes care of the larvae which start independent life when they become about ten centimetre long.

6. *Lepidosteus* or *Lepisosteus*: It is commonly known as **gar-pike** (Fig. 12.5 D). It is found in the rivers of North America and Cuba. It is a primitive fish and has survived for millions of years without any change. It is represented by many species. One species *L. tristoechus* is said to attain a length of about 10 feet. It is a voracious fish feeding on other fishes. The body is elongated and cylindrical, with a long snout bearing sharp teeth for feeding. The exoskeleton consists of closely fitting thick ganoid scales and dermal denticles which are fused to the surface of scales. The dorsal fin is placed near the posterior end. The tail is semiheterocercal. The air bladder is paired and opens dorsally. It functions as lung. *Lepiodosteus* also shows a mixture of primitive and advanced features like *Amia*. Skeleton is completely ossified. During breeding season it moves to deeper parts of lakes and rivers. Eggs are laid on vegetation and stones. The male guards the eggs and larvae. The larvae have a sucker with the help of which they attach to stones. They feed on larvae of crustaceans.

7. *Clarias*: It is commonly known as **mangur**. *C. batrachus* is an Indian species found in fresh water and salt water throughout India. The body is elongated, about one and a half feet long and without scales (Fig. 12.6A). The head is covered with bony plates on the sides and top. It bears four pairs of barbels on the antero-dorsal end. The dorsal fin is long, without spines and running along the whole trunk upto the caudal fin. The anal fin is also long. The caudal fin is more or less rounded and separate from dorsal and anal fins. The pectoral fin is provided with a spine. It has a pair of accessory respiratory organs which are branched, tree-like situated dorsal to the gills. It takes in oxygen from the air. The air bladder is connected with the Weberian ossicles of the ear. *Clarias* is very nourishing fish and is relished as food.

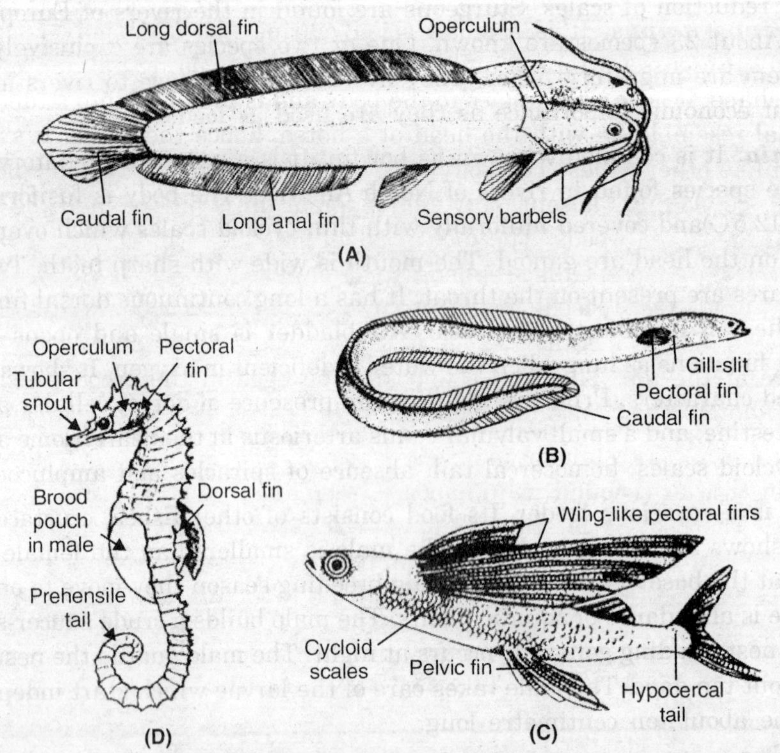

Fig. 12.6 *Bony fishes.* **(A)** *Clarias;* **(B)** *Anguilla;* **(C)** *Exocoetus;* **(D)** *Hippocampus*

8. *Anguilla*: It is commonly known as **eel** (Fig. 12.6B). *A. bengalensis* is an Indian species which is similar to other species found in Europe and America. It lives in marshes. The body is elongated, snake-like, about 3 feet long. The skin appears naked but minute scales are embedded in the dermis. The dorsal fin is long and continuous with the caudal and anal fins. The pectoral fin is small and the pelvic fin is absent. Air bladder is present. The skin functions as accessory respiratory organ both in water and in air. It moves by wriggling movement on damp grass. *Anguilla* is a migratory fish. It moves from river to sea in autumn, spawns in deep water and then die. The females do not have gonoducts and the eggs are discharged through genital pores. The newly hatched larvae are called **leptocephali** (singular **leptocephalus**). They are transparent and so are called **glass eels.** They are pelagic and float on the surface of water. In one to three years they metamorphose to forms called **elevers**. They migrate to rivers and streams, sometimes by crawling on moist grass, and become adults.

9. *Exocoetus* and *Cypselurus*: They are commonly known as flying fish (Fig. 12.6C) because they have large pectoral fins which spread like wings for gliding, after jumping out of water. The pectoral fin has black spots. Some species may have large pelvic fins which also help in gliding. The caudal fin has a large lower lobe which helps in taking a leap outside water. They are capable of moving short distances in air but large-sized fishes can reach an altitude of 1 metre and glide for a distance of 100 metres. The flying fishes are small to medium-sized, about one or one and a half feet in length. The body is covered with cycloid scales. Eyes are large. The dorsal fin is situated opposite to the anal fin. Air bladder is present. It feeds on prawns, small fishes and their larvae. It is an excellent edible fish.

10. *Hippocampus*: It is a small sized marine fish commonly known as **sea-horse** (Fig. 12.6D). It occurs in tropical and temperate seas including Indian Ocean. The head has a peculiar shape. It is at right angles to the body and is produced into a tubular snout, having a superficial resemblance with the head of a horse, hence the name. It swims in a upright position, unlike other fishes. The body is covered by an armour of bony plates. The pectoral fin is small, pelvic fin is absent, only single dorsal fin is present and the tail is prehensile, often coils around sea weeds. The mouth is without teeth and is situated at the tip of the snout. The eyes are large. Gills are made up of many small lobe (**lophobranchs**). Operculum is large and plate-like. It feeds on crustaceans and their larvae which are sucked in from a distance by the tubular mouth. It shows sexual dimorphism. The male has a brood pouch with an aperture at the top. The female deposits the eggs in the brood pouch where the eggs develop and hatch. The young ones come out of the pouch when they are capable of swimming.

11. *Anabas*: It is commonly known as **climbing perch**. It was believed to climb trees, but it is not true. It is found in Indian rivers. *Anabas* (Fig. 12.7A) is a small fish, about 8 inches long. The body is laterally compressed and covered with large overlapping cycloid scales. The dorsal and anal fins are long and made up of two parts : long anterior part with stiff spine and a small posterior part with soft fin rays. The pelvic fin is situated close to the pectoral fin. The operculum bears a row of backwardly directed spines. The lateral line is discontinuous. The gill opening is narrow. The accessory respiratory organs are a pair of **labyrinthine organs** which are formed by much folded plates covered with vascular mucous membrane. Each labyrinthine organ lies in a **suprabranchial cavity** dorsal to the gill chamber. They are used to breathe atmospheric air. They move on land with the help of their pectoral fins and opercular spines. They can live outside water for many days.

12. *Pleuronectus* or *Synaptura*: It is commonly known as **flat fish** (Fig. 12.7B). It is marine fish adapted for living at the bottom. The body is thin, asymmetrical, laterally compressed and flat. Both the eyes are present on the upper side. The upper side is coloured while the lower side is white. The body is covered with imbricate ctenoid or cycloid scales. The mouth is narrow and protractile. The dorsal and anal fins are long and without spines. They cover the whole body. The caudal fin is well developed. The air bladder is absent in adults. It has four pairs of gills. The flat fish swims close to the bottom by undulating movement of the body. It feeds on molluscs. The larvae are pelagic and bilaterally symmetrical. They undergo metamorphosis to give rise to asymmetrical bottom dwelling adults. During metamorphosis pigmentation on the two sides changes and one eye migrates to its final position close to the other eye on the upper side.

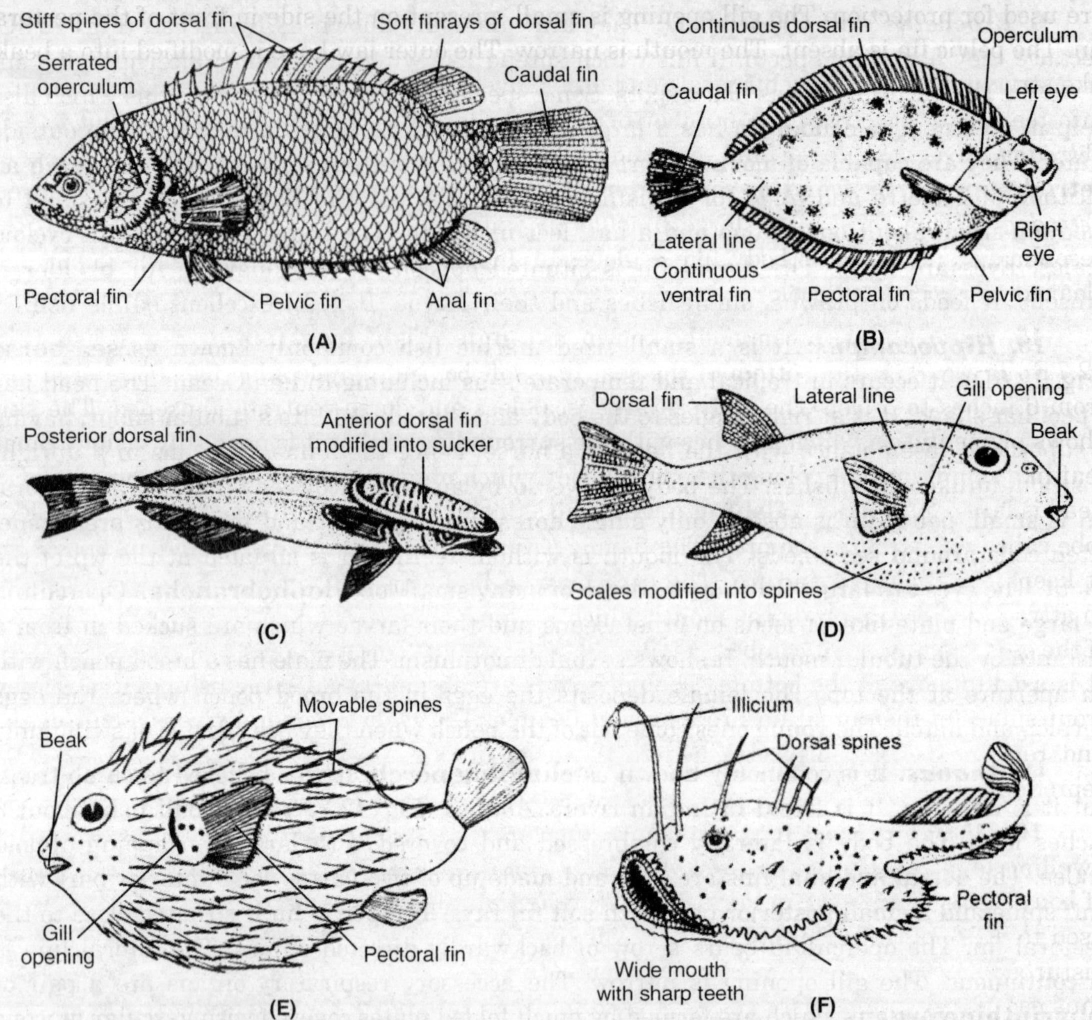

Fig. 12.7 *Bony fishes*. **(A)** *Anabasp;* **(B)** *Pleuronectus;* **(C)** *Echeneis;* **(D)** *Tetrodon;* **(E)** *Diodon;* **(F)** *Lophius*

13. *Echeneis* or *Remora*: It is commonly known as **sucker fish** (Fig. 12.7 C). It is a small marine fish less than 1 feet in length. The head is depressed and bears a peculiar, large, oval **adhesive disc** or **sucker** made up of two rows of transverse lamellae. The sucker is modified anterior dorsal fin. The fish is a good swimmer but generally becomes attached to huge fast swimming fishes by its sucker and enjoy travelling from one place to another. The body is smooth but bears minute cycloid scales. Air bladder is lacking. The dorsal and anal fins are without spines. It feeds on small fishes. It is used by the natives in Africa to catch sea turtles and fishes.

14. *Tetrodon* (globe fish) (Fig. 12.7 D) and ***Diodon* (porcupine fish)** (Fig. 12.7 E): Both these fishes are peculiar species having a globular body. They move slowly with the help of their caudal fin. Body is covered with spines which are modified scales. The spines in *Tetrodon* are small but in *Diodon* they are large and movable as in porcupine. The spines are used for protection. The gill opening is small, present on the side in front of the pectoral fin. The pelvic fin is absent. The mouth is narrow. The outer jaw bone is modified into a beak-like structure for shearing off corals and other invertebrates. The beak is divided by sutures into teeth-like structure. The number of teeth is characteristic—two in *Diodon* and four in *Tetrodon*. Eyes are large and protruding. *Tetrodon* contains a powerful poison called **tetrodoxin**, mostly found in some internal organ. It can be fatal to man when attacked. *Tetrodon* is able to swallow large amount of water or air in a sac-like extension of the oesophagus and swell its body like a globe. In Japan *Tetrodon* is relished as food but it is cleaned by specialized cook.

15. *Lophius*: It is commonly known as **angler fish**, **frog fish** or **devil fish** (Fig. 12.7 F). It occurs at the bottom of the sea. It is soft bodied, ugly-looking fish ranging in size from 6 inches to 3 feet. The body is large, scaleless and dorso-ventrally flattened. The skin shows pigmentation which matches with the surrounding so that it is not easily visible. Small leaf-like skin flaps hang along the edge of body which mimic the leaves of aquatic plants. The head is large and carries a **fishing organ** consisting of a rod called **illicium** and an expanded lobe called **esca** to lure the prey. The fishing organ is modified first ray of spinous dorsal fin. It keeps on moving to and fro. The esca may be luminous or may mimic small organisms to attract the prey towards the mouth which is a wide gape with sharp, long, recurved teeth. If the esca is damaged, it can be regenerated. The pectoral fin is well developed and pediculate. It is used to move at the bottom. The pelvic fins are absent. Besides the fishing organ, a few front spines of the dorsal fin are long and flexible. The male is generally very small in size and remains permanently attached to the head of the female. It is concerned only with reproduction.

16. *Dactylopterus*: It is commonly known as **flying gurnard**. It is a marine bottom dwelling fish. The body is covered with bony plates. The head is also covered by an armour of bony plates. The eyes are large. The pectoral fins are large with free inner rays which are used to walk at the bottom of the sea. They are not used to fly or glide in the air but when disturbed, the fish opens the brightly coloured large pectoral fin to frighten the predators and thus escapes. Two large free spines are present just in front of the dorsal fin. It can produce sound by hyomandibular stridulations.

17. *Batrachus*: It is commonly known as **toad fish**. It is a marine fish found is shallow water in North America. The body is not covered with scales. The head is large, flat and toad-like and bears large protruding eyes. The mouth is large. The dorsal fins are spinous. In some species the spines are hollow and associated with poison gland. Some species are capable of producing a loud sound by vibrating its air bladder.

13

CHAPTER

SUPERCLASS PISCES:

GENERAL ACCOUNT

SKIN AND SCALES

Skin

The skin forms the outer most covering of the body. It not only protects the body from injuries and infections but also performs other functions like respiratory, excretory and osmoregulatory. Many specialized structures like electric organ, poison glands, bioluminiscent organs, etc. are derivatives of skin and they play very important part in the metabolic activities of the body.

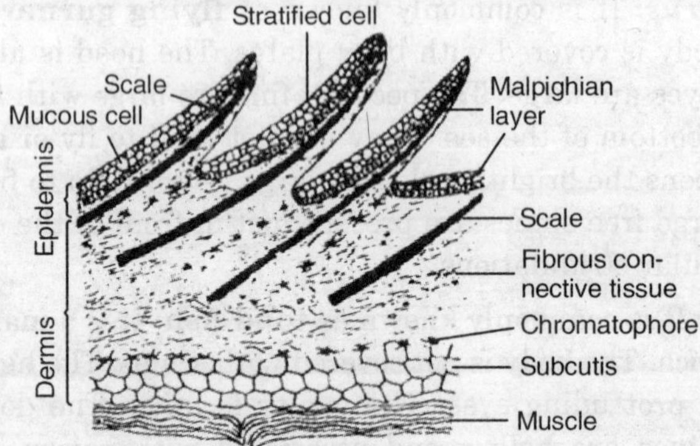

Fig. 13.1 V.S. of skin of teleost

Structure: The skin is made up of two layers: outer **epidermis** and inner **dermis**.

The epidermis is **ectodermal** in origin. It is made up of many layers of flattened cells (Fig. 13.1). The lower layers are made up of columnar cells forming **stratum germinativum**. The cells of this layer are always multiplying to replace the worn out cells.

The dermis is **mesodermal** in origin and consists of (Fig. 13.1) connective tissue, blood vessels, nerves and cutaneous sense organs. It is made up of two layers : the thin upper layer of loose connective tissue is called **stratum spongiosum** and the thick dense lower layer is called **stratum compactum. Chromatophores** are present in the dermis. They give beautiful colours to fishes.

Numerous flask-shaped **mucin**-secreting cells are scattered in the epidermis and dermis. Mucin is a glycoprotein which mixes with water to form a slimy **mucous** covering the whole body. These cells develop in stratum germinativum and migrate to the surface. In general, the fishes which have few scales or are scaleless have large number of mucous cells. The mucous performs many important functions. It lubricates the fish and thus reduces body friction in water while swimming enabling it to swim faster. It is slippery, so helps to escape predators. It protects from microbial infections. Some fishes use it for preparing nest during the breeding season. *Protopterus* (lung fish) uses it to prepare its cocoon for summer sleep.

Some glandular cells are present in the skin which may become modified to form poison glands in some fishes and are used as structures of offense and defence. Exoskeleton in the form of scales, bony plates or denticles are all derivatives of skin and provide protection to the body.

Scales

The skin of all vertebrates, except amphibians, is covered with an exoskeleton in the form of **scales**. The scales are derivatives of skin. They may be derived either from the epidermis (**epidermal scales**) or from the dermis (**dermal scales**). In all terrestrial vertebrates the scales are epidermal in origin, but in fishes they are dermal in origin, except the placoid scales which are derived by the secretory activity of the dermis as well as epidermis. Some fishes lack scales (cat fishes and *Torpedo*) and in a few species the scales or plates are confined to certain regions of the body (*Acipenser, Polyodon*).

Structurally the scales in fishes are of five types:

1. Placoid scales
2. Cosmoid scales
3. Ganoid scales
4. Cycloid scales
5. Ctenoid scales

The cycloid and ctenoid are together called **bony ridge scales**.

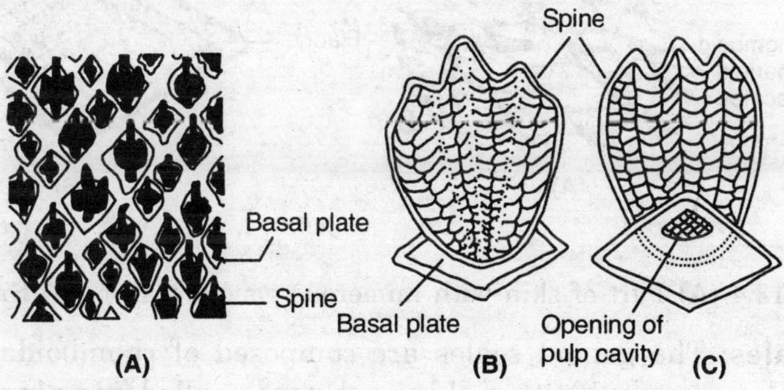

Fig. 13.2 (A) Part of skin with numerous placoid scales; **(B)** Single scale

Placoid scales: Placoid scales or **dermal denticles** are present in cartilaginous fishes. Each scale consists of a diamond-shaped calcareous **basal plate**, embedded in the dermis, from which a **trident spine** projects out of the epidermis (Fig. 13.2). The spines are directed backwards. The calcareous tissue which forms the basal plate is similar to the cement of our teeth. A small opening is present in the centre of the basal plate. It leads into a **pulp cavity** which traverses the spine and is filled with **pulp** containing numerous dentine forming **odontoblast cells**. The spine is made up of a calcareous substance, the **dentine**, externally covered by a hard dense **enamel-like** material. The dentine is relatively spongy and a number of fine branches or **canaliculi**, arising from the pulp cavity, run through it. The scales are set so closely together in the skin that they give it a sandpaper like quality.

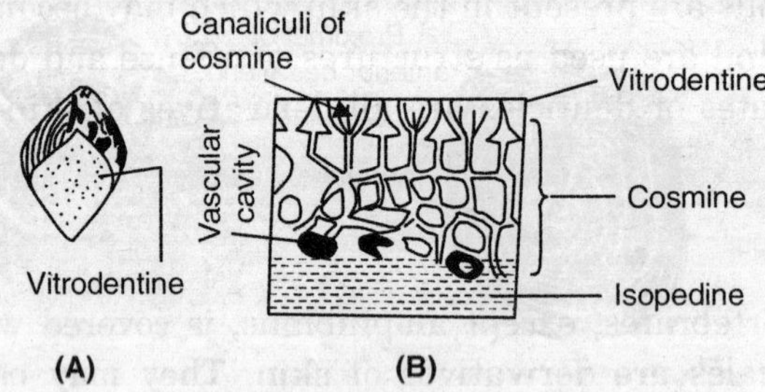

Fig. 13.3 **(A)** Cosmoid scale; **(B)** T.S. of the same

Cosmoid scales: The cosmoid scales (Fig. 13.3) are found in ostracoderms, placoderms and extinct crossopterygians. They are absent in living fishes. The cosmoid scales consist of **three** layers. The **outer layer** formed of cell-less dentine-like substance is called **cosmine**. It contains pulp cavity. On the outer surface the cosmine contains a thin layer of hard glossy substance, the **vitrodentine**. The **middle layer** is the **vascular layer** consisting of **spongy bone**. It contains **vascular spaces** and numerous fine radiating **canaliculi**. The **innermost layer** is **isopedine** comprising several layers of bone lying parallel to one another. It is traversed by channels for blood vessels.

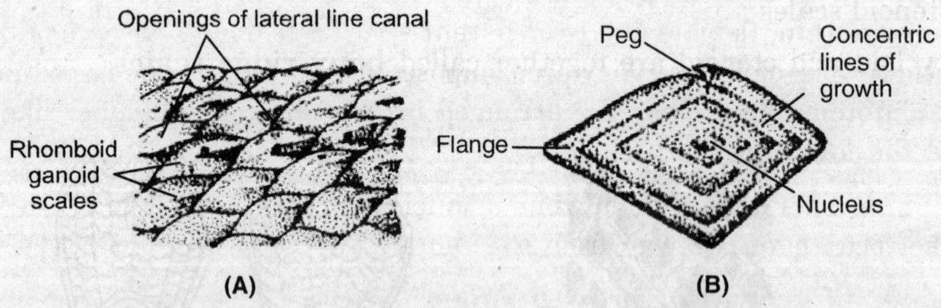

Fig. 13.4 **(A)** Part of skin with numerous gonoid scales; **(B)** Single scale

Ganoid scales: The ganoid scales are composed of rhomboidal or diamond-shaped (Fig. 13.4) bony plates covered with a shiny substance called **ganoine** which gives a shiny lustre to the scales. The scales fit together closely like tiles of a floor and form a protective bony armour on the body. Each ganoid scale consists of **three layers**. The **outermost layer** is made up of cell-less dentine-like substance, **ganoine** arranged is successive lamellae. The **middle layer** is cosmine-like layer traversed by canaliculi. The **innermost layer** is isopedine consisting of layers of bone. The spongy bone or vascular layer is absent. These layers are not universally present and so the ganoid scales are of two types:

1. **Palaeoniscoid ganoid scales** in which all the three layers are present, for example, *Polypterus, Acipenser*.
2. **Lepidosteoid ganoid scales** in which only two layers are present—outer ganoine and inner isopedine, for example, *Lepidosteus*.

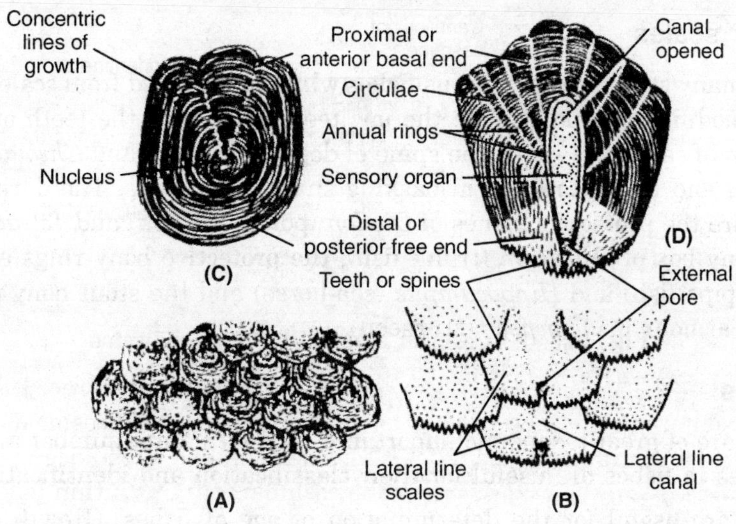

Fig. 13.5 Part of skin with numerous; **(A)** Cycloid scales; **(B)** Ctenoid scales; **(C)** Cycloid; **(D)** Ctenoid

Bony ridge scales: The cycloid (Fig. 13.5 A and C) and ctenoid (Fig. 13.5 B and D) scales are known as bony ridge scales. They are present in majority of teleosteans (bony fishes). They are thin, flexible and transparent structures due to absence of first and middle layers of other types of scales. They are composed of two layers—the isopedine layer and the inner dense, fibrous layer. They are arranged in an overlapping manner like tiles of a roof. They show the following features:

1. **Focus:** This is the central zone or nucleus and is the first part to develop. In some the focus may be away from the centre.
2. **Lines of growth or growth rings:** These are a series of concentric bony ridges around the focus.
3. **Radii:** These are grooves which radiate from the focus to the margin of the scale.
4. **Annuli:** These are annual (yearly) growth marks.
5. **Anterior margin:** The anterior margin lies embedded in the dermis.

6. **Posterior margin:** The posterior margin of cycloid and ctenoid scales is different and is easily recognizable. In cycloid scales it is in the form of eroded bony ridges and in ctenoid scales it is marked by the presence of comb-like teeth or **cteni**.

The origin and development of both types of scales are similar. The cycloid scales are roughly circular in shape being thicker in the centre and thinning out towards the margin. They are present in lung fishes and lower teleosteans such as carps, cods etc. The ctenoid scales are also circular and can be differentiated from cycloid in having a serrated free edge. It is found in modern teleosteans. Some fishes have both types of scales, for example, in *Pleuronectus* ctenoid scales are present on the dorsal surface and cycloid scales are present on the ventral surface.

Derivatives of scales

There are many structures in various fishes which are derived from scales. The following structures are modified placoid scales : the jaw teeth in sharks, the teeth on the elongated saw-like rostrum of saw fish *Pristis*, the spine of dogfish *Squalus* and *Chimaera*, the sting of sting ray *Trygon* and the gill rakers in basking shark *Cetorhinus*. The derivatives of other types of scales are the protective spines of *Diodon* (porcupine fish) and *Tetradon* (globe fish), the complete bony box of *Ostracion* (trunk fish), the protective bony rings around the body of *Syngnathus* (pipe fish) and *Hippocampus* (sea-horse) and the stout bony plates or scutes forming a bony armour in *Acipenser* (sturgeon).

Uses of scales

1. Scales are of great taxonomic importance. The structure, number and arrangement of scales in fishes are useful for their classification and identification.
2. Scales are useful for the determination of age of fishes. Growth of a fish is not constant, it fluctuates with seasons. In summer, when abundant food is available the fish grows rapidly and so the scales also grow fast. In winter, growth slows down due to temperature dependent metabolic activities. The widely separated lines of growth (**wide lines**) indicate rapid growth and the **close lines** indicate slow growth. A series of wide and close lines on a scale corresponds to one annual ring or one year in the life of fish.
3. Scales are useful in the determination of number of spawnings, particularly in migratory fishes like salmon. When salmon migrates from the feeding ground to the breeding ground it does not feed, resulting in calcium deficiency. This causes erosion of scale margins leaving spawning marks. These marks can be counted to determine the number of times the fish has spawned.

FINS OF FISHES

Fins may be defined as firm appendages which are the organs of locomotion and balance. All types of fins are supported by skeletal rods called **fin rays**. In cartilaginous fishes the fin rays are slender, segmented, horny structures called **ceratotrichia** and in bony fishes they are branched, jointed bony structures called **lepidotrichia**.

Types of fins

G.C. Kent (1967) and Hildebrand (1974) have classified the fins into two types:

1. **Median** or **unpaired fins** which include a **dorsal** fin on the back, an **anal** fin on the ventral side behind the anus and a **caudal** fin at the end of the tail.
2. **Lateral** or **paired fins** which include the **pectorals** and **pelvics** which correspond to the forelimbs and hindlimbs, respectively, of terrestrial vertebrates.

Median fins

Dorsal fin: It is located in the middorsal line of the body. Some fishes may have a single dorsal fin (sea-horse *Hippocampus*), some may have double (anterior and posterior dorsal fin as in *Latimaria*) and in some it may be triple (cod fish *Gadus*). It may be reduced (*Notopterus*), elongated (*Clarias*), continuous with caudal and anal fins (*Anguilla*) or divided into many finlets (*Polypterus*).

Modifications of dorsal fin: In some fishes such as *Echeneis* (sucker fish) the first dorsal fin is modified into a **sucker** for attaching to big fishes. In *Lophius* (angler fish) the first spine of the dorsal fin is modified into a **fishing organ** or **bait** to lure the prey. In some bony fishes, for example, *Oncorhynchus* (Pacific Salmon), the fin rays are absent, so the dorsal fin is **fleshy** and is called **adipose fin**.

Anal fin: It is located beyond the anus. It may be long as in *Anabas*, reduced to spines as in *Trichiurus*, absent as in bottom dwellers like *Raja* or may be continuous with the caudal and dorsal fins.

Modifications of anal fin: In male *Gambusia* the anal fin is modified into an intromittent organ called **gonopodium**.

Caudal fin: It is present at the end of the tail. It is often well developed in fishes because it serves as the chief propelling organ during swimming. In some fishes the caudal fin may be reduced or absent, for example, in sea-horse (*Hippocampous*) the tail is prehensile; in sting ray (*Trygon*) and eagle ray (*Myliobatis*) the tail is long, slender and whip-like. The caudal fin is of different shapes in different groups of fishes, so it plays an important role in the identification and classification of fishes.

Types of caudal fin: Different types of caudal fins found in fishes are:

1. Protocercal or diphycercal
2. Heterocercal
3. Homocercal

Protocercal or Diphycercal (first or primary tail). It is probably the most primitive type of tail fin of fishes. It is found in cyclostomes, primitive sharks, living lung fishes, *Chimaera, Latimaria,* larval bony fishes and some deep sea fishes (Fig. 13.6 A and B). In protocercal condition the hind end of the vertebral column is straight and divides the fin into two equal or symmetrical lobes, the **dorsal** or **epichordal lobe** and the **ventral** or **hypochordal lobe**. The term protocercal is used for all types of symmetrical tails whereas the term **diphycercal** (double) is used for secondarily symmetrical tail as in Dipnoi. A modified diphycercal tail occurs in *Latimaria* (Fig. 13.6 C) in which a third lobe called **median**

lobe is present between the epichordal and hypochordal lobe. In *Chimaera, Anguilla* and some deep sea fishes the tail is tapering and symmetrical and terminal part of vertebral column is elongated and straight. This is not truely protocercal and is called **isocercal** or **leptocercal** (Fig. 13.6 D). During development the isocercal tail is formed by the reduction of hypochordal lobe and elongation of the dorsal and anal fins until a continuous fin-fold is established. Such a tail is probably developed secondarily in adaptation to sluggish habits of bottom living and deep sea fishes. In all the fishes the caudal fin passes through the protocercal stage during development.

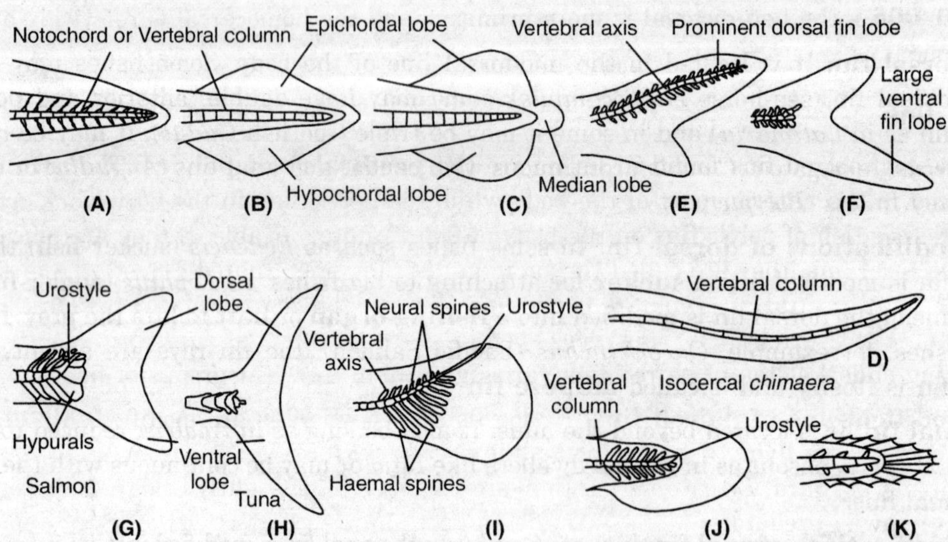

Fig. 13.6 Different types of tail or candal fins in Fishes. **(A)** qand; **(B)** Protocercal or diphycercal; **(C)** Modified diphycercal; **(D)** Isocercal; **(E)** Heterocercal; **(F)** Hypocercal; **(G)** and **(H)** Homocercal; **(I)** and **(J)** Abbreviated homocercal; **(K)** Gephyrocercal

Heterocercal (unequal tail): It is characteristic of Chondrichthyes and some primitive bony fishes. The hind end of the vertebral column is bent upwards and continues almost upto the tip of the much larger dorsal or epichordal lobe making the caudal fin unequal or asymmetrical. It is characteristic of fishes which have a ventral mouth and are bottom feeders. The enlarged dorsal lobe directs the fish towards the bottom (Fig. 13.6 E).

The **reverse heterocercal** condition is found in fossil *Agnattra*, some primitive fishes and flying fish *Exocoetus*. It is known as **hypocercal tail** (Fig. 13.6 F). In this case the vertebral column turns downwards and terminates in the larger ventral or hypochordal lobe. The dorsal lobe is reduced in size. It enables the flying fish to attain maximum speed for gliding when it leaves the water.

Homocercal (equal tail): It is characteristic of most higher bony fishes (Fig. 13.6 G and H). Externally it is symmetrical consisting of equal sized lobes, but internally it is asymmetrical because hind end of the vertebral column is turned upwards and greatly shortened, and does not reach the posterior end of the tail fin. Actually the dorsal or epichordal lobe is reduced and the ventral or hypochordal lobe is greatly developed into a single or two equal and symmetrical lobes. Homocercal tail is characteristic of fishes with a terminal mouth because it propels the fish straight forward.

The homocercal tail is derived from the heterocercal types and so intermediate types are found in many bony fishes. In *Amia*, *Lepidosteus* and *Polypterus* the tail shows intermediate condition between heterocercal and homocercal and is called **abbreviated homocercal type** (Fig. 13.6 I and J). The vertebral column is shortened and withdrawn to the base of the caudal fin, the hypochordal lobe is greatly enlarged to form a single lobe and the epichordal lobe is reduced or vestigeal. In *Fierasfer* the caudal fin disappears completely and the vertebral column becomes reduced or vestigeal forming a **gephyrocercal tail** (Fig. 13.6 K).

The protocercal tail is the most primitive type of tail but of the two major types of tail found in fishes, the heterocercal is more primitive and the homocercal is relatively a recent development.

Paired fins

The two paired fins found in fishes are the **pectorals** and **pelvics**. The pectoral fins are placed in the anterior part of the body while the pelvic fins in the posterior part of the body. The position of pelvic fins on the body of fish is quite variable. If they are placed close to the pectoral fins, they are called **thoracic** and if they are situated close to the anus, they are called **abdominal**. In some the pelvic fins are absent. The paired fins were not present in the ancestral vertebrates. They developed during the course of early fish evolution. The supporting endoskeleton of paired fins varies greatly in different groups of fishes.

Modifications of paired fins. In dog fish the pelvic fins are modified into **claspers** in the male. In the flying fish, *Exocoetus*, the pectoral fins are enlarged and help in gliding. In the flying gurnard, *Dactylopterus*, they are enlarged and brightly coloured and are used to frighten away predators.

Functions of fins

1. Caudal fins are used for forward propulsion during locomotion (**propelling** device).
2. Paired fins help to control the direction of movement during locomotion (**steering** device).
3. Dorsal and anal fins prevent the body from yawing (turning around the vertical axis) and rolling (turning around the longitudinal axis) during swimming (**balancing** device).
4. The paired fins help to maintain equilibrium when the body is at rest.
5. Lung fishes use paired fins as legs for walking on land.

Origin of fins

Median fins: The median fins of fishes are believed to have developed as a result of differentiation of a continuous embryonic fin fold running dorsally along the back upto the tip of the tail and then continued as a ventral fold upto the cloaca. This fold is then supported by a series of cartilaginous rods. This condition is seen in lampreys and is regarded as a primitive condition of the median fin. In modern fishes separate dorsal, ventral and caudal fins are formed by the concentration of radials in certain areas and the degeneration of the fold in the intervening spaces between the fins.

Paired fins: Many theories have been proposed to explain the origin of paired fins but still there is no satisfactory hypothesis.

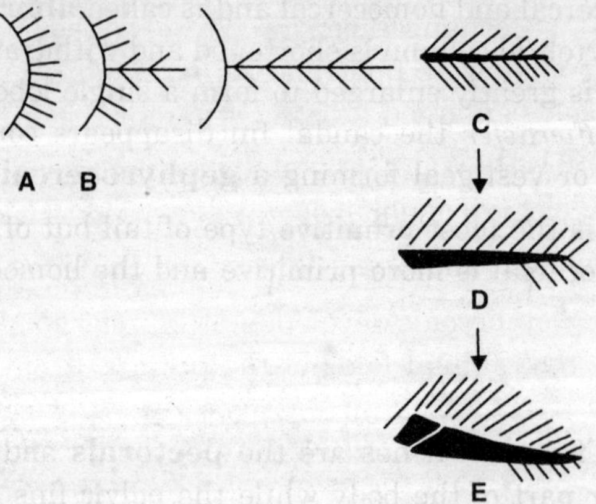

Fig. 13.7 Origin of fin skeleton from gill arch and gill rays as proposed by Gegenbaur

1. **Gill arch theory:** This theory was proposed by Gagenbaur. According to this theory the paired fins are a modification of gill structures, the girdles represent the gill arches and the fins and their skeleton represent the gill septa (Fig. 13.7). Each gill septum is supported by an arch from which rays extend into the lateral part. The median ray is believed to arise first forming an axis and the rays then become arranged above and below it. Such a condition is seen in the fin of *Ceratodus*. The location of pectoral fin behind the last gill arch prompted this theory but it does not explain the position of pelvic fins which lie far away from the last gill arch. It is believed that the location of pelvic fin at the posterior end is due to migration of some of the gill arches that had lost their respiratory function. There are many objections to this theory as embryological and morphological facts do not support it.

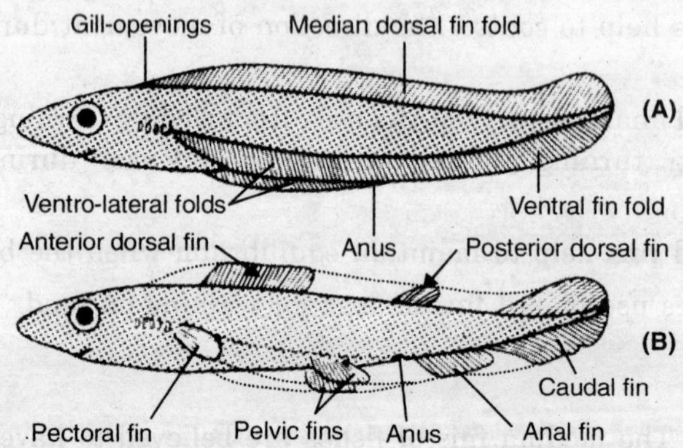

Fig. 13.8 Origin of fin according to fin-fold theory proposed by Bal four

2. **Fin fold theory:** This theory was proposed by Balfour and Thatcher separately, but almost at the same time, and later supported by a number of scientists like Mivart, Wiedersheim, Haswell, Regan, Goodrich, etc. These authors are of the view that the paired fins are basically of the same nature as the median fins and like them, have been derived from paired longitudinal folds (Fig. 13.8).

According to the fin fold theory the ancestral vertebrate had a pair of continuous lateral folds, just like the metapleural folds of *Branchiostoma*, which extended from behind the gills upto the end of the tail. These folds were separate in the proximal part but fused in the caudal region behind the anus forming a ventral fin fold. Besides there was also a dorsal median fold. The continuous folds were hindrance in swimming and undulating movements, so certain portions of these folds enlarged forming different types of fins and the intervening portions disappeared.

There are various embryological and palaeontological evidences which support this theory.

1. The mode of development and basic structure of paired and median fins are similar.
2. In the early stages of embryonic development in some elasmobranchs the paired fins are in the form of longitudinal folds of the body wall having parallel fin rays.
3. In the extinct shark *Cladoselache*, the pectoral and pelvic fins are broad and supported by parallel fin rays indicating origin from longitudinal finfold.
4. In the extinct acanthodians (*Climatius*) a row of spines and finlets are present on either side between the pectoral and pelvic fins which are regarded as remnant of finfolds.

Although there are so many evidences supporting this theory, it is still not widely accepted.

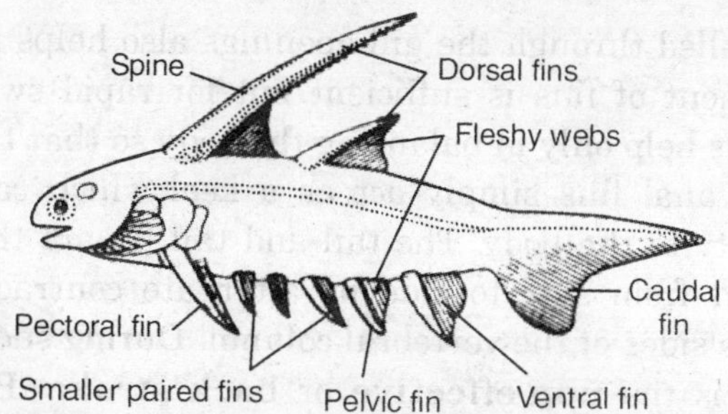

Fig. 13.9 Origin of fin according to fin spines theory

3. Fin Spines theory: This is a recent hypothesis which suggests that primitive fin consisted of metamerically arranged spines with fleshy lobes as in some extinct acanthodian fishes (Fig. 13.9). In the course of development all the spines and lobes have been lost except the anterior pair (pectorals) and the posterior pair (pelvics).

4. Ostracoderm theory: Some ostracoderms had lateral fleshy lobes on either side in the anterior region and these have been regarded as precursors of pectoral fin. Some other ostracoderms had spiny fins on either ventro-lateral sides like the acanthodians. Of these spiny fins some were retained in the anterior (pectoral) and posterior (pelvic) regions and others were lost. From this it was concluded that the paired fins originated from ostracoderm ancestors.

5. Cyclostome theory: In cyclostomes paired appendages are absent and so this goes against the view that ancestral craniates had continuous fin folds. But since cyclostomes are probably degenerate descendants of ostracoderms, it appears that they lost their paired fins which were probably present in ostracoderms.

LOCOMOTION

Fishes are adapted for movement in water. They possess a perfectly stream-lined body which offers least resistance in swimming in water. The fish swims mainly by two methods:

1. Body movement brought about by alternate contraction and expansion of the myotomes,
2. Movement of fins.

Fig. 13.10 Diagram showing use of tail during swimming. a - b is the axis of fish at rest

Water that is expelled through the gill openings also helps in forward progression. For slow swimming movement of fins is sufficient but for rapid swimming body movement is more important, the fins help only in balancing the body so that the fish may not turn upside down. The dorsal and anal fins simply act as a keel which can be lowered or raised as required to give stability to the body. The tail and tail fin are the chief propelling organ in swimming. It is lashed from side to side by alternate contraction and relaxation of the muscles on the opposite sides of the vertebral column. During such movement the tail is first bent to one side, this is the **non-effective** or **back stroke**. By a stroke in the opposite direction the tail is straightened, this is the **effective** or **forward stroke**. Thus by alternate strokes in the left and right sides the fish forces its way through water (Fig. 13.10).

Types of locomotion

Three types of locomotion have been described in different species of fishes:

1. Crangiform
2. Anguilliform
3. Ostraciform

Crangiform: This is the most common type of locomotion in fishes. In this type of locomotion when the fish wants to move forward, the first few myomeres of the anterior end of the body on one side contract so that the head bends to one side. This is followed by a series of alternate contraction and relaxation of successive myomeres, first on one side of the body and then on the other side from the head to the tail end. Thus the body of the fish is thrown into a S-shaped curve during swimming. The forward thrust is due to the pressure of the tail against the surrounding water.

Anguilliform: This is characteristic of eel-like fishes (*Anguilla*) which have long laterally compressed body. They swim actively by undulating movement of the body. The movement is serpentine and the fish looks like a creeping snake. This movement is also initiated by the contraction of first few myomeres on one side so that the anterior part of the body is thrown into a curve which passes backwards in the form of a wave set up by alternate contraction and relaxation of muscle segments on each side of the body. The forward thrust is due to pressure of the body against water. The caudal fin has no role in movement so it is very much reduced or absent.

Ostraciform: This type of locomotion is characteristic of the trunk fish *Ostracion*. The body is enclosed in a hard and rigid bony case and the tail with fan-like tail fin projects behind. In this fish undulating body movements are not possible, the dorsal and anal fins serve as the chief propelling organ for slow movement. Some other fishes like the globe fish (*Tetrodon*), porcupine fish (*Diodon*) and sea-horse (*Hippocampus*) also swim slowly mainly by the propelling action of their dorsal and anal fins.

Other methods of locomotion are jumping, gliding or flying, burrowing, etc. Some fishes like *Salmon*, *Mugil*, *Cyprinus*, etc. jump out of water with a great force either to escape enemy or to avoid an obstacle. The flying fish *Exocoetus* can glide for some distance outside water with the help of its enlarged pectoral fin. In *Cypselurus* both the pectoral and pelvic fins are enlarged for gliding in air. Some fishes burrow in the mud with the help of their swimming movement.

AIR BLADDER

The **air bladder** or **swim bladder** or **gas bladder** is more or less sac-like structure which lies dorsal to the alimentary canal beneath the vertebral column and kidneys but outside the coelom. It arises as an outgrowth from the dorsal or lateral walls of the foregut and opens into the oesophagus by a **pneumatic duct** which is short and wide in lower teleosts and long and narrow in others. In specialised teleosts it may be lost during the course of development. Air bladder is present in almost all bony fishes except a few bottom dwellers in which it has been lost. It is absent in elasmobranchs. The air bladder is filled with a mixture of oxygen, carbon dioxide and nitrogen.

The air bladder varies greatly in structure, position, mode of development and function in different groups of fishes.

Chondrostei: *Polypterus* is one of the most primitive bony fish and it has the most primitive type of air bladder. It is in the form of a bilobed sac with smooth walls. Of the two lobes the left one is smaller and the right one is larger. It opens ventrally into the oesophagus by a single duct (Fig. 13.11A).

Dipnoi or lung fishes: The air bladder is similar in structure to the amphibian lungs. In *Protopterus* and *Lepidosiren* (Fig. 13.11 B) the air bladder is bilobed but the two lobes are equal in size. The walls are thick, highly vascular and alveolar. They open ventrally into the oesophagus by a single duct. In *Neoceratodus* (Fig. 13.11 C) the air bladder is single lobed lying dorsal to the gut but opening ventrally into the oesophagus by a narrow pneumatic duct. The walls are highly vascular and alveolar.

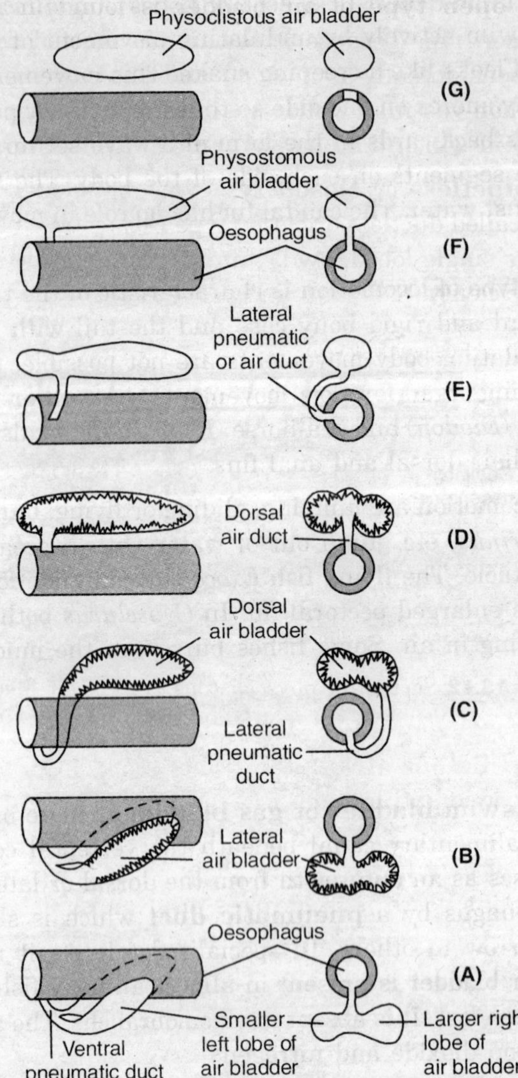

Fig. 13.11 Different types of air bladder in fishes. Left row in L.S. and right row in T.S. **(A)** *Polypterus* and *Clamoichlthyes*; **(B)** *Protopterus* and *Lepidosiren*; **(C)** *Neoceratodus*; **(D)** *Amia* and *Lepidosteus*; **(E)** *Erythrynus*; **(F)** Generalized teleost and *Acipenser*; **(G)** Specialized teleost

Holostei: In *Amia* and *Lepidosteus* (Fig. 13.11 D) the air bladder is single lobed, lies dorsal to the gut and opens dorsally into the oesophagus. The walls are thick, vascular and alveolar. In *Amia* the air bladder is very large and the alveoli are small but more numerous than in *Lepidosteus*. They serve as accessory respiratory organ as well as hydrostatic organ.

Teleostei: The air bladder in higher bony fishes is concerned mainly with hydrostasis and play little or no role in respiration. It is more specialized and is of two types:

(i) Physostomous

(ii) Physoclistous

Physostomous or **open type** of air bladder is found in more generalized teleost (Fig. 13.11 F). It is of open type because it remains connected with the oesophagus by a pneumatic duct which opens dorsally as in *Amia* and *Lepidosteus*. The air bladder is single lobed with smooth and non-vascular walls. An intermediate stage is found in the teleost *Erythrynus* (Fig. 13.11E) in which the pneumatic duct opens laterally into the oesophagus.

Physoclistous or **ductless** or **closed type** of air bladder is characteristic of more specialized teleosts. It is called ductless or closed type because the pneumatic duct is lost (Fig. 13.12). The air bladder is single lobed, oval, thin walled, gas filled sac extending along the whole length of the body cavity.

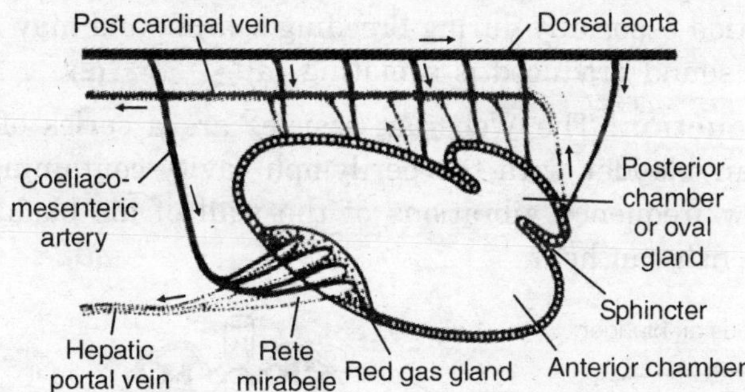

Fig. 13.12 Section of physoclistous type of air bladder showing its structure and blood supply

Structure of physoclistous air bladder: The air bladder is incompletely divided into two chambers by a constriction with a sphincter. The **anterior chamber** is large and the **posterior chamber** (Fig. 13.12) is small and is called **oval gland** (Fig. 13.12). The wall of air bladder is made up of two layers—outer layer of connective tissue is called **tunica externa** and inner layer of smooth muscle fibres and epithelial gas gland is called **tunica interna**. Underlying each gas gland is a network of fine capillaries called **rete mirabele** (plural **retia mirabilia**). The gas gland and rete mirabele together are called **red body** or **red gland** because of its red colour. It is present in the wall of the anterior chamber. The red gland receives blood from coeliaco mesenteric artery and returns blood to the hepatic portal vein. The gas is secreted into the bladder from the blood by the red body. The posterior chamber is thin walled and allows resorption of gases by the blood. It can close and open by the sphincter. When the sphincter is open the oval gland absorbs gas from the air bladder. Secretion and resorption of gases is under the control of the autonomic nervous system.

Functions of air bladder

The air bladder performs a number of important functions.

1. Hydrostasis: The air bladder functions as an important hydrostatic organ in majority of the teleosts. The density of fish flesh is more than that of water surrounding it and the gas in the air bladder makes it equal to that of water. The fish becomes weightless and is kept in equilibrium so that its position is maintained without much effort. Secretion of gas lowers the specific gravity, so the fish rises and resorption of gas increases the specific gravity, so the fish sinks. Thus the fish is able to maintain itself in equilibrium.

2. Respiration: In lower teleosts such as ganoid fishes and lung fishes the air bladder functions as lungs. The fish frequently comes to the surface of water to gulp in air and pass it back to the air bladder for respiration. In physostomous fishes the air bladder functions as accessory respiratory organ. In physoclistous fishes the air bladder can be used for storing oxygen to be used in emergency.

3. Sound production: Air bladder plays an important role in sound production. In some species the wall of air bladder vibrates due to elastic apparatus attached to it, thus producing sound. Sometimes vibrations in the wall of air bladder are produced by the action of special muscles attached directly to the wall. In *Polypterus*, *Lepidosiren*, etc. sound is produced by forceful expulsion of gases from the air bladder. The sound produced in fishes helps in communication especially during breeding season, and may also be used for offense. In the toad fish the sound produced is a mating call.

4. Auditory function: The Weberian ossicles are a series of small bony structures which connect the air bladder with the perilymph cavity containing the internal ear. The ossicles transmit low frequency vibrations of the wall of air bladder to the membranous labyrinth which the fish can hear.

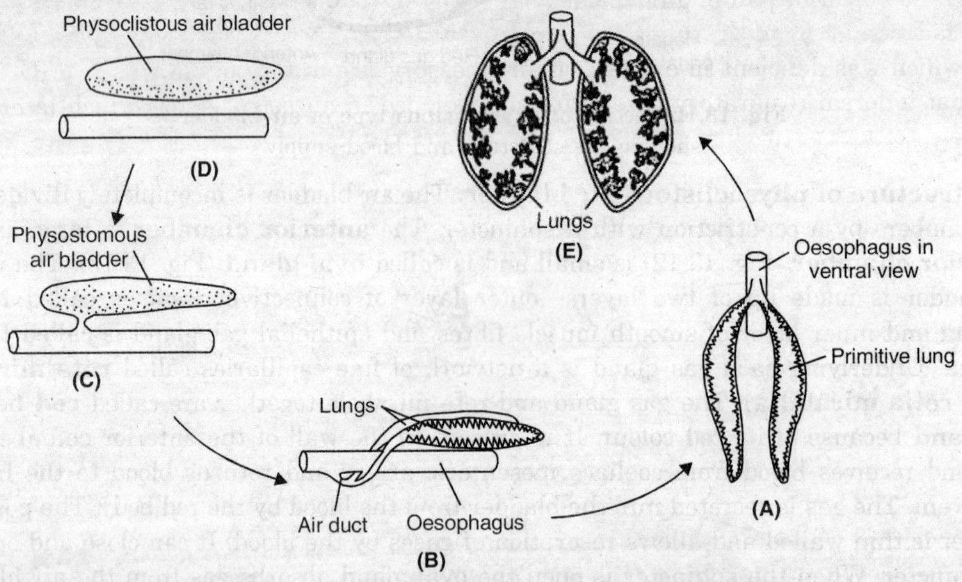

Fig. 13.13 Origin of air bladder and lungs from primitive condition; **(A)** Primitive or generalized lungs (lung fishes); **(B)** Transitional lungs (*polypterus*); **(C)** Physostomous air bladdder (some Primitive tebeosts); **(D)** Physoclistous air bladder (specialized teteosts); **(E)** Lungs of land vertebrates

ORIGIN OF AIR BLADDER AND ITS RELATIONSHIP WITH TETRAPOD LUNGS

The air bladder is generally considered homologous with the lungs of tetrapods in origin, structure and development. This suggests a positive relationship between the two. Probably both arose as pharyngeal pouches. In fact, the air bladder was regarded as the precursor of lungs of tetrapods, but this view was not supported by any evidence. The present view is that lungs are more primitive than air bladder because they are present in primitive teleost (dipnoi and ganoids) and air bladder is the characteristic feature of higher teleosts. It is regarded as a specialized modification of lungs.

From the generalized or primitive lungs three lines emerged (Fig. 13.13).

(i) One line led to the air bladder or swim bladder of most modern teleosts. The connection with the gut shifted from ventrial to dorsal. The intermediate stages like attachment to the gut laterally as in *Erythrynus* support this view.

(ii) Second line without any change in position led to the lungs of land vertebrates. The lining of the lungs developed internal foldings to increase the surface area for respiration.

(iii) They were lost in elasmobrans and same teleosts.

There are differences between the two structures. The air bladder arises from the dorsal wall of the foregut while the lungs arise from the ventral wall of the pharynx. The blood supply of the two is also different. The lungs function as respiratory organs whereas the chief function of air bladder is hydrostasis. Goodrich (1930) is of the view that both these structures develop from posterior pair of gill pouches but Jones (1957) is of the view that the air bladder evolved independently of the lungs. The early evolution of Actinopterygii occurred in swampy waters which was deficient in oxygen, so an accessory respiratory organ was required. It is likely that a dorsal respiratory diverticulum developed in ancestral fishes which later gave rise to air bladder of modern teleosts.

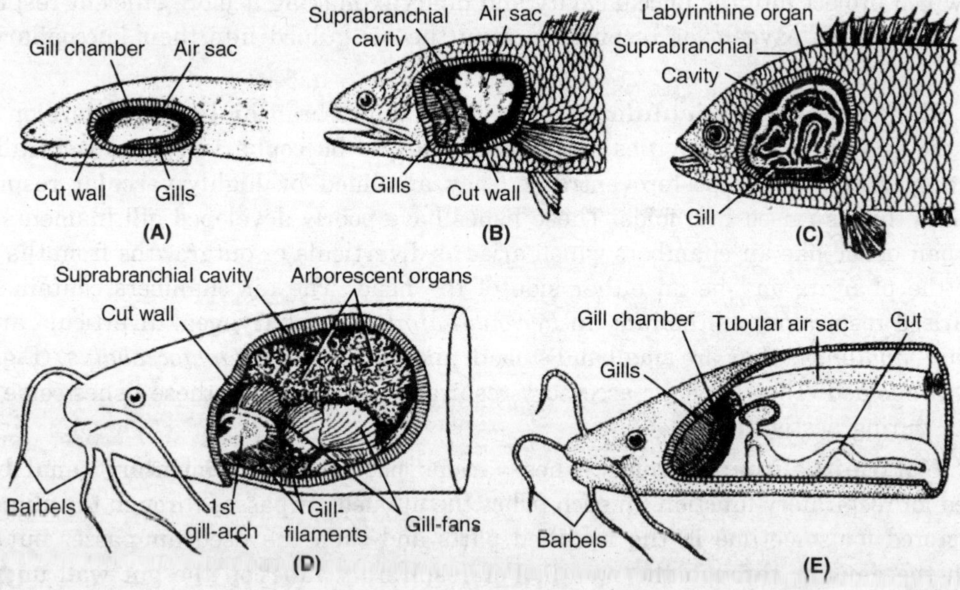

Fig. 13.14 Accessory respiratory organs of some teleosts. **(A)** *Amphipnous*; **(B)** *Channa*; **(C)** *Anabas*; **(D)** *Clarias;* **(E)** *Heteropneustes*

ACCESSORY RESPIRATORY ORGANS

Accessory respiratory organs are special respiratory structures present in addition to gills in some tropical fresh water fishes. They generally develop in response to certain environmental conditions and enable the fish to tolerate oxygen deficiency or to sustain life out of water for short periods. They are rarely present in marine fishes (have been reported in the dogfish *Chiloscyllium*). Accessory respiratory organs are generally present in fishes living in shallow stagnant fresh water or in hill stream fishes, as the streams dry up in summers.

Several types of accessory respiratory organs have developed in different species of fishes.

1. Skin: The skin of eels like *Anguilla*, *Amphipnous cuchia* is moist and highly vascular and serves as an important organ for respiration when they habitually leave water and migrate from one place to another through damp vegetation. In the mudskippers *Periophthalmus* the larvae and embryos breathe through thin, moist and permeable skin before the emergence of gills. In many fishes the median fin folds are vascular and help in cutaneous respiration. In sturgeons and some catfishes the opercular folds are highly vascular and serve as accessory respiratory organ. In *Mastacembelus* the unmodified gills secrete a large amount of mucous which keeps them moist and help in gaseous diffusion for sometime when the fish is out of water.

2. Buccopharyngeal epithelium: In some fishes like *Periophthalmus*, *Boleophthalmus*, *Amphipnous* (Fig. 13.14), *Electrophorus*, etc., the epithelial lining of the buccopharyngeal cavity is richly supplied with blood vessels. It may remain simple or may develop folds or pleats which project into the buccal cavity and pharynx making it more efficient respiratory organ. They obtain oxygen for respiration from the air gulped into their buccopharyngeal cavity.

3. Pharyngeal diverticulum: In some fishes supra-branchial cavities develop in the roof of the pharynx. These cavities open freely into the buccopharynx anteroventrally and into the opercular cavity posteroventrally. They are lined by highly vascular respiratory epithelium that is raised into folds. These fishes have poorly developed gill filaments. They have a pair of sac-like air chambers which arise as diverticula or outgrowths from the dorsal wall of the pharynx and lie on either side of the head. The air chambers contain highly vascularised respiratory epithelium. In *Periophthalmus* the pharyngeal diverticula are very small, in *Amphipnous* they are small and smooth and in *Channa* (= *Ophiocephalus*) (Fig. 13.13 B) they are folded. They serve as accessory respiratory organ when these fishes come out of water or during aestivation.

4. Gut Epithelium: In some fishes various parts of the alimentary canal become modified for respiratory function. In such fishes the inhaled air passes through the alimentary canal, stored for sometime in the modified parts and then the used air passes out either through the anus or through the mouth. For respiratory function the gut wall undergoes certain modifications:

(i) the wall of modified parts becomes thin and transparent due to great reduction of muscles layers,

(ii) the gland cells are absent.

(iii) the epithelial lining becomes highly vascularized.

Intestinal respiration is found in *Lepidocephalus guntea* and *Gobitus*, rectum is modified in *Callichthyes* and stomach in *Misgurus fossilis*.

5. Pelvic fins: In the American lungfish *Lepidosiren*, during the breeding season the pelvic fins of male become enlarged and covered with highly vascular filaments. It is believed that these filaments act as accessory respiratory organ and enable the male to guard the developing eggs and larvae without going up for aerial respiration.

6. Opercular chamber: In *Periophthalmus* and *Boleophthalmus* the opercular chamber becomes modified for aerial respiration. The membrane lining the chamber becomes thin and vascular for exchange of gases. The inhaled air passes through the gill slits into the opercular chamber where it is stored for some time. The opercular chamber bulges out in the form of two little balloons in the hinder end of the head and after sometime when the air passes out through the small external branchial aperture, they collapse.

7. Branchial diverticula: In some specialised air breathing fishes, complex accessory respiratory organs are present in the branchial or opercular chamber. They are in the form of sac-like diverticula called **opercular lungs**, which develop from the dorsal surface of the branchial chamber and lie above the gill. They may contain specialized structures like labyrinthine organ or rosettes to increase the respiratory surface. Some important examples are:

(i) ***Heteropneustes* (= *Saccobranchus*):** In this species four pairs of gill are present but the gill lamellae are reduced in size, four pairs of fans develop on these gill arches. Besides these fans a pair of air sacs (Fig. 13.14E) extend posteriorly from the suprabranchial chamber upto the middle of the caudal region. They are thin walled, long, tubular highly vascular structures lying embedded between the myotomes of the body. The air sacs are modified derivative of the basic gill structure. They receive blood from the fourth afferent branchial vessel of its side. The respiratory lining of air sacs is thrown into folds and ridges and is composed of both vascular and non-vascular areas. The air enters and leaves the air sacs through the pharynx.

(ii) ***Anabas*:** In *Anabas* the air breathing organ consists of a pair of spacious suprabranchial chambers on either side of the skull lying between the first gill arch and the hyomandibular (Fig. 13.14 C). It communicates freely with the buccopharyngeal cavity by an inhalant aperture and with the exterior through the opercular cavity by means of an exhalant aperture. Each chamber contains a special **labyrinthine organ** which develops from the epibranchial segment of the first gill arch and consists of three concentrically arranged bony plates, the margins of which are wavy. The whole labyrinthine organ is a complex structure and is covered by a highly vascular mucous membrane. Thus it serves to increase the area for absorption of oxygen. The air is drawn is through the mouth into the suprabranchial chamber and expelled out through the opercular opening.

(iii) ***Trichogaster*:** The accessory respiratory organ of *Trichogaster* is similar to *Anabas* but simpler in structure. It consists of suprabranchial chamber, labyrinthine organ and respiratory membrane. The air chamber is in the form of a spiral organ consisting of two leaf-like expansions and is composed of loose connective tissue covered by a vascular epithelium. The respiratory membrane lining the air chambers and covering the labyrinthine organ consists of vascular and non-vascular areas.

(iv) ***Clarias*:** The accessory respiratory organ of *Clarias* consists of a pair of suprabranchial chamber, two beautiful rosettes or air trees, four fans and respiratory membrane (Fig. 13.14 D). The suprabranchial chamber lies above the gills. Each chamber contains a beautiful **air tree** or **rosette** called **dendritic** or **arborescent organ**. They are supported by the epibranchials of the second and fourth branchial arches. Each air tree is a highly branched structure supported by cartilaginous internal skeleton and covered by vascular epithelium. Some of the primary gill lamellae of each gill arch are fused to form a fan or gill plate. There are four fans on either side and each consists of vascular and non-vascular areas. The respiratory membrane lining the suprabranchial chamber also consists of vascular and non-vascular areas. Well developed inhalent and exhalent apertures are present in the suprabranchial chamber. The fish periodically comes to the surface of water and gulps in air.

8. Air bladder: In some fishes like *Amia*, *Polypterus*, *Lepidosteus* and in lung fishes the air bladder is most highly evolved and acts as a **lung**. It is truly an accessory respiratory organ. Its walls are highly vascular and alveolar. The gills in these fishes are poorly developed. They utilise atmospheric oxygen for respiration. In *Notopterus* the air bladder has a wide pneumatic duct and is covered by a single layer of epithelium with a network of blood capillaries. It helps in diffusion of gases between the blood and the air contained in the air bladder.

HILL STREAM FISHES

The hill stream fishes can be divided into two groups:

(i) **Temporary inhabitants**

(ii) **Permanent inhabitants**

Temporary inhabitants are those fishes which migrate to hill streams for some specific reason such as spawning. So they do not possess any special modification, for example, *Salmon* and *Hilsa*.

Permanent inhabitants are those fishes which live there permanently, so they have to face the prevailing conditions of the hill streams. These fishes show several important modifications. Some important hill stream fishes are *Garra*, *Balitora*, *Barilius*, *Schizothorax*, *Nemacheilus*, *Barbus*, *Glyptothorax*, *Glyptosternum*, *Pseudoecheneis*.

Conditions in the hill streams

1. **Strength of water current:** The rate of flow of water in the hill streams is very high. It may vary with seasons. In order to withstand the rapid flow of water the fishes of the hill streams develop adhesive organs.

2. **Food:** The food available in the hill streams is sufficient but in the form of algae covering rocks and stones. Other types of vegetation cannot grow in the hill streams because of strong water current. The mouth and jaws of hill stream fishes are modified for scraping algal material.

3. **Depth:** The hill streams are shallow and the water is very clear. So the hill stream fishes have to tolerate strong light during the day.

4. Oxygen in water: Due to rapid flow the oxygen content of water is very high.

5. Rocks and stones: The substratum of hill streams is made up of rocks and stones which provide shelter to small fishes.

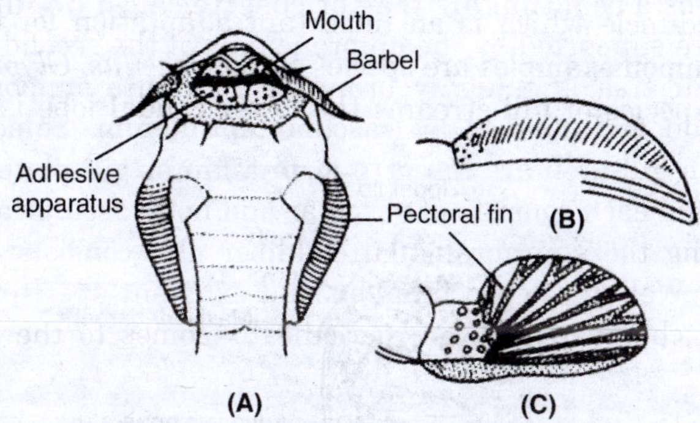

Fig. 13.15 **(A)** *Glyptosternum* showing structure and form of adhesive apparatus; **(B)** Under-surface of pectoral spine, highly magnified; **(C)** Pectoral fin showing reflected inner rays

Adaptations of hill stream fishes

1. The external body form: The body is greatly flattened and may become leaf-like in some highly specialised species like *Balitora*, *Glyptothorax*, *Glyptosternum* and *Pseudoecheneis*. In most of the species the ventral surface is straight and horizontal and the dorsal surface is slightly arched. The head is usually small and semi-circular (Fig. 13.15 A). The fishes are generally small in size so that they can easily find shelter under rocks and stones.

2. Scales: The scales are generally small in size and partly or completely embedded in the skin as in *Schizothorax* and *Nemacheilus*. In some species they are totally absent except in the anal and scapular region. In some species they are present on the dorsal and lateral surfaces and reduced or absent on the ventral surface particularly in the region of the chest which is modified for adhesion to rocks and stones.

3. Paired fins: The paired fins are modified for adhesion. They are shifted outwards and come to lie horizontally along the sides of the body. The outer rays become thick and flat and serve for adhesion. The inner rays increase in number, are directed upwards and are kept in motion (Fig. 13.15 B and C). They also help in respiration. Some special muscles develop in pelvic fins which keep them pressed against the rocks in resting position.

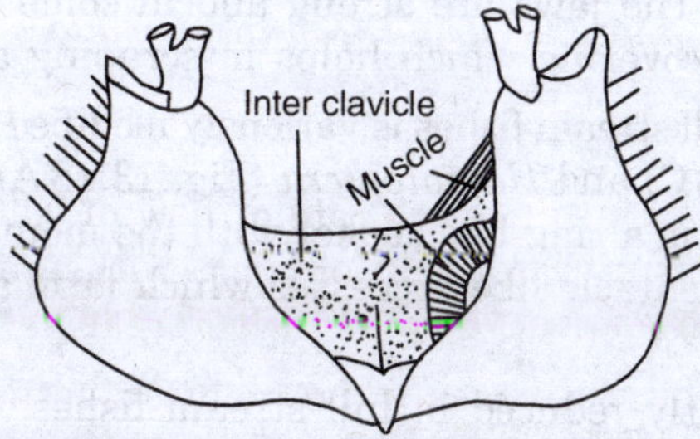

Fig. 13.16 Dorsal view of pectoral girdle of *Glyptothorax*

4. Pectoral and pelvic girdles: The pectoral and pelvic girdles are modified particularly in those species in which fins are used for adhesion. Their shape is modified due to horizontal position of the fins. Keel-like ridges develop on the ventral surface of the inter-clavicular bone for attachment of muscles (Fig. 15.16) which are better developed. In some species like

Glyptothorax and *Pseudoecheneis* the bones of the pectoral girdle are fused to provide more strength.

5. Caudal fin and its peduncle: Most of the hill stream fishes have a long, narrow band-shaped caudal peduncle which is an important adaptation for high altitude and fast flowing water. Some common examples are species of *Nemecheilus*, *Glyptothorax*, *Homaloptera*, *Balitora*, etc. In most species of hill streams the lower caudal lobe is longer and the upper lobe is smaller.

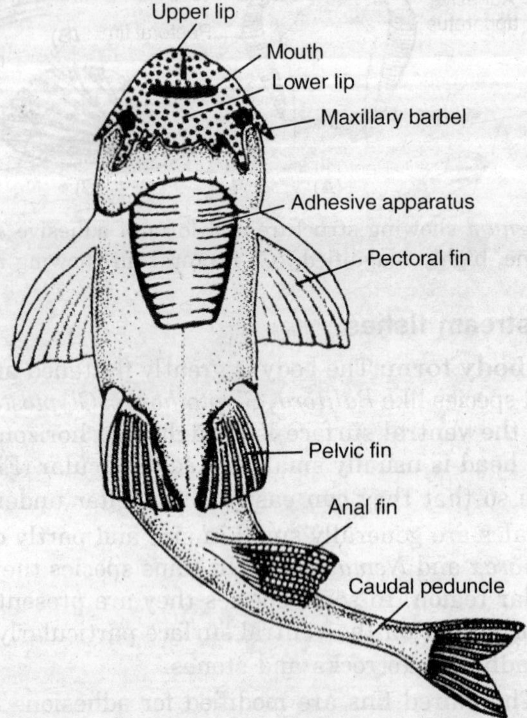

Fig. 13.17 *Pseudoecheneis* showing horizontal paired fins, ventral mouth and fleshy lips forming adhesive sucker and adhesive apparatus

6. Mouth, lips and barbels: The mouth is shifted to the ventral side and is crescentic or semi-circular in shape. The jaws are strong and in some species like *Schizothorax*, they are covered with a horny covering which helps in scraping algal material for feeding.

The lips of different hill stream fishes is variously modified to form a sucker for attachment. In *Nemacheilus* (Fig. 13.18 C) and *Hamaloptera* (Fig. 13.18 A) the lips are thick, swollen and divided in the middle to form a ring-like sucker with the mouth. In *Balitora* the lips are thick and cut into a number of tentacle-like processes which form an effective sucker when pulled outwards.

The barbels are greatly reduced in hill stream fishes. They are short and stumpy in *Balitora* and short, thick and fleshy in *Glyptothorax* and *Pseudoecheneis* (Fig. 13.17).

7. Eyes: The eyes are small and lie close to each other on the upper surface.

8. Gill openings: The gill slits are restricted to the sides of the body as the ventral surface is modified for adhesion. They extend upto the base of the pectoral fin. The restriction of gill slits to the sides may affect respiration but certain factors help in respiration:

(i) Water in hill streams is rich in oxygen.

(ii) Since the gill slits are small the fish is able to retain water in the branchial chamber for a longer time.

(iii) The inner rays of the pectoral girdle are in constant motion, thus they help in respiration. They are richly supplied with blood capillaries for gaseous exchange.

9. Air bladder: The air bladder in hill stream fishes is greatly reduced or absent because it is of no use as a hydrostatic organ. It is not required as accessory respiratory organ because water is well aerated.

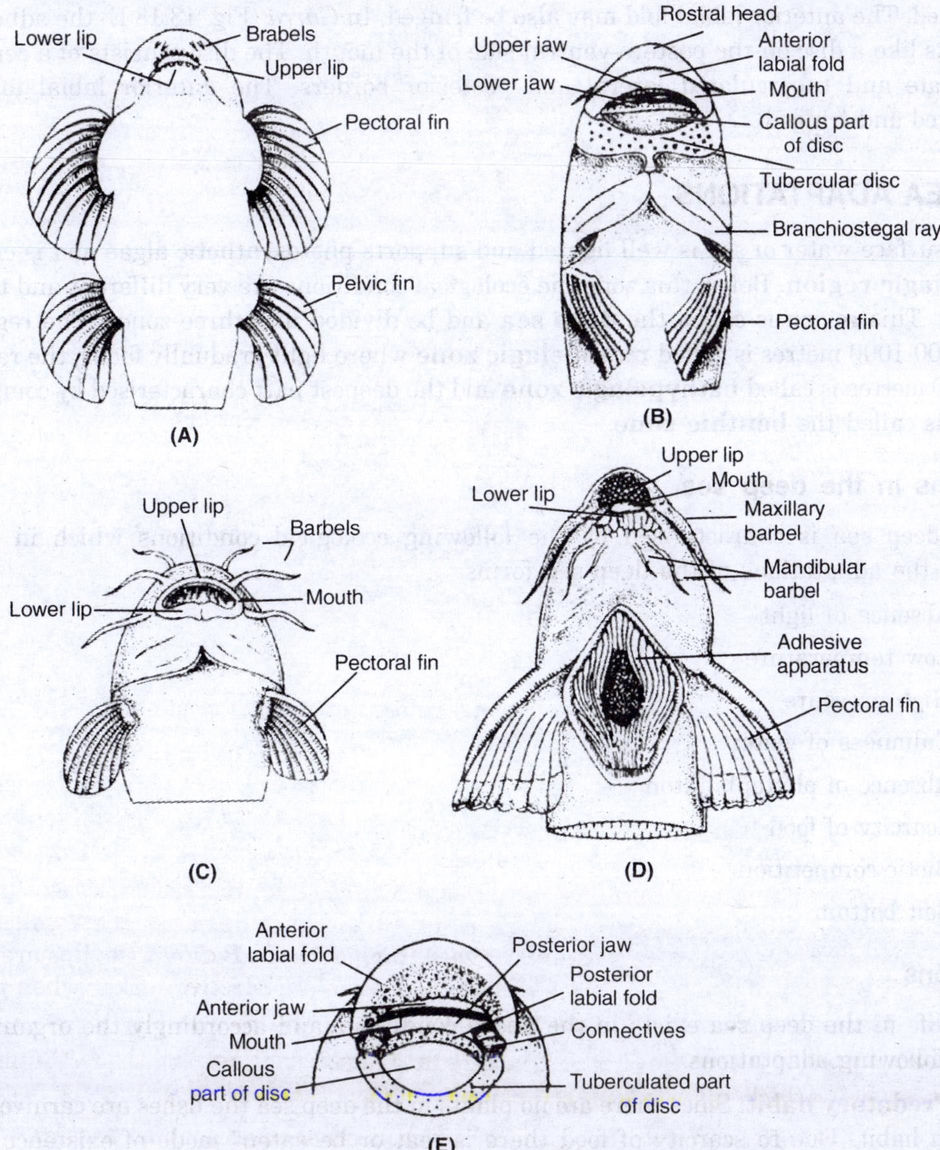

Fig. 13.18 *Adhesive* apparatus of **(A)** *Homaloptera;* **(B)** *Schizothorax;* **(C)** *Nemacheilus;* **(D)** *Glyptothorax;* **(E)** *Garra*

10. Adhesive organ: The skin between the pectoral fins and surrounding the mouth is variously modified in hill stream fishes to form an **organ of adhesion.** In *Glyptosternum,* (Fig. 13.15 A) *Gylyptothorax* (Fig. 13.18 D) and *Pseudoecheneis* (Fig. 13.17) the skin of the ventral side of the thorax between the pectoral fins is thrown into grooves and ridges. Such striated structures can also be seen on the barbels, lateral and ventral sides of the mouth, on the chest and undersurface of the pectoral and pelvic fins. In *Nemacheilus* (Fig. 13.18 C) the mouth aperture is surrounded by thick horny lips which function as a sucker. The base of the pectoral fins is thickend and cushion-like. In *Schizothorax* the jaws are covered by a hard callous plate for scrapping food. Posterior to the plate the skin is thickened and tuberculated. The anterior labial fold may also be fringed. In *Garra* (Fig. 13.18 E) the adhesive organ looks like a disc on the postero-ventral side of the mouth. The disc consists of a central callous plate and tuberculated lateral and posterior borders. The anterior labial fold is tuberculated and fringed.

DEEP SEA ADAPTATIONS

The surface water of sea is well lighted and supports photosynthetic algae and is called the **epipelagic region**. Below this zone the ecological conditions are very different and there is no light. This region is called the **deep sea** and be divided into three zones. The regions between 200-1000 metres is called **mesopelagic zone** where light gradually fades, the region below 1000 metres is called **bathypelagic zone** and the deepest part characterised by complete darkness is called the **benthic zone**.

Conditions in the deep sea

The deep sea is characterised by the following ecological conditions which in turn determine the adaptations of the deep sea forms:

1. Absence of light
2. Low temperature
3. High pressure
4. Calmness of water
5. Absence of phytoplankton
6. Scarcity of food
7. Biotic competition
8. Soft bottom

Adaptations

The life in the deep sea exists in the above conditions and accordingly the organisms show the following adaptations:

1. **Predatory habit:** Since there are no plants in the deep sea the fishes are carnivorous in habit. Due to scarcity of food there is "eat or be eaten" mode of existence.
2. **Large mouth and distensible stomach:** The deep sea fishes have large mouth which enables them to swallow prey larger than themselves (e.g., gulpers, swallowers

and viper fish). Some deep sea fishes have large distensible stomach, (e.g., *Chiasmodon niger* and *Saccopharynx ampullaceus*) and in some dentition is well developed (e.g., *Photostomias* and *Linophryne*).

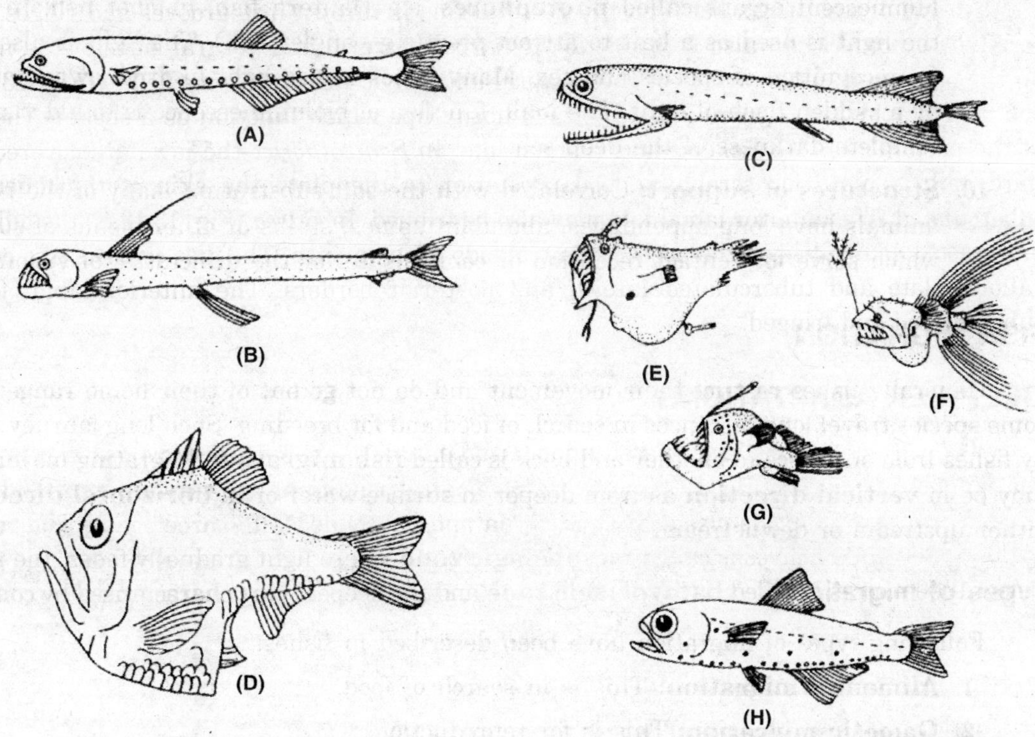

Fig. 13.19 Some deep sea fishes. Mesopelagic fishes. **(A)** Bristle mouth (*Cyclothone*); **(B)** Wiper fish (*Choliordius*); **(C)** Loose jaw (*Malacosteus*); **(D)** Hatchet fish (*Argyropelecus*); **(E)** Linophryne; **(F)** Black devil (*Melanocetis*); **(G)** Fan-fin (*Caulophryne*) anger fisher; **(H)** Lantern fish (*Mycelophium*)

3. **Scavengerous feeding habit:** Most of the deep sea fishes feed on the body of surface forms that sink to the bottom, some even feed on the excreta of the surface animals. As temperature is low in the deep sea, decomposition of bodies falling from above is slowed down so the deep sea fishes enjoy almost fresh diet.
4. **Size of the body:** The body size of many deep sea fishes is small, although gigantic forms are also known. Due to high pressure the body is very much compressed.
5. **Weak endoskeleton:** The skeleton of deep sea fishes is poorly calcified so it is weak, light and flexible. This is caused by low temperature prevailing in the deep sea.
6. **Eyes:** Some deep fishes exhibit greatly enlarged eyes called **telescopic** eyes which are highly effective in visioning lights of very low intensity. Some have very small eyes which are of no use and still others have lost their eyes completely. In those fishes which have small eyes or lost their eyes have well developed sensory organs like antennae. The lateral line system is also well developed.
7. **Pressure sense:** Deep sea fishes have less developed pressure sense. Turbulence in water created by nearby animals can be recognised and depending upon the nature of turbulence, they react by flight or by approach.

8. **Colouration:** The deep sea fishes are brightly coloured and red colour is dominant, other colours are violet, black or brown.
9. **Bioluminescence:** Many deep sea fishes produce their own light by means of luminescent organs called **photophores**, e.g., lantern fish, hatchet fish. In some the light is used as a bait to attract prey (e.g., angler fish). The light is also used for recognition of species and sex. Many fishes utilise light to drive away enemies by a sudden flash of light. The main function of bioluminescence is to aid vision in complete darkness of the deep sea.
10. **Structures of support:** Correlated with the soft substratum many of the bottom animals have long appendages, abundant spines, stalks or other means of support which serve for contact reception or compensate for the difficulties of vision.

FISH MIGRATION

Generally fishes restrict their movement and do not go out of their home ranges. But some species travel long distances in search of food and for breeding. Such long journey taken by fishes from one place to another and back is called **fish migration**. Migrating movements may be in **vertical direction** as from deeper to surface water or in **horizontal direction**, either upstream or downstream.

Types of migration

Following types of migration have been described in fishes:

1. **Alimental migration:** This is in search of food.
2. **Gametic migration:** This is for reproduction.
3. **Climatic migration:** This is to secure more suitable climatic conditions.
4. **Osmoregulatory migration:** This is to control its internal osmotic pressure.

Methods of migration

A fish can make migratory movements by the following methods:

1. **By drifting:** Slow migratory movements in which the fishes are carried passively by water current. It may be directional if all water is moving in one direction.
2. **Random locomotory movement:** These movements are random in direction and may lead to either uniform distribution or aggregation.
3. **Oriented swimming movements:** Fishes swim in a particular direction:
   (i) either towards or away from the source of stimulation, or,
   (ii) at some angle between them and the source of stimulation.

Periodicity of migration

The fishes show great variation in the periodicity of their migration. It varies between different species and even within a species and appears to be due to many biotic and abiotic factors. In most fishes migrations occur at regular intervals and may be daily, monthly, seasonally, yearly, biannually or even longer. The Pacific salmon, for example, may remain in the sea for many years before migrating to the breeding ground. Similarly eels spend several years in fresh water before migrating to sea for breeding.

Patterns of migration

Patterns of migration vary in different species and within the same species. Myers (1949) has described the following patterns of fish migration:

1. **Diadromous**—it is of three types:
   (a) **Anadromous**
   (b) **Catadromous**
   (c) **Amphidromous**
2. **Potamodromous**
3. **Oceanodromous**

1. **Diadromous:** These fishes are truly migratory because they migrate between sea and fresh water.

   (a) Andromous: Fishes which spend major part of their life in sea and migrate to fresh water during the breeding season for spawning are called anadromous fishes, for example Atlantic salmon (*Salmo salar*), Pacific salmon (*Oncorhynchus*) and *Hilsa ilisha*. They travel thousands of kilometre in sea and hundreds of kilometre in rivers to reach their spawning site. After egg laying they return to their feeding place in the sea. Some species die after spawning.

   Salmon. They breed in winter. With the approach of winter both the sexes migrate to fresh water mountain streams reaching the place where they grew up some years back. They stop feeding and their colour changes to dull reddish brown from silver. They make a shallow saucer-shaped nest in the bottom gravel in which the female lays eggs. The Atlantic salmon may breed two or three times in its life. After hatching the larval fishes grow for some time in fresh water and then migrate to the sea where they grow faster as food is abundant.

   (b) Catadromous: Fishes which spend major part of their life in fresh water but migrate to sea for egg laying are called catadromous fishes, for example, the fresh water eel, *Anguilla*. It tarvels thousands of kilometre in rivers to reach their spawning ground in the sea. The adults die after laying eggs.

   *Anguilla***.** The breeding season is autumn. With the beginning of autumn season the colour of the fish changes from yellow to metallic silver. They stop feeding, the alimentary canal shrinks, the gonads enlarge and they migrate to sea. Egg laying takes place in deep waters. The adults die after egg laying. The eggs hatch into small (about 6 mm long) transparent pelagic larvae (Fig. 13.20 A) called **leptocephalia** (singular leptocephalus = small head). They have sharp, needle-like teeth for feeding. They grow into **elevers** or **glass eels** (Fig. 13.20B) during their long return journey. The elevers are about 8 cm. long with cylindrical body. They reach fresh water and feed and grow to become adult yellow eels in some years (Fig. 13.20 C).

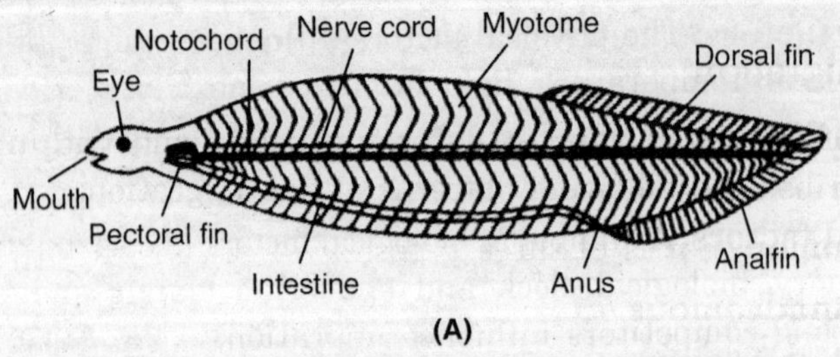

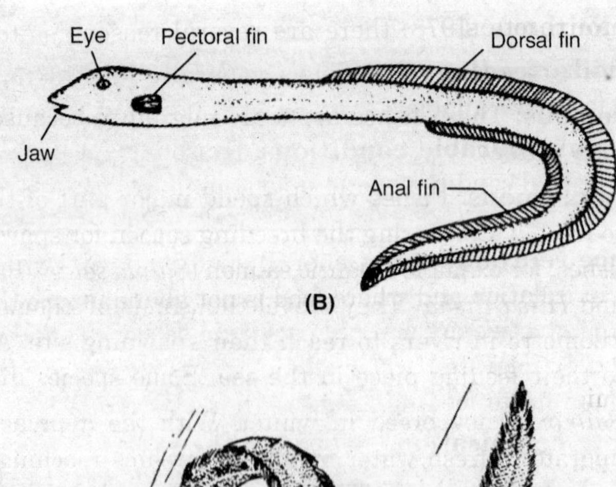

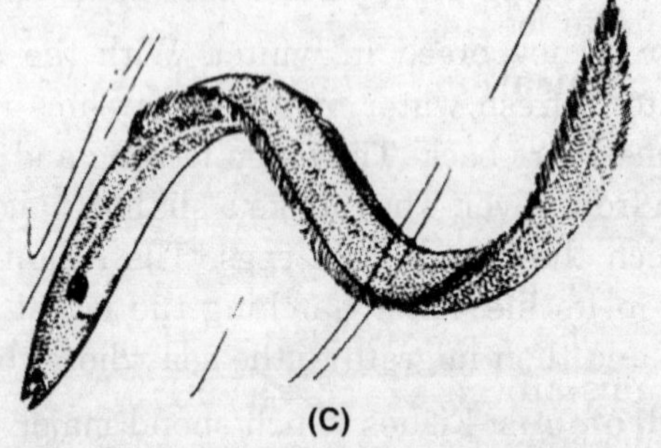

Fig. 13.20 Migration in Anguilla. (A) Leptocephalus; (B) Elever; (C) Adult eel

(c) **Amphidromous:** Fishes in which migration from fresh water to sea or vice versa, not for breeding, but occurs regularly at some stage in the life cycle, are called amphidromous fishes, for example, migration of some gobies.

2. **Potamodromous:** Truly migratory fishes in which migration remains confined to freshwater are called potamodromous. They travel long distances in search of spawning ground and after egg laying return to the feeding place, for example, carps and trouts.

3. **Oceandromous:** Truly migratory fishes which live and migrate in sea are called oceanodromous. They travel long distances in search of spawning ground and after egg laying return to the feeding area, for example, cods, herrings, tunnas and mackerels.

Factors influencing migration

There are a number of factors which influence migration.

1. **Physical factors:** The physical factors influencing migration are bottom materials, depth of water, temperature, light intensity and photoperiod, current, turbidity, etc.
2. **Chemical factors:** Factors like salinity, pH, smell and taste of water greatly influence migration.
3. **Biological factors:** A number of biological factors like sexual maturity, blood pressure, food, memory, biological clock and endocrine glands and presence or absence of predators and competitors influence migration.

Causes of migration

According to Northcote (1978) there are several reasons as to why fishes migrate.

1. **To maximize feeding:** The fishes generally have definite feeding areas where abundant food is available.
2. **To avoid unfavourable conditions:** Seasonal changes may bring about changes in environmental conditions, and migration is a means to escape such unfavourable conditions.
3. **To enhance reproduction:** Generally fishes have separate areas for reproduction which are in isolation and where food is not abundant. So they migrate from feeding areas to reproductive areas in the breeding season. These areas are not suitable for feeding so they migrate back to feeding areas after egg laying. The young fish after hatching migrate to feeding areas.
4. **To promote colonisation.**

Advantages of migration

1. Nikolsky is of the view that migration is an adaptation towards abundance. The breeding or nursery ground may not have sufficient food for the immature and mature members of a large population. Hence fishes have separate breeding and feeding areas. This prevents overcrowding in breeding areas.
2. It may be of some advantage to a species which returns to breed in an area where the environmental conditions are similar to those under which they survived when young.
3. The timing of spawning migration ensures that sufficient numbers of both sexes arrive at a place almost at the same time and thus enhance reproductive success of the group.

PARENTAL CARE

Parental care may be defined as nursing of eggs and young ones by the parents to enhance the chances of their survival. It is an instinctive behaviour. It is rare among fishes. Lack of parental care is associated with the production of large number of eggs so that at least a few, if not all, may survive. Parental care is observed in those species which produce limited number of eggs. Various methods have evolved among fishes to ensure proper development of eggs and young ones by one or both sexes.

1. **Nest building:** A number of fishes show nest building habit to take care of their young ones.

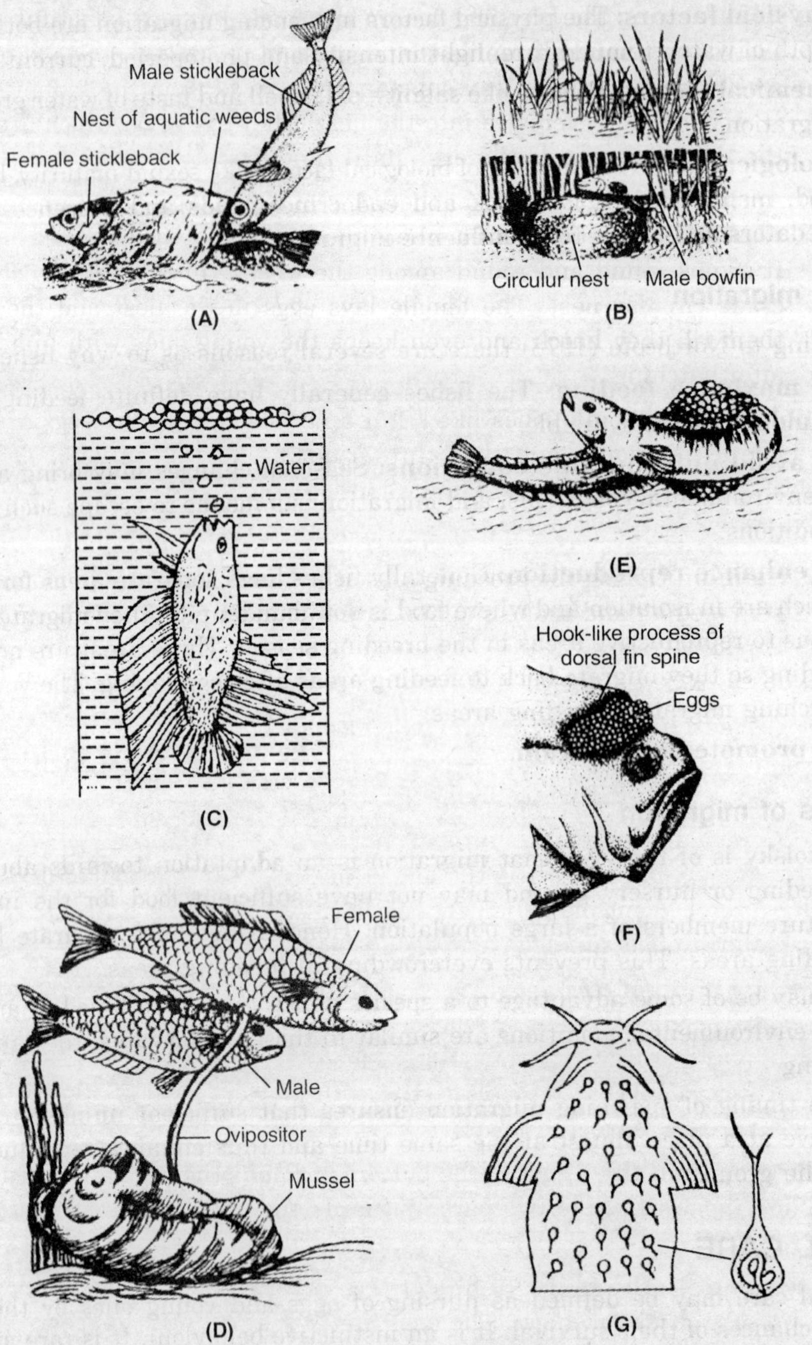

Fig. 13.21 Parental care in fishes. **(A)** Male stickleback (*Gasterosteus aculeatis*); **(B)** Male bowfin (*Amia calva*); **(C)** Bubble nest of *Betta*; **(D)** Female *Rhodeus* laying eggs within the valves of fresh water mussel; **(E)** *Pholis* coiling around the egg mass; **(F)** Male Kurtus, **(G)** Female Platrystachus

(a) Weed nest:

(i) ***Gasterosteus* (Stickleback):** It is a small fresh water fish found in ponds and lakes. During the breeding season the male develops beautiful colouration in the body. It builds a tunnel-shaped nest (Fig. 13.21 A) of dead aquatic plants and binds it with a sticky substance produced from the kidney. By courtship it attracts many females into the nest. They lay eggs, which are fertilized by the male. He guards the eggs until they hatch.

(ii) ***Amia calva*:** It is found in the lakes of North America. During the breeding season it moves to an area where there is abundance of weeds (Fig. 13.21 B). It swims round and round among the weeds till they are crushed to form a crude circular nest. The female lays eggs in the nest and the male guards them till they hatch and even keeps the young ones with him for sometime after hatching.

(b) Bubble nest: In some fishes like *Betta* and *Macropodus* the eggs are collected by the male in its mouth and then thrown out as floating nests of mucous and air bubble (Fig. 13.21 C). The male guards the nest.

(c) Pit nest:

(i) **Salmon:** During the breeding season both sexes migrate to mountain streams, stop feeding and make shallow saucer-shaped pits in the bottom gravel. The female lays eggs in the pit nest and the eggs are fertilized by the male. The fertilized eggs are then covered by a layer of fine gravel.

(ii) **African lungfish *Protopterus*:** The male digs a hole at the base of tall swamp grasses. The female lays eggs in the nest and the male guards the eggs and also aerates the water by his slow body movements.

(iii) **American lungfish *Lepidosiren*:** The male makes a tunnel-shaped burrow in swampy places in which the female lays eggs. The male guards the developing eggs and young ones. Its pelvic fin develop vascularised filaments to aerate the water around the eggs and young ones.

2. **Breeding habit of *Rhodeus*:** It is a small fish found in the rivers of central Europe. When the female is ready for egg laying the oviduct extends out in the form of a long tubular ovipositor which is used to deposit eggs within the valves of fresh water mussel (Fig. 13.21D). The male fertilizes the eggs as they are laid. After hatching the young ones leave the host.

3. **Coiling around the eggs:** In the butterfish *Pholis* the eggs are rolled into a round ball and then one of the two parents guards the eggs by coiling around them (Fig. 13.21E).

4. **Attachment to the body:** In *Kurtus* (Fig. 13.21E) the male develops a bony hook-like process on its forehead. The eggs are grouped in two bunches by filamentous processes of egg membrane and attached to the hook such that one bunch of eggs lies on either side of the head as the male swims in water. It carries the eggs until they hatch.

5. **Integumentary cups:** In the catfish *Platystacus*, during the breeding season the skin of the lower surface of the body of the female becomes soft and spongy. The female presses the belly against the fertilized eggs so that each of them becomes attached to the skin in small integumentary cups (Fig. 13.21 G) by a stalk and remain there till hatching takes place.

6. **Shelter in mouth cavity:** Some species provide protection to the developing eggs by keeping them inside their mouth.

   (i) ***Tilapia*:** The female carries the eggs in its buccal cavity. Even after hatching the young ones take shelter in the buccal cavity of their mother (Fig. 13.22 A). They come out of the mouth, swim about in water very near the mouth and return in case of any danger.

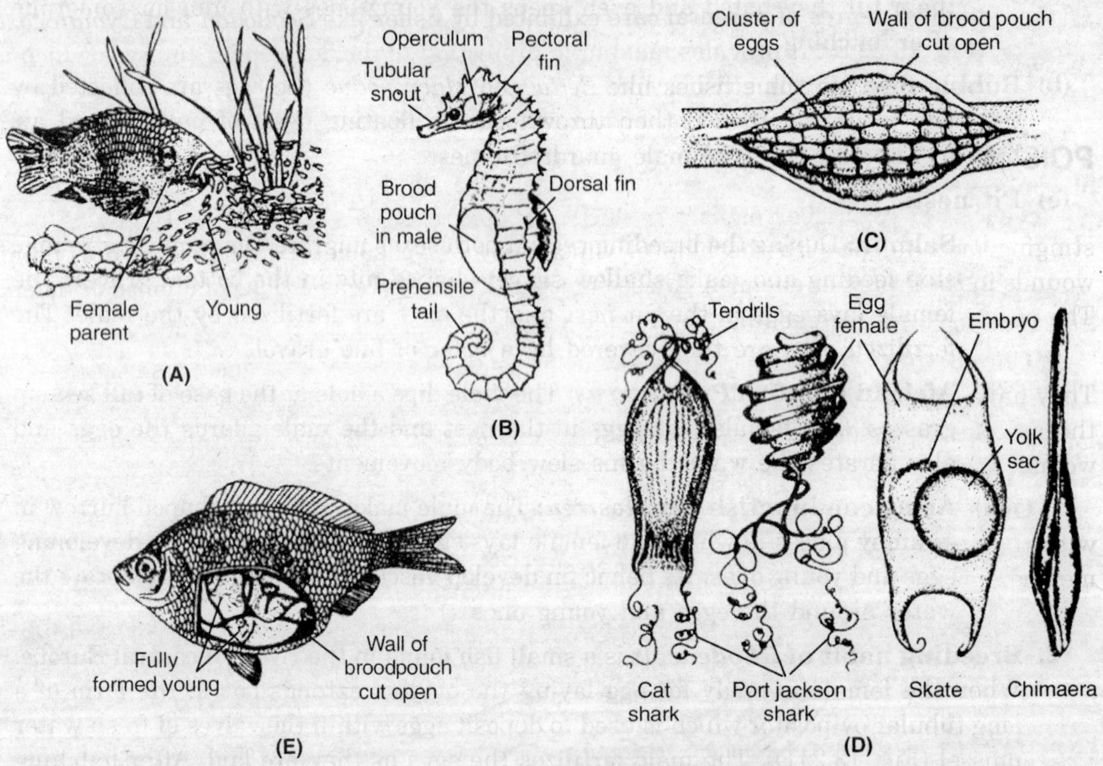

Fig. 13.22 Parental care in fishes. **(A)** Female *Tilapia*; **(B)** Hale *Hippocampus*; **(C)** Wall of brood pouch of *Syngnathus* cut open to show cluster of eggs; **(D)** Different types of mermaids purses; **(E)** Body cavity of *Cymatogaster* cut open to show the fully formed young ones

   (ii) **The catfish *Arius*:** The male carries the eggs and young ones in the buccal cavity and do not feed during this period.

   (iii) **The Brazilian catfish *Loricaria*:** The lower lip of the male is enlarged to form a pouch in which the eggs develop and hatching takes place.

7. **Brood pouch:** In sea horse *Hippocampus* (Fig. 13.22 B) and pipe fish *Syngnathus* (Fig. 13.22 C) the male has a brood pouch on the belly. The female transfers the eggs into the brood pouch where they are retained till hatching.

8. **Mermaids purse:** In some sharks and rays a special type of protective horny egg capsules called mermaids purse are secreted by the shell glands of the oviduct. They are generally rectangular in shape and the four corners are produced into tendril-like elastic filaments (Fig. 13.22 D) with the help of which they attach to sea weeds. The development is completed inside and the young ones come out by rupturing the egg case.

9. **Feeding young ones on mucus:** A unique form of parental care is seen in *Symphrodon discus* where both the parent allow the newly hatched larvae to feed on the mucous covering their body.

10. **Viviparity:** Some fishes are viviparous, that is, development takes place inside the body of the mother and young ones born are similar to the adult. Viviparity is the highest degree of parental care exhibited by fishes like *Scoliodon* and *Cymatogaster* (Fig. 13.22 E). It provides maximum protection to the developing embryos which are nourished by yolk sac placenta.

POISONOUS FISHES

There are a number of fishes which are poisonous. They have poison glands and a stinging mechanism with the help of which they are capable of inflicting severe, painful wounds in their prey or enemies. These structures thus serve as organs of offense or defence. The poison glands are modified epidermal glands.

Chondrichthyes: Among cartilaginous fishes the sting rays and eagle rays are poisonous. They have a spine on their tail. There is no poison gland associated with the spine. Probably the mucous secreted by the gland cells of the skin is poisonous. The spine inflicts painful wounds which may be fatal.

Osteichthyes: Among bony fishes well defined poison gland is usually found in association with the operculum or dorsal fin and in some, the pectoral fin. It is generally found in the members of the families Batrachidae and Siluridae.

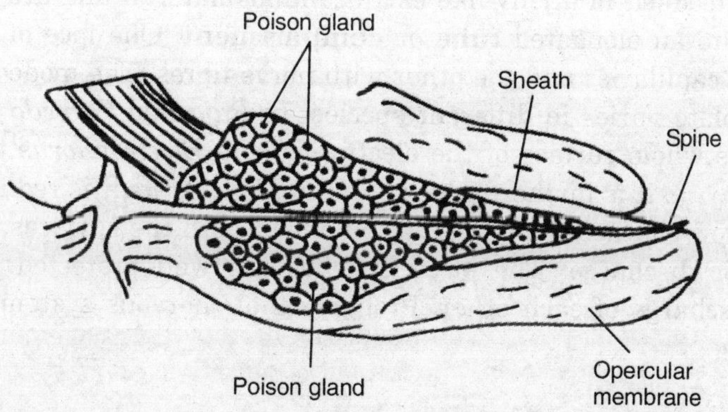

Fig. 13.23 Opercular spines and poison glands of *Trachinus*

In *Trachinus* (Fig. 13.23) the spine of operculum functions as poison gland. It projects posteriorly and bears a groove along its upper and lower surface. A poison gland is present at the base of the groove on both the surfaces but there is no duct. The gland secretes a poisonous fluid into the groove which is introduced into the body through the spine. In some fishes there is no poison groove, instead a complete canal resembling the poison fang of a snake is present. At the base of the canal there is a poison sac which discharges its contents into the canal.

In scorpion fish (*Pterois*) the dorsal spines are deeply grooved. At the base of each groove a poison sac containing poisonous fluid is present. The sac is prolonged into a duct which opens at the tip of the spine.

In the Indian catfish *Heteropneustes* the pectoral fin has a spine with a poison gland at the base. In some Siluridae both the dorsal and pectoral spines are serrated and have a poison sac at the base.

ELECTRIC ORGANS

About 250 fishes, both Chondrichthyes and Osteichthyes, possess electric organ. These fishes are unique in the animal kingdom because they are capable of generating an electric current. Some important examples are :

Chondrichthyes

Torpedo and *Narcine* (electric ray), *Raja* (skate).

Osteichthyes

Gymnarchus, *Mormyrus* (elephant fish), *Electrophorus electricus* (electric eel), *Malapterus*, *Uranoscopus* (= *Astroscopus*) (star gazer)

Structure of electric organ

The position and shape of electric organ is very variable in different species but its microscopic structure is similar in all. Each electric organ is made up of a large number of disc-shaped cells called **electroplates** or **electroplaxes**. These cells are so arranged that all of them face in the same direction in a particular species. Each electroplate is multinucleate and has a transparent cytoplasm, so the electric organ looks like a clear gelatinous mass. The electroplates are embedded in a jelly-like extra-cellular material and are bound together by connective tissue into an elongated **tube** or **compartment**. One face of the electroplate is supplied with blood capillaries and the other with nerve fibres. The mode of contact between nerves and electroplate varies in different species. In *Raja* and *Torpedo* the nerves branch and extend over the whole surface of the electroplate. In *Electrophorus* the area of contact is reduced, in *Mormyrus* it is further reduced and in *Malapterus* it is reduced to a point. The electroplate may show characteristic foldings on one or both the surfaces. This improves the efficiency of electric discharge. The electric fishes are well protected against their own current and the discharge of each other. Probably their nervous system has an unusually high insulation.

Origin of electric organ

A study of development of electric organs reveals that the electroplate is a modified muscle fibre. In *Torpedo* some branchial muscles have lost their original function of moving the gill arches and transformed into electric organ. In electric eel, skate and *Mormyrus* they are derived from some lateral muscles of the tail. In *Malapterus* some body muscles have transformed into electric organ and in star gazer the electric organ has developed from some eye muscles.

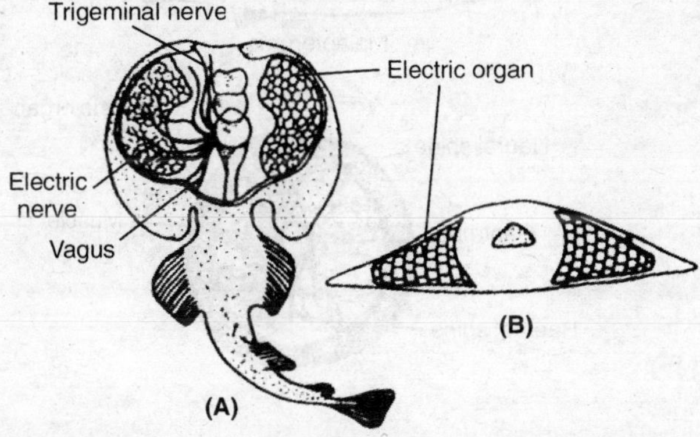

Fig. 13.24 *Torpedo* dissected to show electric organ

Electric organs of *Torpedo* (electric ray): In *Torpedo* there are two large and two small electric organs on each side of the head, the smaller one lies enclosed within the larger one. Each electric organ is kidney-shaped (Fig. 13.24). It has about 45 vertical hexagonal columns having about 400 electroplates. It is innervated by a large nerve originating from a special lobe of the brain. The innervated side is negative and the other side is positive. The current passes from the dorsal or positive side to the ventral or negative side. The discharge is about 30 volts and sometimes upto 50 volts.

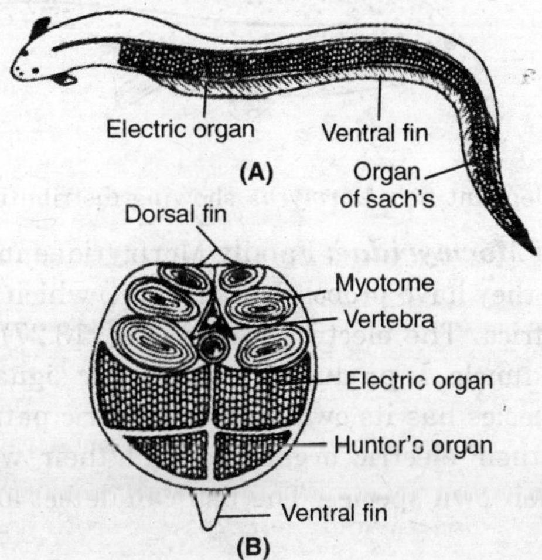

Fig. 13.25. (A) *Electrophorus electricus* dissected to show electric organ; **(B)** T.S. electric organ

Electric organs of *Electrophorus* (electric eel): In *Electrophorus* the electric organs are present on each side of the body and extend along the whole length of the fish. They occupy nearly half the mass of the body. The smaller one present on the ventral side is called **Hunter's organ** (Fig. 13.25). Each electric organ consists of about 70 compartments each containing about 6000 electroplates. It is innervated by nerves arising from the spinal cord. *Electrophorus* is capable of producing a discharge of 500 volts on land and 250 volts in water. The polarity is from head to tail. It can give a severe shock to man.

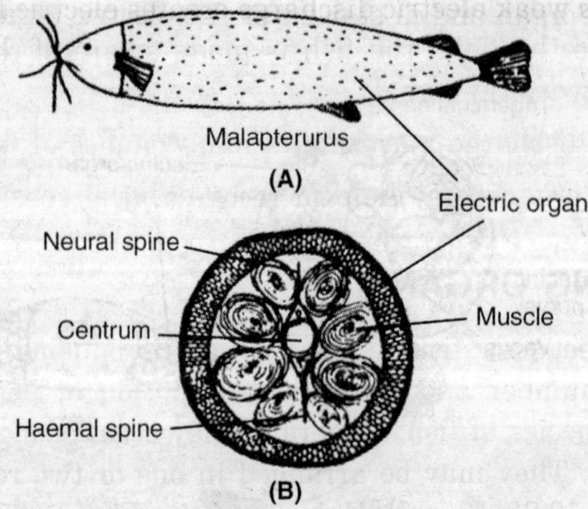

Fig. 13.26 (A) Electric organs in *Malapterus*; (B) Same T.S.

Electric organs of *Malapterus* (African catfish): The electric organs (Fig. 13.26) are located in the skin of the fish forming a thick, semi-transparent covering over the body. Each organ is divided in a number of compartments containing large number of electroplates which are innervated by a nerve from the spinal cord. The nervous end lies towards the tail. The polarity is from tail to head.

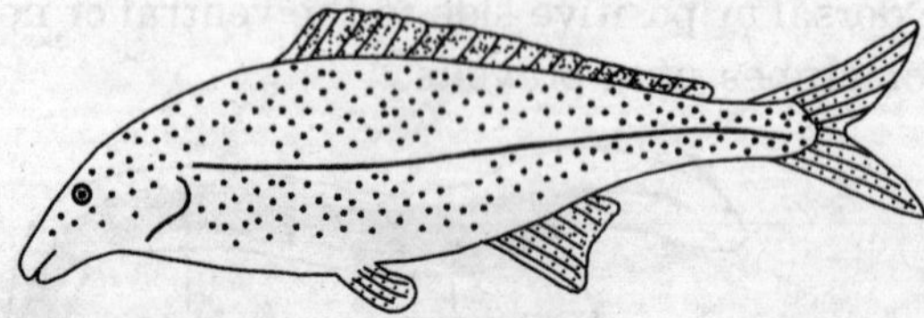

Fig. 13.27 Elephant fish *Mormyrus* showing distribution of electric organ

Electric organs of *Mormyridae*: Family Mormyridae includes 190 species of elephant fishes (so called because they have proboscis-like snout) which live in muddy waters of rivers and lakes of Tropical Africa. The electric organs (Fig. 13.27) in these fishes is a modified muscles of the caudal peduncle. It produces weak electric signals which form an electric field around the fish. Each species has its own distinct electric pattern. These fishes have poorly developed eyes so use their electric organs to find their way, to search their prey and recognize members of their own species. The fish can detect any object breaking the electric field.

Electric organs of *Uranoscopus* (star gazer): They have a pair of electric organs which are in the form of two small oval patches behind the eyes. They can give quite painful electric shock.

Functions of electric organs

1. The primary function of electric organ is defensive, but fishes with powerful electric organs use them both offensively to hunt their prey and defensively against their enemy.

2. In *Mormyrus* weak electric discharge creates electric field around the fish and if any object breaks the field, the fish is made aware of this change. Thus electric field serves as a warning signal.
3. The electric discharge may help in maintenance of territoriality by individual fish.
4. The electric organs may help in recognition of species or even in sex recognition.

LIGHT PRODUCING ORGANS

A number of deep sea fishes possess light producing or luminiscent organs called **photophores**. The number and mode of distribution of photophores on the body varies greatly in different species of fish. Generally they occur along the lateral and ventral sides of the body and head. They may be arranged in one or two rows running from head to tail as in *Scopelus* (Fig. 13.28 A) and *Halosauropsis* (Fig. 13.28 B). In *Opostomias* they are arranged in transverse bands on the body. The photophores are probably specialised gland cells of the epidermis. In *Porichthyes* (toad fish) a very large number of photophores are present along the lateral line. Photophores may also be present on elongated first fin ray of dorsal or pectoral fin, as in the angler fish. Besides photophores few large and more complex organs may also be present in some fishes, for example, one or two **suborbital organs** are present in *Opostomias micripnus* and *Pachystomias microdon*.

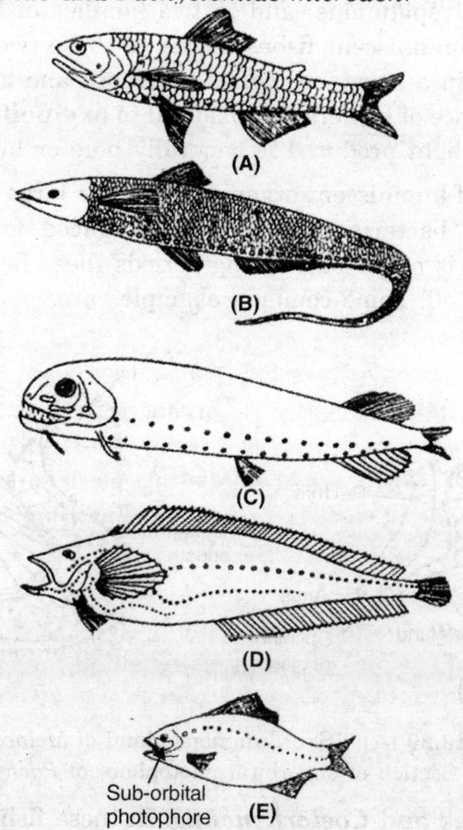

Fig. 13.28 Light producing organs. **(A)** *Scopelus*; **(B)** *Halosauropsis;* **(C)** *Pachystomias*; **(D)** *Porichthyes*; **(E)** *Photoblepharon*, showing suborbital photophore

Structure of photophores

Simple photophores: The simple photophores consist of a series of radially arranged glandular tubules which receive nerve fibres from brain as well as spinal cord.

Complex photophores: The suborbital organ of *Pachystomias* (Fig. 13.29 B) is a complex photophore. It is a cup-shaped structure consisting of several concentric layers. The outermost layer is a black pigment sheath followed by a layer of numerous glandular tubules. A thick layer of light reflecting spicules is present in the cup, the inner margin of which has numerous radial gland tubules. The mouth of the cup is occupied by a lens-like body and the skin forms a covering like iris over it. It is supplied by a branch of the fifth cranial nerve.

Types of luminiscent organs

The light producing organs are of two types:

(i) Self luminous.

(ii) Light produced by symbiotic bacteria.

Self luminous

luminiscent organs can produce their own light. They are found is some elasmobranchs (families Squalidae and Tropidinidae) and in two families of bony fishes (Stomatidae and Scopelidae). Most of the luminiscent fishes belong to these two families. The self luminous luminiscent organs contain a substance called **luciferin** and an enzyme called **luciferase**. The luciferin in the presence of luciferase is oxidised to **oxyluciferin** and in this process light is emitted. In fishes the light produced is generally blue or blue-green.

In the second type of luminiscent organs there are a large number of glandular tubules which secrete symbiotic bacteria. The light is produced for long periods by bacterial luminiscence. Since light is produced for long periods these fishes have mechanical devices to turn the light on and off. Some common examples are:

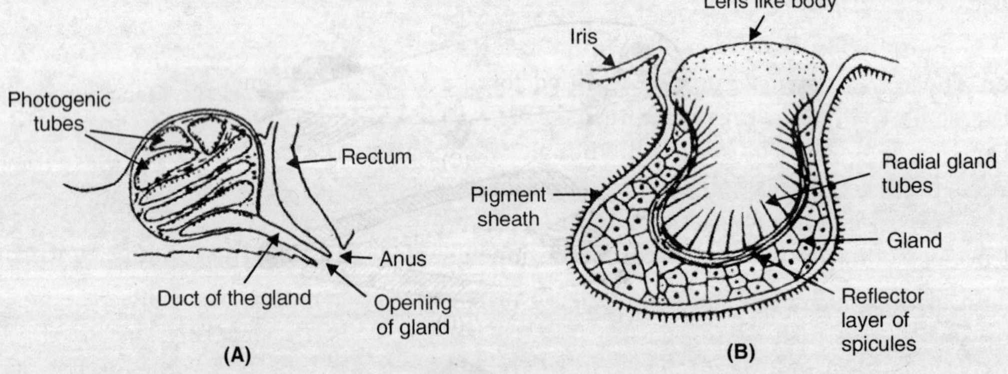

Fig. 13.29 (A) T.S. of luminous gland of *Melacocephalus;*
(B) Section of suborbital photophore of *Pachystomias*

(i) ***Malacocephalus* and *Coelorhynchus*:** In these fishes the gland opens by a duct on the ventral surface of the fish a little in front of the anus. The gland is richly supplied with blood capillaries (Fig. 13.29 A).

(ii) ***Photoblepharon* and *Amalops*:** In these fishes an elongated luminiscent organ is present below each eye (Fig. 13.28 E) and consists of numerous long, parallel, glandular tubes with rich blood supply. In *Photoblepharon* a black tissue can be drawn up like eyelid over the organ, thus cutting off the light.

Control of photoreceptors

The luminiscent organs are under the control of the nervous system and endocrine glands. The cells are activated by efferent fibre and some endocrine glands like the supra-renal gland may also activate it. They are also sensitive to adrenaline.

Uses of luminiscent organs

1. They enable the deep sea fishes to search their prey in the dark.
2. Some fishes use them to attract the prey near the mouth.
3. In some fishes they act as defensive organs. The light from these organs dazzle and confuse the enemy for a moment and the fish escapes.
4. They also help in recognising members of the same species.
5. In *Porichthyes* they act as a warning signal for the predator. This fish has a poisonous spine, so it is avoided by the predator.

SOUND PRODUCING ORGANS

Many fishes are capable of producing characteristic sound with the help of specialised organs. Production of sound is only known in the teleosts, the elasmobranchs do not produce sound. There are two types of sound producing organs in fishes.

(i) **Stridulating organs**

(ii) **Air bladder**

Stridulating organs

In many bony fishes sound is produced by rubbing two bony structures against each other. These bony structures are modified for stridulation. In some species the pectoral fin and in some the dorsal fin is modified for stridulation. In *Cottus scorpius* the preoperculum and in *Dactylopterus* the hyomandibular may be modified for the same purpose. In *Scomber* sound is produced by friction between superior and inferior pharyngeal teeth. The trigger fish *Balistes* produces characteristic drumming sound by stridulation between the post clavicle and a longitudinally grooved area on the inner surface of the corresponding cleithrum.

In *Callomystax*, the sound producing organ is very complex. The neural spines of the third, fourth and fifth vertebrae are fused to form a laterally compressed bony plate. The hinder part of the plate is divided into two with inner margins bearing closely set vertical ridges. This portion receives the first interspinous bone of the dorsal fin. Both the surfaces of interspinous bone are ridged like a file. The vertical movements of sixth and succeeding vertebrae cause the interspinous bone to rub against the ridges of the bony plate producing a characteristic grating sound.

Air bladder

In many teleost sound is produced by vibrations of the wall of air bladder with the help of special muscles. In siluroid fishes there is an **elastic spring mechanism** to vibrate the walls of air bladder. The transverse processes of the fourth vertebra are converted into highly elastic and flexible springs which are bent downwards and backwards and their ends are embedded in the wall of air bladder. Two powerful muscles arise from the occipital region of the skull and are inserted on the anterior face of the two springs. When these muscles contract and relax they cause the wall of the air bladder to vibrate, thus producing a sound. In some species the air bladder is divided internally into a number of compartments by septa. This increases the intensity of sound by the vibration of free ends of the septa.

In other teleosts the air bladder is not connected to any part of the skeleton. The contraction and relaxation of the muscles of the wall of air bladder cause vibratory movements and the gas in the air bladder produces sound. Some examples are *Diodon*, *Tetradon*, *Hippocampus*, *Dactylopterus*, *Ophiocephalus Ostracion*, etc. In the drumfish *Pogonias chromis* most powerful sound is produced by the air bladder. This drumming sound is produced in the breeding season and can be heard at a distance.

Importance of sound production

1. Sound produced by a particular species may serve as recognition sound, thus helping the individuals to aggregate in large numbers. This is of significance during the breeding season as it increases the chances of fertilization.
2. Sound production is useful for those species which have reduced eyes or live in deep sea. It helps the fish to locate the prey or predator and members of its own species.

ECONOMIC IMPORTANCE OF FISHES

Fishes are most numerous of all vertebrates with more than 20,000 species. From economic point of view there are many **beneficial** as well as **harmful** fishes. The economic value of fish is so much that it has led to the development of fish industry which earns lot of revenue for the country through foreign exchange by exporting fish and fish products. In the fish industry no part of the fish is discarded.

Beneficial fishes

Many fishes are beneficial to man because they and their products are very useful.

Fish as food: Fish is used as food since ancient times in all parts. It is a very valuable food because of the following reasons :

1. It is a rich source of proteins and contains all essential amino acids.
2. Its cholestrol content is low.
3. It is good in taste and easily digestible.
4. It is the richest source of elements like iodine and phosphorus and vitamins A and D.

Fishes can be eaten in various ways, either cooked or even raw, pickled, salted, smoked or canned. The fishes deteriorate very rapidly after being caught so they must be consumed immediately or preserved for later use. The simplest method of preservation is to cut open the fish and dry it in air. For future consumption fishes can be preserved by smoking, salting or refrigeration. The fins of sharks are used for preparing soup.

Some common edible fishes are:

Fresh water fishes—*Catla*, *Labeo*, *Cirrhina*, *Wallago*, *Mystus*, *Ophiocephalus*, *Clarias*, *Hilsa*, *Tor tor*, *Schizothorax*, *Barilius*, *Notopterus*, etc.

Marine fishes—*Scoliodon*, *Raja*, *Trygon*, *Somber*, etc.

Fish meal. Fish meal is prepared from scraps obtained from canneries or from whole fish like sardines, mackerels, ribbon fish, etc. which are not relished by man. It is used as artificial food for cattle and poultry. It is prepared by cooking the fishes in water, pressing them to remove water, then drying them in the sun and finally grinding them into powder. The powder can be stored in air light containers and its nutritive value can be retained for a long time. It contains 60% protein and high percentage of calcium phosphate.

Fish manure: Fishes that are not fit for human consumption are dried in sun and then ground into powder which is used as manure. It contains high percentage of nitrogen, phosphate and calcium. **Fish guano** is a rich nitrogen and phosphate fertilizer which is prepared from the material left over after extracting oil from the fish.

Fish oil: Fish oil is of two kinds—**body oil** and **liver oil**.

The oil that is extracted from the whole body is called body oil. It has traces of vitamin A and D. It is obtained from fishes like sardine, herring and salmon. On the basis of iodine content body oil is of 2 types—high iodine content (**drying oil**) and low iodine content (**semi-drying oil**). Body oil is used in painting, varnishing, in preparing soaps and candles and in leather and steel industries.

The oil that is obtained from the liver of certain fishes like sharks, rays, cods and tunas, is called fish liver oil. It is rich in vitamin A and D. It contains 55 to 75% fat and 5 to 10% proteins and is used in preparing medicine.

Fish flour: Fine fish manure of good quality is called fish flour. It is mixed with wheat flour and used for making bread, biscuits, cake and sweets.

Fish protein: It is obtained by removing fat from fish flour and contains 80 to 90% protein. It is a white powder without fishy smell. It is used to increase the protein content of bread, cake, ice cream, etc. It is also used in preparing medicines.

Fish glue and isinglass: Fish glue is prepared from the connective tissue of fish skin, fins and bones. It is used as an adhesive for glass, leather, book binding and furnitures.

Isinglass is a shining powder prepared from the lining of air bladder of fishes. The shine is due to presence of guanine crystals in the lining of air bladder. The bladder is removed, cleaned and dried in the sun and then exported to prepare isinglass. It is used for clearing wine, beer, making edible jelly and for preparing adhesive material.

Fish skin: The skin of fish yields two important products:

(i) **Shagreen:** Skin of sharks along with scales is tanned to produced **shagreen** which was used, in the old world, for casing books, jewel boxes, sword handles, etc. It is also used as an abrasive for polishing wood and ivory.

(ii) **Fish leather:** The thick, rough skin of sharks and rays is tanned with crushed scales to produce **fish leather** which is used for making purses, shoes, bags, belts, boxes, etc. Some tribal people prepared hats, used during war, from skin of porcupine fish *Diodon* and Japanese prepare lantern from the skin of puffer fish *Tetrodon*.

Manufacture of artificial pearls: A substance called guanine is obtained from the silvery scales of some fishes by scrapping. It is used for poishing hollow glass which are filled with wax and used as artificial pearls.

Sports and games: Fishing is a hobby of millions of people all over the world. They catch a number of species like **trout**, **salmon**, **carp**, etc. from fresh waters and *Tarpon* from marine waters.

For decoration: Many people have aquarium, pools and ponds in their homes or office where they keep beautifully coloured fishes for decoration. Breeding and maintaining aquarium fishes has become a very popular business these days. Some common aquarium fishes are gold fish. (*Carassius auratus*), angel fish (*Pterophyllum*), sword tail guppy (*Xiphophorus*), minnow (*Gambusia affinis*) *Trichogaster*, *Tilapia*, Siamese fighter (*Betta splendens*), paradise fish (*Macropodus*), etc. Some of the aquarium fishes are artificially cultured and not found in nature.

Scientific study: Some fishes like *Scoliodonl, Labeo*, etc. are dissected in laboratories for type study. Some are used as experimental animals in various scientific fields like genetics, embryology, animal behaviour and pharmacology.

Biological control: Many fishes like *Chela*, *Puntius*, *Colisa*, *Barilius*, *Rasbora*, *Ambassis*, etc. feed on the larvae of many insects including mosquito. These larvivorous fishes are introduced into water where insect larvae breed. Thus they help in reducing the population of mosquito and other insects.

Harmful fishes

Destructive fishes: The nuisance value of sharks is enormous especially to fisherman. They remove bait from hooks, steal netted fishes and often tear the nett. Some large species often capsize small boats.

Injurious fishes: Many large species of sharks and rays often injure man and occasional death has also been reported due to injuries caused by spines of sting rays. *Carcharodon carcharias,* the great white shark, attacks human beings in Australian waters. Some fishes produce electric shock which may prove fatal to small animals and even man. The flesh of some fish (*Tetrodon*) is poisonous.

Host of parasite: Some fishes act as intermediate host of many human parasites and thus help in the spread of disease.

14

CHAPTER

AMPHIBIA

The classs Amphibia includes animals that partly live on land and partly in fresh water. The Amphibia were the first chordates that lived out of water way back in the Carboniferous period. Although they are regarded as the first land vertebrates, they are not fully adapted for terrestrial life. They spend the early part of their life cycle in water and adult stage on land, hence the name **Amphibia** (**Gr.** *amphi*; dual; *bios*, life). The name Amphibia was given by Linnaeus. The group includes such familiar animals as the frogs and toads (**Anura**), the newts and salamander (**Urodela**) and tropical, burrowing, legless amphibians (**Apoda**). This is the smallest class of vertebrates having nearly 2,500 species. The amphibia occupy an intermediate place between the fish and reptiles. They have descended from some fish-like ancestors on the one hand and are themselves ancestors of reptiles that have given rise to the birds and mammals. The primitive stem Amphibia are called **Labyrinthodontia**.

TYPE FROG

The frog, although highly specialized amphibians, is usually described as a type of the class, because it is easily obtained. It is a nice, clean animal easy to dissect and is admirably suitable for physiological experimentations. The frog has been a favourite object of study for a long time.

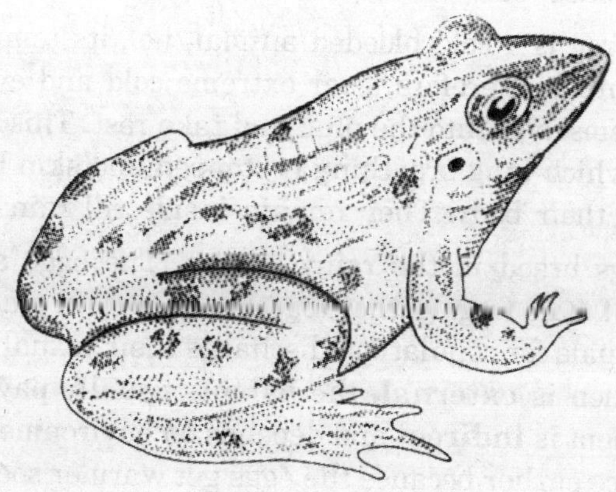

Fig. 14.1 *Frog*–The animal in sitting posture

HABITAT

The frog generally lives in or near water but wanders away from the aquatic home in damp or wet weather. It is obligatory for the frog to live near water because of the following reasons:

(i) to keep the skin moist for cutaneous respiration.

(ii) the eggs are laid in water in which the young hatch out and lead aquatic life for sometime.

(iii) water provides a good means of protection to these defenceless animals.

HABITS

Food: Frog is carnivorous. It feeds chiefly on earthworms, insect, spiders, etc. which it swallows whole. It catches its prey with the help of its sticky protrusile tongue. It can even catch insects on wings.

Water: Frogs do not drink water but absorbs it through the skin which is loosely attached to the body and as such is able to store considerable amount of water.

Shedding of skin: It has been observed that the outer layer of skin, which consists of a very thin transparent membrane, is cast off at certain periods. This skin comes off in large patches, usually the skin covering the toes comes out last. The frogs usually eat the shed skin.

Croaking sound: Frogs produce croaking sound in rainy season. The voice of male is hoarse, deep and loud due to the presence of a pair of distensible vocal sacs on the throat. It is often used as a call to the mate. Female croaks are slower and rarer.

Escape from enemies: When caught in hand it makes violent efforts to escape using its fore-and hindlimbs. At the same time a fluid is poured out to make the already slimy skin more slippery. It also inflates its lungs which helps in slipping. It makes a pain scream when attacked by enemies. Frogs generally hide themselves from their enemies by changing the colour of their skin to match with that of the surrounding. This phenomenon is called **camouflage**. This protective colouration is due to dispersion or concentration of special pigment cells in their skin. Some common enemies of frog are snakes, mongoose, crow, vultures, aquatic birds, fishes, etc.

Hibernation: The frog is a cold blooded animal, i.e., its temperature varies with that of the environment. It cannot, therefore, bear extreme cold and even extreme heat. During intense cold they bury themselves into the mud and take rest. This is known as **hibernation** or **winter sleep** during which lung breathing is stopped and skin breathing continues. They use the energy stored in their body. They remain inside till rain fall.

Breeding: The frogs breed in the rainy season. The eggs are laid in water. During breeding time a number of male frogs gather together at certain places in shallow water, start croaking to attract the female for copulation, discharge their sexual products and then scatter on the ground. Fertilization is **external,** the parents usually pay no attention to the eggs afterwards. The development is **indirect** and depends on environmental conditions. In shallow ponds hatching takes place earlier because the eggs get warmer soon. In deep water hatching may take long. The young one that hatches out is a free-swimming **tadpole** larva which undergoes **metamorphosis** to become adult.

LOCOMOTION

Locomotion in frogs is brought about in two ways:

(i) by **leaping** on land

(ii) by **swimming** in water

Leaping: In ordinary resting position the front part of the body is inclined upwards and is supported by short forelimbs having a peculiar twist (Fig. 14.2A). The posterior part of the body is supported on the hindlimbs which are folded in the manner of letter **Z** and act like a spring in jumping. This is called **sitting** or **squatting posture**. From such a posture the frog jumps by a sudden extension of the hindlimbs throwing its body into the air. The forelimbs act as shock absorbers when the animal returns to the ground. They also help to manipulate the direction for which it adjusts its body before each leap. A frog can jump two to three feet in each leap.

Fig. 14.2 *Frog*–(A) Stages of Leaping

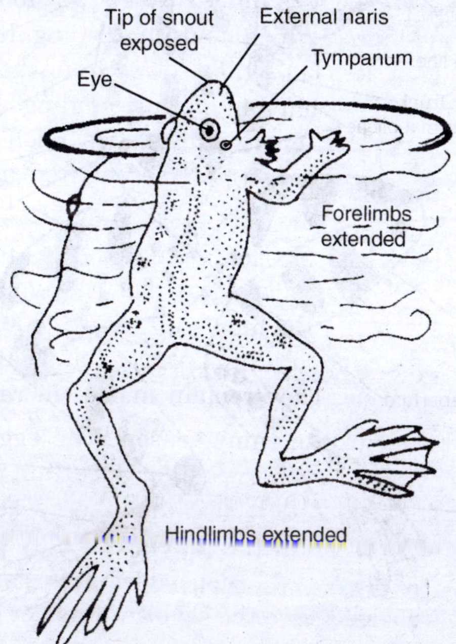

Fig. 14.2 *Frog*–(B) Animal floating in water

Swimming: In swimming also the hindlimbs are used. In the starting position the hindlimbs are folded in the manner of latter **Z**, as in leaping, but as the animal begins to swim the legs are extended quickly (Fig. 14.2 B). As the hindlimbs are long and the toes are webbed, they act like oars enabling the animals to swim. The forelimbs are held back after each stroke. They serve two purposes : (i) they help in propelling the animal to some extent and (ii) in guiding the direction of movement. If the animal does not make strokes in succession, i.e., it does not swim, then it keeps floating with its hindlimbs and body extended in water and the snout carrying the nostrils is exposed. If disturbed in this position, it quickly dives under water.

EXTERNAL CHARACTERS

Shape and size

The body is more or less spindle-shaped with the anterior end pointed and the posterior end somewhat rounded. It is stream-lined and slightly flattened dorsoventrally to help in swimming in water. The frogs range in size from 12 to 18 cm in length and 5 to 8 cm in width. The male is generally smaller and slimmer than the female.

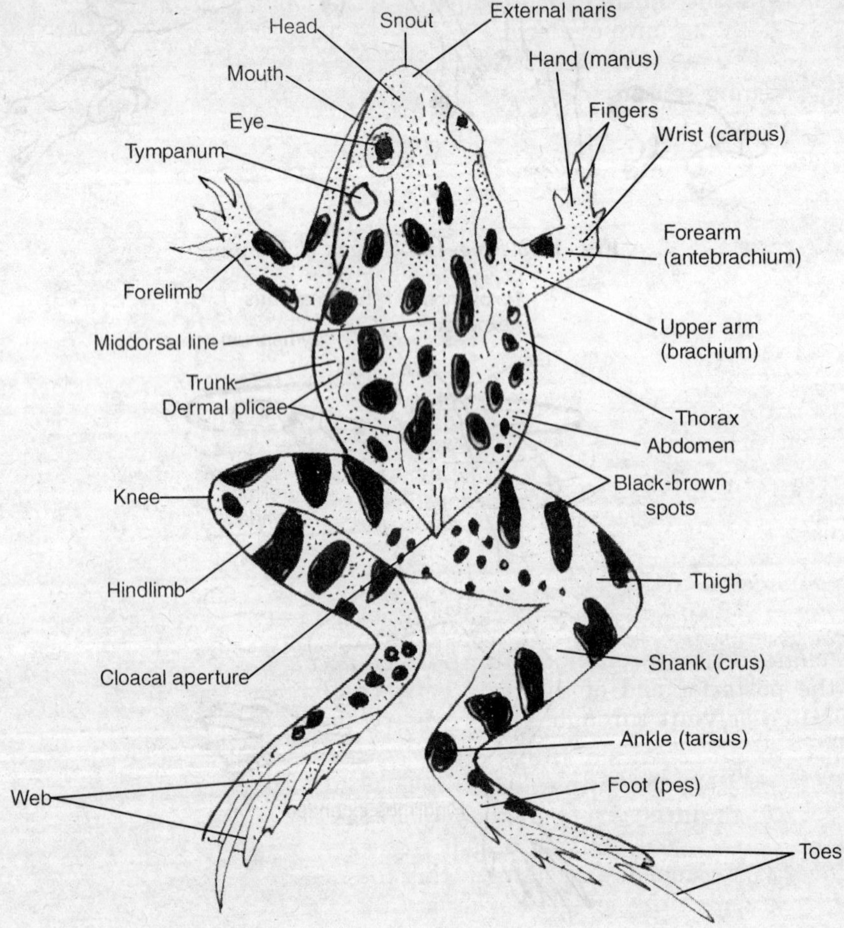

Fig. 14.3 *Frog*–External features (dorsal view)

Divisions of the body

The body is divided into two parts: the **head** and **trunk**. A tail is lacking in the adult stage and there is no neck.

Head: The head is almost triangular in shape joined directly to the trunk, as the neck is absent. The short blunt anterior end is called **snout**. The snout terminates into a wide, transverse opening the **mouth** extending from one ear to the other. The mouth opens into the buccal cavity. Lying above and behind the tip of the snout there are two **external nares** or **nostrils**. Each nostril is guarded by a **valve** externally and opens into the buccal cavity by **internal nares**. The head bears two large, rounded, bulging **eyes** situated dorsolaterally. When pressed downwards they make marked prominence in the buccal cavity. The eyes do not rest on any bone as in higher animals. The eyes are protected by **eyelids**, the upper eyelid is thick, pigmented and capable of very little movements, but the lower eyelid is thin, semi-transparent and capable of free movement. In the eye of frog, a **nictitating membrane** (third eyelid) is also present. It is a thin transparent membrane which arise from the lower eyelid. It covers and protects the eye during swimming and keeps the eye moist in air. Behind the eyes, on either side, there is a conspicuous, nearly circular deeply pigmented, tough membrane called **tympanic membrane** or **tympanum**. A careful observation reveals a small prominence in the centre of the tympanum, this is due to **columella**, a bony rod that lies inside. Frog does not have external ear. In male frog a pair of prominent bluish patches called **vocal sacs** is present on the throat. They act as resonators and intensify the croaking sound during breeding season.

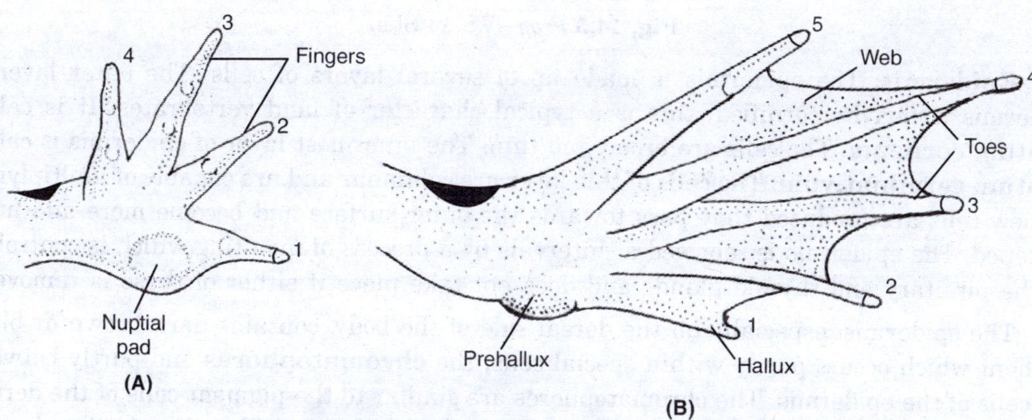

Fig. 14.4 *Frog*–(A) Hand of male showing nuptial pads; (B) Foot showing web

Trunk: The trunk is somewhat oval in shape. Along the mid-dorsal line the back is raised into a characteristic prominent **hump** which is conspicuous when the frog is in sitting posture. At the posterior end of the body between the two legs there is a small circular **cloacal aperture** or **vent** through which urinary, faecal and sex products are discharged.

The trunk bears two pairs of **limbs**. The anterior pair of limbs are the **forelimbs**. They are short and consist of the **upper arm** (**brachium**), **forearm** (**antebrachium**), **wrist** (**carpus**) and **hand** (**manus**). The hand bears four fingers or digits without web. The **thumb** is inconspicuous or vestigeal and can be felt within the skin. In male frogs the base of the first inner finger is thickened, especially during the breeding season, to form **nuptial pads** (Fig. 14.4 A) for grasping the female during copulation.

The posterior pair of limbs are the **hind limbs** or **legs**. They are long, powerful and consist of **thigh**, **shank** (**crus**), elongated **ankle** (**tarus**) and large **foot** (**pes**). The foot consists of five elongated clawless **toes** which are connected by a thin **web** of skin (Fig. 14.4 B). There is a vestigeal sixth digit called **prehallux** which can be felt within the skin.

Skin: The skin of frog is smooth (has no scales, hairs, etc.), moist and loosely attached to the body. It is thicker on the dorsal side than on the ventral. It is composed of two layersv :**epidermis** and **dermis** (Fig. 14.5).

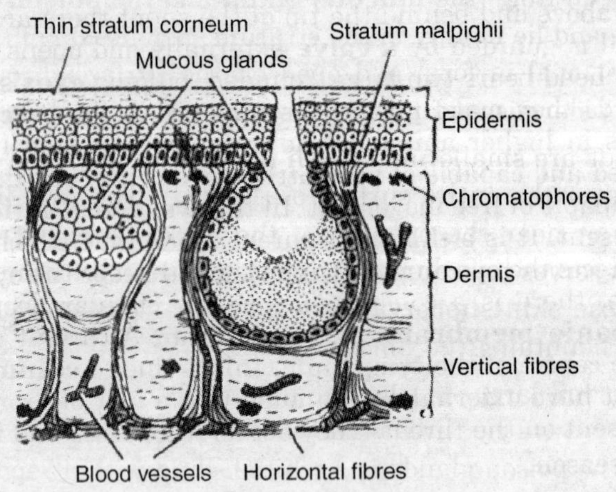

Fig. 14.5 *Frog*–V.S. of Skin

Epidermis: The epidermis is made up of several layers of cells. The outer layer of epidermis is heavily cornified, this is a typical character of land vertebrates. It is called **stratum corneum**. The cells are broad and thin. The innermost layer of epidermis is called **stratum germinativum**. The cells of this layer are columnar and are capable of multiplying. As new cells are produced they pass towards the outer surface and become more and more flattened. The epidermis is renewed at intervals by a process of moulting which is controlled by the pituitary and thyroid glands and does not take place if either of these is removed.

The epidermis, especially on the dorsal side of the body, contains dark brown or black pigment which occurs partly within special cells, the **chromatophores** and partly between the cells of the epidermis. The chromatophores are similar to the pigment cells of the dermis but are less abundant. In certain regions all the cells of the epidermis contain pigment granules. The epidermal pigments appear and disappear in the same region in the course of few months. The main source of colour is located in the pigment cells of the dermis.

Dermis. The dermis is divisible into two layers : an outer **stratum spongiosum** and inner **stratum compactum**. The stratum spongiosum consists of a loose network of fibrous connective tissue richly supplied with lymph spaces and blood vessels. Just beneath the epidermis it forms a thin layer which contains numerous pigment cells. In the deeper parts are embedded the glands. The stratum compactum is made up of dense connective tissue whose fibres run in a wavy course parallel to the surface of the skin. At intervals thin layer is crossed by vertical strands consisting of connective tissue fibres, smooth muscle fibres, elastic fibres, nerves and blood vessels. The vertical strands often extend through the stratum spongiosum into the epidermis. They help in squeezing out secretions of the glands.

Beneath the stratum compactum lies a loose layer of sub-cutaneous connective tissue which is separated by large lymph spaces into two, one beneath stratum compactum and the other above the musculature. The outer layer is vascular and consists of numerous **stellate cells** containing greyish white pigment granules. These cells are abundant on the underside where they produce white colouration characteristic of the under surface.

Glands: The glands of the skin are more developed in amphibians than in fishes. There are two types of glands in frog—the **mucous gland** and the **poison glands**. The glands are of simple alveolar type and lie mainly in the stratum spongiosum of the dermis. Typically the glands are spherical or oval in form and open on the surface through narrow neck which extends through the epidermis and terminates in a triradiate opening at the outer end.

The mucous glands are smaller and much more abundant than the poison glands. They are found practically over the entire surface of the body. They are lined with a single layer of epithelial cells except near the opening of the neck where there are two layers. They produce mucous which serves to keep the skin moist, essential for cutaneous respiration. The slimy mucous makes the skin slippery, hence it escapes enemy. The mucous also prevents evaporation and thus maintains the amount of necessary water.

The poison glands are larger but less abundant and less uniformly distributed over the surface of the body. They are relatively more numerous on the dorsal side and on the hind legs. The secretion of the poison glands is a whitish fluid with a reported burning taste. The poison gland is less developed in frog than in toad, where they are grouped forming **parotid glands**. Comparatively little is known about the property of poison in the frog. In man it produces irritation of eyes and nose. When swallowed it produces nausea.

Some amphibians produce certain secretions with characteristic smell. Their function, perhaps, is to attract the opposite sex.

Colour: The colour of common frog is olive green with dark spots and patches. The hind legs are marked by cross bars of brownish green colour. The ventral surface is yellowish white. The males are usually darker than the females. The colour is produced by three layers of pigment cells : the **melanophores** lying deepest, the **guanophores** lying next and the yellow **lipophores** overlying these. The guanophores are full of granules which by diffraction produce a blue green colour while the lipophores filter out the blue. The skin of frog is able to change colour in harmony with the surrounding. The phenomenon is called **camouflage (protective colouration)**. The change in colour is produced by expansion of pigments in the melanophores under the influence of the pituitary gland. Movement in other chromatophores can also affect the colour, yellow being produced by the disarrangement of guanophores, blue (though rarely) is produced due to absence of lipophores and red by pigments in the lipophores. These colours may contribute to the patterns that the animals present. The colour patterns adopted are usually cryptic or concealing. Many frogs generally expose brightly coloured patches on the thighs when they jump. This presumably serves to startle the attacker, and such colours are called **dynamic** or **startling**. The colour also protects the organs from the effects of light, which may cause contraction when it falls directly on the muscles. Dark colour of adults and eggs may also help in the absorption of heat.

ENDOSKELETON

The skeleton of the larval frog is cartilaginous, but in the adult, most of it is replaced by bone (Fig. 14.6). Such bones are called **cartilage bones** or **replacing bones.** Sometimes bones develop from certain regions of the skin or dermis where there was no pre-existing cartilage, such bones are called **membrane bones**. In some parts the primary cartilaginous skeleton becomes hard due to impegnation by calcium salts, such bones are called **calcified cartilage**. Cartilage may persist on the ends of limb bones to form smooth joint surfaces and in parts of the skull and limb girdles. The skeleton is of two types : the skull, vertebral column and sternum comprise the **axial skeleton** and the limbs and girdles form the **appendicular skeleton**.

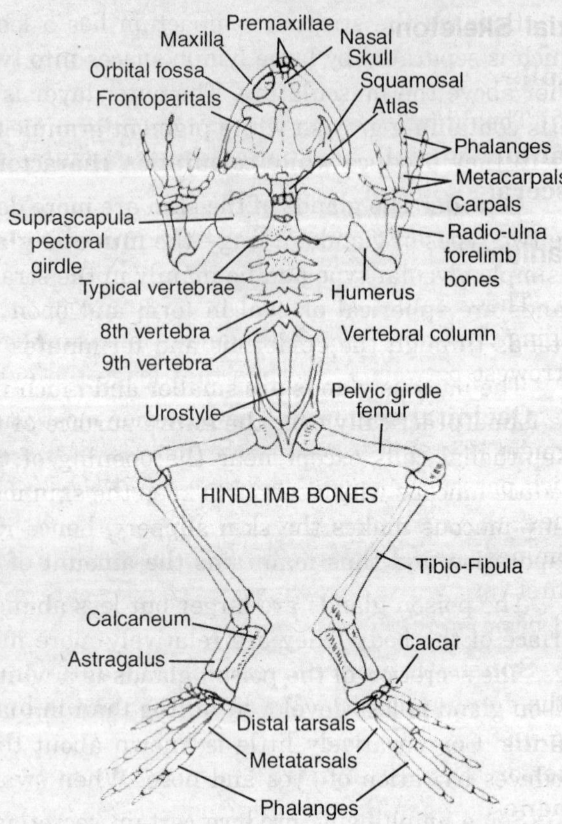

Fig. 14.6 *Frog*–Complete skeleton

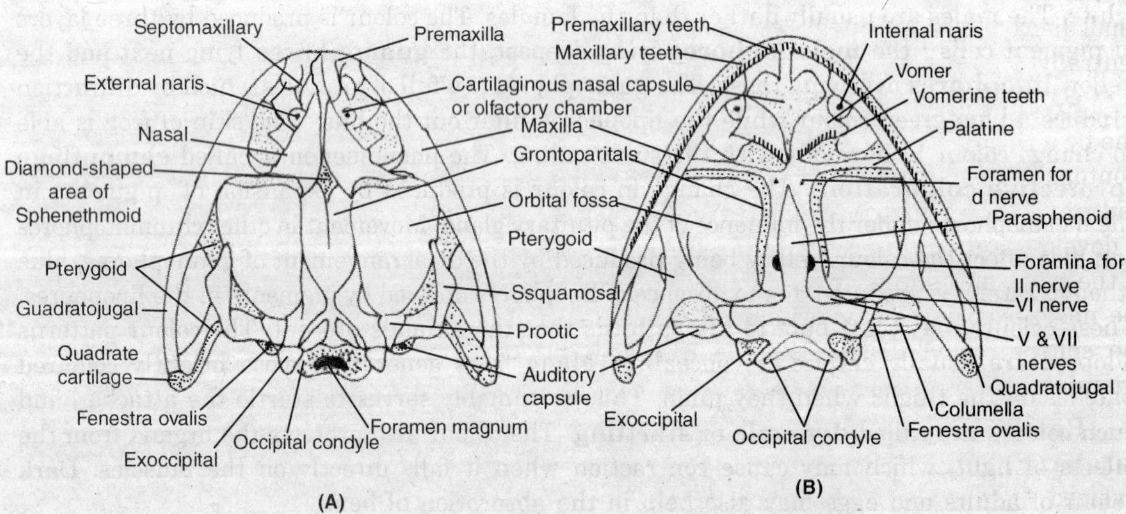

Fig. 14.7 *Frog*–Skull; **(A)** Dorsal view; **(B)** Ventral view

Axial Skeleton

Skull

The broad flat skull consists of three parts: **cranium** or **brain box**, paired **sense capsules** and **visceral skeleton**.

Cranium

The cranium or brain box is a narrow cavity lodging the brain. It is cartilaginous in nature being narrowest between the orbits that lodge the eyes.

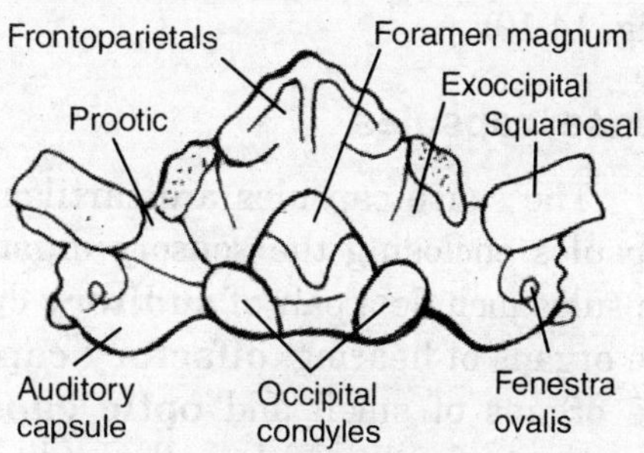

Fig. 14.8 *Frog*–Occipital segment

Occipital segment: The posterior most part of the cranium is formed by the occipital segment (Fig. 14.8). Posteriorly there is a large opening called **foramen magnum** through which passes the spinal cord. On either side of the formen magnum there are two irregular bones, the **exoccipitals**. The posterior ends of the exoccipitals bear two oval, convex prominences, the **occipital condyles** which articulates with the two concavities of the atlas or first vertebra. Externally each condyle bears an exit hole for vagus nerve. The supraoccipital and basioccipital are absent.

Sphenethmoid: Surrounding the anterior end of the cranium is a tubular bone, the sphenethmoid (Fig. 14.9). It is divided by a transverse partition into anterior wide **ethmo-idal region** and posterior **spheno-idal region**. The ethmoidal region is divided by a vertical partition into right and left cavities each enclosing an olfactory sac. The cavity of sphenoidal region encloses the forebrain. The bone is overlapped by the fronto-parietals and nasals dorsally (only a small diamond-shaped area is visible) and parasphenoid ventrally.

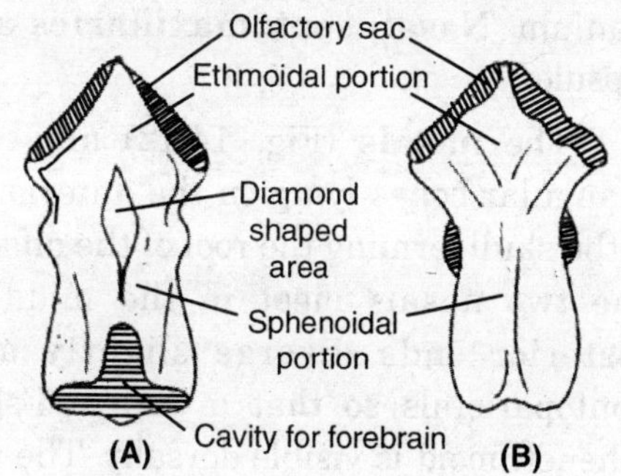

Fig. 14.9 *Frog*–Sphenethmoid
(A) Dorsal view; **(B)** Ventral View

Frontoparietals: These are two membrane bones (Fig. 14.10) united along the middle line by a suture. They are long and flat bones forming most of the roof of the brain case. Each frontoparietal is made up of two bones, the **frontal** and **parietal** which are separate in the early stages of development but fuse afterwards. The anterior end of the two bones slithgly diverge and articulate with the nasal exposing a small diamond shaped area of the sphenethmoid. Posteriorly the fronto-parietals articulate with the exoccipitals.

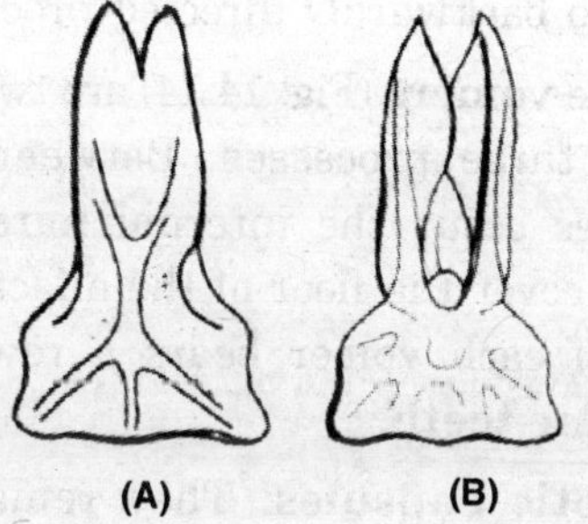

Fig. 14.10 *Frog*–Frontoparietals
(A) Dorsal view; **(B)** Ventral view

Parasphenoid: It is also known as the **parabasal bone**. It is a large dagger-shaped or inverted 'T'-shaped (⊥) membrane bone covering the entire floor of the cranium ventrally. The long arm of T overlaps the spenethmoid bone ventrally and lateral processes of the bone underlie the auditory capsules (Fig. 14.10).

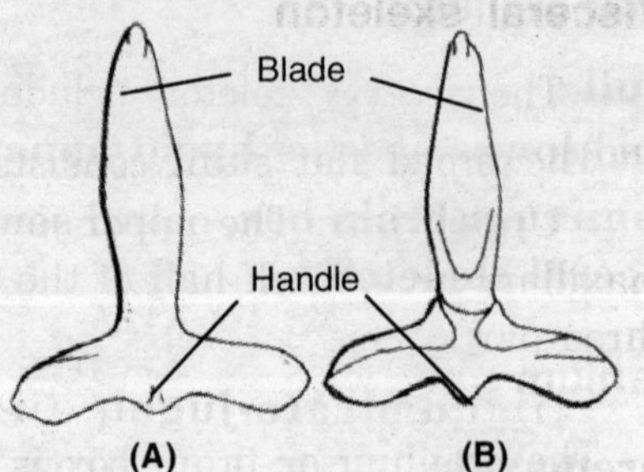

Fig. 14.11 *Frog*–Panasphenoid; **(A)** Dorsal view; **(B)** Ventral view

Sense capsules

The sense capsules are cartilaginous and bony capsules enclosing the sensory organs. The sensory capsules include a pair of **auditory capsules** lodging the organs of hearing; **olfactory capsules** enclosing the organs of smell and **optic capsules** enclosing the organs of sight. Of these the auditory and olfactory are fused with the cranium while optic is not fused.

Auditory capsule: The auditory capsules are cartilaginous and fused with the sides of the posterior end of the cranium. They are formed by an irregular cartilage bone, the **pro-otic** (Fig. 14.15) which lie at the sides and in front of the exoccipitals. Each forms a ring-like lateral projection on either side. On the anterior side of each prootic there is an aperture called **fenestra ovalis** serving as exit for the 5th, 6th and 7th cranial nerves.

Olfactory capsule: The olfactory capsules are fused with the anterior end of the cranium. **Nasal**, **septomaxillaries** and **vomers** are the membrane bones of the olfactory capsule.

The **nasals** (Fig. 14.12) are two large, flat triangular bones lying on the anterior dorsal region of the skull forming the roof of the olfactory capsules. The two nasals meet in the middle line. Their posterior ends diverge slightly and meet the frontoparietals so that a diamond-shaped area of sphenethmoid is visible dorsally. The anterior end of the nasals meet the dorsal process of the premaxillae and partly form the boundary of the external nares.

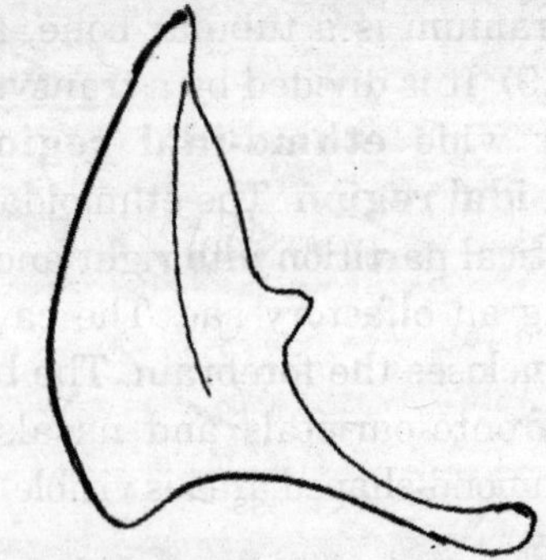

Fig. 14.12 Nasal

The **septomaxillaries** (Fig. 14.13) are a pair of small irregular-shaped bone lying near the anterior end of the nasal. It consists of a broad basal plate with two backwardly directed processes.

The **vomers** (Fig. 14.14) are two triradiate bones having three processes. Between the two outer processes occur the internal nares. Ventrally the vomers cover the floor of the olfactory capsule. The angle of each vomer bears a row of few pointed **vomerine teeth**.

Optic capsules: They remain cartilaginous throughout and are not fused with the cranium. They lodge the eyes.

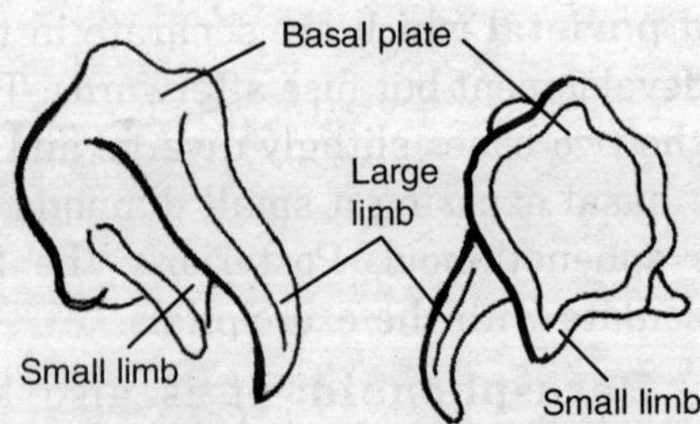

Fig. 14.13 *Frog*–Septomaxillary, **(A)** Dorsal view; **(B)** Ventral view

Visceral skeleton

The visceral skeleton includes the two **jaws—upper** and **lower** and the **hyoid apparatus**.

Upper jaw: The upper jaw is also known as the maxillary arch. Each half of the upper jaw divided into three parts :

(i) **quadrato-jugal**, (ii) **maxilla** and (iii) **premaxilla**.

The **quadrato-jugals** are short, comma shaped bone (Fig. 14.15) devoid of teeth, forming the posterior part of the upper jaw. It joins the maxilla anteriorly by an oblique suture and posteriorly it articulates with the pterygoid and squamosal bones.

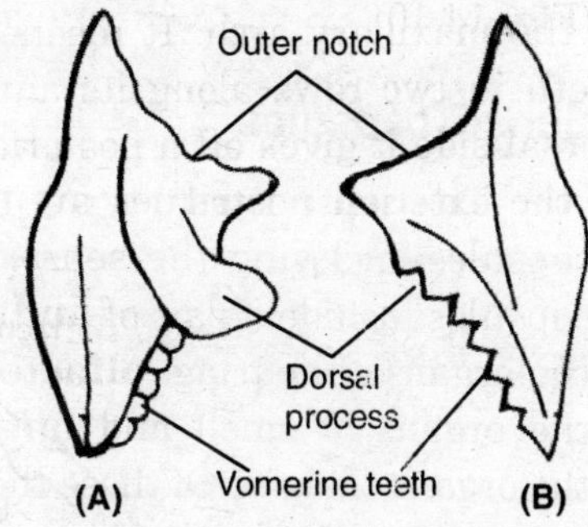

Fig. 14.14 *Frog*–Vomer; **(A)** Dorsal view, **(B)** Ventral view

The **maxilla** (Fig. 14.15) is the largest bone and forms the greater portion of the outer margin of the upper jaw. It is a long and slightly curved bone provided with numerous sharp, pointed, backwardly directed **teeth** throughout its length. Its posterior end is joined with the quadrato-jugal and anterior end with the premaxilla. Dorsally it gives off a **frontal process** overlapped by the nasal.

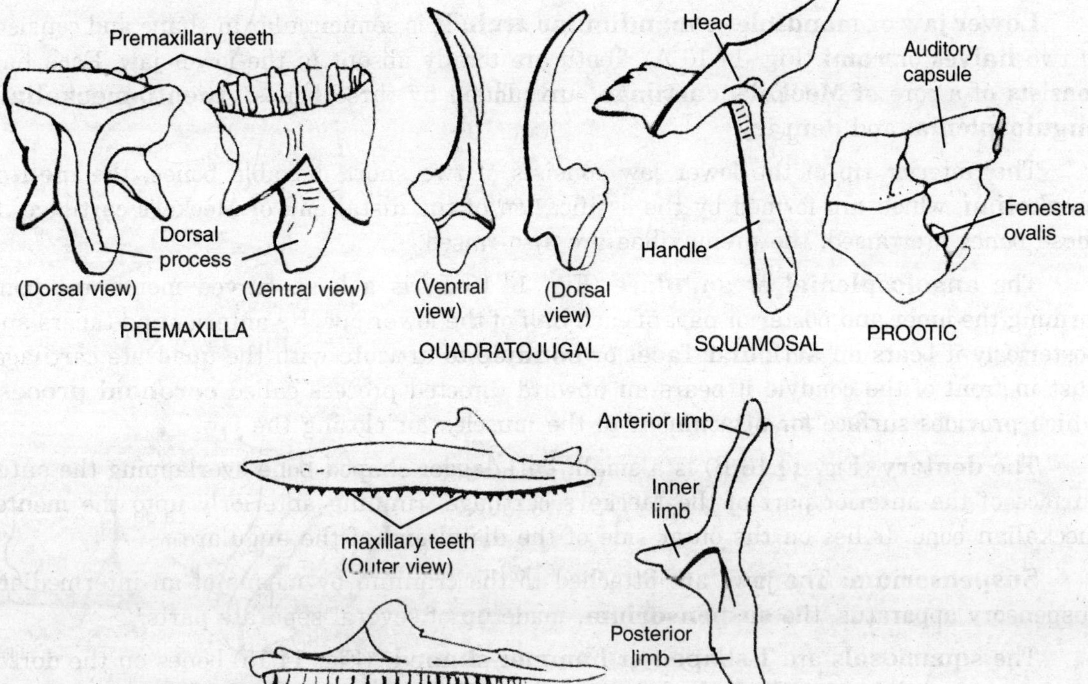

Premaxillary teeth
Dorsal process
(Dorsal view)
(Ventral view)
PREMAXILLA
(Ventral view)
(Dorsal view)
QUADRATOJUGAL
Head
Handle
SQUAMOSAL
Auditory capsule
Fenestra ovalis
PROOTIC
maxillary teeth
(Outer view)
(Inner view)
(MAXILLA)
Anterior limb
Inner limb
Posterior limb
PTERYGOID

Fig. 14.15 *Frog*–Disarticulated bones of upper jaw

The **premaxilla** (Fig. 14.15) is a small irregular bone forming the anterior-most part of the maxillary arch. It meets its fellow along the middle line. It also bears sharp conical teeth in two rows along its anterior margin. Posteriorly it is joined to the maxilla. On its dorsal side it gives off a posteriorly directed process which forms part of the inner boundary of the external nostril.

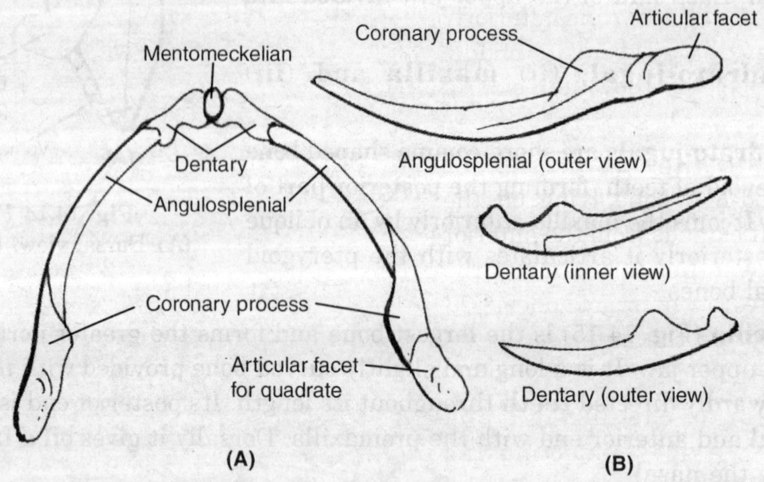

Fig. 14.16 *Frog*–**(A)** Complete mandible; **(B)** Loose bones of mandible

Lower jaw or **mandible** or **mandibular arch:** It is semicircular in shape and consists of two **halves** or **rami** (Fig. 14.16 A). Teeth are totally absent in the lower jaw. Each half consists of a core of **Meckel's cartilage** surrounded by three bones : **mento-meckelian, angulosplenial** and **dentary**.

The anterior tip of the lower jaw consists of two short movable bones, the **mento-meckelian,** which are formed by the ossification of the distal end of Meckel's cartilage. If these bones are raised, the premaxillae are also raised.

The **angulosplenial** or **angulare** (Fig. 14.16 B) is a long, curved membrane bone forming the inner and posterior part of each half of the lower jaw. Its anterior end tapers and posteriorly it bears an **articular facet** or **condyle** to articute with the quadrate cartilage. Just in front of the condyle it bears an upward directed process called **coronoid process** which provides surface for attachment to the muscles for closing the jaw.

The **dentary** (Fig. 14.16 B) is a small, flat, dagger-shaped bone overlapping the outer surface of the anterior part of the Meckel's cartilage, running anteriorly upto the mento-meckalian bone. It lies on the outer side of the distal end of the angulare.

Suspensorium: The jaws are attached to the cranium by means of an intermediate suspensory apparatus, the **suspensorium**, made up of several separate parts.

The **squamosals** are **T-shaped** or **hammer-shaped** (Fig. 14.15) bones on the dorso-lateral side of the posterior end of the cranium. The stem of squamosal remains closely applied on the angle of the jaw. The posterior limb of T articulates with the prootics while the anterior end extends obliquely downwards in front and remains free.

The **pterygoid** is a **triradiate** or **V-shaped** (Fig. 14.15) bone lying ventral to the squamosal. The inner limb of pterygoid runs beneath the long stem of squamosal to connect with the angle of the jaw (quadratojugal and quadrate cartilage), the outer arm runs anteriorly joining the maxilla at about its middle and third limb joins the outer side of prootics.

The **palatine** is a slender transverse rod-like bone lying on the ventral side. It extends from the anterior end of the sphenethmoid and meets the maxilla where the pterygoid meets it. Ventrally it forms the anterior side of each orbit.

The **quadrate cartilage** is a thin rod-shaped cartilage connecting the mandible and cranium at each angle of the mouth. In adult frog it is completely concealed between the squamosal (dorsally) and pterygoid (ventrally). On the inner side it is fused with the auditory capsule and on the outer side with the posterior end of quadratojugal.

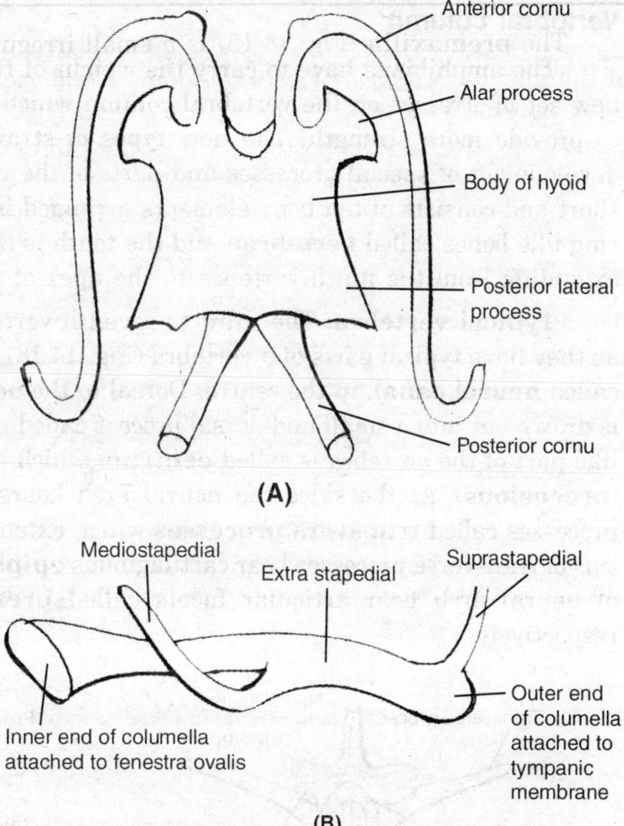

Fig. 14.17 *Frog*–**(A)** Hyoid apparatus; **(B)** Columella auris

Columella auris: Columella auris (Fig. 14.17 B) is partly bony and partly cartilaginous rod-like ear ossicle present in the cavity of each middle ear. Its outer end is attached to the middle of the tympanic membrane and inner end is attached to the fenestra ovalis of the auditory capsule. It serves to conduct sound vibrations to the inner ear.

Hyoid apparatus: In adult frog the gills disappear, so the remains of the branchial arches of the two sides of the hyoidean arch form the **hyoidean apparatus** (Fig. 14.17 A). It lies in the floor of the mouth and provides surface of attachment to the tongue. The hyoid apparatus is cartilaginous. It consists of following parts :

Body of hyoid: It is a thin flat cartilaginous plate more or less quadrangular in outline. It lies beneath the tongue. All its angle give rise to short processes.

Anterior cornua: From the anterior margin of the body of hyoid arise two long slender rods of cartilage called **anterior cornua** or **horns**. Each horn runs upwards and backwards on either side of the neck and joins the prootic bone of the auditory capsules.

Posterior cornua: From the posterior margin of the body of hyoid arises a pair of stout bony diverging processes called **posterior cornua** or **thyrohyals** or **thyroid processes**. They lie on either side of the glottis.

Vertebral column

The amphibians have to carry the weight of the body on four legs. This places entirely new set of stresses on the vertebral column which is largely bony and articulated together to provide more strength. The new types of strain involve new muscle attachments and development of special processes and parts of the vertebrae. The vertebral column of frog is short and consists of ten bony elements arranged in a linear series. The first nine are small ring-like bones called **vertebrae** and the tenth is the **urostyle** which is a long rod-like bone extending from the ninth vertebra to the apex of the pelvic girdle.

Typical vertebra: The third to seventh vertebra of frog are called **typical vertebra** as they have typical parts of a vertebra (Fig. 14.18). Each vertebra is ring-like and has a hole called **neural canal**, in the centre. Dorsal to the neural canal there is a **neural arch** which is drawn out into a small mid-dorsal process called **neural spine**. The ventral thick solid rod-like part of the vertebra is called **centrum** which is concave in front and convex posteriorly (**procoelous**). At the sides the neural arch bears a pair of elongated outwardly directed processes called **transvers processes** which extend almost at right angles to the body. The tips of transverse processes bear cartilaginous **epiphysis**. The anterior and posterior margins of neural arch bear articular facets called **prezygapophysis** and **postzygapophysis** respectively.

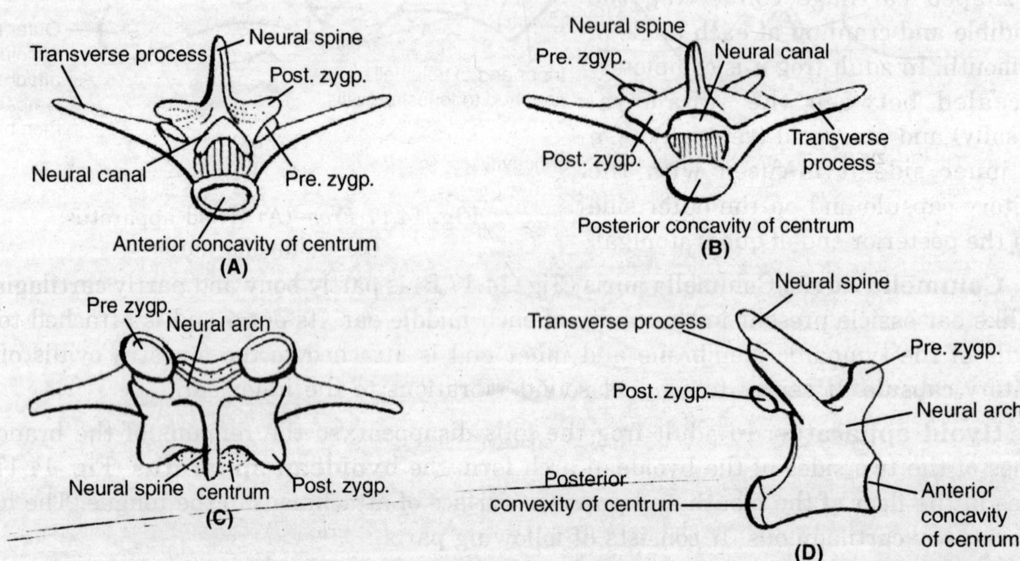

Fig. 14.18 *Frog*–Typical vertebra. **(A)** Anterior view; **(B)** Posterior view; **(C)** Dorsal view; **(D)** Lateral view

First vertebra: The first vertebra or **atlas** (Fig. 14.19) is highly specialized because it has to support the skull on the anterior face. It is a small ring-shaped vertebra devoid of transverse processes and prezygapophysis and with reduced centrum and neural spine. The anterior face of the centrum bears a pair of oval concave facets for the articulation of the occipital condyles. The neutral arch is well developed with postzygapophysis present on the posterior margin.

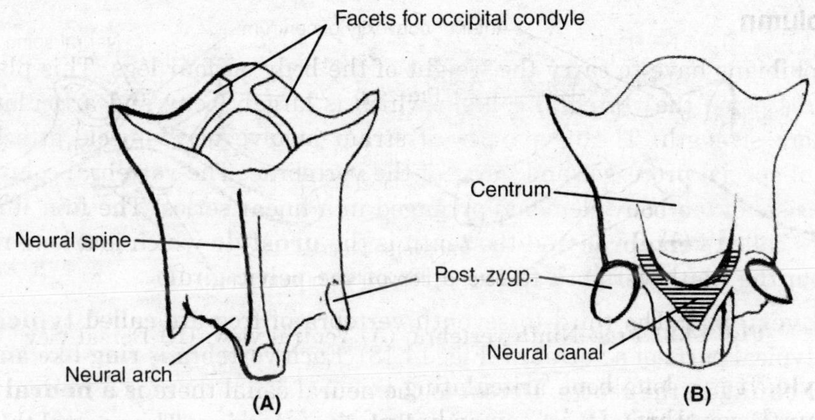

Fig. 14.19 *Frog*–Atlas vertebra, (A) Dorsal view; (B) Ventral view

Second vertebra: It is similar is structure to the typical vertebra but the neural spine is short and the transverse processes are short and broad and flat distally (Fig. 14.20A).

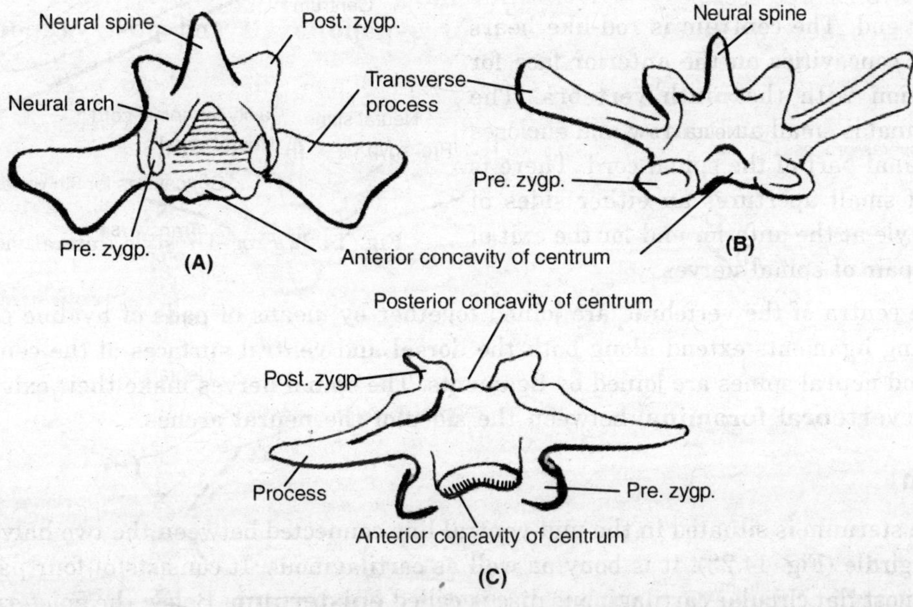

Fig. 14.20 *Frog*–(A) Second vertebra; (B) Fourth vertebra; (C) Eighth vertebra

Eighth vertebra: The eighth vertebra is specialized. The centrum is **amphicoelous** or **biconcave**. The anterior concavity receives the convexity of the seventh vertebra and the posterior concavity receives the convexity of the ninth vetrebra (Fig. 14.20 C).

Ninth vertebra: The ninth or **sacral** vertebra (Fig. 14.21) has a **biconvex centrum**. Anteriorly the centrum has a single convexity which articulates with the concavity of the centrum of the eight vertebra. Posteriorly a pair of convexities are present for articulation with the urostyle. The transverse processes are large, cylindrical and strong and directed backwards articulating with the ilia of the pelvic girdle. Neural spine is reduced, prezygapophysis are well developed and postzyapopysis are absent.

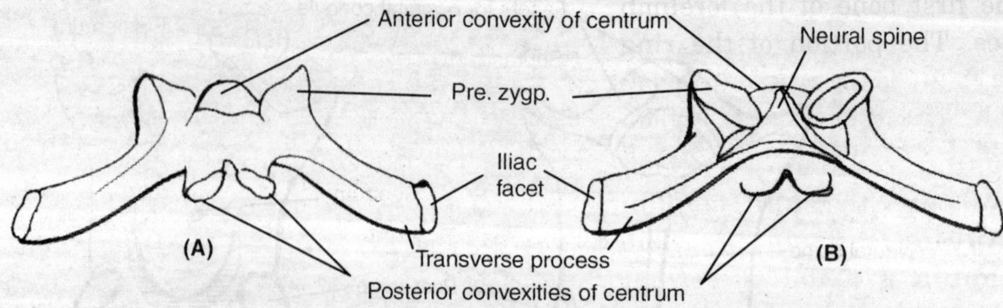

Fig. 14.21 *Frog*–Ninth vertebra, **(A)** Ventral view; **(B)** Dorsal view

Urostyle: It is a long bone aritculating with the ninth vertebra. It is somewhat triangular is shape (Fig. 14.22) with a broad anterior end and a tapering posterior end. On the dorsal side it is elevated into a prominent **keel** or **ridge** which extends nearly up to the posterior end. The centrum is rod-like bears a pair of concavities on the anterior face for articulation with the ninth vertebra. The neural canal is small and narrow and encloses the terminal part of the spinal cord. There is a pair of small apertures an either sides of the urostyle at the anterior end for the exit of the last pair of spinal nerves.

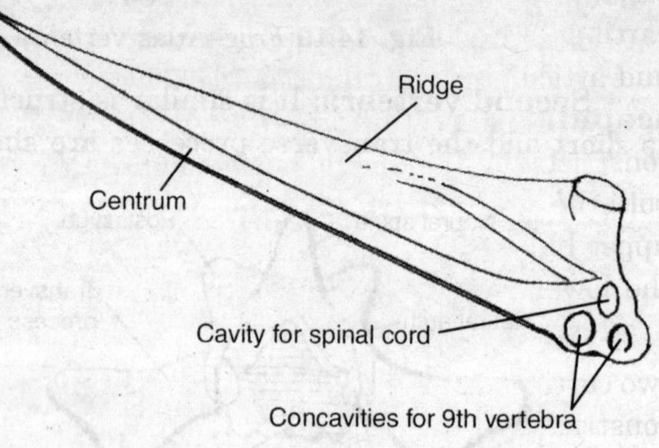

Fig. 14.22 *Frog*–Urostyle (lateral view)

The centra of the vertebrae are joined together by means of pads of hyaline cartilage. Connecting ligaments extend along both the dorsal and ventral surfaces of the centra. The arches and neural spines are joined by ligaments. The spinal nerves make their exit through the **intervertebral foramina**, between the sides of the neural arches.

Sternum

The sternum is situated in the mid-ventral line connected between the two halves of the pectoral girdle (Fig. 14.23). It is bony as well as cartilaginous. It consists of four parts. The anteriormost flat circular cartilaginous disc is called **episternum**. Below the episternum lies a bony rod called **omosternum** which connects it with the clavicles. The **mesosternum** is a cartilaginous rod ensheathed in bone, projecting behind the epicoracoids. The terminal broad cartilaginous plate is called the **xiphisternum**. There are no ribs in frog.

Appendicular skeleton

Pectoral girdle

The pectoral girdle surrounds the anterior part of the body in the manner of a girdle and protects the inner organs, like heart, lungs and liver. It also gives support to the forelimbs. It is made up of two **half rings** (Fig. 14.23) one on each side of the body. Each half ring bears, in the middle of its hinder surface, a cup, the **glenoid cavity** or **fossa** with which the first bone of the forelimb articulates. The portion of the ring above the cavity is the scapular region and that below is the coracoid region.

Scapular region: This region has got two parts: the **suprascapula** and **scapula** (Fig. 14.23). The **suprascapula** is a flat, somewhat rectangular bone which holds the body around the back and covers first four vertebrae dorsally. Its distal part is expanded and made up of calcified cartilage while the lower part is bony and articulates with the scapula. The **scapula** is a stout, oblong bone constricted in the middle but broad at both the ends. Posteriorly it forms the upper half of the glenoid cavity which receives the head of the first bone of the forelimb. At the lower end of the scapula there is a cartilaginous projection, the **acromion**.

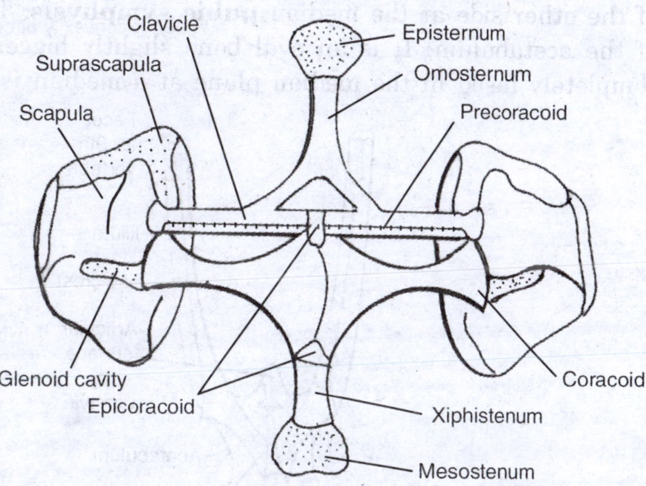

Fig. 14.23 *Frog*–Pectoral girdle and sternum (ventral view)

Coracoid region: This region consists of two bones, the **coracoid** and **clavicle** and two cartilages, the **precoracoid** and **epicoracoids** (Fig. 14.23). The **coracoid** is a stout bone constricted in the middle and inner end broadened. The outer end forms the other half of the glenoid cavity. This bone connects the sternum with the posterior end of the scapula. The **precoracoid** is a slender strip of cartilage running from the anterior edge of the scapula to the sternum. It lies attached to the posterior end of clavicle. The **clavicle** is a slender membrane bone, closely applied to the anterior extremity of the precoracoid. It is separated from the coracoid by a wide gap called **coracoid foramen**. The **epicoracoid** is a thin strip of calcified cartilage which meets its fellow in the middle and lies between the ventral ends of the precoracoid and coracoid.

Pelvic girdle

The pelvic girdle gives support to the hindlimbs. It is a V-shaped structure (Fig. 14.24 A), the apex of which lies at the posterior end of the skeleton and receives the tip of the urostyle, and the two limbs of the girdle diverge anteriorly and receive the two transverse processes of the sacrum. The girdle is made of two similar halves called as **innominatum**. Each half ring is composed of three bones—**ilium**, **ischium** and **pubis**. Almost in the middle of the ring there is a cavity known as the **acetabulum** which receives the thigh bone of the hindlimb.

The **ilium** is a long bone which runs forward and meet the transverse processes of the ninth vertebra. On its dorsal surface there is a thin prominent vertical ridge called **iliac crest**. Posteriorly the two ilia meet at the **iliac symphysis** which lies in the median plane and form the anterior half of the acetabulum. The **pubis** is a triangular hyaline cartilage forming about one-sixth of the lower side of the acetabulum. In older frog it is a calafied cartilage. It lies on the ventral side of the pelvic girdle. It is completely fused with its fellow of the other side at the median **pubic symphysis**. The **ischium** forms the posterior part of the acetabulum. It is an oval bone slightly bigger than the pubis. The two ischia are completely fused in the median plane at a median **ischiatic symphysis**.

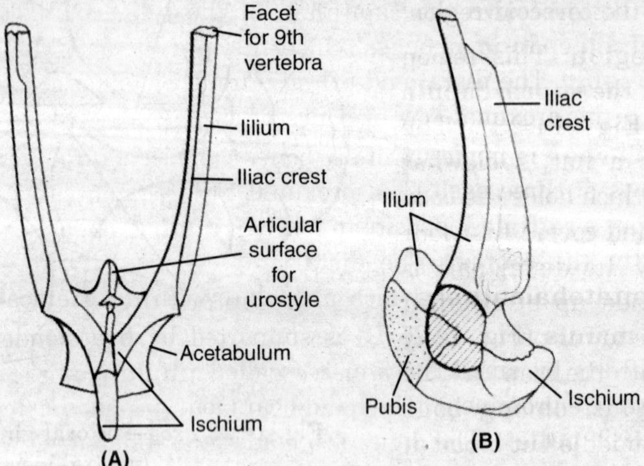

Fig. 14.24 *Frog*–Pelvic girdle, **(A)** Dorsal view; **(B)** One half in lateral view.

The elongation of the two ilia, the reduction of pubis and ischium and fusion of these bones forming an almost circular disc is very remarkable in the pelvic girdle. It helps the animal in leaping.

Forelimb

The forelimb consist of one long bone called the **humerus**, two bones of the fore arm, the **radius** and **ulna** and several bony pieces and rods forming the wrist and fingers.

The **humerus** (Fig. 14.25 A) is the bone of the upper arm. It is a small cylindrical bone with a slightly curved **shaft**. It has a round ball-like head on its proximal end which fits into the glenoid cavity of the pectoral girdle. The head is covered by calcified cartilage. Its lower end has a rounded prominence called **capitulum** or **trochlea**, in the middle. On either side of the capitulum there is a small projection called **condyle** which articulates with the groove of radio-ulna. On the ventral side of the shaft there is a large crest the **deltoid ridge**, starting from the head of the humerus almost to half of its length.

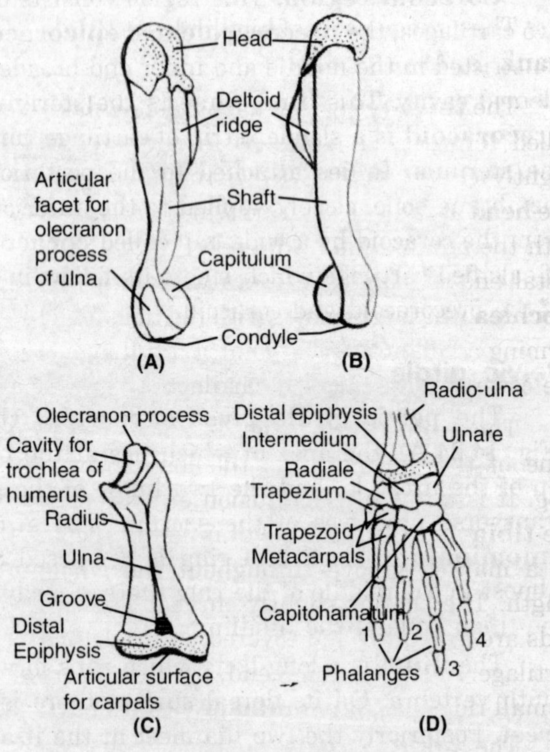

Fig. 14.25 *Frog*–Bones of forelimb, **(A)** Posterior view; **(B)** Anterior view of humerus; **(C)** Radio-ulna; **(D)** Bones of hand

The skeleton of fore arm consists of a compound bone the **radio-ulna** (Fig. 14.25 C) which is formed by the fusion of two bones, the inner **radius** and outer **ulna**. The line of union between the two bone is very distinct especially at the distal end. Proximally the radio-ulna has a concavity to receive the capitulum of the humerus. The ulnar part projects forwards forming an **olecranon process** which fits into the articular facet of the humerus forming the **elbow joint**. The distal end of radio ulna widens and ends in two **epiphyses** for articulation with the proximal row of bones of carpal.

The **wrist** or **carpus** is made up of six bones called **carpals** (Fig. 14.25 D) which are arranged in two rows of three each. The proximal row articulates with the radio-ulna and consists of three bones—**radiale** (opposite to radius), **ulnare** (opposite to ulna) and **centrale** or **intermedium** (in the centre). The distal row consists of three bones—**trapezium**, **trapezoid** and **capitatohamatum**, which articulate with the metacarpals.

The **hand** or **manus** (Fig. 14.25 D) is supported by five slender rod-like bones, the **metacarpals** succeeded by small bony pieces called **phalanges**. The first metacarpal is rudimentary and four are prominent. The hand bears four prominent fingers or digits and one rudimentary **thumb** or **pollex**. The digits are made up of phalanges which vary in number. The first and second digits have two phalanges each and the third and fourth have three phalanges each.

Hindlimb

The various parts of hindlimb are **thigh**, **shank**, **ankle** and **foot**.

The **thigh** consists of a long slender bone called **femur** (Fig. 14.26 A). It consist of a slightly curved **shaft** with a rounded knob-like **head** at the proximal end which articulates with the acetabulum of the pelvic girdle. The distal end is expanded laterally to form the **trochlea** which articulates with the tibio-fibula forming a ball and socket joint. Both the ends are covered with calcified cartilage.

The **tibio-fibula** (Fig. 14.26 B) is the bone of the **shank**. It is the longest bone frog. It is formed by the fusion of two bones the **tibia** and **fibula**. Line of union is shown by a marked groove throughout the whole length. The bone is slightly curved. The two ends are expanded and covered with calcified cartilage. At the proximal end, the tibia bears a small ridge called **cnemial** or **tibial crest**.

The **ankle** or **tarus** consists of two rows of bones called **tarsals** and each row consists of only two bones. The bones of the proximal

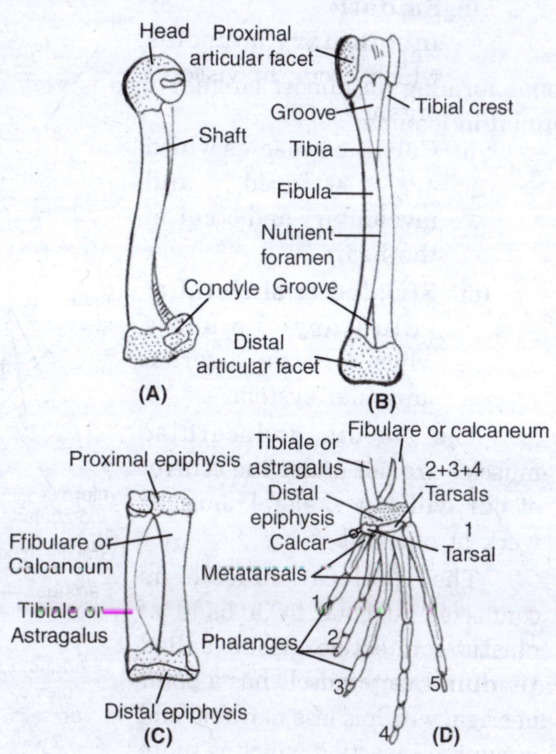

Fig. 14.26 *Frog*–Bones of hindlimb, (A) Humerur, Femur; (B) Tibio-fibula; (C) Same Astralagus calcaneum; (D) Bones of foot.

row are long, and united together at both ends but separated in the middle by a wide gap. The two ends are covered by calcified cartilage. The inner thin and slightly curved bone lying opposite to the tibia is called **tibiale** or **astragalus** and the outer thick and large bone is called **fibulare** or **calcaneum** (Fig. 14.26 C). The distal row of trasals are two very small bones.

The **foot** or **pes** (Fig. 14.26 D) consists of five elongated cylindrical **metatarsal** bones and five **toes** or **digits**. The first and second digits contain two phalanges each, the third and fifth have three phalanges each and the fourth has four. The first digit is called **hallux**. On the inner side of first toe there is an additional digit called **prehallux** or **calcar**.

The extension of the skeleton of foot and webbing of the digits helps in swimming. The length of hind legs of frog gives a good leverage in jumping.

MUSCULATURE

As the ancestral vertebrates moved out of water to land and changed the mode of locomotion, the muscular system also became modified.

In vertebrates three types of muscles are present

(i) **Smooth** or **involuntary muscles** which occur in visceral organs.

(ii) **Cardiac muscles** which are striated and involuntary and occur in the heart.

(iii) **Striated** or **skeletal** or **voluntary muscle** which form the external muscular system.

The smooth and cardiac muscles are not under the control of our will but skeletal muscles work at our will.

The skeletal muscle is connected to bone by a band of elastic connective tissue called **tendon**. Every muscle has a point of origin which is less movable and a point of insertion which is more movable. The muscles are arranged in opposing groups in such a way that when one group contracts, the other group relaxes. Different types of muscles found in frogs are:

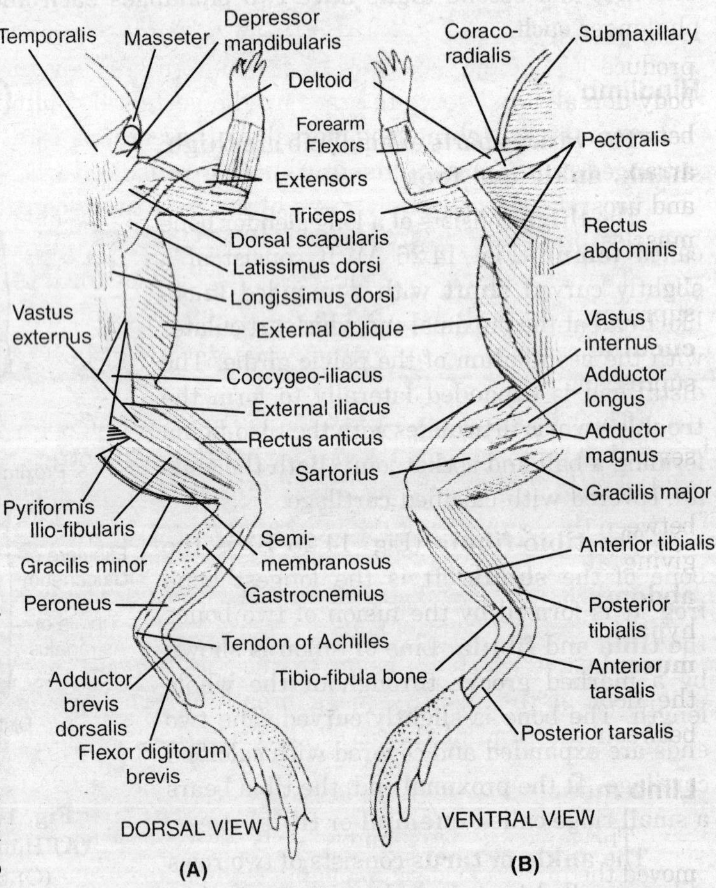

Fig. 14.27 *Frog*–Musculature exposed, **(A)** Dorsal view; **(B)** Ventral view

1. Flexor muscle: This muscle bends one part on another example, **biceps** bends forearm on upper arm.

2. Extensor muscle: This muscle straightens a part, example, triceps straightens the forearm.

3. Abductor muscle: This muscle draws a part away from the axis of body, example, **deltoid muscle** draws arm forward.

4. Adductor muscle: This muscle draws a part towards the axis of the body, example, **latissimus dorsi muscle** draws back the arm.

5. Depressor muscle: This muscle lowers a part, example **depressor mandibular muscle** lowers the jaw to open the mouth.

6. Levator muscle: This muscle raises or elevates a part, example, **masseter muscle** raises the lower jaw to close the mouth.

7. Rotator muscle: This muscle rotates a particular part, example, **pyriformis muscle** rotates the femur.

Trunk muscles

In anurans the dorsal portion of myotomes (**epaxial musculature**) no longer have to produce locomotory effect by lateral flexion, so they have become reduced. They bend the body dorsally and serve to brace up the vertebral column on the sacrum. Short muscles run between the vertebrae and dorsal to these there is a continuous sheet of longitudinally arranged fibres, the **longissimus dorsi muscle**, running from head to the sacral vertebra and urostyle. At the posterior end of the body the **coccygeo-sacratlis** and **coccygeo iliacus** muscles brace the urostyle on the pelvic girdle.

There are a series of muscles that bind pectoral girdle to the axial skeleton. From the suprascapula **rhomboid** and **levator sapulae muscles** run to the vertebrae and skull. The **cucullaris muscles**, derived from the lateral plate musculature, runs from the skull to the suprascapula.

The **hypaxial musculature** is more developed than in fishes. It is differentiated into several parts to support the viscera which needs better support on land than in water. In the mid-ventral region are the longitudinally arranged fibres, **rectus abdominis**, extending between the sternum and pubis. Transverse tendinous fibres interrupt these fibres at intervals giving an appearance of segmentation to them. The **obliqus externus** and **transversus abdominis** are thin sheets of muscle fibres supporting the sides. In the anterior region the hypaxial muscles have become restricted to the throat, where they from the **hyoid musculature** which are meant to raise (**mylohyoid muscle**) and lower (**geniohyoid muscle**) the floor of the mouth during breathing. The **submaxillary muscles** run transversely between the rami of jaw.

Limb muscles

The muscles of the limbs have presumably been derived from the radial muscles that moved the fins. In modern amphibians the limb muscles arise independently in the mesenchyme of the limb. There is no evidence that the limb musculature has been derived from the myotomes. The segmental origin, however, is established by the fact that the limb muscles are innervated by branches of spinal nerves of several segments.

In the **forelimb** there is a sheet of fibres running transversely to the main axis of the body and attached to the sternum and hypaxial muscles at one end and humerus at the other. Within this sheet occur the **deltoideus**, **pectoralis**, **coracoidialis** and **coracobranchialis** muscles. There is a set of muscles in each segment that serves to flex it on the next. The **branchioradialis** flexes the elbow joint, **flexor carpiradialis** and **flexor carpiulnaris** flex the wrist. The **flexor digitorum longus muscle** arise from the medial condyle of the humerus and is inserted on the carpus and terminal phalanges by tendons. The **flexor digitorum brevis muscles** arise from this tendon for insertion on the digits. Two muscles, **latissimus dorsi** and **dorsalis scapulae** run from the middle of the back to the humerus and serve to abduct and draw back the whole limb. The **triceps** are extensors of the elbow. The **extensor carpiradialis** and **extensor carpiulnaris** are extensors of the fingers.

In the hindlimb there are two sets of muscles, the **anterior muscles** which draw the limb forward and flex and adduct its joints and **posterior** muscles which draw it back and extend and abduct it. The anterior muscles of the thigh include the **pectineus** and **adductors** running from the pelvic girdle to the femur. They move the whole limb inwards (adduction). The **sartorius**, **biceps**, **semi-membranosus** and **semi-tendinosus** are two-joint muscles producing flexion at the knee as well as at the hip. The posterior muscles include **gluteus** and **tensor fascia lata** extending from pelvic girdle to femur across the thigh joint. The large **cruralis** (including the **rectus femorals** and **triceps**) runs from girdle and femur to tibia. This is the main extensor of the knee. The extension of knee is an important part of jumping movement of frog. The muscles that flex the ankle joint include the **tibialis anterior** and **paroneus** extending from the femur to the tarsus. The **gastrocnemius** (**plantaris longus**) runs from the femur and is attached to the tarsus through the tendon called **tendon of Achilles**. It is at the back of tibio-fibula and extends the ankel in swimming or jumping. Another muscle called **tiabialis posterior** runs from tibia to the tarsus. There is an elaborate system of small muscles for bending and stretching the toes and abducing them away from each other, so as to expand the circle and help in swimming.

BODY CAVITY

The body cavity or **coelom** of frog is relatively large and is without partition as in higher vertebrates. All the internal organs or viscera are lodged in the coelom. The coelom is lined by a thin transparent membrane called **parietal peritoneum**. The viscera is suspended in the coelom and is covered by coelomic epithelium, the **visceral peritoneum**. In some cases the coelomic epithelium is pulled out into double sheets between which connective tissue is laid down to form mesenteries through which the organs are supplied with blood vessels and nerves. Various organs in the body cavity are slung from the dorsal wall of the body cavity or connected with one another by mesenteries. On the anterior of the main body cavity a small **pericardial cavity** is cut off. This is enclosed by an extension of the peritoneum called **pericardium** and lodges the heart. The coelom is filled with **coelomic fluid** secreted by the epithelium. This fluid acts as a lubricant and enables the viscera to slip easily over one another.

DIGESTIVE SYSTEM

The digestive system includes the **alimentary canal** where the processes of digestion and absorption take place; and the **digestive glands** whose secretions bring about digestion.

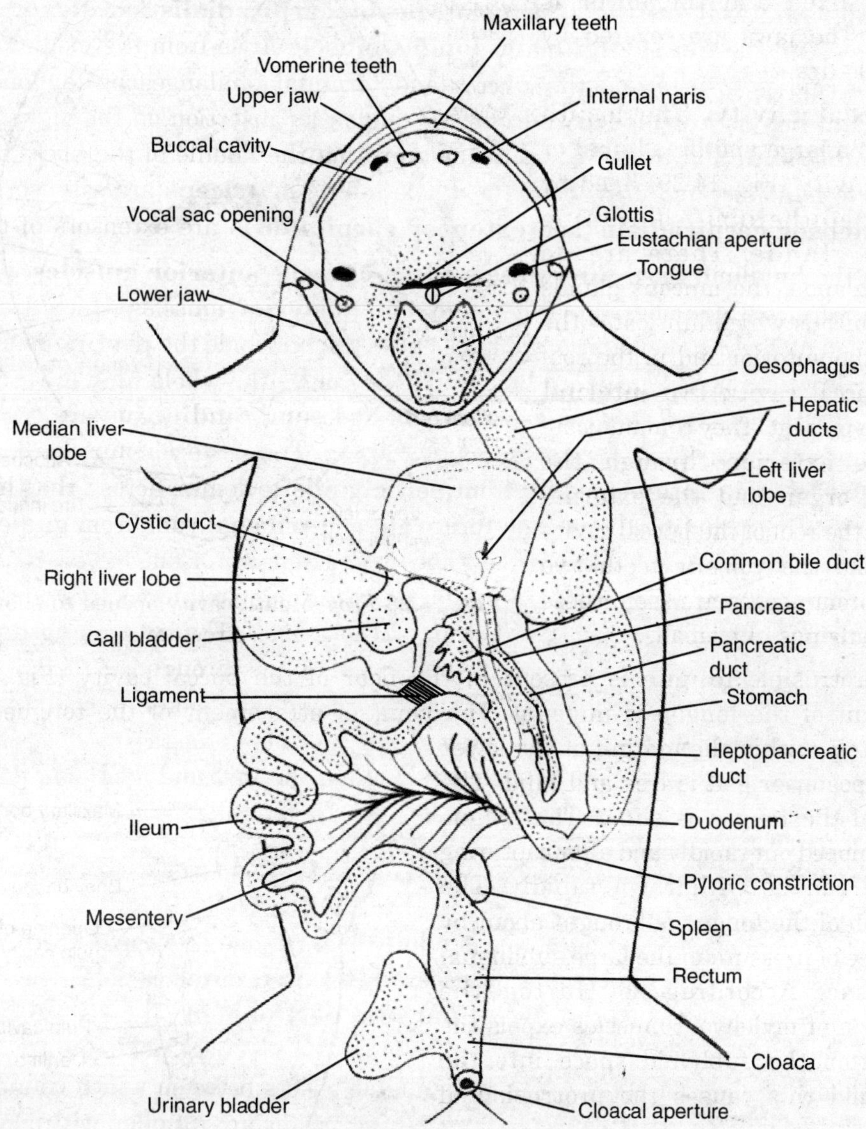

Fig. 14.28 *Frog*–Digestive system

Alimentary canal

The alimentary canal is a long coiled tube of varying diameter and includes the **mouth**, **buccal cavity**, **pharynx**, **oesophagus**, **stomach**, **small intestine**, **large intestine** and **cloaca (Fig. 14.28)**.

Mouth: The mouth is terminal. It is a wide opening extending from one side of the snout to the other. It is bounded by the jaws of which the upper is fixed and the lower is movable. The jaws are covered by immovable lips.

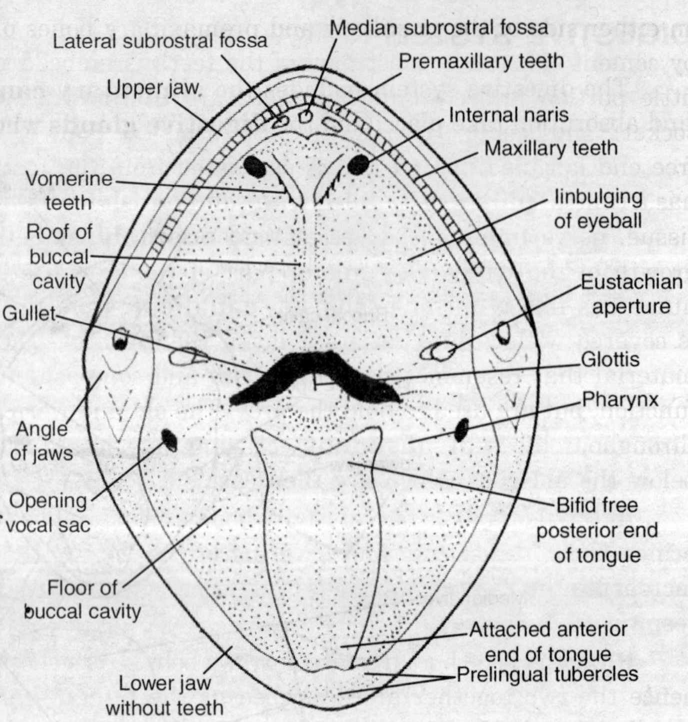

Fig. 14.29 *Frog*–Mouth cavity opened to show internal structure

Buccal cavity: The mouth leads into a large and broad **oral** or **buccal cavity** (Fig. 14.29) lined by ciliated epithelium containing **mucous glands**. There are no salivary glands, the mucous glands secrete mucous which lubricates the food. At the anterior end in the roof of the buccal cavity, two **internal nares** are present. They communicate with the exterior through the olfactory organ and the external nares. In the roof of the buccal cavity just behind the vomeric teeth two marked prominences are seen, these are the bulgings of eyeballs.

A protrusible **tongue** is present in the floor of the buccal cavity (Fig. 14.29). The attachment of the tongue is unusual. The point of attachment of the tongue is **anterior,** immediately within the margin of the lower jaw; the posterior end is free and bifid. The surface of the tongue is slimy. The tongue can be whipped out rapidly and after capturing the prey can be withdrawn rapidly. The protrusion of the tongue is brought about by the change of pressure in the large sublingual lymph sac. According to Hartog the contraction of mylohyoid muscles expels the lymph from the subhyoid space into the tongue and this causes the protrusion of tongue.

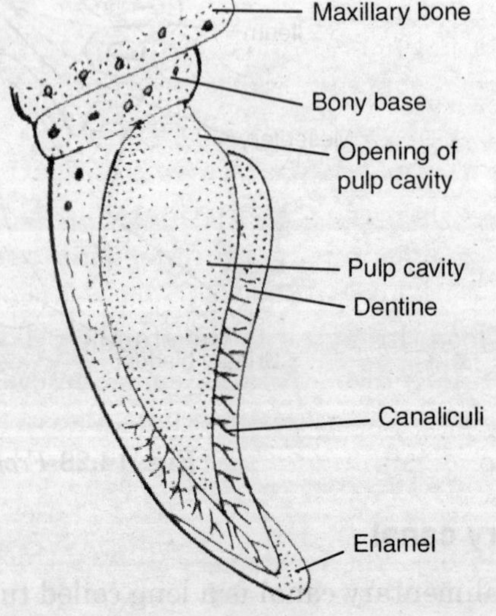

Fig. 14.30 *Frog*–Vertical section of maxillary tooth

Teeth: The upper jaw of frog bears **teeth** (Fig. 14.30) and the lower jaw is without them. The vomers present in the roof of the upper jaw also bear teeth called **vomeric teeth**. The teeth are small, conical and backwardly pointed and occur in a row on either side of the maxillary and premaxillary bones of the upper jaw. They are attached by cement substance. Most part of the teeth is embedded in the mucous membrane, only a little portion projects out. The teeth are **homodont** (similar) and **acrodont** (not set in socket). The basal portion of the tooth forms the **root** by which it is attached to the jaw. The free end is called **crown** and is separated from the root by a transverse furrow. The tooth has a central **pulp cavity** filled with soft nourishing tissue called **pulp** containing connective tissue, nerve fibres, blood vessels and **odontoblast cells** which produce new material for growth of the tooth. The greater part of the crown is composed of **dentine** traversed by numerous fine branching canals or **canaliculi** from the pulp. The upper half of the crown is covered with a hard resistant layer, the **enamel**. The root of the tooth is made up of a material that resembles bone in nature and composition. The teeth are not masticatory in function, but are used to hold the prey. The old and worn out teeth are continuously renewed throughout life (**polyphyodont**), ceasing only in old individuals. New teeth are produced below the old ones which are then cast out.

In the floor of the buccal cavity beneath the tongue lies the body of the hyoid. The hyoid brings about the raising and lowering of the floor of the lower jaw thereby decreasing and increasing the internal capacity of the buccopharyngeal cavity. This arrangement is used in respiration.

Pharynx: The buccal cavity narrows behind as a short pharynx without clear demarcation, hence the two together are described as the **buccopharyngeal cavity**. In the roof of the pharynx open the eustachian tubes by two wide laterally placed **eustachian apertures** which lead into the tympanic cavity of the middle ear. In the floor of the pharynx there is a median slit-like opening called **glottis**, behind the tongue. It leads into the trachea and opens for breathing and closes when the food is being swallowed. In male frog the **opening** of **vocal sacs** lie on either side near the angle of jaws on the floor of the pharynx. The pharynx` opens into the oesophagus through a wide opening called **gullet**.

Oesophagus: The pharynx leads into the oesophagus which is a short, muscular and highly distensible tube. The inner lining is folded longitudinally to allow expansion during passage of food through it to the stomach. The epithelial lining also contains some mucous glands which secretes an alkaline digestive juice. The oesophagus merges with the stomach.

Stomach: The stomach is a dilated thick walled, muscular, slightly curved part of the alimentary canal lying on the left side of the body cavity. It is attached to the dorsal body

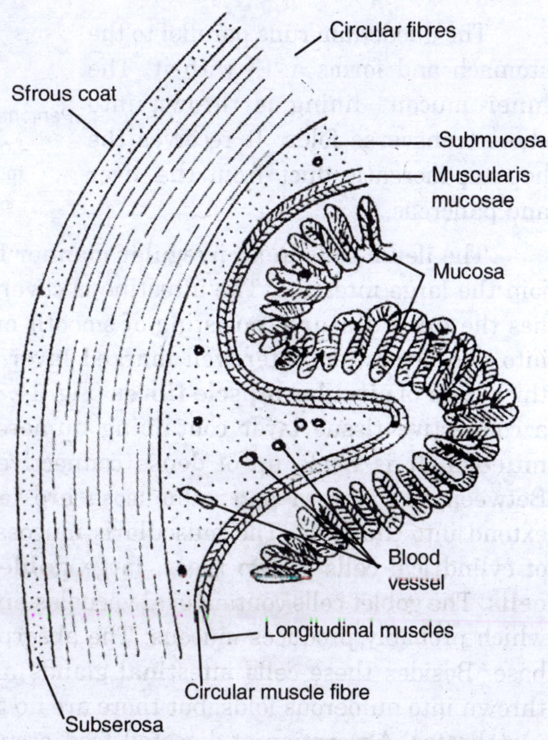

Fig. 14.31 *Frog*–A portion of stomach in T.S.

wall by a mesentery called **mesogaster**. The larger anterior part is called **cardiac stomach** and the narrow posterior end is called **pyloric stomach**. The wall of the stomach is thick and made up of six layers (Fig. 14.31) : (i) the **mucosa** or inner lining has many glands, (ii) the **submucosa** is a network of connective tissue containing blood and lymph vessels and nerves, (iii) the **muscularis mucosa** is made up of both circular and longitudinal bundles of smooth muscle fibres. (iv) the layer of circular muscle fibres which becomes thicker towards the pyloric end, (v) the **subserosa** or outer connective tissue layer, (vi) the **serosa** which is the peritoneum. The glands present in the mucous epithelium are of two kinds : (i) the multicellular **gastric glands** which secrete the enzyme pepsinogen and (ii) the unicellular **oxyntic cells** which secrete hydrochloric acid. The inner lining is thrown into longitudinal folds which allow distension when food is received. The pyloric end of stomach is slightly constricted and opens into the small intestine. The opening is guarded by a **pyloric valve**.

Small intestine: The small intestine is the longest part of the alimentary canal. It is a narrow coiled tube attached mid-dorsally to the body wall by a fold of mesentery. It is divided into two parts which are, however, not clearly demarcated, the short anterior part is called **duodenum** and the remaining part is called **ileum**.

The duodenum runs parallel to the stomach and forms a 'U' with it. The inner mucous lining is thrown into short transverse folds. It receives the hepatopancreatic duct from the liver and pancreas.

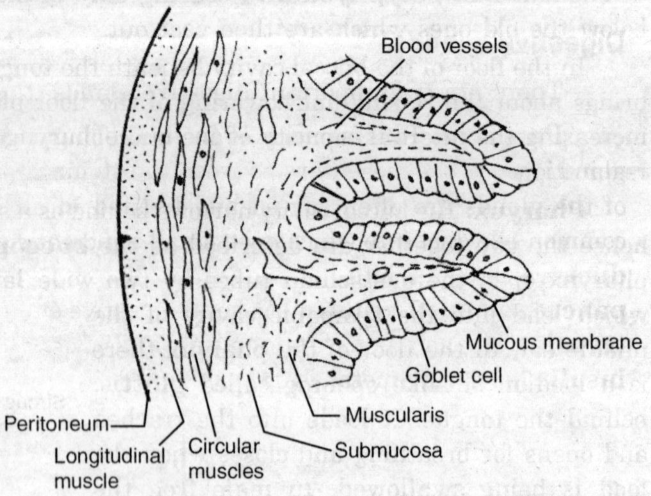

Fig. 14.32 *Frog*–A portion of ileum in T.S.

The ileum coils in an irregular manner forming several loops and widens posteriorly to join the large intestine. The intestine is covered by a thin coat of **peritoneum** within which lies the **musculature** consisting of smooth muscle fibres. The musculature is differentiated into two layers, an outer well marked layer of **longitudinal muscle fibres** and an inner thick layer of **circular muscle fibres** (Fig. 14.32). Within the musculature lies the **submucosa,** a connective tissue layer containing numerous blood vessels. The innermost layer is the **mucosa**. It is made up of dense connective tissue and contains more cellular elements. Between the mucosa and submucosa are large irregular lymph spaces which frequently extend into the folds. The muscularis mucosa is absent. The epithelium of mucous consists of cylindrical cells of two types, large **goblet** or **breaker cells** and ordinary **absorptive cells**. The goblet cells contain oval vacuoles and transparent, more or less, granular substance which probably produces mucous. The absorptive cells are narrow with oval nuclei near the base. Besides these cells intestinal glands are also present in the mucosa. The mucosa is thrown into numerous folds, but there are no true villi, crypts and definite glands as in higher vertebrates. Absorption of digested food occurs in the ileum.

Large intestine: The small intestine leads into the **large intestine** or **rectum**, the transition from one to other is not well marked. It passes straight back and opens into the **cloaca** by an anus guarded by **anal sphincter**. The large intestine is composed of the same layers as the small intestine. The epithelium of mucosa is thrown into numerous low longitudinal folds. Reabsorption of water takes place and faecal matter is stored in the rectum.

Cloaca: The cloaca is a small sac-like structure into which the anus, urinary bladder and urinogenital apertures open. The cloaca opens to the outside by a **cloacal aperture** or **vent** lying at the posterior end of the trunk.

Spleen: A dark red spherical body, the spleen is attached to the mesentery between the ileum and rectum. It is found in initimate association with the alimentary canal but has nothing to do with digestion. It is a well developed lymphoid structure which not only gives rise to new blood corpuscles but disposes off the old ones.

Digestive glands

There are two important digestive glands, the **pancreas** and **liver**.

Pancreas: It is a much branched, elongated, tubular, flattened and yellow coloured gland lying in the mesentery between the stomach and the duodenum. The terminal branches of the glands are often curved and twisted in an irregular manner. It is traversed by the common bile duct into which open small pancreatic duct thus forming the **hepatopancreatic duct** which opens into the duodenum. The pancreas has an **exocrine part** which secretes **pancreatic juice** containing several digestive enzymes (**steapsin** or **lipase**, **amylopsin** and **trypsin**). It also has an **endocrine part** fromed by **islets of langerhans** which produces **insulin** for sugar metabolism.

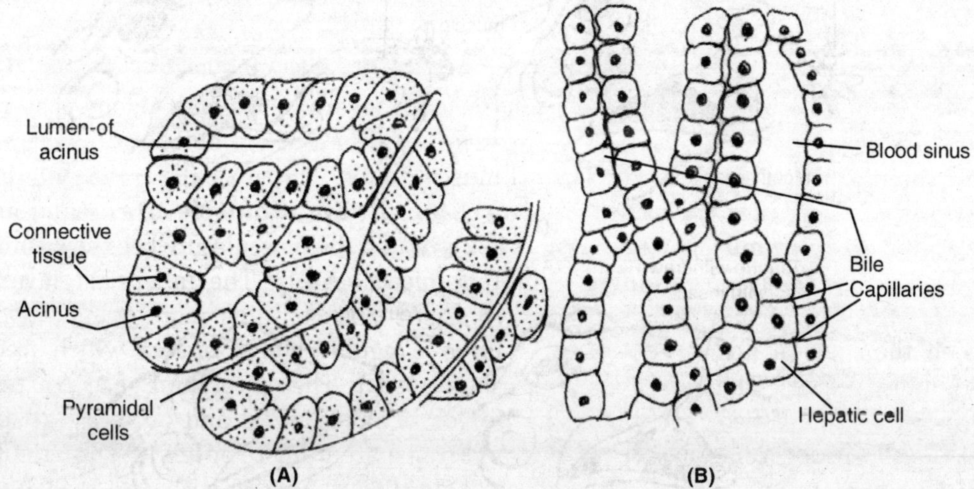

Fig. 14.33 *Frog*–**(A)** T.S. of pancreas, **(B)** T.S. of liver

Liver: The liver is a large reddish-brown gland located close to the heart and lungs and is supported by a fold of mesentery. It is made up of two lobes of which the left is again subdivided into two. Thus there are three lobes, **right**, **left** and **median**.

The liver lobes are made up of several polygonal cells (Fig. 14.33 B) which secrete greenish alkaline juice called **bile**. The bile is stored in a large spherical, thin walled sac the **gall bladder** which lies between the main lobes of the liver. The **cystic ducts** from the gall bladder and **hepatic ducts** from the liver combine to form the common **bile duct** which passes through the substance of the pancreas and receives small pancreatic ducts. The two together form the **hepatopancreatic duct** which opens into the duodenum. Bile has not digestive enzymes, it only emulsifies fats and takes no active part in digestion.

The secretory cells of the liver are cubical or polyhedral in form with large nuclei. They are arranged in parallel longitudinal rows appearing like columns or cords. The cytoplasm contains protein granules, small drops of fats, lumps of glycogen and some pigments. In between the liver cells occur **bile capillaries**. The liver receives arterial blood through hepatic artery and also from the portal system, which includes the anterior abdominal vein from the ventral body wall and the hepatic portal vein from the stomach, intestine pancreas and spleen. All the material, absorbed by the blood from the digestive organs passes through the liver before entering the general circulation. The liver is also supplied with lymph vessels which form perivascular lymph spaces around the capillaries.

Nutrition

Ingestion of food: The frog feeds chiefly on insects, earthworm, spiders, crustaceans, small molluscs, small fish, small frogs as well as tadpoles. It is carnivorous. It swallows its prey whole. The method of capturing prey (Fig. 14.34) remarkable is because it can catch insects on wing. When feeding it opens its mouth and flicks out its tongue and strikes the

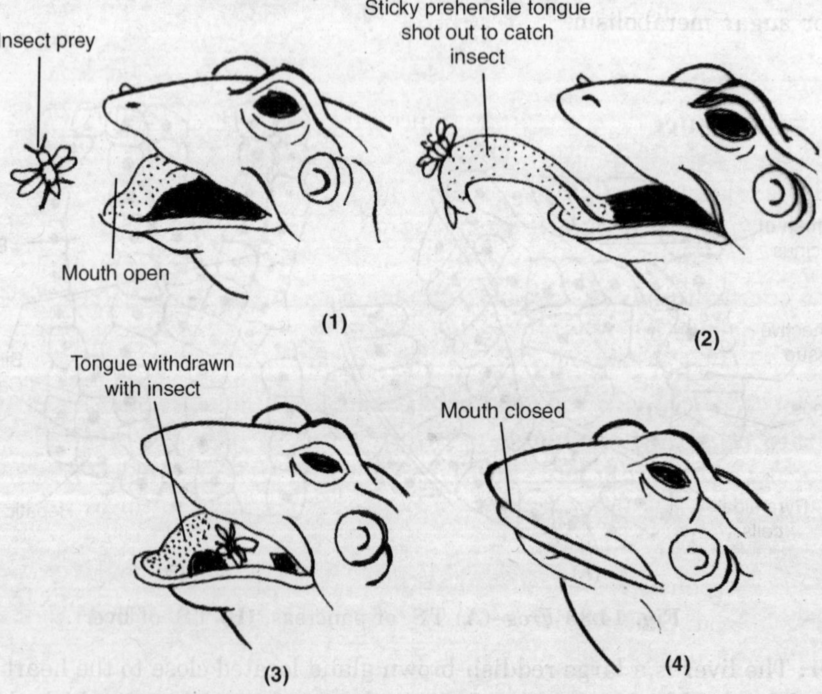

Fig. 14.34 *Frog*–Mechanism of capturing prey

prey with it. The tongue is covered with a viscid, sticky substance so that the prey adheres to it and by a return flash is brought into the back of the pharynx. From here the food is pushed by pharyngeal contraction into the oesophagus. The teeth are used to hold the food during swallowing. The prominences of the eyeball also help in pushing the food inwards. The salivary glands are absent. The mucus secreted by the lining of the buccopharyngeal cavity and oesophagus lubricates the food. From the oesophagus the food is pushed down by a wave of contraction of its muscular walls, called **peristalsis**.

Digestion of food: In the stomach the food is subjected to the action of gastric juice secreted by gastric gland. The gastric juice contains hydrochloric acid and inactive pepsinogen enzyme. Pepsinogen in the presence of hydrochloric acid is converted into active **pepsin** which hydrolyses proteins into **peptones** and **proteoses**. The gastric juice also contains the hormone **gastrin** which activates cells secreting hydrochloric acid. In the stomach the food is retained for 2-3 hours. Here it takes the form of a semidigested thick creamy acidic fluid called **chyme**. When the chyme is formed the pyloric sphincter relaxes and it passes down to the duodenum.

When chyme reaches duodenum it causes the production of a number of intestinal hormones with specific functions. The hormone **enterogastrone** reaches the stomach through blood and stops the production of gastric juice. Another hormone **cholecystokinin** brings about contraction of gall bladder and release of bile juice into the duodenum. The hormones **secretin** and **pancreozymin** induces the pancreas to produce its juice and release it into the duodenum. The secretion of intestinal juice, **succus entericus** is induced by the hormone enterocrinin. The digestion of food in the intestine is completed by three digestive juices :

Bile: Bile is greenish alkaline watery fluid which contains no digestive enzymes, so it takes no active part in digestion. It reduces the acidity of chyme and prepares the fluid for the activity of pancreatic lipase which can act only in alkaline medium. Bile brings about emulsification of fats.

Pancreatic juice: The pancreatic juice is a watery alkaline fluid very rich in enzymes which act best in an alkaline medium. It contains **trypsin** (proteolytic) **amylopsin** (carbohydrate digesting) and **lipase** (lipolytic) enzymes. **Trypsin** is present as inactive **trypsinogen** which is activated by intestinal juice and converts proteoses, peptones and polypeptides into amino acids. **Amylopsin** or **amylase** acts on starch (polysaccharide) and converts it into maltose (disaccharide). **Lipase** or **steapsin** act on emulsified fats converting them into fatty acids and glycerol.

Succus entericus: The intestinal juice also contains several enzymes which act upon all classes of food **Erepsin** converts polypeptides into amino acids. There are a number of carbohydrate digesting enzymes—**maltase** converts maltose into glucose, **sucrase** or **invertase** converts sucrose into glucose and fructose and **lactase** converts lactose to glucose and galactose. Fats are broken down into fatty acids and glycerol by lipase enzyme.

Absorption: The absorptive area of the alimentary canal lies in the small intestine. This area is enormously increased by the development of folds called **villi**. These folds have rich supply of blood. Glucose, fructose and amino acids pass into blood capillaries of the folds, then through the hepatic portal system into the liver. Fatty acids and glycerols pass into lymphatic capillaries or lacteals of the folds. Water, minerals and other nutrients are absorbed directly through the epithelial lining.

Assimilation: The absorbed food undergoes assimilation or is used for liberation of energy during respiration. Glucose is either stored as glycogen in liver or skeletal muscles or converted into fat or incorporated in protoplasm. Fatty acids and glycerol are converted into fats and stored in adipose tissue. Both carbohydrates and fats serve as sources of energy. Amino acids form proteins which are used for growth and repair or may undergo deamination to form urea.

Egestion: As the undigested food passes downwards the excess of water contained in the food residuum is extracted by the walls of the large intestine. It is then stored in the rectum in the form of semi-solid **faeces**. It is ejected or extruded at intervals through the anus into the cloacal chamber and finally to the exterior through the cloacal aperture.

CIRCULATORY SYSTEM

The circulatory system of amphibians is constructed on a plan different from that of the fish. Due to the appearance of lungs the single circuit heart of the fish undergoes special modifications. The blood flows in two circuits. It goes from the heart to the lungs and after oxygenation returns to the heart to be pumped back to different parts of the body. Thus the heart has two separate auricles and there is a mechanism in the ventricle and the truncus arteriosus to avoid mixing of oxygenated and deoxygenated blood. The blood vascular system of frog is closed and includes the following parts : (i) blood, (ii) heart, (iii) arterial system, (iv) venous system (v) lymphatic system. The circulatory system is concerned with the transport of gases, nutrients, hormones, wastes products, etc.

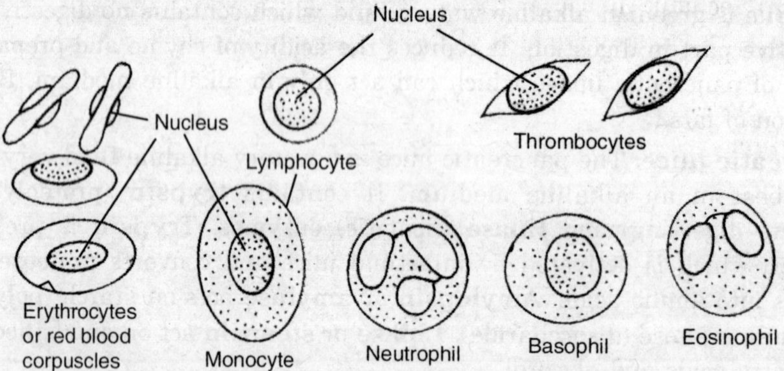

Fig. 14.35 *Frog*–Different types of blood cells

Blood: The blood is a liquid connective tissue. It presents the general structure consisting of a clear liquid called **plasma** in which various types of cells called **corpuscles** are suspended (Fig. 14.35).

Plasma forms major part of the blood. It contains about 90% water in which various soluble substances like mineral salts, absorbed food, waste materials, hormones, etc. are dissolved.

The blood corpuscles are of three types:

(i) **Red blood cells** (RBC) or **erythrocytes**

(ii) **White blood cells** (WBC) or **leucocytes**

(iii) **Platelets** or **thrombocytes**.

The **erythrocytes** are oval in shape and have an oval nucleus in the centre. Seen on edge they appear flattened and slightly bulging in the centre. They are 14 μ by 23 μ in size. There are about 250,000 to 450,000 RBC per cubic millimeter of blood. The cytoplasm of the cell contains haemoglobin.

The **leucocytes** vary both in size and form. The cytoplasm is nearly colourless but often granular. The form of nucleus varies being usually spherical in smaller leucocytes. There are about 5,000 to 7,000 WBC per cubic millimeter of blood. They have independent power of locomotion and are amoeboid. They are phagocytic ingesting bacteria and other foreign particles. They may pass out of delicate walls of the capillalries.

The **thrombocytes** are small nucleated spindle-shaped cells playing an important role in clotting of blood.

The blood cells are produced mainly in the spleen and bone marrow. The old worn out blood cells are destroyed in the spleen.

Heart: The heart is situated in the anterior part of the body cavity, ventral to the liver and protected by the pectoral girdle. It is somewhat triangular in shape, broad anteriorly and narrow posteriorly. The heart lies enclosed within a thin transparent sac called **pericardium**, whose cavity is completely cut off from the coelom, although continuous with it in early development. The pericardium is made up of two layers, the **parietal pericardium** forming the outer wall of the sac and the **visceral pericardium** which closely invests the heart. The narrow space between the two layers is filled with a watery **pericardial fluid** which protects the heart from mechanical shocks, friction and injury.

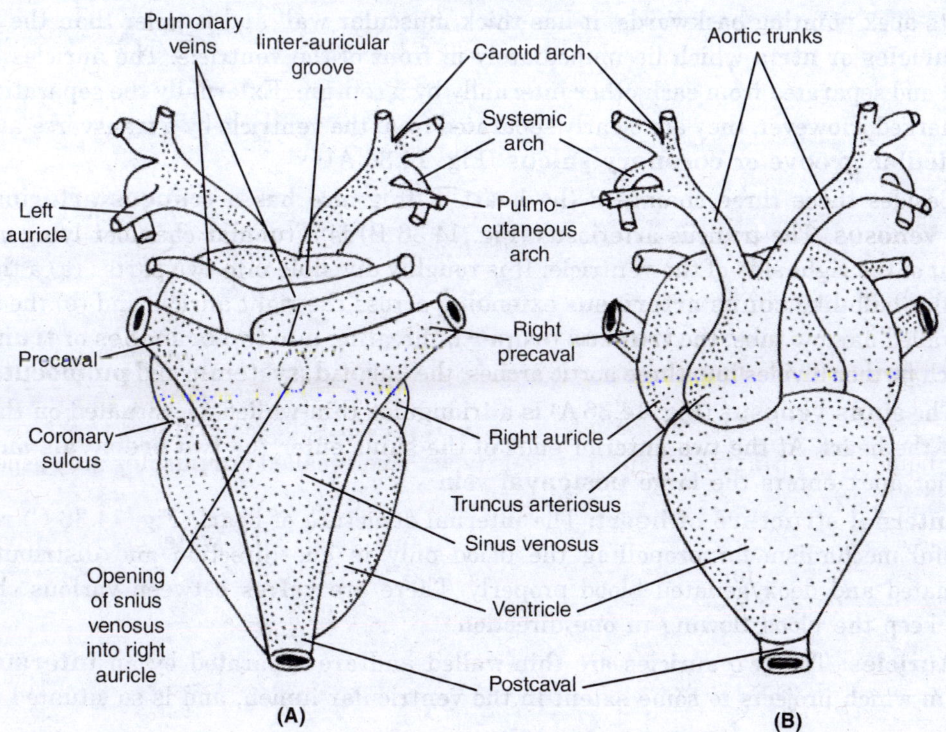

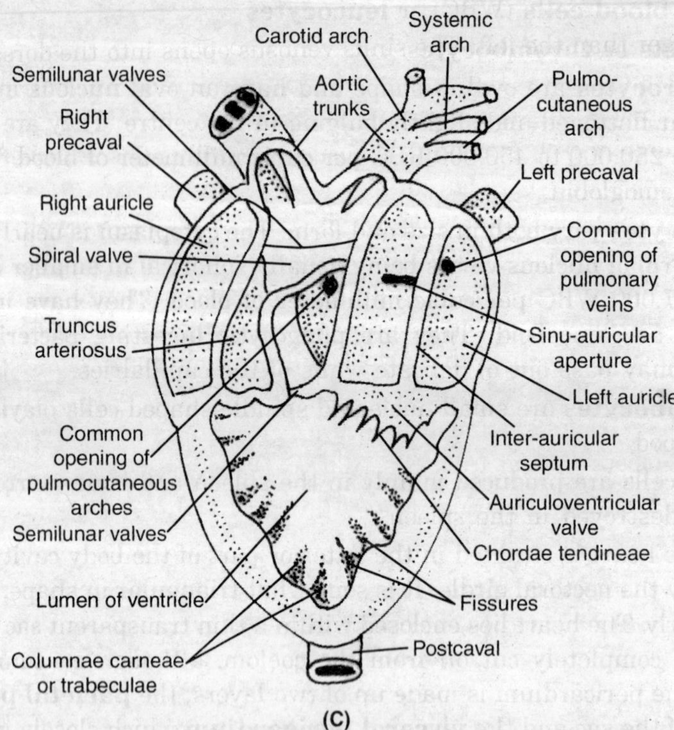

Fig. 14.36 *Frog*–Structure of heart. **(A)** Dorsal view; **(B)** Ventral view; **(C)** Internal structure

Chambers of heart: The heart of frog is three chambered : (i) single conical **ventricle**, with its apex pointing backwards, it has thick muscular wall and is paler than the rest; (ii) two **auricles** or **atria** which lie immediately in front of the ventricle. The auricles are thin walled and separated from each other internally by a septum. Externally the separation is not well marked. However, they are clearly separated from the ventricle by a transverse **auriculo-ventricular groove** or **coronary sulcus** (Fig. 14.36 A).

Besides these three chambers the heart of frog also has a **truncus arteriosus** and **sinus venosus**. The truncus arteriosus (Fig. 14.36 B) is a tubular chamber lying ventrally in front of the right side of the ventricle. It is roughly divisible into two parts : (a) a thickened muscular bulb-like **conus arteriosus** extending across the right auricle and (b) the anterior thin-walled narrow tube, the **truncus proper** bifurcating into two **branches** or **trunks** each of which further divides into three aortic arches, the **carotid**, **systemic** and **pulmocutaneous**.

The **sinus venosus** (Fig. 14.36 A) is a triangular thin-walled sac situated on the dorsal side of the heart. At the two anterior ends of the sinus enter the two **precavals** and at the posterior apex enters the large **postcaval** vein.

Internal structure of heart: The internal structure of heart (Fig. 14.36 C) reveals a beautiful mechanism for propelling the blood only in one direction and distributing the oxygenated and deoxygenated blood properly. There are valves between various chambers which keep the blood flowing in one direction.

Auricles: The two auricles are thin-walled and are separated by an **interauricular septum** which projects to some extent in the ventricular lumen, and is so situated that the right auricle is larger than the left. The sinus venosus opens into the dorsal wall of the right auricle by a large transverse oval opening, the **sinu-auricular aperture** which lies close to the **inter-auricular** spetum. The aperture is guarded by a pair of flap-like **valves** which allow the blood to flow in from the sinus venosus and prevent back flow. Slightly anterior to the sinu-auricular aperture, in the left auricle, just near the spetum there is a small **opening of the pulmonary vein**. It has no valves. The vein perforates the wall obliquely hence the pressure caused by the contraction of the auricles serves to close the opening. Both the auricles open into the ventricle by a large opening the **auriculo-ventricular aperture** which is divided by the inter auricular septum. The opening is guarded by 4 **valves**, two large ones on dorsal and ventral edges and two small ones on either side. Small fibres called **chordae tendinae** arise from the ventricular wall to be inserted into the free edges of the valves to prevent them from being pushed back into the auricles when the ventricle contracts and thus keep the blood flowing from the ventricle into the auricles.

Ventricle: In frog the ventricle is not divided into two chambers as in birds and mammals. Its has thick muscular and spongy wall. Its inner surface bears number of fissure like chambers separated from each other by muscular partitions or ridges called **columnae carnae** or **trabeculae**. They receive the blood coming from the auricles and prevent the mixing of the two samples.

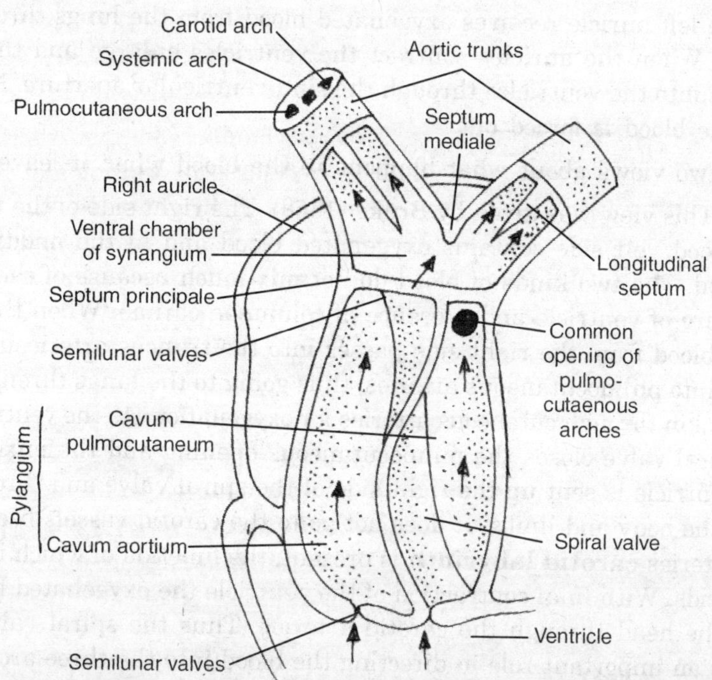

Fig. 14.37 *Frog*–Internal structure of truncus arteriosus in ventral view

Truncus arteriosus: The opening of the ventricle into the truncus arteriosus is guarded by three (four, according to Sharma, 1957) semilunar valves. They are in the form of pockets, open anteriorly and prevent back flow of blood from the truncus into ventricle. In the truncus there are three more pocket-shaped semilunar valves placed anteriorly. This set of valves divides the truncus into two parts (i) the long basal thick-walled **conus arteriosus** or **pylangium**, (ii) the distal short thin-walled **bulbous aorta** or **synangium** (Fig. 14.37). The cavity of pylangium is incompletely divided into two by a membranous fold the **spiral valve**, that arises in the right ventral side. This traverses the chamber obliquely. The left dorsal chamber is called **cavum pulmocutaneum** and the right ventral chamber is called **cavum aorticum**. The spiral valve is attached by its dorsal edge to one of the semilunar valve in front and its ventral edge hangs freely in the cavity of pylangium. Anterior to the spiral valve in the pylangium, there is a common aperture of the two pulmocutaneous arches. The separate openings of carotid and systemic arches lie in the synangium (Sharma, 1957, has described a common aperture for the carotid and systemic arches in frog). All the apertures are guarded by **valves**. Distally the synangium bifurcates into two **branches** or **trunk** which further divides into three arches—**carotid**, **systemic** and **pulmocutaneous**.

Working of heart: The heart keeps on beating throughout life. Contraction of heart is called **systole** and relaxation of heart is called **diastole**. The heart pumps the blood to all parts of the body through **arteries** and receives blood from different parts of the body through **veins**. The different chambers of heart contract in a regular sequence and the valves present prevent the back flow of blood. When sinus venosus contracts the auricles expand, the right auricle receives deoxygenated blood from the body tissues through the sinu-auricular aperture and the left auricle receives oxygenated blood from the lungs through the common pulmonary vein. When the auricles contract the ventricles enlarge and the blood from the two auricles pass into the ventricles through the sinu-ventricular aperture. Next the ventricle contracts and the blood is forced out.

There are two views about what happens to the blood when it leaves the ventricle.

Old view: This view was given by Brüke (1858). The right side of the ventricle contains deoxygenated blood, left side contains oxygenated blood and in the middle region there is some mixed blood. The two kinds of blood do not mix much because of their viscous nature and spongy nature of ventricle and presence of columnae carnae. When the ventricle starts contracting the blood from the right side passes into the truncus arteriosus and directed by the spiral valve into pulmocutaneous arteries, thus going to the lungs through the pulmonary arteries and the skin through cutaneous arteries for oxygenation. As the ventricular contraction proceeds the spiral valve closes the pulmocutaneous opening and the mixed blood from the middle of the ventricle is sent up the right side of the spiral valve and through the systemic vessels goes to the body and limbs. It does not go to the carotid vessels because at the bases of the carotid arteries **carotid labyrinth** is present, the function of which is to prevent blood flow in the carotids. With final contraction of the ventricle the oxygenated blood from the left side goes into the head through the carotid arteries. Thus the spiral valve in the truncus arteriosus plays an important role in directing the blood into the three arches. Although the ventricle is undivided yet a perfect double circulation is possible in this way.

But now this hypothesis is not regarded valid but described as a "fantasy arising from incorrect deductions". This theory assumes that the blood returning from the lungs is highly oxygenated and that the blood returning from the remaining parts of the body is deoxygenated as such is sent directly for oxygenation. This is not altogether correct for the frog uses its lungs for breathing only under certain circumstances. While the frog is at rest the air is continually pumped into and out of the buccal cavity, through the nostrils, it is forced into the lungs only rarely. The respiration goes on in the buccal cavity, the exchange of gases taking place in the moist mucous membrane of the buccal cavity. Gaseous exchange also occurs in the moist skin (**cutaneous respiration**). Naturally the blood returned from the skin (**cutaneous vein**) and from the buccal cavity (**jugular vein**) is richer in oxygen content than the blood returned from the lungs by **pulmonary veins**. While in water only cutaneous respiration is possible and the frogs normally spend a considerable time under water. Under these conditions the blood from the lungs will be poorest in its oxygen content. Naturally the blood returned from the skin will contain more oxygen. These facts created doubt in the old theory.

The frog's heart is now regarded as a single circuit heart. As these animals breathe mainly by the skin and mucous membrane of the buccal cavity, assisted by lung breathing, the oxygen obtained by the lungs is supplementary and is distributed along with the oxygen received from the buccal cavity to all parts of the body.

Modern view: Some authors like Vandervael (1933) and Foxon (1953) have established that the blood passes up on both sides of the spiral valve in the truncus arteriosus simultaneously, and that the spiral valve does not change its position during contraction of ventricle. The blood is completely mixed in the ventricle and flows simultaneously through the three arches to all parts of the body. Thus according to this view the inter auricular septum and the spiral valve of the truncus arteriosus are of no importance in frogs. DeLong (1962) stated that the carotid arches receives highly oxygenated blood, the systemic receive mixed blood and pulmocutaneous arches receive least oxygenated blood. The exact mechanism, however, is not well understood.

It seems that the spiral valve mainly exists for mechanical reasons. It prevents the truncus arteriosus from closing altogether and also provides a degree of mechanical support to the truncus which ensures that the main contraction is in its diameter and not in its length. The carotid labyrinth which is believed to help in making the selection of blood, has a different function. It acts as a sensory organ detecting pressure changes. The presence of two auricles also cannot be explained according to the above view. In some tailed amphibians the atruim is undivided into which arterial and venous blood is poured. There are some points, however, that need further investigations.

What is the use of pockets in the ventricular walls? It is clearly understood that the muscles of the heart require plentiful supply of oxygen to carry on their work. For this work the heart has a coronary supply which is very poor of frog. There is only a single small coronary artery which does not supply the muscles of the ventricle. The muscles get their supply of oxygen from the blood that passes through it. As the ventricular walls need more oxygen in response to the great pumping work they have to carry on, its oxygen assimilating surface has been increased by the appearance of pockets.

Arterial system

The arterial system begins with the truncus arteriosus which, after its bifurcation, splits into three pairs of arteries which are symmetrically disposed on either side of the middle line.

Of these the anterior **common carotid** goes to the head, the **systemic** to the body and viscera, and the **pulmocutaneous** to the lung and skin.

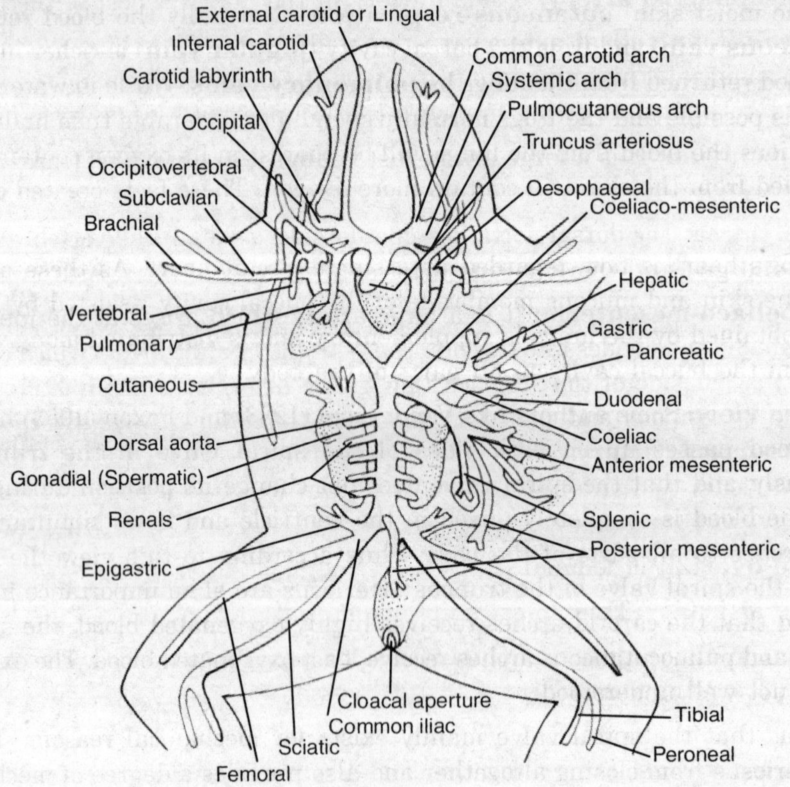

Fig. 14.38 *Frog*–Arterial system

Common carotid: It is a short vessel running forward but soon bifurcates into two:

(a) External carotid: It is a small inner branch which runs forward to the tongue as such it is also called the **lingual**. En route it supplies branches to the thyroid gland and muscles of the hyoid.

(b) Internal carotid: It is the large outer branch. At the base where the common carotid divides there is an oval enlargement called **carotid gland** or **labyrinth** which consists of a network of blood capillaries. It is probably a sense organ and controls the blood pressure in the vessel beyond. The internal carotid runs forward and gives rise to a **palatine artery** which supplies the roof of the mouth, a little farther it gives rise to the **cerebral carotid** that enters the cranium in the region of the orbit and supplies the brain, and the third branch, the **ophthalmic** passes onward and supplies the eye and neighbouring tissues.

Systemic arch: It is the longest arch. The two systemic arches pass outward and curve around the oesophagus to join to form a median **dorsal aorta,** which proceeds backward beneath the vertebral column. Each systemic arch gives rise to following arteries:

(a) **Laryngeal artery:** It is a small artery, just near the origin of systemic arch, supplying the larynx and certain muscles of the hyoid.

(b) **Oesophageal artery:** A small artery in the dorsal aspect of the oesophagus.

(c) **Occipitovertebral artery:** It is a small artery just behind (or common with) the oesophageal. This splits into two, a **vertebral** extending along the vertebral column and an **occipital** giving branches to the upper and lower jaws, orbit and nose.

(d) **Subclavian:** It is a large artery just behind the occipito-vertebral and supplies the arm.

Dorsal aorta: The dorsal aorta lies mid-dorsally beneath the vertebral column. It runs posteriorly and gives off the following arteries:

(a) **Coeliaco-mesenteric:** It is a large artery which arises at the junction of the two systemic and divides into two branches—(i) the **coeliac** artery which gives off branches to the left side of the stomach (left gatric artery), a right **gastric artery** to right side of stomach and pancreas and **hepatic artery** to the liver; (ii) the **arterior mesenteric artery** to the spleen (**spleenic artery**), intestine (**intestinal artery**) and cloaca and anterior portion of rectum (**rectal artery**).

(b) **Gonadial artery:** A pair of small arteries supplying the gonads (**spermatic** in male and **ovarian** in female).

(c) **Renal artery:** Four to six pairs of small arteries supplying the kidneys.

(d) **Posterior mesenteric artery:** Near the posterior end of the dorsal aorta arises the posterior mesenteric artery, a small vessel supplying the posterior portion of the rectum.

(e) **Iliac arteries:** Posteriorly the dorsal aorta divides into two large iliac arteries supplying the hind limb. Each iliac as it enters the legs given off a branch, the **epigastric** to the ventral body wall and a **rectovesical** to the rectum and urinary bladder. Finally the iliac bifurcates into two, a **femoral** to the skin and muscles of the anterior part of the thigh and to the hip and the **sciatic** which enters the limb along with the sciatic nerve and supplies the lower leg.

Pulmocutaneous arch: It runs outward and upwards giving off posteriorly a **pulmonary artery** to the lung and a great **cutaneous branch** to the skin and various other structures. Of these the **auricularis** supplies the tympanum, thymus, lower jaw, pharynx and hyoid, etc., the **dorsalis** supplies the skin of the back, and the **lateralis** supplies the skin of the side of body.

Venous system

The venous system of frog includes the following veins:

(i) **Pulmonary veins**

(ii) **Caval veins**

(iii) **Renal portal system**

(iv) **Hepatic portal system**

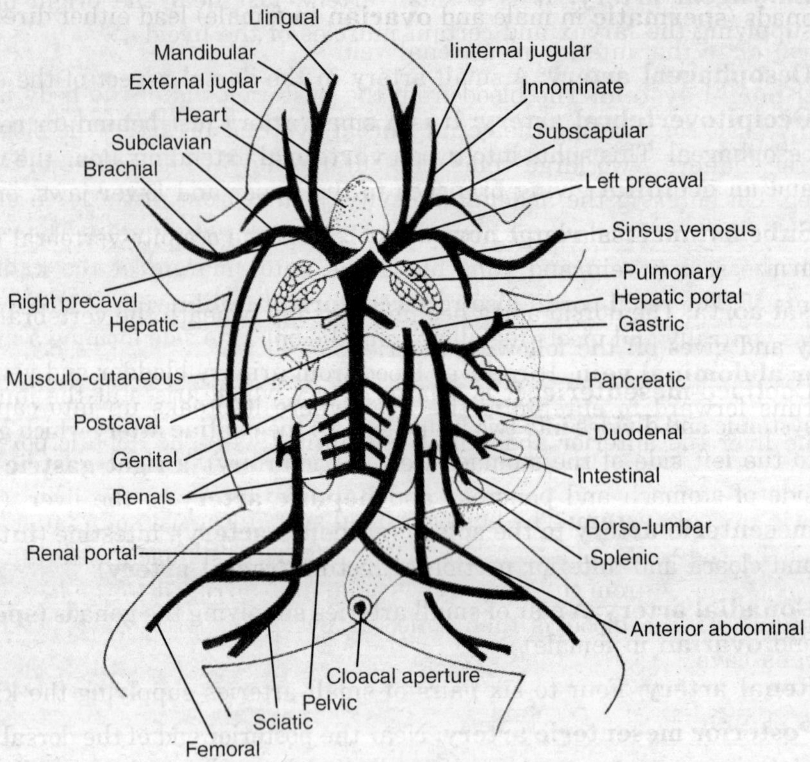

Fig. 14.39 *Frog*–Venous system

Pulmonary veins: The oxygenated blood from the lungs is collected by the left and right pulmonary veins which unite to form the common pulmonary vein opening into the right auricle dorsally.

Caval veins: The deoxygenated blood from various parts of the body is collected by a pair of anterior vena cavae and a single posterior vena cava, all opening into the sinus venosus.

(a) Anterior vena cavae or precavals: The precavals collect blood from the anterior part of the body. Each anterior vena cava is formed by the union of three branches:

(i) **External jugular:** It receives blood from the tongue (**lingual**), hyoid, thyroid and outer margin of lower jaw (**mandibular**) and floor of the mouth.

(ii) **Innominate:** It receives the **internal jugular** bringing blood back from the brain and other parts of the head, and the **subscapular** bringing blood from the shoulder.

(iii) **Subclavian:** It receives the **branchial vein** from the forelimb and the large **cutaneous vein** collecting blood from the side of the body and the head.

(b) Posterior vena cava or postcaval: It arises between the kidneys and runs forward ventral to the dorsal aorta. Before opening into the posterior end of the sinus venosus it receives a pair of **hepatic veins** from the liver. Four to six **renal veins**

collect blood from each kidney and pour it into the post caval. The veins from the gonads (**spermatic** in male and **ovarian** in female) lead either directly to the vena cava or to the anterior most renal vein.

Renal portal system: The blood from the posterior end of the body is collected by a set of blood vessels constituting the **renal portal system**. It carries the blood to a capillary system in the kidneys. Two large veins, the outer **femoral** and inner **sciatic** collect blood from the leg. On entering the abdominal cavity, just in front of the base of the thigh, the femoral divides into a dorsal **renal portal** and a ventral **pelvic vein**. The renal portal vein unites with the sciatic vein and runs along the outer margin of the kidneys and sends branches into it. The renal portal also receives **dorso-lumbar** vein from the body wall. The pelvic passes ventrally and meets its fellow from the opposite side forming a **median ventral** or **anterior abdominal vein**. It receives blood from urinary bladder and ventral abdominal wall and runs forward to enter into the liver where it breaks up into capillaries. Before entering the liver the anterior abdominal vein unites with the hepatic portal system by a small loop.

Hepatic portal system: The hepatic portal system consists of a large **hepatic portal vein** formed by the union of several branches from the stomach, intestine, pancreas and spleen. It collects blood from the alimentary canal and carries it to the liver where it breaks up into capillaries. The blood is then collected by the hepatic veins and poured into the posterior vena cava.

Lymphatic system

The lymphatic system includes the **lymph**, **lymph vessels**, **lymph hearts**, **lymph spaces** and **spleen**.

Lymph: Although the blood is conveyed to the tissues by the blood vessels, it does not come directly in contact with body cells and tissues. The walls of fine blood vessels or capillaries are very thin and certain portions of the fluid plasma carrying materials like plasma proteins and white blood cells escape through them and bathe the tissues. This fluid is called **lymph** or **tissue fluid**. It is a colourless fluid because it lacks red blood cells. It has less concentration of calcium and phosphorus than in blood. It is a medium of exchange between the blood and tissue. It carries food and oxygen to the cells and takes away waste substances from the cells.

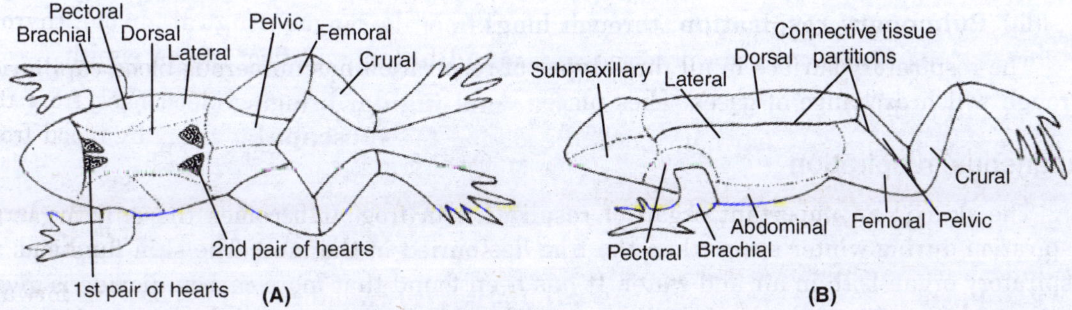

Fig. 14.40 *Frog*–(A) Lymph heart and sinuses in dorsal view; (B) Lymph sinuses in lateral view

Lymph or lymphatic vessels: Lymph vessels are fine thin-walled vessels of various sizes forming a network. The **lymph capillaries** unite to form larger vessels which open into the venous system thus returning lymph back to blood from which it originates. The lymph capillaries form a network in all the body organs except the nervous system. The lymph vessels are made up of connective tissue fibres with some unstriped muscles and lined with endothelial cells. They are provided with valves that prevent back flow of lymph.

Lymph hearts: Where the lymphatic vessel join the venous system small pulsating vessels called **lymph hearts** are present. They slowly drive lymph into veins. In frog there are two pairs of lymph hearts. The anterior pair is situated below the scapula and opens into scapular vein and the posterior pair is situated at the sides of the tip of urostyle and opens into the femoral vein.

Lymph spaces: In frogs some lymph vessels become enlarged to form large **lymph channels** or **sinuses**. There are large subcutaneous spaces present beneath the skin so it is loosely attached to the underlying muscular body wall. At intervals the skin is attached to body muscles by sheets of tissue called **septa** so that the space is divided into chambers on the dorsal, lateral and ventral surfaces.

Spleen: The spleen is a spherical dark red gland present in the mesentery between the ileum and rectum. It contains the largest mass of lymphatic tissue in the body. It produces antibodies, erythrocytes and phagocytic lymphocytes and also disposes off the old erythrocytes.

VOCAL AND RESPIRATORY ORGANS

In the vertebrate animals the vocal and respiratory organs are intimately associated owing to the fact that the production of sound is caused by the expulsion of air from lungs. Except the sound made by a few fishes the voice makes its first appearance in the amphibians. In the urodeles, the lowest division of Amphibia, the voice is feebly developed or entirely absent, but attains its maximum in some of the anurans.

Types of respiration

In frog respiration takes place by three methods:

(i) **Cutaneous respiration**, through moist surface of skin.

(ii) **Buccopharyngeal** or **buccal respiration**, through lining of buccopharyngeal cavity.

(iii) **Pulmonary respiration**, through lungs.

The respiratory surface in all three types of respiration has numerous blood capillaries through which exchange of gases takes place.

Cutaneous respiration

The skin is an important organ of respiration in frog. It becomes the only organ of respiration during winter sleep when the frog lies buried in the mud. The skin functions as respiratory organ both in air and water. It has been found that more carbon dioxide is given out through the skin than through the lungs. The skin has a very rich blood supply and is permeable to gases. It is always kept moist, for exchange of gases through the skin is possible under moist condition only. The gases first dissolve in moist surface and then diffuse into blood. The skin is kept moist by secretion of mucous. It provides a large space that is in contact with air for exchange of gases.

Buccal respiration

The mucous lining of the buccal cavity is richly supplied with blood capillaries where exchange of gases takes place. During buccal respiration the mouth remains permanently closed and the nostrils remain open. The floor of buccal cavity is alternately raised and lowered and the air is drawn into and expelled out of the buccal cavity through the open nostrils. The glottis remains closed during the entire process so that the air may not enter and leave the lungs.

Pulmonary respiration and sound production

Vocal apparatus: The sound producing organs of frog are located in a thin-walled chamber or box called **larynx** or **laryngo-tracheal chamber** (Fig. 14.41) situated just below the pharyngeal cavity at the beginning of the entrance of the lungs. The larynx opens into the pharynx by a slit-like **glottis** and by a pair of very small tubes called **bronchi** into the lungs. It is held between the stout thyroid processes of the hyoid apparatus to which it is attached by muscles and connective tissue. Its walls are supported by a pair of **arytenoid cartilages** and a **cricoid cartilage**. The arytenoid cartilages are a pair of semilunar valves which rest upon the cricoid cartilage. Their upper edges form the lateral margins of glottis. They provide attachment to muscles which open and close the glottis. The cricoid cartilage consists of a slender ring surrounding the larynx and lying in nearly the same place as the thyroid process of the hyoid, to which it is closely attached.

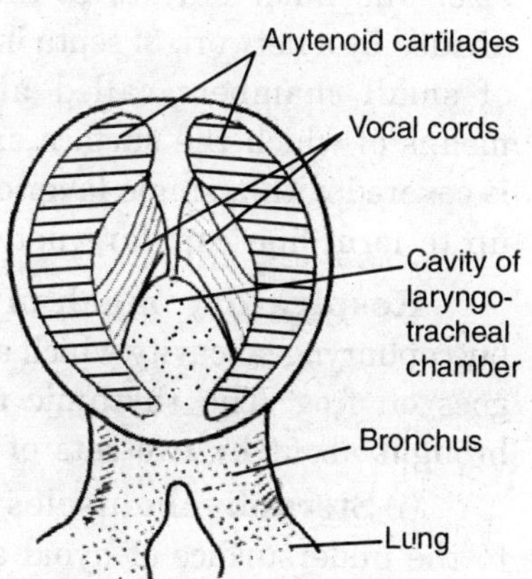

Fig. 14.41 *Frog*–Laryngeo-tracked chamber opened to show vocal cards

The true sound producing organs consists of a pair of elastic bands, the **vocal cords** extending longitudinally across the larynx. Sound is produced by expulsion of air from the lungs which set the free edges of the vocal cords in vibration. Vibrations in sound are caused by altering the tension on the cords through the action of laryngeal muscles. The vocal apparatus of male frog is much larger than that of the female.

The males of many species of *Rana* possess a pair of **vocal sacs** situated at the side of the pharynx. These sacs are out-pocketings of the pharyngeal wall which extend backward between the skin and the body. They communicate with the mouth by small openings in the floor in front of the angle of the lower jaw. The vocal sacs are distended during croacking by the pressure of air in the buccal cavity. They serve as resonators to re-enforce the sound produced by the vocal cord. They are absent in females.

Organs of respiration: The chief organs of pulmonary respiration are a pair of **lungs**. The lungs are not very well developed. The frog uses its lungs only when the need for oxygen is great. The ordinary requirement is fulfilled by cutaneous and buccal respiration. The air enters and leaves the lungs via **respiratory tract**. The respiratory tract consists of external nostrils, nasal chambers, internal nostrils, buccopharyngeal cavity, glottis, laryngo-tracheal chamber and two bronchi.

Lungs: The lungs (Fig. 14.42) are ovoid thin-walled sacs suspended freely inside the peritoneal body cavity one on either side of the heart. The outer surface of lungs is covered by a fold of peritoneum. They are capable of great distension and are inflated through the glottis. When the air is let out of the lungs, they shrivel to an inconspicuous size. The inner surface of the lungs is divided by a network of septa into a series of small chambers called **alveoli**, by means of which the surface exposed to air is greatly increased. The inner surface of alveoli is covered with a single layer of epithelial cells richly supplied with blood vessels which break up to form fine capillary network where exchange of gases takes place.

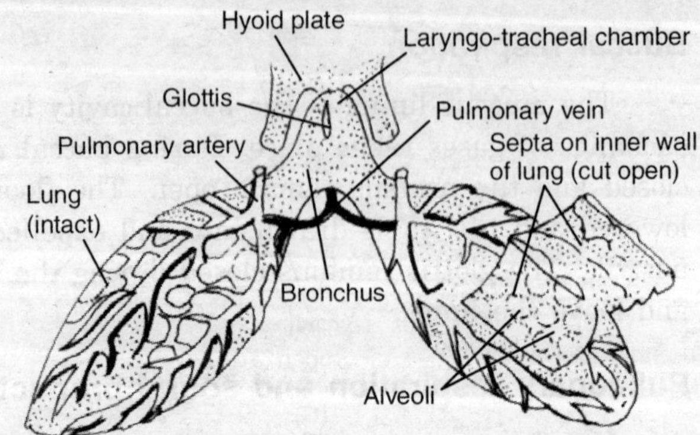

Fig. 14.42 *Frog*–Respiratory organs in dorsal view. Right lung partly cut open to show the alveoli

Respiratory mechanism: The air is drawn in by the action of the floor of buccopharyngeal cavity which actually acts as a buccal force pump. Ordinarily buccal respiration goes on frog. The rhythmic up and down movements of the floor of the buccal cavity is brought about by two sets of muscles:

(i) **Sternohyal muscles** attached at the lower end to the sternum and at the upper end to the undersurface of hyoid apparatus,

(ii) **Petrohyal muscles** attached below to the upper surface of the hyoid apparatus and above to the squamosal bone of the skull.

During buccal respiration the glottis remains closed and no air passes into the lungs. The air is forced into the lungs during lung breathing. There are two steps in pulmonary respiration: (i) **inspiration**, (ii) **expiration**.

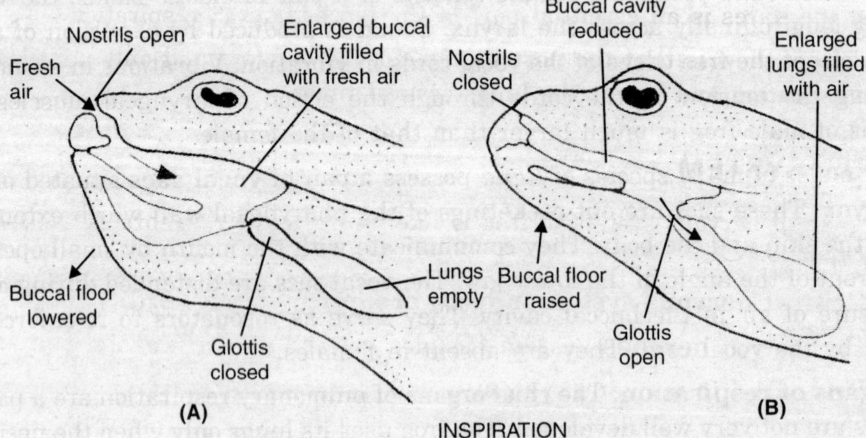

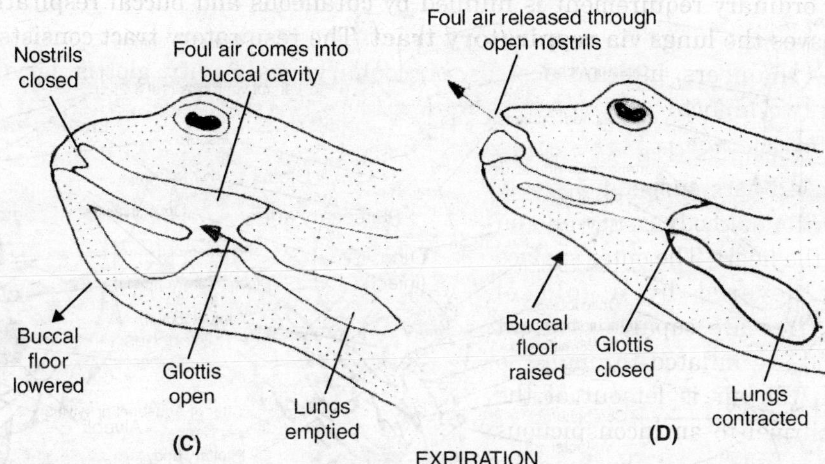

Fig. 14.43 *Frog*–Mechanism of breathing, **(A)** and **(B)** Inspiration **(C)** and **(D)** Expiration

Inspiration: The hyoid apparatus and the floor of buccopharyngeal cavity are lowered (Fig. 14.43 A and B) as in buccal respiration and air enters the cavity through the nostrils (the glottis and mouth remain closed). The floor of the buccal cavity is lowered by the contraction of sternohyal muscles. The glottis now opens and the premaxillae bones of upper jaw are raised and the external nares close. Now the petrohyal muscles contract raising the hyoid apparatus and the floor of buccal cavity. As a result the volume of the buccal cavity is reduced and air is forced into the lungs through the glottis. The drawing in of air into the lungs is called **inspiration**.

Expiration: Inspiration is followed by another process called expiration. When the lungs are filled the glottis closes and air is retained in the lungs for some time (Fig. 14.43 C and D). Expiration is brought about by the contraction of the muscles of the body wall aided by the elasticity of the walls of the lungs. With the external nares close the floor of the buccopharyngeal cavity is lowered and the air from the lungs is drawn into the cavity through the glottis. Finally the glottis is closed, the external nares open and the floor of the buccal cavity is raised forcing the air out. During expiration the lungs are emptied. The arrangement for closing the nares is an essential part of the mechanism of respiration.

The recent view is that the nostrils remain open always and only mixed air from the buccal cavity enters and leaves the lungs several times.

NERVOUS SYSTEM

The plan of nervous system of frog is like that of other vertebrates. It is composed of three sub-divisions.

(i) **Central nervous system** includes **brain and spinal cord**.

(ii) **Peripheral nervous system** includes **cranial and spinal nerves**.

(iii) **Sympathetic nervous system** includes **sympathetic nerves and ganglia**.

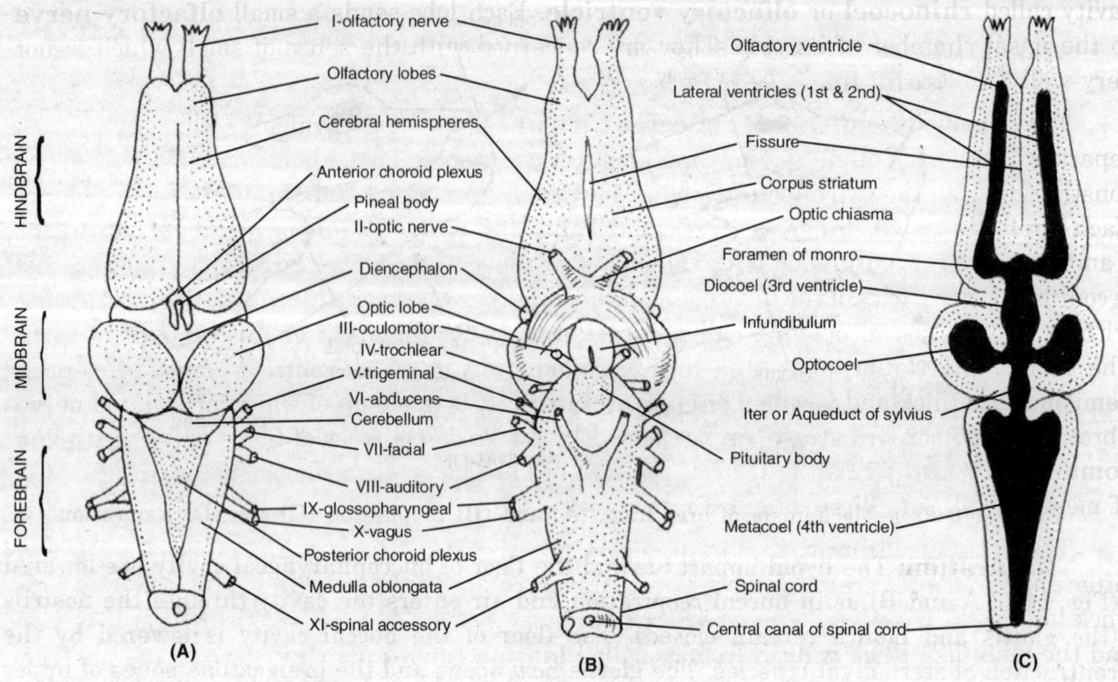

Fig. 14.44 *Frog*–Structure of brain. (A) Dorsal view; (B) Ventral view; (C) Longitudinal section showing ventricles

Central nervous system

Brain

The **brain** of frog is elongated, somewhat flattened, bilaterally symmetrical, white coloured structure lying in the cranial cavity of the skull. It is surrounded by two membranes called **meninges**. (singular, meninx) made up of connective tissue. The inner thin, pigmented, vascular membrane closely applied to the brain is called **piamater**. The outer tough fibrous membrane lining the interior of the cranial cavity is called **duramater**. The space between the two membranes is called **subdural space** which is filled with a clear, watery shock absorbing **cerebrospinal fluid**. The inner cavities of brain are called **ventricles**. They are continuous with each other and are also filled with cerebrospinal fluid. The brain can be divided into three parts (Fig. 14.44 A):

(i) **Forebrain** or **prosencephalon**

(ii) **Midbrain** or **mesencephalon**

(iii) **Hindbrain** or **rhombencephalon**

Forebrain: The forebrain is the largest part of the brain. It consists of a pair of **olfactory lobes**, two **cerebral hemispheres** and a **diencephalon**.

The **olfactory lobes** are the anterior most part of the forebrain. They are small elongated lobes pushed in front of the cerebral hemispheres, of which they seem to be a continuation. They are fused together in the median plane but each lobe contains a separate cavity called **rhinocoel** or **olfactory ventricle**. Each lobe sends a small **olfactory nerve** to the nasal chamber of its side. They are concerned with the sense of smell which is not very well developd in frog.

The **cerebral hemispheres** or **cerebrum** are long, narrow, oval and smooth structures separated from each other by a median longitudinal groove. They are demarcated by a slight constriction from the two olfactory lobes. They are narrow in front and slightly wider behind. Each cerebral hemisphere has a large cavity called **lateral ventricle** or **paracoel** (ventricles I and II) which is continuous with the cavity of the olfactory lobe. Posteriorly the paracoel is continuous the cavity of the diencephalon called **III ventricle** or **diocoel**. They communicate with each other by an opening called **interventricular foramen** or **foramen of Monro**. The roof of the cerebrum is thin and is called **palluim**. The ventrolateral side of each hemisphere is thick and is called **corpus striatum**. It is made up of white medullated nerve fibres. The two corpora strata are connected by a transverse tract of fibres called **anterior commisure** which forms the floor of the fore brain. The cerebral hemispheres are the seat of memory, will, intelligence and consciousness.

The **diencephalon** or **thalamencephalon** is a short, rhomboidal, unpaired structure situated behind the cerebral hemispheres. Its cavity is called **III ventricle** or **diocoel**. The thick lateral walls of the diocoel are called **optic thalami**, the floor is called **hypothalamus** and the plate-like thick wall in front is called **lamina terminals** which marks the anterior boundary of the diencephalon. On its dorsal surface the diencephalon possesses the **anterior choroid plexus** which is a network of blood capillaries. It also supports the **pineal body** dorsally. On its ventral surface it bears a X-shaped structure called **optic chiasma** formed by two optic nerve which cross each other. Just posterior to the optic chiasma there is a hollow bilobed projection called the **infundibulum** which bears a small oval, non-nervous, glandular **hypophysis** or the **pituitary body**.

Midbrain: The midbrain is the broadest part of the brain. It consists of a pair of large and well developed **optic lobes** or **corpora bigemina** and **crura cerebri**. The optic lobes are two prominent rounded protuberances on the dorsal side and project sidways so that they are visible from below. Each optic lobe has a cavity called **optocoel** or **optic ventricle** which communicates with the fourth ventricle behind through a narrow cavity called **iter** or **aqueduct of Sylvius**. The iter communicates with the third ventricle in front. The **crura cerebri** are two thick longitudinal bands of nerve fibres which run longitudinally beneath the optic lobes connecting the diencephalon and medulla.

Hindbrain: The hindbrain consists of **cerebellum** or **metencephalon** and the **medulla oblongata** or **mylencephalon**.

On the antero-dorsal region of the hindbrain there is a narrow transversely running ridge of nervous matter called the **cerebellum**. This is very small and reduced in frog. It controls muscular contraction and equilibrium.

The **medulla oblongata** is simply a widening of the spinal cord. It is a small, somewhat triangular structure. Its cavity is called **metacoel** or **IV ventricle** which communicates with the iter in front and posteriorly it is continuous with the central canal of the spinal cord. On its dorsal surface there is a reddish brown triangular area called **posterior choroid plexus.** Medulla oblongata controls involuntary actions like heart beat, respiration, etc.

Spinal cord

The **spinal cord** of frog presents typical structures. It is a short, somewhat flattened structure which extends posteriorly from the medulla oblongata through the foramen magnum and neural canal of the vertebral column. It terminates in a tapering narrow thread, the **filum terminale** which extends into the urostyle. The spinal cord presents two enlargements during its course. The anterior enlargement is present in the branchial region and is known as **branchial** or **cervical swelling** and nerves for arms arise from it. The posterior enlargement is present anterior to the filum terminale and is known as **lumbar** or **sciatic swelling** and nerves for hind legs arise from it. The spinal cord is traversed throughout its length by two median longitudinal grooves, one on the dorsal side (**dorsal fissure**) and one on the ventral side (**ventral fissure**). The spinal cord encloses a **central canal** which is in continuation with the ventricles of the brain. Spinal nerves arise by two **roots**, a **dorsal** and **ventral** at the sides of the cord. The internal structure of the cord shows that it is made up of mainly **ganglion cells** and **nerve fibres**. The central part of the cord consists of **gray matter** chiefly of nerve cells and non-medullated nerves surrounding the central canal. At the sides it is produced to form dorsal and ventral **horns** or **cornua**. The gray matter is surrounded by **white matter** consisting mainly of medullated fibres running mostly longitudinally. Spinal cord controls reflex actions.

Peripheral nervous system

The peripheral nervous system includes the **cranial nerves** arising from the brain and the **spinal nerves** arising from the spinal cord.

A nerve may be **afferent** with **sensory nerve fibres** or **efferent** with **motor nerve fibres** or **mixed** (both sensory and motor fibres). Sensory nerve fibres carry impulses from the receptors to the central nervous system and the motor nerve fibres carry impulses from the central nervous system to the effector organs. In the body there are four systems of nerve fibres : **somatic sensory** and **somatic motor**; and **visceral sensory** and **visceral motor**.

Cranial nerves

In frog there are ten pairs of **cranial nerves** which emerge from the cranium through various foramina and supply different organs of the body. They are numbered I to X (Fig. 14.45).

I. Olfactory nerve: It is the first cranial nerve which arises from the olfactory lobes by two roots. Each nerve passes through a small foramen in the ethmoid bone and is distributed to the walls of the nasal chamber. It is a sensory nerve and is concerned with smell.

II. Optic nerve: It is the second cranial nerve which arises from the diencephalon and after crossing in the chiasma, emerges from the side of the cranium and innervates the eyes. It is also sensory in nature and is concerned with vision.

III. Occulomotor nerve: It is a small motor nerve arising from the crura cerebri and innervates four muscles of the eyeball (rectus superior, rectus inferior, rectus medialis and obliquus inferior). It also give a branch to the ciliary ganglion.

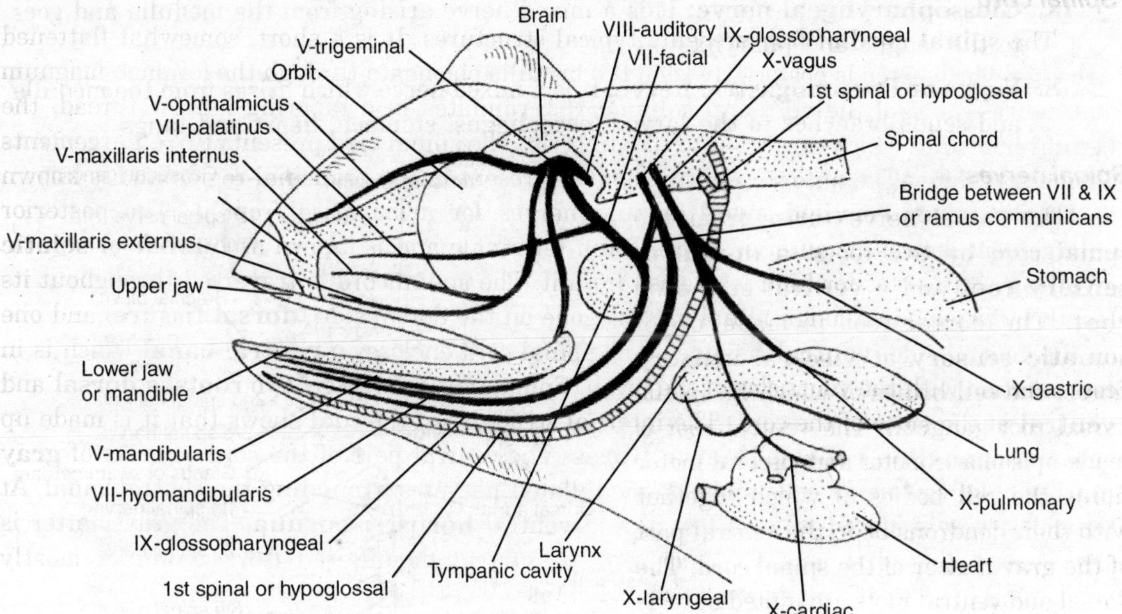

Fig. 14.45 *Frog*–Cranial nerves

IV. Trochlear or pathetic nerve: It is a small motor arising from the dorsal side of the brain between the optic lobes and cerebellum and supplying the superior oblique muscle of the eye.

V. Trigeminal nerve: This is the largest cranial nerve. It arises from the sides of the anterior end of medulla by a pair of roots which unite to form a large **prootic ganglion** before leaving the skull. From this ganglion arise two branches, the **ophthalmic** and the **maxillo-mandibular**, which leave the skull by a foramen in the anterior part of the pootic bone. The ophthalmic runs along the dorsal side of the orbit and is distributed to the skin of the anterior part of the head. The maxillo-mandibular nerve runs onward behind the eye and divides into the **maxillary superior** which supplies the upper jaw and the **mandibular** which supplies the lower jaw. The maxillary is a somatic sensory nerve while the mandibular is a visceral motor nerve.

VI. Abducens nerve: This is the sixth cranial nerve which arises from the ventral side of the medulla. It is a motor nerve. It enters the orbit and innervates the external rectus and retractor bulbi muscles of the eye ball.

VII. Facial nerve: It arises from the medulla close behind the third and divides into two branches the **palatine** to the roof of the buccal cavity (palate) and the **hyomandibular** which divides into the **hyoidean** and **mandibular** going to the tongue and the muscles of the lower jaw respectively. It is a mixed nerve.

VIII. Auditory nerve: It is a somatic sensory nerve which arises from the medulla and goes to the inner ear.

IX. Glossopharyngeal nerve: It is a mixed nerve arising from the medulla and goes to the tongue and pharynx.

X. Vagus or Pneumogastric nerve: It is a mixed nerve which arises from the medulla and sends branches to the larynx, oesophagus, stomach, heart and lungs.

Spinal nerves

The spinal nerves arise from the spinal cord by two roots, **a dorsal** or **sensory root** and a **ventral** or **motor root**. The dorsal root is made up of **somatic sensory** or **visceral sensory** fibres, the cell bodies of which lie in the dorsal root ganglion. The ventral root is made up somatic motor and visceral motor fibres the cell bodies of which together with their dendrons lie in the ventral part of the gray matter of the spinal cord. The dorsal and ventral roots are united outside the vertebral column to form a spinal nerve which emerges through a small intervertebral foramen. Thus the spinal nerves are mixed in nature. In frog white chalky masses called **calcareous bodies** or **gland of Swammerdam** are present which serve as reserve of calcium.

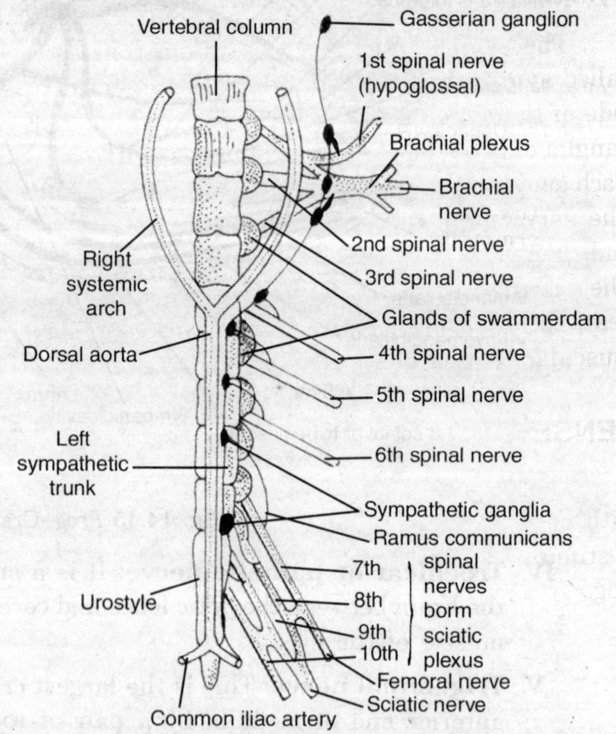

Fig. 14.46 *Frog*–Spinal nerves and sympathetic trunk of the left side

The adult frog possesses only ten pairs of spinal nerves (Fig. 14.46), the number in tadpole is much larger, the posterior ones degenerate with the loss of tail. At the point of their origin each spinal nerve gives off 3 branches : a short **ramus dorsalis** to the dorsal skin and muscles, a large **ramus ventralis** to the skin and muscles of ventral side and a very short **ramus communicans** to the sympathetic ganglion.

The **first** spinal nerve emerges between the first and second vertebrae and its principal branch, the **hypoglossal nerve** innervates the tongue and several muscles attached to the hyoid. The **second** nerve is quite large and receives a branch of the first anteriorly and **third** spinal nerve posteriorly forming the **branchial plexus**. The **fourth**, **fifth** and **sixth** nerves are small and are distributed mainly to the skin and muscles of the wall of abdomen. The seventh, eighth and ninth pass almost directly backward and anastomose forming the **sciatic** or **lumbosacral plexus**. Before merging into the sciatic plexus the seventh gives rise to the **ilio-hypogastric** branch to the muscles of the abdomen. From the plexus arises the **sciatic nerve** and goes into the leg on each side. Proximal to the knee each sciatic divides into the **tibialis** and **peroneus**. The tibialis gives branches to the gastrocnemius, tibialis posticus and numerous muscles of the planter surfaces of the foot. The peroneus innervates the peroneus muscle, the tibialis anticus and the muscles on the extensor surface of the foot. The small **tenth** nerve receives a branch of the ninth and the ischio-coccygeal plexus which gives branches to the bladder, cloaca, oviducts and posterior lymph hearts.

Sympathetic nervous system

The sympathetic nervous system consists of two thread-like strands or nerve cords called **sympathetic trunks** (Fig. 14.46) which runs beneath the vertebral column on either side of the dorsal aorta. Each nerve cord has ten **sympathetic ganglia** and corresponding ganglia of both the cords are connected with each other by small **transverse commissures**. Each ganglion is connected with the spinal nerve by a small nerve called **ramus communicans**. The nerve cords run anteriorly, enter the skull and join the vagus ganglion of tenth nerve and then run forward and join the gasserian ganglion of fifth nerve and finally terminates. The system controls many internal functions not under voluntary regulation such as heart beat, secretion of digestive juices, muscular movements in the stomach and intestine and muscular tone of blood vessels.

SENSE ORGANS

The sensory system includes **organs of special sense** or **receptors**. They are supplied with nerve fibre and so they convey the stimulus to the central nervous system. Each kind of stimulus affects a particular sense organ. As in man, there a five types of sense organs in frog.

1. **Organs of touch** or **tangoreceptors**
2. **Organs of taste** or **gustatoreceptors**
3. **Organs of smell** or **olfactoreceptors**
4. **Organs of sight** or **photoreceptors**
5. **Organs of hearing and balance** or **stato-acoustic receptors**

Organs of touch or tangoreceptors

In the skin of frog numerous minute **tactile organs** or **tangoreceptors** are present under the **epidermis**. These are supplied with sensory nerve endings. The nerve endings never reach the cells of the outermost layer of epidermis. The tactile organs are sensitive to many types of stimuli such as touch, temperature, chemical, humidity, light, pain, etc.

Organs of taste or gustatoreceptors

The organs of taste are present in the form of **taste buds** on the tongue and roof of the buccal cavity. Each taste bud is made up of group of barrel-shaped columnar cells some of which are **neurosensory** or **receptor** or **taste cells** and others are **supporting cells**. The neurosensory cells are slender and elongated, their free ends bear numerous delicate taste hairs projecting above the surface while the opposite ends are innervated with sensory nerve fibres. The supporting cells are larger and their free ends are without sensory fibres. The taste cells are innervated by nerves from VII and IX cranial nerves.

Organs of smell or olfactoryreceptors

The olfactory organs include a pair of **nasal** or **olfactory sacs** situated in the olfactory capsule of the skull. Each nasal sac-communicates with the outside by the **external nares** and with the buccal cavity through the **internal nares**. The nasal sacs are lined by epithelial cells. Some of the cells are **neurosensory cells** or **olfactoreceptors** while other cells are **supporting cells**. The olfactoreceptor cells are provided with a tuft of fine sensory hairs on their outer free ends projecting into the lumen of the olfactory sac and nerve fibres on their inner ends. The nerve fibres unite to form the **olfactory nerve** which carry stimuli about smell to the brain. The epithelial lining also has **mucous secreting** glands which keep the nasal epithelium moist.

Organs of sight or photoreceptors or eyes

The organs of sight are a pair of **eyes** situated in the orbital fossae one on either side of the head. They are similar to the eyes of other vertebrate in general structure and function. The eyes are covered with two **eyelids**, the **upper** and **lower eyelids**. The **upper eyelid** is thick, narrow fold extending from the upper edge of the orbit. It has no power of independent movement, but may be raised or lowered by the movements of the eyeball. The **lower eyelid** is thick and capable of movement. There is a thin, semilunar, transparent **nictitating membrane** which is not a separate membrane but upper thin and transparent part of the lower lid and lies folded beneath it when the eye is uncovered. It covers the eye when the frog is in water or underground. The movement of the nictitating membrane is brought about by two types of muscles the **retractor bulbi** and **levator bulbi** muscles. When the retractor bulbi muscles contract the eyeball is drawn deep into the orbits, the eyelids come closer and the nictating membrane is unfolded to cover the eyes. When the levator bulbi muscles contract the eyeballs protrude, the eyel ds are separated and the nictitating membrane is folded back to its original position.

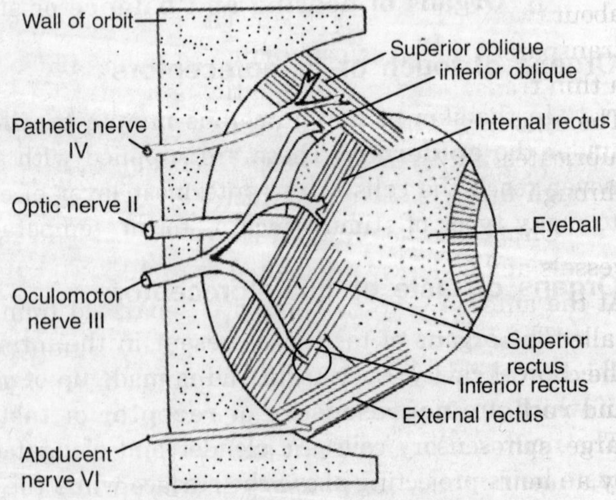

Fig. 14.47 *Frog*–Muscles and nerve supply of the eye

Muscles of the eyeball: The eyeball is moved by a set of six extrinsic muscles. There are four **rectus muscles** (**anterior, posterior, superior** and **inferior rectus muscles**) and two **oblique muscles external** and **internal** (Fig. 14.47). The four rectus muscles rotate the eyeball forwards, backwards, upwards and downwards respectively. They arise from the inner wall of the eyeball and are inserted round the equator of the eye ball at various points. The superior oblique muscles bring about the rotation of the eyeball along the axis between the orbit and cornea and the superior rectus bring about opposite movement. These muscles arise from the front part of the orbit. The frog can protrude its eyeball with the help of levator bulbi muscles and withdraw it deep into the orbit with the help of retractor bulbi muscles.

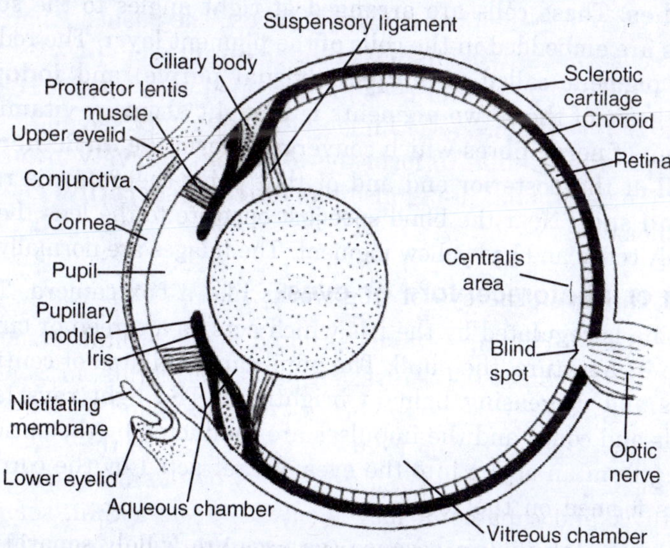

Fig. 14.48 *Frog*–V.S. of the eye

Structure of eye: The outer one-third part of the eye is exposed and is transparent while the remaining two-third part is concealed and is opaque. The wall of the eyeball is made up of three concentric coats or layers : the outermost layer is **sclerotic**, the middle layer is **choroid** and the innermost layer is **retina**.

Sclerotic: It is the outer thick protective layer made up of fibrous connective tissue and cartilage. It is actually the part of optic capsule which has not fused with the skull and forms about two-third of the eyeball. The front one-third part of the sclerotic is raised, circular and transparent and is called **cornea**. It allows light to enter the eye. The cornea is covered with a thin transparent membrane the **conjunctiva** which is a transparent continuation of skin. The eyes have **Harderian glands** at their lower inner angle. The secretion of these glands lubricates the eyes and keep them moist. Excess secretion is drained into the nasal sac through fine naso-lacrymal duct.

Choroid: The **choroid** is present beneath the sclerotic. It is richly supplied with blood vessels and is darkly pigmented to exclude all light except that entering through the cornea. At the anterior end the choroid is separated from cornea forming a pigmented circular disc called **iris**. In the centre of the iris there is an opening called **pupil**. At the base of the iris the choroid thickens to form a ring like **ciliary body**. Attached to the iris there are **circular** and **radial muscles** which can reduce and increase the pupil. Behind the pupil there is a large, spherical transparent and crystalline **lens** enclosed in a **lens capsule**. It is suspended by a membranous **suspensory ligament** attached to the margin of the lens and the ciliary body. The iris divides the hollow of the eyeball into two chambers. The small anterior chamber between the cornea and lens is the **aqueous chamber** filled with a watery fluid called **aqueous humour**. The large posterior chamber behind the lens is called the **vitreous chamber** and is filled with a jelly-like substance called **vitreous humour**.

Retina: It is the innermost highly sensitive layer restricted to the posterior part of the eye behind the ciliary body. It is made up of several layers. The outermost layer lining the choroid is the **pigmented layer**. Beneath this layer are present two types of sensory cells, the **rods** and **cones**. These cells are arranged at right angles to the surface of retina and their narrow ends are embedded in the cells of the pigment layer. The rods and cones contain a photosensitive pigment called **rhodopsin** (visual purple) and **iodopsin** (visual violet) respectively. Synthesis of these two pigments is brought about by vitamin A. The innermost layer is made of up of nerve fibres which converge to form the **optic nerve**. The optic nerve leaves the eyeball at the posterior end and at that point there are no rods and cones. This area is called **blind spot**. Near the blind spot just opposite to the lens there is a **yellow spot** which contains only cones and has yellow pigment. The images are normally formed in this area.

Working of the eye: The eye of frog works like a box camera. The amount of light falling on the retina is regulated by the iris which acts as a screen or curtain, admitting the rays only through its aperture, the pupil. The pupil can be dilated or contracted to allow more light in darkness and decreasing light in brightness. The light rays falling on the retina stimulate the rods and cones and the impulses are conducted by the optic nerve to the brain. Light rays coming from an object into the eyes are refracted by the cornea and lens and an inverted image is formed on the retina.

Frog has **monocular vision** because the eyes are widely separated. The eyes of frog has no power of **accommodation** for viewing both near and distant objects as human eyes have. The lens cannot either be changed in form or brought nearer to or farther from the retina, so that only those objects are in focus which are at a particular distance from the eye. Images of objects at other distances are not clearly outlined resulting in imperfect vision. In air the frog is **myopic** or short sighted. It can clearly see those objects which are near at hand. The myopia of frog is evidently an advantage to the animal in securing food. It is well known that frogs can perceive movement much more than other forms and for this purpose indefinite and blurred images and sufficient.

Organs of hearing and balancing or ears: The **ears** are the statoacoustic organs or organs of hearing and balancing in frog. There is no external ear in frog. The ear consists of only two parts, **middle ear** and **internal ear**.

Middle ear: The **ear drum** or **tympanic membrane** is situated postero-laterally on the head just behind the eye. It is a dark, flat, circular patch of skin tightly stretched over a ring like cartilage called **annulus tympanicus**. The cavity of the middle ear closed by the ear drum is called **tympanic cavity**. It contains air and communicates with the pharynx through the Eustachian tube. The air causes the eardrum to vibrate without being damaged. The tympanic cavity is crossed by a slender club-shaped bone and cartilage the **columella auris**, the outer end of which is attached to the eardrum and the inner end to a small cartilaginous nodule, the **stapedial plate** (Fig. 14.49). The bony partition between the

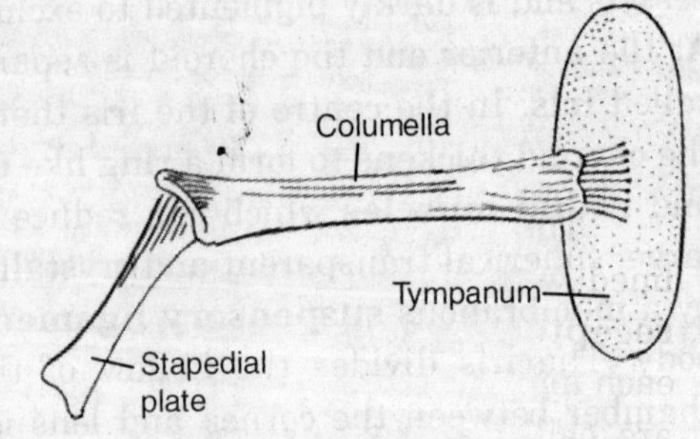

Fig. 14.49 *Frog*–Columella auris

tympanic cavity and the auditory capsule has a small oval aperture called **fenestra ovalis** which is covered by a membrane. The stapedial plate is also fused with the fenestra ovalis.

Internal ear. The inner ear lies within the **auditory capsule** (Fig. 14.50) formed by the prootic and **exoccipital** bones of the skull. Its cavity is filled by a clear protective watery fluid called **perilymph** in which the **inner ear** or **membranous labyrinth** floats. It is a sac-like structure divisible into two parts : a large dorsal oblong **utriculus** and a small ventral oval **sacculus**. From the posterior side of the sacculus arise two small irregular pockets called **lagena** (or **cochlea** of higher vertebrates) and **pars basilaris** (now regarded as a part of lagena). Above these lies another protuberance, the **pars neglecta** which comes from the utriculus. The membranous labyrinth gives rise to three **semicircular canals** (Fig. 14.50) embedded within the cartilage of the auditory capsule. There is an **anterior** and a **posterior vertical canal** and a **horizontal canal**, all three at right angles to each other. At one end each canal has an enlargement called **ampulla**. From the inner side of the sacculus arises a narrow tube the **ductus endolymphaticus** which is filled with a milky fluid, the **endolymph**. In the endolymph are suspended many small particles of calcium carbonate called **otoliths**. The membranous labyrinth is innervated by fibres of auditory or VII cranial nerve.

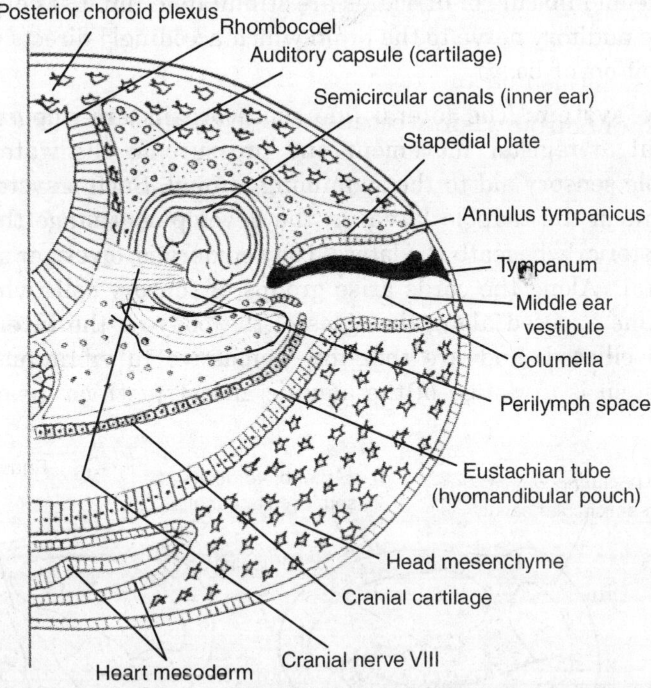

Fig. 14.50 *Frog*–Section of head passing through the ear region

The wall of the membranous labyrinth is made up of fibrous connective tissue and is lined with cubical epithelium. The epithelial lining is modified at certain places to form **receptors** or **sensory patches** or **acoustic spots** or ridges. There is one acoustic spot in each ampulla and one each in sacculus, utriculus and lagena. The acoustic spots of ampullae are called cristae and those of sacculus, utriculus and lagena are called maculae. Each acoustic spot is made up of supporting cells and tall sensory hair cells with stiff sensory hair-like process at their free end.

Working of ear: The ear of frog performs two functions hearing and equilibrium of the body.

The frogs do not have well developed hearing mechanism. Instaed of cochlea they have a small patch of sound receiving tissue, the lagena and sacculus. Frogs usually respond to croaks, this proves that they can hear, but they cannot discriminate between tones of various pitch. In hearing the sound waves strike upon the tympanum and produce vibrations which are transmitted through the columella and stapedial plate to the perilymp of the auditory capsule. From the perilymph the vibrations are transmitted to the endolymph of the inner ear. These vibrations stimulate the sensory cells of the sacculus and lagena giving rise to impulses which on being transmitted to the brain through the auditory nerve, produce sensation of sound.

The utriculus and the three semicircular canals along with their ampullae are concerned with the maintenance of equilibrium of the body. It there is any change in the position of the body, the endolymph and otoliths start movement and exert pressure due to the which the cristae of ampullae and macula of utriculus are stimulated. Impulses of change in position are transmitted by the auditory nerve to the brain which accodingly directs the concerned muscles to correct the position of body.

Lateral line system: The lateral line sense organs are important structures which enable the animal to register movement and pressure in the water around them. This provides a valuable sensory aid to the swimming animal. Such a system is **absent** in adult frog, but is present in the tadpole larva. In the development stage the tenth cranial nerve sends a growth posteriorly beneath the lateral body epidermis, on either side. It runs posteriorly to the tip of the tail. Along the cords arise groups of sensory cells which grow through the epidermis to become exposed along the sides of the body as the **lateral line system**. The exposed cells are ciliated, and are therefore sensistive to vibrations in the surrounding medium. This system is a vestige of the aquatic ancestry, which disappears by the time of metamorphosis.

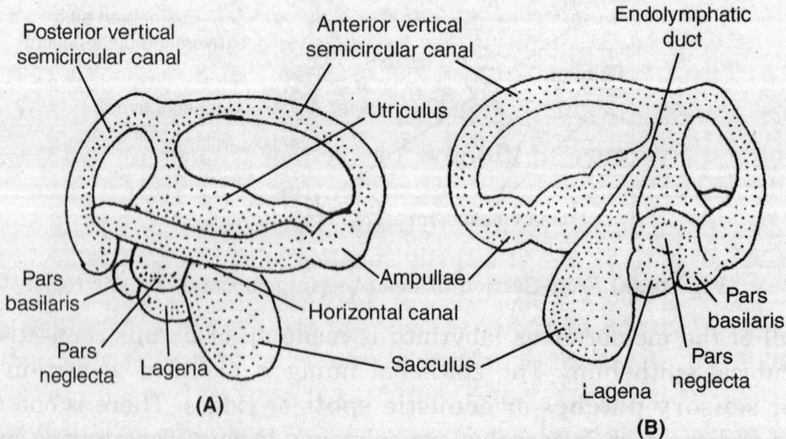

Fig. 14.51 *Frog*–Membranous labyrinth, **(A)** Outer view; **(B)** Inner view

EXCRETORY SYSTEM

The excretory system of frog comprises a pair of **kidneys** and **ureters** and an **urinary bladder**.

Kidneys

The kidneys are the main organs of excretion. In the tadpole the kidneys are **pronephric** but in the adult they are **mesonephric** in nature. The kidneys are a pair of compact, oval, flattened, dark red bodies lying in subvertebral lymph spaces above the coelom, one on either side of the vertebral column. They are covered by a fold of peritoneum on the ventral surface only. The ventral surface is flatter than the dorsal. The outer surface is smooth and convex and the inner surface is straight with notches. The ventral surface of each kidney has an elongated yellow coloured **adrenal** or **suprarenal gland** of endocrine function. At the anterior end of each kidney finger-like **fat bodies** and **reproductive organs** (**testis** in male and **ovaries** in female) are attached. From the outer edge of the posterior end of each kidney arises the **urinary** or **mesonephric** or **Wolffian duct**, the **ureter**.

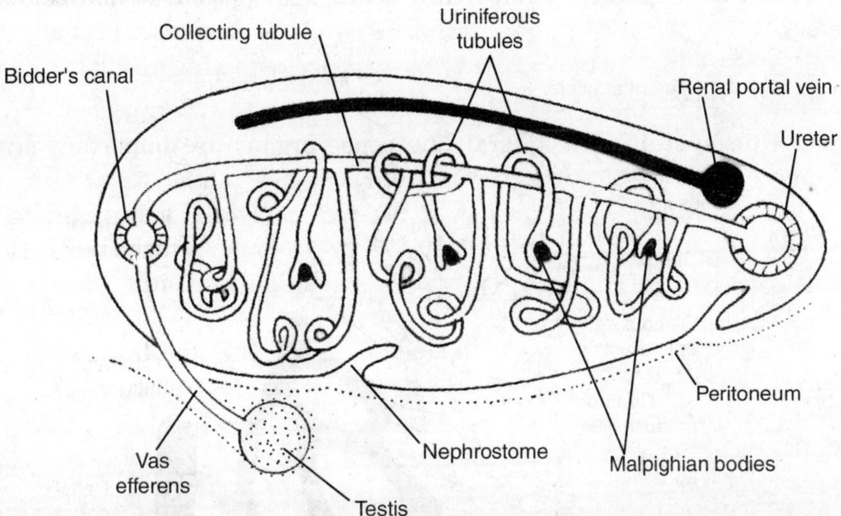

Fig. 14.52 *Frog*–Diagrammatic T.S. of kidney showing the arrangement of renial tubules and collecting tubule

Histological structure of kidney: Each kidney is made up of a large number (about 2000) of coiled microscopic tubules called **uriniferous** or **renal tubules** or **nephrons** (Fig. 14.52), bound together by connective tissue. These tubules are the main functional units of the kidneys. The kidneys are also richly supplied with blood vessels and their capillaries. Each tubule has a double-walled cup-shaped **Bowman's capsule** at the proximal end. It enclosses a network of blood capillaries called **glomerulus** (Fig. 14.53) which receives blood from an **afferent arteriole** of the **renal artery**. From the glomerulus blood is collected by **efferent arteriole** joining the **renal vein**. The Bowman's capsule and the glomerulus are together known as the **Malpighian body** or **renal corpuscle**. At the other end each renal tubule opens into a **collecting tubule** which runs transversely along the dorsal margin of the kidney. The collecting tubule opens into the Bidder's canal towards the inner side and ureters towards the outer side. On the ventral surface of each kidney there are numerous ciliated funnel–shaped **nephrostomes** which drain wastes from the body cavity and discharge into the renal veins. In tadpoles the nephrostomes are connected with the renal tubules.

Ureters

The ureters are urinary or mesonephric or Wolffian ducts which arise from the posterior outer edge of each kidney and pass backward to open into the dorsal side of the cloaca. The opening of the two ureters lie side by side on two papillae. In male frog the ureters dilate, just posterior to the kidneys, to form a **vesicula seminalis** in which sperms are stored. Thus in male frog ureters function as **urino-genital ducts** since they discharge urine are well as sperms.

Urinary bladder

It is a large, bilobed, thin-walled, transparent elastic sac which opens into the ventral side of the cloaca by a sphinctered aperture which lies opposite to and below the openings of the ureters.

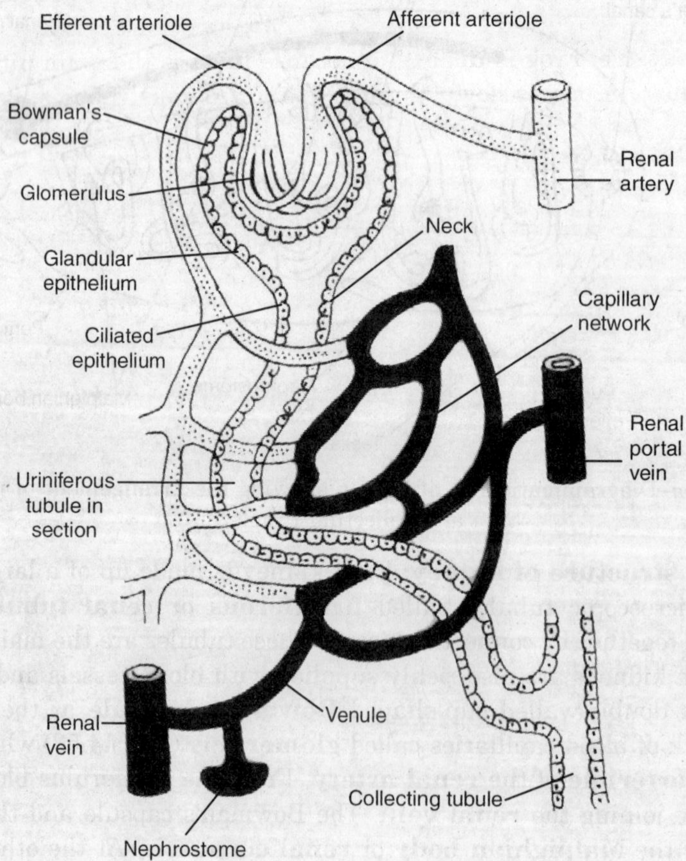

Fig. 14.53 *Frog*–Section of uriniferous tubules with its blood capillaries

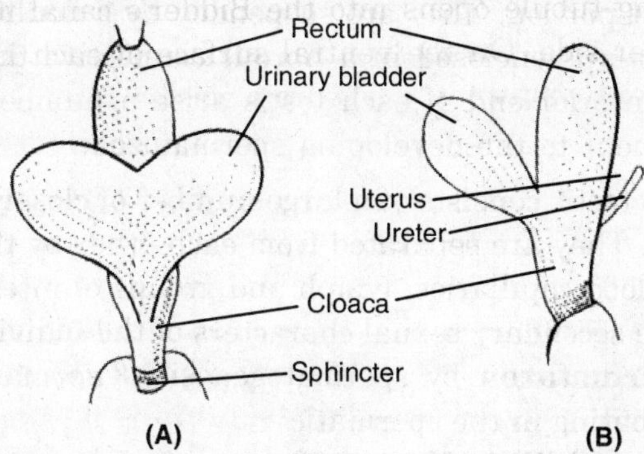

Fig. 14.54 *Frog*–Urinary bladder, **(A)** Ventral view; **(B)** Lateral view

Physiology of Excretion

During the processes of life certain substances are produced which are of no use to the organism. These substances must be got rid off to maintain life. The process of removal of waste substances is known as excretion and this is done by the kidneys. The blood brings these substances to the kidneys by renal artery. Ultrafiltralion occurs in the glomerulus, as a result nitrogenous wastes (mainly **urea**), water, inorganic salts, glucose, etc. are filtered into the Bowman's capsule. Frog is **ureotelic** because **urea** is the main nitrogenous waste. As the glomerular filtrate passes down through the uriniferous tubules, useful substances such as glucose, amino acids, inorganic chlorides and some water are selectively reabsorbed. The remaining harmful substances such as urea, some salts and water form urine which passes through the collecting tubules into the ureters and goes to the cloaca. It is stored in the urinary bladder for sometime and when the bladder is full it is discharged through the cloacal aperture.

REPRODUCTIVE SYSTEM

The sexes are separate. The adults show sexual dimorphism. The function of the reproductive system is production of sex cells and transporting them outside the body. The sex cells are produced by the **ovaries** in the female (ova) and **testes** in the male (sperms). The sex products are carried to the outside in the two sexes by very different methods.

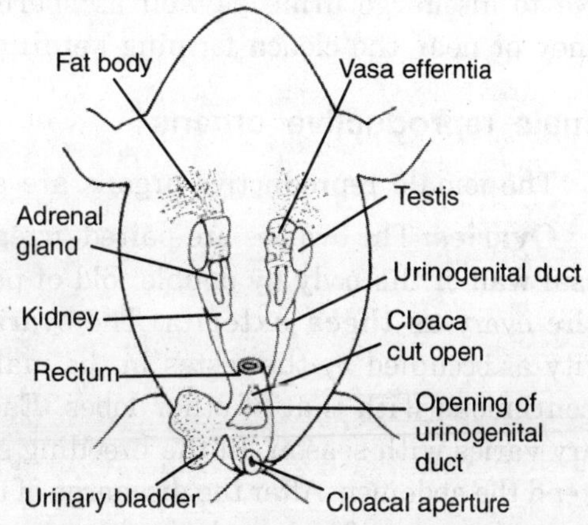

Fig. 14.55 *Frog*–Male urinogenital system

Male reproductive organs

The male reproductive organs are a pair of **testes, vasa efferentia** and two **urinogenital ducts** (Fig. 14.55).

Testes: The testes are whitish or yellowish, cylindrical or ovoid organs lying ventral to the kidneys attached to the outer edge of its anterior most portion. They are suspended by a double fold of peritoneum called **mesorchium**. Each testis is enclosed in a fibrous connective tissue capsule called **tunica albuginea**. Near the anterior end of each testis arise a number of finger-like **fat bodies** which provide nourishment to the developing spermatozoa.

Histologically each testis consists of a large number of closely packed oval shaped coiled **seminiferous tubules**. They are separated from each other by thin partitions of connective tissue which contains blood capillaries, lymph and groups of intestitial cells which secrete a hormone influencing the secondary sexual characters of the individuals. The epithelial lining of tubules produce **spermatozoa** by spermatogenesis. The mature **sperms** are found in bundles (Fig. 14.56A) floating in the spermatic fluid. Each sperm (Fig. 14.56B) is about 0.03 mm in length, has an elongated **head** with a **nucleus** and a rounded **acrosome** at the anterior end, a short **middle piece** containing centriole and mitochondria and an elongated filamentous **tail**.

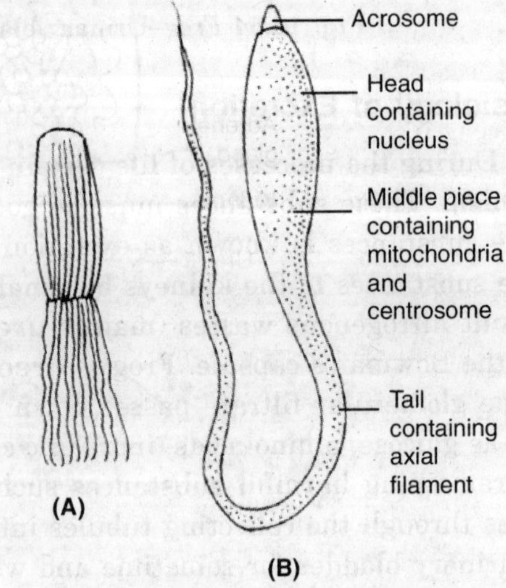

Fig. 14.56 *Frog*– **(A)** Sperm bundle; **(B)** Amature sperm

Vasa efferentia: From each testis arise a number (10 or 12) of slender tubes called vasa efferentia which leave the inner margin of the testis and enter the kidney. They are connected with the seminiferous tubules of the testis. Inside the kidney they are connected with the anterior uriniferous tubules which contain spermatozoa during the breeding season. They open into the Bidder's canal of the kidney which is connected to the ureters through the collecting tubule. The sperms pass through the substance of the kidney into the urniogenital duct.

Urinogenital duct: In male frog the ureters function as the urinogenital duct as they serve to discharge urine as well as sperms. The ureters are dilated just posterior to the kidney or near the cloaca forming **seminal vesicles** which stores sperms until required.

Female reproductive organs

The female reproductive organs are a pair of **ovaries** and **oviducts** (Fig. 14.57).

Ovaries: The ovaries are paired, irregular by folded, multilobed, organs attached to the dorsal wall of the body by double fold of peritoneum, the **mesovarian** which surrounds the entire ovary as **theca externa**. The ovaries occupy the same relative position in the body cavity as occupied by the testes in the male. Each lobe of the ovary is hollow and its cavity is continuous with that of other lobes. Each ovary may contain 7-12 lobes. The size of the ovary varies with season. In the breeding season they may fill the body cavity and may even distend the abdomen. After the discharge of the ova the ovaries become so small and pigmentless that it becomes difficult to locate them.

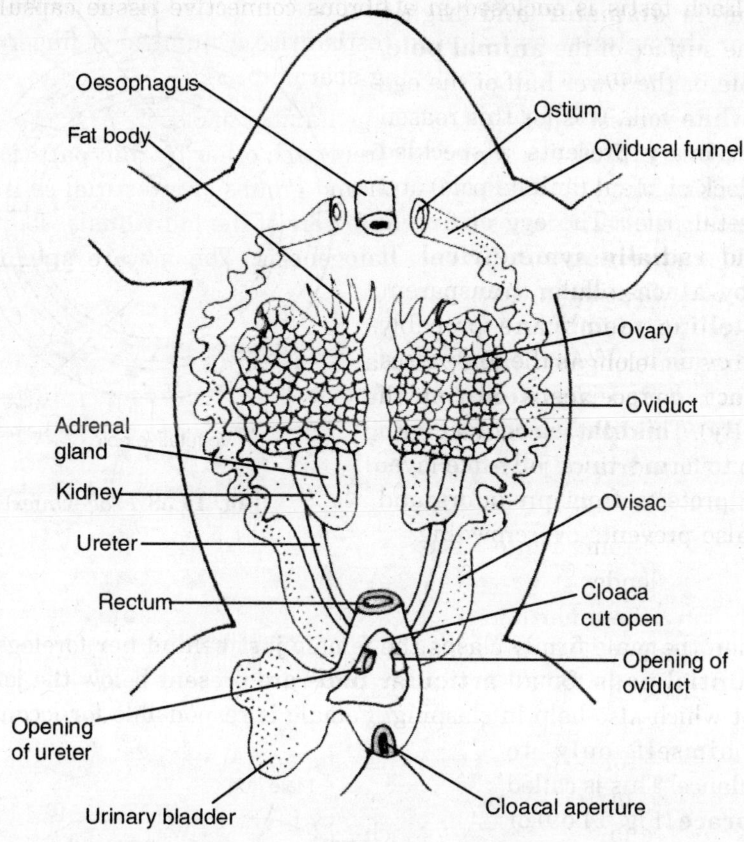

Fig. 14.57 *Frog*–Female urinogenital system

Histologically each ovary contains innumerable dark rounded **ova** in different stages of development. The ova arise from the germinal epithelial lining of the ovary by oogenesis. The mature ova or eggs are released into the body cavity by the rupture of the ovarian wall. The number of eggs varies from species to species (2,000 in *Rana pipiens* and 20,000 in *R. catesbiana*). The ova are moved by the action of cilia of the peritoneum to the ostium of the oviduct. In front of each ovary finger-like **fat bodies** are attached but less closely than in the male. They function as store-house of nutriment.

Oviducts: On the side of each ovary lies a greatly convoluted, white coloured tube, the oviduct, suspended by a double fold of peritoneum. The oviducts run from below the lungs to the cloaca, where they open separately on papillae just in front of the openings of the ureters. Anteriorly each oviduct opens by a slit-like **aperture** or **ostium** with ciliated and highly elastic wall. At the posterior end just before opening into the cloaca each oviduct widens to form a thin walled highly distensible **ovisac**, sometimes also called **uterus**. The eggs collect in the ovisac before being laid.

The mature ovum (Fig. 14.58) is spherical about 1.75 mm is diameter and highly pigmented on the surface of the **animal pole**. The **vegetal pole** or the lower half of the egg is white due to white yolk. It is for this reason that the mature ovary presents a speckled appearance of black pigment (animal pole) and white yolk (vegetal pole). The egg of frog is **telolecithal** and **radially symmetrical**. It is surrounded by a non-cellular transparent two layered **vitelline membrane** probably derived from the ovum itself. As the egg passes down the oviduct, it receives a coating of **albumen** (or **jelly**). This coat, on contact with water, swells up to form a thick jelly-like layer which probably protects from predators and infections and also prevents overcrowding.

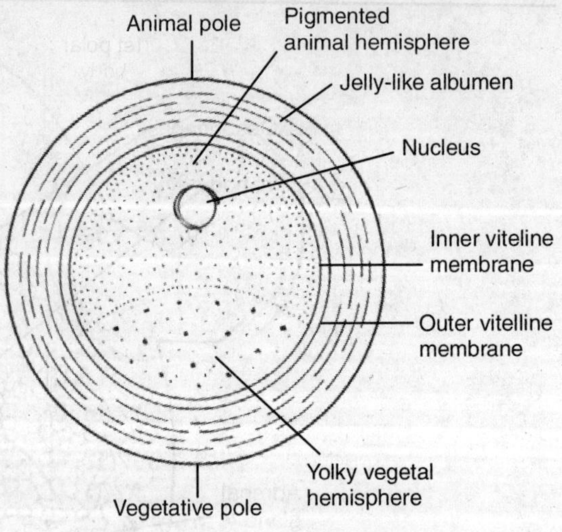

Fig. 14.58 *Frog*–Unfertilized egg

Copulation

In copulation the male firmly clasps the female just behind her forelegs by his forelegs and enlarged **nuptial pads**. Small **articular pads** are present below the joints of all digits of hand and foot which also help in clasping. Female is responsible for locomotion while the male exerts himself only to maintain the balance. This is called **amplectic embrace** (Fig. 14.59) or **amplexus**. Amplexus occurs in frogs and toads but not in urodeles. The duration of amplexus is several days after which the eggs are laid (oviposition). As the eggs pass out through the cloaca the male deposits the sperms over them (insemination). The eggs should be inseminated soon after they are shed because the life of eggs is very short and therefore quick fertilization is essential.

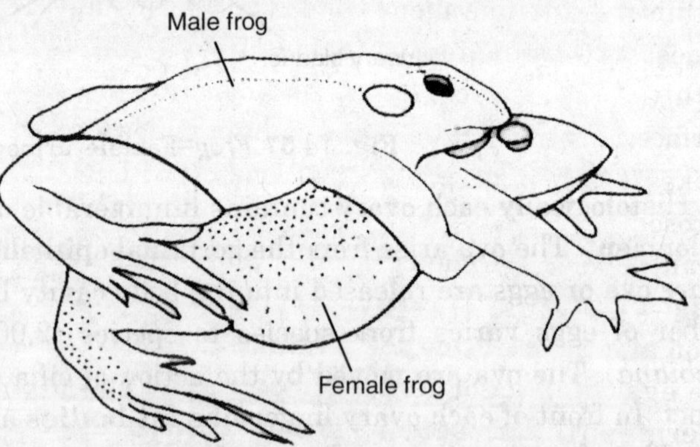

Fig. 14.59 *Frog*–Amplectic embrace

Fertilization

In frog fertilization is external. Ovulation is brought about under the influence of some pituitary hormone. The **first maturation division** occurs at the time of ovulation and the **first polar body** is extruded (Fig. 14.60 A). The egg is now called the **primary oocyte**. It is at this stage, the egg is released in the body cavity. The **second maturation division** begins soon after first is over but is not completed till the egg is activated by the sperm.

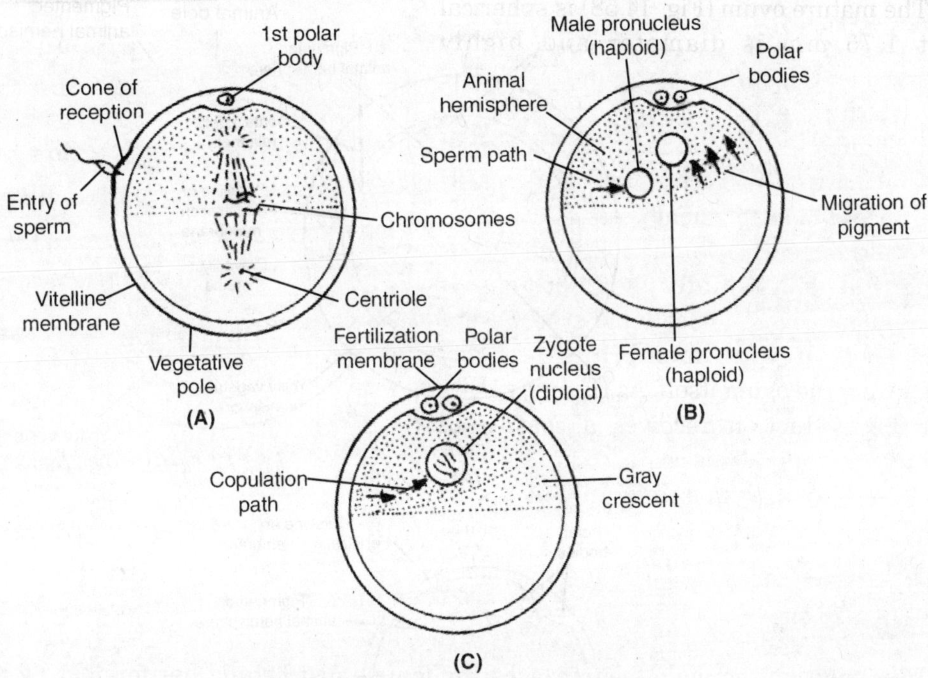

Fig. 14.60 *Frog*–Stages in fertilization

The eggs are laid in the **secondary occyte stage** with the first polar body present within the vitelline membrane. The sperms always penetrate the egg near the animal pole almost at the equator, by making its way through the jelly. After one sperm makes contact with the egg surface, others do not enter the jelly. At the point of entry of sperm the egg surface is raised into a small papilla called **cone of reception**. The sperm head generally makes a direct perpendicular contact with the vitelline membrane and with the help of acrosome works its way through it and the tail is left behind. The sperm takes several minutes to enter. The entry of sperm brings about the extrusion of the **second polar body (Fig. 14.60 B)**. It is extruded through a small pit that appears close to the first polar body. Soon after penetrating the sperm head (now known as the **male pronucleus**) moves towards the **egg nucleus** or **female pronucleus** and the passage is known as the **penetration** or **sperm path**. This becomes evident because the spermatozoon carries some pigment granules from the surface layer along with it in the form of a dark streak. The male and female pronuclei are haploid. Frequently the egg nucleus moves away with the result that the sperm path is diverted to meet the egg nucleus. This diverted course of the sperm path is called the **copulation path** (Fig. 14.60 C) because it results in the fusion of the two nuclei to form a zygote with diploid nucleus (diploid number of chromosome in frog is 26). By the time the two nuclei fuse, the two sperm centrosomes are ready to form the division spindle for the first cleavage.

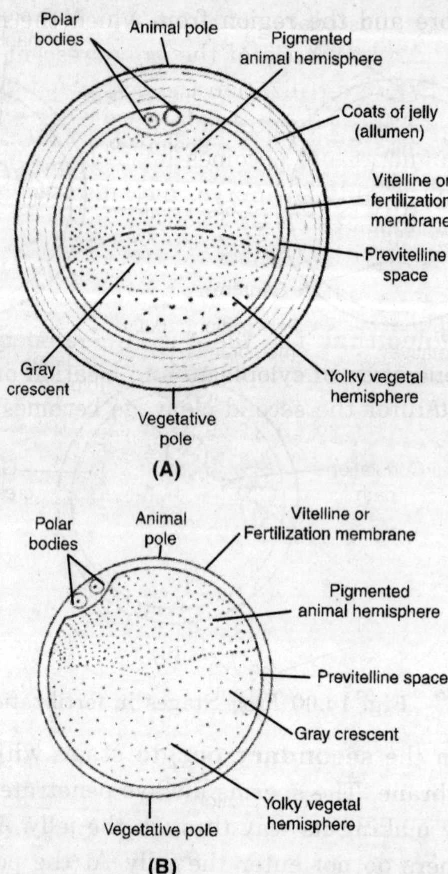

Fig. 14.61 *Frog*–**(A)** Fertilized egg in dorsal view; **(B)** Zygote without jelly in lateral view

Changes in the egg: Soon after the penetration of the sperm some changes take place in the egg. There is an almost immediate loss of water from the egg, so the vitelline membrane gets separated from the plasma membrane of the egg and a space called **perivitelline space** (Fig. 14.61) is formed between the two. It is filled with a fluid which is released by exocytosis from zygote cytoplasm. The egg rotates in the fluid in such a way that the black pigmented animal hemisphere comes to lie uppermost and the heavier yolk laden vegetal hemisphere is lowermost. The vitelline membrane is now known as the **fertilization membrane** which most probably prevents polyspermy (Wolf et al., 1976). At the completion of maturation the jelly swells up to its maximum as it imbibes more water. Ultimately the jelly becomes arranged in three layers, the thickness of which is more than the diameter of the egg.

Following the invasion by the spermatozoon the egg substance undergoes many changes. There is a streaming of protoplasm towards the animal pole and of yolk towards the vegetal pole. The minute pigment granules on the surface of the egg which appear motionless in the unfertilized egg show violent activity, and this involves loss of pigment from the surface of the egg. As a result of this, between the animal and vegetal poles, an area appears which is gray in colour. This area is crescentic in shape, hence called the **gray crescent**, and it establishes the plane of embryonic symmetry. The gray crescent is entirely a surface phenomenon and has nothing to do with the fusion of gametes. It marks the position of the

dorsal lip of the blastopore and the region from which the mesoderm and notochord will later develop. Thus with the appearance of the gray crescent the orientation of the future embryo can be stablished. Before fertilization the egg is radially symmetrical about the egg axis passing from the animal to the vegetal poles. With the appearance of the gray crescent the egg loses its radial symmetry and becomes bilaterally symmetrical. They gray crescent marks the dorsal surface of the future embryo, the animal pole marks the anterior end and the vegetal pole, the posterior end.

Cleavage

The egg of frog is **telolecithal**, i.e., there is large amount of yolk concentrated at one pole, opposite to the concentration of cyloplasm and location of the nucleus. The cleavage is **holoblastic**, i.e. **total** and after the second cleavage becomes unequal. The first **cleavage**

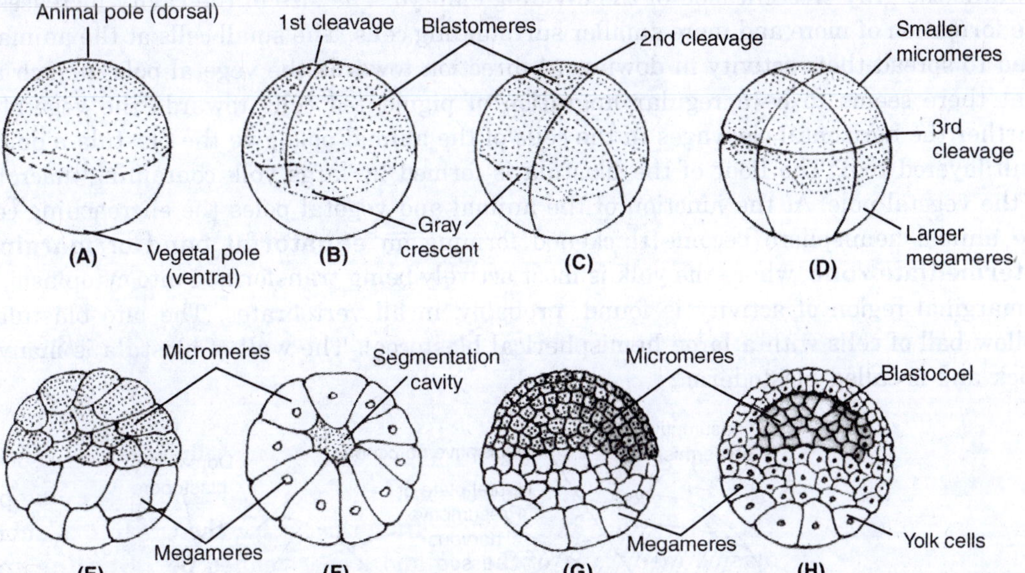

Fig. 14.62 *Frog*–Cleavage. **(A)** Zygote; **(B)** 2-cellstage; **(C)** 4-cell stage; **(D)** 8-cell stage; **(E)** Early blastula; **(F)** V.S. of early blastula; **(G)** Late blastula; **(H)** V.S. of late blastula

begins two and a half to three hours after fertilization and is **vertical** or **meridional** (Fig. 14.62 B). It begins as a small depression near the centre of the animal hemisphere and gradually extends vertically in the form of a groove up to the vegetal pole. It divides the zygote into two identical cells. The **second cleavage** (Fig. 14.62 C))begins about an hour after the first. The furrow of this cleavage also begins at the centre of the animal hemisphere and is at right angles to the first. It is also **vertical**. It divides the zygote into four equal **blastomers** which are not qualitatively identical because any two contain material from the gray crescent. The **third cleavage** begins about thirty minutes after the second. The cleavage plane is **horizonal** and slightly above the equator (Fig. 14.62 D). It produces eight distinct cells of two types : the four upper, small, pigmented blastomerers at the animal pole are called **micromeres** and the four lower, large yolk-containing blastomeres at the vegetal pole are called **macromeres**. The **fourth cleavage** follows twenty minutes after the third and tends to be vertical. The cleavage rate is accelerated with each division. The subsequent divisions become so irregular that it is not possible to trace out any plan of procedure. The segmentation is more rapid in the micromeres at the animal pole than in the macromeres because of the large quantity of yolk present in the vegetal hemisphere. This solid, ball-like multicellular embryo is called **morula** by some writers but morula stage is absent in frog.

Blastula formation

The eight cell stage of the developing embryo shows the beginning of a cavity. This is the result of segmentation, so it is called the **segmentation cavity**. Since this stage is known as **blastula**, the cavity is also called **blasotocoel** (Fig. 14.62 E and F). It enlarges with each of the early cleavages and is filled with albuminous fluid arising from the surrounding fluid. The blastocoel appears in an eccentric position above the level of the equator, slightly towards the gray crescent side of the dividing embryo. The size of the cavity increases with the formation of more and more smaller surrounding cells. The small cells at the animal pole tend to spread their activity in downward direction towards the vegetal pole in such a way that there seems to be a regular migration of pigmented cells towards the vegetal pole. Further the horizontal cleavages in the cells of the animal pole give the blastula a double or multilayered roof. The floor of the blastula is formed by large yolk containing macromeres of the vegetal pole. At the junction of the animal and vegetal poles the encroaching cells of the animal hemisphere become thickened forming an **equatorial band** or **marginal** or **intermediate zone**, where the yolk is most actively being transformed into cytoplasm. Such a marginal region of activity is found, probably, in all vertebrates. The late blastula is a hollow ball of cells with a large hemispherical blastocoel. The wall of blastula is many cells thick and is called blastoderm.

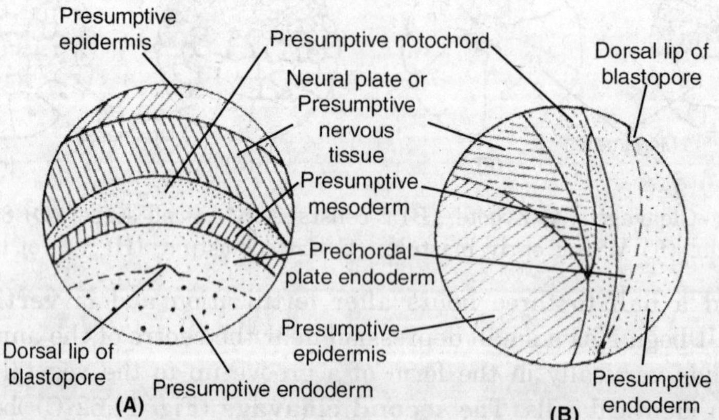

Fig. 14.63 *Frog*–Presumptive areas, **(A)** portero-dorsal view; **(B)** side view

Fate maps in frog

The cells of the blastula do not present much distinction in their make up except in the degree of pigmentation, size and yolk content. There is no apparent indication as to what part the cells will play in the formation of the future embryo. Many scientists have succeeded in demonstrating the ultimate fate of different cells even at the blastula stage. They have used many experimental devices, such as vital staining, and constructed maps, known as **fate maps**, showing the ultimate fate of various areas. This is called the prospective or presumptive fate of different regions. Thus a fate map is simply a topographical surface mapping of the blastula. The surface of blastula can be divided into three regions (Fig. 14.63):

1. A large area of small deeply pigmented cells around the animal pole.
2. A light pigmented marginal or intermediate zone or equatorial band around the equator.
3. A region of large cells around the vegetal pole.

The blastula is stained with various harmless dyes and the coloured areas are watched. The animal hemisphere has two areas. The single outer layer of cells forms the epithermal layer which forms the epithelium of the integument and lining of the nervous system. In this region material for nose, ear and foregut may be found. The inner layer of cells give rise to the **neuroblast** which forms the nervous system and sense organs, so it is designated as the **nervous layer**. It also has a small localized area which will take part in the formation of eyes of the embryo. The animal hemisphere region is designated as the **presumptive ectoderm** on the fate map. The endoderm arises partly from the marginal zone area and largely from the vegetal hemisphere which is designated as the **presumptive endoderm**. It gives rise to the alimentary canal, endodermal lining of the mouth, gill region and pharynx. In the marginal zone there is a large area on the dorsal side of the blastula which forms the **notochord**. On both sides of the notochord area in the marginal zone, material for **mesoderm** is present. It gives rise to the segmental muscles of the body. The lateral and ventral parts of the marginal zone give rise to the mesodermal lining of body cavity, kidney, etc.

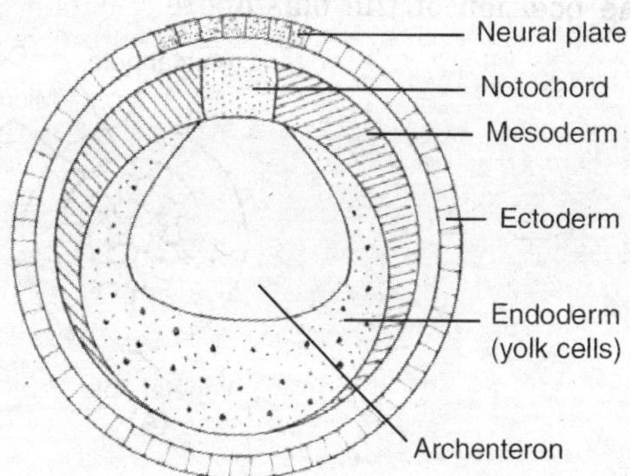

Fig. 14.64 *Frog*–T.S. gastrula showing primary germinal layers

Gastrulation

Gastrulation is the embryological term for the complex cell movements which occur in almost all animals at the cleavage period. These movements carry those cells, which will form the future internal organs, from their superficial position in the blastula to approximately in their definitive position inside the embryo. As a matter of fact during gastrulation several complicated types of activities go on within, rather simultaneously, but the ultimate result is that the presumptive ectoderm comes to enclose the presumptive notochord, mesoderm and endoderm (Fig. 14.64). A remarkable series of mass migration of cells known as **formative movements** take place. In essential, gastrulation involves a rearrangement of materials already present. The changes involved in gastrulation include the following process:

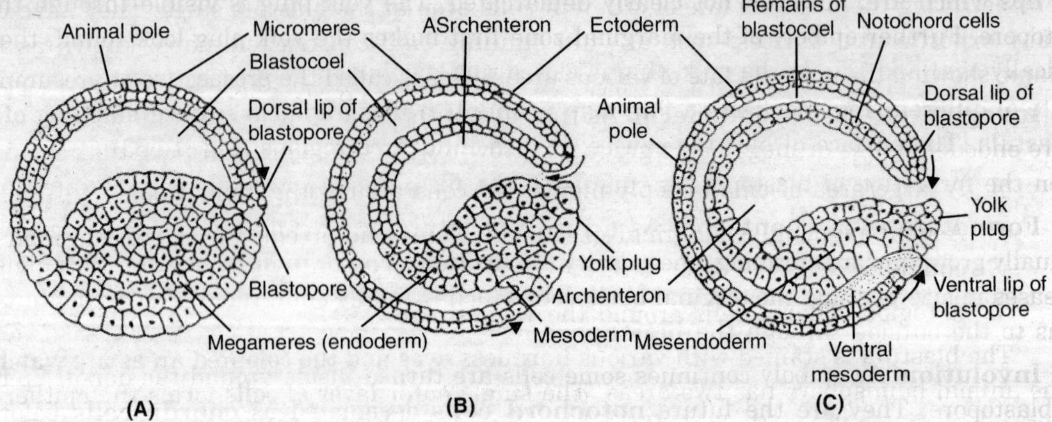

Fig. 14.65 *Frog*–Stages of gastrulation in median sagittal section

Epiboly: During epiboly there is progressive growth of the pigmented micromeres over the light coloured yolky megameres of the vegetal hemispheres until the pigmented cells completely cover the megameres except a small circular patch called the **yolk plug** marking the position of the blastopore.

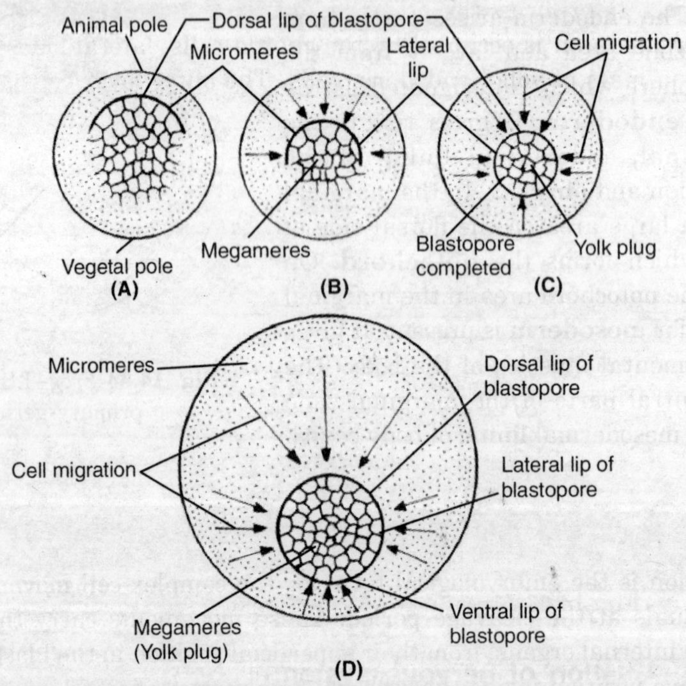

Fig. 14.66 *Frog*–Stages in the formation of blastopore

Formation of blastopore: When epiboly begins a small crescent-shaped groove appears on the postero-dorsal margin of the blastula. Its anterior pigmented margin is called the **dorsal lip** of the blastopore and the backwardly projecting horns form the **lateral lips** of the blastopore. As epiboly continues the lateral lips approach each other to form the **ventral lip** of the blastopore. Thus the crescentic groove becomes a complete circular blastopore with four lips which are, however, not clearly demarcated. The yolk plug is visible through the blastopore. Further epiboly of the marginal zone first makes the yolk plug look round, then vertically oval and finally the yolk plug is closed and disappears.

Invagination of the endoderm: As the cells of the animal pole continue to grow, the future endoderm cells or megameres gradually grow towards the blastopore and sink inside. When the formation of blastopore is completed the future endoderm cells become internal.

Formation of archenteron: As a result of inturning of cells the crescentic groove gradually grows inwards forming a new cavity the **archenteron** or **primitive gut**. It gradually increases in size pushing the original blastocoel which is finally obliterated. The archenteron opens to the outside through the **blastopore**.

Involution: As epiboly continues some cells are turned inside around the dorsal lip of the blastopore. They are the future **notochord cells** designated as **chorda cells** before differentiation. These cells lie beneath the **neural plate cells**. The future **mesoderm cells** roll inside the lateral lips of the blastopore and are arranged on either side of the chorda cells between the ectoderm and endoderm. The mesoderm of ventral side rolls over the ventral lip of the blastopore.

A completed gastrula has a fully developed archenteron opening to the outside through the blastopore which later forms the proctodaeum. The blastocoel is completely obliterated. The roof of the archenteron is occupied by the chorda cells, lateral sides by mesoderm cells and the floor has a mass of yolky endoderm cells. The outer layer consists of ectoderm and neural plate cells.

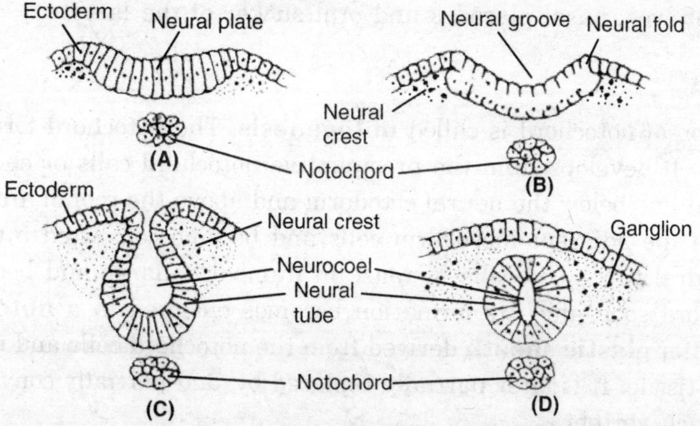

Fig. 14.67 *Frog*–Stages in the formation of neural tube

Neurogenesis (formation of nervous system)

Neurogenesis begins towards the end of gastrulation . The presumptive **neural plate area**, which extends as a **median band** from the dorsal lip of the blastopore in the anterior direction, becomes thickened froming a mid-dorsal **medullary** or **neural plate.** Soon the lateral margins of the neural plate thicken to form **neural** or **medullary folds**. This is followed by the appearance of a longitudinal depression, the **neural groove**, in the medullary plate. The embryo becomes elongated and this stage is called **neurula**. The medullary folds grow towards each other and fuse in the midline. Thus the open groove is converted into a closed **neural tube** with a central cavity called **neurocoel**. The wide anterior end of the neural tube forms the brain and the narrow posterior end forms the spinal cord. Anteriorly the neural tube remains open for sometime by an opening called **neuropore**. It soon closes and the anterior end of the neural tube enlarges to form the **primary cerebral vesicle** which soon constricts to form **fore**-, **mid**-, and **hind-brain vesicles**. The posterior opening is the blastopore. At the posterior end the medullary folds merge into the sides of blastopore and as the folds meet in the middle they form the primitive streak. The blastopore is reduced to a slit called proctodaeum which marks the site of future anus. The archenteron now opens to the outside by way of the blastopore, but opens into the posterior end of the neurocoel. The canal connecting the archenteron and the neurocoel is called the **neurenteric canal**. Later the neurenteric canal closes and there is no connection between the neural tube and the enteron.

As the neutral folds fuse the neural tube separates from the dorsal ectoderm which becomes a continuous sheet of cells above the mid-dorsal line. The neurocoel is lined with ciliated and pigmented ectoderm and occurs as the much reduced **central canal** of the spinal cord and the brain. On the lateral margins of the neural plate, some ectoderm cells which do not take part in the formation of the neural tube appear as a semicircular elevated **ridge** or **crest** called **neural crest** or **sense plate**. It contains materials which will give rise to parts of the autonomic system, dorsal root ganglia of cranial and spinal nerves, visceral arches, lens of eyes, nasal placodes and oral sucker of the larva.

Notogenesis

Formation of notochord is called **notogenesis**. The notochord forms the axial skeleton of the animal. It develops from the presumptive notochord cells or chorda cells lying along the mid-dorsal line below the neural ectoderm and above the roof of archenteron. These cells separate from the adjacent mesoderm cells and become converted into a **cylindrical rod** below the neural tube. The cells increase in number, enlarge and become vacuolated. The entire notochord soon after its formation becomes enclosed in a **notochochordal sheath** made up of outer **elastic sheath** derived from the notochord cells and inner **fibrous sheath** of connective tissue. It is later partially replaced by, and partially converted into material of centrum of each vertebra.

Differentiation of mesoderm and formation of coelom

In the frog embryo the mesoderm is in the form of sheets of blocks without any cavity (unlike *Branchiostoma* in which it first appears as hollow mesodermal pouches) extending on either side of the notochord in the form of an inverted horse shoe around the archenteron and the yolk. The uppermost part of the mesoderm that lies on either side of the notochord is called **segmental** or **vertebral plate** and the more ventral portion is called the **lateral plate mesoderm**. Soon this mesoderm gets differentiated into three distinct regions : the **dorsal somite mass** or **epimere** close to the notochord, the intermediate cell mass or

mesomere and the most lateral thinner region, the **hypomere** or **lateral plate**. The mesoderm now forms a complete layer. At this stage the embryo is **tiploblastic**, i.e., all the three germ layers, ectoderm, endoderm and mesoderm are formed.

Epimere and formation of somites: The epimere of each side divides transversely into blocks of cells called **somites**. The first somite appears on the anterior end just behind the pharynx and the others follow in sequence towards the posterior end of the embryo. In all 45 pairs of somites are formed in 6 mm long tadpole and out of this 34 pairs are formed in the tail. A narrow cavity called the **myocoel** develops eccentrically within each somite. The outer wall of somite, towards the ectoderm in thin and is called **dermatome**. It gives rise to the dermis of the skin. The inner wall of somite, towards the notochord, is called **myotome**. It gives rise to the skeletal muscles of the body. The part of the somite near the lower edge of myotome is called **sclerotome**. It gives rise to the vertebral column and some parts of the skull. Gradually the myocoel disappears.

Mesomere: The upper part of the remaining mesoderm on either side of the archenteron differentiates into kidney forming tissue. This band of mesoderm is called **mesomere** or **nephrotome**. It retains its connection with the hypomere a little longer, but finally separates from it. It is also divided into a series of segments. The nephrotomes soon develop a cavity called **nephrocoel**. They later give rise to tubular outgrowths on the sides which move backward finally uniting into a **longitudinal duct**. This is the beginning of the **pronephros** or head kidney with tubules as excretory agents and longitudinal duct as pronephric duct opening into the cloaca.

Hypomere and development of coelom: The remaining part of the mesoderm is called the **hypomere** or **lateral plate mesoderm**. The lateral plate mesoderm on both sides of the archenteron splits dorsoventrally into two layers : outer **parietal** or **somatic** layer and inner **visceral** or **splanchnic** layer. The somatic mesoderm in conjunction with the adjacent ectoderm forms the **somatopleurae** that gives rise to the skin, its blood vessels and connective tissue. The splanchnic mesoderm in conjunction with the gut endoderm forms **splanchnopleurae** which gives rise to the lining epithelium, muscles and blood vessels of the intestine. The space between the somatic and splanchnic layers forms the **body cavity** or **coelom** (a **schizocoelom**). One cavity appears on each side. Eventually the two coelom grow towards each other and meet ventrally in the floor of the archenteron to form a single coelomic cavity.

Development of heart

The heart and the surrounding pericardium develop from the mesoderm lying below the pharynx. At the end of neurulation the free edges of the mesodermal mantle gradually grow towards the middle line below the pharynx. At this stage some loose cells are derived from the free edges of the growing mesodermal mantle. These are called **endocardial cells**. They accumulate below the pharynx and become arranged in the form of two thin tubes which soon fuse to form a single thin walled tube called **endocardium**. It forms the endothelial lining of the cavity of the heart. The lumen of the tube becomes the cavity of the heart. The endocardial tube bifurcates at both ends, the two branches at the anterior end represent **ventral aortae** and the two branches at the posterior end are called **vitelline veins**.

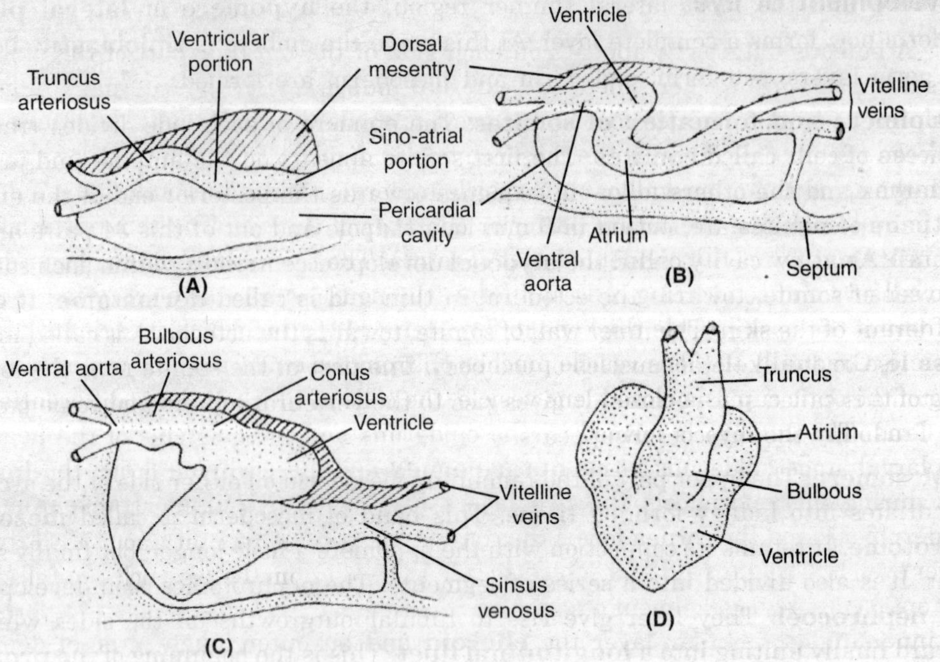

Fig. 14.68 *Frog*–Stages in the development of heart

When the endocardial tube is being formed, the growing edges of the mesodermal mantle meet and fuse below the endocardial tube. The floor of the pharynx induces a split in the mesodermal sheet as a result of which a pair of space develop in it. The two space grow towards each other but remain separated by a median strip called **ventral mesocardium**. It soon disappears and the two spaces become continuous and form the pericardial cavity. The outer mesodermal wall represents **somatic mesoderm** and the inner wall is called **splanchnic mesoderm**. The somatic mesoderm forms the lining of the pericardial cavity and is called the **pericardium**. The portion of the splanchnic mesoderm surrounding the endocardium is called **epimyocardium**. It later splits into a thin outer layer called **epicardium** and a thick inner layer called **myocardium**.

Inside the pericardium the heart lies longitudinally as a simple straight tube. Its two ends remain attached to the pericardium by dorsal mesocardium. The simple tubular heart elongates rapidly and the middle myocardium grows in thickness. The different regions of the tube grow differently in thickness and constrictions appear to form different chambers. The posterior most part of the tube differentiates as **sinus venosus** which receives a pair of **vitelline veins**. It opens into a thin walled **atrium** which is followed by a thick walled **ventricle**. The later leads to the **bulbus arteriosus** which opens into the **truncus arteriosus**. The attached anterior and posterior ends of the tube do not allow it to grow in length in the antero-posterior direction. As a result the straight tube becomes folded in the form of a reversed S-shaped tube. The posterior part (atrium) of the original tube folds to become dorsal and anterior in relation to the more anterior part, the ventricle, of the tube. Similarly, the original anterior part (ventricle) becomes ventral and posterior.

Development of eye

A bulbous outgrowth appears on each side of the anterior part of the embryo directly dorsal to each oral sucker. These are external indications of the internally enlarging **optic vesicles**. The optic vesicles continue to grow towards the epidermis. A pressure is developed between the two, as a result the mesenchyma is displaced and the vesicles come in contact with the epidermis. The wall of optic vesicle undergoes invagination forming a double walled **optic cup** enclosing the future posterior chamber of the eye. The thin outer layer develops pigments and forms the pigmented coat of retina. The thick invagninated layer differentiates into sensory and conducting cells and forms the neurosensory layer of retina. The epidermis in front of the optic vesicle thickens to form **lens placodes** which invaginate to form a **lens vesicle**. Gradually the lens vesicle pinches off from the epidermis and comes to lie within the ring of the optic cup as a closed lens vesicle, the cells of which undergo differentiation to form the **lens**. The **cornea** is formed by the epidermis and mesenchyma of the head region. In the larval stage epidermis forms **outer cornea** and mesenchyma forms the **inner cornea** but during metamorphosis the two layers fuse to form a single transparent layer. The epidermis around the cornea forms the conjuntiva. It remains continuous with the upper and lower eyelids. The **choroid** and **sclerotic** coats develop from the mesenchyma cells surrounding the optic cup. As development progresses the mesenchymal coat becomes differentiated into an inner highly vascular layer the **choroid** and an outer tough layer of densely fibrous connective tissue called **sclera** or **sclerotic coat**.

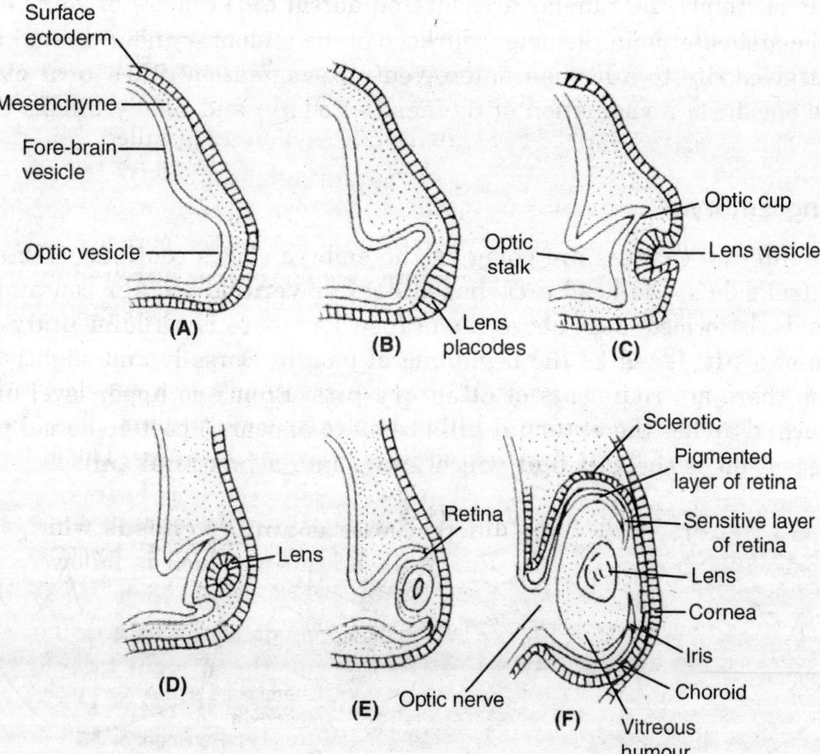

Fig. 14.69 *Frog*–Stages in the development of eye

After the formation of lens the pigmented retinal epithelium at the rim of the optic cup and the choroid coat grow over the edges of the lens and form the pigmented **iris**. The outer epithelium of iris is formed from the choroid mesoderm. The opening in front of the lens is called **pupil**. Smooth musle fibres develop within the iris and control the size of the pupil. The **ciliary body** develops from the non-neural part of the sensory retina on the inner side of the rim of the optic cup. The mesodermal cells of the choroid form the **ciliary muscles**. The **suspensory ligaments** emerge from the ciliary body, become attached to the lens capsule and suspend the lens.

Formation of gut

The anterior portion of the archenteron expands both ventrally and laterally beneath the notochord and infundibulum of the brain forming the **foregut** and its derivatives. The large cavity of the foregut is the **pharynx** which gives rise to **visceral pouches**. The most anterior of these is the **mandibular arch** which is associated with the development of the muscles of the jaws. The **hyoid arch** is the next lying posterior to the mandibular arch. Between these develop the **hyomandibular pouch** that never breaks through to form the cleft. The next three are the **first**, **second** and **third branchial pouches**. These extend outward towards the ectodermal invagination (visceral grooves). The foregut also gives rise to a medioventral posteriorly directed outgrowth, the **liver diverticulum**, which extends a short distance underneath the yolk, and gives rise to the **liver**, **gall bladder** and **bile duct**. The **midgut** is simply the tubular archenteron dorsal to the mass of yolky endoderm. The portion of the archenteron in the neighbourhood of the temporary neural canal is the hindgut. The foregut gives rise to a median antero-ventral evagination called **oral evagination**. It meets the stomodaeal invagination of the head ectoderm and finally breaks through as the mouth.

Prehatching embryo

As the internal changes are going on the embryo shows continuous external changes. On the ventral side of the head a U-shaped **oral** or **ventral sucker** containing mucous or cement glands, is formed. Just above the oral sucker there is a rudimentary **stomodaeum** in the form of a **pit**. It marks the beginning of mouth. Dorsally and slightly lateral to the stomodaeum there are rudiments of **olfactory pits**. From the upper level of the first and second branchial arches the **external gills** begin to appear. A postero-dorsal projection near the proctodaeal pit is the **tail bud** which grows into a **postanal tail**.

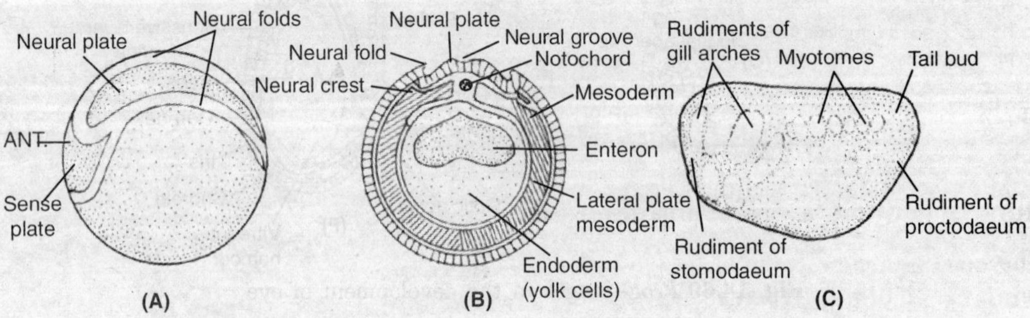

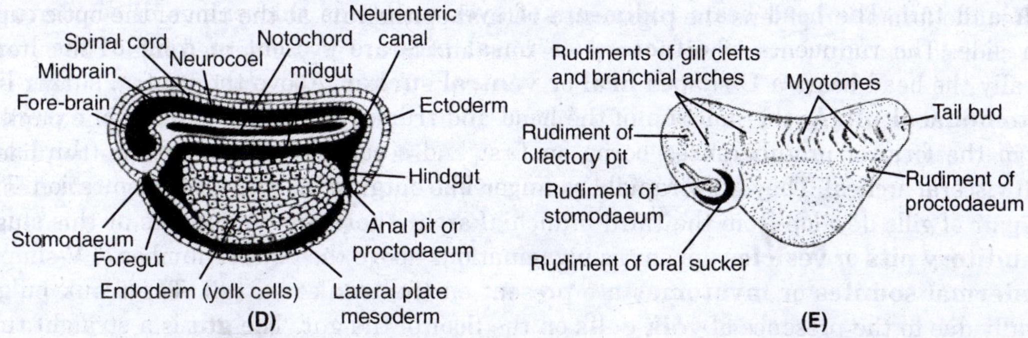

Fig. 14.70 *Frog*–Neurula stage. (A) Early neurula inside view; (B) T.S. early neurula; (C) Late neurula in side view; (D) T.S. late neurula; (E) Prehatching larva

Hatching: The embryo hatches when it is about 6 mm long (about two weeks after fertilization). The interval between fertilization and hatching depends upon temperature of the environment. It is believed that hatching is accomplished with the help of temporary glands in the region of the sucker. Probably these glands secrete an enzyme under the influence of which the jelly is dissolved providing space for the escape of the larva. Even after hatching the larvae are seen attached to empty jelly capsules, but the larvae do not eat jelly. It nourishes itself from the yolk still present in the cells of the floor of the archenteron. Before hatching the embryo shows constant swimming movements because of the cilia on its epidermis. The embryo also shows considerable muscular movements which continue even after hatching. Occasionally the larva may coil and exhibit S-shaped body contractions which are exclusively muscular.

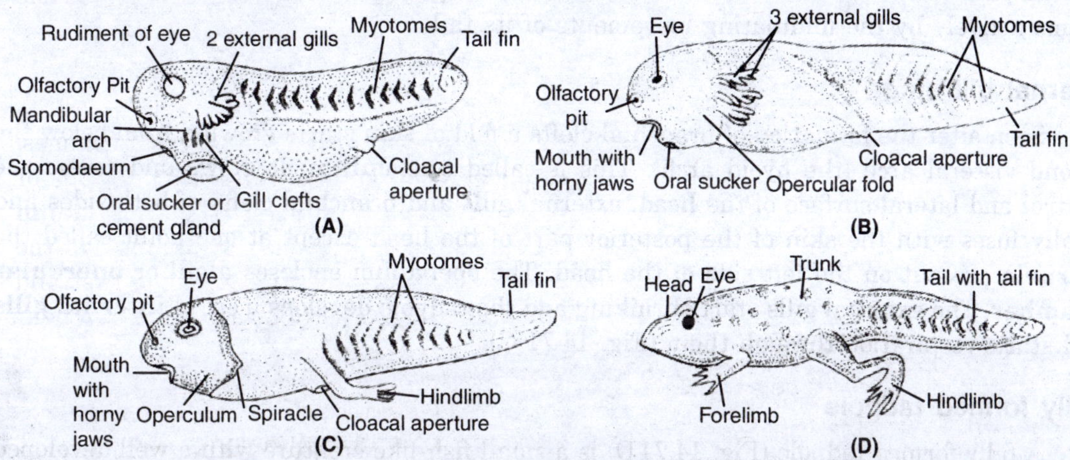

Fig. 14.71 *Frog*–(A) Newly hatched tadpole; (B) External gill stage; (C) Internal gill stage; (D) Fully formed tadpole

Structure of tadpole larva at the time of hatching

The newly hatched larva of frog is called **tadpole** (Fig. 14.71 A). It is a small, blackish, free-swimming fish-like creature about 6 mm in length. The body is divisible into **head**,

trunk and **tail**. The head bears rudiments of **eyes** visible as thickenings of ectoderm on either side. The rudiments of **olfactory** or **nasal pits** are present in front of the head. Ventrally the head bears a U-shaped **oral** or **ventral sucker**. Above the ventral sucker lies the **stomodaeal pit**. At the junction of the head and trunk there are two pairs of **external gills** in the form of protuberances borne on first and second branchial arches (third and fourth visceral arches). They grow rapidly as finger-like outgrowths and start respiration. The third pair of gills develop from the third branchial arch. There are no gill slits at this stage. The **auditory pits** or **vesicles** appear as invaginations above the gills. A number of **V**-shaped **mesodermal somites** or **myotomes** are present on the trunk and tail. The trunk bulges ventrally due to the presence of **yolk cells** on the floor of the gut. The **gut** is a straight tube which opens to the outside through a **cloacal aperture** present ventrally at the base of the tail. The heart is a **S**-shaped tube. The functional kidneys are a pair of **pronephros**. The **tail** is short and without tail fin. The **epidermis** of the larva bears **cilia**.

External gills stage

A few days after hatching the stomodaeum develops connection with the pharynx and the **mouth** is formed. The mouth develops jaws and lips which acquire **horny papillae** or **teeth** with the help of which the tadpole starts feeding on algae and other aquatic vegetation. The **external gills** (Fig. 14.71 B) grow longer, become highly vascular and exchange of gases takes place through them. They remain functional for a short time and later replaced by **internal gills**. Four vertical slits appear on either side of the head below the external gills. These are called the **gill slits** or **clefts** or **branchial clefts**. The pharynx communicates with the exterior through them. The tail becomes longer and acquires **dorsal** and **ventral caudal fins** which are continuous around the tail. The fins are without **finrays**. The larva swims actively by the undulating movements of its tail.

Internal gill stage

Soon after the formation of branchial clefts a fold of skin starts growing from below the second visceral arch (the hyoid arch). This is called **operculum**. It grows and covers the ventral and lateral surface of the head, external gills and branchial clefts of both sides and finally fuses with the skin of the posterior part of the head except at one point called the **spiracle** present on the left side of the head. The operculum encloses a gill or **opercular chamber**. The external gills start shrinking and the tadpole develops a set of **internal gills** and starts respiration through them (Fig. 14.71 C).

Fully formed tadpole

A fully formed tadpole (Fig. 14.71D) is a small fish-like creature with a well developed tail, tail fin and tail muscles which help in locomotion. It has eyes, nostrils, mouth, spiracle, a very long spirally coiled intestine and cloacal aperture. The oral sucker has disappeared. For respiration water enter through the mouth, passes through the pharynx, branchial clefts, over the internal gills in the opercular chamber and finally to the outside through the spiracle. In later stage **lungs** develop from the pharynx and tadpole respires both by gills and lungs. For lung respiration it frequently comes to the surfaces to gulp in air. The **lateral line sense organs** develop. **Mesonephric kidneys** becomes functional. At the end of larval life, the **limb buds** start appearing, the **hind limb buds** appear first while **forelimb buds** appear a little later and remain covered by operculum for sometime. The left forelimb emerges through the spiracle while the right one through the operculam.

Metamorphosis

After leading an active free swimming existence for some time (about two or three weeks) the tadpole undergoes a series of rapid changes transforming it into a terrestrial tail-less adult frog with completely different morphology and physiology often for life in a different habitat. The beginning of metamorphosis is indicated by frequent trips of tadpole to water surface to gulp air for breathing. It stops feeding. Metamorphosis is completed in a few days. During metamorphosis some structures undergo reduction and finally disappear, some become functional and some structures which occur and function in the tadpole as well as the adult get transformed or remodelled in order to meet the requirements of the adult mode of life.

Following are the major changes which take place during metamorphosis:

1. The lung-type or pulmonary respiration takes over the internal gill phase.
2. The gill clefts are closed and the operculum falls off.
3. The larval skin is shed.
4. The horny jaws are lost and replaced by true jaws which are first of cartilage and then of bone.
5. The mouth widens and the snout assumes the adult shape.
6. The lateral line sense organs disappear.
7. The tail, tail fin and muscles are lost by regre-ssion.
8. The two pairs of limbs become long and functional. They develop joints and toes. The hindlimbs arise first just before metamor-phosis and the forelimbs emerge just before coming out of water.
9. The middle ear and tymp-anic membrane develops.
10. Thc cycs bulge on the dorsal surface of the head. They develop eye-lids and nictitating mem-brane.
11. The tongue develops from the floor of the mouth.

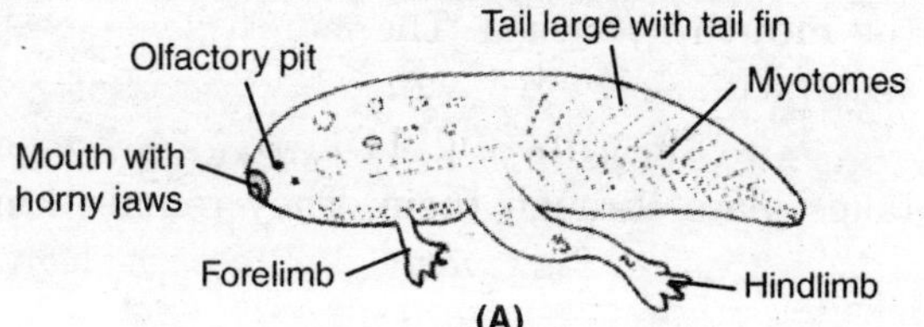

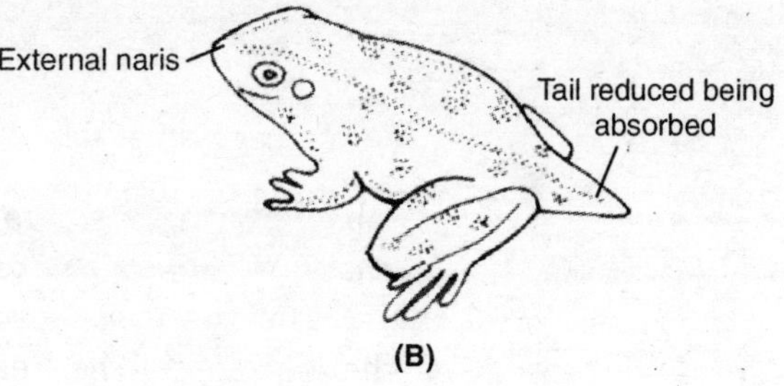

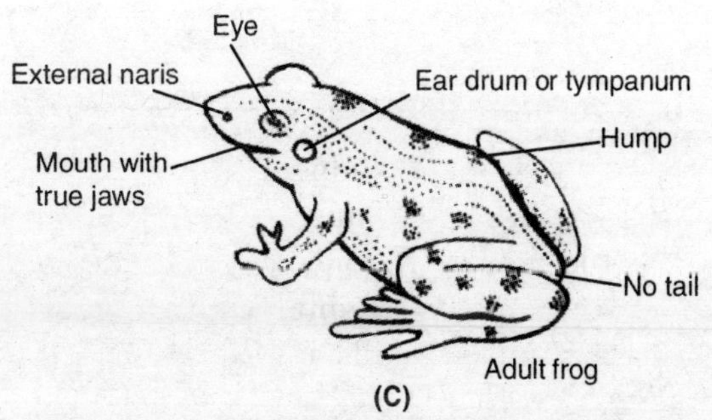

Fig. 14.72 *Frog*–Stages in metamorphosis

12. In response to the change in diet the long and coiled intestine shor-tens and straightens.
13. The kidneys become mesonephric.
14. The brain gets differentiated.
15. The heart becomes three-chambered.
16. The skin thickens and becomes glandular. The surface layer becomes keratinized and characteristic pigmentation pattern develops.
17. Some endocrine glands function actively and the definitive gonads appear.

With all these changes a frog in miniature is formed and leaves water.

15

CHAPTER

AMPHIBIA:

GENERAL CHARACTERS AND CLASSIFICATION

The transition from aquatic to terrestrial mode of life was one of the most important events in the evolutionary history of vertebrates. The amphibians made the first attempt but could not become completely terrestrial. The class **Amphibia** (**Gr.** *amphi*; dual; *bios*; life) include, animals that live partly in fresh water and partly on land. Both in structure and function they stand between the fishes and reptiles.

GENERAL CHARACTERS

1. The amphibians are cold blooded, carnivorous, tetrapod vertebrates.
2. Skin is moist, smooth or rough and glandular. Scales absent, in some forms scales concealed in the skin.
3. Paired fins absent, instead there are two pairs of limbs, some forms are limbless. Toes four or five or fewer, hindlimbs with webbed toes.
4. Skeleton largely bony. Notochord not persistent. Ribs, if present, not attached to the sternum. Skull with two occipital condyles.
5. Two nostrils connected to the mouth cavity are present. They are valved to exclude water.
6. Eyes often with movable eyelids.
7. Auditory apparatus with a single rod-like ossicle, the columella, fitting into fenestra ovalis.
8. Mouth large, usually with small homodont teeth in both upper and lower jaws. Tongue is often protrusible.
9. Heart is three chambered (two auricles and one ventricle) with one or three pairs of aortic arches. Sinus venosus present. Renal and hepatic portal system well developed. Red blood cells oval, with nucleus.
10. Respiration by lungs, skin or mouth lining, all together or separately, larva respires with the help of gills, some adults also use gills for respiration.
11. Vocal cords present in frogs and toads.

12. Excretory organs are a pair of mesonephric kidneys. Ureters open into the cloaca. Urinary bladder large. Excretion ureotelic.
13. Brain with ten pairs of cranial nerves.
14. Lateral line sense organs present in larva and some aquatic adults.
15. Sexes separate. No copulatory organ in males. Females mostly oviparous. Gonoducts open into cloaca.
16. Eggs with gelatinous covering, laid in water. Fertilization external.
17. Development indirect. Larva is called tadpole. Larva aquatic.

CLASSIFICATION

The class Amphibia includes the living salamanders (urodela), frogs and toads (Anura), the limbless caecilians (Gymnophiona) and many fossils forms of Carboniferous and Permian time. About 2,500 living species are known. The amphibians have been classified by different authors but now-a-days the classification given by G.K. Noble (1924) is generally followed. He divided the amphibians into six orders, three extinct and three living. The extinct amphibians are placed in subclass **Stegocephalia** and the living ones in subclass **Lissamphibia**.

Subclass I: STEGOCEPHALIA

1. Extinct amphibians, lived from Devonian to Triassic.
2. Endoskeleton massive.
3. Skull with a solid roof of membrane bones which were deeply sculptured.
4. Exoskeleton of dermal scales and bony plates.

Fig. 15.1 *Eryops*

Order 1. Labyrinthodontia

1. Large crocodile or salamander-like aniphibians known as stem amphibians (oldest known Tetrapoda).
2. Existed from Carboniferous to Triassic period.
3. Skull large and completely roofed with bones.
4. Teeth large, with greatly folded dentine in both the jaws.
5. Many species aquatic, some terrestrial.

Example: *Eryops*

Order 2. Phyllospondyli

1. Small salamander-like amphibians.

2. Lived from carboniferous to Permian period.
3. Head large and flat.
4. Vertebrae tubular, with well developed transverse processes with stout ribs.
5. Spinal cord and notochord lodged in a single cavity.
6. Fingers four and toes five.
7. Regarded as ancestors of modern Salientia and Urodela.

Example: *Branchiosaurs* (*Ichthyostega*).

Order 3. Lepospondyli

1. Small salamander-like or eel-like amphibians.
2. Lived from Carboniferous to Permian period.
3. Vertebrae cylindrical, each composed of a single piece. Neural arch and centrum continuous. Ribs articulate with the vertebral column intervertebrally.
4. Regarded as ancestors of modern Gymnophiona.

Examples: *Dipocaulus*, *Lysorophus*.

Subclass II: LISSAMPHIBIA

1. Includes all living amphibians
2. Exoskeleton absent
3. Teeth small and simple.

Order 1. Gymnophiona (**Gk.**, *gymnos*, naked; *ophioneos*, serpant-like) or **Apoda** (**Gk.**, *a,* without; *podos*, foot).

1. Primitive burrowing amphibians with elongated eel-type body, commonly known as caecilians.
2. Burrows in surface soil or in damp places near streams.
3. Limbs and limb girdles absent.
4. Tail short or absent.
5. Eyes small, functionless and covered by skin.
6. Tympanum and tympanic cavity absent.
7. Numerous minute dermal scales embedded in the skin in transverse rows.
8. Males with protrusible copulatory organ.

Examples: *Ichthyophis*, *Uroaeotyphlus*.

Order 2. Urodela (**Gk.**, *uro,* tail; *delos*, visible) or **Caudata** (**L.**, **cauda**, tail; **ata**, bearing)

1. Lizard-like amphibians with a well developed tail which helps in progression.
2. Limbs weak, almost of equal size.
3. Skin without scales.
4. Tympanic membrane and tympanic cavity not developed.
5. Teeth present in both the jaws.
6. Gills may be present or absent in adults.

7. Male without copulatory organ.
8. Larvae resemble adults; aquatic.

Suborder 1. Cryptobranchoidea

1. Most primitive, permanently aquatic form.
2. Skin soft and flabby.
3. Eyelids absent.
4. Gills absent in adults.
5. Premaxillary spine short.
6. Fertilization external.

Examples: *Cryptobranchus*, *Megalobatrachus*.

Suborder 2. Ambystomatoidea

1. Adults live on land.
2. Eyelids present.
3. Premaxillary spine large.
4. Teeth arranged in transverse rows across the posterior margin of the vomer.
5. Fertilization internal.
6. Exhibit the phenomenon of neoteny or paedogenesis.

Example: *Ambystoma*.

Soborder 3. Salamandroidea

1. Eyelids presents or absent.
2. Teeth arranged on palate and prevomers.
3. Respiration by gills or lungs, in some by skin.
4. Vertebrae opisthocoelous.
5. Fertilization internal, usually viviparous.

Examples: *Triton* and *Triturus* (newts), *Salamandra* (salamander), *Amphiuma* (congo eel), *Plethodon*.

Suborder 4. Proteidae

1. Aquatic, bottom dwelling forms with depressed body.
2. Permanent larval forms with three pains of external gills and two pairs of gill slits in adults.
3. Tail with caudal fin.
4. Eyelids absent.
5. Teeth present.
6. Skull cartilaginous, without maxillae.

Examples: *Proteus*, *Necturus* (mud puppy).

Suborder 5. Meantes

1. Aquatic, permanent larval forms.
2. Forelimbs small, hindlimbs absent.

3. Eyelids absent.
4. External gills well developed, persistent.
5. Jaws with horny covering.

Examples: *Siren* (mud eel), *Pseudobranchus.*

Order 3. *Salientia* (**L.**, *saliens*, leaping) or **Anura** (**Gk.**, *an*, without, *aura*, tail)

1. Body short and broad, with short forelimbs and long hindlimbs.
2. Head and trunk fused, neck and tail absent.
3. Mouth wide opening. Mandibles without teeth.
4. Eyelids well developed, nictitating membrane present.
5. Tympanum present.
6. Skin loose and scaleless.
7. Gills present only in the larval stage. Adults respire by lungs.
8. The vertebral column is short, ribs absent or reduced.
9. Fertilization external. Developement indirect.

Suborder 1. Amphicoela

1. Vertebrae amphicoelous, ribs free.
2. Two relict tail muscles present.
3. Fertilization internal.

Examples: *Leoplema*, *Ascaphus.*

Suborder 2. Opisthocoela

1. Vertebrate opisthocoelous.
2. Ribs free.

Examples: *Alytes* (mid-wife toad), *Bombinator*, *Pipa*, *Xenopus.*

Suborder 3. Procoela

1. Vertebrae procoelous. Urostyle with two condyles.
2. No free ribs.

Examples: *Bufo* (common toad), *Hyla* (tree frog).

Suborder 4. Anomocoela

1. Vertebrae procoelous or amphicoelous.
2. No free ribs.
3. Sacral vertebrae fused with urostyle.
4. Upper jaw with teeth.
5. Skin rough.

Examples: *Pelobates*, *Scaphiopus.*

Suborder 5. Diplasiocoela

1. Vertebrae one to seven procoelous, eighth amphicoelous and ninth convex anteriorly and with two condyles posteriorly.

2. Ribs absent.
3. Pectoral girdle fused with sternum.

Examples: *Rana* (common frog), *Rhacophorus* (flying frog).

REPRESENTATIVE TYPES

Ichthyophis

Ichthyophis is commonly known as **caecilian**. It is found in tropical countries like Sri Lanka, Philippines, Borneo, Mexico to Argentina and India. It lives in burrows in moist ground. In feeds on small invertebrates. The animal is worm-like and slender, about 30 cm long. The body is divided into head, trunk and tail. The tail is very much reduced. The body is covered with smooth, slimy, transversely ringed skin, with small calcified scales, arranged in transverse rows, embedded in it. Squirt glands are present in the skin which discharge an irritating fluid. The head bears a pair of small, functionless eyes covered with skin, nostrils and a small protrusible sensory tentacle between the eye and nostril. The tympanum and tympanic cavity are absent. The skull is compact and bony. The limbs and limb girdles are absent. The vertebrae are amphicoelous. The right lung is elongated while the left is reduced. The anus is subterminal. The cloaca of male functions as eversible copulatory organ. Fertilization is internal. The female coils around the eggs and takes care of them till they hatch.

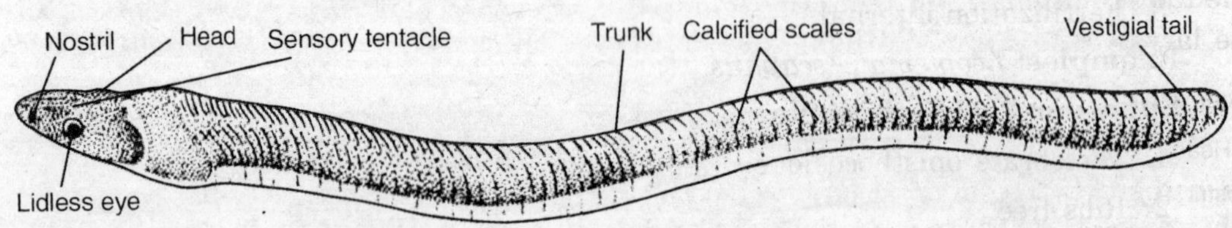

Fig. 15.2 *Ichthyophis*

Necturus

Necturus is commonly known as **mud puppy**. It is an aquatic salamander found in the lakes and rivers of U.S.A. It crawls on the bottom and feeds on small fishes and invertebrates. *Necturus* is about 30-40 cm long, rust-coloured with blackish spot. The body is divided into head trunk and tail. The head bears small eyes without eyelids, a wide mouth and nostrils. The forelimbs and hindlimbs are short and weak and adapted for crawling on the bottom of rivers and lakes. The tail is laterally compressed and provided with a tail fin. It is the main organ of progression. The typanum and tympanic cavity are absent. Behind the head on each side there are three bushy red-coloured external gills and two gill slits. They are the main respiratory organ. The lateral line system is well developed. The eggs are laid in nests attached by stalks to underside of stones. The female guards the nests. The adult animal is a permanent neotenic larval stage because metamorphosis could not be induced even after administration of thyroxin.

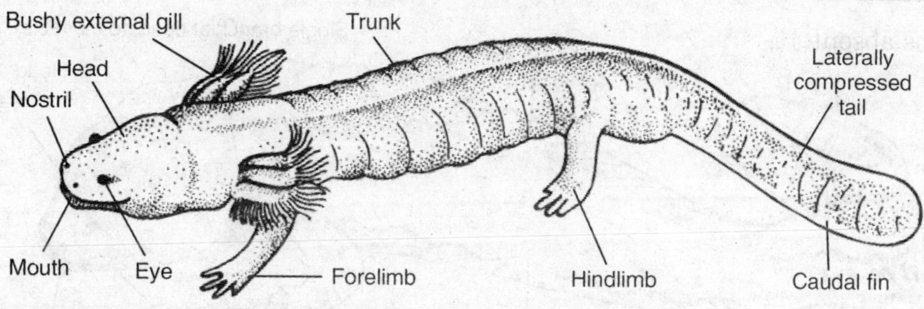

Fig. 15.3 *Nectures*

Proteus

Proteus is commonly known as **olm** or **cave salamander**. It inhabits deep and limestone caves of Europe, America and Germany. It is a neotenic adult with three pairs of red coloured branched external gills and two pairs of gill slits. The lungs are also present. It often comes to the surface, gives out bubbles of air and engulfs fresh air into the lungs. Body is about 20-25 cm long, eel-like, covered with unpigmented skin and divided into head, trunk and tail. The eyes are rudimentary, sunken deep in the head, covered by opaque skin and without eyelids. The limbs are poorly developed with two digits in the hindlimb and three in the forelimb. The skull is cartilaginous. Both jaws have teeth. The vertebrae are amphicoelous. The tail is flattened and compressed and provided with caudal fin. It helps in locomotion. If the larvae are kept in red light they develop functional eyes and pigmented skin.

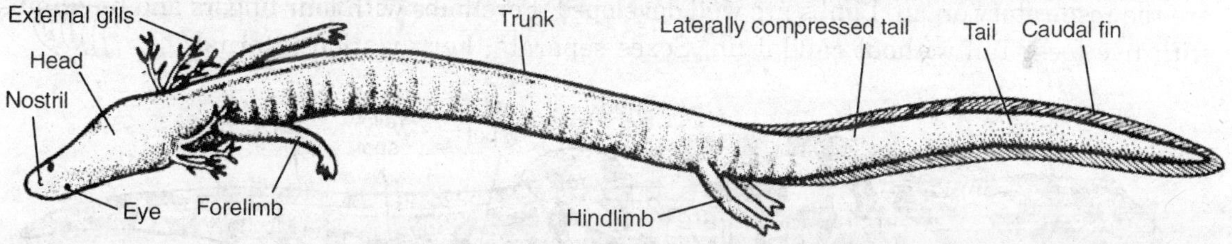

Fig. 15.4 *Proteus*

Cryptobranchus

Cryptobranchus is commonly known as **hell blender**. It is found in America, Japan, Eastern Asia and New York. It inhabits running water with rock slabs and feeds on small worms, small fishes and crustaceans. The body is elongated, depressed, about 60 cm long and divided into head trunk and tail. The head is broad and flattened so it gives it a ferrocious look. It bears nostrils and a pair of small eyes without eyelids. The trunk and tail are provided with a fleshy fold of skin. The trunk bears two pairs of limbs with four digits in forelimb and five in hindlimb. The tail is laterally compressed with a caudal fin. Gills are absent but a single gill slit or spiracle is present which acts as excurrent opening for the water taken into the mouth during buccal respiration. The female lays about 300 eggs arranged in strings. Fertilization is external. The larvae hatch is November and take several years to mature.

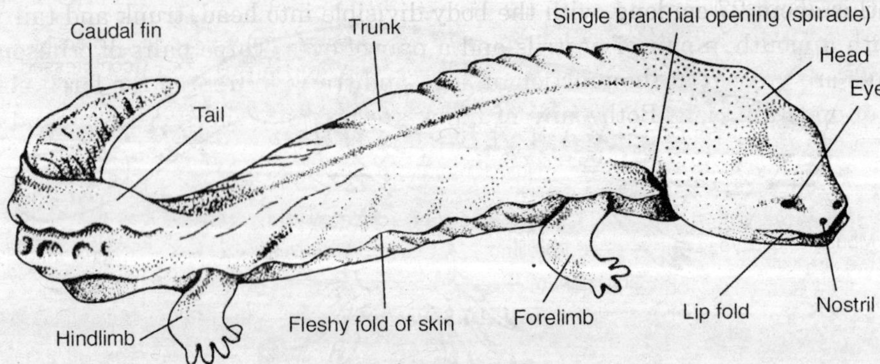

Fig. 15.5 *Cryptobranchus*

Ambystoma

Ambystoma is found in North America, Central Mexico and USA. The adults are terrestrial and larvae are aquatic. *A. tigrinum* or **tiger salamander** presents the classic example of neoteny and paedogenesis in Amphibia. The larvae fail to metamorphose into adult, retain 3 pairs of external gills and become sexually mature. This stage is known as **axolotl**. Another species of *Ambystoma* is *A. maculatum* which is commonly known as **spotted salamander**. The body is lizard-like, with transverse grooves on the sides and about 18-20 cm long. The skin is poisonous with yellow and orange spots all over the body. The head is depressed, with a large mouth, a pair of eyes with eyelids and a pair of poison glands called **paratoids**. Lungs are the respiratory organ. Limbs are well developed. Forelimbs with four fingers and hindlimbs with five toes. Tail without caudal fin. Sexes separate. Fertilization internal.

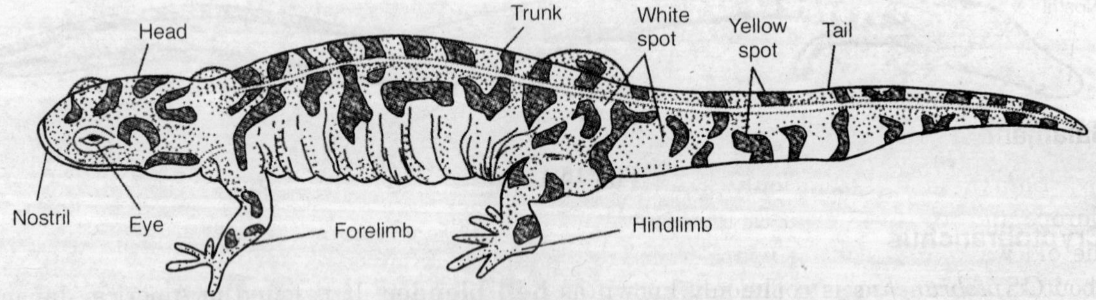

Fig. 15.6 *Ambystoma*

Axolotl larva

At one time axolotl was considered an adult and a separate genus *Siredon* but later it was found to be the larva of *Ambystoma*. It shows neoteny and paedogenesis. Due to deficiency of thyroid hormone or lack of iodine in water, the larvae fail to metamorphose into adult, retain three pairs of external gills (neoteny) and become sexually mature (paedogenesis) to produce young ones. Other conditions favouring neoteny are abundance of food, oxygen supply and low temperature. The axolotls can be induced to metamorphose by injecting thyroid hormone. Metamorphosis may become difficult if the larva grows older.

Axolotl is about 27 cm long, with the body divisible into head, trunk and tail. The head is broad with a mouth, a pair of nostrils and a pair of eyes. Three pairs of crimson coloured external gills are present at the junction of head and trunk in front of the forelimbs and has four pairs of open gill cleft. Both pairs of limbs are present. The tail has a caudal fin.

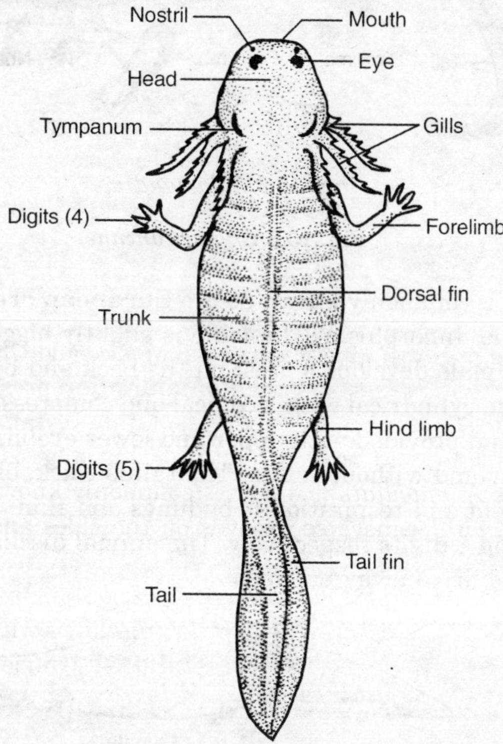

Fig. 15.7 *Axolotl larva*

Salamandra

Salamandra is commonly known as **fire salamander**. It is found in Eastern Asia, Europe and North America. It is commonly found under logs, stones, cracks and crevices of the old walls. The body is lizard-like. The males are shorter than the females, measuring about 12-15 cm in length. The skin is brilliantly black with irregular yellow patches on the back and limbs. The eyes are provided with movable eyelids. A pair of large paratoid glands are present behind the head. The tail is subcylindrical, without caudal fin. The limbs are well developed and lift the body well above the ground. The gills and gill clefts are absent and lungs are present. Some species like *Salamandra salmandra* are viviparous. Fertilization is internal. The male lays sperms in a capsule called **spermatophore** which is picked up by the female by her cloacal lips for internal fertilization. The female gives birth to young larvae which complete their development in water.

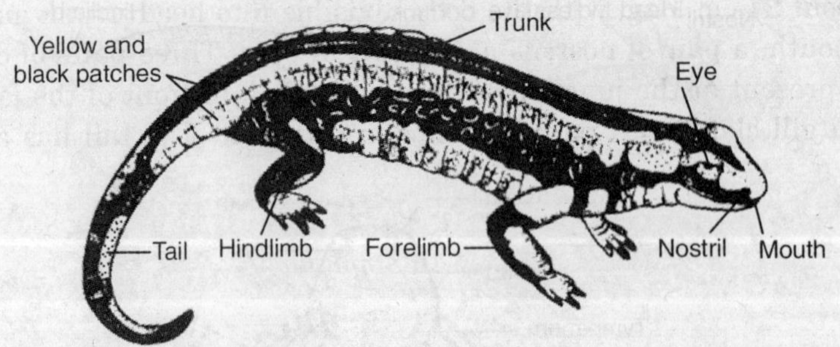

Fig. 15.8 *Salamandra*

Taricha (= Triturus)

Taricha or *Triturus* is commonly known as the **European crested newt**. It is terrestrial and shows marked sexual dimorphism. The male is slightly bigger than the female. During the breeding season the male develops a crest on the back and becomes brilliantly coloured. The body is elongated and cylindrical with a conical and compressed head. The mouth is wide and the eyes are small and provided with upper and lower eyelids Tympanum is absent. The skin is smooth and slimy and without scales. The tail is thick, elongated and provided with caudal fin. Gills are absent and respiration is by lungs and skin. The fore and hindlimbs are well developed with 4 and 5 digits respectively. The animal discharges a venomous secretion from its dorsal glands to drive away enemies.

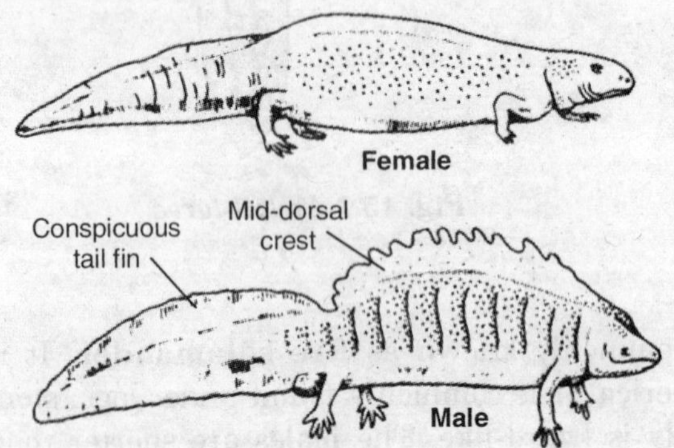

Fig. 15.9 *Triturus*

Amphiuma

Amphiuma is commonly known as **congo eel**. It is generally found in swamps and rice fields is North America. It feeds on molluscs, cray fish and small fishes. The body is elongated eel-like, about one meter long and covered with smooth black skin. The eyes are small functional but without eyelids. The limbs are very poorly developed, having 2 or 3 digits and do not help in locomotion. The adults have no external gills but have a pair gill slits. Lungs are present and trachea is long. Eggs are laid in water and female coils around the egg till they hatch.

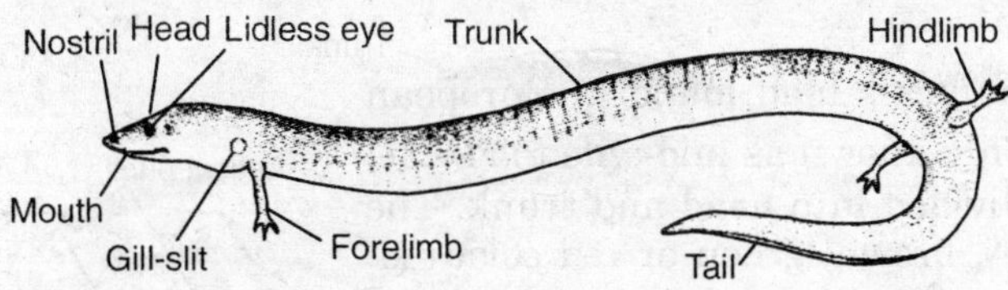

Fig. 15.10 *Amphiuma*

Siren

Siren is commonly known as **mud eel**. It inhabits burrows, muddy ditches and ponds in the United States of America. It is a permanent neotenic form. The body is elongated, eel-like, about 75 cm in length and covered with minute papillae. Eyes small, without eyelids. Jaws without teeth, with a horny covering. Forelimbs small, with four digits, hindlimbs absent. It has three pairs of external gills and one pair of gill slits. A well developed vomerò-nasal organ or Jackobson's organ is present. Fertilization is probably external.

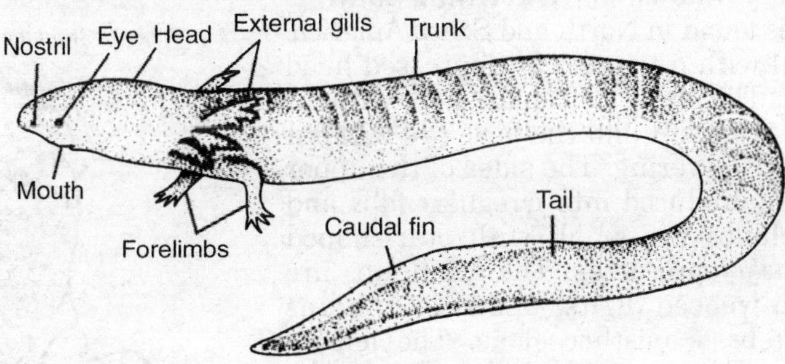

Fig. 15.11 *Siren*

Bufo

Bufo, the **common toad,** is found in all parts of the world except Australia. It is represented by about 300 species. It is terrestrial, nocturnal and enters water only to breed and spawn. During day time it hides under logs, stones or in burrows. They feed on worms, snails and insects by trapping them with their sticky tongue. The skin is rough, dry and warty and is protective in function. Body is divided into head and trunk. The head has a mouth, large eyes with eyelids, nostrils and tympanum. Behind the tympanum there is a large poison secreting **parotid gland**. They have no teeth on the jaws. The limbs are short, provided with horny tips and poorly developed webs. Vertebrae procoelous. Eggs are pigmented and laid in a gelatinous string. Young toads mature in many years. The parotid gland of toad produces two toxic substances—**bufotalus** and **bufogus**. When disturbed the toads urinate and produce foul unpalatable secretions from their skin glands, so they are avoided by their enemies.

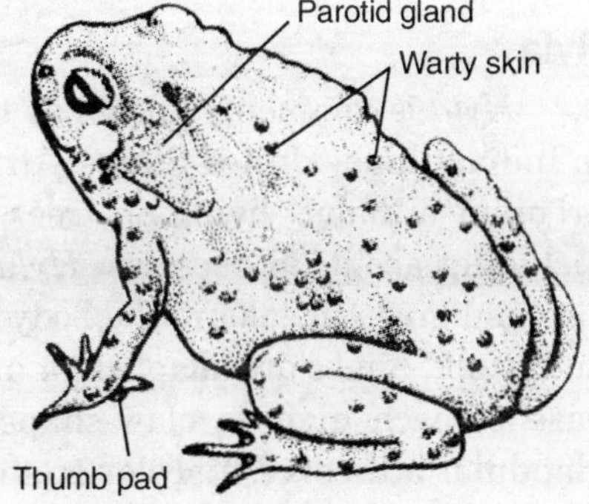

Fig. 15.12 *Bufo*

Alytes

Alytes is an amphibious toad found in European countries. It is commonly known as mid-wife toad. The body is 5-8 cm long, divided into head and trunk. The skin is warty and grey, brown, green or red coloured. The head has a pair of protruding eyes and large tympanum. The tongue is rounded and non-protrusible. The lower jaw has no teeth Males without vocal sacs. Vertebrae opisthocoelous and ribs present throughout life. The pupil is vertical, Fertilization is external. The male shows parental care. It strokes the cloaca of female with its toes for sexual stimulation. The female lays eggs in strings which the male winds around its posterior body and thighs. The male stays in damp places until the tadpoles hatch and enter water.

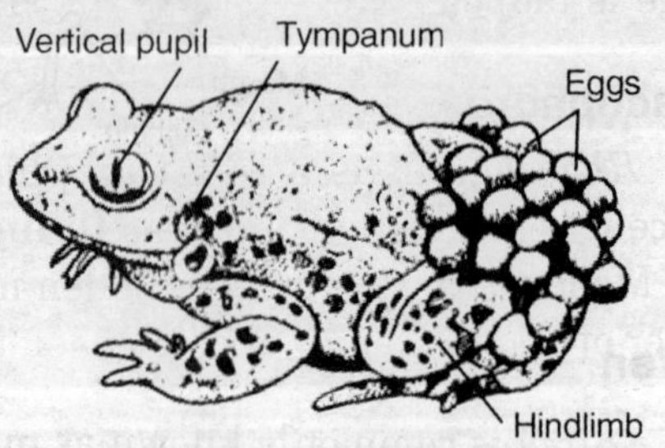

Fig. 15.13 *Alytes*

Pipa

Pipa, commonly known as **Surinam toad** is strictly aquatic and is found in North and South America. It is an ugly animal with a triangular depressed head and a large trunk. The eyes are small and without eyelids. The tongue is absent and the jaws are without teeth but have a horny covering. The sides of the upper part of the snout are produced into irregular faps and tenatcles. The forelimbs are small with star-shaped papillae having sensory processes. The hindlimbs are well developed with webbed digits. The skin contains poison glands. Vertebrae opisthocoelous. The female shows parental care. During the breeding season the skin of female becomes soft and pitted. Fertilization is external. The male presses the fertilized eggs on female's back. Each egg is enclosed in a little pocket or vascular pouch covered with operculum. Development is completed in the pouch and young toads come out of the pouch in about three months. The females changes its skin after the escape of young ones.

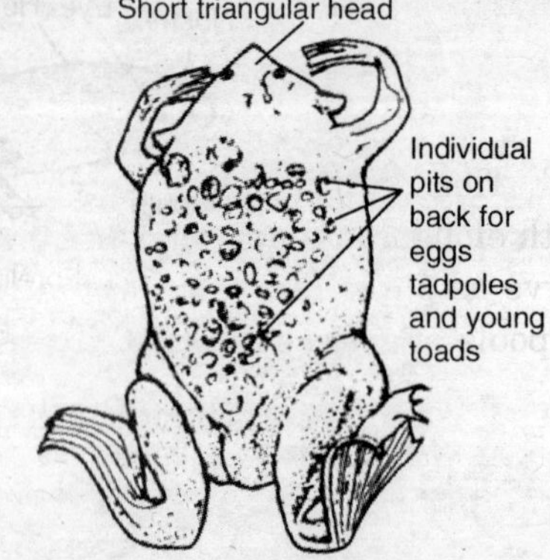

Fig. 15.14 *Pipa*

Hyla

Hyla is commonly known as **tree frog**. It is found in India, China, United States, Africa and Canada. It is arboreal in habit, living on trees. It is a large genus including about 350 species. *Hyla* is about 3-8 cm in size and green in colour. The body is divi ded into head and trunk. Limbs are adapted for arboreal life. Terminal base of each digit is claw-shaped and swollen into glandular **adhesive discs** or **cushions** which are used to climb trees. Eyes well developed with horizontal

Fig. 15.15 *Hyla*

pupil, tympanum present. The vocal sacs are large so their voice is very loud. Upper jaw bear teeth and lower jaw is without teeth. Fertilization external. Eggs are laid in water and larval stage is tadpole.

Rhacophorus

Rhacophorus is a tree frog which is able to glide from a higher level to a lower one, hence it is commonly known as **flying frog**. It is found in Africa, South Eastern Asia, Japan and Madagascar. It remains hidden under stones or on trees and come out at twilight. It has power of colour change. Body slender, female is larger than male. The belly narrows posteriorly.

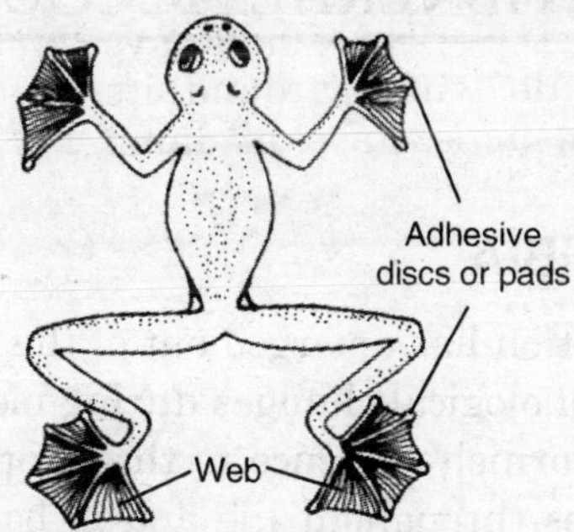

Fig. 15.16 *Rhacophorus*

Head broad and somewhat conical. Tympanum and eyes with eyelids present. Limbs with elongated digits bearing terminal adhesive discs. Webs develop between the digits which serve as parachute and help in gliding. Eggs are laid in gelatinous foam over shallow water of pools and rice fields. Many species of *Rhacophorus* show parental care.

16 CHAPTER

AMPHIBIA:
GENERAL ACCOUNT

EMERGENCE OF AMPHIBIA

The amphibian organization has emerged out of the piscine (fish-like) pattern is almost an established fact. The morphological changes during metamorphosis and the anatomy and palaeontology of both forms furnish evidence to this. Both are cold blooded, respire by gills in early life and in some cases throughout life and in both the air bladders serve as lungs. To prevent dessication both lay eggs in water and almost always larval development is completed in water. The anatomical resemblances between the two groups are so close that it led Huxley to put both Amphibia and Pisces under **Ichthyopsida**. Moreover, recent reserаches in palaeontology leave no doubt that the Amphibia have originated from some fish-like ancestor.

In an attempt to explore the likely fish-like ancestor of Amphibia it must be borne in mind that the modern Amphibia cannot be compared to modern fishes because the modern forms have become highly sepcialized in response to the demands of modern environments. This is a fact that the amphibians have not descended from the modern fish.

In the Carboniferous age there were three distinct orders of Amphibia, as revealed by fossil history. These were **Labyrinthodontia**, **Lepospondyli** and **Phyllospondyli**. The first two were present not only in abundance, but presented a variety of forms in the lower Carboniferous, which indicates that they must have originated earlier, at least in the Devonian and possibly in the Silurian times. Another important point that limits the search for the tetrapod ancestor is the fact that the amphibians arose from ancestors in the fresh water, because all the fossil records that establish the relationship are obtained from the fresh water deposits.

First amphibian: The earliest known amphibian was *Ichthyostega*. It belonged to the order **Labyrinthodontia** and subclass **Stegocephalia**. It apeared in the Devonian period of the Palaeozoic era about 300 million years ago. It flourished during the Carboniferous period which was the age of amphibians and disappeared in the Triassic period of Mesozoic era.

Probable ancestor: During the Devonian period three major groups of bony fishes were contemporaries of Labyrinthodontia—**Dipnoi**, **Actinopterygii** and **Crossopterygii**. The ancestor of amphibians must be one of these.

At the first thought it strikes that the dipnoans were the ancestors of Amphibia. This is because of certain obvious resemblances between the modern amphibians such as the urodeles and the modern dipnoans. Both breathe, to a large extent, by lungs and the distribution of blood vessels to and from these organs have close resemblance. Other similarities may be found in certain features of skull, brain, urinogenital system and early development. These similarities led some authors to conclude that the urodeles originated from the Dipnoi. The most primitive dipnoans may have possessed some of the urodele resemblances but they also possessed a number of distinctive features which would preclude them from the direct ancestorship of modern Amphibia. Further study revealed that the similarities between the two groups were due to convergent evolution. It was concluded that Dipnoi is definitely an evolutionary side-line and cannot be considered as ancestor of Amphibia.

The Actinopterygii do not come close to the ancestral position because of the absence of internal nostrils and lack of a large fleshy fin from which a land-limb might develop. It also possesses many specialized features. This leaves the Crossopterygii or the lobe-finned fishes as the probable ancestor of amphibians. During the Devonian the Crossopterygians evolved in two lines : Rhipidistia and Coelocanthini. The rhipidistian crossopterygians are regarded to be on the main line of evolution of amphibians. They are represented by two genera: *Osteolepis* and *Eusthenopteron* which show many morphological similarities with early amphibians.

1. The bones of the skull and jaws are similar.
2. The pentadactyl amphibian limb developed from the lobe of the crossopterygian fin.
3. The position of external and internal nares is similar.
4. The teeth in both are of labyrinthodont type, i.e., the enamel has number of infoldings.
5. Both used air bladder as lungs when they moved from one pond to another.

The most primitive amphibian *Ichthyostega* had features of both the groups and represents a transitional stage.

TERRESTRIALIZATION IN AMPHIBIAN

Terrestrialization refers to gradual transformation from aquatic life to land life. In this process the organism loses aquatic adaptations and develops new adaptations for terrestrial life. From protozoa to fishes origin and evolution occurred mainly in water. It was the ancestor of Amphibia who made the first attempt for terrestrialization. The amphibian started terrestrialization but they were not fully successful. They still remained chained to water so they are described as modified fishes adapted for life on land. Terrestrialization was, however, completed by the next group, the Reptilia.

During terrestrialization some remarkable changes have occurred in the morphology, anatomy and physiology of Amphibia. Some striking modifications are:

1. Respiration: The gills were replaced by lungs. The gill respiration was lost and pulmonary respiration developed. With this transition internal nares developed. This helped in breathing when the mouth remained closed. The skin also became modified for cutaneous respiration.

2. Desiccation: Desiccation is a major problem of land life. In order to avoid desiccation the early amphibians remained close to water and frequently returned to water. Some early amphibians developed tough skin to resist desiccation.

3. Gravity: Gravity is a strong factor influencing life of animals living on land. The first amphibian developed a strong backbone, with powerful girdles at two points on the vertebral column and strong limbs to withstand gravity.

4. Muscles: The amphibians developed strong muscles to lift the body above the ground.

5. Locomotion: The paired fins were replaced by strong paired limbs which became the chief locomotory organ.

6. Circulatory system: The auricle became divided into right and left halves which helped to separate oxygenated and deoxygenated blood. Out of the six pair of aortic arches of fishes, three pairs (first, second and fifth) disappeared, the third pair became the **carotid artery**, fourth pair **systemic aorta** and the sixth pair **pulmonary aorta**.

7. Sense organs: The lateral line sense organs were of no use for land life so they disappeared. Besides internal ear the amphibians developed a middle ear with a bone called **columella auris**. This intensified hearing. The eyes developed **eyelids** which were used for protecting and cleaning the cornea in air.

8. Reproduction: Land animals must either go back to water to reproduce or they must develop methods for protecting the eggs on land. Though amphibians have developed many terrestrial adaptations they could not solve the problem of reproduction. So they have to return to water for laying eggs and early development also takes place in water. Thus they have remained chained to water.

Causes for transition from water to land

What caused the Crossopterygians to leave their aquatic habitat and come to land? There are many different views from different zoologists. A few are being discussed here.

1. Preadaptation: Preadaptation refers to the possession of characters which will be useful to the possessor in a new environment. The crosspterygians already had a number of adaptations for land life. For example, the air bladder helped them to breath on land, the lobed fins helped them to walk on land. These features helped the crossopterygians to make an excursion on land.

2. Drought condition: According to Barrel (1916) the Devonian was a period of seasonal drought, so the crossorpterygians were forced to search new fresh water lakes and streams. In order to find suitable water bodies they had to cross dry land. They used their lobed fins to cross the land which in due course of time got converted into pentadactyl limb.

3. Desire for water: Romer (1958) rejected the above view. According to him the early amphibians and reptilians inhabited water until the Pennsylvian period. It was the desire for more water that caused the crossopterygians to move from one pond to another.

4. Enemies: Some scientists are of the view that enemies in water forced the crossopterygians to leave water and move to land.

5. Favourable habitat: Since land provided abundant food, oxygen, shelter and breeding places, it attracted the crossopterygians from their aquatic environment which met with periodic dryness.

Adaptive Radiation

The modern amphibians have undergone elaborate adaptive radiation. Once they come on land they started diverging in different habitats. They exhibit various types of adaptations.

1. Aquatic adaptations: The newts and salamanders, with a few exceptions, have returned to water and exhibit many secondary aquatic adaptations.

1. Body is elongated and eel-like.
2. They respire with the help of gills.
3. They have developed branchial circulation.
4. Fins present on the trunk and tail.
5. Skull feebly ossified.

Some common examples are *Cryptobranchus Amphiuma*, *Siren*, *Proteus* and *Necturus*.

2. Terrestrial adaptations: Frogs and toads are simple terrestrial amphibians. The frogs have to return to water to keep their skin moist. The toads, on the other hand, have wart-like projections which prevent evaporation of water from the skin. Since they move freely on land some modifications associated with their leaping habit are:

1. Elongation of hindlimbs and the pelvic girdle.
2. Shortening of the vertebral column.
3. Development of powerful muscles.

The other terrestrial amphibians show the following adaptations:

3. Cave adaptations: Some amphibians like *Proteus*, *Typhlomolge* and *Typhlotriton* are permanent cave dwellers and they possess the following features:

1. Body is elongated and cylindrical.
2. Eyes reduced or degenerate.
3. Limbs reduced or absent.
4. External gills present for respiration.
5. Skin soft, non-pigmented or darkly pigmented.

4. Burrowing adaptations: The caecilians like *Ichthyophis* and *Urotyphlus* and some anurans like. *Cacopus*, *Pelobates* and *Engystoma* live in burrows. The caecilians show the following burrowing adaptations:

1. Body elongated, worm-like and without limbs.
2. Skull bony and compact.
3. Eyes reduced, covered and functionless.

The burrowing anurans possess the following features:

1. Snout protruding.
2. Mouth narrow.
3. Eardrums hidden.
4. Limbs and snout modified for digging.

5. Volant adaptation: *Rhacopharus* (fying frog) is the only flying amphibian. The digits of the limbs are connected by a web. The webbed feet act as parachute and help the animal to leap for 6-10 metres in the air from one tree to another.

6. Arboreal adaptation: A number of amphibians live on trees. Some common species are *Hyla Phyllomedusa*, *Chiromantis* and *Rhacophorus*. They possess the following structural modifications:

1. Adhesive or sucking disc or pad is present at the tip of each digit. They help in attachment.
2. An opposable finger or toe is present.

PARENTAL CARE IN AMPHIBIA

Parental care is a complex behaviour shown by parents to take care of the offspring. It is well developed in Amphibia where extreme modifications in structure and behaviour are observed. There are several devices for protection during the early stages of development and they fall under two major categories:

I. Protection by means of nests or nurseries.

II. Direct nursing by the parents.

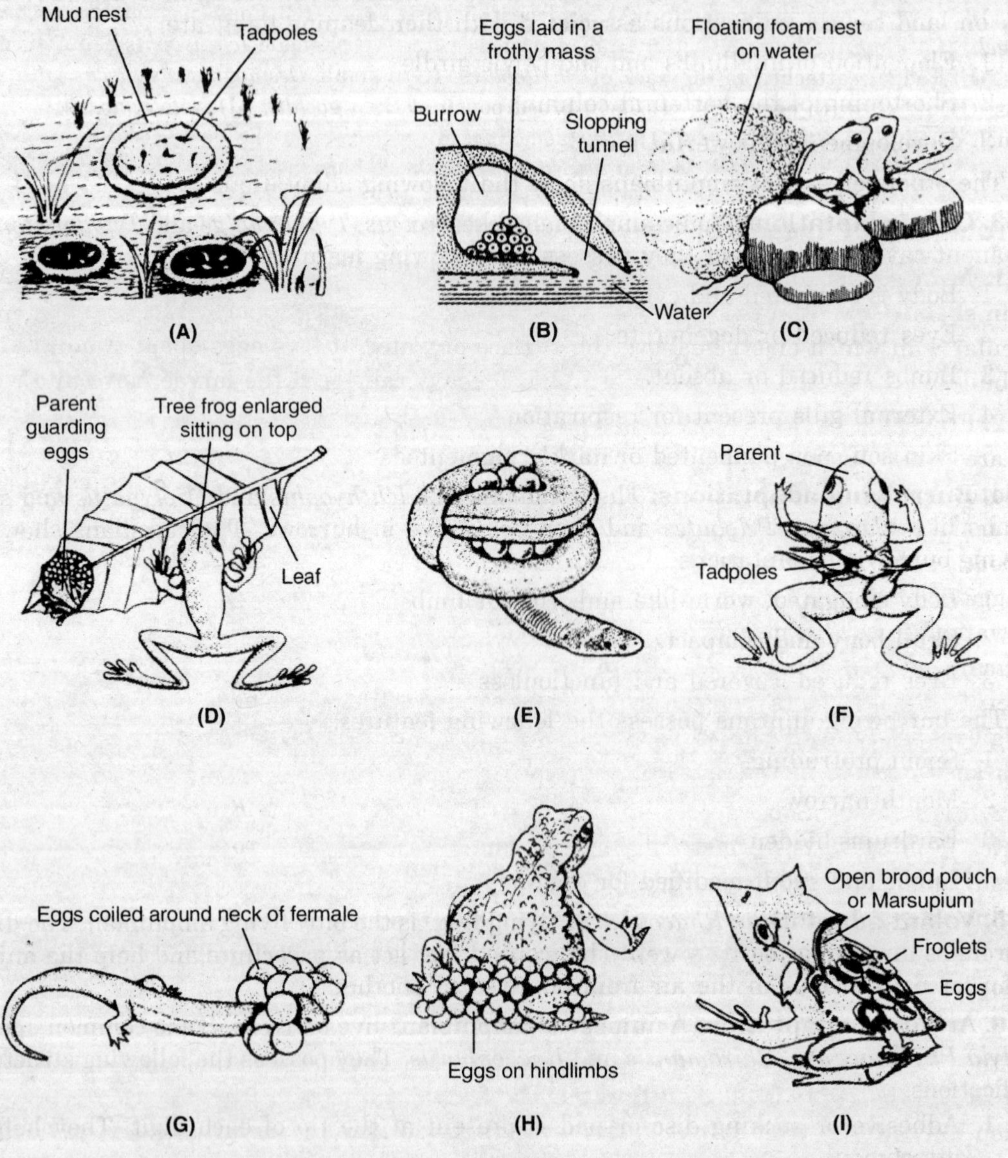

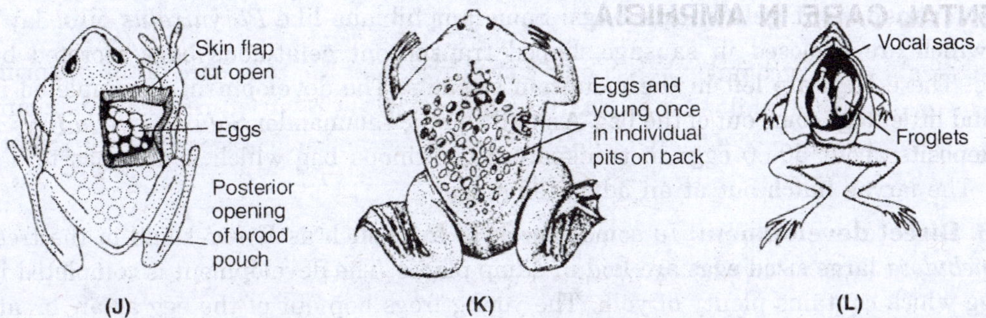

Fig. 16.1 Parent care in Amphibia. **(A)** Mud nest of *Hyla Faber;* **(B)** Foam nest of *Rhacophorus schlegeli;* **(C)** Floating foam nest; **(D)** Leaf nest of tree frog; **(E)** Female *Ichthyophis* coiling around eggs; **(F)** Tadpole attached to the back of *Phyllobates*; **(G)** Female *Desmognathus fuscus* with egg mass; **(H)** *Alytes* with egg mass; **(I)** Open brood pouch of *Hyla goeldii;* **(J)** *Nototrema* with brood pouch cut partially to show eggs; **(K)** *Pipa* showing dermal cells on the back; **(L)** *Rhinoderma drawini* with froglets in vocal sacs

I. Protection by means of nests or nurseries

1. Mud nests: The Brazilian tree frog *Hyla fabre* (Fig. 16.1 A) makes a basin-shaped nest in shallow water on the border of pond. The female scoops 3-4 inches of mud and builds a circular wall which emerges above the surface of water. It lays eggs about two days after nest building. Later the nests are destroyed by heavy rains and the larvae move into water.

2. Foam nests: Some amphibians make foam nests with their mucous secretion and the eggs are laid in them. In the Japanese flying from *Rhacophrous schlegli* (Fig. 16.1 B) the copulating male and female bury themselves in damp earth near pond or flooded rice fields. They make a hole or tunnel a few inches above water. The female producs a secretion from the cloaca which is beaten into froth or foam in which the eggs are laid. The male impregnates the eggs. During rainy season when the eggs hatch the larvae are carried down the tunnel into water for further development. The female of South American tree frog *Leptodactylus mystacinus* beats the mucous secretion into foam and fills it in holes near water. It lays eggs in them and when the tadpoles hatch they are carried into water. The female of many species of frog beat the mucous into a frothy mass with her hind legs. It lays eggs in the frothy mass which keep floating in water (Fig. 16.1 C). When the tadpoles hatch they drop into water.

3. Leaf nests: Some species of tree frogs like *Phyllomedusa hypochondrales* of South America, *Rhacophorus malabaricus* of India and *Chiromantis* of Tropical Africa lay eggs on leaves (Fig. 16.1 D) and branches of aquatic plants overhanging water. They fold the leaves and glue their margins with some cloacal secretion. The eggs are laid in these leaf nests and are protected from egg predators. The larvae after hatching fall into water and complete their development.

4. Tree nests: The tree frog *Hyla resinfictrix* makes nest in shallow cavities in trees. It lines the cavity with beewax which is collects from the nest of some stingless bee. During the rainy season the cavity is filled with water. The female lays eggs in it and the young ones also develop in it after hatching. In *Autodax* (order Urodela) the female lays 10-20 eggs in a hole in trees or sometimes in dry holes in ground. The female or both parents remain with the eggs and developing young ones protecting them from enemies and also providing them moisture.

5. Transparent gelatinous bags: Some amphibians like *Phrynixalus biroi* lay large eggs which are enclosed in sausage-shaped transparent gelatinous bags secreted by the female. These bags are left in the mountain streams. The development is completed in the eggs and little frogs come out of the bag. A small aquatic salamander *Salamandrella Keyserlingi* also deposits about 50-60 eggs in transparent gelatinous bag which it attaches to aquatic plant. The larvae hatch out at an advanced stage.

6. Direct development: In some species of frogs such as *Hylodes* and in the tree frog *Hyla nebulosa* large sized eggs are laid in damp places. The development is completed inside the egg which contains plenty of yolk. The young frogs hop out of the egg as air breathers. The red backed Salamander *Plethodon cinereus* also shows direct development. The young ones that hatch out are miniature adult.

7. Defending in territories: Many species of frog maintain definite territories to protect the eggs. For example, the male green frog *Rana clamitans* attacks any intruder in its territory. The male *Mantophyrne robusta* sits over and holds the elastic gelatinous bag containing developing eggs. Some species of tree frogs sit near or on top of egg mass until they hatch and if the guarding frog is removed the eggs dry up and die.

II. Direct nursing by the parents

1. Coiling around the eggs: In some caecilians like *Ichthyophis glutinosa* (Fig. 16.1 E) and *Hypogeophis* the female lays large eggs in a hole in damp soil. She coils her body around the bunch of eggs and protect them from enemies or dessication until hatching. The congo eel *Amphuima* also shows similar behaviour. In the salamander *Plethoclon* the eggs are laid in a small package of about five each beneath stones or rotten log and the female coils round them. In *Megalobotrachus maximus* it is the male which coils round the eggs and defined them till hatching.

2. Transferring tadpoles into water: The small frogs like *Phyllobates*, *Dendrobates Pelobates* and *Arthroleptis* of South America and Tropical Africa lay eggs on the ground. The tadpoles on hatching adhere to the body (Fig. 16.1 F) of one of the parents with their sucker-like mouth and carried to water.

3. Eggs carried by the parents: In some amphibians the eggs are glued to the body of one of the parents and carried from one place to another and on hatching the tadpoles are released in water. The female *Desmognathus fuscus*, (Fig. 16.1 G), the dusky salamander carries the eggs around her neck. The female Srilankan tree frogs, *Rhacophorus reticulatus* carries the eggs on its belly. The male *Alytes obstetricans* (Fig. 16.1 H), the midwife toad entangles the eggs around its hindlegs until they hatch.

4. Eggs carried on the back in pouches: In many frogs and toads the eggs are carried on the back either exposed or in pouches. In the Brazilian tree frog *Hyla goeldii* (Fig. 16.1I) the posterior part of the back of female serves as open brood pouch or marsupuim in which it carries the eggs which develop into miniature frogs before leaving the mother's back. In *Nototrem* (Fig. 16.1J) the eggs are covered on the back of the female by skin forming a single brood pouch opening posteriorly by a small aperture in front of the cloaca. The pouch develops only during the breeding season.

In the aquatic Surinam toad, *Pipa americana* (Fig. 16.1 K) the eggs are carried on the back of the mother. The skin of the back of female becomes thick, soft, gelatinous and vascular. The fertilized eggs are pressed on the back. The skin grows around the eggs until each is enclosed in a dermal cells which is finally covered by a lid most probably formed by a secretion from the glands of skin. There are about 100 eggs on its back. Complete metamorphosis occurs inside the dermal cell and young ones leap out.

5. Organs as brood pouch: One of the most remarkable modes of nursing has been observed in a small frog *Rhinoderma darwini*. The fertilized eggs (atleast two) are sheltered and develop in the large vocal sacs of male (Fig. 16.1 L). The young ones emerge out as miniature frog. The female of West African tree frog *Hylambates breviceps* carries the eggs in her mouth cavity. The male *Arthroleptis* carries the developing larvae in its mouth. One unique example of breeding inside an organ of the body is of the Australian frog *Rheobatrachus silus*. The female keeps the eggs in her stomach and the fully developed froglets are expelled through the mouth.

6. Viviparity: Some amphibians are viviparous, i.e., they give birth to young ones. The female retains the eggs in the oviducts and produce fully developed young ones of small size. Some common examples are East African toads *Pseudophryne vivipara* and *Nectophryne tornieri, and Salmandra salmandra* which give birth to about ten to fifty young ones and *S. atra* which produces one or two young ones. In Gymnophiona many viviparous species have been reported, for example *Dermophis*, *Typhlonectes*, etc.

NEOTENY AND PAEDOGENESIS

Neoteny (**Gr.,** *neos*, youth; *tenio*, to extend) refers to the retention of larval characters in the adult, for example, retention of larval gills in some adult urodeles. **Paedogenesis** or **paedomorphosis** refers to precocious development of reproductive organs in the larval body, as in some salamanders.

The two terms, neoteny and paedogenesis were first proposed by Kollmann. They are often used as synonyms by many scientists, but actually they are two very distinct evolutionary processes. In some aquatic larval urodeles, the two processes seem to overlap and it is difficult to distinguish between them. The animals may either be considered as adult which have retained certain larval characters due to failure of meltamorphosis (neoteny) or they may be considered as larvae which have become sexually mature (paedogenesis).

The classic example of neoteny and paedogenesis in Amphibia is the **axolotl**. It was one of the earliest examples of neoteny to be discovered. Axolotl is the larva of *Ambystoma* (salamander) which lives in high attitudes of Mexico (*A. mexicana*) and Colorado (*A. tigrinum*). The name axolotl has been derived from the Aztec Indian words, *atl* meaning water and *xolotl* meaning servant (thus axolotl means servant of water). The axolotl has been known for centuries now. Its history dates back to the 16th century when the Spanish settlers used it as food. Wagler in 1830 for the first time named it *Siredon pisciformis* thinking that it is related to the perribranchiate Amphibia *Siren* which retains gills permanently and retain the form of larva throughout life and do not undergo metamorphosis.

Under normal conditions *Ambystoma* passes through the aquatic larval stage and then undergoes metamorphosis to transform into air-breathing adults. Under certain circumstances the larva fails to undergo metamorphosis and becomes sexually mature. This stage is called axolotl. It has a tail fin, three pairs of external gills and four pairs of gill openings. It breeds several times in a year.

In 1865 some individuals were brought to France in Paris. There they started to breed and cruiously enough their young went through all the changes characteristic of metamorphosis. They lost their gills and fin, shed their larval skin, and emerged as animals different from larvae. The axolotl were finally identified as sexually mature larvae of *Ambystoma*.

Factors affecting neoteny

The causes of neoteny are not clearly understood but it is a known fact that certain environmental factors affect metamorphosis. The absence of these factors causes failure of metamorphosis and favours neoteny.

1. Abundance of food in the aquatic environment is the cause of retention of larval characters (Gadow 1903).
2. According to Shufeldt if water is cold it inhibits the secretion of thyroxin (a hormone which induces metamorphosis in amphibians). The low temperature inhibits metamorphosis and favours neoteny (Huxley 1929).
3. Lack of iodine (a component of thyroxin hormone) in water causes failure of metamorphosis and retention of larval characters.
4. Salinity of water favours neoteny (Weismann).

This has been proved by the fact that drying up of swamps, lack of food and increase in temperature induce axolotls to undergo metamorphosis. Treatment of axolotls with thyroxin or thyroid stimulating hormone (TSH) induce them to transform into air breathing adult salamanders.

Genetic basis of neoteny

Many scientists, on experimental basis, have shown that **genes** are responsible for neoteny. Genes that bring about metamorphosis are simply suppressed in neotenic forms. They are not absent because occasionally these forms metamorphose into adults. According to Etkin and his coworkers (1968) the level of **thyroid hormone** is kept low in the body by genetic mechanism during early premetamorphic phase. They also established that **prolactin hormone** acts an an inhibitor of metamophosis and its level is kept high at this time. Probably the structural genes responsible for the synthesis of thyroid hormone are **swtiched off** by some operator genes whereas the genes responsible for the synthesis of prolactin are **switched on**. In such condition the **hypothalamus** becomes sensitive to the level of thyroxin in the blood. As a result the **neurosecretory apparatus** of the hypothalamus produces a **thyrotropin releasing factor** (**TRF**) which stimulates the **anterior lobe** of the **pituitary gland** to produce **thyroid stimulatory hormone** (**TSH**) which triggers the production of thyroid hormone. This is followed by a fall in the level of prolactin and metamorphosis starts. From this it is clear that poor secretion of thyroid hormone and lack of responsiveness of larval tissue to thyroid hormone favours neoteny.

Types of neoteny

Three types of neoteny have been described by Kollmann.

1. Partial neoteny: When metamorphosis is delayed due to certain ecological or physiological changes in the environment, for example, overwintering tadpole larvae of *Hyla arborea*, *Rana esculenta*, *Bufo vulgaris*, etc.

2. Intermediate neoteny: When the larva matures sexually but undergoes metamorphosis when the environmental conditions are favourable, for example, axolotl.

3. Total neoteny. When the animal retains larval characters throughout life as well as becomes sexually mature. Even treatment with thyroxine hormone does not induce metamorphosis, for example, *Necturus*, *Siren*, *Proteus*.

Significance of neoteny

Different scientists tried to explain neoteny in different ways. De Beer considered it to be a simple evolutionary process. Weismann regarded it to be retarded evolution or atavism or reversion of ancestral condition. He was of the view that originally amphibians were aquatic, gills breathing and limbless animals and larval features indicate ancestral phylogenetic stage. The neotenic forms were considered to be the most primitive living amphibians. Now the neotenic forms are considered to be secondary sepcialization, an adaptation of advantage.

ECONOMIC IMPORTANCE

Although all amphibians are of some economic importance to man but frogs and toads are of great economic value.

1. Enemies of insects: Frogs and toads destroy a large number of insects and maintain a balance in nature. They destroy insect pests of many crops and help the farmers. Pack (1922) has reported a case where the toad was of real value in fighting an outbreak of sugar-beet webworms. In the paddy fields of China, Japan and India frogs have been found to destroy a large number of insects harmful to the crop. Frogs eat eggs of mosquito and thus prevent the spread of malaria.

2. As food: Many species of frogs like *Rana pipiens*, *R. catesbeiana* and *R. palustris* are consumed as food in many parts of the world. Usually frog's leg form a staple food. Attempts have been made to establish frog farms but with no success. In China frogs and toads are dried and sold as food. In America *Necturus* and axolotls and in Japan giant salamanders are used as food.

3. As study material: Frogs and salamanders are universally used in laboratory for the elementary course of Biology. In USA *Necturus* is also dissected. They are used for research in pharmacology and physiology. The larvae of amphibians serve as excellent material for the study of many problems of embryology and endocrinology.

4. Medicinal value: In China toads are used in preparation of many medicines. The skin of toads secretes a digitalis like substance which when injected into human body increases blood pressure.

5. Poisonous amphibians: Many amphibians produce irritating poisonous substance. The toad *Bufo marinus* produces a poisonous secretion (**Gifotoxin**) which is fatal to dogs and cats. Some tribal people in South America use poison secreted by the frog *Dendrobates* to poison their arrows. *Salamandra salamandra* is a poisonous amphibian which shows warning colouration.

6. Other uses: In Japan and other countries skin of toad is used as a source of leather. It is used for making delicate purses. Skin of some frogs is used as glue in book binding. In USA bull frogs are used for recreation in jumping frog contest. Frogs are also used as baits in fishing.

Many superstitions have been attached to frogs and salamanders. In parts of South China, the brilliant *Polypedates dennysi* is worshipped by a cult and carried in a chair by faithful members of the group (Pope, 1931). A toad shaped amulet from near the river Nile is dated at 3400 B.C. and a Chinese sacrificial vessel in the form of a toad belongs to an ancient dynasty (2205 to 1122 B.C.). These indicate the role frogs and toads played in religion. Aesop (560 B.C.) included frogs among his fables and Aristophanei (448-380 B.C.) dramatized their croaking choruses.

GYMNOPHIONA

The **Gymnophiona** or **Apoda** are vermiform amphibia without limbs and limb girdles. The body is covered with numerous small scales embedded in the skin and arranged in transverse rows. The eyes are small functionless and covered by skin. The tympanum and tympanic cavity are absent.

Distribution

The Gymnophiona shows discontinuous distribution. They mostly inhabit the tropical and subtropical regions of Central and South America, Africa, India and Malaya Archipelago. Three genera found in South India are *Ichthyophis*, *Uraeotyphlus* and *Gegenophis*.

Habit and Habitat

The Gymnophiona burrow in surface soil in damp places near streams. A few species are aquatic. They mostly feed on worms and other small invertebrates.

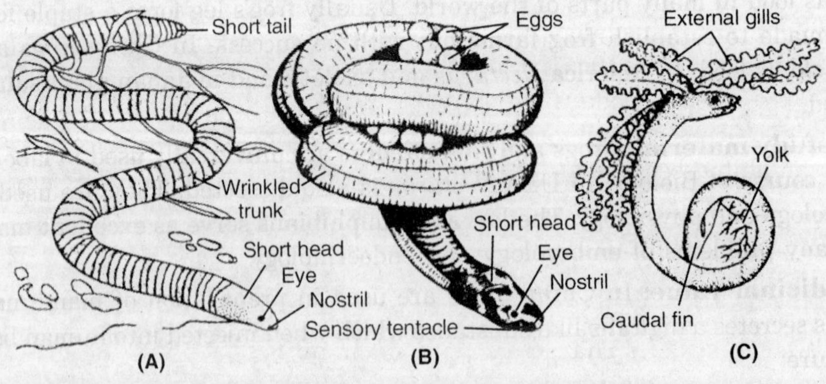

Fig. 16.2 Ichthyophis glutinosa. **(A)** Male; **(B)** Female; **(C)** Larva with 3 pairs of external gills

General characters

1. They have long slender worm-like body with much reduced tail. Since the tail is reduced the vent or cloaca is almost terminal.
2. The skin is smooth, slimy, with many minute calcified scales embedded in it and arranged in transverse rows.
3. The eyes are small, concealed by skin and functionless, hence the common name **blind worms**. Though the eye is reduced it has all the usual parts.
4. A pair of retractile sensory tentacle is present in a pit behind the nostril on either side of the head. It helps the animal to feel the way.
5. There is no tympanum and tympanic cavity.
6. The skull is compact and has a complete bony roof.
7. The jaw suspensorium is formed by the squamosal and quadrate.
8. The maxilla is fused with the palatine to form maxillopalatine.
9. They may have as many as 250 vertebrae. The vertebrae are amphicoelous. The ribs do not meet to form a sternum. The notochord is persistent.
10. Limbs and limb girdles absent, hence the name **Apoda**.
11. The lungs are asymmetrical, the right being much larger than the left.
12. There are only two pairs of aortic arches, the systemic and pulmonary which are connected by **ductus Botalli.** The carotid arises from the systemic.
13. The males have protrusible copulatory organ.
14. Fertilization is internal.
15. The eggs are large and yolky, usually laid in masses in moist ground near water.
16. The female *Ichthyophis* shows parental care (Fig. 16.2 B) by coiling around the eggs till they hatch. The larval stage is passed inside the egg. It has three pairs of external gills which are shed on hatching (Fig. 16.2 C).
17. The South American aquatic species, *Typhlonectes* is viviparous or ovoviviparous.

Phylogenetic position

There are no fossil evidences to determine the exact ancestry of Apoda. Based on the vertebral structures it was thought that the apodans evolved from the Lepospondyli. According to Colbert (1969) the Apoda should not be considered as the direct descendents of the Lepospondyli. He considers all modern amphibians (Anura, Urodela and Apoda) under a single group Lissamphibia because all of them show certain similar characters.

AFFINITIES OF APODA

Resemblances

Apoda shares many characters with other amphibians. These includes structure of heart, presence of conus arteriosus, the form of brain, the structure of urinogenital system and the absence of tympanic cavity.

Differences

The Apoda differs from other amphibians in having dermal scales embedded in the skin, absence of limbs, presence of ductus Botalli, presence of copulatory organ and absence of gill-breathing larva.

Peculiar characters

The Apoda shows certain peculiar characters like absence of limbs and limb girdles, reduced eyes, corotid arising from the systemic and presence of retractile tentacles.

17

CHAPTER

REPTILIA:

TYPE UROMASTIX

The **lizards** have been defined as reptiles with movable quadrate bones, the right and left halves of the mandibles united by suture, a transverse anal opening, paired copulatory organs, a cloacal bladder and with the anterior end of the brain case never completely closed. The body, in majority, is covered with horny epidermal scales, usually possessing four well-developed (sometimes degenerate) limbs, and have eyes with movable eyelids. Majority of lizards are terrestrial, some (*Chameleon* and the flying lizard, *Draco*) are entirely arboreal and some (*Dibamus*) are almost entirely subterranean. The terrestrial forms have depressed body, the arboreal forms have compressed body and subterranean forms have elongated and cylindrical body. Here *Uromastix*, the **spiny tailed lizard**, has been described as a type of this group. The description, however, can be safely used for the study of other lizards such as the ordinary wall lizard *(Hemidactylus)* or the monitor lizard (*Varanus*).

DISTRIBUTION

The genus *Uromastix* is represented by 11 species distributed in the deserts of North Africa and North-Western Asia. In India it is represented by only one species, *U.hardwickii*, found in North-Western India (Uttar Pradesh, Punjab, Haryana, Rajasthan) and Gujarat. It has also been reported from Sind and Kathiawar. The specimens differ slightly from those of India but the variation is not sufficient to warrant racial distinction. In Hindi it is popularly known as '**sanda**'.

Habits and habitat

Uromastix inhabits sandy tracts where vegetation is scanty. It is also known as **sand lizard** because of its sandy habitat. The lizard is widely distributed over the area in which it occurs. *Uromastix* is a burrowing lizard. Each one lives in its own burrow, four to eight feet deep. The burrow is made in a zigzag manner. The size of animal can be determined by the size of the burrow. The entrance of the burrow is closed by a small heap of sand when the animal is inside. The burrows are used for shelter, protection from enemies and for egg laying. The lizard is diurnal. It never leaves its burrow until the sun is up and after feeding returns to the burrow during the heat of the day and carefully closes the entrance. Sometimes empty burrows may be occupied by some snake. *Uromastix* is herbivorous. It feeds on grass, flowers, fruits and succulent leaves. Sometimes it eats insects.

Defensive mechanism

Uromastix is calm, gentle and non-agressive in nature. It can be tamed easily in captivity. It is slow in its movement. When disturbed it tries to frighten away the disturber by lifting its head in a peculiar way and opening its mouth at the same time. *Uromastix* protects itself from snake or other predators in a very interesting way. When danger is perceived it quickly enters the burrow keeping 3–4 inches of its tail outside and shaking it violently. If the predator still attacks the tail, it is hurt badly by the horny spines on the sides of the tail. This finally drives the enemy away. On the other hand, if the snake can get into the hole before the lizard, then it becomes defenceless and is caught by the head by the snake and eaten.

Some tribes in Rajasthan eat it as food and its flesh is said to be excellent and sweet like chicken. The tail is considered to be a delicacy. They catch the lizard by digging out the burrow or by filling the burrow with water, thus forcing the lizard to wiggle out. Sometimes they frighten the lizard by making a hissing snake-like sound. The lizard rushes to the burrow with their tail projecting out and lashing violently. The tribal people catch them with their tail and pull them out of their burrow.

Breeding

Uromastix hibernates in winter to avoid cold season. During this period it feeds on the accumulated body fats. It breeds in spring when it comes out after hibernation. The male fights for the female. There is a period of courtship before copulation which takes place during the day when it is sunny and warm. Fertilization is internal. The female lays about 15 eggs in its burrow. The eggs are large, measuring about 20–30 mm in size and dusty white in colour. Development is direct. The newly hatched young ones resemble the parents.

EXTERNAL FEATURES

Size, shape and colour

The body is massive, elongated, about 20–30 cm long and depressed. In colour it is yellow brown with dark spots above and whitish below. There is a large blackish spot on the front of each thigh. The throat often has black spots.

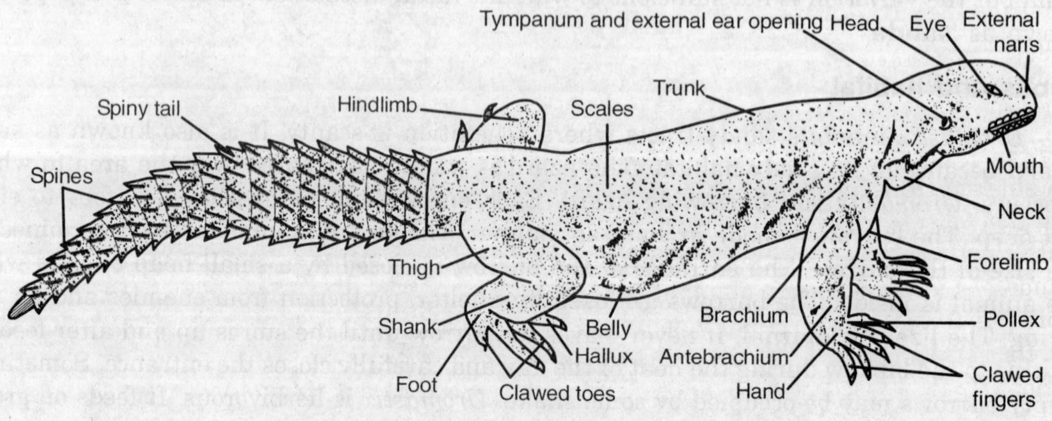

Fig. 17.1 *Uromastix hardwickii*–External features

Division of the body

The body is divided into four distinct regions: **head, neck, trunk and tail** (Fig. 17.1).

Head: The head is small, depressed and somewhat triangular in shape. The anterior end is produced into a blunt **snout** having a pair of large apertures, the **external nares** or **nostrils**. At the time of digging the burrow the external nares are covered by a prominent swelling of its posterior mucous membrane which acts like a **valve.** The **mouth** is a wide slit-like aperture running along the anterior border of the head. It is bound by immovable **lips** covered with scales. The lower lip is circular while the upper lip has a sharp edge and overlaps the lower lip. The **eyes** are small, situated dorso-laterally on each side, almost in the middle of the head. The **pupil** is rounded and the **iris** is light orange in colour. Each eye is provided with upper and lower movable and opaque **eyelids**, the lower eyelid is more mobile than the upper. The eyelids completely close the eyes. A transparent thin third eyelid, the **nictitating membrae** is also present. It lies folded at the anterior corner of the eye, spreads over the cornea and cleans it when required. It *Gecko* both the eyelids are fused and form a transparent covering over the eye as in snakes. Behind and below each eye there is a small oval sunken pit known as the **external auditory meatus** or **external ear** opening. Anterior border of the ear opening is feebly denticulate. Below the auditory meatus lies a brownish oval patch of skin the **tympanic membrane** or **tympanum**.

Neck: The neck is very short and connects the head and trunk. The skin of the sides of the neck is very loose with more or less distinct transverse **folds**. When the neck is stretched the folds disappears.

Trunk: The trunk is broad and elongated, rough and convex dorsally and smooth and flat ventrally. The skin on the sides of the trunk is loose and folded. The trunk is supported by two pairs of short, strong and muscular pentadactyle **limbs**. The forelimbs are slightly shorter than the hindlimbs. The **forelimbs** arise from the trunk just behind the neck. Each consists of the **upper arm** or **brachium**, the **fore arm** or **antebrachium** and the **hand** or **manus** bearing five **digits** or **fingers** ending in a curved and pointed **claw**. The first digit or **pollex** is the smallest and the fourth is the largest. The **hindlimbs** are situated at the posterior end of the trunk more towards the ventral side. Each has a proximal **thigh** or **femur**, middle **shank** or **crus** and distal **foot** or **pes** bearing five **toes** with horny **claws**. The first toe is called **hallux** and is smallest and the fourth toe is largest. A few **spines** are present near the knee joint. The under surface of the manus and pes in *Gecko* bears leaf-like adhesive pads, the surface of which is divided by a series of transverse plates which help them to climb smooth vertical surface. In *Uromastix* and *Varanus* the limbs are used for digging.

A slit-like transverse opening is present mid-ventrally at the junction of the trunk and tail. On the ventral surface of the thigh 12–18 minute **preanofemoral pores** (Fig. 17.2) are present. They are arranged in a single row from the cloaca to the knee joint. Each pore is surrounded by a complete ring of minute **scales** of which one is larger than others. The pores are more prominent in males than in females. They are openings of simple tubular ectodermal glands, the secretion of which form temporary spines which help in holding the female during copulation.

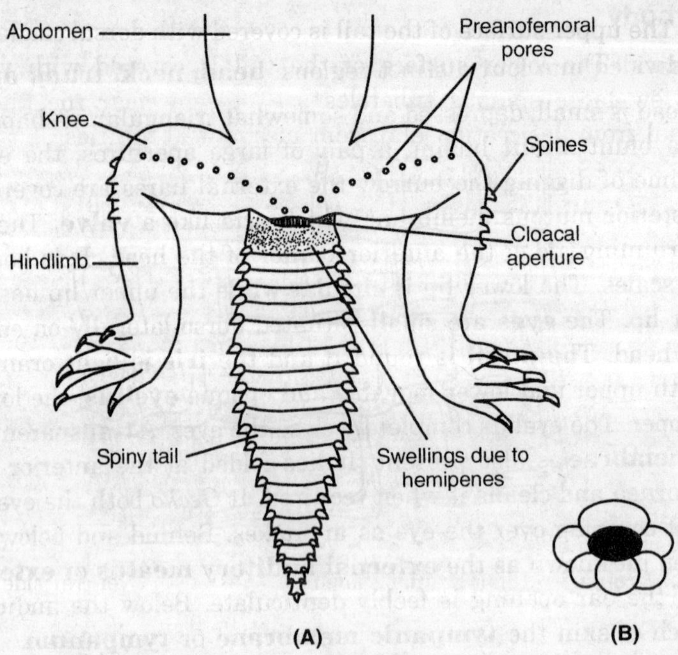

Fig. 17.2 *Uromastix*–**(A)** Ventral view of posterior part of the body of male showing cloacal aperture and femoral pores; **(B)** One femoral pore enlarged, surrounded by scales of which one is quite large

Tail: The tail is elongated, dorso-ventrally compressed, thick in front and gradually tapering at the posterior end. Dorsally the tail is covered with a cross series of enlarged squarish **spinose scales**. The lower surface of the tail is covered with squarish scales about as large as the ventrals. It is due to the presence of spiny scales on the tail it is called **spiny tailed lizard**. The tail is used as an organ of defence. Some lizards break off their tail (**autotomy**) very easily to keep the attention of their enemies engaged to it while they escape. The tail regenerates later on. In *Uromastix* caudal autotomy is absent.

EXOSKELETON

The body of *Uromastix* is covered by horny epidermal scales which form the exoskeleton. The scales are periodically shed off (**ecdysis**) and replaced by a new set of scales formed beneath the old ones. In majority of lizards it comes off in flakes but in those which have a much elongated body it may be shed in one piece. The scales are formed from thickened areas of the integument which grow up and backward and become cornified. They usually have a free posterior border and more or less overlap each other. The upper **head scales** are unequal, smooth or obtusely keeled, large on the snout and small over the outer part of the upper eyelids. The scales on the cheek are oval, 12 to 14 upper labials more or less denticulated on their free margin. The **trunk scales** show simple arrangement (Fig. 17.3). The dorsal scales (**dorsals**) are numerous small and smooth and arranged in cross series. The ventral scales (**ventrals**) are subquadrangular in shape and as large as or larger than the largest dorsals. The tail is covered with whorls of spinose scales larger than those of trunk and head. The upper surface of the tail is covered with dorsal caudals arranged in about 20 transverse rows. The lower surface of the tail is covered with ventral scales called **subcaudals**. A few large spinous tubercles are present near the knee joint. The scales protect the animal from desiccation and from mechanical injuries.

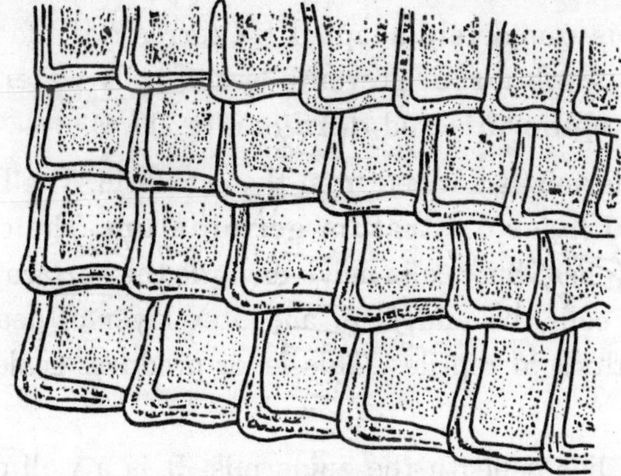

Fig. 17.3 *Uromastix*–Photographs showing scales on the ventral side of the body

The claws of the digits are also modified scales formed from the stratum corneum of the epidermis. Each claw is made up of a dorsal horny plate called **unguis** and a ventral plate called **subunguis**.

In some reptiles such as *Varanus salvator* osteoderms occur. These are ossification of the cutis in the form of bony plates. These are absent in *Uromastix*.

Sexual dimorphism

Sexual dimorphism in *Uromastix* is not well marked. However, the two sexes differ in the following characteristics:

1. The male is slightly smaller in size and darker in colour than the female.
2. The preanofemoral pores are more prominent in male than in female.
3. During the breeding season the abdomen of female becomes broader. In male a pair of **copulatory sacs** or **hemipenes** appear as nodule-like swelling on the ventral side on the base of the tail (absent in female). They open by a pair of apertures into the proctodaeum of the cloaca.

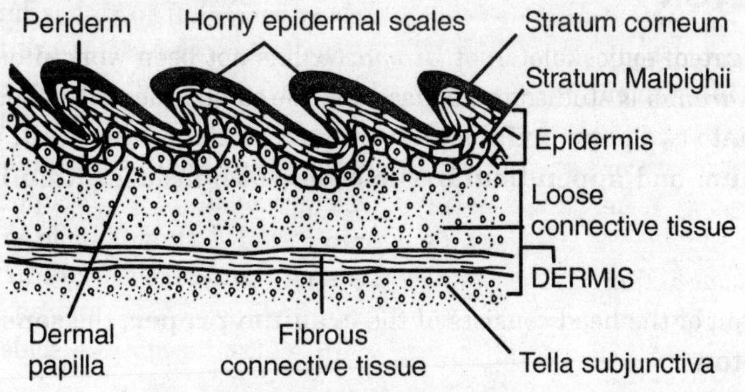

Fig. 17.4 *Uromastix*–V.S. of skin

BODY WALL

The bodywall includes the **skin** or **integument**, **muscles** and **peritoneum**.

Skin

The skin of reptiles is completely adapted for terrestrial life. It is dry, covered with scales and without glands. Such a skin prevents loss of body water. Histologically the skin is made up of two layers: **epidermis** and **dermis** (Fig. 17.4)

Epidermis: The epidermis is derived from the ectoderm. The basal or innermost layer is known as the **germinal layer** or **stratum germinatum**. The cells of this layer divide actively and produce new cells. These cells move towards the surface, flatten and accumulate **keratin**. The keratinized cells die and form an outer layer of dead cells called **stratum corneum** which is quite thick in reptiles. This layer produces scales which form the outer protective exoskeleton.

Dermis: The dermis lies beneath the epidermis. It is a well developed thick layer of dense fibrous connective tissue composed of muscle fibres, nerves, blood vessels and chromatophores. There are no glands in the skin except a few **femoral glands** which are found on the ventral surface of the thigh. In males these glands secrete a substance which hardens into temporary spines which help to hold the female during copulation.

Muscles

A layer of muscles is present beneath the skin. Histologically there are three types of muscles: (i) **smooth** and **involuntary muscles** found in the alimentary canal; (ii) **striated** or **skeletal** and **voluntary muscles** associated with the bodywall and appendages and (iii) **cardiac muscles** found in the heart. The muscles play an important role in locomotion and in supporting the body.

Peritoneum

A thin layer of closely fitting mesodermal cells lie beneath the muscles and lines the coelom. It is known as the peritoneum or coelomic epithelium.

ENDOSKELETON

The structure of endoskeleton of *Uromastix* has not been worked out in detail. Usually the skeleton of *Varanus* is studied in its place and the same is described here. The endoskeleton can be divided into two parts: **axial skeleton** which includes the **skull, vertebral column, ribs** and **sternum** and **appendicular skeleton** which includes **limb bones** and **girdles.**

SKULL

The skeleton of the head consists of the **cranium proper**, the **sense capsules** and the **visceral skeleton**.

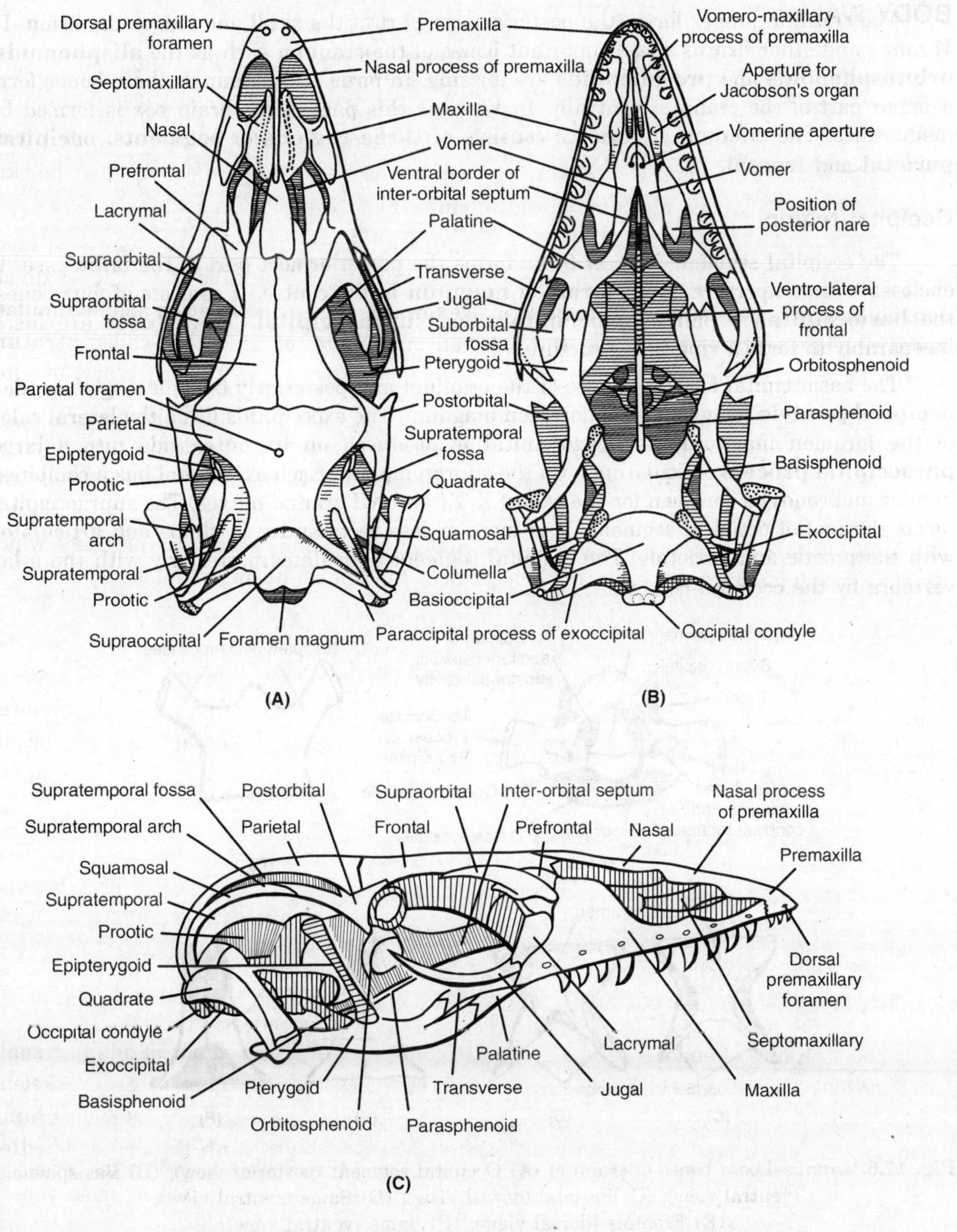

Dorsal premaxillary foramen
Premaxilla
Nasal process of premaxilla
Septomaxillary
Maxilla
Nasal
Vomer
Prefrontal
Ventral border of inter-orbital septum
Lacrymal
Palatine
Supraorbital
Transverse
Supraorbital fossa
Jugal
Suborbital fossa
Frontal
Pterygoid
Parietal foramen
Postorbital
Parietal
Supratemporal fossa
Epipterygoid
Quadrate
Prootic
Supratemporal arch
Squamosal
Supratemporal
Columella
Prootic
Basioccipital
Supraoccipital
Foramen magnum
Paraccipital process of exoccipital
Vomero-maxillary process of premaxilla
Aperture for Jacobson's organ
Vomerine aperture
Vomer
Position of posterior nare
Ventro-lateral process of frontal
Orbitosphenoid
Parasphenoid
Basisphenoid
Exoccipital
Occipital condyle
(A)
(B)
Supratemporal fossa
Postorbital
Supraorbital
Inter-orbital septum
Nasal process of premaxilla
Supratemporal arch
Parietal
Frontal
Prefrontal
Nasal
Premaxilla
Squamosal
Supratemporal
Prootic
Epipterygoid
Dorsal premaxillary foramen
Quadrate
Occipital condyle
Palatine
Lacrymal
Septomaxillary
Exoccipital
Basisphenoid
Pterygoid
Transverse
Jugal
Maxilla
Orbitosphenoid
Parasphenoid
(C)

Fig. 17.5 *Varanus*–Skull. (A) Dorsal view; (B) Ventral view; (C) hateral view

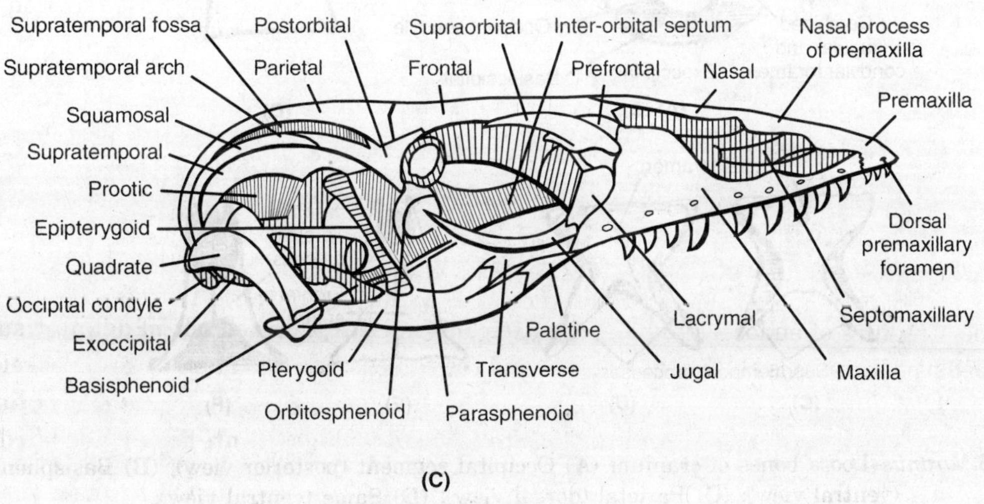

Cranium

The cranium proper forms the posteriormost part of the skull and lodges the brain. In *Varanus* and other lizards some important bones of the cranium such as the **alisphenoids, orbitosphenoids** and **presphenoids** are missing. In birds and mammals these bones form a major part of the cranium ventrally. In *Varanus* this part of the brain box is formed by membranes. The cranium of *Varanus* consists of **three regions** or **segments: occipital, parietal** and **frontal.**

Occipital region

The occipital segment (Fig. 17.6 (a)) forms the posteriormost part of the brain case. It encloses a large aperture called **foramen magnum** in the centre. It consists of four bones: the **basioccipital,** a pair of **exoccipitals** and **supraoccipital.** These bones are fused inseparably to form a ring-like structure.

The basioccipital forms the base of the cranium and posteriorly bears a single rounded **occipital condyle** lying below the foramen magnum. The exoccipitals lie on the lateral sides of the foramen magnum. Each exoccipital is produced on its outer side into a large **paroccipital process** articulating with the supratemporal. Each exoccipital has a combined jugular and condylar foramen for the exit of X, XI and XII cranial nerves. The supraoccipital forms the roof of occipital segment. It is more or less rectangular in shape and articulates with the prootic and parietals. The occipital segment articulates posteriorly with the atlas vertebra by the occipital condyle.

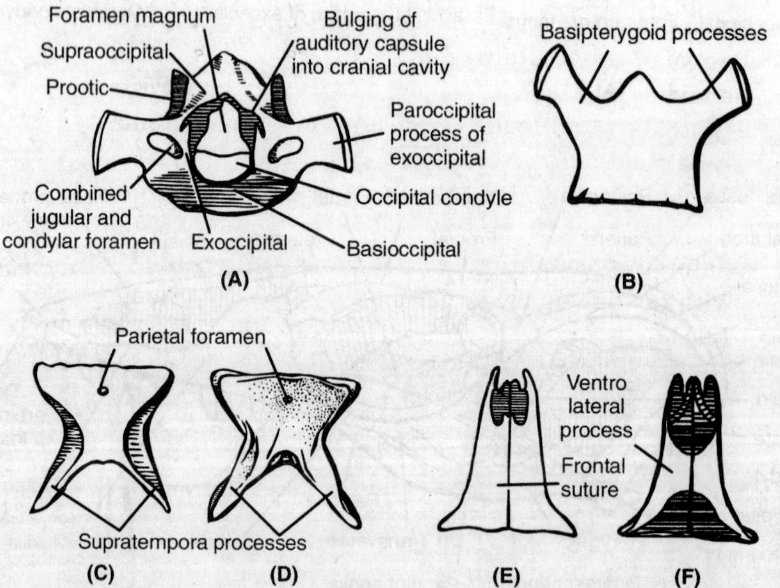

Fig. 17.6 *Varanus*–Loose bones of cranium **(A)** Occipital segment (posterior view); **(B)** Basisphenoid (ventral view); **(C)** Parietal (dorsal view); **(D)** Same (ventral view); **(E)** Frontals (dorsal view); **(F)** Same (ventral view)

Parietal region

The parietal region consists of two bones: the **basisphenoid** (Fig. 17.6 B) and the **parietal** (Fig. 17.6 C and D). The basisphenoid is a broad, flat somewhat rectangular bone present on the ventral side and forming the base of the parietal region of the cranium. Anteriorly it bears a pair of lateral processes called **basipterygoid processes**. Each process bears a cartilaginous pad at its tip and articulates with the pterygoid of its side. Posteriorly it articulates with the anterior margin of the basioccipital. Laterally it remains free. At its anterior margin it bears a median, knob-like process which receives the hinder end of the parasphenoid (which is very much reduced in *Varanus*).

The parietal is a large, flat bone on the dorsal side of the cranium. It forms the roof the cranium in the parietal region. It is formed by a pair of bones completely fused in the adult. It is broad in front and narrow behind. A **parietal foramen** is present in the centre on the mid-dorsal line. Anteriorly the parietal articulates with the frontal by a prominent suture. Each postero-lateral angle is produced behind into a prominent **supratemporal process** which articulates with quadrate, squamosal, supratemporal and paroccipital process of its side. The supratemporal processes form the outer margins of the post-temporal fossae. Posteriorly the parietal is not joined with the supraoccipital but there is a distinct gap filled with fibrous tissue which is removed during cleaning. The lateral sides are uncovered as the alisphenoid bones are absent in *Varanus*.

Frontal region

The frontal region consists of three bones: a pair of **frontals** (Fig 17.6 E and F) and a **parasphenoid**. The roof of the frontal region of the cranium is formed by a pair of frontals that are fused in the middle line by a prominent **frontal suture**. Posteriorly the frontals are separated from the parietal by a distinct **coronal suture**. Each frontal is a large triangular bone lying on the dorsal side, narrow anteriorly and broad posteriorly and produced into a ventro-lateral process which unite along the mid-ventral line. It remains free on its outer lateral margins as orbitosphenoid bones are absent in *Varanus*. Frontals articulate anteriorly with nasals, antero-laterally with prefrontals, postero-laterally with postorbital, behind with parietal and below with the parasphenoid. The parasphenoid is a long, narrow, rod-like bone in the mid-ventral line.

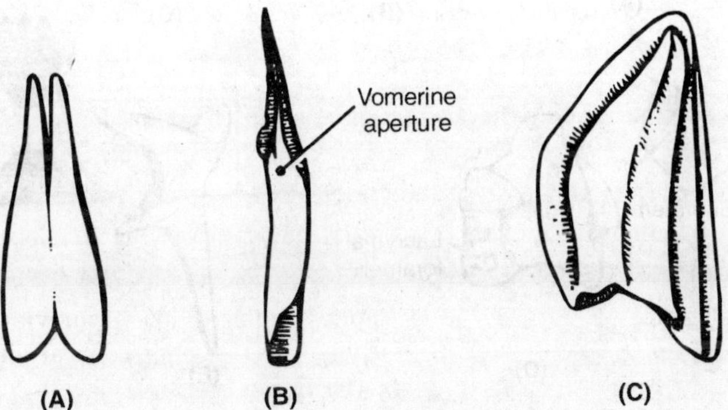

Fig. 17.7 *Varanus*–Bones of olfactory capsule, **(A)** Nasals (dorsal view); **(B)** Vomer (Ventral view); **(C)** Septomaxillary (dorsal view)

SENSE CAPSULES

The sense capsules protect and lodge the special sensory organs. There are three pairs of sense capsules: **olfactory, auditory** and **optic.**

Olfactory capsules: The olfactory capsules lodge the organs of smell. They lie side by side on the antero-lateral side of the cranium. Each capsule is made up of three bones: **nasal, vomer** and **septomaxillary**.

The **nasals** (Fig. 17.7 A) are a pair of small slender bones which cover the olfactory capsule dorsally. Each nasal is a flat, triangular bone narrow anteriorly and broad posteriorly. The two nasals are fused along the middle line to form a single bone. The nasals articulate in front with the nasal process of the premaxillae and the two septomaxillaries and posteriorly they articulate with the frontals.

The two **vomers** (Fig. 17.7 B) are elongated rod-shaped bones forming the ventral surface of the olfactory capsule. They are fused with each other anteriorly and separated posteriorly. They form the inner margin of the posterior nares. Each vomer is perforated by a **vomerine aperture** in the centre. It articulates in front with the vomero-maxillary process of the premaxillae and behind with the palatine.

The **septomaxillary** (Fig. 17.7 C) is an irregular flat bone lying one on either side of the nasal process of the premaxillae.

Auditory capsules: The auditory capsules lodge the organs of hearing. They are situated on the postero-lateral side of the cranium. Each auditory capsule consists of three bones: **prootic, opisthotic** and **epiotic**. In *Varanus* the epiotic is fused with the supraoccipital and the opisthotic is fused with the exoccipital, only prootic remains separate. It is a small irregular bone lying just outside the supraoccipital.

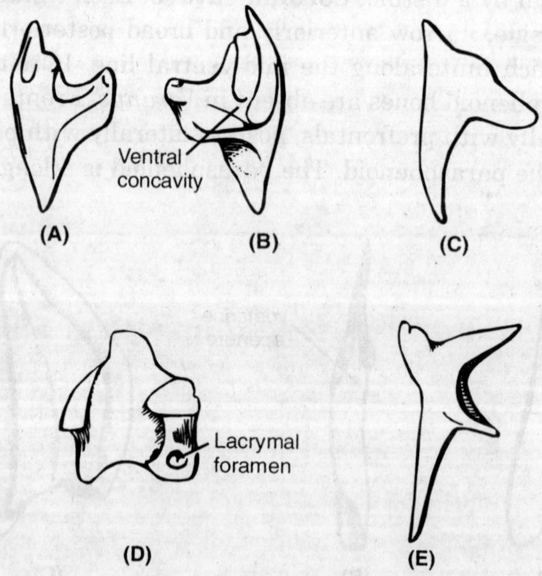

Fig. 17.8 *Varanus*–Bones of optic capsule, **(A)** Prefrontal (outer view); **(B)** Same (inner view); **(C)** Supraorbital (outer view); **(D)** Lacrymal; **(E)** Post frontal

Optic capsules or orbits: The optic capsules lodge the organs of sight. The two orbits lie one on either lateral side in the middle of cranium. They are large in size and separated from each other by a thin longitudinal vertical partition, the **interorbital septum.** Each orbit is made up of five bones: **prefrontal, supraorbital, lacrymal, postfrontal** or **postorbital** and **jugal.**

The **prefrontal** (Fig. 17.8 A and B) is a small somewhat triangular bone with a deep cup-like concavity on its ventral side. It lies between the frontal and maxilla and forms the anterior boundary of the orbit. Posteriorly it articulates with the supraoccipital and ventrally with the lacrymal.

The **supraorbital** (Fig. 17.8 C) is a small triangular bone overhanging the anterior part of the orbit. Its broad base articulates with the prefrontal while the free pointed apex is directed posteriorly towards the postfrontal.

The **lacrymal** (Fig. 17.8 D) is a small squarish bone lying on the anterior boundary of the orbit. It has a small aperture for the lacrymal duct.

The **post-orbital.** (Fig. 17.8 E) or **post-frontal** lies behind the frontal and forms the postero-dorsal boundary of the orbit. It is irregular in shape and has four processes. The two inner processes form a deep notch which lodges a part of both the parietal and the frontal. The posterior process is the largest and articulates with the squamosal to form the supra-temporal arcade. The antero-lateral process remains free and forms the posterior boundary of the orbit.

The **jugal** is a small slender curved bone forming the postero-ventral boundary of the orbit. A wide gap is present between the jugal and the post-orbital.

VISCERAL SKELETON

It includes the **jaws (lower** and **upper**), **suspensorium** and **hyoid apparatus**.

Upper jaw: The upper jaw is closely fused with the cranium. It consists of two halves or **rami** which are fused anteriorly but diverge posteriorly. Each half is made up of two sets of bones: the **outer** and **inner**. The outer set is made up of **premaxilla, maxilla** and **jugal** and the inner set is made up of **pteryoid**, **palatine**, **transverse** and **epipterygoid**.

The **premaxillae** (Fig. 17.9 A and B) of the two halves are fused into a single median bone which forms the anterior extremity of the snout. Its dorsal surface bears a pair of **premaxillary foramina** while its ventral surface bears 6–8 small conical teeth along its margin. Posteriorly it bears three processes: a single median elongated dorsal **nasal process** which articulates with the anterior end of fused nasals, and a pair of wing-like ventral vomero-maxillary processes which articulate with the vomer. Anteriorly the premaxillae articulate with the maxilla.

The **maxilla** (Fig. 17.9 D and E) forms the major part of the anterior half of the upper jaw on either side. It is a long irregular bone lying behind the premaxilla. The outer surface of maxilla has a number of small pores called **maxillary foramina.** Its main body is called **alveolar part**. It bears 8–10 small conical and slightly pointed **pleurodont teeth**. On the inner side the palatine process is present, but is poorly developed, so a gap is present between the maxilla and palatine. The maxilla articulates anteriorly with the premaxilla, laterally with the palatine, prefrontal, supraorbital and lacrymal and behind with the jugal and transverse.

The **jugal** (Fig. 17.9 C) is a small, slender, curved bone behind the maxilla. It also borders the orbit ventrally. It articulates anteriorly with the maxilla and lacrymal, inwardly with the transverse and remains free posteriorly.

The **pterygoid** (Fig. 17.9 F and G) is a pair of large, elongated irregular bone present in the roof of the mouth. At the anterior end it has a **palatine process** and at the posterior end it has a **quadrate process**. Anteriorly it articulates with the palatine on the inner side and the transverse on the outer side. Posteriorly it articulates with the basipterygoid process of the basisphenoid on the inner side and the quadrate on the outer side.

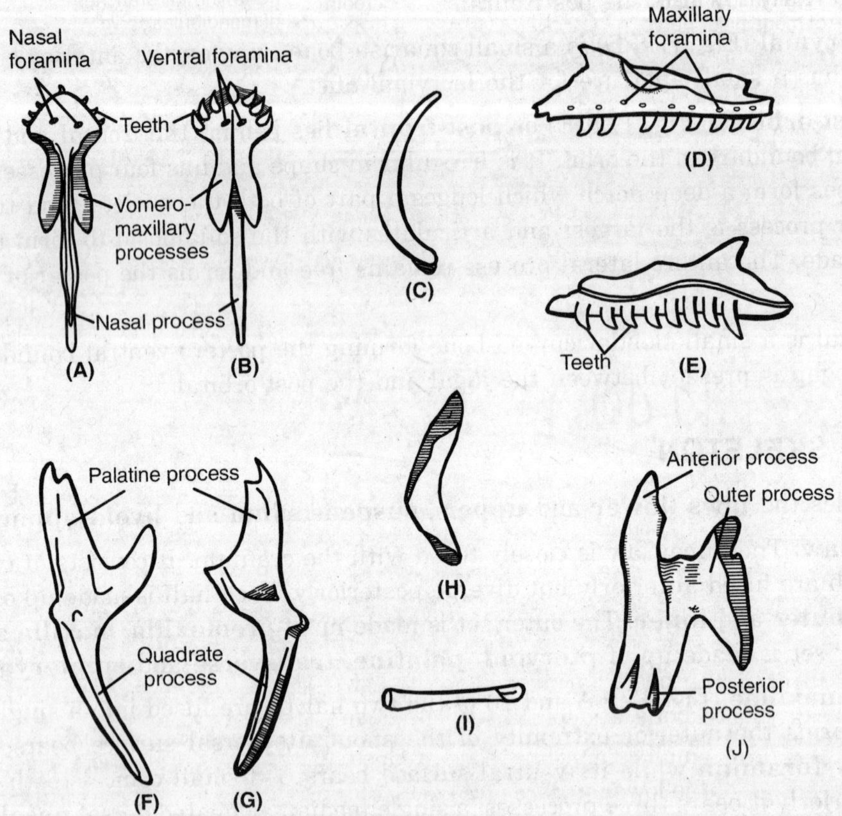

Fig. 17.9 *Varanus*–Loose bones of upper jaw, **(A)** Premaxillae (dorsal view); **(B)** Same (ventral view); **(C)** Jugal; **(D)** maxilla (outer view); **(E)** Same (inner view); **(F)** Pterygoid (ventral view); **(G)** Same (dorsal view); **(H)** Transverse; **(I)** Epipterygoid; **(J)** Palatine (ventral view)

The **palatine** (Fig. 17.9 J) is a pair of small, flat and somewhat irregular bones present in the roof of the mouth. Each palatine has 3 processes: the elongated **anterior process** which articulates with the vomer, the small **posterior process** which articulates with the pterygoid and the broad **outer process** which articulates with the maxilla and transverse. The palatines also form the posterior boundary of internal nares.

The **transverse** (Fig. 17.9 H) or **transpalatine** or **ectopterygoid** is a small, slightly curved bone present on the inner side of the jugal. It connects the pterygoid with the jugal and maxilla. It also forms the floor of the orbit.

The **epipterygoid** (Fig. 17.9 I) also known as **columella crani** is a slender bony rod extending almost vertically between the pterygoid and prootic.

The **septomaxillary** is a pair of small irregular bone lying dorsally above the vomer, one on either side of the nasal process of the premaxillae. It articulates laterally with the maxilla and posteriorly with the nasal.

Lower jaw or mandible: The lower jaw consists of two halves or **rami** (Fig. 17.10) united anteriorly but diverge posteriorly. Each ramus is made up of six bones: **articular**, **angular**, **supra-angular, coronoid, splenial** and **dentary**, around an axial **Meckel's cartilage**.

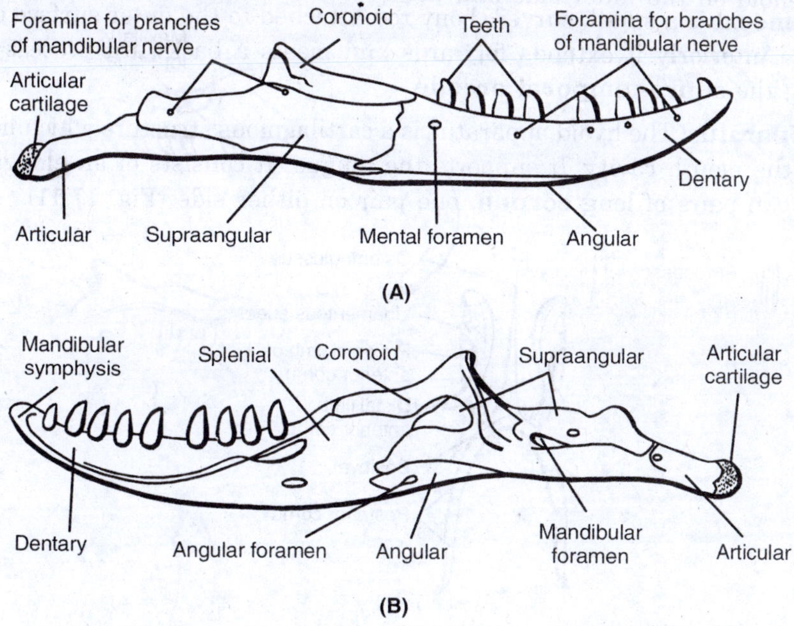

Fig. 17.10 *Varanus*–Lower jaw, **(A)** Outer view; **(B)** Inner view

The **articular** is the posteriormost bone of each ramus. Dorsally it bears an articular surface for the quadrate bone and posteriorly it terminates into an **articular cartilage**.

The **angular** is a small, splint-like bone wedged in between the articular and dentary, covering the lower half of the lower surface of the articular. It has a pore called **angular foramen**.

The **supra-angular** is a flat, elongated, somewhat rectangular bone lying in the middle of the ramus covering the upper half of the outer surface of the articular. It has a pair of pores called **mandibular foramina** for the mandibular nerves.

The **coronoid** is small, somewhat triangular bone present in the middle of the ramus on the dorsal side. Just behind the last tooth it gives off a prominent upwardly directed **coronoid process.**

The **splenial** is a flat, irregular, membranous bone lodged in a groove on the inner face of the dentary.

The **dentary** is the largest bone, constituting almost the anterior half of each ramus. It is united with its counterpart by a **suture**. The dorsal surface of a dentary bears 8–10 small **mandibular teeth** which are of pleurodont type. Its outer surface bears numerous **foramina** for mandibular nerves.

Suspensorium: The bones which help in the suspension of the lower jaw with the cranium and the upper jaw form the suspensorium. The two bones are **quadrate** and **squamosal**.

The **quadrate** is a small, thick, rod-like bone attached obliquely to the outer postero-lateral side of the cranium. Anteriorly it articulates with the quadrate process of the pterygoid and the articular bone of the lower jaw. Posteriorly it articulates with the squamosal, supratemporal, supratemporal process of the parietal and paroccipital process of the exoccipital.

The **squamosal** is a small, curved, bony rod attached to the outer surface of the supra-temporal bone. Anteriorly it extends forwards and meets a backward process of the post-frontal forming the **supra-temporal arcade**.

Hyoid apparatus: The hyoid apparatus is a cartilaginous structure which lies embedded in the floor of the mouth cavity. It supports the tongue. It consists of an elongated body or **basihyal** and two pairs of long **cornua,** one pair on either side (Fig. 17.11).

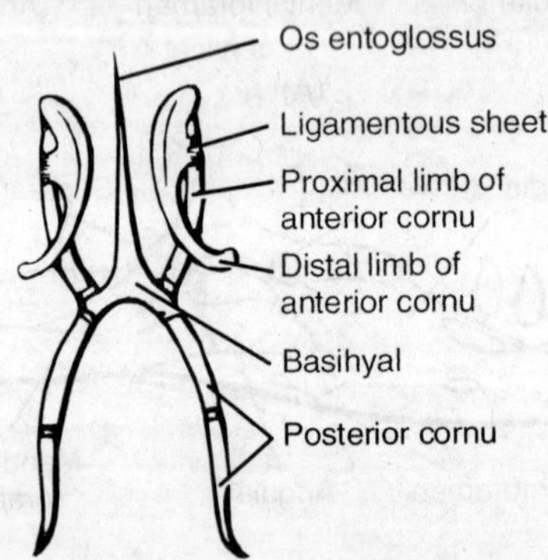

Fig. 17.11 *Varanus*–Hyoid apparatus

The **basihyal** or **body proper** is an elongated median cartilaginous rod, broad basally and tapering anteriorly into the **process lingualis** or **os entoglossus**.

The **anterior cornua** are elongated cartilaginous rod attached to the body proper ventrally, running upwards and then curving downwards. Each cornua is made up of two segments— **proximal** and **distal limbs.** The two segments are joined together by a **ligamentous sheet**. The free end of the distal limb is flattened.

The **posterior cornua** are also cartilaginous rods arising from the posterior end of the body proper and running backwards. It is also made up of two segments.

VERTEBRAL COLUMN

The vertebral column of *Varanus* consists of variable number of vertebrae, the number ranges from 83 to 85. It can be divided into four regions: the **cervical** consisting of 8 vertebrae, the **thoracolumbar** containing 22 vertebrae, the **sacral** having only 2 vertebrae and the **caudal** region with 51–53 vertebrae.

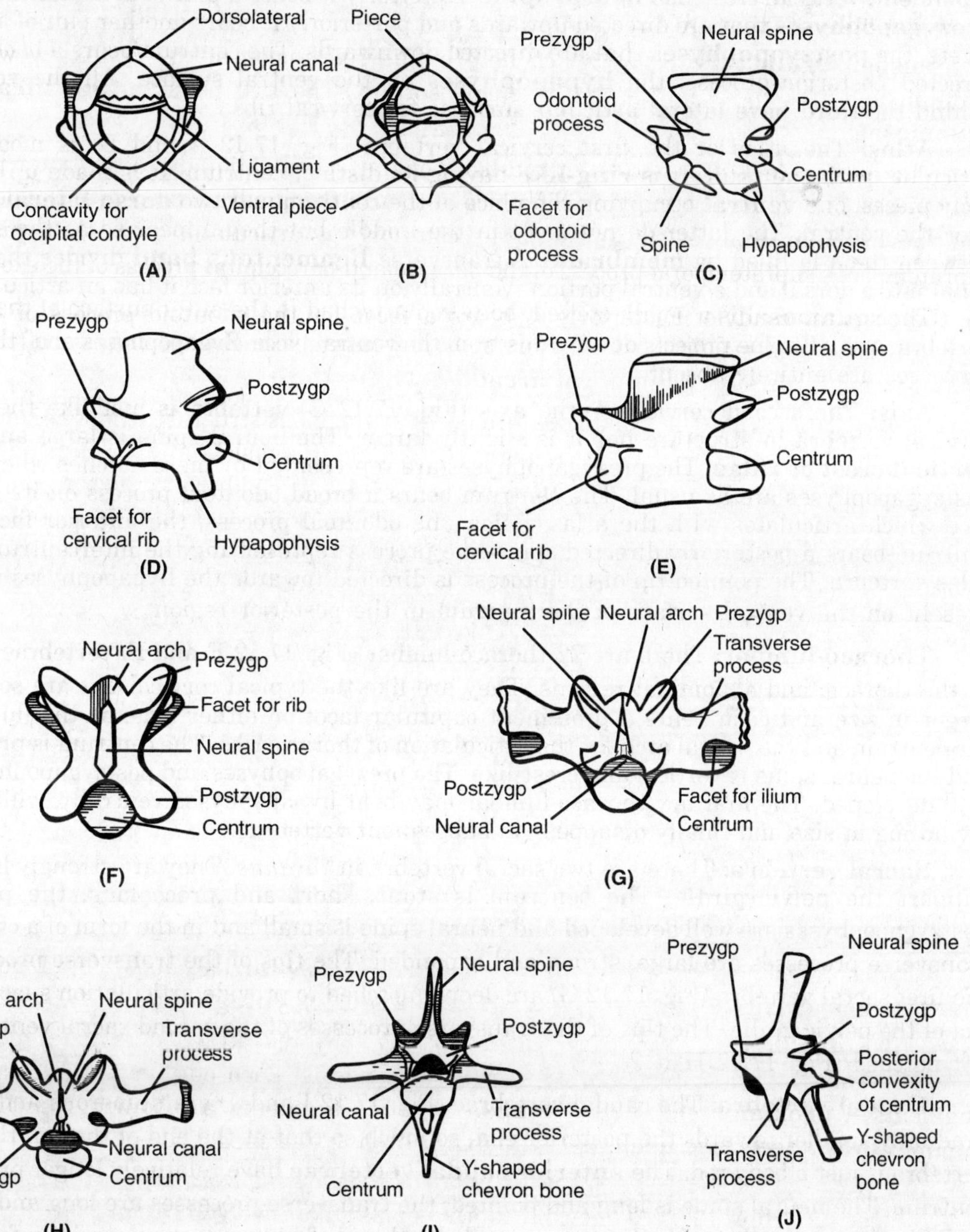

Fig. 17.12 *Varanus–Vertebrae.* (A) Atlas (anterior view); (B) Same (Posterior view); (C) Axis (lateral view); (D) Typical cervical (lateral view); (E) Thoraco-lumbar (lateral view); (F) Same (dorsal view); (G) First sacral (dorsal view); (H) Second sacral (dorsal view); (I) Caudal (posterior view); (J) Same (lateral view)

Cervical vertebrae: A typical cervical (Fig. 17.12 D) vertebra consists of a stout **centrum** which is hollow in front and convex behind, i.e., **procoelous**. The **neural arch** bears a prominent, vertical, crest-like **neural spine.** Anteriorly it bears a pair of articular facets, the **prezygapophyses** that are directed upwards and posteriorly it bears another pair of articular facets, the **postzyapophyses** that are directed downwards. The centrum bears a backwardly directed posterior process, the **hypapophyses** on the ventral surface. All the vertebrae behind the third have lateral articular surfaces for cervical ribs.

Atlas: The atlas or the first cervical vertebra (Fig. 17.12 A and B) is modified to articulate with the skull. It is **ring-like** having no distinct centrum. It is made up of three bony pieces, one **ventral** occupying the place of the centrum and two **dorso-lateral** arching over the ventral. The latter do not meet in the middle but they appear so as the space in between them is filled by **membrane**. A transverse **ligamentous band** divides the neural canal into a dorsal and a ventral portion. Ventrally on its anterior face it has an articular facet for the occipital condyle. Posteriorly it bears a facet for the odontoid process of the axis vertebra. A small spine projects downwards from the ventral piece. Zygapophyses and transverse processes are entirely absent.

Axis: The second cervical or the axis (Fig. 17.12 C) vertebra is just like the typical cervical vertebra in structure but it is slightly larger. The neural spine is large and like a **vertical crest** or **ridge.** The prezygapophyses are represented by mere notches whereas the postzygapophyses are as usual. The centrum bears a broad odontoid process on its anterior face which articulates with the atlas. Below the odontoid process, the anterior face of the centrum bears a posteriorly directed spine-like process representing the intercentrum of the atlas vertebra. The pointed tip of the process is directed towards the hypapophyses which is present on the ventral surface of the centrum in the posterior region.

Thoraco-lumbar: There are 22 thoraco-lumbar (Fig. 17.12 E and F) vertebrae present in the thoracic and abdominal regions. They are like the typical cervical but are somewhat larger in size and each bears a prominent capitular facet on either side, at the junction of the centrum and the neural arch for the articulation of thoracic ribs. The centrum is procoelous and the neural spine is vertical and crest-like. The prezygapophyses and postzygapophyses are well developed. The first few thoraco-lumbar may bear hypapophyses ventrally, which go on dwindling in size and finally disappear in subsequent vertebrae.

Sacral vertebra: There are two sacral vertebra in *Varanus*. They are strongly built and support the pelvic girdle. The centrum is stout, short and procoelous, the pre- and postzygapophyses are well developed and neural spine is small and in the form of a crest. The transverse processes are large, strong and expanded. The tips of the transverse processes of the first sacral vertebra (Fig. 17.12 G) are deeply notched to provide articulation space for the ilia of the pelvic girdle. The tips of the transverse processes of the second sacral ventebra are not notched (Fig. 17.12 H).

Caudal vertebra: The caudal vertebrae (Fig. 17.12 I and J) are numerous and become gradually smaller towards the posterior end, so much so that at the end of the tail the whole vertebra is just a bony rod. The **anterior caudal vertebrae** have relatively longer procoelous centrum. The neural spine is long and pointed, the transverse processes are long and slender and the pre-and postzygapophyses are well developed. The most characteristic feature of anterior caudal vertebra is the presence of Y-shaped **chevron bone** attached to the ventral surface of the centrum. The two limbs of Y articulate with the centrum enclosing the haemal canal for the passage of haemal artery and vein. The **posterior caudal vertebrae** differ from anterior in lacking chevron bone. All the processes are reduced and at the end of the tail it is represented merely by a rod-like centrum.

Ribs

Except the first three cervical vertebrae, all precaudal vertebrae bear ribs. The ribs are slender curved rods. They are single-headed (**unicephalous**) and articulate with the capitular facets of the vertebrae. The **cervical ribs** are short rods, without cartilaginous part, and are not connected with the sternum. The **thoracic ribs** (Fig. 17.13 A) are relatively larger. Each rib has a dorsal bony **vertebral part** connected with the vertebra and a ventral cartilaginous **sternal part**. The first-three thoracic ribs are connected with the sternum by their sternal parts (**sternal ribs**). The remaining thoracic ribs are short and remain free ventrally. Their free ends carry a little cartilaginous rod representing dwindling sternal ribs.

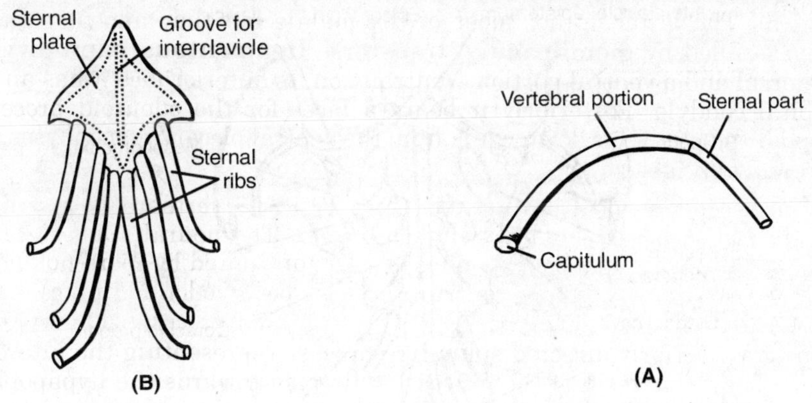

Fig. 17.13 *Varanus*–**(A)** Thoracic rib; **(B)** Sternum (ventral view)

STERNUM

The **sternum** or **breast bone** (Fig. 17.13 B) consists of a rhomboidal plate of cartilage embedded in the ventral thoracic wall. Anteriorly the plate has a **groove** on its mid-ventral surface into which fits the elongated stem of T–shaped cartilaginous **interclavicle**. On the small anterior end of the T rest slender curved rods, the **clavicles**, one on either side. The antero-lateral borders of the plate articulate with the coracoids and epicoracoids of the pectoral girdle and its postero-lateral border bears two facets for sternal ribs. The posterior end of the plate bears two sternal ribs.

PECTORAL GIRDLE

The **pectoral** or **shoulder girdle** (Fig. 17.14) is situated at the anterior end of the trunk. It consists of two similar halves lying on either side of the interclavicle. Each half is made up of the following parts:

The **supra-scapula** is a flat distally expanded plate made of cartilage articulating internally with the scapula. Its distal border is free and slightly curved inwards.

The **scapula** is an oblong bony plate constricted in the middle. Its outer end articulating with the supra-scapula is broder than its inner end which articulates with the coracoid. At the junction of the scapula and the coracoid along the ventral side a **glenoid cavity** is present for the articulation of the head of humerus. From the anterior end of the scapula just opposite to the glenoid cavity arises a completely ossified process, the **mesoscapula**.

The **coracoid** is a stout flat bone constricted in the middle and broadly expanded at its inner end. Two large **gaps** or **fenestrae** divide each coracoid into a narrow anterior portion, the **procoracoid**, a narrow middle portion, the **mesocoracoid** and a broad posterior **coracoid proper** which articulates with the antero-lateral border of sternum. The **epicoracoid** is an irregular-shaped cartilage starting from the inner end of the coracoid and running along the inter-clavicle up to its end.

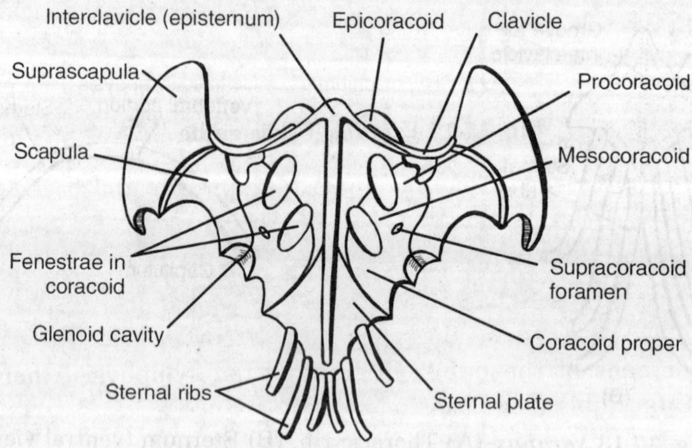

Fig. 17.14 *Varanus*–Pectoral girdle

The **interclavicle** or **episternum** is a T-shaped or anchor-shaped investing bone, the stem of which is placed longitudinally in between the two halves of the pectoral girdle and the cross-piece or the two lateral arms of the T are situated a little in front of the scapula. The posterior end of the stem is closely applied to the ventral surface of the sternum, reaching up to its posterior end.

The **clavicle** is a small, narrow, flat and curved bone articulating with the anterior surface of the arms of the interclavicle. The outer ends articulate with the junction of scapula and suprascapula. The inner ends of the clavicle remain separated in the middle.

The pectoral girdle along with the ribs and sternum protect the heart and lungs. It provides articulation to the bones of the forelimbs.

PELVIC GIRDLE

The **pelvic** or **hip girdle** (Fig. 17.15) is situated at the posterior end of the trunk. It consists of two similar halves, the **ossa innominata** which meet in the middle line by a vertical **ligament**. Each half is a triradiate structure made up of three bones – **ilium, pubis** and **ischium.** The bones are solid, hard and meet each other but do not fuse. On the outer side at the point where the three bones meet there is a concave articular surface, the **acetabulum,** into which the head of the femur fits. At the base of the depression of the acetabulum the lines of fusion of the bones can be seen clearly.

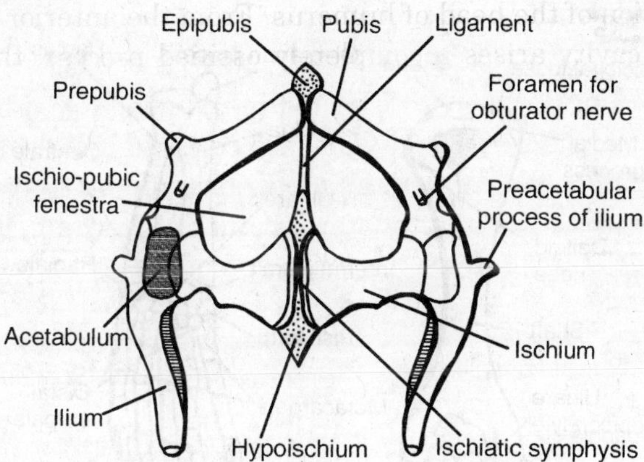

Fig. 17.15 *Varanus*–Pelvic girdle

The **ilium** is a compressed rod-like bone passing upwards and backwards to articulate with the sacral vertebra. It forms nearly one-third part of the acetabulum and on its outer side just in front of the acetabulum it is produced into a **pre-acetabular process.**

The **pubis** is a flat and somewhat curved bone. It passes downwards from the acetabulum to meet its fellow of the other side, along the midventral line, at the **pubic symphysis**. Between the anterior ends of the pubis, in front of the symphysis, there is a nodule of cartilage, the **epipubis**. Pubis also forms one-third part of the acetabulum. Near its fusion with the ilium and ischium, each pubis has an oval foramen for the obturator nerve. Just external to and slightly posterior to the foramen arises a small rod-like process, the **prepubis** which is directed outward.

The **ischium** is a flat and slightly curved bone which runs inwards from the acetabulum and meets it fellow of the other side at the **ischiatic symphysis.** The **hypoischium** is a small piece of calcified cartilage in between the two ischia, at the anterior side. It is also called **os cloaca** as it passes backwards from the ischiatic symphysis and supports the ventral wall of the cloaca. A wide space is enclosed between the ischia and pubes. It is divided into two lateral **ischio-pubic fenestrae** by a median ligament which may not be visible in dried specimens as it is often removed while cleaning.

The pelvic girdle protects the kidneys and provides articular surface for the bones of thc hind limbs.

FORELIMB

The forelimb shows the usual patten of pentadactyle limb (Fig. 17.16). It consists of humerus, radius and ulna and bones of the fore-foot.

The upper arm consists of a single bone, the **humerus** which is made up of an elongated **shaft** with both the ends greatly expanded. The head lies at the proximal end and fits into the glenoid cavity of the pectoral girdle. A **medial process** is present at the proximal end and a prominent **deltoid ridge** is present below the head. The distal end has a pulley-like articular surface, the **trochlea** which is divided into **radiale** and **ulnare epicondyles** for articulation with the radius and ulna respectively.

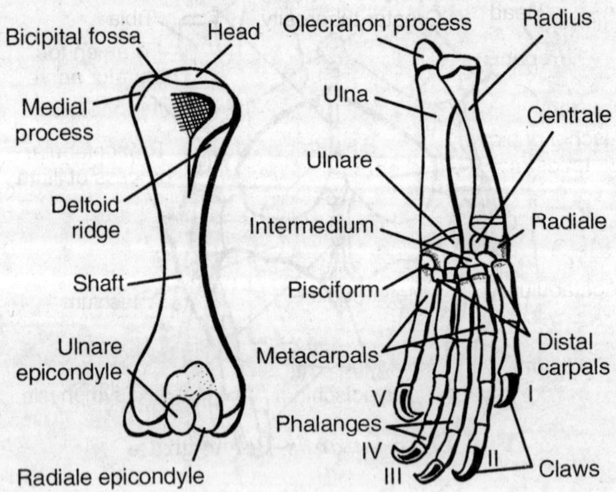

Fig. 17.16 *Varanus*–Bones of forelimb

The forearm is made up of two bones, **radius** and **ulna** of which the radius is slender and composed of a **shaft** and two **epiphyses.** Distally it has a concave articular facet for the carpus. The ulna is stouter, its proximal end is produced into a upwardly directed **olecranon process**, while its distal end bears a convex articular surface for the carpus.

The **carpus** or **wrist** consists of ten polyhedral bony pieces or **carpals** arranged in three rows. The proximal row consists of three bones, the **radiale, ulnare** and **intermedium,** the central row of one, the **centrale** and the distal row of five small pieces. A **pisiform** is attached to the distal epiphysis of the ulna on its post axial side.

The **digits** are five each consisting of a **metacarpal** bone and varying number of **phalanges.** The first digit or **pollex** has two phalanges, the second three, the third four, the fourth five and the fifth has three only. The terminal phalanx of each digit bears a horny **claw**.

HINDLIMB

The hindlimb consists of femur, tibia and fibula and bones of hind foot (Fig.17.17).

The **femur** is a stout bone of the thigh. It is made up of a **shaft** and two **epiphyses.** The anterior or proximal epiphysis develops a rounded **head** that fits into the acetabulum. Near the head on the preaxial side there is a prominence called **lesser trochanter** while on the postaxial side there is a minor **greater trochanter**. The distal extremity is pulley-shaped having two **condyles** and a **tuberosity** for the articulation of tibia and fibula.

The **shank** or **crus** consists of two bones, the **tibia** and **fibula**. The tibia is a stout somewhat curved bone having a longitudinal ridge, the **cnemial crest** along the dorsal surface. The proximal end bears two concave facets for articulation with the condyles of femur. The fibula is a slender bone the proximal end of which articulates with the tuberosity of femur. The distal end of both tibia and fibula articulates with the tarsus.

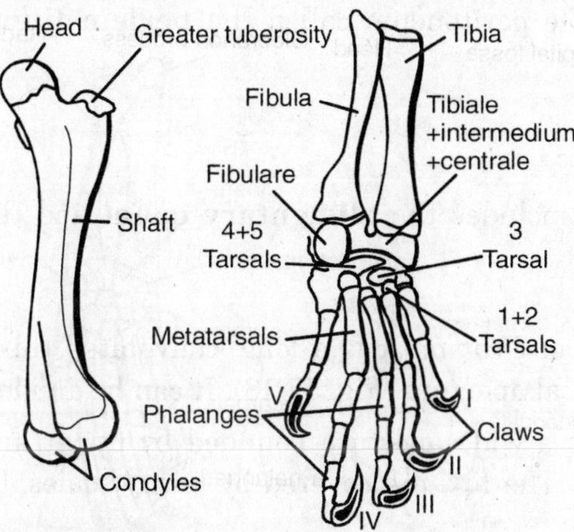

Fig. 17.17 *Varanus*–Bones of hindlimb

The **pes** or **tarsus** consists of only five **tarsal bones**, two in the proximal row and three in the distal row. The two tarsals of the proximal row are suturally fused with the tibia and fibula. The bones of the distal row are cuboidal and of different sizes and articulate with the **metatarsals** of the toes. The number of **phalanges** in each toe is variable, the first has two, the second three, the third four, the fourth five and the fifth three. The terminal phalanx of each toe bears a horny **claw**.

BODY CAVITY OR COELOM

The coelom is a large cylindrical cavity lined by coelomic epithelium known as **peritoneum.** It is divided into two parts: a small anterior **pericardial cavity** lodging the heart and a large **pleuro-peritoneal** or **perivisceral** or **abdominal cavity** lodging the viscera which includes alimentary canal, digestive glands, lungs, kidneys, fat bodies, etc.

Mesenteries

The digestive tract is attached, for the most of its length to the median dorsal line of the coelom by the **dorsal mesentery**. In the pyloric region of the stomach it is missing. The portion of the dorsal mesentery supporting the stomach is called **mesogaster**. The lung is attached to the mesogaster by a short mesentery. A **ventral mesentery** exists between the viscera and body wall. One part of it forms the long mesentery extending between the median ventral line of the body wall and the median line of the ventral face of the liver. This is the **falciform ligament** of the liver. It contains a number of blood vessels which pass from the ventral body wall into the substance of the liver. In the free posterior margin of the falciform ligament the ventral abdominal vein crosses from the body wall to the liver. A **gastrohepatic ligament** or **omentum** extends from the anterior part of the stomach to the dorsal face of the liver. The **hepatoduodenal ligament** joins the duodenum and liver and encloses the greater part of pancreas. The part of the mesentery from the urinary bladder to the midventral line of the body wall is called the **median ligament.** Each gonad has a mesentery: **mesovarium** in the female, **mesorchium** in the male and **mesotubarium** of the oviduct.

An anterior extension of the peritoneum called the **pericardium** encloses the pericardial cavity.

DIGESTIVE SYSTEM

The digestive system includes the **alimentary canal** and the **digestive glands.**

Alimentary canal

The alimentary canal of *Uromastix* is a long, convoluted tube beginning at the mouth and terminating in the cloacal aperture (Fig. 17.18). It can be divided into the following parts:

Mouth: The mouth is a wide aperture bounded by upper and lower **jaws** covered by fleshy and immovable **lips**. The lips are covered by horny scales. The mouth opens into the buccal cavity.

Buccal cavity: The buccal cavity is narrow in front and broad behind. It is lined with mucous membrane. It contains **tongue, teeth** and **internal nares**.

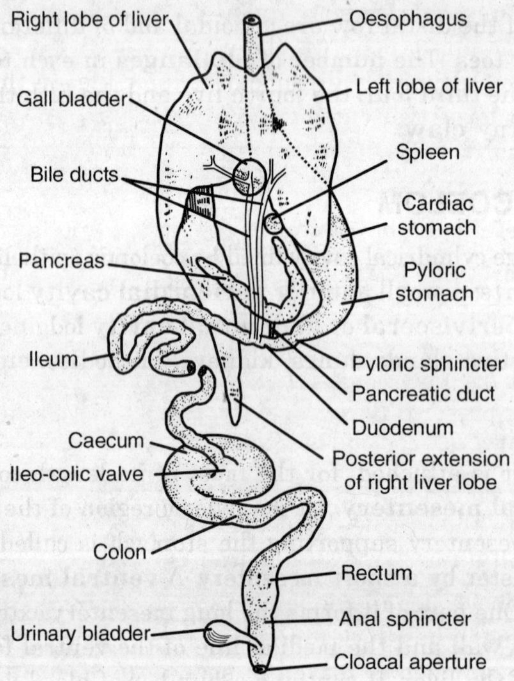

Fig. 17.18 *Uromastix*–Digestive system

The tongue is a well developed, elongated, fleshy structutre present in the floor of the buccal cavity. The anterior end is narrow, free, bifid and protrusible. The posterior end is broad and widely bifurcated. The tongue is covered with numerous glandular papillae having taste buds. It has a mid-dorsal groove (Fig. 17.19).

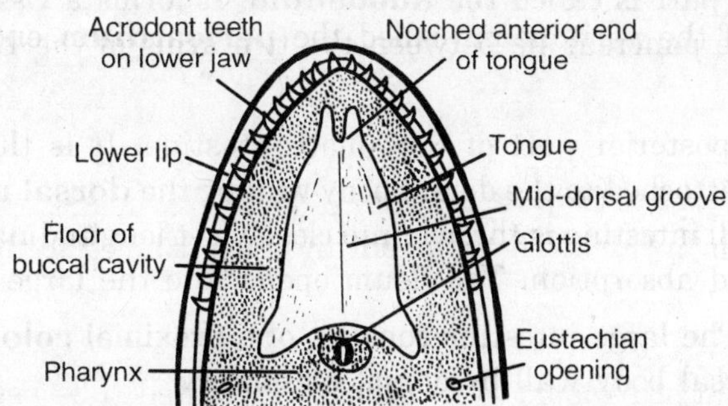

Fig. 17.19 *Uromastix*–Floor of buccal cavity showing the position of tongue

The teeth are present on both the upper and lower jaws. In the upper jaw they are present on premaxilla and maxilla and in the lower jaw on the dentary. The teeth are acrodont, *i.e.,* they are attached to the parapet or edge of the jaw bones.

In some lizards the teeth are usually divided into incisors, canines and molars. In *Uromastix* the young at birth has four incisors in each upper and lower jaw, but as growth proceeds the two central upper incisors are replaced by single large, cutting tooth, a downward prolongation of the premaxillary bone, the lateral incisors remaining intimately attached to the central tooth and may ultimately disappear in later life. In the lower jaw the incisors of each mandible unite to form two cutting surfaces. There are no proper canines and the anterior molars are usually worn down and show a more or less toothless gap.

The nasal chambers communicate with the buccal cavity by a pair of small apertures, the **internal nares** present in the roof of latter near the anterior end.

Pharynx: Posteriorly the buccal cavity opens into the pharynx. The lining of the pharynx is thrown into longitudinal folds. In the floor of the pharynx, between the posterior bifurcation of the tongue, lies a median longitudinal slit, the **glottis** which leads into the trachea or wind pipe. On each lateral side of the pharynx near the jaw angle a small aperture of the eustachian tube is present. Posteriorly the pharynx leads into the oesophagus through a wide aperture the **gullet** which is distensible and opens only when the food is swallowed.

Oesophagus: The oesophagus is a long, narrow, muscular tube which passes straight down the neck dorsal to the trachea and opens into the stomach. It has great power of distension. Its inner wall is thrown into longitudinal folds.

Stomach: The stomach lies in the left half of the body cavity. It is a long, tubular sac wider than the oesophagus. It has thick muscular walls. It is suspended by a fold of peritoneum called **mesogaster.** The stomach is curved in such a way that its inner side is concave and outer side is convex. It can be divided into two parts: the anterior part is the **cardiac stomach** and the posterior part is the **pyloric stomach**. The end of pyloric stomach is marked by a muscular constriction called **pyloric sphincter** or **valve**. The inner lining of the stomach is thrown into a number of longitudinal folds which are more developed in the cardiac part and less developed in the pyloric part.

Small intestine: The stomach is followed by the small intestine which is a long, narrow and coiled tube. It is divided into two parts:

The small anterior part is called the **duodenum**. It forms a U-shaped structure with the pyloric stomach. The pancreas lie between the two arms of the U. It receives bile and pancreatic duct.

The **ileum** is the posterior part of the small intestine. It is the longest part of the alimentary canal and is attached to the dorsal body wall by the **dorsal mesentery**. The inner mucous lining of the small intestine is thrown into closely-set longitudinal folds which increase the area of secretion and absorption. The ileum opens into the large intestine.

Large intestine: The large intestine consists of a proximal **colon** and distal **rectum**. It is attached to the dorsal body wall by dorsal mesentery.

The colon is the thin walled anterior portion of the large intestine. A blind pouch called **caecum** arises from the anterior part of colon. An **iliocolic valve** is present internally at the junction of the ileum and colon. The function of colon is formation of faeces and absorption of water.

The rectum is a short, tubular, thick–walled posterior part of the large intestine. It stores faeces. It opens into the cloaca. A sphincter is present between the two.

Cloaca: The cloaca is internally divided into three chambers: **coprodaeum, urodaeum** and **proctodaeum**. The coprodaeum is the anterior chamber and receives the rectum. It is constricted into several chambers. The urodaeum is the middle chamber and receives the ureters and gonoducts dorsally and the urinary bladder ventrally. The wall of urodaeum has a sphincter which regulates the flow of urine. The proctodaeum is the posterior chamber. It opens to the outside by the **cloacal aperture** which is a transverse slit present at the junction of the trunk and tail. The cloaca absorbs water from urine as well as faeces. This is a terrestrial adaptation as it serves to conserve water. The urine and faeces are converted into semisolid dehydrated **castings** or **pellets** which are egested through the cloacal aperture.

Digestive glands

In *Uromastix* there are five glands associated with the alimentary canal. They are **salivary glands**, **gastric glands**, **liver**, **pancreas** and **intestinal glands**.

Salivary glands: Keith is of the view that there are a few salivary glands which open into the buccal cavity and probably secrete mucus.

Gastric glands: The gastric glands are microscopic simple or branched structure present in the mucous membrane of the stomach. They discharge gastric juice into the cavity of the stomach.

Liver: The liver is a bilobed (Fig. 17.18), dark red gland present behind the heart and between the two lungs. It is connected with the stomach by a thin membranous fold, the **gastro-hepatic omentum**. The right lobe is larger, narrower and longer extending posteriorly up to the right gonad. The left lobe is short and broad lying ventral to the cardiac stomach. A small third lobe or **dorsal lobe** is found attached to the dorsal side of the left lobe. A small rounded **gall bladder** is present on the ventral side of the posterior end at the point of union of the two lobes. A **cystic duct** arising from the dorsal side of the gall bladder, a **hepatic duct** arising from the right liver lobe and a **hepatic duct** arising from the dorsal lobe join together to form the **first bile duct** which opens into the duodenum. A **second bile duct**

originates from the left liver lobe and opens independently into the duodenum. The liver secretes bile which is stored in the gall bladder.

Pancreas: The pancreas (Fig. 17.18) is a narrow whitish diffuse gland lying in the loop formed between the duodenum and the pyloric stomach. It secretes pancreatic juice which is discharged into the duodenum by a **pancreatic duct** which originates from the posterior end of the pancreas and opens into the duodenum just in front of the opening of the bile duct.

Intestinal glands: The intestinal glands are microscopic glands present in the mucous membrane of the small intestine. They discharge intestinal juice into the lumen of the intestine.

BLOOD VASCULAR SYSTEM

The blood vascular or circulatory system of *Uromastix* shows many changes that are associated with the adoption of air breathing habit. It is of closed type. It includes the **heart, arterial** and **venous systems** and **blood**.

Heart

The heart is a reddish coloured triangular organ situated midventrally in the anterior part of the thoracic cavity just below the sternum. It is longer than broad. The broad base is directed forwards and the pointed apex backwards. The heart is enclosed in a two layered, thin, transparent sac, the **pericardium.** The outer layer of pericardium is called **parietal pericardium** and the inner layer is **visceral pericardium.** A narrow **pericardial cavity** is enclosed between the two layers. It is filled with a watery **pericardial fluid** which allows free movement of the heart and also protects it from shocks and mechanical injuries. The apex of the heart is attached to the pericardium by a small, well developed cord, the **gubernaculum cordis.** The wall of the heart is made up of three layers: the outermost **tunica adventia**, the middle **tunica media** and the innermost **tunica intima,** made up of cardiac muscles.

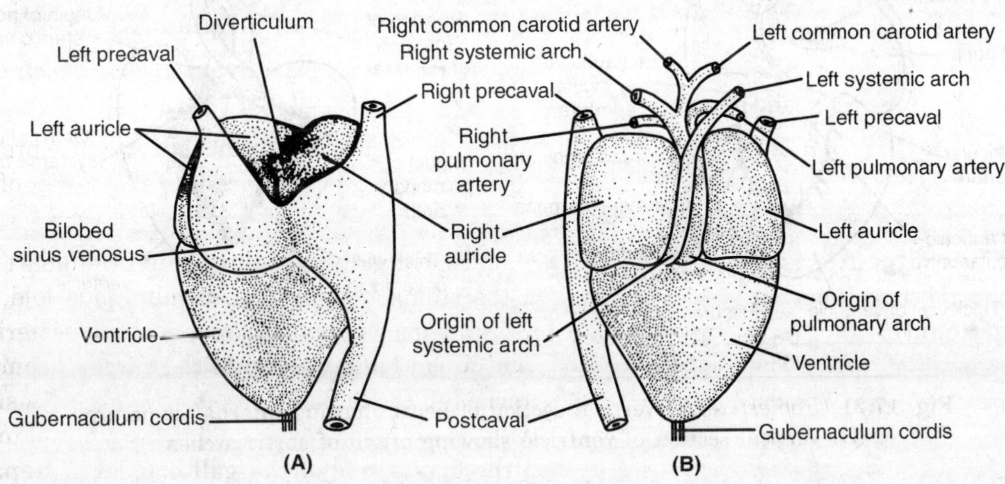

Fig. 17.20 *Uromastix*–Structure of heart. **(A)** Dorsal view; **(B)** Ventral view

The heart consists of four chambers: a **sinus venosus**, two **auricles** or **atria** and a **ventricle.**

Sinus venosus: The sinus venosus (Fig. 17.20) is a large, thin-walled, bilobed chamber present on the dorsal side of the heart lying transversely above the auricles but closely attached to the right auricle. It opens into the right auricle through an oval **sinu-auricular** or **sinu-atrial opening** which is not guarded by any valve, but the lips are thick and muscular and kept closed except when the blood is forced through it into the right auricle. At the anterior ends the sinus venosus receives two **precaval veins** and the posterior apex receives the large **postcaval vein**. They collect deoxygenated blood from the body and open by three separate apertures into the sinus venosus. Externally a deep groove divides the sinus venosus into a larger right and a smaller left lobe.

Auricles: The two auricles (Fig. 17.20) are thin-walled chambers externally marked by a faint longitudinal **interauricular groove.** They are externally separated from the ventricle by a transverse **auriculo-ventricular groove** or **coronary sulcus.** The right auricle is larger than the left. From the anteromedial border of the right auricle, on the dorsal side, a small **diverticulum,** of unknown function, is given off. The auricles open into the ventricle by **auriculo-ventricular aperture.** Internally the two auricles are separated by a thin, muscular and vertical partition, the **inter-auricular septum**. The anterior end of the septum remains attached to the wall of the auricles while the posterior end reaches the auriculo-ventricular aperture dividing it into right and left parts (Fig17.21A). The opening is guarded by **auriculo-ventricular valve** attached to the posterior free end of the interauricular septum and consisting of two **semilunar flaps,** one guarding the right opening and the other left, The inner lining of the wall of auricles is raised into a network of muscular ridges called **musculi pectinati.**

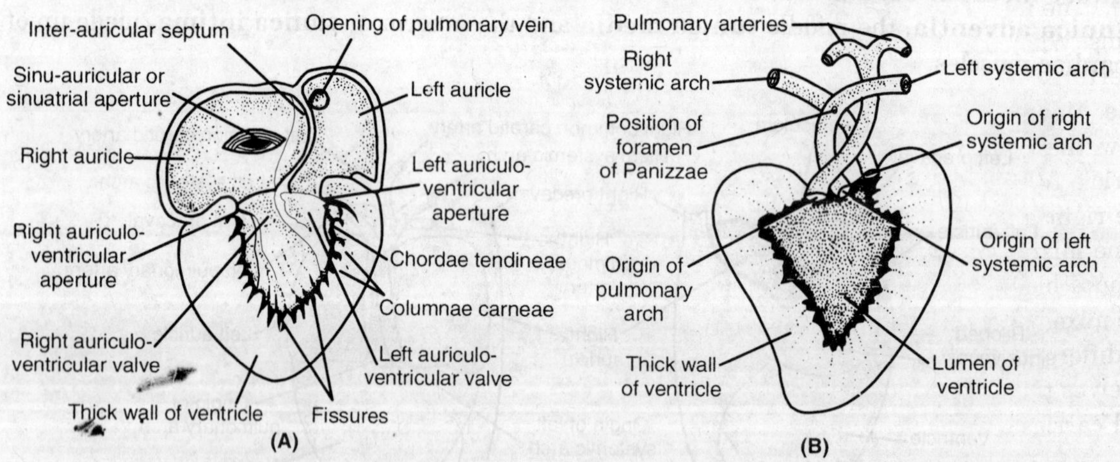

Fig. 17.21 *Uromastix*–**(A)** Vertical section of heart showing internal structure; **(B)** Vertical section of ventricle showing origin of aortic arches

The common pulmonary vein bringing oxygeneated blood from the lungs opens into the left auricle on the anterodorsal surface, close to the interauricular septum by a small circular aperture. In *Uromastix* a forward lip-like outgrowth of the dorsal wall of the auricle guards the pulmonary aperture. In most of the lizards, however, the opening is not guarded by any valve.

Ventricle: The ventricle (Fig. 17.20) is a conical, thick-walled, muscular portion of the heart posterior to the auricles. The apex of the ventricle is somewhat rounded and connected with the wall of the pericardium by a well developed ligamentous cord, the **gubernaculum cordis.** The right lateral surface of the ventricle is slightly concave while the left is evenly convex. The inner surface of its wall is thrown into a number of ridges, the **columnae carnae**, as a result its cavity is greatly reduced. Each auriculo-ventricular aperture is guarded by a single semilunar auriculo-ventricular valve which allows the flow of blood only in one direction, i.e., from the auricle to the ventricle. The free edges of the valves are connected to the inner wall of the ventricle by firm cords called **chordae tendinae**. A prominent **muscular flap** or **interventricular septum** arises from the ventral wall of the ventricle on the left side near the apex and extends horizontally forwards towards the right. It divides the cavity of the ventricle into a large dorsal part, the **cavum dorsale** on the left and a small ventral part, the **cavum pulmonale** on the right. The two parts are continuous with one another over the free margin of the ridges. The left side of the ventricle receives oxygenated blood from the left auricle and the right side receives deoxygenated blood from the right auricle. The chordae tendinae keep the blood of the two sides partially separated.

The three aortic arches arise independently from the ventricle (Fig. 17.21 B). The **pulmonary arch** arises from the cavum pulmonale while the systemic arches arise from the cavum dorsale. A semilunar valve is present at the base of each arch. It allows the blood to flow from the ventricle to the arch only.

Course of circulation

The sinus venosus receives deoxygenated blood from the whole body by the three caval veins. When it contracts the blood is forced into the right auricle through the sinu - auricular aperture. Owing to the spongy nature of the ventricle this blood is well retained in the right side. Meanwhile the left auricle receives oxygenated blood from the lung via pulmonary vein. Now the two auricles contract simultaneously forcing the oxygenated blood from the left auricle to the left side of the ventricle i.e., cavum dorsale and the deoxygenated blood from the right auricle to the right side of the ventricle, i.e., cavum pulmonale. There is, however some mixing of arterial and venous blood in the ventricle. When the ventricle contracts the impure blood from the cavum pulmonale goes to the lungs through the pulmonary arch and the mixed and pure blood from cavum dorsale goes into the systemic arches and is distributed to different parts of the body. The arrangement of systemic arches is such that the head receives the purest blood, as it has greater oxygen requirement and the dorsal aorta receives and carries mixed blood.

Arterial system

The **arteries** are blood vessels which carry pure blood from the heart to various organs of the body. All the arteries together form the **arterial system**. There are three aortic arches, the two **systemic** and a **pulmonary** (Fig. 17.22), which arise separately from the ventricle.

Pulmonary arch: The pulmonary arch originates from the right ventral side (the cavum pulmonale) of the ventricle, so it is fully visible in the ventral view of the heart. It runs forwards and divides into right and left **pulmonary arteries** which curve backwards and enter the right and left lungs respectively. They are the only arteries which carry deoxygenated blood to the lungs for purification.

Systemic arches: Both the systemic arches come off separately from the ventricle. The right systemic arch arises from the left side of the ventricle and carries oxygenated blood while the left systemic arch arises from the right side and carries mixed blood. The left systemic is fully visible in the ventral view of the heart whereas the right systemic in the dorsal view. The two systemic arches cross each other just in front of the heart. At the point of contact they communicate with each other through a **foramen of Panizzae.** The two systemic arches then curve backwards around the oesphagus and finally meet each other posterior to the heart in the mid-dorsal line forming a median vessel, the **dorsal aorta**. Before forming the dorsal aorta the right systemic arch gives off two important branches: **innominate** and **common subclavian** (Fig. 17.22).

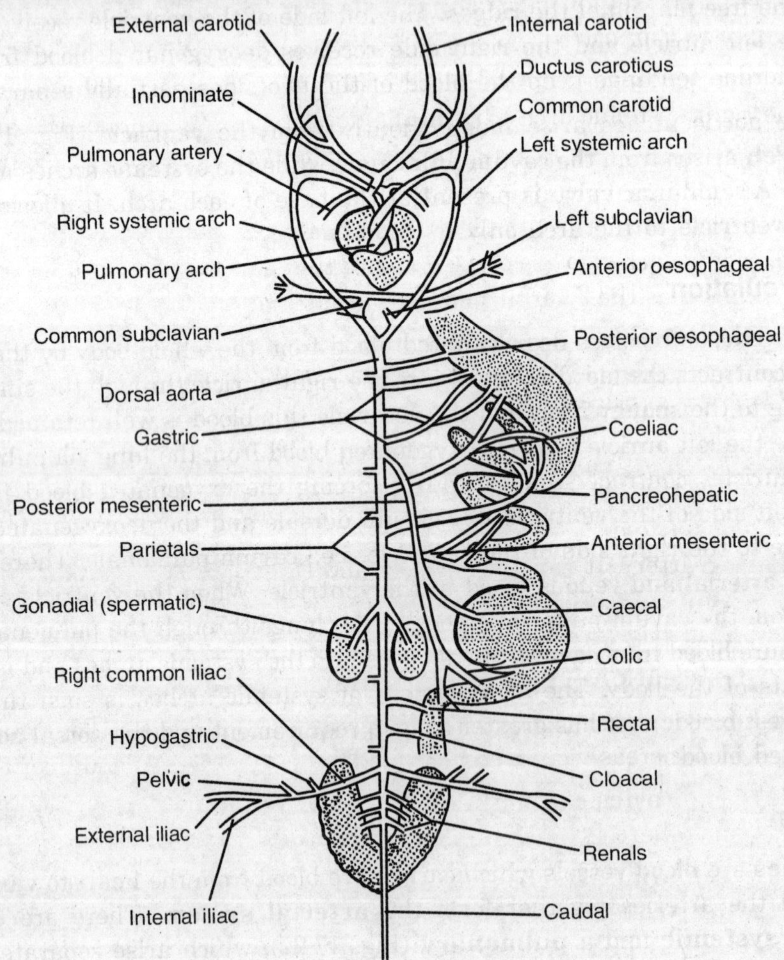

Fig. 17.22 *Uromastix*–Arterial system

Innominate: Before it starts curving to the right side, the right systemic arch gives off a very short artery called **innominate** or **carotis primordia** (**primary carotid**) which soon bifurcates into right and left **common carotid artery**. Both the common carotids run almost parallel to the systemic arch for a little distance, then turn forwards and divide into two branches, the **external** and **internal carotids.** The external carotid artery carries oxygenated blood to thyroid, oesophagus, trachea, floor of buccal cavity, larynx, tongue, hyoid and its muscles, etc. The internal carotid artery is a fairly large vessel which supplies oxygenated blood to the face, jaws and brain. The internal corotid of both sides is connected with the systemic arch of its side by a small vessel, the **ductus caroticus**. Since the right systemic arch gives rise to the carotid artery, it is also known as the **carotico-systemic arch**.

Common subclavian: The common subclavian arises from the right systemic arch a little distance before its union with the left systemic arch to form the dorsal aorta. It soon divides into right and left subclavian artery running on either side to supply the forelimbs, shoulder girdle and its muslces. Each subclavian divides into three arteries: (*i*) **scapular artery** to the scapular region, (*ii*) **coracoid artery** to the coracoid region and (*iii*) **brachial artery** to the arms.

The left systemic arch gives off a small branch, the **anterior oesophageal,** just before its union with the right systemic arch to form the dorsal aorta. It supplies the anterior part of the oesophagus.

Dorsal aorta: The dorsal aorta (Fig. 17.22) is formed by the union of the right and left systemic arches. It runs straight backwards, mid-dorsally beneath the vertebral column, giving off a number of branches to various organs of the body. It goes straight up to the tail region where it is known as the **caudal artery**. Various branches of the dorsal aorta are as follows:

1. **Posterior oesophageal artery:** It is the first artery arising from the dorsal aorta just after its formation. It supplies the posterior part of the oesophagus.
2. **Gastric arteries:** There are three gastric arteries supplying the stomach. They are small, unpaired and named as **anterior, middle** and **posterior** gastric arteries.
3. **Posterior mesenteric artery:** It is a large median artery arising a little behind the middle gastric artery. It runs posteriorly and supplies the large intestine by three branches: (*i*) **caecal** (caecum), (*ii*) **colic** (colon) and (iii) **rectal** (rectum).
4. **Coeliac artery:** It is a large unpaired vessel which supplies the stomach (**gastric**), pancreas (**pancreatic**) and spleen (**spleenic**).
5. **Pancreo-hepatic artery:** It is a fairly big vessel that divides into three branches supplying the pancreas (**pancreatic**), duodenum (**duodenal**) and liver (**hepatic**).
6. **Anterior mesenteric** or **intestinal artery:** It is an unpaired artery supplying the small intestine where it breaks up into four or five large branches.
7. **Genital or gonadial arteries:** These are a pair of arteries supplying the gonad. In male they are called **spermatic artery** and in female **ovarian artery.**
8. **Rectal artery:** An unpaired artery passing down along the dorsal wall of the rectum.

9. **Cloacal artery:** It is a small unpaired artery supplying the lateral wall of cloaca and the extreme posterior end of rectum.
10. **Common iliac arteries:** In the region of the sacral vertebrae arise a pair of common iliac arteries which supply the hindlimbs. Each artery breaks up into the following branches: (*i*) **hypogastric artery** to the urinary bladder, (*ii*) **pelvic artery** to the pelvic girdle; (*iii*) **external iliac artery** to the muscles of the thigh and running as **femoral** in the distal region, and (*iv*) **internal iliac artery** to the lower part of the leg.
11. **Renal arteries:** Three pairs of renal arteries arise at the level of the kidneys and supply blood to it.
12. **Caudal artery:** The posterior continuation of the dorsal aorta into the tail forms the caudal artery. It passes through the haemal canal of the caudal vertebrae.
13. **Parietal arteries:** Along its course the dorsal aorta gives off about 15 pairs of dorso-lumbar or parietal arteries at regular intervals. They supply the vertebral column, dorsal trunk and caudal muscles.

Venous system

The **veins** collect deoxygenated blood from different organs of the body and return it to the heart. All the veins together form the **venous system.** The venous system of *Uromastix* can be studied under the following heads:

1. **Pulmonary veins**
2. **Caval veins**
3. **Hepatic portal system**
4. **Renal portal system.**

Pulmonary veins: There is a single thin-walled pulmonary vein (Fig. 17.23) of moderate size which is actually formed by the union of several veins from both the lungs. It leaves the left lung anteriorly and runs along the ventrolateral side of the trachea. It passes dorsally to the sinus venosus and opens into the left auricle close to the interauricular septum. It is the only vein which carries oxygenated blood to the heart.

Caval veins: All the blood from different parts of the body is returned to the heart through three large venous trunks which enter the sinus venosus. These are the **right** and **left precaval veins** or **anterior venae cavae** and the single **post caval vein** or **posterior vena cava** (Fig. 17.23).

Precaval veins: There are two precaval veins—the right and the left, which collect venous blood from the anterior part of the body. Each precaval is formed by the union of four veins:

(i) **Common jugular vein**
(ii) **Subclavian vein**
(iii) **Intercostal vein**
(iv) **Laryngo-tracheal vein.**

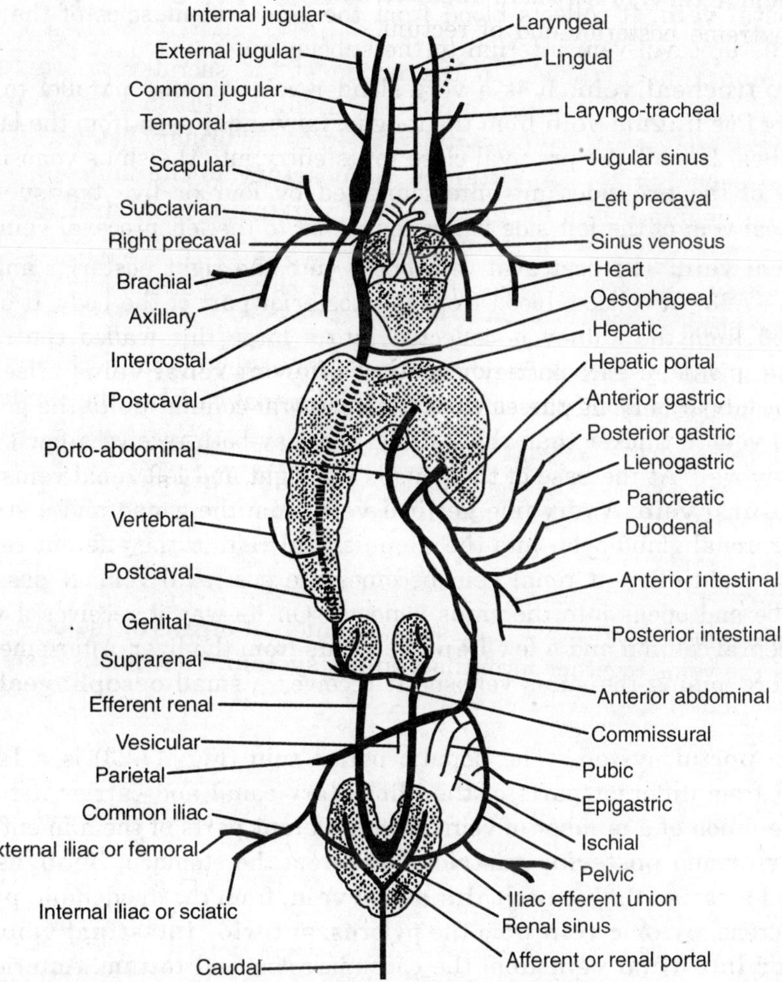

Fig. 17.23 *Uromastix*–Venous system

Common jugular vein: It is formed by the union of **external** and **internal jugular veins** (Fig. 17.23) at the level of the tympanum. The external jugular vein passes forward by the side of the trachea and receives blood from the upper jaw by a **maxillary vein** and from the lower jaw by a **mandibular vein.** The internal jugular vein receives blood from the brain by a **cerebral vein** and from the orbital region by an **orbital vein.** The common jugular vein then runs back into the neck where it receives a **temporal vein** from the auditory region. At the base of the union of the common jugular vein and temporal vein there is a large **jugular sinus**. The left precaval receives blood only from the internal jugular vein and the subclavian, the external jugular vein is absent (Bhatia, 1929).

Subclavian vein: It is a large vein that joins the posterior end of the jugular sinus. It collects blood from the forelimbs by three small veins: *(i)* **Scapular** from the shoulder region, *(ii)* **Brachial** from the arms and *(iii)* **Axillary** from the armpits and thorax.

Intercostal vein: It collects blood from the skin and muscles of the chest and runs forwards to the precaval vein internal to the subclavian.

Laryngo-tracheal vein: It is a very slender vein running parallel to the trachea. It receives a very fine **lingual vein** from the tongue, **laryngeal vein** from the larynx and a vein from the trachea. It joins the precaval close to its entry into the sinus venosus. The laryngo-tracheal vein of the two sides are interconnected by four or five **transverse loops.** The laryngo-tracheal vein of the left side is not connected to the left precaval vein (Bhatia, 1929).

Postcaval vein: The postcaval vein opens into the right posterior angle of the sinus venosus (Fig. 17.23). It collects blood from the posterior part of the body. It originates in the kidneys. Blood from the kidney is collected into a large thin-walled central **renal sinus** present in the middle of each kidney. A pair of **efferent renal veins** arise from the renal sinus and run forwards along the sides of the vertebral column up to the gonads. The right efferent renal vein is thicker than the left. On the way both receive a few small veins from the dorsal body wall. At the base of the gonads the right and left renal veins join each other by a **commissural vein**. A very fine **genital vein** from the gonad and a **suprarenal vein** from the suprarenal gland open into the commissural vein or the efferent renal vein of that side. Only the right efferent renal vein becomes the postcaval vein. It passes through the right liver lobe and opens into the sinus venosus. On its way it receives a **vertebral vein** from the vertebral column and a few **hepatic veins** from the liver. After emerging out of the liver and before joining the sinus venosus it receives a small **oesophageal vein** from the oesophagus.

Hepatic portal system: The hepatic portal vein (Fig. 17.23) is a large vein which collects blood from different parts of the alimentary canal and carries it to the liver. It is formed by the union of a number of veins from different parts of the alimentary canal. They are the **anterior** and **posterior gastric veins** from the stomach, **lienogastric vein** from the spleen and a part of the stomach, **duodenal vein,** from the duodenum, **pancreatic vein** from the pancreas, **pyloric vein** from the pylorus, **anterior intestinal vein** from the ileum and **posterior intestinal vein** from the caecum, colon and rectum. Anteriorly the hepatic portal vein receives a branch of the anterior abdominal vein to form the **portoabdominal vein** which enters the left liver lobe. The blood from the liver is collected by the **hepatic veins** which open into the postcaval vein.

Renal portal system: The venous blood from the tail is collected by the **caudal vein** which starts from the tip of the tail and runs ventral to the caudal artery through the haemal canal of the caudal vertebrae receiving on its way a number of branches from the tail muscles (Fig. 17.23). It then enters the kidneys and splits up into two **afferent renal** or **renal portal veins** running forwards over the ventral surface of the kidneys. In the matrix of the kidneys the renal portal veins break up into fine capillaries. Each renal portal vein receives a **cloacal vein** from the side of the cloaca and a **rectal vein** from the dorsal region of the rectum. Anteriorly each renal portal vein receives a fairly large vein, the **common iliac,** from the hindlimbs, somewhere in the middle of the kidney. The two are connected by a small **transverse connection**, the **iliac-afferent union.** The common iliac is formed by the union of two branches: the **external iliac** or **femoral vein** collecting venous blood from the thigh and the **internal iliac** or **sciatic vein** from the lower leg. The **pelvic veins** arise antero-laterally from the common iliacs. The two pelvic veins meet each other mid-ventrally in front of the kidneys to form the **anterior abdominal vein.** During its course each pelvic vein receives several small **parietal veins** from the posterior muscles, **vesical vein** from the bladder, **ischial vein** from the ischium, **pubic vein** from the pubic region and **adipose vein** from fat bodies. The anterior abdominal vein is double at places indicating a primitive feature. At the anterior end the two components of the anterior abdominal vein remain separate, one enters the left liver lobe directly while the other joins the hepatic portal vein to form the **porto-abdominal vein** before entering the left liver lobe. The blood from the liver is collected by hepatic veins which open into the postcaval vein.

Blood

The blood of *Uromastix* presents the same structure as that of frog. It consists of a fluid matrix or **plasma** in which different types of blood **cells** or **corpuscles** float. The **red blood cells** or **erythrocytes** are oval, biconvex, nucleated and contain haemoglobin. The **white blood cells** or **leucocytes** vary both in size and form. The cytoplasm is colourless and the nucleus is usually spherical. They have independent power of locomotion and are amoeboid. The red blood cells of reptiles have a longer life than that of mammals. Their haemoglobin concentration and oxygen carrying capacity is almost half of mammals. The reptiles are **cold blooded** or **poikilothermic** animals.

RESPIRATORY SYSTEM

In reptiles exchange of gases cannot take place through the skin (cutaneous respiration) as in frog because it is dry, tough and horny. The modern reptiles do not, at any time of life, breathe by gills. On the walls of temporary pharyngeal clefts minute papillae have been observed which have been interpreted as vestiges of gill filaments and no functional gills occur in repitilian embryo. In adult reptiles respiration is entirely **pulmonary,** i.e., by **lungs,** except in some cases, such as in sea turtles where buccopharyngeal respiration takes place. The respiratory system of *Uromastix* consists of two parts: **respiratory tract** and **respiratory organs** (Fig. 17.24).

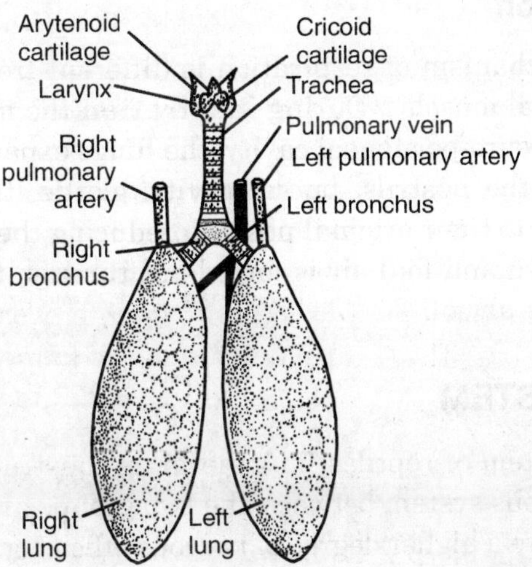

Fig. 17.24 *Uromastix*–Respiratory system

Respiratory tract

The respiratory tract begins from the **external nares** present on the tip of the snout. The external nares lead into small sacs called **nasal chambers** which open into the buccal cavity by a pair of **internal nares.** On the floor of the buccal cavity a small, median, longitudinal slit called **glottis** is present. It leads into a small box-like chamber, the **larynx** (Fig. 17.24) which is supported by a pair of **arytenoid cartilages** present in the antero-lateral wall and a median **cricoid cartilage.** The larynx is provided with two pairs of muscles: the inner **musculus compression laryngis** which serves as a sphincter of the glottis and the outer **musculus dilator laryngis** which serves as a dilator of the glottis. The vocal cords are absent in *Uromastix.* The larynx leads into a long cylindrical tube, the **trachea** which runs backwards through the neck ventral to the food pipe. The wall of trachea is provided with complete rings of cartilage, the **tracheal rings,** which prevent it from collapsing. The posterior end of the trachea bifurcates to form two small and narrow tubes, the **bronchi** which are also supported by complete cartilaginous rings. Each bronchus enters the lung of its side through its anterior end. It does not branch into bronchioles but directly form **alveoli** inside the lungs.

Respiratory organs

The respiratory organs of *Uromastix* are a pair of **lungs** (Fig. 17.24). They are elongate, fusiform, hollow sacs of orange colour lying on either side of the heart, dorsal to it and extending posteriorly along the outer margin of the liver lobes. The wall of lungs is thin and elastic. The inner surface is raised into a network of fine **ridges** or **septa** enclosing small air sacs or **alveoli.** The alveoli have a rich supply of blood capillaries. Gaseous exchange takes place in the alveoli. The anterior part of the lungs has more alveoli and is more vascularised than the posterior part, so the anterior part is more respiratory than the posterior part. The posterior part serves as a reservoir for residual air.

Mechanism of respiration

In *Uromastix* the mechanism of respiration is different from that of frog. It is affected by ribs and their inter-costal muscles. During **inspiration** the muscles pull the ribs outward increasing the size of the pleuro-peritoneal cavity, the lungs expand and fresh air from outside enters the lungs through the nostrils, buccal cavity, glottis, trachea and bronchi. During **expiration** the ribs return to their original position reducing the size of the pleuro-peritoneal cavity, the lungs are pressed and foul air is forced out through the same passage. Exchange of gases takes place in the alveoli.

NEUROSENSORY SYSTEM

The neurosensory system of reptiles, in general, is constructed on the amphibian plan. The differences in the nervous system between the two groups are not apparent anatomically, but reptiles however, indicate a higher degree of nervous efficiency. One of the major differences is the absence of the lateral line system in reptiles. This system is peculiarly adapted to aquatic mode of life and as reptiles are purely terrestrial, this system is gone. In amphibians its traces are present. The absence of this system is fully compensated by improvement in other sensory organs. With the appearance of a horny layer some other integumentary sense organs are lost.

NERVOUS SYSTEM

As in other vertebrates the nervous system of *Uromastix* consists of the following parts:

1. Central nervous system
2. Peripheral nervous system
3. Autonomic nervous system.

Central nervous system

The central nervous system includes the **brain** and **spinal cord.**

Brain

The brain of *Uromastix* is quite simple (Fig. 17.25), just like that of frog. It lies inside the cranial cavity. It is surrounded by two **membranes** or **meninges:** the outer membrane is **duramater** and the inner is **piamater.** In between the two membranes **arachnoid fluid** is present, it protects the brain from mechanical injuries. Histologically the brain is made up of two types of nervous tissues: the **grey matter** composed of nerve cells and **white matter** composed of nerve fibres. In the forebrain the grey matter lies external to the white matter while in the midbrain and the medulla oblongata this arrangement is reversed. The brain consists of the usual three divisions: forebrain, midbrain and hindbrain.

Forebrain: The forebrain is the anterior part of the brain. It consists of **olfactory lobes, cerebral hemispheres** and **diencephalon**.

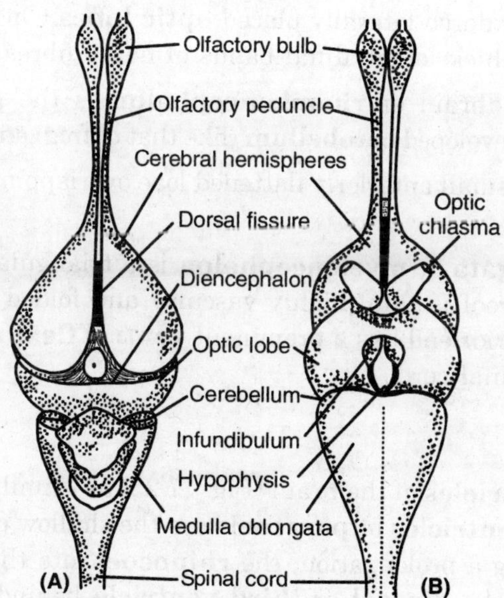

Fig. 17.25 *Uromastix*–Structure of brain, **(A)** Dorsal view; **(B)** Ventral view

The two **olfactory lobes (telencephalon)** are the anterior-most part of the forebrain. In *Uromastix* they are smaller in size than that of frog. Each olfactory lobe consists of two parts: **olfactory peduncle** and **olfactory bulb.** The olfactory peduncle is a narrow and elongated stalk-like structure dilated in front into a small nodule-like **olfactory bulb** from which the olfactory nerve originates.

The **cerebral hemispheres** are a pair of oval bodies narrow in front and broad behind. They are closely applied to each other but clearly demarcated by a mid-dorsal groove called **median longitudinal fissure**. The posterior end of the olfactory peduncle is attached to the narrow anterior end of the cerebral hemisphere. The roof of the cerebral hemispheres is called **pallium**. It is thinner than the floor and lateral walls which together constitute the **corpus striatum**. The corpora striata of the two hemispheres are connected by a transverse **anterior commissure** above which lies the **hippocampal commissure** connecting the hippocampal region of both the hemispheres.

The **diencephalon** is a small rounded area between the cerebral hemispheres and the midbrain. It is hardly visible from the dorsal side. Its lateral walls are thickened forming **optic thalami.** The roof is thin, non-nervous and highly vascular forming the **anterior choroid plexus**. Behind the optic thalami there is a transverse band of nerve fibres called the **posterior commissure.** From the dorsal surface of the diencephalon arises a median outgrowth the **pineal apparatus.** It consists of two parts: anterior **parietal organ** or **parietal eye** and posterior **pineal body** or **epiphysis.** In most lizards the parietal eye is a eye-like structure. The thick floor of diencephalon is called **hypothalamus.** The **infundibulum** arises from the ventral surface of diencephalon behind which occurs the **pituitary body**. The two optic nerves cross each other forming the **optic chiasma** which lies in front of the infundibulum.

Midbrain: The **midbrain** or **mesencephalon** is the middle part of the brain. It consists of a pair of large, rounded, dorso-laterally placed **optic lobes.** On the ventral surface of the optic lobes there are two thick longitudinal bands of nerve fibres called **crura cerebri.**

Hindbrain: The hindbrain or **rhombencephalon** is the posteriormost part of the brain. It consists of poorly developed **cerebellum** (like that of frog) and the **medulla oblongata**.

The **cerebellum** is a small anteriorly flattened lobe overlapping the antero-dorsal surface of the medulla oblongata.

The **medulla oblongata** or **myelencephalon** is a triangular structure broad in front and tapering behind. Its roof is thin, highly vascular and folded and forms the **posterior choroid plexus**. Its posterior end has a prominent **ventral flexure** where it passes into the anterior portion of the spinal cord.

Cavities of the brain

The **cavities** or **ventricles** of the brain (Fig. 17.25) are similar to those of the frog. The first and second **lateral ventricles** or **paracoels** are the shallow cavities within the cerebral hemispheres, each sending a prolongation, the **rhinocoel** into the olfactory bulb anteriorly and communicating with the **diocoel** or **third ventricle** behind by a small aperture, the **foramen of Monro**. The diocoel is the laterally compressed cavity of the diencephalon. The

optocoel is the cavity of each optic lobe. Both the optocoels are connected by a narrow passage, the **iter**. The **metacoel** or the **fourth ventricle** is the cavity of the medulla oblongata. It is connected with the diocoel through the iter.

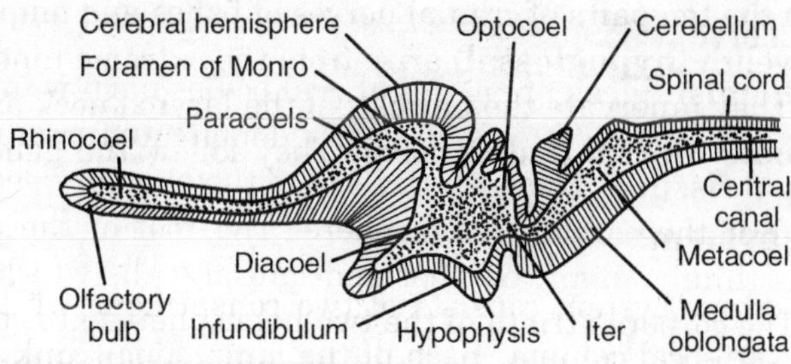

Fig. 17.26 *Uromastix*–Sagittal section of brain showing ventricles

The brain of reptiles shows a relative increase in the size of the cerebral hemispheres and the olfactory bulbs are reduced. The increase in the size of the hemispheres indicates an increase in instinctive activities. The young animals have not to be taught. The animal automatically and inevitably carries out the activities which are normal and characteristic of the species. The behaviour of reptiles is certainly on a higher level than that of amphibians in general.

The different parts of the brain perform different functions. The olfactory lobes control the sense of smell. The cerebral hemispheres are the seat of intelligence, memory, emotions and will. The diencephalon acts as a relay station in conveying the impulses to the cerebral hemispheres and as integrating centre for the autonomic nervous system. The optic lobes are concerned with the sense of sight. The cerebellum controls voluntary movements and also maintains equilibrium. The crura cerebri carry sensory impulses from the medulla oblongata and spinal cord to the optic thalami and then to the cerebral hemispheres. The medulla oblongata controls involuntary movements.

Spinal cord

The spinal cord is a white coloured, dorsoventrally flattened tube running through the neural canal of the vertebral column. It has the typical vertebrate structure but closely resembles that of frog. A very shallow groove, the **dorsal fissure** extends along the mid-dorsal line of the cord and a prominent **ventral fissure** extends along the mid-ventral line. The spinal cord encloses a narrow central cavity, the **central canal** which opens anteriorly into the cavity of the medulla oblongata but is closed posteriorly. It is filled with the cerebro-spinal fluid. The meninges covering the brain are continuous over the spinal cord.

In section the spinal cord consists of characteristic H-shaped **grey matter** with prominent **dorsal** and **ventral cornua** or **horns** and **white matter** which surrounds the grey matter. The grey matter is mainly composed of nerve cells and non-medullated nerve fibres. The **dorsal** and **ventral roots** of the spinal nerves arise from the dorsal and ventral horns of grey matter. The white matter is divided into **dorsal, ventral** and **lateral funiculi.** It consists of medullated nerve fibres. The spinal cord controls reflex activities.

Peripheral nervous system

The peripheral nervous system consists of twelve pairs of **cranial nerves** from the brain and sixteen pairs of **spinal nerves** from the spinal cord. First ten pairs of cranial nerves correspond with the ten pairs of cranial nerves of fishes and amphibians. The eleventh (**accessory**) and the twelfth (**hypoglossal**) arise from the extreme hind end of the brain and behind the tenth pair. They innervate the muscles of the larynx, neck and tongue. The spinal nerves arise by two **roots, dorsal** and **ventral**. They follow the general pattern of frog.

Autonomic nervous system

The autonomic nervous system consists of two conspicuous white trunks that extend along each side of the vertebral column. Each of the autonomic trunk is connected with the spinal nerves on the same side of the body by a number of ganglionic enlargements. The autonomic nervous system carries on the usual sensory and motor functions. It helps to regulate involuntary reactions.

SENSE ORGANS

The sense organs of *Uromastix* have not been studied. The following description applies to the reptiles in general and to *Lacerta* in particular. Not much is known about the organs of touch and taste.

Olfactory Organs

In reptiles, on the whole, the organs of taste are lacking. This is compensated by better development of **nasal** or **olfactory organs** (Fig. 17.27) which are concerned with the sense of smell. The nasal organs are a pair of **olfactory sacs** or **nasal chambers** which communicate with the exterior by a pair of **external nares** and internally with the mouth cavity by a pair of **internal nares.** The epithelial lining of the nasal chambers is smooth and occupied by specialised olfactory sensory cells. In some reptiles the surface area is increased by folding. Each olfactory cell is a narrow elongated cell with a nucleus in the middle, fine sensory processes at the free end and a fine nerve fibril at the other end.

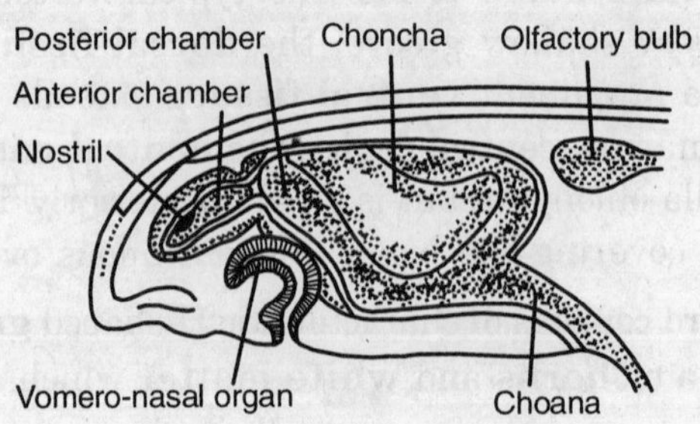

Fig. 17.27 *Uromastix*–Sagittal section of nead showing nasal chambers and Jacobson's organ

In reptiles, the nasal cavity is provided with an accessory olfactory organ called **vomero-nasal organ** or **organ of Jacobson** (Fig. 17.27). It develops as a ventral hollow outgrowth from each nasal cavity forming a pair of small pouches lying in the anterior region of the roof of the mouth and each opening into the oral cavity by a narrow **nasopalatine duct.** The two ducts open close together just behind the front teeth and anterior to the internal nares. The Jacobson's organs are vascular and lined with olfactory epithelium. They are innervated by branches of olfactory and trigeminal nerves. The Jacobson's organs are best developed in lizard, snakes and *Sphenodon* but absent in adult crocodiles. Probably they serve to smell the food when taken into the buccal cavity. The scent particles are introduced into them by the tongue tip. They also help in recognition of enemies, in locating the member of opposite sex, etc.

Eyes

The eyes of reptiles are more advanced than those of amphibians due to terrestrial mode of life. They are large, globular and lodged into spacious orbits. There are two movable **eyelids** covered with small scales. The lower eyelid is more movable than the upper. The eyelids protect the eyes from dust, heat and rain. A semitransparent third eyelid, the **nictitating membrane** is present in the anterior corner of the eye. It can be pulled rapidly over the cornea. It protects and cleans the eye. Two glands are present in the eye. The **harderian gland** present on the anterior side of the eye ball lubricates the nictitating membrane and the **lacrymal gland** present on the posterior side of the eyelid keeps the eyes moist and clean.

The structure of eyeball resembles that of the frog. It is made up of usual three layers: **fibrous tunic**, **choroid** and **retina** (Fig. 17.28).

The fibrous tunic is thick and tough and is divided into two distinct regions: the small transparent anterior portion is called the **cornea** and the large opaque posterior portion is called **sclerotic.** The cornea bulges forwards. A thin transparent epithelial layer called **conjunctiva** is fused with the outer surface of the cornea. The sclerotic is cartilaginous and is supported from outside by a ring of 11 bony **sclerotic plates** or **ossicles**.

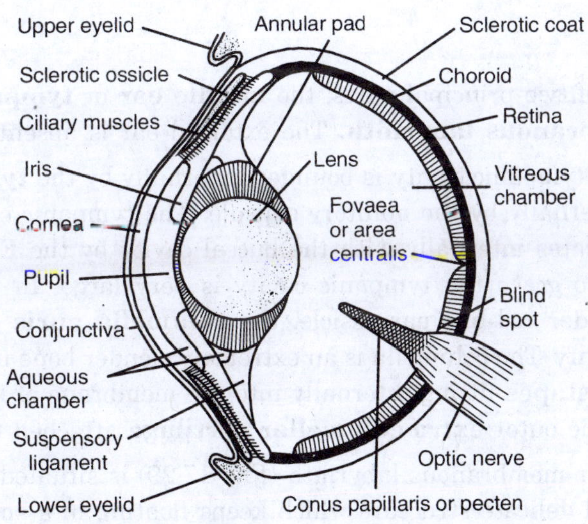

Fig. 17.28 *Uromastix*–Vertical section of eye

The choroid is thin, vascular and pigmented and lies in close contact with the sclerotic. It serves to darken the eye. At the junction of the sclerotic and cornea, the **ciliary body** is present. Just in front of the cavity of the eye **iris** is present. It is perforated in the centre by an aperture, the **pupil.** The iris possesses circular and radial muscles which reduce and enlarge the pupil respectively. The iris also contains pigments which givè it the characteristic light orange colour.

The retina lies inside the choroid. The portion of the retina lying in contact with the choroid is known as the **optical part**. It is thick and sensitive and contains more **cones** than **rods**, thus there is a good daylight vision and colour perception but little night vision. Some yellow **oil droplets** are found in this part to protect it from excessive exposure by filtering the light at the blue end of the spectrum. In lizards the cones are of unique type and are called **double cones**. The remaining part of the retina is thin and non-sensory. The part that lines the ciliary body is known as the **ciliary part** and that lines the iris is the **iridial part**. At the junction of ciliary and optical parts the retina is irregular and is called **ora serrata**. The optic nerve arises from the inner surface of the retina at a point called **blind spot** as no image is formed in this part. It emerges after piercing the three coats at the back of the eye and joins the brain. The part of the retina where distinct image is formed lies opposite to the centre of the lens and is called the **yellow spot** or **area centralis**. Usually a cushion-like, vascular, pigmented process called **pecten** projects from the blind spot into the vitreous chamber. It probably provides nourishment to the retina.

The **lens** is a biconvex transparent body lying behind the iris. It is more convex behind than in front. A ring of soft tissue called **annular pad** is present around the lens. Well-developed **ciliary muscles** and **ciliary pocesses** arise from the ciliary body and form **suspensory ligaments** attached to the annular pad. They keep the lens in position. The lens and suspensory ligaments divide the cavity of the eye into two compartments: the anterior small **aqueous chamber** filled with a watery fluid, the **aqueous humour** and the large posterior **vitreous chamber** filled with **vitreous humour**.

Ear

The ear consists of two principal parts: the **middle ear** or **tympanic cavity** and the **internal ear** or **membranous labyrinth.** The external ear is absent.

The middle ear or tympanic cavity is bounded externally by the **tympanic membrane** or **tympanum** and internally by the auditory capsule. The tympanic cavity is a fluid-filled cavity which communicates internally with the buccal cavity by the **Eustachian tube**. In large species like *Gecko gecko* the tympanic cavity is very large. In *Uromastix* also it is relatively large. A slender rod-like ear ossicle, the **columella auris** extends horizontally across the tympanic cavity. The columella is an extremely slender bone made up of two parts: an inner bony rod, the **stapes** fitting internally into the membrane of fenestra ovalis of the auditory capsule and the outer **extra-columellar cartilage** attached to the tympanum.

The internal ear or membranous labyrinth (Fig. 17.29) is situated within the auditory capsule. It is a soft and delicate structure which keeps floating in a small space filled with a fluid called **perilymph**. It is made up of a bag-like structure, the **vestibule,** the upper cylindrical part of which is known as **utriculus** and the lower large rounded part is the **sacculus**. Three **semicircular canals,** one **horizontal** and two **vertical,** one end of each is enlarged to form an **ampulla,** open into the utriculus. The sacculus gives rise to a small flat lobe called **pars lagena** which represents the cochlea of mammalian ear. It contains a special organ of hearing, the **papilla basilaris** which corresponds with the organ of corti in mammals. A narrow tube, the **ductus endolymphaticus** leads upwards from the sacculus and ends blindly over the hindbrain. According to some workers another sensory area, the **macula lagenae** is present near the apex of the duct, but its function is not known.

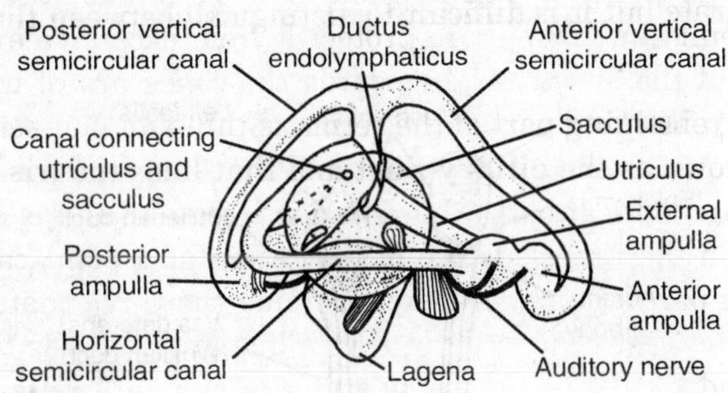

Fig. 17.29 *Uromastix*–Internal ear (outer view)

URINOGENITAL SYSTEM

In *Uromastix,* as in other vertebrates, the **excretory** and **reproductive** systems are closely associated and are together described as the **urinogenital system.**

EXCRETORY SYSTEM

The excretory system is similar in both the sexes. It includes a pair of **kidneys**, a pair of **ureters** and a **urinary bladder.**

Kidneys: The kidneys (Fig. 17.30) are **metanephric**. They are dark red, irregular-shaped bodies lying in the posterior part of the abdominal cavity in close contact with the dorsal body wall. Ventrally they are covered with **peritoneum**. Each kidney is made up of two lobes: a broad **anterior** and a narrow **posterior lobe.** The anterior lobes are free and apart while the posterior lobes are united together so that the kidneys appear V-shaped. A pad of connective tissue is present between the posterior lobes of the two kidneys. Histologically each kidney is made up of **connective tissue** containing blood capillaries, nerve fibres and muscle fibres. Numerous narrow coiled **uriniferous tubules** lie embedded in the connective tissue. All the tubules open into a common urinary duct, the **ureter**.

Ureters: From the ventral surface of each kidney arises a narrow, delicate metanephric duct or ureter (Fig. 17.30). Both the ureters run backwards and open into the middle chamber (urodaeum) of the cloaca. The opening is situated on a small papilla.

Urinary bladder: The urinary bladder is a median thin-walled sac opening into the urodaeum of cloaca ventrally (Fig. 17.30).

Physiology of excretion: The reptiles are uricotelic, that is, they excrete uric acid. The excretory waste consists of uric acid and urate particles which are not much soluble in water. They leave the kidneys in the form of yellowish urine which passes through the ureters into the urodaeum and from there into the urinary bladder where it is stored. Water is reabsorbed from the cloaca and urinary bladder so that urine is discharged in semi-solid state containing mainly uric acid.

REPRODUCTIVE SYSTEM

The sexes are separate but it is difficult to distinguish between the two sexes from mere external appearance.

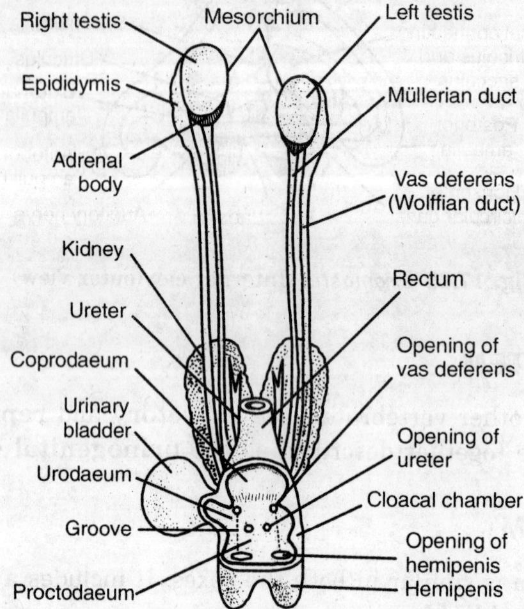

Fig. 17.30 *Uromastix*–Male urinogenital system

Male

The male reproductive organs include a pair each of **testes, vas deferens** and **copulatory sacs** or **hemipenes.**

Testes: The testes (Fig. 17.30) are two white oval bodies situated in the body cavity quite ahead of the kidneys. Both testes are not on the same level. The right testis is anterior to the left and is situated just behind the right lobe of the liver. Each testis is suspended by a fold of peritoneum, the **mesorchium** that keeps it attached to the bodywall. The **adrenal gland** is attached to each testis posteriorly. Internally each testis is made up of a number of coiled **seminiferous tubules**, the epithelial lining of which produce **spermatozoa**.

Vasa deferentia: The seminiferous tubules of each testis unite to form bigger tubes called **vasa efferentia.** These unite to form a thick convoluted mass, the **epididymis** lying close to the outer posterior surface of the testis. The epididymis continues behind as a narrow convoluted tube the **vas deferens** or **sperm duct** or **Mullerian duct** (Fig. 17.30). It runs posteriorly along the dorsal body wall and opens into the urodaeum of the cloaca independent and in front of the opening of the ureter. According to Das (1960) the vestigeal **Mullerian duct** runs parallel to the vas deferens of each side from the testis to the urodaeum.

Copulatory sacs or **hemipenes**. The hemipenes (Fig. 17.30) are a pair of eversible hollow sacs lying midventrally under the skin behind the cloacal aperture at the base of the tail. Normally they lie contracted and appear like small swellings but when erected they become cylindrical. They are made up of erectile tissue. Each hemipenis consists of a proximal hard **stalk** or **pedicel** and a distal soft **head** or **glans**. A longitudinal **groove** or **sulcus** runs to them from the opening of the vas deferens for the conveyance of the sperm. During copulation only one of the organ is inserted but which one is immaterial. This, however, depends upon the side the male happens to be during copulation. The erection of the hemipenis depends partly on muscular action and partly engorgement of blood sinuses in the tissue of the hemipenis.

Female

The female reproductive organs are a pair of **ovaries** and a pair of **oviducts.**

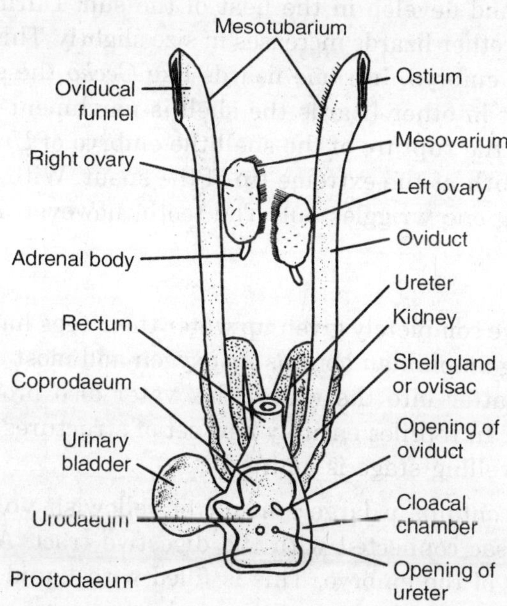

Fig. 17.31 *Uromastix*–Female urinogenital system

Ovaries: The ovaries (Fig. 17.31) are irregularly oval-shaped, yellowish-white organs situated in the posterior part of the body cavity, anterior to the kidney (not as much as the testis). The right ovary is somewhat anterior to the left. A fold of peritoneum called the **mesovarium** keeps the ovary attached to the dorsal wall of the body cavity. The outer surface of each ovary is raised into rounded protuberances or **follicles,** each containing an **ovum**. The ova are in different stages of development.

Oviducts: The oviducts or **Müllerian ducts** (Fig. 17.31) are two long, thin-walled and much folded tubes independent of the ovaries. A fold of peritoneum, the **mesotubarium** keeps the oviduct of each side attached to the dorsal body wall. The wide anterior end of each oviduct is called **oviducal funnel**. It opens into the body cavity by a longitudinal opening called **ostium** which is directed outwards. The margin of ostium is ciliated. The two oviducts run posteriorly and as they pass over the ventral surface of the kidney each dilates to form a **shell gland** or **ovisac** where the ova are stored and provided with egg shell. The terminal part of the oviduct is called **vagina**. It opens dorsally into the proctodaeum of the cloaca anterior to and independent of the opening of the ureter.

Copulation and fertilisation

During the breeding season, that is, March and April, the male and female undergo courtship followed by copulation which occurs on a warm and sunny day between 10 a.m. to 3 p.m. The male becomes agile and agressive and fights for female. During copulation the hemipenes of the male are everted and one is inserted into the cloaca of the female to transfer the sperms. Fertilisation is internal. Most lizards are oviparous. The female *Uromastix* lays about 10–15 shelled eggs in burrows. Development is direct.

Eggs: The eggs of *Uromastix* are large measuring about 20–30 mm in size. The eggs are dirty white in colour and develop in the heat of the sun. During incubation period the egg of *Uromastix* and some other lizards increases in size slightly. This is because of absorption of moisture and growth of embryo. In some lizards like *Gecko* the shell contains calcareous deposits and is brittle, but in other lizards the shell is parchment-like and contains only a small amount of lime. For the rupture of the shell the embryo of *Uromastix* is provided with a sharp calcareous **egg-tooth** at the extreme tip of the snout. With the help of this the shell is ruptured and the young one wriggles out. The tooth, however, is shed after birth.

Development

Since the reptiles have completely given up water their eggs have developed complicated structures. The developing amphibian obtains its oxygen and most of its food from the water and excretes its waste matter into the water; the water also protects it from drying and against mechanical injury. In reptiles entirely new set of structures have developed for these functions as the water dwelling stage is omitted.

The egg of reptiles contains a large amount of yellowish **yolk**, which as the embryo grows, is contained in a sac connected with the digestive tract. A large sac, the **amnion**, develops around the body of the embryo. This is filled with liquid and protects the embryo against injury and desiccation. From the posterior end grows out a tubular sac, the **allantois**, in which the waste matter of the body is deposited. The entire egg is protected by a firm **shell** which is porous. Beneath the shell lies a membrane, the **chorion**, richly supplied with blood vessels. This acts as a respiratory surface taking in oxygen and giving off carbon dioxide. These structures satisfy all the needs of the developing organism.

The cleavage in reptiles is **meroblastic**. In such eggs most of the egg mass consists of inert yolk, while the protoplasm is confined to a small area capping the yolk at the animal pole. Cleavage and blastula formation are confined to this clear protoplasmic area, the yolk does not cleave at all. The first and second cleavages pass through the centre of the animal pole, but they are not complete cleavages of the egg. Additional equatorial divisions separate a cluster of cells at the pole from more peripheral members. Further subdivisions, rather irregular, form a **blastodisc** or **blastula**. The cell mass tends to be slightly lifted off the underlying yolk, the space formed below is considered to be equivalent to the **blastocoel**. The blastula, however, is a flat sheet and not a sphere. The potential embryonic regions are distributed over the flat blastodisc as in the shark. Over the periphery, the disc is tightly adherent to the yolk. In one region (the future posterior end of the embryo) the margin of the disc is for a time free growing lip. At this end, as in sharks, cells from the dorsal surface roll inward to the under surface of the disc and extend forward forming a thin sheet of endoderm. Thus a two layered disc is formed. This is the primary act of **gastrulation.** The free posterior margin of the disc is comparable to the dorsal lip of the blastopore.

A different procedure is adopted for mesoderm formation. On the dorsal surface of the disc the **primitive streak** shortly appears, extending forward from the region of the blastoporal lip. It consists of parallel ridges with a grove between them. At the front end of the groove develops a pit which extends forward and downward as a canal to pierce the roof of the underlying gut cavity. The primitive streak is a region of crucial activity in the formation of the embryo. On the dorsal surface of the disc there is a steady movement of cells medially and posteriorly into the margin of the primitive streak. In the walls of the streak the sheets of cells move downward and then fan outward to interpose themselves between ectoderm and endoderm. Anteriorly, the cells taking part in this movement push forward from the pit region to form the notochord, laterally the material moves outward to form the somites and other mesodermal structures. Obviously the primitive streak (although not an open blastopore) functions as a blastopore in forming the mesoderm. The primitive streak persists for a considerable period during embryonic development. The development of reptiles, in other ways, is similar to that of the birds.

18 CHAPTER

REPTILIA:
GENERAL CHARACTERS AND CLASSIFICATION

The reptiles are the first vertebrates adapted for life on dry land. They flourished and attained maximum development during the Mesozoic era. By the end of the Triassic most of the orders were established and some became extinct. The evolution of reptiles was rapid. They have radiated towards all kinds of habitats, that is, land, water and air. It is now definitely known that the reptiles arose from the amphibians and in turn gave rise to birds and mammals. The characters of reptiles are, in fact, a combination of characters found in fishes and amphibia on one hand and in birds and mammals on the other hand. The reptiles are cold blooded animals which breath by lungs throughout their life and have body covered with scales. In winter they become dormant and hibernate in crevices. The class name **Reptilia** refers to the mode of movement (**L.**, *reptum*, to creep or crawl). The study of reptiles is known as **Herpetology (Gr.**, *herpeton*, reptile).

GENERAL CHARACTERS

1. Reptiles are cold blooded, terrestrial or aquatic, creeping or burrowing animals.
2. Skin is dry and without any glands.
3. Body is provided with an exoskeleton of scales, shields, plates or scutes.
4. Body is bilaterally symmetrical and divided into four regions: head, neck, trunk and tail.
5. Limbs two pairs (**Tetrapoda**), pentadactyle, with clawed digits. Limbs absent in snakes and some lizards.
6. Endoskeleton bony. Skull articulates with the vertebral column by a single median occipital condyle. (monocondylic).
7. Mandibles made up of many small pieces and articulate with the cranium by the quadrate bone.
8. A characterstic T-shaped interclavicle is present.
9. Mouth terminal, simple, conical teeth present on the jaws. In turtles teeth are replaced by horny beaks.
10. Alimentary canal terminates in a cloacal aperature.
11. Reptiles are carnivorous.

12. Respiration is by lungs.
13. Heart with two auricles and incompletely divided ventricle, four chambered in crocodiles. The red blood cells are oval and nucleated.
14. Brain is more developed than that of amphibians. Twelve pairs of cranial nerves present. Lateral line system absent. Jacobson's organ present in the roof of the mouth.
15. Kidneys metanephric. Excretion uricotelic.
16. Sexes separate. Males with a pair of copulatory sacs or hemipenes.
17. Fertilization internal. Eggs laid on land. Development direct. Young ones resemble the adults.
18. Embryonic membranes (amnion, chorion, allantois and yolk sac) are formed during development.
19. No parental care.

CLASSIFICATION

The class **Reptilia** can be divided into five subclasses on the basis of presence or absence of certain openings in the temporal region of the skull (Fig. 18.1). It is represented by about 7,000 living species placed in 4 out of the 16 orders that flourished during the Mesozoic era. Rest of the orders have become extinct.

Subclass I: ANAPSIDA

Skull roof solid, temporal opening absent (Fig. 18.1 A).

Order 1. Cotylosauria

1. Extinct. Carboniferous to Triassic.
2. Resemble Labyrinthodont amphibians.
3. Limbs feeble and project laterally.
4. Jaws and teeth poorly developed.

Examples: *Seymouria*, *Camptorhinus*.

Order 2. Chelonia (Gr., *chelone*, turtle) or Testudinata (L., *testudo*, turtle)

1. Permian to recent
2. Body short, broad and oval, encased in a shell of rounded dorsal carapace and flat ventral plastron.
3. Limbs pentadactyle with clawed and/or webbed digits, or paddle-like.
4. Thoracic vertebrae and ribs usually fused with the carapace.
5. Skull solid, quadrate immovable, sternum absent.
6. Jaws without teeth, covered with horny sheath.
7. Cloacal apereture is longitudinal.
8. Copulatory organ single.
9. Marine forms are **turtles**, freshwater forms **terrapins** and lands forms **tortoises.**

Examples: *Chelone*, *Chrysemys*, *Trionyx*, *Testudo*, *Dermachelys*.

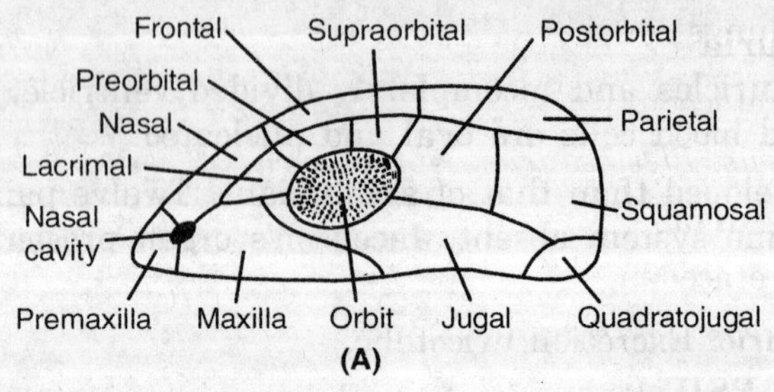

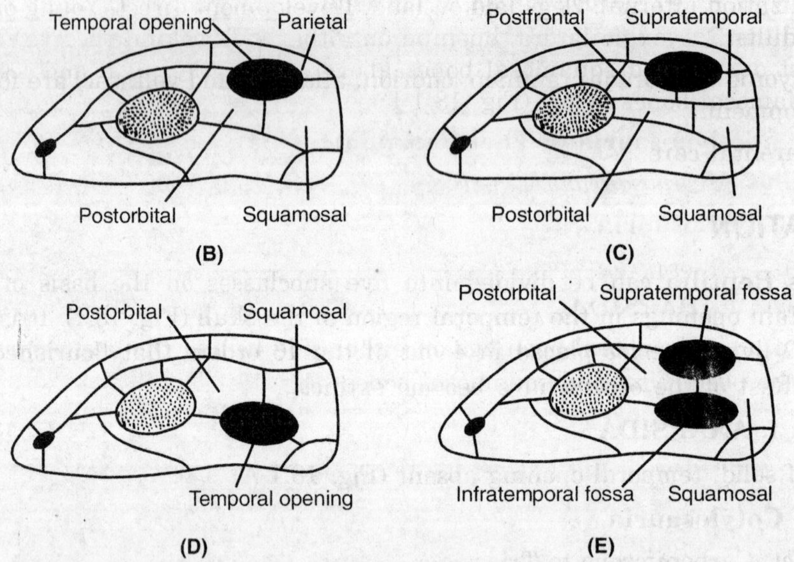

Fig. 18.1 Different types of reptilian skull in lateral view. (A) Anapsida; (B) Euryapsida; (C) Parapsida; (D) Synapsida; (E) Diapsida

Subclass II: EURYAPSIDA

1. Extinct.
2. Skull with single dorso-lateral temporal opening on either side behind the eye, bounded below by post-orbital and squamosal bones and above by the parietal bone (Fig. 18.1 B).

Example: *Plesiosaurus*.

Subclass III: PARAPSIDA

1. Extinct. From Mesozoic to Cretaceous.
2. Skull with single dorso-lateral temporal opening on either side behind the eye, bounded below by supratemporal and postfrontal bones and above by the parietal bone (Fig. 18.1 C).

Order 1. Protosauria

From Permian to Triassic.

Examples: *Araeoscelis*, *Tanystropheus*.

Order 2. Plesiosauria

From Triassic to Cretaceous.

Examples: *Pilosaurus, Plesiosaurus, Nothosaurus.*

Order 3. Ichthyosauria

From Triassic to Cretaceous.

Examples: *Mixosaurus, Ichthyosaurus.*

Subclass IV: SYNAPSIDA

1. Extinct. From Carboniferous to Permian.
2. Skullwith a single lateral opening on either side behind the eye, bounded by the post orbital and squamosal bones above and by the jugal and lower part of the squamosal bones below (Fig. 18.1 D).

Order 1. Pelycosauria (= Theromorpha)

Examples: *Varanosaurus*, *Dimetrodon.*

Order 2. Mesosauria

Example: *Mesosaurus.*

Subclass V: DIAPSIDA

Skull with two temporal openings on either side behind the eye separated by a bar formed by the postorbital and squamosal bones (Fig. 18.1 E).

Superorder I: LEPIDOSAURIA

1. Two temporal vacuities present.
2. Humerus bone with two foramina.

Order 1. Eosuchia

Extinct. Permian to Eocene.

Examples: *Youngina*, *Prolacerta.*

Order 2. Rhyncocephalia (**L.,** *rhynchos*, snout + **Gr.,** *kephale*, head).

1. Lizard-like burrowing reptiles with elongated body.
2. Body covered with granular scales, a row of mid dorsal spines present.
3. Limbs pentadactyle, with clawed digits.
4. Vertebrac amphicoelous. Numerous abdominal ribs present. Ribs single headed and with uncinate process.
5. Parietal foramen with vestigeal pineal eye present.
6. Quadrate bone immovable. Teeth acrodont.
7. Cloacal opening transverse slit.
8. Copulatory organ absent in male.

Examples: *Sphenodon punctatum*, only living species.

Order 3. Squamata (**L.,** *squama*, scale)

1. Includes lizards and snakes.
2. Body covered with horny epidermal scales and shields.
3. Limbs clawed, absent in snakes and some lizards.

4. Quadrate movable. Vertebrae procoelous. Ribs single-headed. Teeth acrodont or pleurodont.
5. Cloacal aperature transverse.
6. Copulatory organ paired and eversible.

Suborder 1. Lacertilia or **Sauria**

1. Includes lizards. Jurassic to recent.
2. Terrestrial, burrowing or arboreal.
3. Limbs and girdles usually well developed.
4. Skull with single occipital condyle. Jugal bone present.
5. Eyelids movable. Nictitating membrane, ear openings and tympanum present.
6. Maxillae, palatines and pterygoids fixed. The two rami of mandible fused in front. Premaxillae bear teeth.
7. Shows caudal autotomy with regeneration.
8. Sternum present.
9. Urinary bladder present.

Examples: *Hemidactylus*, *Calotes*, *Chaemeleon*, *Uromastix*, *Draco*, *Phrynosoma*, *Varanus*, *Heloderma*, *Gecko*, *Mabouia*, *Iguana.*

Suborder 2. Ophidia or **Serpentina**

1. Includes snakes. Cretaceous to recent.
2. Body elongated, slender and narrow.
3. Limbs and girdles absent. Pelvic girdle is present in some snakes like *Boa* and *Python*.
4. Terrestrial or aquatic, burrowing or arboreal.
5. Skull with triple occipital condyle. Jugal bone absent.
6. Eyelids immovable. Nictitating membrane, ear openings and tympanium absent.
7. Maxillae, palatines and pterygoid freely movable and help in biting mechanism. The two rami of the mandible joined by an elastic ligament and can be widely stretched to swallow a large prey. Premaxillae without teeth.
8. Sternum absent.
9. Do not show caudal autotomy.
10. Urinary bladder absent.
11. Tongue bifid and protrusible.

Examples: *Python*, *Naja*, *Typhlops*, *Eryx*, *Boa*, *Vipera*, *Hydrophis*, *Crotalus*, *Bungarus*.

Superorder II: ARCHOSAURIA

1. The upper temporal vacuity of the skull closed but both temporal arches present.
2. Humerus with single foramina.

Order 1. Pseudosuchia (= Thecodontia)

Extinct. Dominant in the Triassic.

Example: *Euparkeria*.

Order 2. Saurischia

Extinct. Triassic to Cretaceous.

Examples: *Brontosaurus*, *Diplodocus*.

Order 3. Ornithischia

Extinct. Triassic to Cretaceous.

Examples: *Iguanodon*, *Stegosaurus*.

Order 4. Pterosauria

Extinct. Jurassic to Cretaceous.

Example: *Pteranodon*.

Order 5. Crocodilia (Gr., ***krokodeilos*****, crocodile)**

1. Large-sized, aquatic, carnivorous reptiles with a long laterally compressed tail having denticulated dorsal crest.
2. Body covered with transverse rows of bony plates and epidermal scales.
3. Limbs short and powerful, forelimbs with five and hindlimbs with four clawed and webbed digits, modified for swimming.
4. Quadrate fixed, parietal foramen absent. Maxillae, palatines and pterygoid united along the midde to form a pseudopalate.
5. Teeth thecodont.
6. Vertebrae procoelous. Sternum present.
7. Ribs two-headed. Thoracic ribs with uncinate process. Abdominal ribs present.
8. Heart four chambered (2 auricles and 2 ventricles).
9. Lungs well developed and spongy.
10. Cloacal aperture longitudinal.
11. Male with single median, erectile and grooved penis.

Examples: *Crocodylus*, *Alligator*, *Gavialis*.

ORDER RHYNCOCEPHALIA

Rhyncocephalia (L., ***rhynchos*****, snout, Gr.,** ***kephale*****, head)** is an order of class Reptilia in which the front part of the head forms an overhanging beak. All the memberes of this

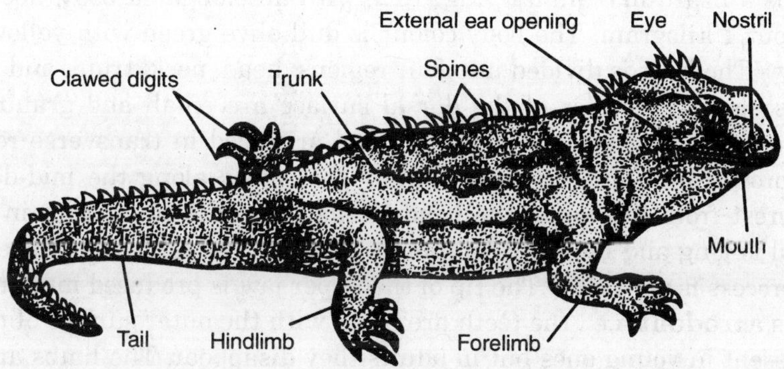

Fig. 18.2 *Sphenodon punctatum*

group became extinct in the Mesozoic era and today it is represented by a single species *Sphenodon punctatum* found only in New Zealand, hence it exhibits isolated distribution. Due to increase in population of man and other carnivorous mammals it has now become restricted to some small neighbouring islands in the Bay of Plenty, North Island. It has survived in its present home in isolation due to absence of natural enemies.

Sphenodon is a primitive lizard and is regarded as the last living witness of by gone ages. It has two pairs of vacuities in the temporal region of the skull, so it is included in the subclass Diapsida. It is popularly known as **tuatara** or **hatteria**.

Fossil rhyncocephalians

The rhyncocephalians appeared during the Cretaceous period of the Mesozoic era and spread throughout the world. One family Rhyncosaurs flourished during the Triassic period and some members attained considerable size, for example, *Scaphonyx* was about 2 feet high and weighed about 200 pounds. After Triassic period the rhyncocephalians became very much restricted. The genus *Homoeosaurus* of the Jurassic period was very similar to the *Sphenodon* of today which indicates that the rhycocephalians have not changed much since their origin.

Living fossil

Sphenodon is one of the oldest living reptile. It retains many primitive characters and has not undergone much change. Again in the course of time all its relatives have become extinct. It is, therefore, often called a **living fossil.** It is also on the verge of extinction, so it is strictly protected by law.

Habit and habitat

Sphenodon lives in small holes or burrows of petrels (sea birds), but it is very fond of water and lies submerged in water for several hours. Hence it is semiaquatic. It is carnivorous and feeds largely on worms, insects, spiders, molluscs, frog, occasionally on lizards and baby sea birds. It is a sluggish animal nocturnal in habit. It crawls slowly and stops frequently to look around. During the day it basks in the sun. In captivity its lifespan is almost 50 years.

External features

Sphenodon is a lizard-like animal (Fig. 18.2) with an elongated body, about 2 feet long and weighing about 1 kilogram. The body colour is dull olive green with yellow spots above and whitish below. The body is divided into four regions: head, neck, trunk and tail. The body is covered with scales. The scales of the dorsal surface are small and granular while the ventral scales are large squarish and plate-like and arranged in transverse rows. The eyes are large, dark brown and with a vertical pupil. The scales along the mid-dorsal line are modified into a crest from the top of the head to the tip of the tail except in the region of the neck. The tail is long and laterally compressed. If lost, the tail can be regenerated as in lizards, but the process is very slow. The tip of the upper jaw is produced into an overhanging beak. Dentition is **acrodont**, i.e., the teeth are fused with the outer surface of the jaw bones. The teeth are present in young ones but in adults they disappear. The limbs are pentadaclyl with five clawed digits. At the junction of the trunk and tail a transverse slit-like cloacal aperture is present. The copulalory organ is absent in male *Sphenodon.* It lays ten eggs in holes in the ground. The eggs hatch in thirteen months.

Endoskeleton

The **skull** is typical diapsidan type with two pairs of temporal fossae— a supratemporal fossa behind the orbit and an infratemporal fossa lateral to the supratemporal fossa. An extra-temporal vacuity, the post-temporal fossa bounded by post-temporal bar is also present in the skull. The lower jaw is made up of six bones on each side and the dentaries are united by ligament. The quadrate is immovable.

The **vertebral column** is divided into five regions: cervical, thoracic, lumbar, sacral and caudal. The vertebrae are **amphicoelous** with persistent notochord. A proatlas vertebra is present between the skull and atlas. The articulation of atlas and axis helps in side to side movement of the head. The caudal vertebrae have a chevron bone (a Y-shaped haemal arch). They are divided by a septum.

The ribs are present in the neck, thorax and lumbar region. All the ribs are single headed, attached to the vertebrae and bear a backwardly directed cartilaginous uncinate pocess. In the thoracic region they are attached to the sternum and in the abdominal region they are embedded in the muscles of the body wall. A **sternum** is present. The anterior end of the sternum is attached to the coracoids and interclavicles of the pectoral girdles and the thoracic ribs are attached to the posterior end of the sterum.

The **pectoral girdle** has a T-shaped median interclavicle, two clavicles which are joined to the interclavicle. The scapula is fused with the procoracoid and has an acromian process. The suprascapula is cartilaginous and has a clavicular process. The **pelvic girdle** is similar to that of lizard but the two ilia are erect. A large ischiopubic foramen is present between ischium and pubis. A cartilaginous hypoischium is present behind ischium and the two pubes form pubic symphysis.

Circulatory system

The heart is typically reptilian, but the three main aortic arches instead of emerging independently from the ventricle come out by a short common trunk. The arterial and venous systems show certain interesting points of similarity with that of Urodela.

Parietal eye

One interesting feature of *Sphenodon* is the presence of **parietal** or **pineal** or **third eye**. It is lodged in the parietal foramen of the skull. On the forehead there is a rosette of scales. The central scale is transparent and the parietal eye lies beneath it. The parietal eye develops as an outgrowth from the roof of the diencephalon. It is made up of lens and retina and is connected to the brain by a nerve. It shows a tendency towards degeneration. The actual role of pineal eye is not known. It was probably functional in extinct forms. Some lizards possess similar pineal eye but have no nerve supply.

Affinities

Sphenodon shows affinities with dinosaurs, crocodiles, Chelonia, Lacertilia and Amphibia.

Affinities with Dinosaurs

Similarities

1. Skull with two pairs of vacuities.
2. Quadrate immovable.
3. Ribs have uninate process.
4. Abdominal ribs present.

Dissimilarities

1. Ribs single headed in *Sphenodon* but double headed in dinosaurs.
2. Teeth acrodont in *Sphenodon* but thecodont in dinosaurs.
3. Clavicle and interclavicle present in *Sphenodon* but absent in dinosaurs.
4. Parietal eye present in *Sphenodon* but absent in dinosaurs.

Affinities with Crocodilia

Similarities

1. Quadrate bone immovable.
2. A proatlas vertebra present between the skull and atlas vertebra.
3. Skull diapsidan type.
4. Ribs with uncinate process.
5. Caudal vertebrae have chevron bone.
6. Abdominal ribs present.

Dissimilarities

1. Teeth acrodont in *Sphenodon* but thecodont in crocodile.
2. Nasal opening double in *Sphenodon* but single in crocodile.
3. Vertebrae amphicoelous in *Sphenodon* but procoelous in crocodile.
4. Clavicle present in *Sphenodon* but absent in crocodile.
5. Pecten absent in *Sphenodon* but present in crocodile.
6. Copulatory organ absent in *Sphenodon* but present in crocodile.

Affinities with Chelonia

Similarities

1. Quadrate immovable.
2. Caudal ribs fused with caudal vertebrae.
3. A process from parietal reaches the squamosal.
4. Urinary bladder present.
5. Pecten is absent.

Dissimilarities

1. Sternum present in *Sphenodon* but absent in Chelonia.
2. Cloacal aperture transverse in *Sphenodon* but longitudinal in Chelonia.
3. Copulatory organ absent in *Sphenodon* but present in Chelonia.
4. The oviduct opens dorsally in *Sphenodon* but ventrally in Chelonia.

Affinities with Lacertilia

Similarities

1. The general body plan is same in both.
2. A proatlas is present.
3. Caudal vertebrae separated by septum and have a chevron bone.
4. Some lizards like *Gecko* have amphicoelous vertebrae.
5. Vertebrae with persistent notochord.
6. Ribs are single headed.
7. Skull diapsidan.
8. Cloacal glands present.
9. Parietal organ present.
10. Oviducts open dorsally.

Dissimilarities

1. An extra vacuity is present in *Sphenodon*.
2. Quadrate immovable in *Sphenodon* but movable in lizard.
3. The rami of jaws are joined by ligament in *Sphenodon* but a symphysis is present in lizard.
4. The vertebrae are amphicoelous in *Sphenodon* but procoelous in all lizards except *Gecko*.
5. Abdominal ribs and uncinate process is present in *Sphenodon* but absent in Lacertilia.
6. Conus arteriosus present in *Sphenodon* but absent in Lacertilia.
7. Pecten absent in *Sphenodon* but present in lizard.
8. Copulatory organ absent in *Sphenodon* but present in Lacertilia.

Affinities with Amphibia

1. In *Sphenodon* the three aortic arches arise from a short common trunk representing the conus arteriosus of the amphibians.
2. In *Sphenodon* ductus arteriosus and ductus caroticus are present, a character which occurs in urodele amphibians and nowhere else.
3. *Sphenodon* also remains immersed in water for long periods like the amphibians.

Unique features of *Sphenodon*

1. Presence of overhanging beak.
2. Presence of parietal eye.
3. Absence of copulatory organ.
4. The eggs hatch in 13 months.

Primitive features of *Sphenodon*

1. Dentition acrodont and homodont.
2. Vertebrae amphicoelous with persistent notochord.
3. Presence of abdominal ribs.
4. Presence of conus arteriosus (amphibian feature).
5. Retension of ductus caroticus and ductus arteriosus (amphibian feature).

Conclusion

Sphenodon has many primitive features and also shows similarities with Amphibia, but it shows many reptilian characterstics. Due to its similarities with lizards it was earliar placed with the lizards. On the basis of anatomical differences between *Sphenodon* and lizards Gunther in 1860 placed it in a separate order, Rhycocephalia. Inspite of its primitive ness *Sphenodon* is neither degenerate nor an ancestral form. It has remain unchanged since the time of its origin.

ORDER CHELONIA

The order **Chelonia** or **Testudinata** include all living anapsids in which the skull has no temporal openings. The Testudinata is the oldest name available for this group and was given by Betesch in 1788 while Chelonia was used by Macartney in 1802. The order Chelonia is represented by **turtles, tortoises** and **terrapins** (Fig. 18.3) which are described as **'reptiles**

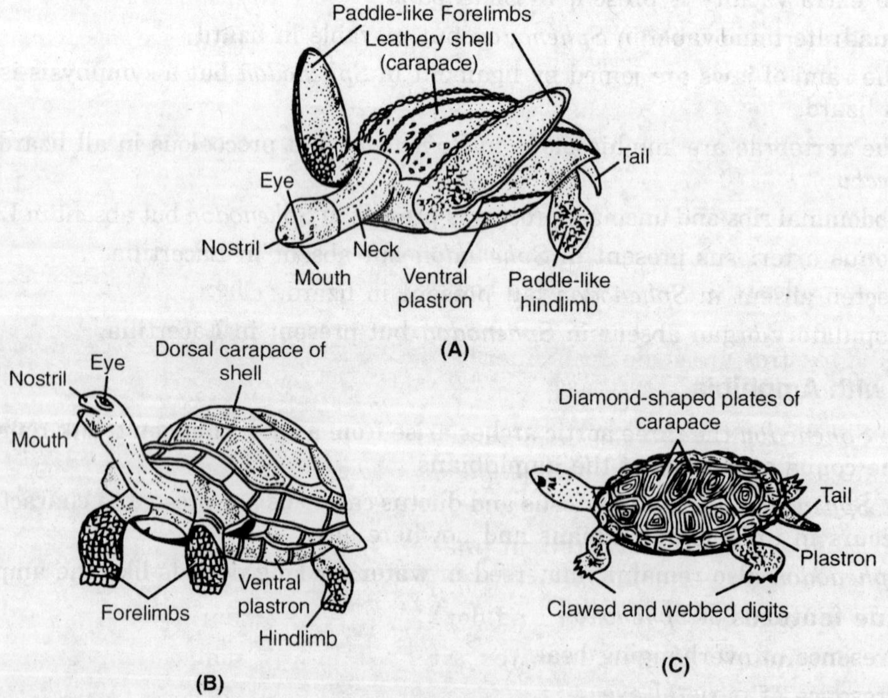

Fig. 18.3 *Chelonians*–**(A)** *Dermochelys coriacea* (leather back turtle); **(B)** *Testudo abingdoni* (giant tortoise); **(C)** *Malacolemys* (diamond back terrapin)

in boxes' as their body is enclosed in a box-like exoskeleton of dorsal **carapace** and ventral **plastron.** There is much inconsistency in the use of the words turtle, tostoise and terrapin. The wholly terrestrial forms are called tortoises, the marine forms are called turtles and terrapin is used for some fresh water forms which are edible. The chelonians are ancient reptiles which appeared in the Triassic and have remained unchanged for almost 175 million years.

Habit and habitat

The chelonians are widely distributed throughout the tropical and temperate zones. They are found in sea, freshwater and marshy land. They are found at the bottom of the sea but can swim successfully. They crawl slowly on land. They come out of water to lay eggs on land. Being cold blooded they hibernate during winter and aestivate in summer. They can survive for long periods without food by lowering their metabolic activity. In times of danger the animal withdraws its head, neck, limbs and tail inside the shell.

Shape and size

The body is more or less elliptical and dorsoventrally flattened. The size is quite variable. Some may be as small as 15 cm. The largest living marine turtle is leatherback turtle, *Dermochelys coriacea,* which is about 2.5 metres long and weighs 600 kg. The largest land form is the giant tortoise, *Testudo abingdoni*, of the Galapagos Island. It may weigh more than 100 pounds. The green edible sea turtle, *Chelonia mydas* is 1.5 metre long and weighs 200 kg.

Body

The body is short and broad with a prominent head, a long retractile neck, trunk and a short tail. The head is small and retractile, with a short and pointed or rounded snout projecting beyond the lower jaw. A pair of nostrils is placed close together on the tip of the snout. The eyes are lateral and have three eyelids. The tympanic membrane is present at the back of the jaw.

Shell

The body of chelonians is enclosed in a bony shell which is their characterstic feature. It consists of a dorsal, convex, shield-like carapace and a ventral, flat, plate-like plastron. Both the carapace and plastron are made up of inner bony plates and outer horny epidermal shields or scutes.

The earapace is made up 51 inner **bony plates** formed by the **ossification** of the **dermis.** Some of them are attached to the ribs and vertebrae. The different plates (Fig. 18.4 A) forming the inner surface of the carapace are:

1. Eight median **neural** or **vertebral plates** fused with the neural spines of second to ninth thoracic vertebrae.
2. A large **nuchal plate** anterior and dorsal to the first neural plate.
3. Three **postneural** or **pygal plates** are present just behind the eighth neural plate. The first two pygal plates, just behind the eighth neural plate, are called **suprapygal** plates. They are attached with the eighth costal plate. The third pygal joins the posterior marginal plates.

4. There are eight **costal plates** on each side of the neural plates. They are firmly united with the ribs below them. and by their inner and outer borders with the neural and marginal plates, respectively.
5. There are eleven **marginal plates** on each side forming the margin of the carapace.
6. An unpaired **postero-median marginal plate** may also be present. It is usually absent or vestigeal.

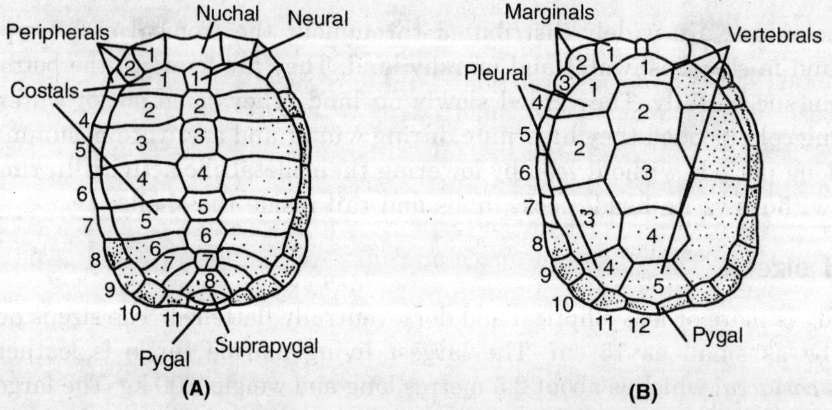

Fig. 18.4 *Carapace*–**(A)** Bony plates; **(B)** Epidermal scules

Externally the carapace is covered by horny **epidermal shields** (Fig. 18.4 B) which are formed by **cornification** of the **epidermis.** They are arranged in a regular fashion. They do not correspond either in position or number with the underlying bony plates. They include:

1. A median row of five **vertebral shields**
2. Four **costal shields** on either side of the vertebrals.
3. A row of twelve **marginal shields** on either side of the costals.
4. An anterior median marginal called **nuchal shield** present in front of the first vertebral and a posterior median marginal called **pygal** or **supracaudal shield** present behind the fifth vertebral.

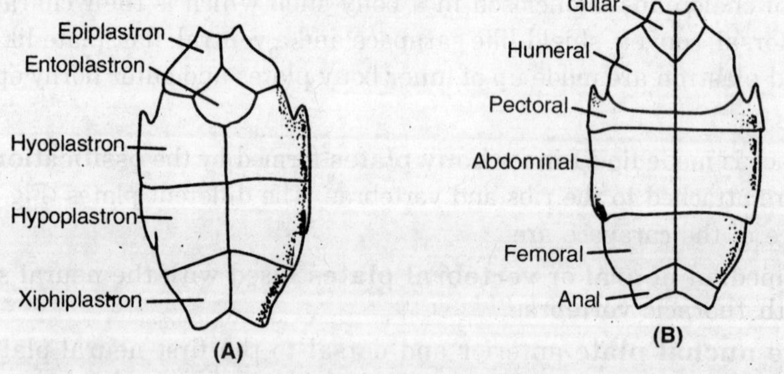

Fig. 18.5 *Plastron*–**(A)** Bonyplates; **(B)** Epidermal scules

The plastron is attached to the marginal plates of the carapace by bridges. It is made up of nine inner **bony plates** (Fig. 18.5 A): a median **entoplastron,** a pair each of **epiplastron,**

hyoplastron, hypoplastron and **xiphiplastron** begining from the anterior end. The entoplastron is embedded between the epiplastron and hypoplastron and is regarded as homologous with the interclavicles. The epiplastron is regarded as homologous with the clavicle and the remaining are modifications of abdominal ribs.

The outer surface of plastron is made up of twelve **epidermal shields** (Fig. 18.5 B). They are arranged in pairs and named **gular, humeral, pectoral, abdominal, femoral** and **anal shields** from the anterior end. In some an **intergular** (sometimes paired) may be present.

The shell of chelonians has brought several important changes in the internal organisation of the animal. The supporting function of the vertebral column is taken over by the solid carapace, as such the muscles attached to the vertebral column are lost or modified. The shoulder girdle has come to lie within the ribs instead of outside them, and therefore, the humerus lies in a horizontal plane with the forearm almost at right angles to it. This is the most disadvantageous arrangement but this is so to extricate the limbs from the shell. The lungs are encased between the carapace and plastron, and are limited still further in their movement by being attached by their dorsal surface to the inner lining of the shell.

Limbs

The limbs are short, pentadactyl and clawed. They are modified according to their habitat (Fig. 18.3). In the marine forms they are modified into flippers or paddles for active swimming. In fresh water forms the digits are webbed and in terrestrial forms the limbs are short and elephant-like to bear the body weight.

Skeleton

The skull is without temporal opening (anapsid condition). The roofing is complete in *Chelone* but in all others it is emarginated for the attachment of muscles of jaw and neck. The jaws are without teeth which are replaced by horny beak-like structure.

The vertebral column consists of eight cervical, ten thoracic, two sacral and reduced caudal vertebrae. The neck vertebrae show remarkable diversity in their mode of articulating with one another (such variation occurs in no other group of reptiles). The thoracic and sacral vertebrae are fused to the carapace.

The girdles are greatly modified and lie encircled within the ribs. The sternum and sternal ribs are absent. The ribs are single headed. The thoracic ribs are fused with the carapace. The abdominal ribs are absent. The cervical ribs are reduced.

Digestive system

The mouth is terminal. The tongue is non-protrusible. The digestive glands, liver and pancreas, are present. The rectum opens into the cloaca which is a longitudinal aperature. The chelonians are herbivorous, carnivorous or piscivorous.

Respiratory system

The respiratory organs are a pair of lungs which are large spongy sacs attached to the inner surface of the carapace both dorsally and laterally. Expansion by the usual method is therefore impossible. It is brought about partly by the protrusion and retraction of the neck and limbs and partly by the muscular action of the throat. Breathing is achieved by the contraction of modified abdominal muscle and the movement of pectoral girdle and hyoid apparatus. Additional respiratory mechanism occur in some aquatic forms. In the Australian turtle cloacal respiration occurs. The lining of cloaca is provided with numerous projections and is highly vascularized.

Circulatory system

The heart is typically reptilian consisting of two auricles and a thick walled ventricle. Sinus venosus is thick walled and lies dorsal to the auricles. The renal and hepatic portal systems are present.

Excretory system

As in other reptiles the excretory organs are a pair of metanephric kidneys. The excretory waste product is mainly uric acid excreted as solid white substance.

Reproductive organ

Male with single copulatory organ or penis. The chelonians are oviparous. The eggs are laid in nests on the ground. They are large, spherical and yolky encased in a calcareous shell.

Phytogenetic status of chelonia

The chelonians show a combination of primitive and specialized features:

Primitive features

1. Absence of temporal vacuity in the skull (anapsid skull). The roof of the skull is completely ossified. The skull is emarginated for attachment of muscles.
2. Pectoral girdle with clavicle, interclavicle and cleithra.

Specialized features

1. Body encased in a shell which protects from enemies.
2. Teeth replaced by horny beak-like structure.
3. Neck vertebrae show remarkable variability in their mode of articulating with each other. Neck retractile.
4. Limbs webbed or modified into flippers or paddle for swimming and heavy and elephant-like in land forms for supporting body weight.
5. The vertebrae and ribs are fused with carapace.
6. The girdles lie within the ribs.
7. The respiratory system is specialized to suit the body structure.

The specialized features of chelonians are not found in any other group of living reptiles. The phylogenetic relationships of chelonia are not very clear. The earliest turtles had all the specialized features present in the present forms. Hence, chelonian organisation evolved in its full form without any intermediate connecting forms.

ORDER OPHIDIA

The order Ophidia includes snakes. The snakes represent a more modern reptilian group than the lizards. They are a few Cretaceous forms which are possibly primitive snakes, but may be advanced lizards. The snakes have not been found before the Eocene, and the more evolved snakes appeared fairly late in the Tertiary. The snakes are considered late comers. They are near the peak of evolutionary development today. The structural organisation of snakes is greatly specialized.

Body shape

The body is generally long, slender and cylindrical with no clear distinction into head trunk and tail.

Limbs

The limbs and limb girdles are absent in most snakes except boas and pythons in which vestiges of hind limbs and their girdle are found. The limbs are lost most probably due to their burrowing habit. Some people are of the view that it is easy to move through deep grass if the limbs are absent. Snakes are known to "swim" through deep grass. Another view is that snakes have evolved as swimming eel-like forms. The majority of the ancestral snakes were burrowers. Most of the lizards that have become limbless are burrowers and most of the living shakes today are burrowers.

LOCOMOTION

All species of snakes are able to swim. They swim by lateral undulations of the body. On land they move by following methods:

1. **Lateral undulations or serpentine or curvilinear locomotion:** The body is thrown into a series of longitudinal and horizontal curves which follow the same pathway.(Fig. 18.6 A).

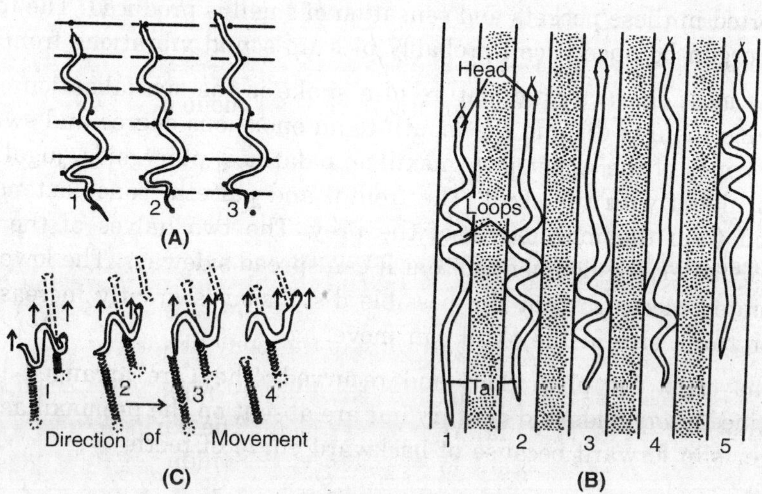

Fig. 18.6 Locomotion in *snake*–**(A)** Serpentine; **(B)** Concertina; **(C)** Side winding

2. **Rectilinear motion:** This type of movement is shown by snakes which have heavy body. The body undulates dorso-ventrally from head backward and the snake moves foward in a straight line. The body is pulled forward in sections by well developed cutaneous muscles that originate on the ribs and insert on the broad ventral scales. The ribs act as internal legs and the scales like setae of an earthworm.
3. **Concertina motion:** This type of movement is used in narrow places like burrows of rodents. The tail is anchored to the ground and the body is drawn into a number of S-loops. The body is pushed forward by straightening the loops and then drawing the tail forward (Fig. 18.6 B).
4. **Sidewinding:** This type of locomotion is used by snakes living in sand, for example, *Crotalus cerastes*. It moves by a curious looping of the body at right angles to the direction of travel (Fig. 18.6 C).

Eyes: In snakes the eyes are completely sheathed and protected externally by stiff transparent covering formed from the fused and immovable upper and lower eyelids. The sclerotic layer has no scleral ossicles. The retinal cells have no coloured oil droplets. Accomodation is brought about by the displacement of lens and not by change in the curvature of lens.

Ears: In snakes the external ear opening, tympanic membrane, eustachian canals and middle ear are lost, most probably due to burrowing habit. Therefore they cannot hear air borne sound. The ear of snakes includes an internal ear and a columella auris. The inner end of columella auris plugs the fenestra ovalis while the outer end is attached to the middle of the quadrate bone. They are able to detect ground vibrations. The sound waves travel along the quadrate, columella auris, fenestra ovalis and finally reach the perilymph of the ear.

Tongue: The tongue of snakes is long, slender and deeply forked. It usually protrudes, even when the mouth is closed, through the opening formed by the notches in the jaws. It serves as an organ of smell. The tongue when extended picks up from the air tiny smell producing particles and is pulled back. In the roof of the front part of the mouth there are certain pockets that can pick up sensation of smell. When pulled back the forked tips of the tongue are inserted in these pockets and sensation of smell is produced. The tongue also acts as an auditory organ because it can probably pick up sound vibrations from the ground.

Skull: The most characterstic feature of a snake is the specialization of its skull and jaws which enable them to distend the mouth to an enormous degree and swallow a prey of large size. The bones of the upper jaw (maxillae, palatines, pterygoids, jugal and quadrate) are lossely articulated with the skull. The frontal and parietal bones extend downward to shield the brain from the movements of the prey. The two halves of the lower jaw are connected only by an elastic ligament so that it can spread sideways. The lower jaw is broken into flexibly jointed pieces so that the possible distension is greatly increased. Except the brain case everything else in the skull can move.

Teeth: The teeth are pleurodont and recurved. They are arranged in rows on the maxillae, palatines, pterygoids and dentary but are absent on the premaxillae. The prey once seized can never slip forward because of backward curve of teeth.

Feeding mechanism

The snakes do not chew their food but swallow it whole. This is possible because of the following structural modification:

1. The jaw bones are loosely articulated with the skull and the two rami of the lower jaw are connected by an elastic ligament. Therefore, the mouth can be expanded many times the diameter of the snake.
2. The pectoral girdle and sternum are absent so the throat and body are also distensible.
3. The glottis is present far anterior in the floor of the mouth opening just behind the lower front teeth, thus breathing is not affected while swallowing.
4. The cartilages of trachea keep it open for breathing while swallowing.
5. The recurved teeth prevent the sized prey from slipping forward and also helps in pushing it back through the mouth and oesophagus into the stomach. It is digested leisurely in the stomach.

Internal organization

The elongation of body has resulted in reduction in its diameter. This has caused rearrangement of internal organs. The left lung is reduced or absent. The gall bladder lies posterior to liver the right kidney and gonad lie anterior to the left.

Poison apparatus

All poisonous snakes have a poison apparatus (Fig. 18.7) in their head (absent in non-poisonous snakes). It is made up of three parts:

(1) Poison glands, (2) Poison ducts, (3) Poison fangs.

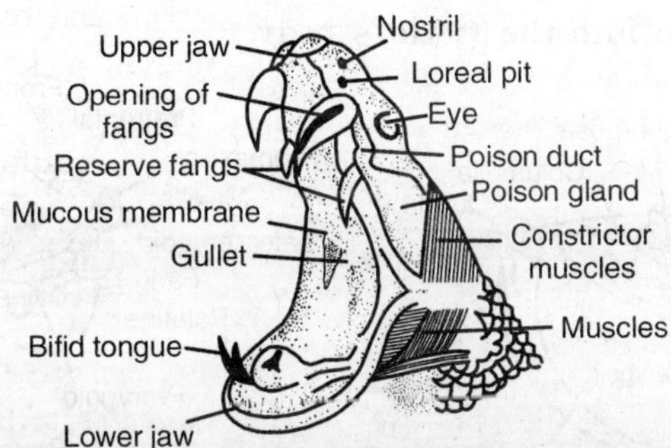

Fig. 18.7 Mouth of poisonous snake open to show the poison apparatus

Poison glands are a pair of sac-like structures situated one on either inner side of the upper jaw, below and behind the eyes. They are probably modified superior labial or parotid glands. The size and shape of poison glands vary with species being small and oval in sea snakes and long and tubular in vipers. The poison glands are held in position by **ligaments**. The anterior ligament attaches its anterior end with the maxilla and the posterior ligament with the quadrate. Each gland is surrounded by fibrous connective tissue and a fan-shaped **constrictor muscle** called **temporal** or **masseter**. It stretches during biting and forces the poison from the gland to the duct.

A narrow **poison duct** connects each poison gland to the poison fang. It opens at the base of the fang.

The **poison fangs** are a pair of modified maxillary teeth. They are long, recurved, sharp and pointed needle-like structures injecting the poison into the body of the victim. A pair of reserve poison fangs is also present. When the functional fang is broken or damaged it is replaced by the reserve fang. Based on their structure and position the fangs are of three types:

1. **Proteroglyphous** (*proto*, first *glyph*, hollowed): It is found in cobra, krait, coral snake and sea snake. The fangs are relatively short, situated at the anterior end of the maxillae and permanently erect. An open groove runs along the anterior face of the fang to conduct saliva into the wound.
2. **Opisthoglyphous.** (***opistho***, behind): It is found in *Dryophis* and snakes of family Colubridae. The fangs are small, and located at the back of maxillae and grooved along its posterior border.
3. **Solenoglyphous** (*solen,* pipe): It is found in vipers and rattle snakes. It is a large fang located in front of each maxilla. Its base is covered by a sheath containing a few reserve fangs. A central canal runs through the fang and opens at the tip. It serves as a passage for the poison. The fangs are movable and remain folded when not in use. They lie close to the roof of mouth when it is closed.

Biting mechanism

Since the bones of the skull and jaws in poisonous snakes are movably articulated, they allow a considerable degree of adjustment during swallowing. In snakes with solenoglyphous type of fangs the biting mechanism includes two steps:

1. Erection of fang.
2. Injection of poison into the victim's body.

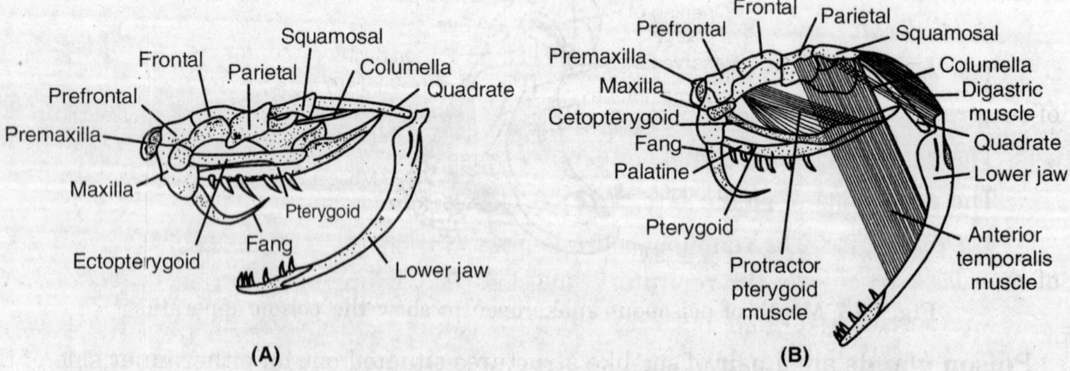

Fig. 18.8 Biting mechanism. **(A)** At rest, mouth closed and fang lies against the roof; **(B)** In striking position mouth open and fang erect

When a snake strikes a series of movements occur in a sequence (Fig. 18.8).

1. Contraction of digastric muscles lowers the mandible, thus opening the mouth.
2. The lower end of the quadrate thrusts forward and this pushes the pterygoid forward. This is also aided by the contraction of sphenopterygoid muscles.

3. This pushes the transverse or ectopterygoid upwards causing the maxilla bearing fangs to rotate through 90° at the hinge joint with lacrimal.
4. As a result the fangs become vertically erect and are ready to strike.
5. The stretching of constrictor muscles around the poison gland forces the poison through the poison duct into the groove of the fang to be injected into the victim.
6. After injecting the poison the mouth is closed by the contraction of temporal muscles and the above movements are reversed. The fangs are withdrawn into the mouth. The vertical fangs rotate and become horizontal.

Snake venom

Snake venom is a clear sticky liquid of faint yellow or green colour secreted by the poison gland. It is tasteless, odourless and acidic in reaction. It is a complex mixture of enzyms and specific toxins and is a good digestive juice. It is fatal only when mixed with blood. It can be swallowed without causing any damage if there is no cut in the mouth or alimentary canal and pass out with faeces. It can be precipitated with reagents like silver nitrate and potassium permaganate and can be dried and stored indefinately without losing its venomous properties. It is soluble in water, salt solution or glycerine. Snake venom may contain both neurotoxic and haemotoxic factors but in some one factor may become dominant.

Effect of snake venom

The venom of different snakes has its own effect. Degree of virulence differs in different snakes. The effect of poison on the victim depends on the amount of poison injected and the degree of virulence of the poison. There are two types of effects of snake venom.

1. **Neurotoxic,**
2. **Haemotoxic.**

The venom of krait, cobra and sea snake is **neurotoxic.** It causes death by paralysis of respiratory muscles and asphyxia.

The venom of vipers is haemotoxic. It causes tissue destruction and haemorrhage.

The symptoms of snake bite of three most common species of poisonous snakes are:

1. **Cobra bite:** The venom of cobra is most virulent. It is neurotoxic. It causes paralysis of muscles particularly the resiratory muscles. The symptoms are:

(i) Piercing pain and burning sensation causing numbness of the bitten part which becomes blue or black.

(ii) Giddiness and vomiting.

(iii) Weakness in legs.

(iv) High pulse rate.

(v) Speechlessness.

(vi) Drooping of saliva and eylids and contraction of pupils.

(vii) Death within few hours due to failure or respiration or heart.

2. **Krait bite:** Krait is the most common poisonous snake. It is very deadly because the amount of poison injected is more than three times than that of cobra. The symptoms are similar to those of cobra except unbearable abdominal pain due to internal haemorrhage. Death occurs within 6–24 hours due to destruction of RBC and paralysis of trunk and limbs.

3. **Viper bite:** The venom is mainly haemotoxic. The symptoms are:

(i) Local swelling and discolouration of the bitten part and acute burning pain.

(ii) A red fluid oozes out of the wound due to massive tissue destruction (necrosis). This frequently requires amputation.

(iii) Pupils dilate.

(iv) Pulse rate increases.

(v) Profuse vomiting.

(vi) Person loses conciousness.

(vii) Death due to paralysis of vaso-motor centres and profuse bleeding.

Antivenin

The best way to cure snake bite is to inject **antivenom serum** or **antivenin** into the body of victim. It counteracts the snake venom. Different antivenins are required for different snake bite because of differences in the nature of the venoms.

Preparation of antivenin: An antivenin is prepared by injecting a horse with gradually increasing doses of the snake venom until the horse becomes fully immunized to any amount of venom injected. Now enough antibodies are formed in the blood of horse to neutalize the effect of that snake venom. The blood serum is collected and preserved. This is antivenom serum or antivenin.

First aid treatment

First aid treatment should be given soon after snake bite before medical treatment can be secured. It includes:

1. A toniquet or ligature should be tightly tied near the bitten part towards the heart by any suitable piece of cloth. This prevents circulation of poison in the body. The torniquet should be loosened for a few minutes in between.
2. Wound should be washed with clean water.
3. Wound should be cooled with ice to slow down blood circulation.
4. Poison may be sucked out by a suction pump and not by mouth.
5. A cut 1 cm deep should be made with a sterilized knife or blade so that some poison may let out with blood.
6. The victim may be given hot milk, tea or coffee.
7. The victim should be shifted to hospital for treatment with an antivenin injection.

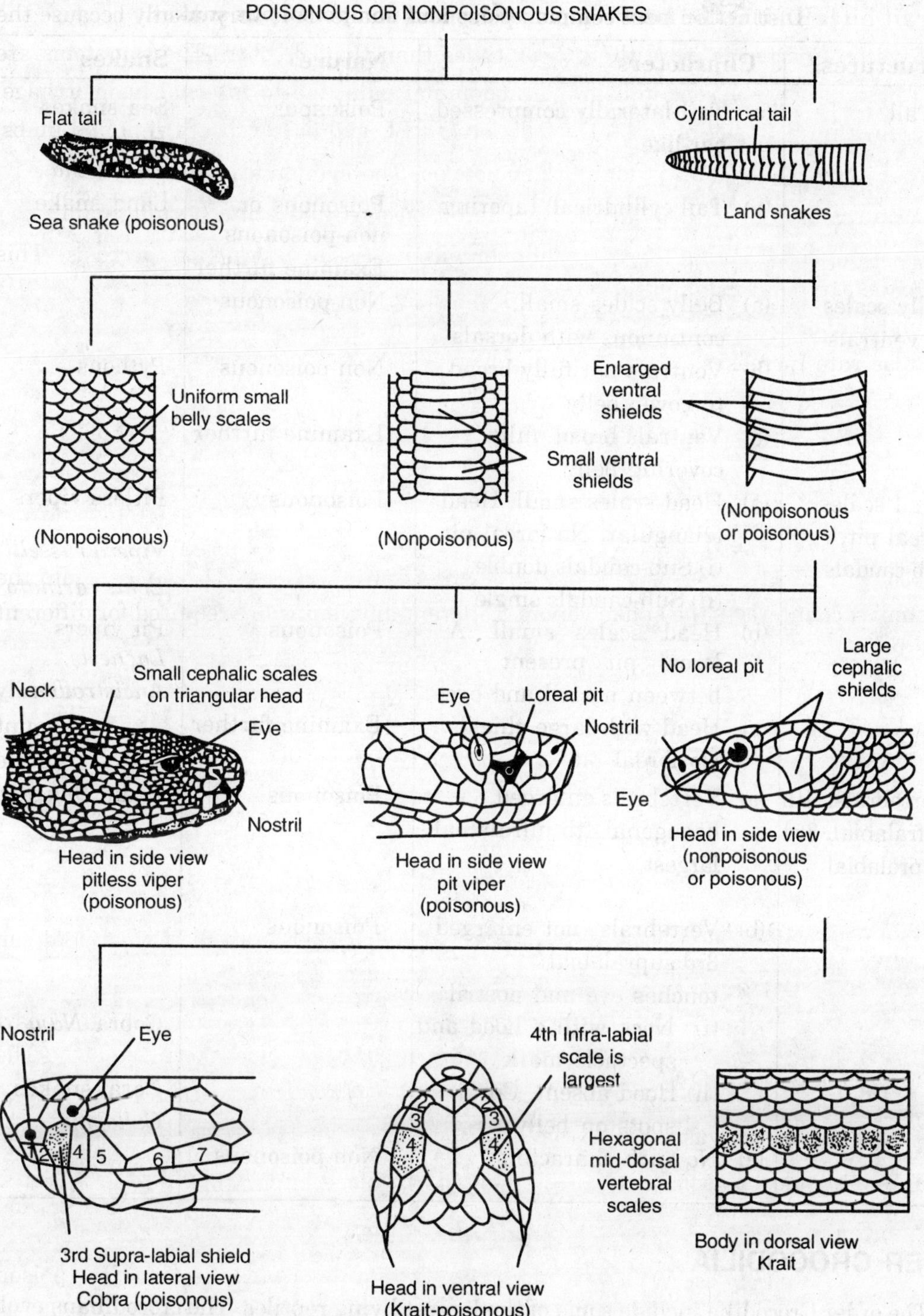

Fig. 18.9 Identification of poisonous and non poisonous snakes

Distinction between poisonous and non-poisonous snakes

Structures	Characters	Nature	Snakes
1. Tail	(a) Tail laterally compressed, oar-like	Poisonous	Sea snakes *Hydrophis, Enhydrina*
	(b) Tail cylindrical, tapering	Poisonous or non-poisonous Examine further	Land snake
2. Belly scales or ventrals	(a) Belly scales small, continuous with dorsals	Non-poisonous	
	(b) Ventrals not fully broad to cover belly	Non-poisonous	Pythons
	(c) Ventrals broad, fully covering belly	Examine further	
3. Head scales, loreal pit, sub-caudals	(a) Head scales small. Head triangular. No loreal pit	Poisonous	Pitless vipers
	(i) Sub-caudals double	"	*Vipera russelli*
	(ii) Sub-caudals single	"	*Echis carinata*
	(b) Head scales small. A loreal pit present between nostril and eye	Poisonous	Pit vipers *Lachesis, Ancistrodon*
	(c) Head with large shields. No loreal pit	Examine further	
4. Vertebrals, 4th infralabial, 3rd supralabial	(a) Vertebrals enlarged, hexagonal 4th infralabial largest	Poisonous	Krait, *Bungarus*
	(b) Vertebrals not enlarged. 3rd supralabial touches eye and nostril	Poisonous	
	(i) Neck with a hood and spectacle mark	"	Cobra, *Naja*
	(ii) Hood absent. Coral spots on belly	"	Coral snakes, *Callophis*
	(c) No such characters	Non-poisonous	

ORDER CROCODILIA

The order Crocodilia include some of the largest living reptiles. The crocodilians evoloved during the Cretaceous period of Mesozoic era. Some large sized marine crocodiles existed during late jurassic period. They had paddle-like feet and lobed tail.

Habit and habitat

The crocodilians are carnivorous, lizard-like creatures living mainly in tropical and subtropical fresh-water bodies. Some species are found in salt-water or estuaries. Thy swim actively with the help of their powerful laterally compressed tail. The webbed limbs also help in swimming. They walk on land with the help of their strong limbs.

Today the order is represented by 21 living species included in 3 families.

Family Alligatoridae: It includes two genera: *Alligator* and *Caiman*. There are two species of *Alligator:* **Chinese alligator** found in the rivers of China and **American alligator** (*A. mississipiensis*) found in south eastern parts of North America.

Family Crocodylidae: The members of this family are widely distributed in the Indo-Pacific region and in Northern Australia. They occur in swamps, estuaries, mangroove and rivers. The fresh-water crocodile is commonly known as **maggar.** It is about 5 metres long. The Americal species is marine. The largest species is the salt water or estuarine crocodile, *Crocodylus porosus* of south east Asia. It is 8–9 metres long.

Family Gavialidae: It is represented by a single species *Gavialis gangeticus,* commonly known as gharial. It lives in the river system of Ganga and Brahmaputra of Northern India and Burma. It attains a length of 8 metres.

General Characters

The **body** is long, stout and depressed. The **head** is large and elongated into a prominent **snout.** The snout is blunt in crocodiles and alligator (Fig. 18.10 A and B), but long and

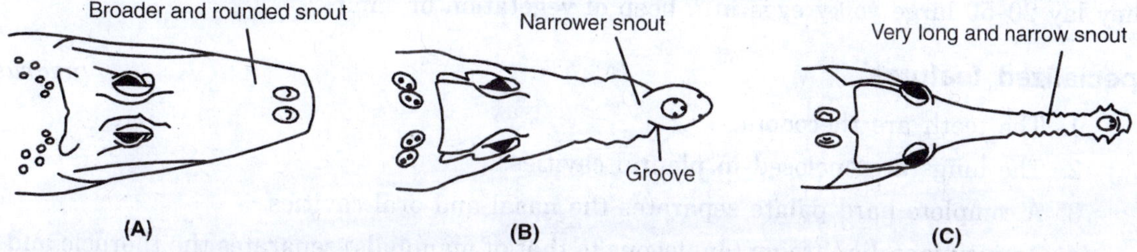

Fig. 18.10 Shape of snout, **(A)** *Alligtor*; **(B)** *Crocodylus*; **(C)** *Gavialis*

pointed in gharial (Fig. 18.10 C). The **nostrils** are situated at the tip of the snout and can be closed by **valves** operated by special set of muscles: **longitudinal dilator muscles** for closing and **constrictor muscles** for opening it. The **eyes** are situated dorso-laterally. They are covered by transparent **nictitating membrane.** The ear opening, **tympanum** is covered by scaly movable flaps of skin. The nostrils, eyes and ears are placed high on the head so that they remain exposed even when the animal is immersed in water, thus breathing, sight and hearing remain unaffected. While swimming the eyes, ears and nostrils remain protected. The **internal nares** are situated in the pharynx. A **flap** present on the hind end of the tongue acts as **intranarial epiglottis** and closes the wind pipe at the back so that the animal can open its mouth to feed in water and at the same time breathe with its nostrils above water. The **limbs** are short but strong, pentadactyl with clawed and webbed digits. They help in swimming as well as for locomotion on land. The tail is large and laterally compressed and serve as a powerful swimming organ. The cloaca is a longitudinal slit.

The **skin** is thick and leathery and covered with horny epidermal **scales**. The scales on the back and belly are rectangular and supported by dermal **bony plates**. Dorsally the tail bears a **crest** of broad plates which are double in the basal half but single in the distal half.

Two pairs of **musk** or **stink glands** are present in crocodiles, one pair on the lower jaw in the throat region and the other near the cloaca. The secretion of these glands has a musky odour, and helps in mating.

The **jaws** are powerful with a single row of teeth for catching and holding the prey. The teeth are **thecodont** i.e., they are set in sockets and have persistent pulp.

Skull is highly sculptured with persistent sutures. Interorbital septum is well developed. Quadrate is large and immovable. A **secondary palate** formed by maxillae, palatines and pterygoids is present and separates the nasal and oral cavities. The **vertebrae** are **amphicoelous** or **procoelous.** A **proatlas** is present, there are only two sacral vertebrae. The caudal vertebrae has cheveron bone. Sternal and abdominal ribs are present, sternal ribs have uncinate process.

The respiratory organs are a pair of **lungs** enclosed in pleural cavities. An incipient **diaphragm** is present between thoracic and abdominal cavities.

The **heart** is four chambered: two **auricles** and two **ventricles.** A complete **interventricular septum** divides the ventricle. The left and right **aortic arches** cross each other and communicate by **foramen of panizzae**, but there is no mixing of blood between the two arches. Crocodilians are warm blooded.

The male has a simple but grooved copulatory organ. The crocodilians are oviparous. They lay 20–50 large yolky eggs in a heep of vegetation or sand.

Specialized features

1. The teeth are thecodont.
2. The lungs are enclosed in pleural cavities.
3. A complete hard palate separates the nasal and oral cavities.
4. An incipient diaphragm (analogous to that of mammals) separates the thoracic and abdominal cavities.
5. The heart is four chambered.

REPRESENTATIVE TYPES

Testudo

Testudo (Fig. 18.11 A) is commonly known as **land tortoise.** It is widely distributed in Asia, Africa and Europe. It is terrestrial, diurnal and feeds on small worms and insects, some are herbivorous. It can live without water for long periods but sometimes in hot weather, it goes into water. It has a life span of 100 years or more. *Testudo abingdoni* found in the islands of Indian ocean and Galapagos island of Pacific ocean is commonly known as the **giant land tortoise**. It weighs about 100 kg and lives for 200–400 years. The body is enclosed in an oval shell made of dorsal convex **carapace** and ventral flat **plastron** which are joined on the sides by skin. The head is mounted on a retractile neck and the tail is short. The limbs are massively built with clawed digits, each with two phalanges only. They are not webbed and are adapted for walking on land. The head, neck, limbs and tail can be withdrawn completely within the shell.

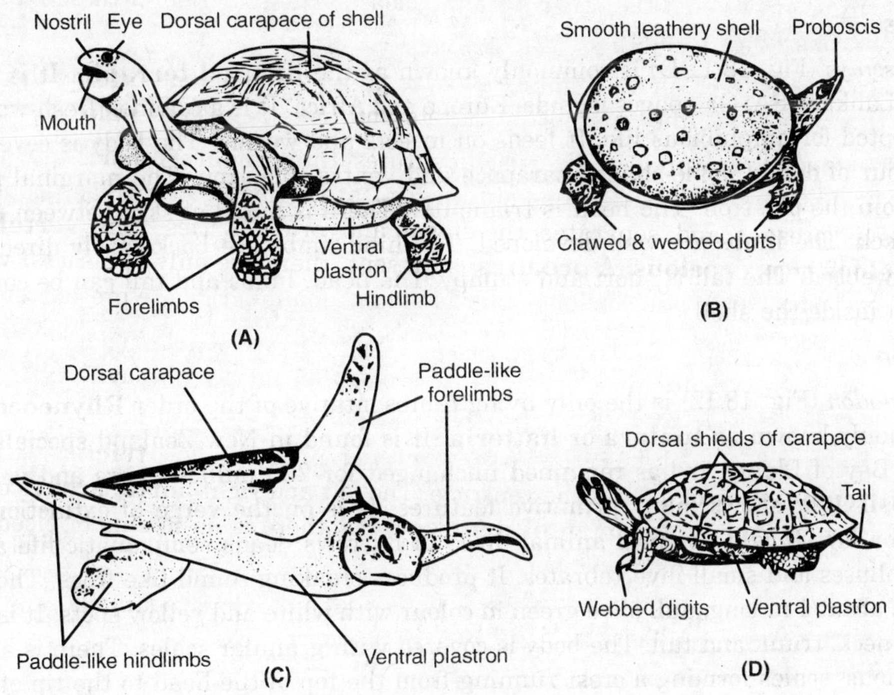

Fig. 18.11 Chelonians–(A) *Testudo abingdon;* (B) *Trionyx spiniferus;* (C) *Chelonemydas;* (D) *Chrysemys*

Trionyx

Trionyx (Fig. 18.11B) is commonly known as **tortoise** or **soft river terrapin.** It is found in the rivers of Asia, Africa and North America. The shell is soft, smooth and leathery and without shields. The skin is soft and leathery. The head is pointed and with greenish or blackish long streaks. The lips are fleshy. The carapace is fused with the vertebral column and ribs. Dorsal surface is olive above and yellowish below. The feet are broadly webbed and only three inner digits are clawed. *Amyda* is also a soft shell turtle found in Indian rivers in which the ribs extend beyond the costal plates of carapace. *Trionyx gangeticus* is found in the rivers of North India and *T. leithi* in the rivers of South India.

Chelone

Chelone (Fig. 18.11 C) is commonly known as **green turtle**. It is edible and its flesh is very delicious. It is found in Indian, Pacific and Atlantic oceans. It is a large turtle measuring about 110 cm, and weighs about 90 kg. The body is divided into head, neck, trunk and tail. The head and neck are not completely retractile. The jaws have denticulate edges. The eyes are well developed and provided with eyelids and nictitating membrane. The tail is short. The shell is rigid. The carapace is heart-shaped and covered with smooth bony shields. The plastron is joined to the carapace by ligaments. The limbs are paddle-like and adapted for swimming. Only first digit of the forelimb is clawed while the digits of the hindlimbs are clawed and webbed. It feeds on sea weeds and also eats fishes. During the breeding season the turtle comes to the shore for egg laying. It lays about 200 eggs.

Chrysemys

Chrysemys (Fig. 18.11 D) is commonly known as the **painted terrapin**. It is found in India, Sri Lanka, USA, Galapagos islands, Europe and Africa. It is a common fresh-water form but is adapted for amphibious life. It feeds on insects and worms. The body is covered by a bony armour of dorsal dome-shaped carapace and ventral plastron. The marginal plates of carapace join the plastron. The head is triangular. There is a fold of skin between the head and the neck. The limbs are well developed. The hind limbs are backwardly directed. The digits are webbed. The tail is short and stumpy. The head, limbs and tail can be completely withdrawn inside the shell.

Sphenodon

Sphenodon (Fig. 18.12) is the only living representative of the order **Rhyncocephalia.** It is commonly known as **tuatara** or **hatteria**. It is found in New Zealand specially in the islands of Bay of Plenty. It has remained unchanged for 200 million years and is called a **living fossil** as it shows many primitive features. It is on the verge of extinction and is protected by law. It is a noctural animal, lives in burrows, leads semiaquatic life and eats insects, molluses and small invertebrates. It produces croaking sound like frogs. The body is lizard-like, about 1 m long, dull olive green in colour with white and yellow spots. It is divided into head, neck, trunk and tail. The body is covered with granular scales. There is a median row of spinous scales forming a crest running from the top of the head to the tip of the tail except in the neck region. The lower surface has transverse row of large squarish plates. The eyes are photosensitive. A distinct **parietal** or **pineal eye** with lens, retina and nerve connection with the brain is present. It is sensitive to light. The tail is thick and laterally compressed. The limbs are pentadactyle, with clawed digits and adapted for walking. The anal aperature is transverse. The copulatory organ is absent in male. The breeding season is November to February. It lays about ten eggs with hard shell in holes in the ground. The eggs hatch in 13 months.

Hemidactylus

Hemidactylus (Fig. 18.12A) or *Gecko* is commonly known as **wall lizard** or **house lizard**. It is distributed worldwide but chiefly found in houses in India, Sri Lanka, China and Africa. It is nocturnal in habit and is adapted to creep on walls. It is a fast runner. It hibernates in winters under wood, logs or in crevices in the wall. It feeds on insects and small invertebrates. The body is about 25 cm long, slender and covered with minute smooth scales. Some lizards have black or dark grey patches on the dorsal surface while others have dark grey dorsal surface. The ventral surface is yellowish white. The head is triangular and contains nostril, vertical ear opening and eyes with immovable eyelids. The tongue is short, sticky and protrusible. The tail is short and can break off at unossified regions between caudal vertebrae. This is called **caudal autotomy.** The broken tail wriggles for sometime to distract

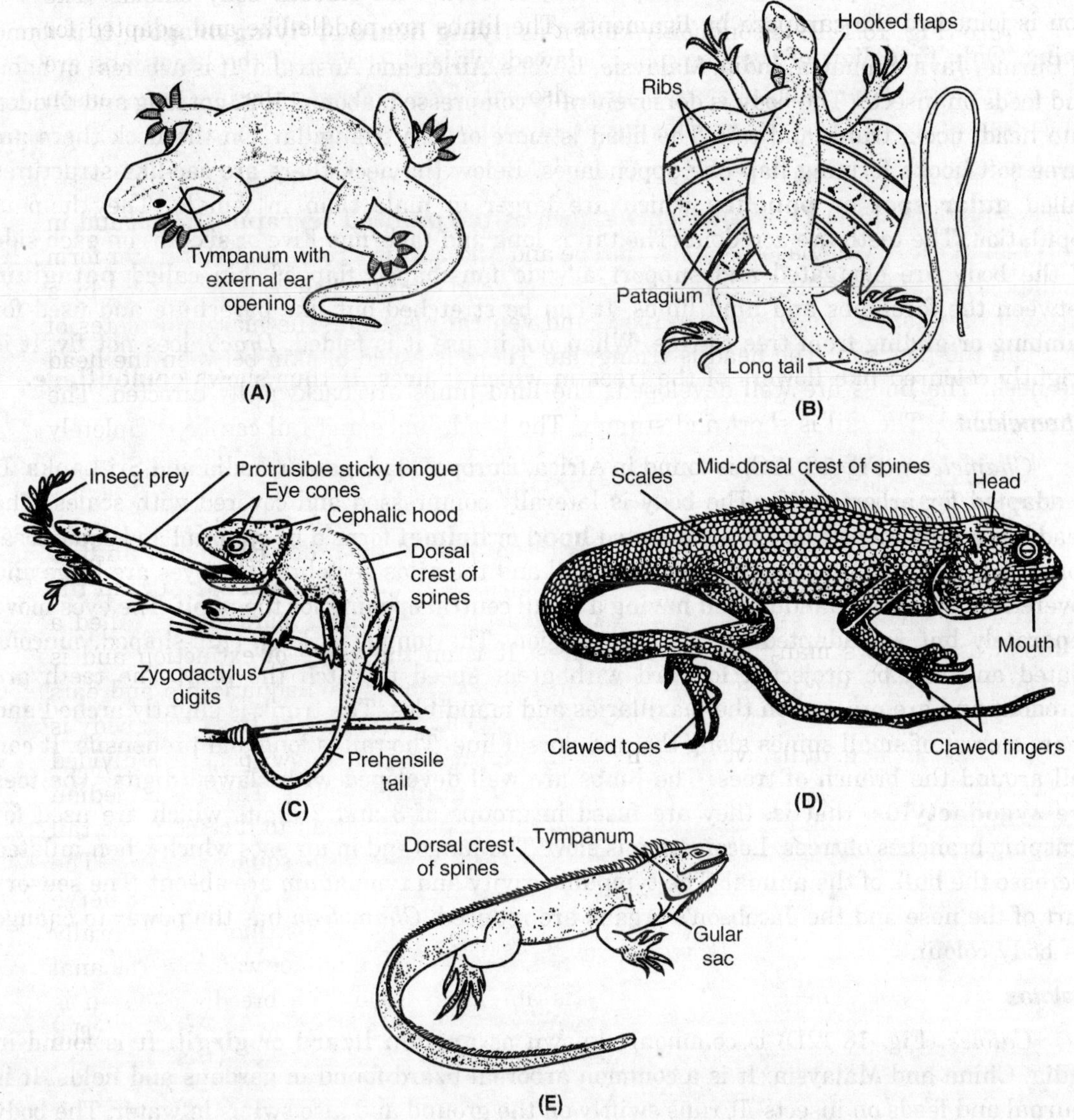

Fig. 18.12 Lizards–(A) *Hemidactylus;* (B) *Draco;* (C) *Chameleon;* (D) *Calotes;* (E) *Iguana*

the enemy and the lizard escapes. The new tail regenerates but is different from the old one because the lost vertebrae are not regenerated. The limbs are well developed with clawed digits. The digits are dilated and the undersurface is provided with adhesive pads which are in the form of double series of ridged lamellae which work under the principle of vacuum. These enable the lizard to creep or walk on the wall or ceiling and other smooth surfaces. Many wall lizards produce sound. The male has femoral pores on the thigh. The eggs are round and the egg shell is hards. In other lizards the egg shell is leathery.

Draco

Draco (Fig. 18.12B) is commonly known as **flying lizard** or **flying dragon**. It is found in Burma, Java, Sumatra, India, Malaysia, Europe, Africa and Australia. It is arboreal in habit and feeds on insects. The body is dorsoventrally compressed, about 15–22 cm long and divided into head, neck, trunk and tail. The head is more or less triangular. On the neck there are three soft hooks forming flap-like appendages. Below the neck there are sac-like structures called **gular sacs** or **pouches** which are larger in male than in female. They help in copulation. The teeth are acrodont. The tail is long and tapering. Five or six ribs on each side of the body are elongated and support a wide fan-shaped flap of skin called **patagium** between the forelimbs and hind limbs. It can be stretched out like parachute and used for climbing or gliding from tree to tree. When not in use it is folded. *Draco* does not fly. It is brightly coloured like flowers of the trees in which it lives. It thus shows **camouflage.**

Chameleon

Chameleon (Fig. 18.12C) is found in Africa, Europe, Madagascar, India and Sri Lanka. It is adapted for arboreal life. The body is laterally compressed and covered with scales. The head has a prominent backwardly directed **hood** or **helmet** formed by occipital and squamosal bones. A proatlas is present between the skull and the atlas vertebra. The eyes are large and covered with a thick glandular lid having a small central opening for the pupil. The eyes move separately but are adapted for binocular vision. The tongue is long, club-shaped, mucous coated and can be projected forward with great speed to catch the prey. The teeth are acrodont and are present on the maxillaries and mandibles. The trunk is slightly arched and bears a crest of small spines along the mid-dorsal line. The tail is long and prehensile. It can coil around the branch of trees. The limbs are well developed with clawed digits. The toes are **zygodactylus**, that is, they are fused in groups of 3 and 2 digits which are used for grasping branches of trees. Locomotion is slow. The lungs end in air sacs which when inflated increase the bulk of the animal. The tympanic cavity and tympanum are absent. The sensory part of the nose and the Jacobson's organs are reduced. *Chameleon* has the power to change its body colour.

Calotes

Calotes (Fig. 18.12D) is commonly known as **garden lizard** or **girgit.** It is found in India, China and Malaysia. It is a common arboreal lizard found in gardens and fields. It is diurnal and feeds on insects. It runs swiftly on the ground and also swims in water. The body is slender, elongated about 35 cm long, brightly coloured, with a greenish tinge and covered by rough, horny, backwardly directed epidermal scales of unequal size. The throat is dark red in colour so it is also known as **blood sucker.** The head is pyramidal in shape containing a pair of oval nostrils, eyes with movable eyelids and nictitating membrane and external ear opening covered with tympanum. Lips are absent. On the head and trunk is present a crest of movable spinous scales pointed backwards, they gradually reduce in size. The tail is long tapering and whip-like. It does not break. The limbs are pentadactyle and the digits are clawed. The cloacal aperture is a transverse slit. The most common Indian species is *Calotes versicolor* which has the capability to change colour from normal olive green to yellow. This is controlled by the hormones of the pituitary gland. The breeding season is May to November. It lays eggs in crevices or in debris.

Iguana

Iguana (Fig. 18.12E) is an arboreal lizard inhabiting South and Central America and West Indies. The body is laterally compressed and about 2.5 m long. The tail is long and tapering. Along the mid-dorsal line from the neck to the tip of the tail there is a crest of long spines which keep on decreasing in size gradually and in the tail they are very small. On the throat there is an expansible fold of skin with spinous margin, which play an important role in courtship and frightening away enemies. It is adapted for active running and climbing. It feeds on leaves and fruits and sometimes on insects and small invertebrates. Another iguanid, *Amblyrhynchus* is the only marine lizard found in Galapagos islands. It eats sea weeds. It is often seen basking in sun on the shore. It is used as food by the native people.

Phrynosoma

Phrynosoma (Fig. 18.13A) is the **horned lizard** which has wrongly been named as the horned toad. It is found in western half of the United States and Central America. It is

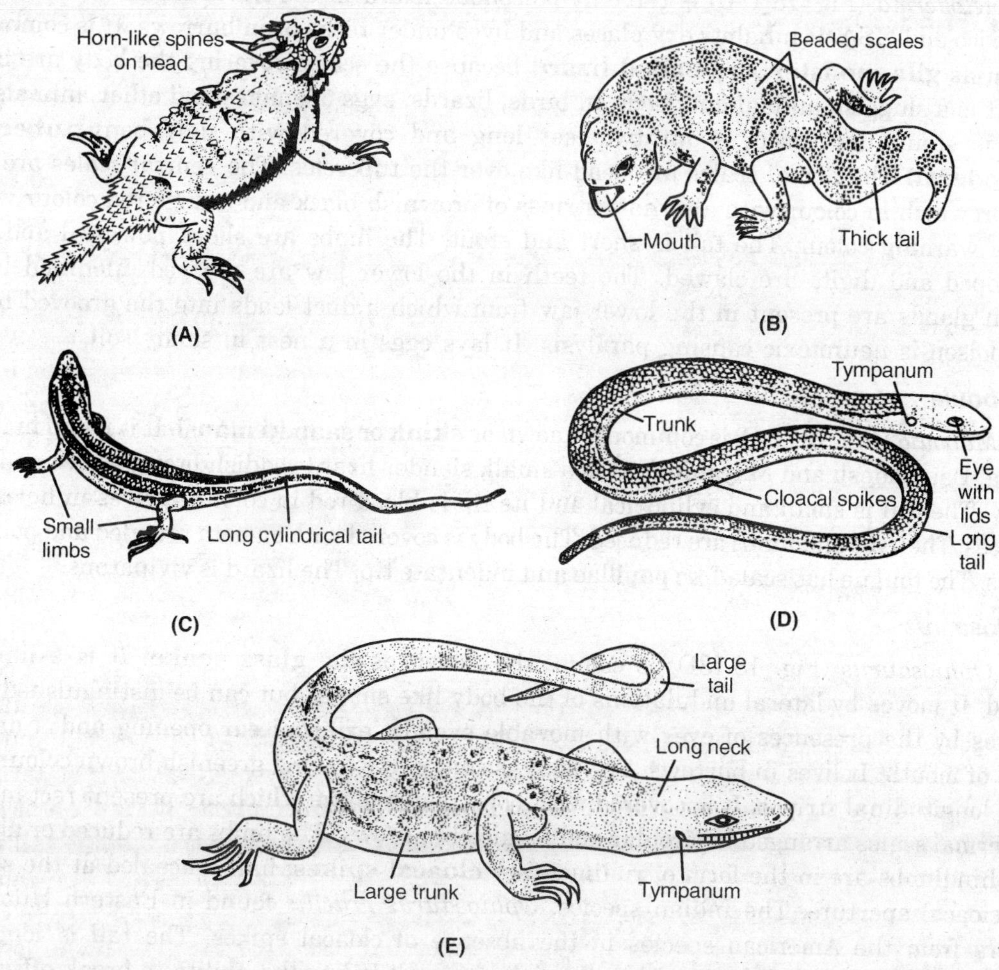

Fig. 18.13 Lizards–**(A)** *Phrynosoma*; **(B)** *Heloderma;* **(C)** *Mabouia*; **(D)** *Ophiosaurus*; **(E)** *Varanus*

adapted for desert life. It lives on dry ground and sandy places. It can live without water for a long time. It is diurnal and feeds on ants and other small insects. In the night it burries itself in sand keeping the head and horns exposed. The body is broad, flattened and covered with spiny scales. The head bears a prominent **helmet**. The scales of the head region are enlarged. There are five enlarged spines on each side of the head and a pair of enlarged horn-like processes below the helmet giving it a grotesque appearance. The undersurface is covered with small keeled scales. The tail is short and bears numerous small and large spines. The dorsal surface is yellowish grey or black or brown in colour. The sides of the lower jaw projects in the form of prominent ledges and protected by a row of small spines. The tongue is short and non-protrusible. The teeth are pleurodont. A row of pores are present on the ventral surface of the thigh in both sexes. The limbs also bear spines. The digits are clawed and adapted for digging sand. *Phrynosoma* is a harmless animal.

Heloderma

Heloderma (Fig. 18.13B) is the only poisonous lizard in the world found in the deserts of Mexico and USA. It inhabits dry places and lives under rocks or in burrows. It is commonly known as **gila monster** or **beaded lizard** because the scales covering the body are bead-like. It is a sluggish animal and feeds on birds, lizards, eggs of snakes and other animals. Its body is stout, depressed, about two feet long and covered with ugly bony **tubercles (osteoderm)** The dorsal scales are bead-like over the tubercles. The ventral scales are flat. It is brownish in colour with alternate stripes of brownish black and orange red colour which act as warning colour. The tail is short and stout. The limbs are short, powerful and well developed and digits are clawed. The teeth in the lower jaw are grooved. Modified labial poison glands are present in the lower jaw from which a duct leads into the grooved tooth. The poison is neurotoxic causing paralysis. It lays eggs in a nest in sandy soil.

Mabibouia

Mabouia (Fig. 18.13 C) is commonly known as **skink** or **saap ki mausi**. It is found in India, China, Bangladesh and Mayanmar. It is a small, slender lizard reddish brown above and pale below. The tail is small and cylindrical and its tip is blood red in colour. It lives in holes and crevices. The eyes and limbs are reduced. The body is covered with smooth rounded and polished scales. The tongue has scale-like papillae and indentate tip. The lizard is viviparous.

Ophiosaurus

Ophiosaurus (Fig. 18.13D) is commonly known as the **glass snake**. It is a limbless lizard. It moves by lateral undulations of the body like snakes but can be distinguished from snakes by the presences of eyes with movable eyelids, external ear opening and a narrow gape of mouth. It lives in burrows. The body is snake-like having greenish brown colouration with longitudinal stripes. It is covered with bony plates above which are present rectangular epidermal scales arranged in circular and longitudinal rows. The limbs are reduced or absent. The hindlimbs are in the form of rudimentary **cloacal spikes**, half concealed at the side of the cloacal aperture. The Indian species, *Ophiosaurus gracilis* found in Eastern Himalaya differs from the American species in the absence of cloacal spikes. The tail is long and constitutes two-third of the total length of the animal. It has the ability to break off its tail when seized. The broken tail then breaks up into a number of small pieces that wriggles about and the lizard escapes. Other limbless lizards are *Rhineura* (worm lizard) and *Anquis fragilis* (blow worm or blind worm). They are also found in Europe, America and Asia.

Varanus

Varanus (Fig. 18.13E) is commonly known as **monitor lizard**. It is found in Africa, South Asia and Australia. It lives in burrows or on trees and is semiaquatic. It is a good climber as well as a good swimmer. It is active at night and feeds on tortoises, squirrel and dead animals. The body is about 60–90 cm long and covered with smooth small scales. It is brownish in colour with darker spots which act like warning colour. The head is triangular in shape, the neck is long, the trunk is large and the tail is long and laterally compressed. The tongue is long, protrusible and deeply bifid. The teeth are large, pointed and pleurodont. The limbs are stout with clawed digits.

Varanus komodoensis is the largest living lizard in the world found in the islands of Malaya Archipelago. It is about 2.5 m long and weighs about 100 kg. It mainly preys upon pig, goat, deer, etc. but can also kill adult water buffalo.

Non-poisonous snakes

Typhlops

Typhlops (Fig. 18.14A) is commonly known as **blind snake** because its eyes are vestigeal and covered by scales. It looks like earthworm hence it is also known as **worm snake.**

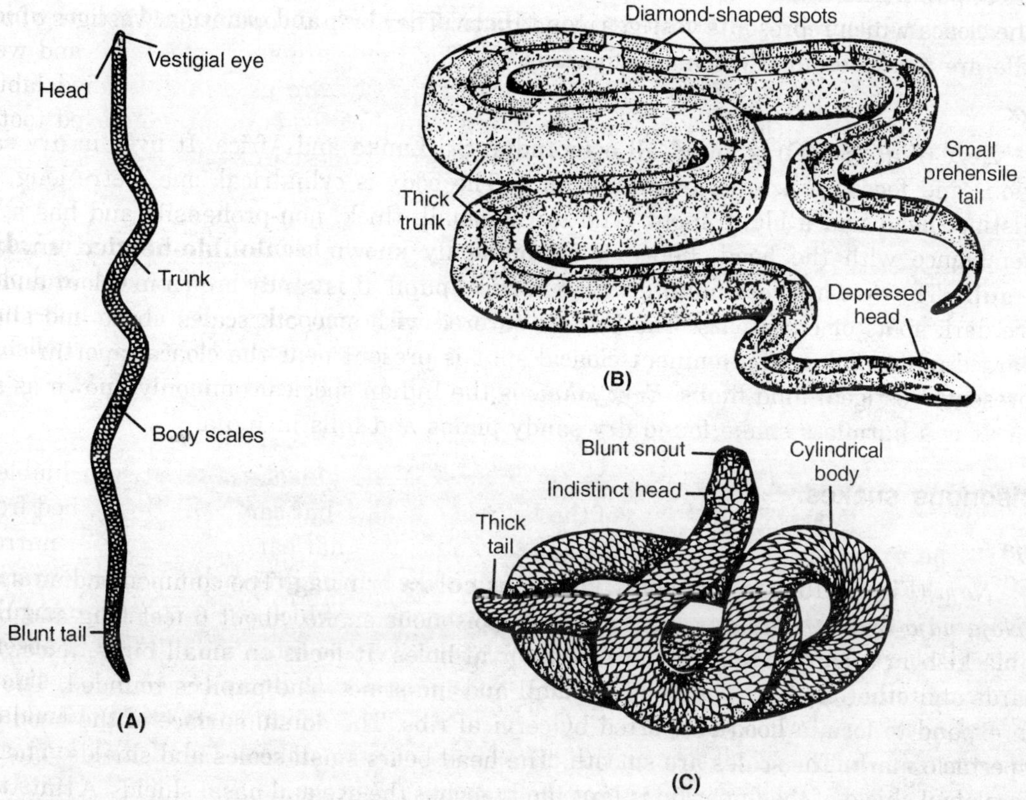

Fig. 18.14 Non piosonous snakes–**(A)** *Typhlops vermicularies;* **(B)** *Python;* **(C)** *Eryx johnii*

It is a small burrowing snake found in the tropical and subtropical regions of the earth. It feeds on small insects, larvae and termites. The body is elongated, cylindrical and covered with small, overlapping, semicircular scales. Its various species range in length from 7 inches to 2 feet. The head is small with blunt snout and covered with large shields. A few teeth are present only in the upper jaw, lower jaw without teeth. The mouth is non-expansible and the tongue is slender, bifid and protrusible. The tail is short and blunt. The left lung is reduced. It has vestiges of pelvic girdle. It is a harmless snake. It is oviparous.

Python

Python (Fig. 18.14B) commonly known as **ajgar** is the largest non-poisonous snake. It has a world-wide distribution except New Zealand. *Python reticulatus* of South East Asia is the largest species measuring about 10 m or more in length. The Indian species *P. molurus* is about 6 m long. It is a huge animal weighing 110 kg. It lives on trees in forests and is nocturnal in habit. It feeds on birds and mammals. The prey animal is constricted in its body coils and killed due to suffocation and then swallowed slowly. *Python* is a sluggish animal with a stout body covered with small scales arranged in many rows. The dorsal surface is brown with dark grey rhomboid spots while the ventral surface is grey with yellowish brown spots. Head is distinct from neck. The anterior half of the head bears shields and the posterior half bears small scales. A brown spear-like mark is present over the head. The eyes have a vertical pupil. The tail is short and prehensile. The adult has a **cloacal spur** on either side of the cloaca which represents vestigeal hind limbs. They help is copulation. Vestiges of pelvic girdle are also present. *Python* is oviparous.

Eryx

Eryx (Fig. 18.14C) is found all over India, Sri Lanka and Africa. It lives in dry sandy regions and feeds on lizards, frogs and mice. The body is cylindrical, one metre long, with indistinct head and a blunt snout. The tail is small, thick, non-prehensile and has a false resemblance with the head, hence it is commonly known as **double-headed snake** or **dumuhi**. The eyes are very minute with a vertical pupil. It is sandy brown in colour and may have dark spots or cross bars. The body is covered with smoooth scales above and slightly enlarged plates below. A prominent cloacal spur is present near the cloacal aperture which represents vestigeal hind limbs. *Eryx johnii* is the Indian species commonly known as **sand boa**. It is a harmless snake found dry sandy plains and hills in India.

Poisonous snakes

Naja

Naja (Fig. 18.15A) is commonly known as **cobra** or **naag**. The common Indian species is *Naja naja* or *N. tripudians.* It is a deadly poisonous snake, about 6 feet long and brown or blackish in colour. It lives under stones or in holes. It feeds on small birds, mice, frogs, lizards and other snakes. The head is small and indistinct. The pupil is rounded. The neck can expand to form a hood supported by cervical ribs. The dorsal surface of the hood bears a spectacle mark. The scales are smooth. The head bears small scales and shields. The third supralabial shield of the upper lip is large and touches the eye and nasal shields. A tiny wedge-shield is present on the undersurface of the 4th and 5th lower labials. Sub-caudal shields are arranged in two rows. Fangs are small, erect and non-movable. A large poison gland enclosed in a fibrous capsule is present on each side and a duct from it runs into the fang. When annoyed it raises the anterior part of its body about one foot above the ground, spreads the hood, produces hissing sound and bites making several wounds. The venom is neurotoxic and fatal. Its enemies are mangoose which kills it and jungle fowl which destroys its eggs.

Naja bungarus or *N. harnnah* is the largest (about 4 metre long) and most poisonous snake and is known as **king cobra**. It is also known as **hamadryas**. It mainly eats monitor lizard and also other snakes including cobra.

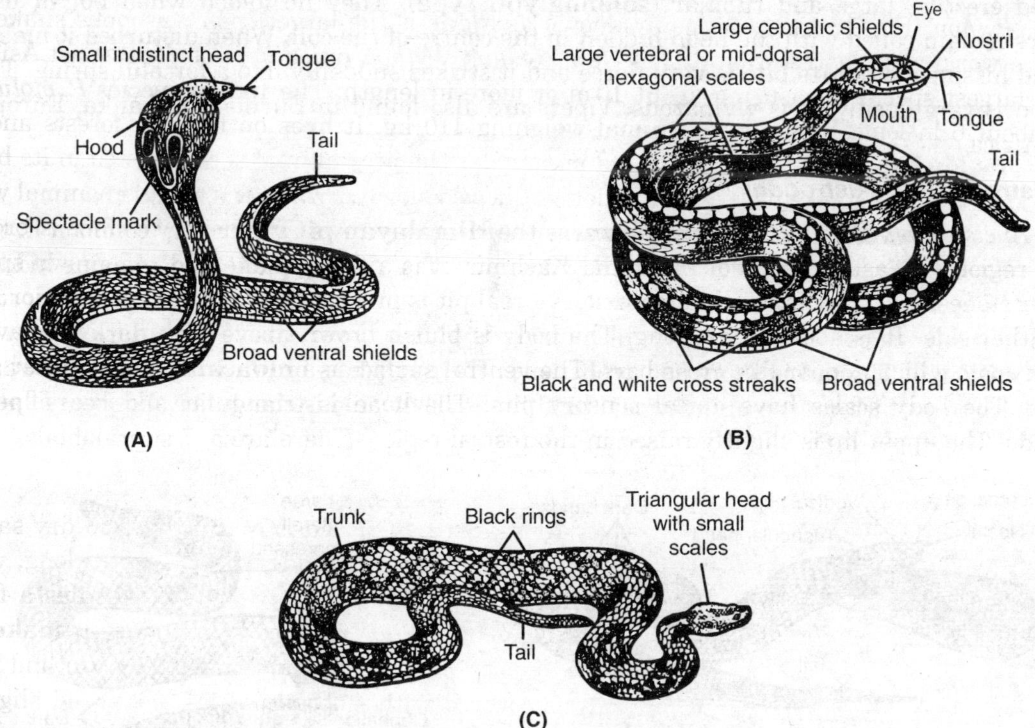

Fig. 18.15 Poisonous snakes–**(A)** *Naja naja*; **(B)** *Bungarus;* **(C)** *Vipera russelli*

Bungarus

Bungarus (Fig. 18.15B) or **krait** as it is commonly known has several species in India and Malaysia. It lives in crevices of walls or under logs and stones. It is nocturnal and feeds on smaller snakes, lizards, toads, mice, etc. It is a shy snake and attacks only when disturbed. It is about 3 feet long. The body is slender. The dorsal surface is pale brown with dark bluish black markings having white cross bars, the ventral surface is white. The fangs are small. The scales are smooth. The backbone is ridged having a central row of enlarged scales. The ventrals are large and hexagonal, fourth infralabial is the largest and subcaudals are in a single row. Krait is oviparous and the female incubates the eggs. The common krait, *B. coeruleus* is a slow moving snake but is deadly poisonous. The venom is almost three times more poisonous than that of cobra. *B. fasciatus* or the **banded krait** has alternate broad black and yellow bands. It is found in northern India and Indonesia.

Vipera

Vipera russelli (Fig. 18.15C) or **Russel's viper** is the largest Indian **pitless viper**. It is about 5 feet long, nocturnal in habbit. The head is large, distinct, flat, triangular and covered with small scales. The nostrils are lateral in position and large. The body scales are keeled. The tail is short and tapering. It is pale brown above and yellow below. The upper surface bears three longitudinal rows of large black rings appearing like chains, hence the name **chain viper**. The head and body are separated by a constriction, the neck. The head shows yellow coloured 'A' mark. The sub-caudals are arranged in two rows. The fangs are paired erectile, large and tubular (**solenoglyph type**). They lie folded when not in use. Vipers remain coiled with the head hidden in the centre of the coil. When disturbed it makes a loud hissing sound. Its bite is very fierce and it strikes suddenly like a forceful spring. The poison is very deadly. It is viviparous. Vipers are also found in Burma, Sri Lanka, Europe, and Africa.

Ancistrodon (= Agkistrodon)

Ancistrodon (Fig. 18.16A) *himalayans* is the **Himalayan pit viper** very common in the hilly regions of eastern parts of India and Kashmir. It is a timid snake and remains hidden under stones, etc. Its poison is haemotoxic. A loreal pit is present between the eye and nostril on either side. It is about 2 feet long. The body is bluish brown above with dark brown or black spots which appear like cross bars. The ventral surface is brown with black and white spots. The body scales have apical sensory pits. The head is triangular and bears large shields. The upper lip is slightly raised in the rostral region. There are 5–7 supralabials, the

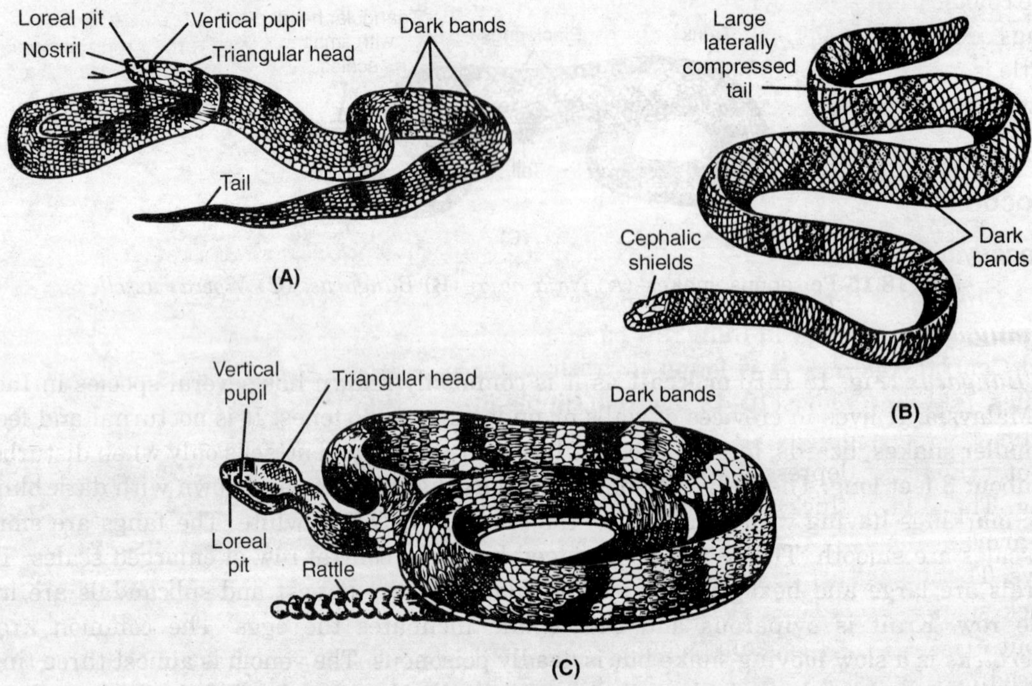

Fig. 18.16 Poisonous snakes–**(A)** *Ancistrodon*; **(B)** *Hydrophis*; **(C)** *Crotallus*

3rd and 4th are enlarged but do not touch the eye. The two postocular, three preocular, and the supraocular are enlarged sideways to form a shield over the eye. The eyes are big with golden iris and vertical pupil. The tail is cylindrical and ends in a long spine-like scale. The sub-caudals are in two rows. As in pitless viper the fangs are well developed, sheathed and of solenoglyph type. It is viviparous.

Hydrophis

Hydrophis (Fig. 18.16B) is commonly known as **sea snake.** It occurs in Bay of Bengal, Malay Archipelago and tropical parts of Indian and Pacific oceans. The body is elongated about 2 metres in length. The head and neck are slender while the tail is laterally compressed, oar-like and helps in swimming. It looks like eel but can be easily recognised by its tail. The general body colour is dark olive green with yellow cross bars above and white below. The head is small and covered by large shields. The eyes are small with round pupil. Nostrils are valvular and lie at the tip of the snout to allow breathing while in water. The ventral scales are small. The 3rd and 4th supralabials touch the eyes. There are 14–18 maxillary teeth behind the fangs. It is viviparous and comes out of water for egg laying. Its venom is neurotoxic.

Crotalus

Crotalus (Fig. 18.16 C) is found in America and Mexico and is easily recognized by the presence of a rattle at the end of the tail, hence it is commonly known as **rattle snake.** It is nocturnal and feeds on small mammals. The body is 2–3 metres long. The head is triangular, distinct from neck and has small scales. A sensory pit is present between the eye and nostril. The colour of body is greyish borwn with dark bands and beautiful pigmentation pattern. The fangs are paired long, movable and erectile. They are folded backwards when not in use. The rattle is made up of 10–12 hollow segments. During locomotion the rattle produces characterstic rattling sound. Before striking the rattle vibrates producing sound which serves as a warning. It is viviparous.

Crocodilia

Crocodylus

Crocodylus (Fig. 18.17) is commonly known as **mugger.** It is found in India, Africa, Australia and Central America. It is found in fresh water bodies. It makes a long (10–13 metres) tunnel in water. It is very dangerous to man. The body is stout, cylindrical, depressed and about 4–6 metres long. The skin is thick, leathery and covered by an armour of bony scutes arranged in transverse rows. The upper surface is dark olive green with black spots. The head is elongated into a blunt snout. The eyes, ear openings and nostrils are placed high up on the head so that they remain exposed even when the whole body is immersed

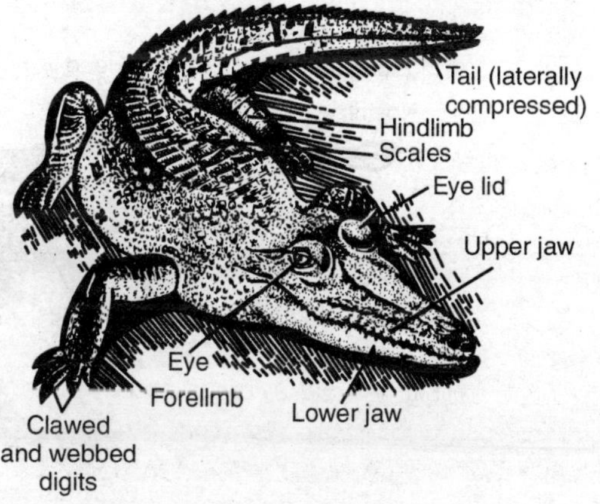

Fig. 18.17 *Crocodylus palustris*

in water. When inside water the nostrils are covered by a valve, ear openings by a flap of skin and eyes by nictitating membrane. The jaws are long powerful and bear numerous bluntly conical and unequal teeth. The tongue is non-protrusible. The tail is long, heavy and laterally compressed. It serves as a powerful organ of swimming. Dorsally it bears a crest of broad plates, double in basal half and single distally. The limbs are short and stout, with 5 fingers and 4 toes. The heart is four chambered. It is oviparous.

Alligator

Alligator (Fig. 18.18A) is found in China and North America. It inhabits shallow water and on slight disturbance burries in sand. Superficially it resembles *Crocodylus.* The body is about 3–4 metre long and covered with thick leathery skin. The exoskeleton is in the form of bony scutes. The dorsal bony scutes do not articulate with each other. They are arranged in 6–8 longitudinal rows. The ventral scutes have very little ossification. The general colour is greyish or dark brown above and yellowish below. The head is broad ending in a bluntly rounded snout. The jaws are large powerful and bear numerous bluntly conical and unequal teeth. The first and fourth mandibular teeth fit into the pits of the upper jaw. The mandibular symphysis is short and extends up to fourth or fifth tooth. The tongue is non-protrusible. The eyes and nostrils exposed when in water. The ear opening is protected by a flap of skin. The nasal bone divides the nasal aperture. The tail is long and laterally compressed. The limbs are short and strong with clawed and webbed toes. The heart is four chambered and bladder is absent. Eggs are laid on land in nests.

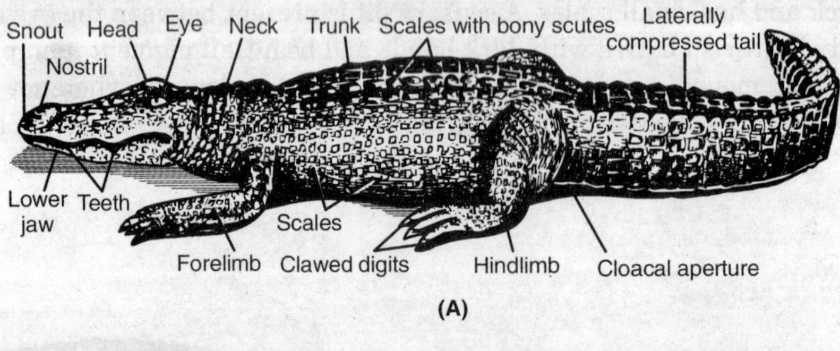

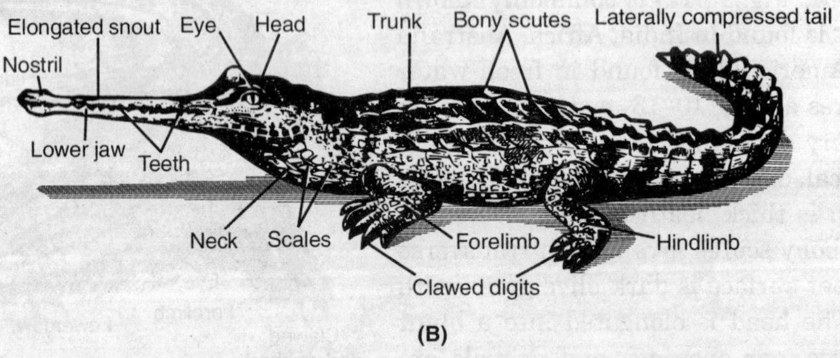

Fig. 18.18 (A) *Alligator mississipiensis;* **(B)** *Gavialis gangeticus*

Gavialis

Gavialis gangeticus (Fig. 18.18B) is commonly known as **ghariyal** or **naka**. It is found in India in the river Ganga. *Gavialis* is also found in Mayanmar and Malaysia. It feeds only on fishes because of narrow throat. It is not dangerous to man. The body is elongated, about 5 metres long. The head is large and produced into a narrow and much elongated snout with nostrils at its tip. The eyes are present on top of the head. The colour is dark olive brown with dark markings above. The body is covered with bony and epidermal horny scales. The jaws are powerful and rimmed with numerous conical and unequal teeth. The mandiubular symphysis is very long, extending up to the level of the last tooth. Fore and hindlimbs are pentadactyle with claws and web. The tail is powerful and laterally compressed. Heart is four chambered and bladder is absent. The ears are covered by a flap of skin.

19

CHAPTER

REPTILIA:
GENERAL ACCOUNT

ORIGIN OF REPTILES

Reptiles are a group of cold blooded creeping animals successfully adapted for life on land. Although amphibians are considered as the first land vertebrates, the reptiles are the first successful land vertebrates completely independent of water. They do not have any fear of dehydration.

The amphibians were first vertebrates to emerge on land but they were still dependent on water. They frequently visited water bodies to keep their skin moist. They laid eggs in water and their early life, that is, the tadpole larva, was aquatic and had gills for breathing.

The success of reptiles on land was because:

1. They had dry skin covered with an exoskeletion of **scales** or **bony plates** to prevent loss of water.
2. Evolution of shelled (**cleidoic**) or **amniotic eggs** which can be laid on land.
3. Development of extra–embryonic membrane, the **amnion** around the developing embryo (absent in Amphibia). The amnion encloses **amniotic cavity** filled with **amniotic fluid** in which the embryo develops.

The oldest amniotic eggs were found from the lower permian sediments in North America.

Origin from amphibia

The reptiles originated about 250 million years ago in the **Carboniferous period** of the **Palaeozoic era**, almost 50 million years after the origin of amphibians. The primitive reptiles most probably originated from some primitive **Labyrinthodont Amphibia** which flourished through the Carboniferous and Permian periods and finally became extinct in the Triassic. The transition was so gradual that it was difficult to ascertain whether the fossils were of advanced Amphibia or primitive reptile. It is presumed that the reptiles had a polyphyletic origin, that is, they arose along a number of independent lines.

Stem reptiles (cotylosauria)

It is now certain that the reptiles have originated in the Palaeozoic era from Labyrinthodont Amphibia, but there is no agreement with regard to the **Protoreptilia** (first reptile). The extinct progenitors of the reptiles belong to the order Cotylosauria of the subclass **Anapsida**. They have been referred to as the **stem reptiles** because they have primitive and unspecialized characters. They have no vacuity in the temporal region. The first traces of these were found in the carboniferous period in deposits formed in the great coal swamp, when the amphibians were at the peak of their development. Some cotylosaurs regarded as the probable ancestors of reptiles are as follows.

Seymouria

According to the most accepted view *Seymouria* (Fig. 19.1) belonging to the order Cotylosauria is the ideal stem reptile. It was found in the lower Permian deposits of the town Seymour in Texas (USA) about 250 million years ago. *Seymouria* was about 2 feet long, aquatic, lizard-like animal with a thick body. The head was small and pointed and the nostrils were dorsally placed. The limbs were short and pentadactyle and the tail was short.

Fig. 19.1 *Seymouria*

It possessed characters both of reptile and Amphibia. Romer (1962) considered *Seymouria* as a reptile-like amphibian and Colbert (1969) considered it a connecting link between Amphibia and primitive reptile.

Amphibian characters

1. Skull flat with little ossification.
2. A prominent otic notch for accomodation of ear drum present.
3. Single occipital condyle in the skull.
4. Intertemporal bone present behind the eye.
5. Palate primitive and incomplete.
6. Teeth folded (labyrinthine); also present on vomers and palatines.
7. Head region with traces of lateral line canal.
8. Vertebrae not properly differentiated.
9. Neck short so pectoral girdle lies behind the head.
10. Only one pair of sacral ribs.

Reptilian characters

1. Skull anapsid type and moncondylic.
2. Interclavicle of pectoral girdle long.
3. Neural arches of vertebrae were broad.
4. Ilium of pelvic girdle expanded for attachment of muscles used for walking.
5. Pelvic girdle attached to vertebral column by sacral vertebrae.

6. Limbs pentadactyle, muscular and arise mid ventrally. The arrangement of toe bones similar to early reptiles and the phalangal formula was reptile-like 2 : 3 : 4 : 5 : 3 or 4.

Conclusion

When *Seymouria* was living the reptiles were already present for about 50 million years. So it cannot be ancestor of reptile. Romer treated *Seymouria* as a reptile and placed it under the order **Cotylosauria** in Reptilia. Other workers treated it as an Amphibia and placed it in the order **Seymouriamorpha** in Amphibia. However, now it is regarded as a **connecting link** between **Labyrinthodontia** (Amphibia) and **Cotylosauria** (Reptilia).

Contemporary to *Seymouria* are **Didectomorphs** and **Captorhinomorphs** which are very primitive reptiles but had their own marked specialization. The didectomorphs were more tortoise like and were on the road to chelonians while the captorhinomorphs were more lizard like and were on the road to typical reptiles and mammal - like reptiles. Because of their primitive skull features both of them have been put under Cotylosauria, the stem reptiles.

Limnoscelis

It is a captorhinomorph cotylosaur. Its fossils were found from late Carboniferous or early Permian in New Mexico. In its general organisation it seems to be the first undoubted reptile (Romer 1946) and probably may be taken to be nearer to the protoreptile as it was also too late an arrival in the geological time scale.

Limnoscelis is about 5 feet long including its tail which is almost as long as the body. The legs were very short and stumpy. Like *Seymouria* it was also aqualic. It had many reptilian features like:

1. The skull was of anapsid type and laterally compressed.
2. The otic notch is absent but its disappearance is indicated by a mark at the back of the skull.
3. The premaxillary teeth were large and overhung the front teeth in the lower jaw. This is a characterstic feature of early reptiles but not seen in amphibians.

Diadectes

It is a didectomorph cotylosaur, a contemporary of *Limnoscelis*, but represents a different line of evolution. *Diadectes* showed the following features:

1. The otic notch was present at the back of the skull.
2. Dentition specialized. The front teeth were chisel-shaped and back teeth were broad with ridged crowns.

The earliest large land reptile were probably the **pareiasaurs** which were related to *Diadectes*. They appeared in late permian and were herbivorous. They had a crest of bony plates along the back which was most probably used as a defence against the carnivorous mammal - like reptiles.

From the above discussion it is evident that the cotylosaurs were the probable ancestors of reptiles. Many workers are of the view that during the Permian period, the cotylosaurs separated into two divergent evolutionary groups:

1. **Therapsida:** It is represented by captorhinomorph cotylosaur which gave rise to mammal - like reptiles and later evolved into mammals.
2. **Sauropsida:** It is represented by didectomorph cotylosaur which evolved into remaining reptiles and birds.

It is, however, not clear at what stage of evolution these two groups separated.

TERRESTRIAL ADAPTATIONS IN REPTILES

The reptiles are the first vertebrates which have become completely adapted for life on dry land both as adult and during embryonic development. They have achieved this by undergoing certain modifications in their **organisation** and their **eggs**. During the transition from Amphibia to Reptilia following changes occurred which enabled them to survive and diversify on land:

1. Skin and **exoskeleton:** Since land animlas live independent of water they require a mechanism to conserve their body water. The skin of reptiles has become **dry, thick** and **rough**. It is covered by **scales, shields, plates, scutes**, etc. which prevent evaporation of water from the body surface.

2. Pentadactyle limbs: The land animals require well developed limbs to move on land. The reptiles have pentadactyle limbs with clawed digits which enable them to walk on land. The limbs are modified according to their habit.

3. Skeleton: The land animals require a well developed skeleton to support their body weight and help them in locomotion. In reptiles the skeleton is strongly developed. The skull is fully ossified. The vertebral column and the girdles supporting the limb are strongly developed.

4. Respiratory organ: The cutaneous and gill respiration of amphibians is replaced by complete **pulmonary respiration**. The respiratory organs are a pair of **lungs** lying in the thoracic cavity. The thoracic muscles which bring about expansion and contraction of the thoracic cavity are well developed and play a very important role in breathing.

5. Circulatory system: The circulatory system has become modified so that there is no mixing of arterial and venous blood. For this the ventricle is partially or completely divided. The aortic arches have independent origin from the heart.

6. Excretory system: The kidneys have become **metanephric.** Excretion is **uricotelic**, *i.e.*, the excretory waste is mainly uric acid.

7. Sense organs: Since land animals are in direct contact with their surrounding, they have well developed eyes and ears. In some like snakes, the ears are absent due to their burrowing habit. They can perceive ground vibrations. Reptiles possess a very special sense organ, the Jacobson's organ which lies in the roof of the mouth. It is a chemoreceptor and is able to detect smell. It helps to locate the prey or members of the opposite sex.

8. Internal fertilization: In amphibians fertilization was external because they bred in water. In reptiles fertilization became internal, an adaptation to terrestrial existence.

9. Copulatory organ: For internal fertilization the male developed a copulatory organ, the penis.

10. Development: The development became direct by omitting the larval stage in the life history.

11. Foetal membranes: The developing embryo is enclosed within four types of foetal membranes, namely **chorion**, **amnion**, **allantois** and **yolk sac**.

The chorion is a thin membrane lying just beneath the calcareous shell.

The amnion forms amniotic cavity filled with amniotic fluid in which the embryo develops. It provides the aquatic environment for the developing land animal.

The allantois is highly vascular and helps in respiration. The allantoic cavity stores end products of metabolism and acts as urinary bladder. The yolk sac provides mourishment to the developing embryo.

12. Eggs: The reptiles laid their eggs on land. The eggs were protected by an outer hard calcareous **shell**, hence they were called **cleidoic eggs**. The developing embryo was surrounded by a membrane, the **amnion** (absent in amphibians), hence eggs of reptiles were called **amniotic eggs** (Fig. 19.2). The amnion encloses a cavity called **amniotic cavity** which was filled with a fluid called **amniotic fluid** in which the embryo develops. Hence the reptilian eggs did not depend on water for development. The amniotic fluid is described as the artificial swimming pool of the developing embryo.

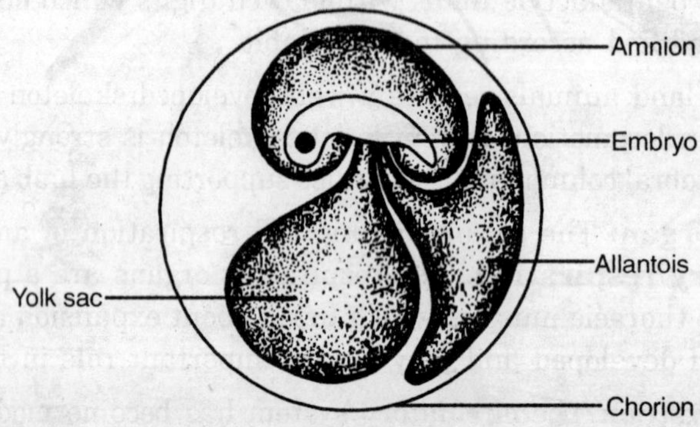

Fig. 19.2 Amniotic egg

ADAPTIVE RADIATION AND EVOLUTION

The evolution of a single group of animal in different directions invading different habitats with different types of adaptations is called **adaptive radiation** or **divergent evolution**. It generally results from competition for food and living space. Among vertebrate groups, the reptiles have shown greatest adaptive radiation during the course of their evolution.

The first or most primitive reptiles were the **Cotylosauria**. They are called **stem reptiles** because they possess primitive characters and have given rise to all other reptilian orders. The skull of Cotylosauria had complete roofing, that is, there was no vacuity or fossa in the temporal region of the skull, hence they were placed in the sub-class **Anapsida**. From this generalized condition the reptiles have evolved along several lines.

Temporal fossae and arcade. The region of skull behind the orbit is known as **temporal region**. In the skull of reptiles the temporal region contains one or two cavities called **temporal fossae**. The fossae are bordered by a set of bones which form an arch or bar called **arcade**. The fossae provide surface for attachment to temporal muscles which help in the movement of lower jaw. The fossae play an important role in the classification of reptiles.

Evolution of reptiles

The reptiles have evolved along several lines from the stem reptiles but the five main lines based on skull roofing are:

1. **Anapsid line**
2. **Synapsid line**
3. **Euryapsid line**
4. **Parapsid line**
5. **Diapsid line.**

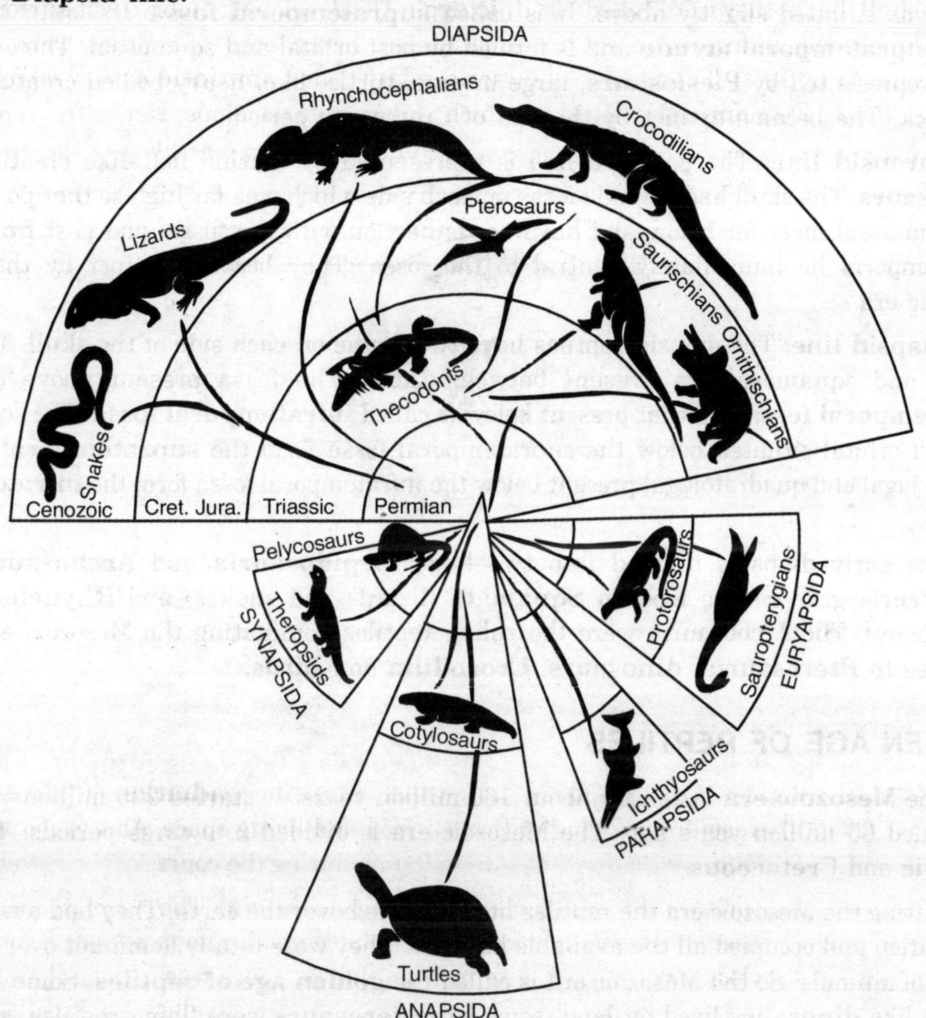

Fig. 19.3 Chart showing evolution of reptiles

Anapsid line: The anapsid line has a skull resembling the stem reptiles. There is no fossa in the temporal region of the skull. Among the living reptiles, the chelonians (turtles and tortoises) have retained anapsid skull. They arose as an early offshoot of stem reptiles in the Triassic period and have remain unchanged since 160 million years.

Synapsid line: The synapsid line is represented by the mammal-like reptiles or **Synapsida**. The early **Pelycosauria** or **Theromorpha** were similar to the cotylosaurs. The later **Therapsida** were more mammal-like. The temporal region of the skull had a single vacuity or fossa called **infratermporal fossa**, on each side behind the orbit. The fossa was bounded above by the post-orbital and squamosal and below by the quadratojugal and jugal. The lower bones together form the **infratemporal arcade**. These reptiles became extinct in the Jurassic period after giving rise to mammals. This type of skull is not found in any living reptile.

Euryapsid line: The euryapsid reptiles also had a single fossa on each side of the skull, but it was situated slightly above. It is called **supratemporal fossa**. Its lower border is called **supratemporal arcade** and is formed by post-orbital and squamosal. The euryapsid line is represented by **Plesiosaurs**, large marine, turtle - like, heavy bodied creatures with long neck. The became extinct by the end of Cretaceous period.

Parapsid line: The parapsid line is represented by marine fish - like creatures, the **Icthyosaurs**. The skull had a single fossa on each side which was far high so that post orbital and squamosal meet far below and have no connection with the fossa, and post frontal and supratemporal lie immediately ventral to the fossa. They became extinct by the end of Mesozoic era.

Diapsid line: The diapsid reptiles have two fossae on each side of the skull. The post-orbital and squamosal are present between them. The fossa present above is called **supratemporal fossa** and that present below is called **infratemporal fossa**. The squamosal and post-orbital situated below the supratemporal fossa form the **supratemporal arcade** and the jugal and quadratojugal present below the infratemporal fossa form the **infratemporal arcade**.

The early diapsids divided into two lines: **Lepidosauria** and **Archosauria**. The Lepidosauria gave rise to modern **Squamata** (lizards and snakes) and **Rhynchocepalia** (*Sphenodon*). The Archosauria were the ruling reptiles dominating the Mesozoic era. They gave rise to **Pterosauria**, **dinosaurs**, **Crocodilia** and **birds**.

GOLDEN AGE OF REPTILES

The **Mesozoic era** lasted for about 160 million years. It started 225 million years ago and ended 65 million years ago. The Mesozoic era is divided into three periods: **Triassic, Jurassic** and **Cretaceous**.

During the Mesozoic era the reptiles literally ruled over the earth. They had a worldwide distribution and occupied all the available habitats. They were totally dominant over all other groups of animals. So the Mesozoic era is called the **golden age of reptiles**. Some Mesozoic reptiles like **dinosaurs** lived on land, some like **pterosaurs** were flying reptiles, and some like **ichthyosaurs** were adapted for aquatic life.

The major groups of Mesozoic reptiles are:

1. **Thecodont**
2. **Dinosaurs** (terrible lizards)
3. **Pterosaurs** (flying reptiles)
4. **Marine reptiles**
5. **Mammal - like reptiles.**

Thecodont

The Thecodont reptiles appeared in the begining of the Triassic period and became extinct by the end of Triassic.

Example: *Erythrosuchus.*

Dinosaurs (terrible reptiles)

The word **dinosaur** was proposed by Sir Richard Owen. It means **terrible reptiles**. The dinosaurs appeard in the middle of Triassic period, dominated the earth up to the Cretaceous period and finally became extinct by the end of Cretaceous period.

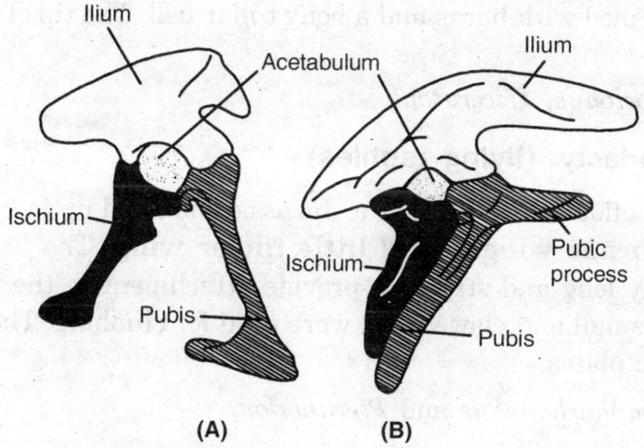

Fig. 19.4 Structure of pelvis. **(A)** Order Saurischia; **(B)** Order Ornithischia

The dinosaurs include two orders:

1. **Saurischia** (reptile-like dinosaurs)
2. **Ornithischia** (bird-like dinosaurs).

Saurischia means reptile- like pelvis, that is, the arrangement of bones of the pelvic girdle is similar to that of reptiles. The pelvis was a **triradiate** (Fig. 19.4A) structure in which the pubic bone extends down and forward below the iluim and the ischium extends down and backward. The early Saurischians were small, bipedal, carnivorous creatures like their thecodont ancestors. They gave rise to two groups:

Suborder 1. THEROPODA. Small to large, bipedal, carnivorous saurischians.

Examples: *Ornithomimus*, *Allosaurus*, *Tyrannosaurus*.

Suborder 2. SAUROPODA. Giant, heavy-bodied, herbivorous saurischians.

Examples: *Apatosaurus (Brontosaurus)* and *Diplodocus*.

Ornithischia evolved from some thecodonts in the Triassic period. The arrangment of bones of pelvic girdle (Fig. 19.4B) is similar to that of birds (ornithischia means bird - like pelvis). The pubic bone has rotated backward so that it is parallel to the ischium. The pelvis is a tetraradiate structure. Its four processes are formed by anterior and posterior extensions of the iluim, prepubis and closely placed ischium and pubis. Ornithischian dinosaurs had no teeth in the jaws. The upper portion of the skull and jaws were modified into a beak. Most of them were quadrupedal.

The ornithischian dinosaurs have been divided into four suborders:

Suborder 1. ORNITHOPODA. Most primitive bipedal creatures.

Examples: *Camptosaurus, Iguanodon.*

Suborder 2. STEGOSAURIA. Possessed peculiar armour of bony plates and spines along the middle of the back.

Example: *Stegosaurus.*

Suborder 3. ANKYLOSAURIA. Head and entire back covered with bony plates.

Example: *Ankylosaurus*.

Suborder 4. CERATOPSIA. Horned dinosaurs. They were of the size of *RHINOCEROS*. *They* had large head armed with horns and a bony collar frill. The tip of the upper jaw formed a small hooked beak.

Examples: *Monoclonus*, *Triceratops*.

Pterosaurs or pterodactyl (flying reptiles)

The flying reptiles flourished during the Jurassic and lived up to late Cretaceous period. They possessed **membrane wings** called **little finger wing**. The little finger or the fifth digit became extremely long and strong to provide attachment to the membrane wing. The other four digits were small and clawed and were used for climbing. The eyes were large and provided with sclerotic plates.

Examples: *Rhamphorhynchus* and *Pteranodon.*

Marine reptiles (fish-like reptiles)

Some reptiles of the Mesozoic era returned to water. Their feet were modified into paddles for swimming.

Examples: *Ichthyosaurus* and *Plesiosaurus.*

Mammal-like reptiles

The synapsid line was represented by mammal - like reptiles. It is divided into three suborders:

Suborder 1. PELYCOSAURIA. Considered to be the first synapsid as it showed most primitive synapsid characters.

Examples: *Varanosaurus, Dimetrodon.*

Suborder 2. THERAPSIDA. Advanced mammal - like reptiles appeared in early Jurassic. They showed teeth differentiation into incisors, canines, premolars and molars, many had two occipital condyles. secondary palate present, dentary bone enlarged, had double jaw articulation. Theriodonts are specialized theraspids.

Examples: *Cynognathus*, *Tritylodon*.

Suborder 3. ICTIDOSAURS. Advanced theriodonts, are connecting link between advanced theriodonts and primitive mammals.

Example: *Diarthrognathus*.

Some mesozoic reptiles

Ornithomimus

Ornithomimus (Fig. 19.5A) is a saurischian dinosaur. It closely resembled ostrich so it is often called **ostrich-dinosaur**. It was about two or three metres long. The hind limbs were long and slender and feet were bird-like. The fore limbs were short and the hand had only three digits. It was dominant during the late Cretaceous period. It ate the eggs of other animals. It was the fastest dinosaur.

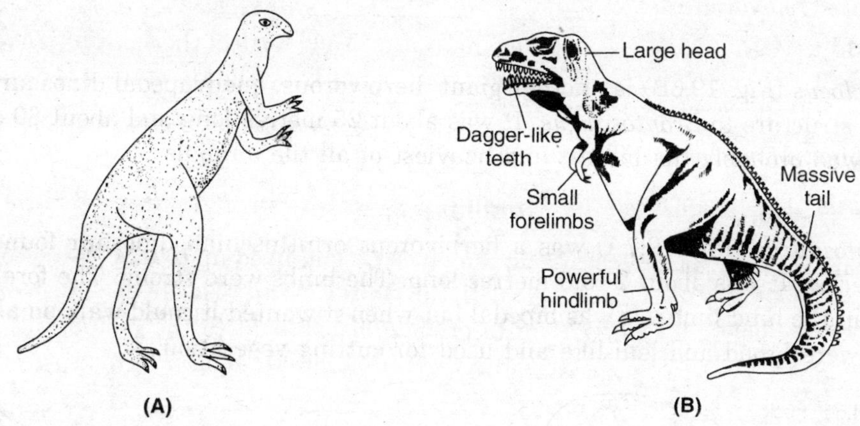

Fig. 19.5 Mesozoic reptiles. **(A)** *Ornithomimus;* **(B)** *Tyrannosaurus*

Tyrannosaurus

Tyrannosaurus (Fig. 19.5B) was the largest flesh-eating bipedal saurischian dinosaur which weighed about 4 quintals. It was about 15 meters long and 5 metres tall. The head was about 1 metre long. It lived during the Cretaceous period about 68 million years back. The fore limbs were extremely short but the hind limbs were strong and massive. The feet had three forward toes and one backward toe. The jaws were long with large dagger-like teeth. Each tooth was about 15 cm long. It was a carnivore adapted to hunting other large reptiles.

Apatosaurus (= Brontosaurus)

Apatosaurus (Fig. 19.6A) was a giant sauropod saurischian dinosaur that lived in the Jurassic period. It was nearly 23 metres long and weighed about 30–60 tons. It was a quadrupedal dinosaur with strong limbs. The fore limbs were slightly shorter than the hind limbs. The feet were broad with large foot pads. The neck and tail were long. The vertebrae had strong spines to provide surface for attachment of muscles. The head was comparatively small and the nostrils were placed at the tip of snout. It lived near swamps and was amphibious. The jaws were weak and armed with sharp, cutting, leaf-shaped teeth adapted to eat soft vegetation (herbivorous).

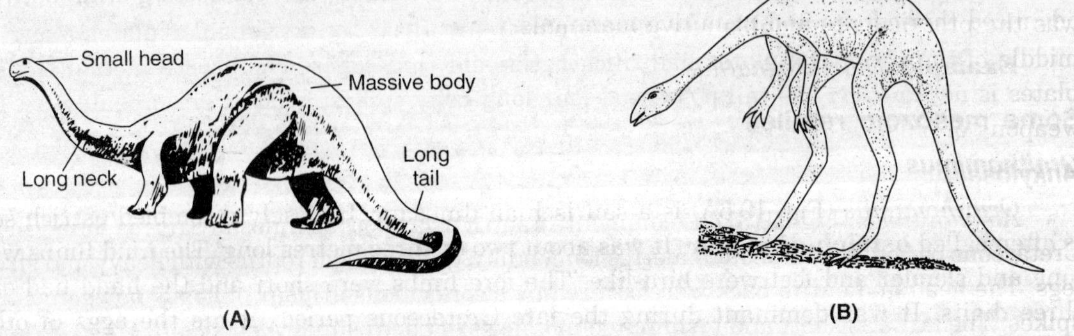

Fig. 19.6 (A) *Apatosaurus;* **(B)** *Diploclocus*

Diploclocus

Diploclocus (Fig. 19.6B) is another giant, herbivorous, quadrupedal dinosaur similar in habits and structure to *Brontosaurus*. It was about 25 metres long and about 80 quintals in weight. It was probably the largest and heaviest of all the animals.

Camptosaurus

Camptosaurus (Fig. 19.7A) was a herbivorous ornithischian dinosaur found in upper Jurassic period. It was about 2 to 6 metres long. The limbs were strong. The forelimbs were shorter than the hind limbs. It was bipedal but when it wanted it could walk on all four legs. The teeth were broad and leaf-like and used for cutting vegetation.

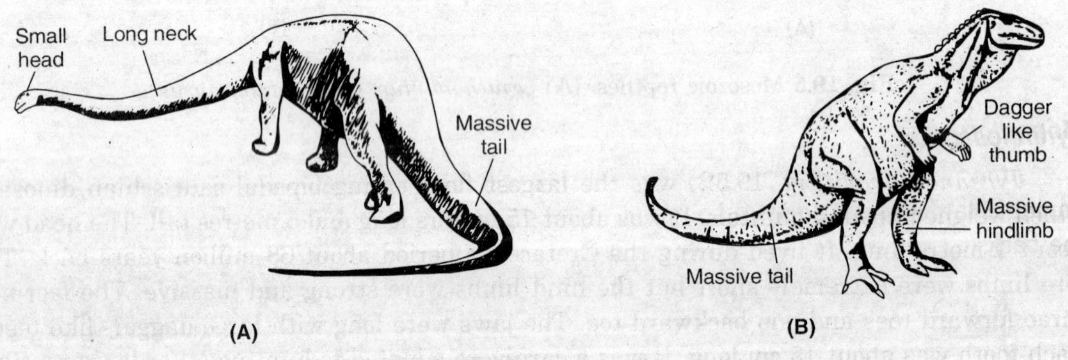

Fig. 19.7 (A) *Camptosaurus;* **(B)** *Iguanodon*

Iguanodon

Iguanodon (Fig. 19.7B) was a terrestrial, herbivorous, bipedal ornithischian dinosaur found in Triassic and lower Cretaceous period. It lived in herds. It was about 9 metres long, heavy-bodied creature with horny beak. The hind limbs were longer and stronger than the fore limbs. The feet had only four clawed toes. The thumb of hand was enlarged into a sharp spike which was most probably used as a weapon of defense.

Stegosaurus

Stegosaurus (Fig. 19.8A) is commonly known as **plated dinosaur**. It was a quadrupedal ornithischian dinosaur about 6 metres long. The limbs were strong and feet broad. The head was comparatively small and the back was arched. The most striking feature of this animal was the presence of a double row of alternately arranged bony, triangular plates along the middle of the back. The plates had thick bases and thin edges. The exact function of these plates is not known. The tail posseseed four long bony spikes, which were probably used as weapons of defense.

Ankylosaurus

Ankylosaurus (Fig. 19.8B) is commonly known as armoured dinosaur. It was found in the Cretaceous period. It was about 6 metres long, bulky, quadrupedal reptile with heavy and strong legs. The head and entire back were covered by a continuous armour of heavy polygonal bony spikes. The tail was short and terminated in a mass of bone that formed a club. The teeth were small and weak suggesting that *Ankylosaurus* was herbivorous and ate soft vegetation.

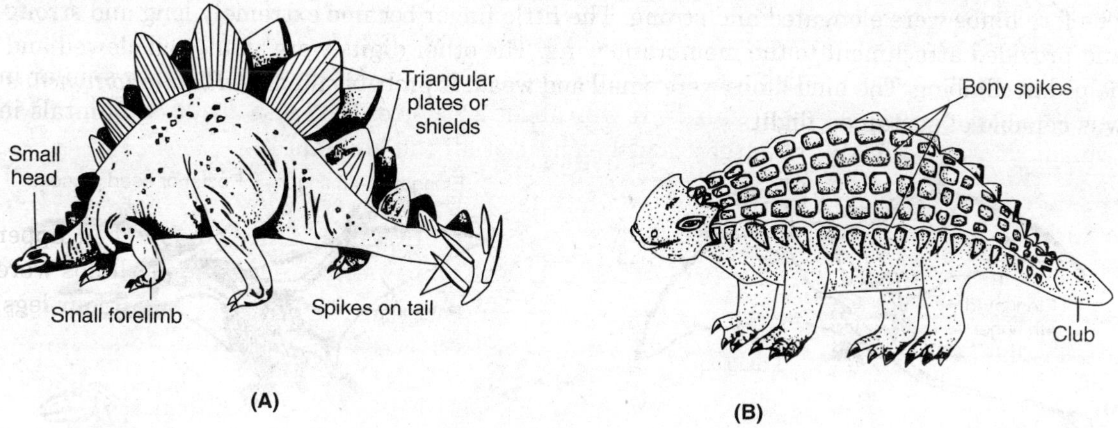

Fig. 19.8 (A) *Stegosaurus;* **(B)** *Ankylosaurus*

Monoclonius

Monoclonius (Fig. 19.9A) resembled the present day rhinoceros (mammal) in size and in having a large **nasal horn** on the front of the skull. It was quadrupedal. It had a collar frill at the back of the head. The limbs were short, stout and strong. The tail was short and tapering.

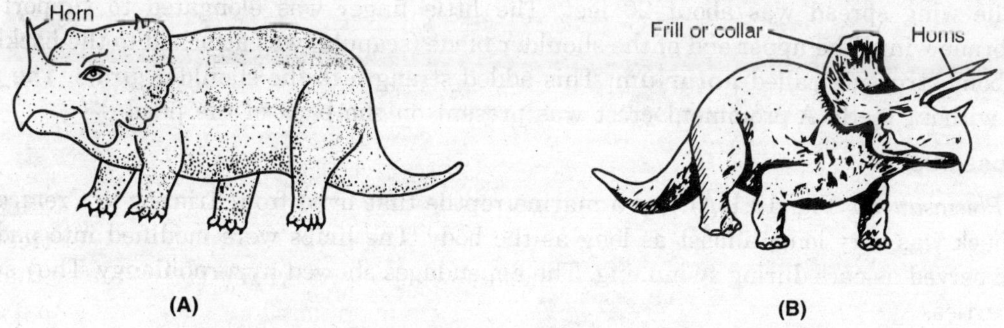

Fig. 19.9 Mesozoic reptiles. **(A)** *Monoclonius;* **(B)** *Triceratops*

Triceratops

Triceratops (Fig. 19.9B) is commonly known as **horned dinosaur**. It possessed three horns, one **nasal horn** on the front of the skull and two **brow horns**, one above each eye. It was of the size of *Rhinoceros*. It also possessed a large perforated **frill** at the back of the head which was formed by backward growth of parietal and squamosal bones. Primarily the frill provided surface for attachment to the muscles of the neck and secondarily it provided protection to the neck and shoulder region. The horns served as weapons of defense. *Triceratops* was dominant in the Cretaceous period and were found in large numbers in North America. It was herbivorous, upland-dwelling dinosaur.

Rhamphorhynchus (Long-tail pterosaur)

Rhamphorhynchus (Fig. 19.10A) was a **flying reptile** found in the Jurassic period. It was about 2 feet long and the wing spread was about 2 feet. It had the characterstic archosaurian skull. The jaws were elongated and had long pointed teeth. The neck was long and flexible. The tail was very long, almost twice the length of the body and ended in a steering rudder. The fore limbs were elongated and strong. The little finger became extremely long and strong and provided attachment to the membrane wing. The other digits were small and clawed and helped in climbing. The hind limbs were small and weak. It probably ate fish. *Rhamphorhynchus* was capable of sustained flight.

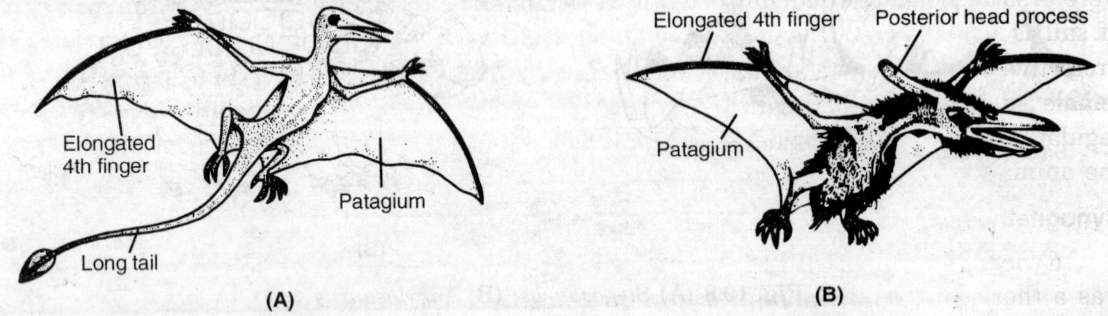

Fig. 19.10 (A) *Rhamphorhynchus;* **(B)** *Pteranodon*

Pteranodon *(short-tail plerosaur)*

Pteranodon (Fig. 19.10B) is the last of the flying reptile. The body and tail were small but the wing spread was about 20 feet. The little finger was elongated to support the membrane wing. The upper end of the shoulder blade (scapula) was attached to the backbone by a bony structure called **notarium**. This added strength to the shoulder girdle. The jaws were without teeth. A prominent crest was present on the back of the head.

Plesiosaurus

Plesiosaurus (Fig. 19.11A) was a marine reptile that lived from Triassic to Cretaceous. The neck was very long, almost as long as the body. The limbs were modified into paddles which served as oars during swimming. The appendages showed hyperphalangy. They swam like-turtles.

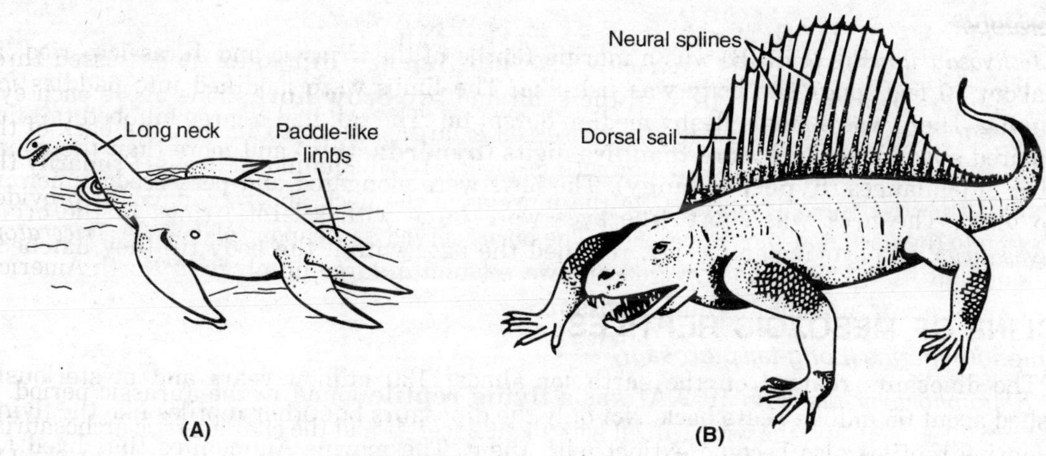

Fig. 19.11 (A) *Plesiosaurus;* (B) *Dimetrodon*

Dimetrodon

Dimetrodon (Fig. 19.11B) was a carnivorous mammal - like reptile of the Permian times. It was about 9 feet long and had dagger-shaped biting teeth. Its most striking feature was the presence of a huge, sail on the back in the middle region. The neural spines of the vertebrae were greatly elongated and supported by a web of skin to form the sail. The actual function of sail is not known but a number of functions have been attributed to it. The sail was protective as it made the animal look big thereby frightening the enemies. It was absent in female so helped to distinguish between the two sexes. It was probably a temperature regulating device as it provided a large area of skin surface for warming up or cooling down the animal.

Cynognathus

Cynognathus (Fig. 19.12A) was an advanced mammal-like reptile. Phylogenetically it was a theriodont reptile ancestral to the mammals. It showed differentiation of teeth. The canines were large indicating that it was predatory. The scapula was mammal-like. The jaw suspensorium was formed by quadrate bone of the skull and angular bone of the lower jaw.

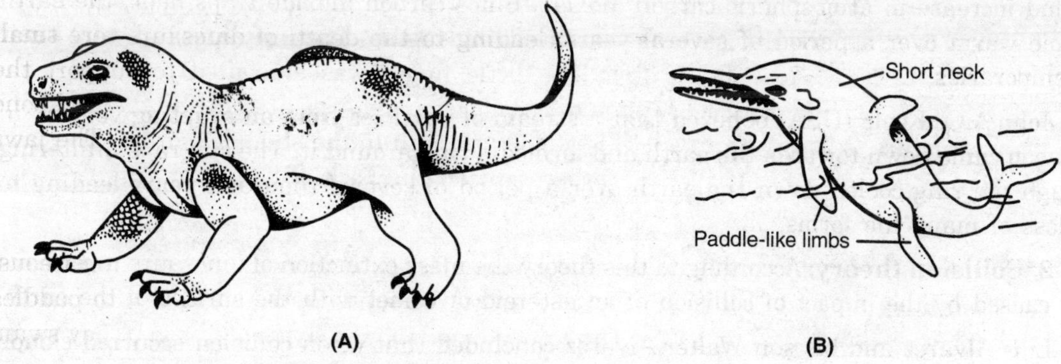

Fig. 19.12 (A) *Cynognathus;* (B) *Ichthyosaurus*

Icthyosaurus

Icthyosaurus (Fig. 19.12B) was a marine reptile of the Triassic and Jurassic period. It was about 10 feet long. The body was fish-like. The limbs were modified into paddles for swimming. There was a single fleshy median dorsal fin. The tail had a large biltobed tail fin. The paired appendages had more than five digits (**hyperdactyly**) and more than the usual number of phalanges (**hyperphalangy**). The jaws were elongated and possessed numerous sharp teeth. The neck was short. The eyes were large with sclerotic rings in the orbit. *Ichthyosaurus* was ovoviviparous, i.e., retained the eggs within the body till they hatched.

DECLINE OF MESOZOIC REPTILES

The dinosaurs roamed on the earth for almost 130 million years and mysteriously vanished about 65 million years back. Not only the dinosaurs but other reptiles like the flying and marine reptiles also became extinct with them. The marine Ammonites that lived for almost 330 million years and the Trilobites that lived for 230 million years also became extinct almost at the same time. Scientists are still puzzled over why so many animals disappeared so suddenly. Some fossil clues suggest that these mass deaths occurred within a short period of one or two years. Various factors like large size, food poisoning, inter and intraspecific competition, epidemic, catastrophism, environmental change, racial senescence etc. have been suggested for their sudden extinction, but none of them are very satisfactory. Different views have been put forward by different scientists regarding their sudden extinction. The two main theories are:

1. Eruption theory: This theory claims that widespread **volcanism** could be the real cause of mass extinction.

Charles B. Officer et.al. (Nature 1987) suggested that volcanic eruptions led to large-scale emission of sulphur dioxide into the atmosphere. The sulphur dioxide destroyed the ozone layer, so ultraviolet rays reached the earth surface killing animals and plants. Large clouds of dust and smoke blocked the sun rays, as a result photosynthesis stopped and earth cooled down. The dinosaurs died due to lack of food and cold atmoshperic temperature. When it rained sulphur dioxide dissolved in water producing acid rain which destroyed marine life.

D. M. McLean (USA) suggested that major volcanic outbursts resulted in loss of marine life and increase in atmospheric carbon dioxide. Since carbon dioxide traps heat, the earth became warm over a period of several years, leading to the death of dinosaurs and their contemporaries.

John A.O. Keefe (USA) believed that a stream of particles from an erupting volcano in the moon came down towards the earth and formed a ring around it. The sun rays infiltering through the ring cooled down the earth over a period of several thousand years leading to the loss of many life forms.

2. Collision theory: According to this theory the mass extinction of dinosaurs must have been caused by the impact of collision of an asteroid or comet with the surface of the earth.

Luis Alvarez and his son Walter Alvarez concluded that when collision occurred clouds of dust entered the atmosphere and stayed for several years blocking sunlight. As a result the earth became cool and dark killing many plants and animals.

Kenneth J. Hsu et al. believed that when collision occurred some poisonous substance like **iridium** was introduced into water bodies and killed the planktons and other life forms. When planktons, which use carbon dioxide, died the concentration of carbon dioxide in the atmosphere increased making the earth warm. This might have led to the extinction of dinosaurs.

The sudden extinction of dinosaurs and thier contemporaries still needs to be analysed and is open for speculation and reasoning.

ECONOMIC IMPORTANCE OF REPTILES

The reptiles are not of great economic importance but they are valued for thier **skin**. The skin of reptiles, chiefly snakes and lizards, is now extensively used for production of **leather** for high-priced shoes, purses, bags and other fancy articles. The wearing quality of reptile leather is greater and they present an infinite variety of pattern. Moreover, the reptile leather can be finished off in any colour. Among the lizards the skin of *Varanus* is very much in demand. The skin of *Uromastix hardwickii* is not of much value but the skin of *U.asmussi* is very much liked abroad. India imports it from Baghdad.

Other economic value

1. Study material. Lizards like *Uromastix*, *Calotes* and *Hemidactylus* serve as good study material and are dissected in laboratories in different parts of the country.

2. As food. The flesh of some lizards is considered a delicacy in same parts of the world. The tail muscles of *Uromastix* is eaten by some tribal people.

3. Religious importance. In Hindu religion the snakes are associated with Lord Shiva and therefore they are worshipped. On naag panchami people in many parts of India offer milk to cobra (naag).

4. Medicinal importance. The quacks sell oil extracted from the fat bodies of *Uromastix* as a cure for impotance and also as a massage oil. The powdered faeces of *Uromastix* mixed with kajal has been claimed to cure corneal opacity.

5. Source of living. Many poor villagers make the snake dance to the tune of flute and earn money for their living.

6. Decoration. The chelonians (turtles and tortoises) are kept in aquarium for decoration purpose.

20
CHAPTER

CLASS AVES

The class **Aves** (**L.** *avis*, bird) includes birds. Of all the animals, birds are the best known and most easily recognized. The most characteristic feature of bird is the presence of feathers. The study of birds is known as **Ornithology** (**Gr.** *ornis*, birds). The birds show significant advance over lowly organised vertebrates in having an insulated body covering, colourful plumage, complete separation of venous and arterial blood in the heart, regulated body temperature, ability to fly, highly developed voice, hearing and sight, migratory habit and specialized care of the young. In all these features they can only be compared with mammals from which they differ in the type of body covering, in the ability to fly and in the manner of reproduction.

The birds have existed for over 130 million years. Although they have existed as a separate class of animal they have retained their basic structure to a remarkable degree. There are many ways in which the birds are adapted to their surrounding. The parts of the world where birds live are rather limited. Though birds have been seen in practically every area of the world from pole to pole and at heights more than that of Mount Everest, yet they have never been found to penetrate more than about 11 feet into the earth or about 100 feet below the surface of the sea. Every year some part of their lives has to be spent in contact with land, for birds cannot build nests at sea.

TYPE COLUMBA

The **common rock pigeon**, *Columba livia* is quite familiar bird best suited for type study because of its convenient size, harmless nature and availability in large numbers. More than 500 species of pigeon are known. They are of two types: **wild** and **domestic**. The wild variety is known as **blue rock pigeon** or **rock dove** or *Columba livia*. There are many domestic varieties which differ from each other in size, colouration, arrangement of feathers, etc. They have descended from the wild variety by artificial selection.

Distribution

The blue rock pigeon is widely distributed in Asia, Europe, and Sourth Africa. It is especially abundant in Palestine in Israel. It has several subspecies. In India two subspecies are found. One subspecies is *Columba livia neglecta*. It is large, pale coloured and occurs upto 13,000 feet on the Himalayas. The other subspecies is *Columba livia intermedia*. It is small, dark and occurs throughout the plains except in some of the wettest area.

Habitat

The blue rock pigeon lives is fissures and holes in rocks forts, old buildings and side walls of wells. In towns and cities it lives in big buildings, godowns, grain markets, temples, mosques, churches, railway stations, etc. They never nest on trees.

Habits

The pigeons are harmless, inoffensive and timid in nature, so they are considered a symbol of peace, harmony and happiness all over the world. They fly in flocks and roost together. They lead a **monogamous life**, that is, one male and female partner live together throughout life. Breeding season occurs throughout life. The solitary couple makes a simple nest of small dry sticks in any place to protect the eggs and young ones from rain and sun. The pigeons show courtship behaviour followed by copulation, Fertilization is internal. The eggs are laid in nest and incubated by the warmth of the parent's body. Hatching takes place in about 15 days. The pigeons show parental care. The young ones are nourished by the parents by fatty curd-like secretion called **pigeon's milk** secreted in their crop. They show homing instinct. The pigeons produce a characteristic sound which resembles the syllables "gooter goon", they do not sing, chirp or screech.

Food: The pigeons are vegetarians. They eat seeds of fruits, grains and pulses. Sometimes they eat insects, snails and slugs, probably mistaken for seeds or grains. In morning and evening they collect in flocks, leave their place of rest to plunder nearby fields.

Locomotion: The pigeons have a pair of powerful wings adapted for swift and strong flight. On land they walk rapidly on their two legs. Such type of walking is called **bipedal gait**. When disturbed they rise suddenly by striking the ground with their wings producing a crackling sound.

EXTERNAL FEATURES

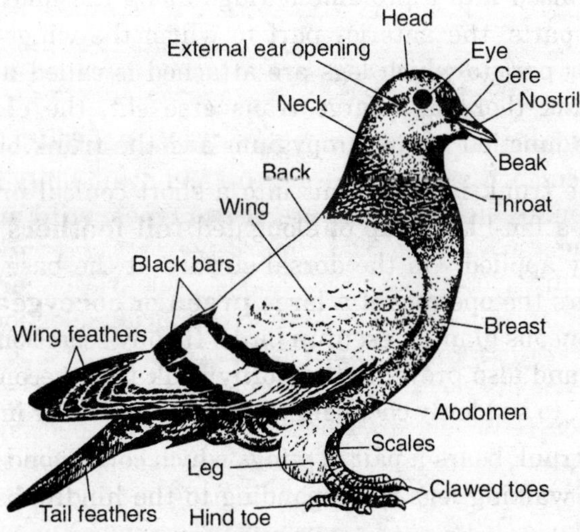

Fig. 20.1 *Columba*–External features

Shape, size and colour

The body is spindle-shaped or fusiform and stream-lined, well adapted for swift and direct flight. It offers minimum resistance to air. It is about 33cm long. The body is slaty grey in colour with glistening, metallic green and purple sheen on the upper breast and round the neck. The feet are crimson red or dark, soles are pale and claws horny. The eyes are orange-red or orange-yellow in colour. The bill is black. The wings have two black bars.

Divisions of the body

The body is bilaterally symmetrical and divided into head, neck, trunk and tail (Fig. 20.1). The entire body is covered with feathers which overlap each other and are directed backwards.

Head: The head is small, elongated antero-posteriorly, rounded and mobile. Anteriorly the head possesses a short, prominent and pointed **beak** or **bill** formed by the elongated upper and lower **jaws** covered by horny sheath called **rhamphotheca**. There are no teeth in the jaws. The terminal wide mouth is bounded by the horny beak. At the base of the upper beak there are two slit-like **external nares** or **nostrils.** Each nostril is dorsally overhung by a small whitish patch of soft and sensitive skin called **operculum** or **cere**. On either lateral sides the head bears two large, rounded **eyes** provided with upper and lower eyelids and a transparent third eyelid, the **nictitating membrane** which can be drawn across the eyeball from the anterior corner of the eye. Below and behind each eye there is an external ear opening or auditory aperture which leads into the tympanic cavity closed by tympanic membrane.

Neck: The neck is long, cylindrical and sharply marked out from the head and trunk. It is very mobile and helps in handling food. It compensates the forelimbs which have become modified into wings.

Trunk: The trunk is compact, plump and spindle-shaped. The boat-shaped **carina** or **keel** of the sternum is raised into a prominent ridge along the midventral line. The trunk can be divided into two parts: the anterior part to which the wings are attached is called **thorax** and the posterior part to which legs are attached is called **abdomen**. Towards the posterior end of the trunk there is a large transverse slit, the **cloacal aperture**. It is situated ventrally at the junction of the uropygium and the trunk.

Tail: Posteriorly the trunk is drawn out into a short conical projection, the **true tail** or **uropygium**. It bears a fan-like group of elongated **tail feathers** or **rectrices** to which the term tail is generally applied. On the dorsal surface at the base of the tail a knob-like papilla is present. It bears the opening of a large **preen** or **coccygeal**, or **oil** or **uropygial gland**. It is the only cutaneous gland present in birds. Its oily secretion is used for lubricating or dressing the feathers and also prevents the horny beak from becoming brittle. The tail is used as a rudder in flight, to suddenly check flight and as a balancer in walking and perching.

Appendages: The trunk bears a pair of wings which correspond to the forelimbs of any land vertebrate and two walking legs corresponding to the hindlimbs.

Wings: The wings are powerful organs of flight. They are attached high on the back to the anterior region or thorax of the trunk. They are elongated, flattened and distally pointed with its longitudinal axis at right angles to the trunk. At rest they remain folded close to the body in the form of "Z". They are extended during flight. The wings show three typical divisions of the forelimb, the **upper arm** or **brachium**, **fore arm** or **antebrachium** and **hand** or **manus**. The parts of hand are closely bound together by skin. All the bones can be identified by feeling them through the skin. The hand bears three imperfectly marked and clawless digits of which the second is the largest. Between the upper arm and the fore arm on the anterior side extends a fold of skin called the **alar membrane** or **prepatagium**. A similar fold exists between the posterior portion of the upper arm and the trunk. This is the **postpatagium**. Wings provide broad surface for support in flight. Their movement serve to alter the manner and direction of flight.

Legs: The legs or hindlimbs arise somewhat anteriorly from the trunk to balance and support the entire weight of the body at rest. Each leg has a short and muscular **thigh** closely bound to the trunk, a long and muscular **shank** extending from the knee downwards and backwards and a **foot** clearly divisible into a **proximal** portion the taro-metatarsus and four digit. The digits are long and clawed. The first digit is called **hallux** and is directed backward. The remaining three digits are directed forward (Fig. 20.3). The legs although small in size are efficient organs of locomotion. The pigeons walk on the ground with considerable rapidity. The legs also help in **perching**.

Skin: The skin of pigeons (like most birds) is dry, loose, thin and hard. The looseness of skin has a definite advantage as it enables the flight muscles to move freely. It is made up of usual two layers: the **epidermis** and the **dermis** (Fig. 20.2), both of which are reduced to a delicate thinness.

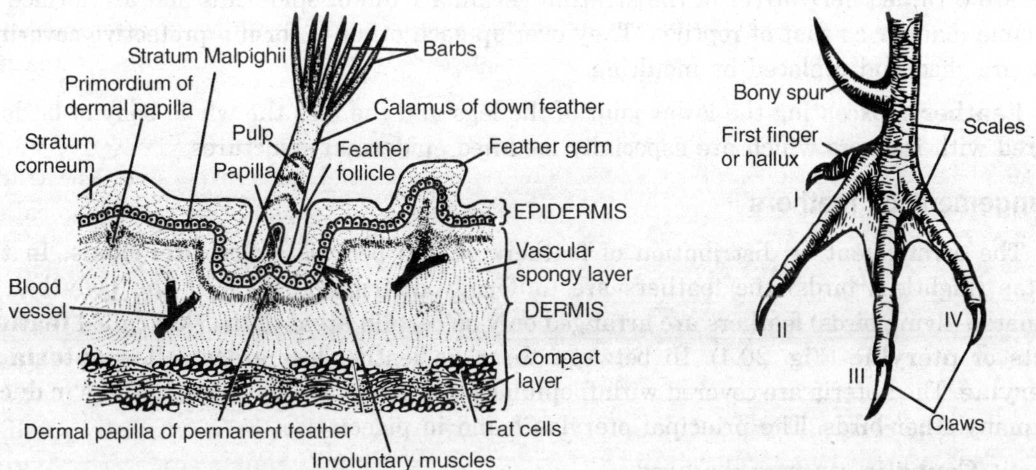

Fig. 20.2 *Columba*–V.S. of skin

Fig. 20.3 Leg of male fowl showing exoskeletal structures

Epidermis: The epidermis is several layers thick. The outermost layer is the **epithelium**. It is made up of a single row of flattened delicate cells. The middle layer is **stratum corneum**. It is made up of flat cornified cells and is horny and protective. It is modified to form several exoskeletal structures. The innermost layer is **stratum Malpighii** or **germinativum**. It is made up of large, cylindrical and actively dividing cells. It gives rise to cells of outer and middle layers.

Dermis: The dermis is mesodermal in origin. It consists of muscle fibres, nerve fibres, blood capillaries and connective tissue.

Pigments occur only in scales and feathers. Both dermis and epidermis have no pigment cells. There are no glands in the skin except the preen glands.

EXOSKELETON

The pigeon has well developed exoskeletal structures which are purely epidermal in origin. They provide protection to the body from temperature fluctuation and radiations and also protects from mechanical injury. The various exoskeletal structures are **claws** and **beak, spur**, **scales** and **feathers**.

Claws and beak: Claws and beak are horny exoskeletal derivatives and have already been described. They are variously modified in different birds. The beak is used for taking in food, for cleaning feathers and for fighting. The claws (Fig. 20.3) help in walking and perching.

Spur: Spur (Fig. 20.3) is a horny projection of tarsometatarsus present in the male of some species of birds. It is a sharp, pointed structure covered by a horny epidermal sheath. It is used for fighting. Spur may also occur on the carpometacarpus bone of the wing in certain birds.

Scales: The scales are found on the legs and feet (Fig. 20.3) and at the base of the beak. They are cornified derivatives of the stratum germinativum of epidermis and are formed in the same manner as that of reptiles. They overlap each other to form a protective covering. They are shed and replaced by moulting.

Feathers: Excepting the lower joint of the legs and the toes the whole body of birds is covered with feathers which are especially modified epidermal structures.

Arrangement of feathers

The arrangment or distribution of feathers on the body is called **pterylosis.** In the Ratitae (flightless birds) the feathers are uniformly distributed all over the body but in Carinatae (flying birds) feathers are arranged only on certain areas of the skin called **feather tracts** or **pterylae** (Fig. 20.4). In between these lie feather-less areas called **apteria** or **apterylae**. The apteria are covered with filoplumes in pigeons and by down feathers in ducks and many other birds. The principal pterylae found in pigeon are:

(i) **Cephalic** covering the head

(ii) **Cervical** on the neck.

(iii) **Humeral** across the upper arm or humerus.

(iv) **Spinal** extending from the neck to tail along the vertebral column.

(v) **Ventralis** a double tract with one branch running along each side of the breast.

(vi) **Wings** or **alar** composed of remiges with their coverts.

(vii) **Caudal** composed of rectrices with their coverts.

(viii) **Femoral** or **lumbar** spreading obliquely on the outer side of the thigh.

(ix) **Crural** covering the shank of legs.

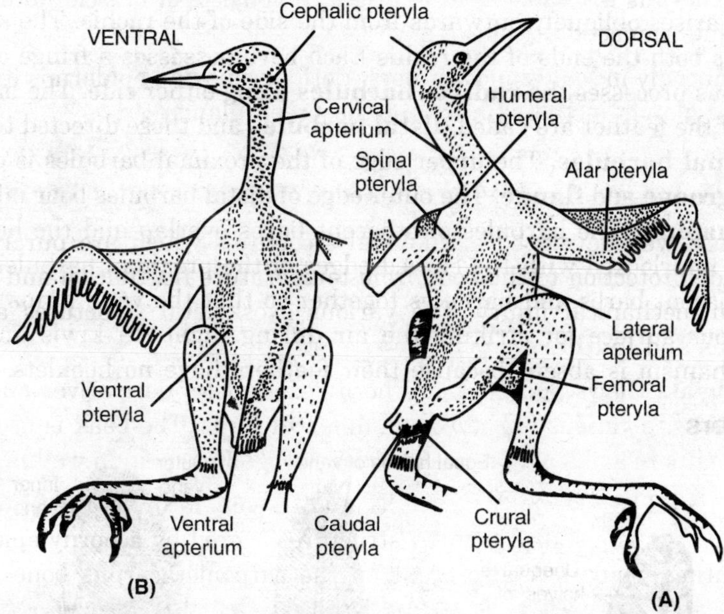

Fig. 20.4 *Columba*–Feather tracts in young birds, **(A)** Dorsal view; **(B)** Ventral view

Structure of a typical feather

The typical structure is presented by the contour feather from the general body surface or flight feather from the wings or tail. A typical feather consists of a **central axis** or **main stem** and an expanded distal portion the **vexillum** or **vane**.

Axis: The central axis or scapus is divided into a basal portion, the **calamus** and a distal portion, the **shaft** or **rachis** (Fig. 20.5A).

The **calamus** is the hollow, tubular, semi-transparent stalk of the feather. Its base is inserted into a pit or epidermal follicle of the skin from which muscle fibres pass to the feather to allow individual movement. At the base of the calamus there is a small aperture, the **inferior umbilicus** into which fits a small conical nutritive **feather** or **dermal papilla** from the dermis. It supplies nutrients and pigments to the developing feather. A second extremely minute aperture is situated at the junction of the quill and vane on the ventral surface. It is the **superior umbilicus**. In pigeon and many other birds the superior umbilicus is covered by a small tuft of soft feather called **after-shaft** or **hyporachis**. They have to be scrapped off by a scalpel to see the superior umbilicus.

The **rachis** or **shaft** is the distal solid continuation of the quill tapering towards the tip. It forms the longitudinal axis of the vane. It differs from the calamus in being stiff, solid, opaque, roughly quadrangular and filled with closely packed pith cells. A narrow **umbilical groove** runs along the ventral surface of the rachis throughout its length.

Vane: The vane or vexillum is the expanded membranous portion of the feather developed around the rachis which forms its longitudinal axis. The rachis divides the vane into two unequal lateral halves. The vane is apparently like a sheet formed by a series of numerous (about 600) narrow, parallel and closely spaced delicate thread-like structure the **barbs** or **rami**. Each barb arises obliquely outwards from the side of the rachis. The size of the barbs decreases towards both the ends of the rachis Each barb possesses a fringe of small delicate oblique filamentous processes the **radii** or **barbules** along either side. The barbules directed towards the tip of the feather are called **distal barbules** and those directed towards the base are called **proximal barbules.** The upper edge of the proximal barbules is deeply curled or rolled forming a **groove** and **flange**. The outer edge of distal barbules bear minute **barbicels** or **hooklets** or **hamuli**. The barbules of adjacent barbs overlap and the hooklets of distal barbules become interlocked with the curved edges of the proximal barbules. The hooklets, thus hold the adjacent barbs and barbules together so that the vane forms a flexible, firm, flat and continuous surface for striking the air during flight. In kiwis and ostriches the interlocking mechanism is absent because their feathers have no hooklets.

Kinds of feathers

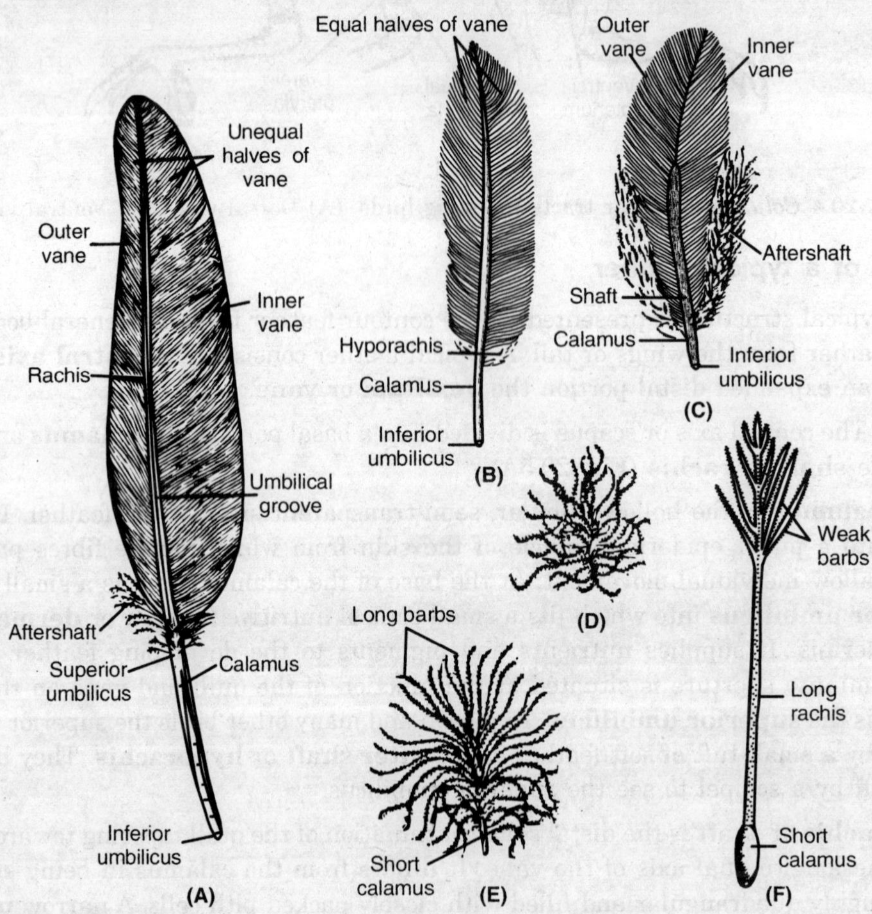

Fig. 20.5 *Columba*–Different types of feathers. **(A)** Typical wing feather; **(B)** Tail feather; **(C)** Contour feather; **(D)** Nestling down feather; **(E)** Permanent down feather; **(F)** Filoplume

In pigeons the feathers are variously modified to serve different function. They are of following types:

1. Quill feathers: The quill or flight feathers are the largest feathers of birds. They have a strong shaft and barbules with an interlocking arrangement. They are classified as follows:

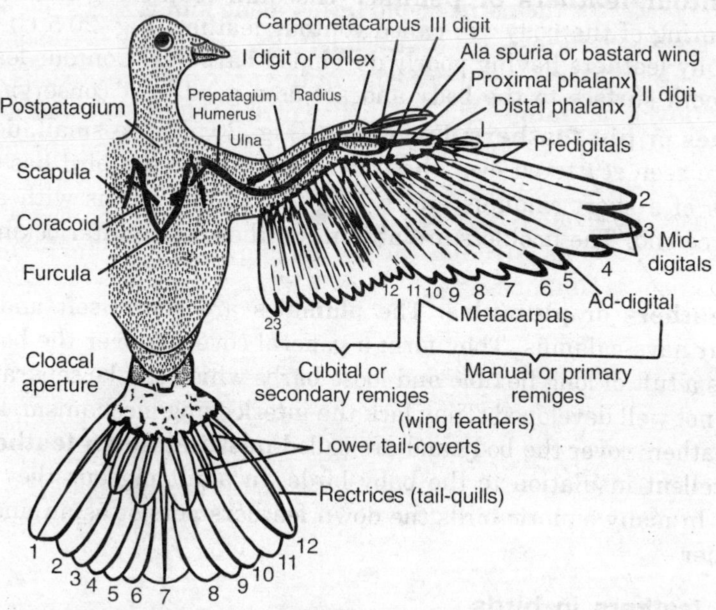

Fig. 20.6 *Columba*–Arrangement of flight feathers (ventral view)

(i) **Remiges:** The remiges (singular **remix**) are the quill feathers of wings (Fig. 20.5 A). The inner or posterior half of the vane is much broader than the outer or anterior half. In pigeons there are remiges attached to the hinder border of each wing. They are divided into **primaries, secondaries** and **tertiaries.** The **primaries** or **mannuals** are attached to the bones of the manus (Fig. 20.6). In all there are eleven primaries and are known according to their position. Six feathers attached to the middle metacarpal are called **metacarpal quills**, one attached to the phalanx of the post axial digit is called **ad-digital**, two attached to the proximal phalanx of the middle digit are known as **mid-digital** and two attached to the distal phalanx of the middle digit are called **pre-digital quills**. One of the pre-digital quills is smaller. The quills that are attached to the ulna are known as **secondary quills or cubitals** (Fig. 20.6). The tertiaries are attached to the humerus and elbow. A small but distinct tuft of feathers called **ala spuria** or **bastard wing** is attached to the pollex or thumb.

(ii) **Rectrices** (singular **retrix**) or **tail feathers** (Fig. 20.5B): There are twelve rectrices in pigeon which are arranged in semicircle or fan-like manner arround the uropygium (Fig. 20.6). The two sides of the vane are almost equal in size. They act as a brake in alighting and as a rudder in vertical or lateral steering.

(iii) **Coverts:** Small quill feathers that cover the bases of the quill feathers of tail and wings are called **coverts**. The coverts covering the bases of wings are called **upper and under wing coverts** and those covering the bases of tail are **upper and under tail coverts.** The interspaces between quills of flight and tail feathers thus present a continuous area to oppose the buoyancy of air.

(iv) **Contour feathers or pennae:** The quill feathers which form the general covering of the body are called contour feather (Fig. 20.5 C). They are small woolly feathers having poorly developed barbules. Contour feathers provide a smooth surface to the body and act as a good heat conserving layer.

2. Filoplumes or **pin feathers:** Filoplumes (Fig. 20.5F) are small, delicate, hair-like feathers which are seen scattered over the body surface after the bird has been plucked. A filoplume consists of a short calamus and a long thread-like rachis with a few barbs and barbules at the free end. The hooklets are absent. So there is no interlocking arrangements of the barbules.

3. Down feathers or **plumules:** The plumules are short, soft and wooly feathers without rachis but have calamus. They form a general covering over the body. The calamus is short and bears a tuft of long flexible and loose barbs which can be separated (Fig. 20.5 E). The barbules are not well developed. They lack the interlocking mechanism. In newly hatched birds the down feathers cover the body and are called **nestling down feathers** (Fig. 20.5 D). They provide excellent insulation to the baby birds. In adult pigeons they are replaced by contour feathers. In many aquatic birds the down feathers persist as an under-coat beneath the contour feather.

Other kinds of feathers in birds

Powder down feather: These feathers are simply down feathers the terminal end of which does not develop beyond the early stage and continually break down into powdery fragments. The powder is used by the bird for cleaning the plumage. Powder down feathers are usually aggregated into patches and sometimes in scattered tracts and tufts. These feathers are absent in pigeon.

Rictal bristles: Rictal bristles are stiff hair-like feathers present around the rectus (base of the bill) of birds such as flycatchers and goat suckers. Each rictal bristle has a short calamus, a long, slender, stiff and pointed rachis with a few rudimentary barbs at its base. They are not found in pigeon.

Tactile feathers or vibrissae: These are present at the root of beak or round the eyes. It is best developed in nocturnal birds but absent in pigeon.

Development of feathers

Feathers are produced usually in several generations, first the **nestling feathers (neossoptiles),** then one or more generations of **juvenile feathers** which may be of various types such as the **perifiloplumes, preplumules** and **prepennae,** finally the **adult feathers** or **teleoptiles** are produced. Like other epidermal structures feathers are moulted either seasonally or at a certain stage in the life cycle, a new generation being produced from the old papillae. Most birds moult in autumn, some again in the spring. The development of feather in birds is very much similar to the development of scales and hairs in reptiles and mammals respectively.

Development of down feather

The feather is first identified as a pimple-like elevation, the **dermal papilla** (Fig. 20.7 A), on the surface of the delicate and semitransparent skin of five or six day old embryo. It is produced by dermal cells and is externally covered by epidermis, As the papilla grows, it becomes cone-shaped. This is the **feather germ** (Fig. 20.7 D). It grows outwards as a

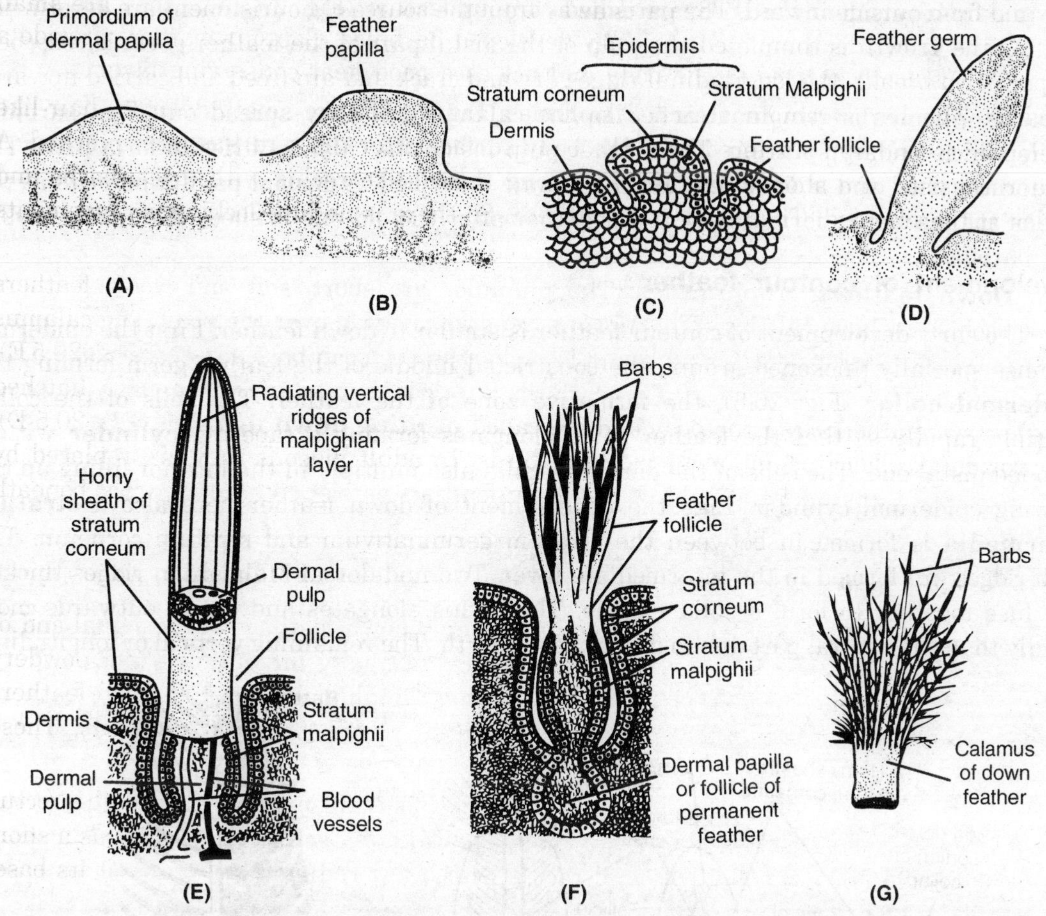

Fig. 20.7 *Columba*–Development of feather (down) **(A)** Primordium of dermal papilla; **(B)** Feather papilla; **(C)** Same in V.S.; **(D)** Feather germ; **(E)** Later stage in V.S.; **(F)** Down feather in follicle—V.S., **(G)** Mature down feather

tapering epidermal cylinder fitted with mesodermal core. The soft central core of the feather germ contains blood capillaries, nerve fibres and connective tissue and is called the **feather pulp**. It feeds the growing feather. As the feather germ elongates a pit called **feather follicle** (Fig. 20.7C) develops around its base. The germinative layer covering the distal part of the feather germ gives rise to an **intermediate layer** of cells between itself and the stratum corneum. A series of vertical radiating ridges, the **barb ridges** (Fig. 20.7E) appear on the inner surface of the distal third of the feather germ. These elongate and extend along the full length of the germ. On either side of the base of the barb ridge some cells produced by the intermediate cell layer become arranged in tangential rows. These are the **primordia of barbules**. The lower part of the feather germ remains intact, becomes uniformly thickened and forms the **calamus**. The horny epidermal layer or stratum corneum forms a transparent and coherent cylindrical sheath called **periderm** around the growing feather. The **cornification** or **keratinization** of the elements progress from the apex of the feather germ towards its base and from outside inward. The parts away from the source of nourishment are keratinized first. As the growth is completed the pulp of the distal part of the feather germ is resorbed, the periderm breaks, the longitudinal ridges dry and crack and are freed and carried upwards as barbs. The barbs remain attached to the calamus but they spread out. The barbules develop by secondary splitting. The feather pulp inside the calamus dries to form **pith**. The periderm is shed and a down feather or plumule is formed. A dermal papilla persists at the base of the feather to form a new down feather after the old one is shed off during ecdysis.

Development of contour feather

The early development of contour feather is similar to down feather. First the epidermis becomes specially thickened around the constricted middle of the feather germ forming the **epidermal collar** (Fig. 20.8), the formative zone of the feather. The cells of the collar multiply rapidly so that the feather germ elongates forming a tapering **cylinder** with a rounded distal end. The cells of the dermal papilla also multiply in the interior filling up the growing epidermal cylinder. Like the development of down feather here also a **stratum intermedia** is formed in between the stratum germinativum and stratum corneum. The barb-ridges are formed in the intermediate layer. Two mid-dorsal Malpighian ridges thicken and fuse together to form a solid **rachis**. The rachis elongates and grows outwards more rapidly than the ventral part due to differential growth. The remaining vertical or longitudinal

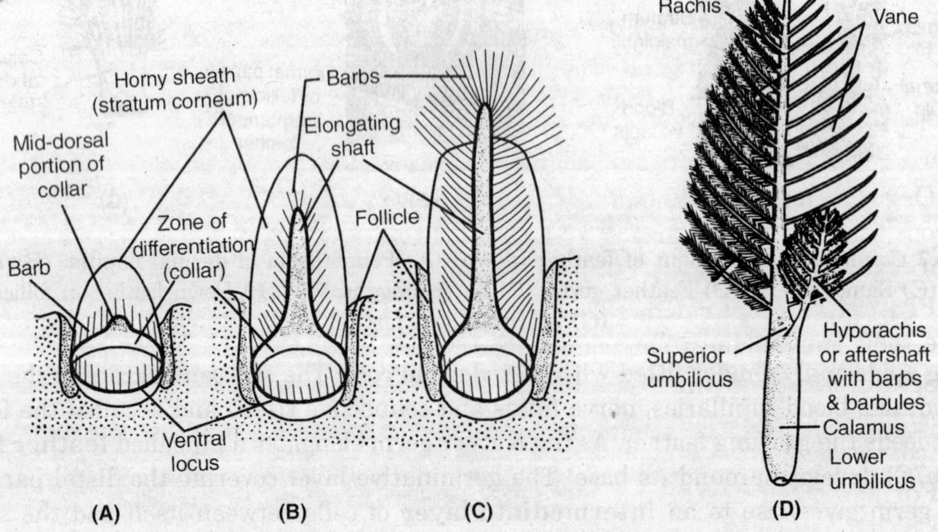

Fig. 20.8 *Columba*–Development of contour feather from down feather stage (schematic). **(A–C)** Differentiation and elongation of rachis; **(D)** Young contour feather

ridges become detached from inside the periderm and migrate on either side of the rachis and become the barbs. As the rachis elongates it carries the barbs up with it. New barbs appear below. The barbules arise later from both sides of the barbs. The fusion and differential growth of two ventral barbs results in the formation of **after-shaft** or **hyporachis**. The periderm splits and dries, the developing feather splits along the mid-ventral line and the barbs flatten out to form the **vane**. The part of the shaft which remains in the skin forms the calamus. The original basal aperture of feather germ becomes the **inferior umbilicus** while the **superior umbilicus** develops later. The feather pulp inside the calamus dries to form the pith.

Moulting or ecdysis

In birds feathers are shed and replaced by a process called **moulting** or **ecdysis**. It occurs after a time gap of six weeks. A dermal papilla present at the base of the feather follicle forms new feather. In some bird moulting is seasonal, for example, peacock but in most birds (for example, pigeon) it is gradual and occurs throughout the year.

Colouration

The feathers of pigeon and other birds show varied and brilliant colouration. The colour of feathers is due to certain pigments deposited during growth. The colours may change or fade out. The colouration in birds is for concealment, recognition and sexual selection.

Functions of feathers

The feathers serve a variety of important functions, some of which are discussed below.

1. Protection: The feathers form a light weight, impervious, flexible, durable and water proof body covering which protects the underlying skin from mechanical and physiological injuries.

2. Heat retention: The retention of body heat is one of the most important physiological function of feathers. The birds have extremely high but constant body temperature between 104° to 112° F. The feathers help to maintain constant body temperature. The plumage forms a non-conducting covering with numerous dead air spaces useful as insulation. In cold weather the feathers fluff out, thus increasing insulation and preventing loss of heat from the body. In warm weather the feathers are held close to the body and allow some heat to escape.

3. Organs of flight: The feathers are light, elastic structures which make the weight of the body light. The wing and tail feathers with airtight linkages between barbules and hooks form surfaces to support the body during flight.

4. Protective colouration: The feathers of birds are usually brilliantly coloured and resemble their surrounding. This camouflage protects them from their enemies.

5. Sexual dimorphism: The colour of feathers of most birds serve as secondary sexual character and help in recognition of sex and courtship display. Usually the male is more brightly coloured than the female.

6. Formation of nest: Some birds use feathers for making their nests. This helps to conserve heat around the developing nestling.

7. Natal covering: The down feathers form a covering around the new born baby bird, thus providing them protection.

8. Other uses: The powdery secretion of powdery down feathers of some birds like herons, bitterns, etc. is used for cleaning the plumage and keeping them free from ectoparasites.

The plumage of birds have served widespread use. The feathers are used for stuffing mattresses and pillows. From time immemorial feathers have been used for arrow making, head-gear and for ornamental wear. In ancient times feathers were used as pen for writing. The feathers used for writing were made from quill feathers of some birds. Some bird feathers are even used today for special kind of writing and drawing. Feathers of some birds like *Rhea* are extensively used for dusters.

ENDOSKELETON

The skeleton of bird is, in general plan, the same as that of other vertebrate animals. One of the chief characteristics of the birds' bones is their lightness, an adaptation to flight. The birds may have large bones but they are not compact. Many bones have great cavities with no marrow, some may contain prolongations of air sacs.

The skeleton of fowl is being described here because it is large in size and convenient for study. Essentially, the **axial skeleton** includes the cranium made up of many bony elements, vertebral column, sternum and ribs and the **appendicular skeleton** attached to the axial includes girdles and limb bones.

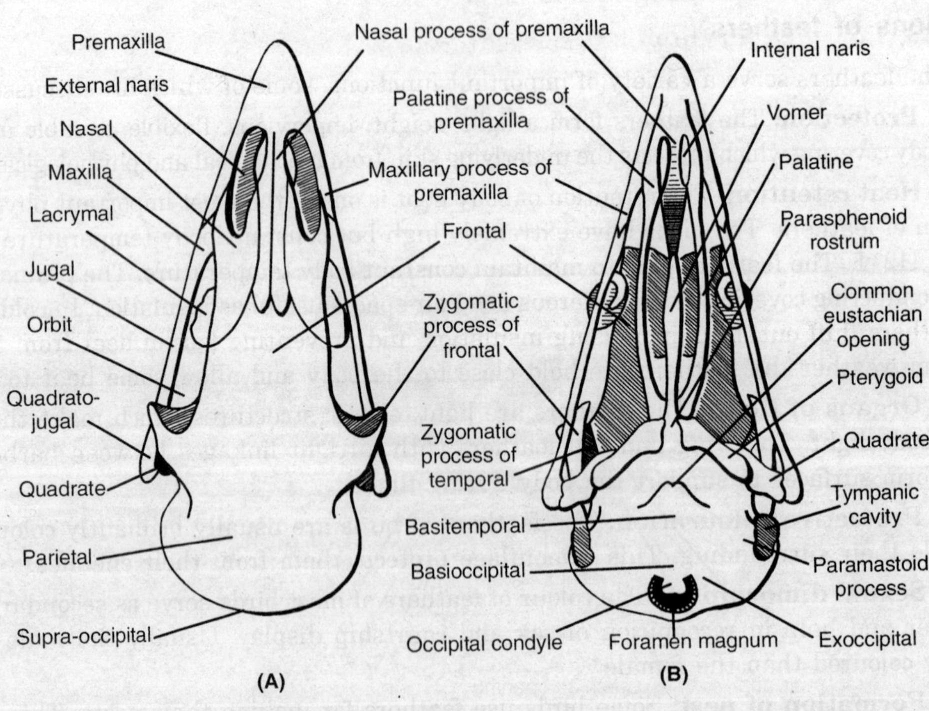

Fig. 20.9 *Fowl*–Skull, **(A)** Dorsal; **(B)** Ventral

SKULL

The **skull or cranuim** (Fig. 20.9 and 20.10) is the endoskeleton of the head of bird. Due to aerial mode of life the skull of bird shows several characteristic features:

1. The number of bones is reduced.
2. The bones are pneumatic, so it is very light in weight.
3. The sutures have disappeared and the bones are fused so intimately that it becomes difficult to identify them separately.
4. The teeth are absent and the jaw bones are modified to form a toothless (edentulus) **horny beak** which acts as a limb for catching food.
5. The **orbits** are large to accomodate large eyes. A very thin membranous inter-orbital septum separates the two orbits. The enlargement of orbits restricts rest of the brain to the back of the cranium.
6. The cranium is large to accomodate the large and well developed brain. The part of the brain which deals with smell is very much reduced and that which deals with coordination of movement and balance is large.
7. Skull is **monocondylic**, *i.e.*, it has only one occipital condyle.
8. Jaw suspensorium is **autostylic**.
9. Palate is of **schizognathous** type.
10. **Foramen magnum** faces downwards.

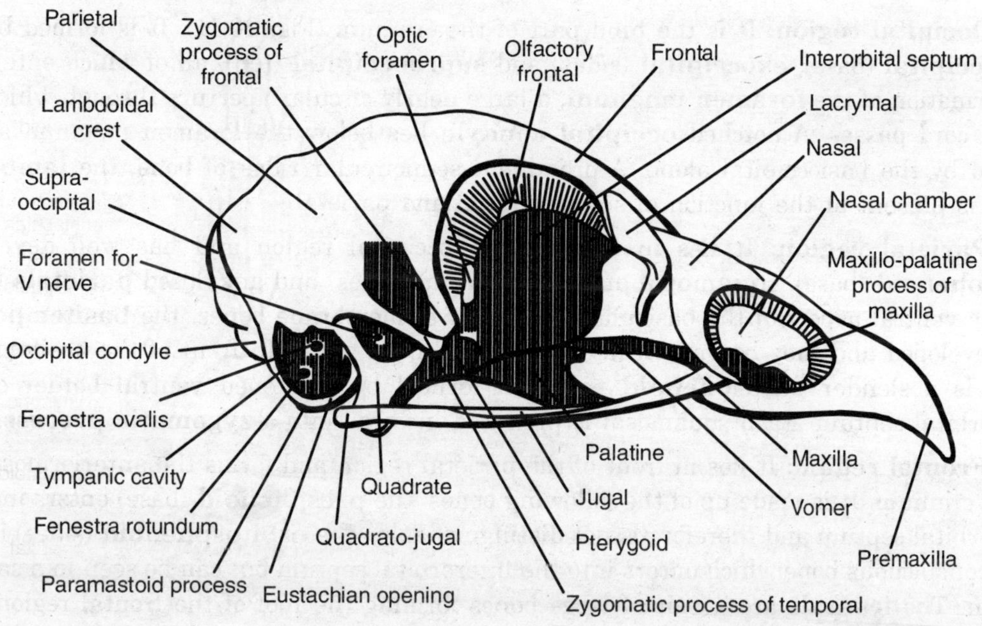

Fig. 20.10 *Fowl*–Skull, lateral view

Regions of skull

The skull as usual consists of three parts: **cranium**, **sense capsules** and **visceral skeleton**.

Cranium

The **cranium** or **brain case** forms the broad, posterior and median box-like part of the skull. It lodges the brain. The outlines of component bones of the cranium can be followed in the skull of young bird. It is as usual divided into three regions: **occipital, parietal** and **frontal**.

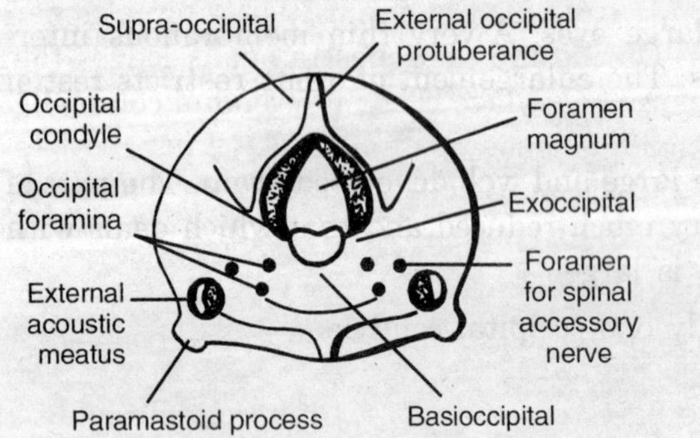

Fig. 20.11 *Fowl*–Occipital segment (posterior view)

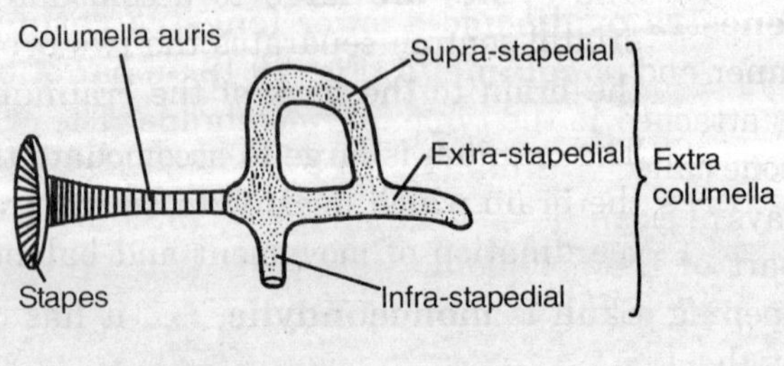

Fig. 20.12 *Fowl*–Columella aurcs

Occipital region: It is the hind part of the cranium (Fig. 20.11). It is formed by the **basioccipital** (base), **exoccipital** (sides) and **supraoccipital** (top), all of which enter into the formation of the **foramen magnum**, a large nearly circular aperture through which the spinal cord passes. A notched **occipital condyle** lies below the foramen magnum and is formed by the basioccipital alone. A prominent semicircular ridge of bone, the **lamboidal crest** is present at the junction of supraoccipital and parietal.

Parietal region: It lies in front of the occipital region and has well developed **basisphenoid** (base), **squamosal** and **alisphenoid** (sides) and ankylosed **parietals** (roof). On the ventral aspect of the basisphenoid, a pair of membrane bones, the **basitemporals**, are developed and they become firmly fused to it in the adult. In front of the basitemporal there is a slender **parasphenoid rostrum** formed by thickened ventral border of the interorbital septum. Each squamosal is produced inwards into a **zygomatic process**.

Frontal region: It lies in front of the parietal region and forms the anteriormost part of the cranium. It is made up of the following bones: the **presphenoid** (base) enters into the interorbital septum and therefore is not distinguishable. The **orbitosphenoid** (sides) is also an inconspicuous bone which enters into the interorbital septum but can be seen in a sagittal section. The **frontals** are a pair of large bones forming the roof of the frontal region. The broad posterior end of each frontal is drawn out into a strong **post-orbital** or **zygomatic process** which joins the zygomatic process of the squamosal.

Sense capsules

The **sense capsules** lodge and protect the special sensory organs. There are three pairs of sense capsules: **auditory** (hearing organs), **optic** (organs of sight) and **olfactory** (organs of smell).

Auditory capsules

A pair of auditory capsules are present one on either lateral sides of the occipital region. They lodge the internal ears. Each auditory capsule is formed by the **prootic bone**. A large cup-shaped **tympanic cavity** is present on either lateral side of the hinder end of the skull. It is formed by the squamosal above and basitemporal below. The **tympanic membrane** is attached just with its prominent outer tip. Within the tympanic cavity almost in the middle there is an oval depression, the **fenestra recess**. It is divided by a partition into an upper **fenestra ovalis** and a lower **fenestra rotunda**. During life the fenestra ovalis lodges the inner end of **columella** (Fig. 20.12), a bar of bone and cartilage. The outer end of columella is attached to the tympanic membrane. The columella is made up of a slender inner rod-like bone called **columella auris** and outer triradiate cartilage called **extracolumella**. The three rays of extracolumella are **suprastapedial**, **extra-stapedial** and **infra-stapedial**. The inner part of the columella auris lodged in the fenestra ovalis is called **stapes**. The posterior opening of the eustachian canal is a funnel-shaped opening at the ventral and anterior end of the tympanic cavity. The eustachian canal passes forwards and inwards between the basitemporal and basisphenoid and opens on the base the skull close to the middle line.

Optic capsules or orbits

The **orbits** are a pair of large spaces separated from each other by a narrow, longitudinal **interorbital septum** which is formed mainly by the mesethmoid along with presphenoid, parasphenoid rostrum and orbitosphenoid. Dorsally the orbit is bounded by the frontal and pre-frontal, anteriorly by the frontal and lacrimal and posteriorly by post-orbital process of frontal and alisphenoid. The orbit is incomplete ventrally. The lacrimal is a large bone produced below into a slender curved process. The interorbital septum has many openings. The **foramen** for the exit of **olfactory nerve** is a median opening at the anterior end of the cranium between the interorbital septum and frontal bone. The **optic foramen** is a large median hole in the orbitosphenoid lying in the posterior part of the orbit. The two orbits communicate with each other and with the cranium through this foramen. The foramina for the exit of third and fourth cranial nerves are small holes towards the outer side of the optic foramen.

Olfactory capsules

In birds the sense of smell is poorly developed so the olfactory lobes of the brain are very much reduced in size. The **olfactory capsules** or **nasal sacs** are relatively smaller in size and remain cartilaginous throughout life. The following bones are found in association with the olfactory capsule. The **nasals** are a pair of lamellar bones which cover the sides and roof of the olfactory capsule. They remain separated from each other by the dorsal nasal processes of the premaxillae. Each nasal is a thin Y-shaped bone. The single posterior process of Y meets the frontal and also articulates with lacrimal. The two anterior processes bear a deep notch in between. They form the upper and lateral boundaries of the external nares and unite with the processes of the premaxillae. There are two vomers but they unite into a single bone which is extremely small and found in continuation with the parasphenoid rostrum at the base of the nasal chambers. In pigeon the vomers are not present. The **ecto-ethmoid** or **turbinals** are poorly developed. The inter-nasal septum is a cartilaginous continuation of interorbital septum. All these structures are entirely cartilaginous and are removed while cleaning the skull.

Visceral skeleton

The **visceral skeleton** includes the upper and lower jaws, suspensorium and hyoid apparatus. The jaws are without teeth.

Upper jaw or maxillo-palatine apparatus

The upper jaw forms two bony **arcades: outer** and **inner**, which remain widely separated but are fused anteriorly and meet posteriorly. The **outer** or **infraorbital arcade** or **suborbital bar** forms the border of the mouth. It is made up of following bones: **premaxilla, maxilla, jugal** and **quadrato-jugal**. The **inner arcade** or **palato-pterygo quadrate bar** forms the roof of the buccopharyngeal cavity. It is made up of **palatine, pierygoid** and **quadrate.**

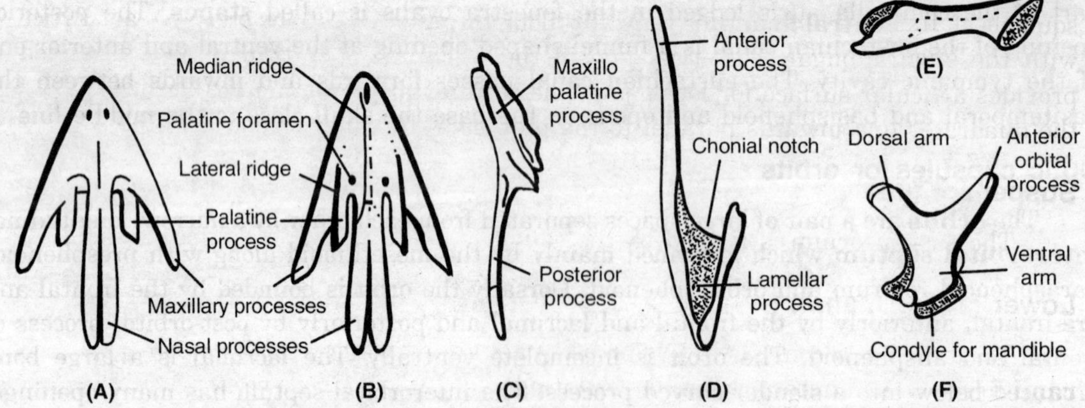

Fig. 20.13 *Fowl*–Loose bones of upper jaw, **(A)** Premaxilla (dorsal view); **(B)** Same (ventral view); **(C)** Maxilla; **(D)** Palatine; **(E)** Pterygoid; **(F)** Quadrate

Premaxilla: It is the anterior most bone of outer arcade of the upper jaw forming practically the whole of the upper beak. It is a large, stout, triradiate bone (Fig. 20.13 A and B) formed by the fusion of its fellow of the opposite side. It is produced backwards into three pairs of processes: the **nasal**, **maxillary** and **palatine**. The large slender ascending **nasal process** runs backwards along the inner side of the nasal bone completing the upper boundary of the external nares. The two outer **maxillary processes** are shorter than the nasal processes and they run backwards and slightly outwards forming the anterior part of the upper jaw. The inner **palatine processes** are the smallest and they run backwards and meet the palatines. They form the anterior part of the palate.

Maxilla: It is a slender bony rod (Fig. 20.13C) lying behind and to the inner side of the maxillary process of the premaxilla. Anteriorly each maxilla is expanded into a thin spongy horizontal lamina, the **maxillo-palatine process** with which the palatine is attached ventrally. The two maxillae do not meet in the middle line so the bony palate remains incomplete with a median cleft. The post-zygomatic process of the maxilla forms the anterior half of the outer arcade.

Jugal: It is a slender, rod-like bone which forms the middle upper border of the suborbital bar. Anteriorly it joins the maxilla and posteriorly meets the quadratojugal.

Quadratojugal: It is a slender rod forming the posterior half of the arcade. Posteriorly it is thickened and articulates with the outer side of the distal end of the quadrate.

Palatine: It is a slender, horizontal rod (Fig. 20.13 D) which forms most of the anterior portion of inner arcade of upper jaw. Anteriorly it is attached with the ventral aspect of the processes of premaxilla and maxilla and posteriorly it is expanded into a broad **lamella** the sides of which articulate with the parasphenoid rostrum and its posterior end articulates with the pterygoid.

Pterygoid: It forms the middle part of the inner arcade of the upper jaw. It is a short and stout, rod-shaped bone (Fig. 20.13 E) set somewhat obliquely and form the posterior roof of the mouth cavity. It articulates behind with the inner surface of ventral arm of the quadrate and in front with palatine and parasphenoid rostrum.

Quadrate: It is a stout triradiate or **Y**-shaped bone (Fig. 20.13 F) lying slightly above the anterior border of the tympanic cavity. Its **dorsal arm** movably articulates with the squamosal. Its **ventral arm** runs ventrally along the outer side of the pterygoid and articulates with the quadratojugal. The inner end of the ventral arm is transversely expanded and provides articular surface for lower jaw. The third or **anterior arm** or **orbital process** of the quadrate runs inwards parallel to the pterygoid. Its anterior end lies free.

Suspensorium

The suspensorium is formed by the squamosal and the quadrate bones of the skull.

Lower jaw or mandible

The lower jaw consists of a pair of long, curved and laterally compressed **halves** or **rami** firmly united anteriorly and articulating with the quadrates posteriorly, (Fig. 20.14). It appears like a single V-shaped bone. Each ramus is made up of five bones, one **replacing**

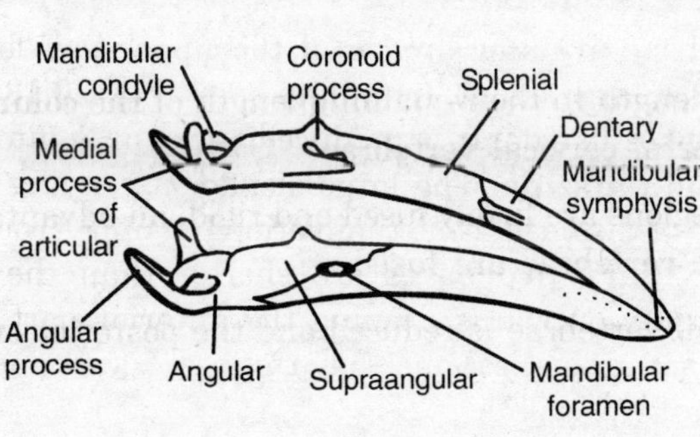

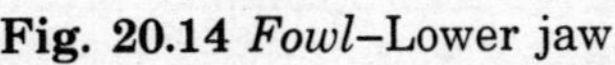
Fig. 20.14 *Fowl*–Lower jaw

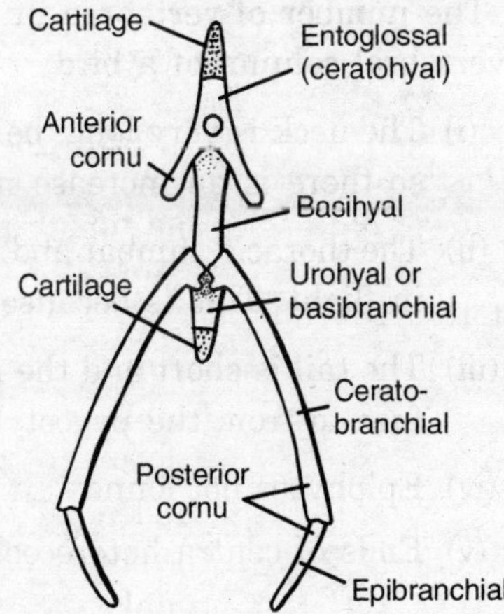

Fig. 20.15 *Fowl*–Hyoid apparatus

bone–articular and four **investing bones–angular**, **supra-angular**, **dentary** and **splenial**, all developing around an original cartilaginous bar the **Meckel's cartilage.** All the bones have the same general relations as in the lizard. These bones can be studied in the mandible of young bird only because in the adults they are fused intimately.

The **articular** forms the expanded posterior end of each ramus. Dorsally it bears a concave **mandibular condyle** for the quadrate and is produced backward and inwards into processes. The **angular** is a slender bone lying beneath the articular along the inner and lower border of the posterior third of the ramus. The **supra-angular** forms the upper part of posterior third of the ramus. It bears a small **coronoid process**. **Splenial** is a thin splint-like bone lying along the inner surface in the middle part of the mandible. **Dentary** is a large bone forming the anterior part of each ramus. Anteriorly it fuses with its fellow of the opposite side at a **symphysis.** Posteriorly it extends below the angular and supra-angular up to the articular. A **mandibular foramen** is present between the dentary and the posterior bones. There are no teeth on the dentary.

Hyoid apparatus

The hyoid apparatus lies in the floor of the mouth and supports the tongue. It is made up of bones and cartilage. The **median body,** is arrow shaped and made up of three bones placed end to end. The anterior piece is known as **entoglossal** or **ceratohyal,** it bears a cartilaginous process in front. The middle bone is **basihyal** and the posterior piece is **urohyal** or **basibranchial**, it has a cartilage behind. A pair of small **anterior cornua** projects backwards from the entoglossal. A pair of long backwardly directed **posterior cornua** or **thyrohyals** arise from the junction of the basihyal and urohyal. Each posterior cornua is made up of two pieces, a long **ceratobranchial** and a short **epibranchial** connected by cartilage.

VERTEBRAL COLUMN

The number of vertebrae in birds is most variable. There are many characteristics of the vertebral column of a bird.

(i) The neck is very long, being equal in length to the remaining length of the column, so there is an increase in the number of cervical vertebrae

(ii) The thoracic, lumbar and abdominal regions are firmly fused and rigid, an advantage in flight. This is because the thoracic vertebrae are fused.

(iii) The tail is short and the number of tail vertebrae is reduced and the posterior ones fuse to from the pygostyle.

(iv) Epiphyses not found.

(v) Ends of centra heterocoelous or saddle shaped.

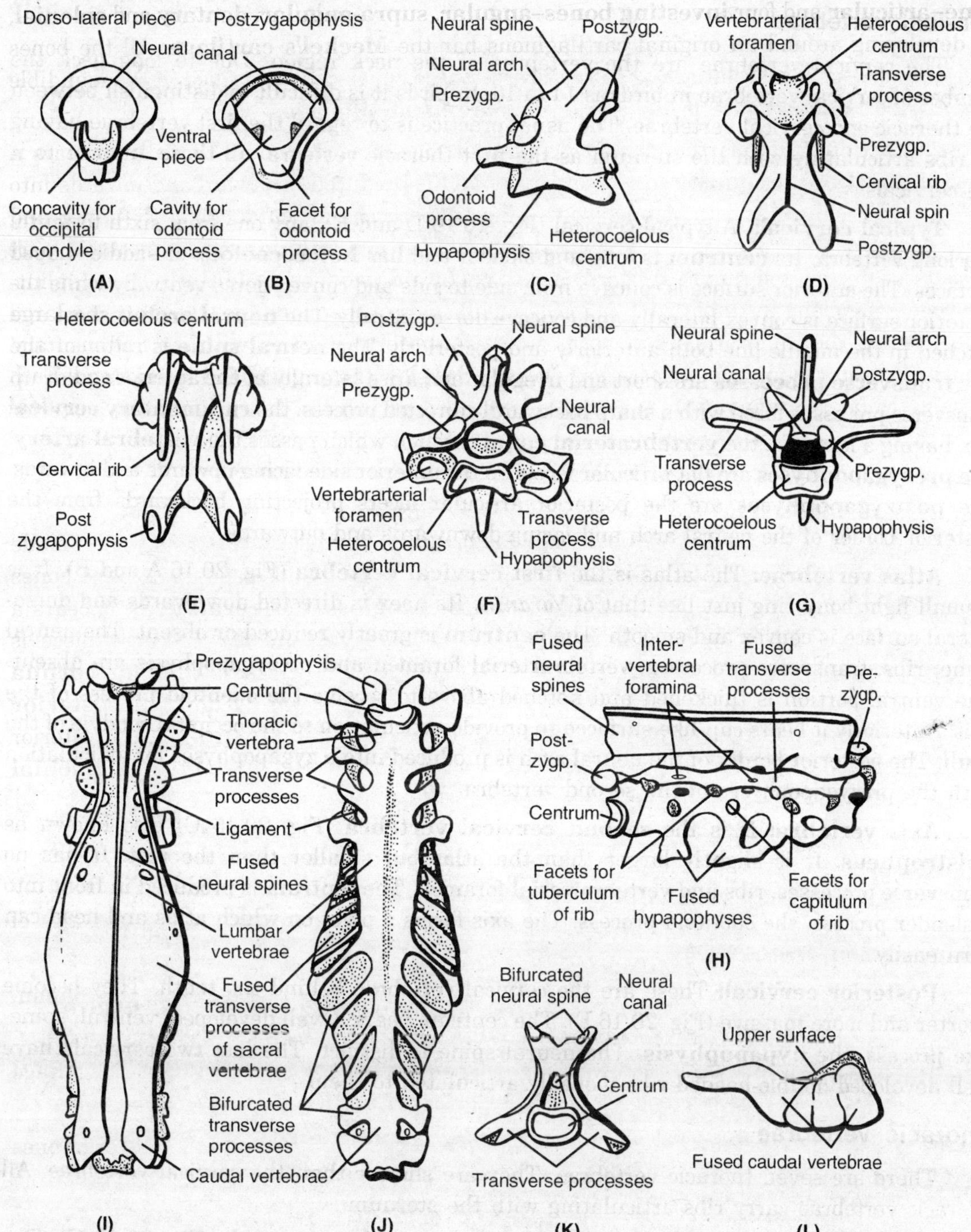

Fig. 15.16 *Fowl*–Vertebrae; (A) Atlas (anterior view); (B) Same (posterior view); (C) Axis (lateral view); (D) Typical cervical (dorsal view); (E) Same (ventral view); (F) Posterior cervical (anterior view); (G) Free thoracic (anterior view); (H) Fused thoracic (lateral view); (I) Synsacrum (dorsal view); (J) Same (ventral view); (K) Free caudal; (L) Pygostyle (lateral view)

Cervical vertebrae

The cervical vertebrae are the vertebrae of the neck region. Due to long neck the number of cervical vertebrae in birds is 14 to 16. In birds it is difficult to distinguish between the thoracic and cervical vertebrae. The usual practice is to regard the first vertebrae having its ribs articulating with the sternum as the first thoracic vertebra, all those anterior to it are cervicals.

Typical cervical: A typical cervical (Fig. 20.16 D and E), any one from sixth to tenth, is a long vertebra. Its **centrum** is long and slender and has **heterocoelous** or saddle-shaped surfaces. The anterior surface is concave from side to side and convex dorso-ventrally while the posterior surface is convex laterally and concave dorsoventrally. The **neural arch** is short and notched in the middle line both anteriorly and posteriorly. The **neural spine** is rudimentary. The **transverse processes** are short and irregular and arise laterally at the anterior end. Each transverse process is fused with a sharp backwardly directed process, the **rudimentary cervical rib**, having a foramen, the **vertebraterial canal** through which passes the **vertebral artery**. The **prezygapophyses** are flat articular facets on the anterior side facing upwards and inwards. The **postzygapophyses** are the posterior articular facets projecting backwards from the posterior border of the neural arch and facing downwards and outwards.

Atlas vertebrae: The atlas is the **first cervical vertebra** (Fig. 20.16 A and B). It is a small light bony ring just like that of *Varanus*. Its apex is directed downwards and dorso-lateral surface is convex and smooth. The **centrum** is greatly reduced or absent. The neural spine, ribs, transverse processes, vertebrarterial foramen and prezygapophysis are absent. The ventral portion is thickened and notched above to receive the odontoid process of the axis. Anteriorly it bears cup-like surfacea to provide articulation to the occipital condyle of the skull. The posterior border of the neural arch is produced into a zygapophyses for articulation with the prezygapophyses of the second vertebra.

Axis vertebra: It is the **second cervical vertebra** (Fig. 20.16 C) also known as **epistropheus.** It is slightly larger than the atlas but smaller than the rest. It has no transverse processes, ribs and vertebrarterial foramen. The centrum is produced in front into a slender process, the odontoid process. The axis forms a pivot on which atlas and head can turn easily.

Posterior cervical: These are the cervical vertebrae behind the tenth. They become shorter and more massive (Fig. 20.16 E). The centrum bears a well developed, ventral, spine-like process, the **hypapophysis.** The neural spine is distinct. The last two cervicals have well developed double-headed ribs movably articulated to them.

Thoracic vertebrae

There are seven thoracic vertebrae. They are shorter than the cervical vertebrae. All thoracic vertebrae carry ribs articulating with the sternum.

Free thoracic: The **first** and **sixth** thoracic vertebrae are free (Fig. 20.16 G). The **centrum** is heterocoelous and has a vertebral process, the **hypapophysis** for the attachment of flexor muscles of the neck. The neural spine is elongated, neural arch and neural canal are present. The pre- and postzygapophyses are well developed. Each vertebrae carries a double headed thoracic rib. The transverse processes bear facets for attachment of the ribs.

Fused thoracic: The second to fifth thoracic vertebrae are completely fused (Fig. 20.16 H), so their neural spine, transverse processes and hypapophyses become confluent and form a prominent compressed fenestrated plate. They bear facets for articulation of thoracic ribs.

Synsacrum

It is a composite bone (Fig. 20.16 I and J) formed by the fusion of 16 vertebrae of different regions. It includes 1 thoracic, 6 lumbar, 2 sacral and 7 caudal vertebrae. The bones of the synsacrum are so intimately fused that it becomes difficult to distinguish them clearly. The first vertebra of the synsacrum is the last **thoracic vertebra**. It containts a pair of free thoracic ribs and is without hypapophysis. About half a dozen **lumbar vertebrae** are firmly fused together and with the last thoracic vertebra. They have stout transverse processes, which abut against the ilia. The neural spines of these vertebrae are fused to form a vertical crest of bone which is continuous in front with that of the last thoracic, and fused along its edge with the dorsal edges of the ilia of the pelvic girdle. The two **sacral vertebrae** follow the lumbars. These have broad centra fused together and their transverse processes are united to form bony plates which support the ilia along their outer margins. The sacral ribs are fused with their transverse processes. The anterior **caudal vertebrae** come next. They are seven in number. The centra of anterior four or five are laterally compressed and in the hind-most two or three are dorso-ventrally compressed. The transverse processes, except in the last, are **double,** that is, made up of two elements. The **dorsal elements** fuse to form a bony plate continuous with those of the vertebrae in front. The **ventral elements** of first four or five vertebrae are rod-like and are regarded as ribs, as they ossify independently of the centra. In the last one or two vertebrae they are smaller. Between the bodies of these vertebrae and the ilia are deep hollows, which lodge lobes of the kidneys.

Caudal vertebrae

The caudal region is short consisting of a few **free** or **movable vertebrae** and a **pygostyle.**

Free caudals: There are four or five small free caudal vertebrae with a centrum, transverse processes, neural canal and bifurcated neural spine (Fig. 20.16 K).

Pygostyle: The last caudal vertebra is called **pygostyle** or **ploughshare** (Fig. 20.16 L). It is a vertical laterally compressed plate formed by the fusion of four or more of the hind-most caudal vertebrae. It has a flat dorsal border and convex ventral border. It provides attachment to caudal muscles and supports rectrices.

STERNUM

The sternum or breast bone (Fig. 20.17 A) of fowl is well developed and one of the most characteristic part of the skeleton. It is broad, flat, unsegmented, boat-shaped bone covering not only the thorax but a great part of the abdomen. It is composed of the following parts:

Metasternun: It is the body proper or main part of the sternum. It is in the form of a broad bony plate concave dorsally and convex ventrally. The dorsal surface bears holes through which air sacs communicate with its cavity.

Keel: The ventral surface of body proper is produced into a triangular ridge or crest called **keel** or **carina.** Its anterior border is thicker and produced ventrally into a **carinal process** which provides surface for the attachment of the great pectoral muscles of fight.

Manubrium: The **manubrium** or **rostrum** or **presternum** is a small plate-like structure projecting vertically from the anterior end of the ventral surface of the metasternum. It has a small notch or foramen.

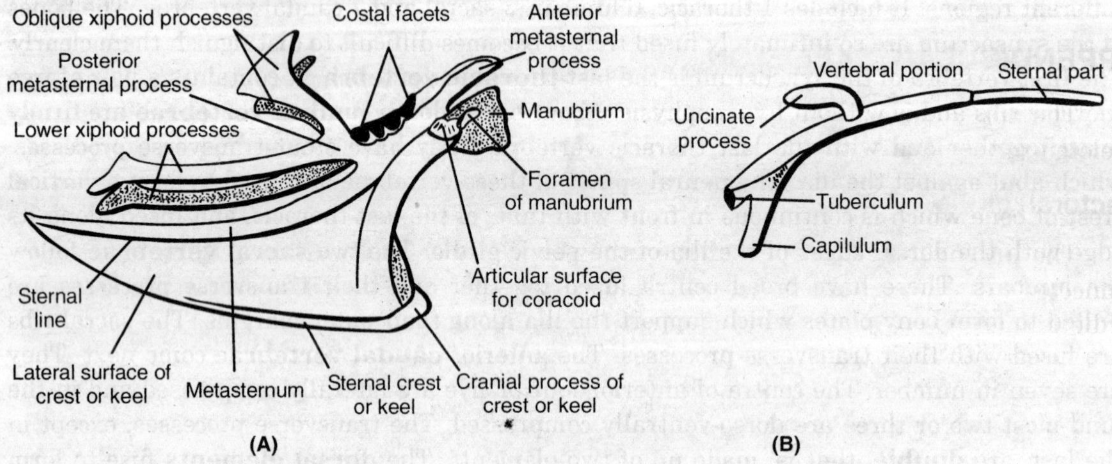

Fig. 20.17 *Fowl*–**(A)** Sternum; **(B)** Thoracic rib

Metasternal processes: These are a pair of large processes on the sides of the anterior end of the body of the sternum. Anterior metasternal process is smaller and is called **costal process**. The posterior metasternal process is larger and called **metostea** or **xiphoid process**. It divides into two flattened branches, the upper branch is called **oblique xiphoid process**. Its free end is broad and terminate behind the last sternal rib. The lower **xiphoid process** runs downwards and backwards and ends freely.

Costal surfaces: The costal surfaces are four or five cup-like depressions in the dorso-lateral border of the anterior end of the body of the sternum. The lower end of sternal ribs articulate with them.

Coracoid grooves: Lying at the base of the manubrium, on either lateral side, there are deep transverse coracoid grooves communicating with each other through a hole. These provide articular facets for the coracoid bone.

RIBS

There are two types of ribs in birds: **thoracic** and **cervical**.

Cervical: The last two cervical vertebrae bear two double-headed, movably articulated cervical ribs.

Thoracic: There are seven pairs of thoracic ribs (Fig. 20.17 B) in fowl. Each rib is made up of two portions: a dorsal flat and curved vertebral part attached to the vertebra and a ventral straight part attached to the sternum. The two parts are united in the middle at a right angle. The vertebral part of a thoracic rib is double headed or **bicephalous,** that is, it consists of two processes—the lower **capitulum** attached to the centrum and the upper **tuberculum** attached to the transverse process. The vertebral part of each thoracic rib except first and last, bears a backwardly directed process called **uncinate process** which overlaps with the next rib. This increases the strength of the thoracic wall which is essential for flight. The first, second and seventh thoracic ribs are without sternal part so they do not articulate with the sternum.

APPENDICULAR SKELETON

The appendicular skeleton comprises of the two girdles–**pectoral** and **pelvic** and the skeleton of the **limbs**.

Pectoral girdle

The **pectoral** or **shoulder girdle** is made up of two similar halves. They are firmly connected with the sternum to provide strong support to the wings. Each half of the pectoral girdle is made up of three bones: **coracoid**, **scapula** and **clavicle**.

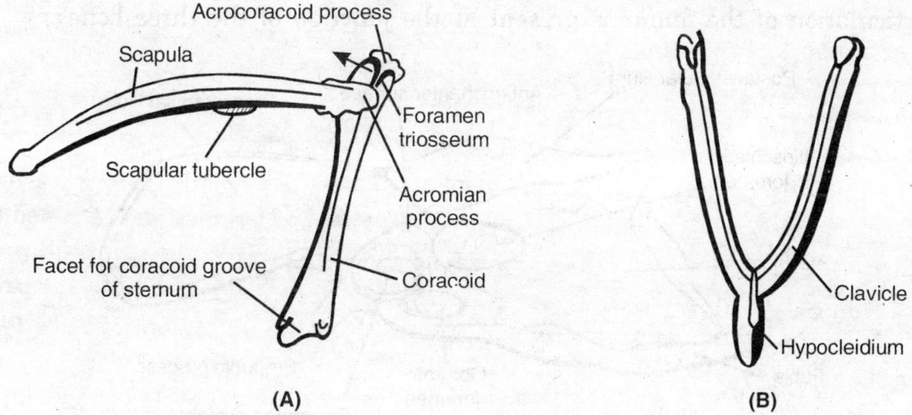

Fig. 20.18 *Fowl*–**(A)** Pectoral girdle (right half, outer view); **(B)** Furcula

Coracoid: The coracoid is a straight and stout bone (Fig. 20.18A) directed ventralwards. It articulates with the coracoid groove on the antero-lateral edge of the sternum. The ventral end also bears a **foramen** leading into an air cavity. The dorsal end of the coracoid is produced into a hook-like **acrocoracoid process** which meets the clavicle. The inner side of the coracoid below the acro-coracoid process is connected with the scapula. The coracoscapular angle is less than a right angle. The outer surface of the coracoid bears a cup-shaped depression which forms major part of the glenoid cavity. The head of humerus fits into the glenoid cavity.

Scapula: The **scapula** or **shoulder blade** (Fig. 20.18 A) is the strongest bone of the shoulder girdle. It is a long flattened sabre-shaped bone lying along the dorsal side of the thoracic framework, almost parallel to the vertebral column. Its expanded anterior end is firmly united with the coracoid by a ligament. On its outer surface it bears a shallow cup-like depression forming a part of the glenoid cavity. The inner end is produced into an **acromion process** to which the clavicle is attached.

Clavicle: The clavicle is a slender, slightly curved bone. Its expanded upper end is connected with acromian process of scapula and acro-coracoid process of the coracoid. Thus a circular gap called **foramen triosseum** is left between the three bones. The pectoralis minor muscle which raises the wing passes through this foramen and is inserted upon the head of the humerus. Ventrally the two clavicles are fused to form a small laterally compressed, rounded plate called **hypocleidium** which is connected with the rostrum of the sternum by a ligament. The **V-shaped** bone thus formed is called **furcula** or **merrythought bone** or the **wish bone** (Fig. 20.18B). It forms a spring-like connection between two halves of the shoulder girdle.

Pelvic girdle

Pelvic girdle is quite large and made up of two halves. Each half is known as **os innominatum** or **innominate bone** (Fig. 20.19). Each half presents the typical triradiate structure. The **ilium**, **ischuim** and **pubis** are the three common components. All the elements of each side are perfectly fused with the synsacrum in between forming a more or less shield-shaped structure covering the organs of the body cavity from above. As usual the **acetabulum** for the articulation of the femur is present at the junction of the three bones.

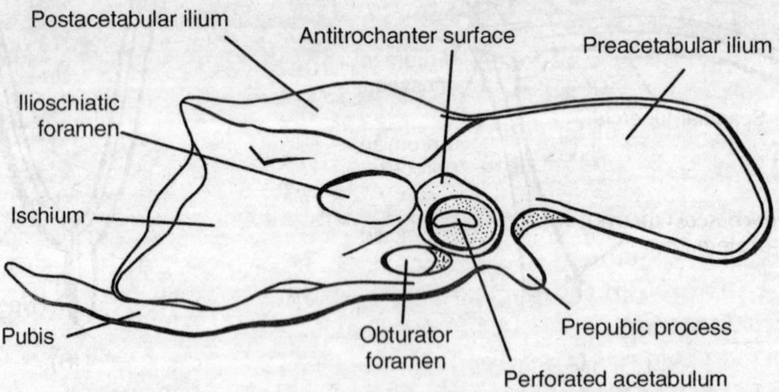

Fig. 20.19 *Fowl*–Pelvic girdle (right half, outer view)

Ilium: It is an expanded bone which divides into two almost equal parts: a **preacetabular part** extending anterior to the acetabulum and a **postacetabular part** posterior to the acetabulum. The inner border of ilium is fused with the synsacrum. The outer surface of preacetabular part is concave and of postacetabular part is convex. The ilium forms dorsal part of the acetabulum. On the outer surface above the acetabulum there is a projection, the **anti-trochanter** that articulates with the great trochanter of the femur.

Ischium: It is a broad almost vertical bone directed backwards parallel to the hinder end of the ilium to which it is attached posteriorly. It forms one-third part of the acetabulum. Just behind the acetabulum it encloses a large oval **ilio-ischiatic foramen**.

Pubis: It is a slender bone directed backwards parallel to the outer and ventral margin of the ischium with which it is often fused. It forms the ventral part of the acetabulum. In front of the acetabulum the pubis projects forwards as a small blunt **preacetabular** or **pre-pubic process**. Behind the acetabulum it is separated from the ischium by a slit-like opening the **obturator foramen**. The free posterior end of pubis extends for some distance beyond the ischium.

Forelimbs

The forelimb forms the skeleton of the wing and presents the structure of a typical pentadactyl limb.

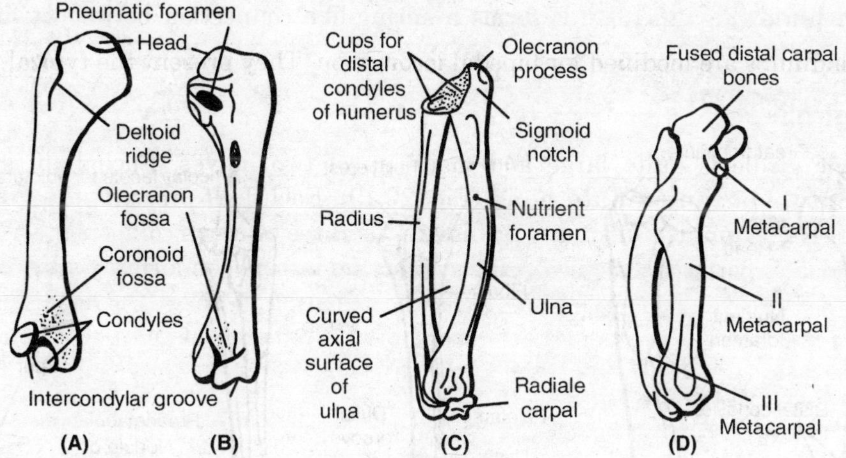

Fig. 20.20 *Fowl*–Bones of forelimbs. **(A)** Humerus (anterior view); **(B)** Same (posterior view); **(C)** Radius and ulna; **(D)** Carpometacarpus

Humerus: It is the bone of upper arm. It is a stout, elongated and slightly curved bone (Fig. 20.20 A and B) expanded at both the ends. The proximal end is smooth and convex and is called **head.** It fits into the glenoid cavity of the pectoral girdle. The head is bordered by two **tuberosities** or **tubercles**. The preaxial or small tuberosity has a prominent, short **deltoid ridge** or **crest.** The postaxial or larger tuberosity has a large **pneumatic foramen** communicating with its air cavity. The distal trochlear end of humerus bears a pair of convex articular surfaces or **condyles** for articulating with the bones of the forearm.

Radius and ulna: These are the bones (Fig. 20.20C) of the forearm. The **radius** is a short, slender and straight bone proximally articulating with the outer or preaxial condyle of trochlea of humerus and distally with the carpus. The **ulna** is longer, stouter and more curved than the radius. A wide space is left between the two bones. Its proximal end articulates with the inner or postaxial condyle of the trochlea of humerus and is projected into a small blunt **olecranon process**. Distally it articulates with the radius and carpus.

Carpals: The wrist or carpus consists of only two bones in the adult (in young birds the distal carpal row consists of three pieces which later fuse with the metacarpus). The smaller preaxial bone is **radiale,** it articulates with the radius. The larger postaxial bone is **ulnare,** it articulates with the ulna.

Carpometacarpus: The manus or palm in the adult consists of a single elongated bone, the **carpometacarpus** (Fig. 20.20D) formed by the fusion of three metacarpals with the distal row of carpals. The **first metacarpal** is represented by a small stumpy projection on the preaxial side. It is fused with the second. The **second metacarpal** is stronger and more or less straight bone. The **third metacarpal** is thinner and slightly curved rod on the postaxial side of the second with which it is fused at both ends.

Phalanges: There are three clawless **digits** or **fingers** present on the three metacarpals. The **first digit** or **pollex** consists of a single phalanx, the **second digit** or **index finger** has two phalanges, the **third digit** has only one phalanx.

Hindlimbs

The hindlimbs are modified for bipedal locomotion. They present the typical pentadactyl structure.

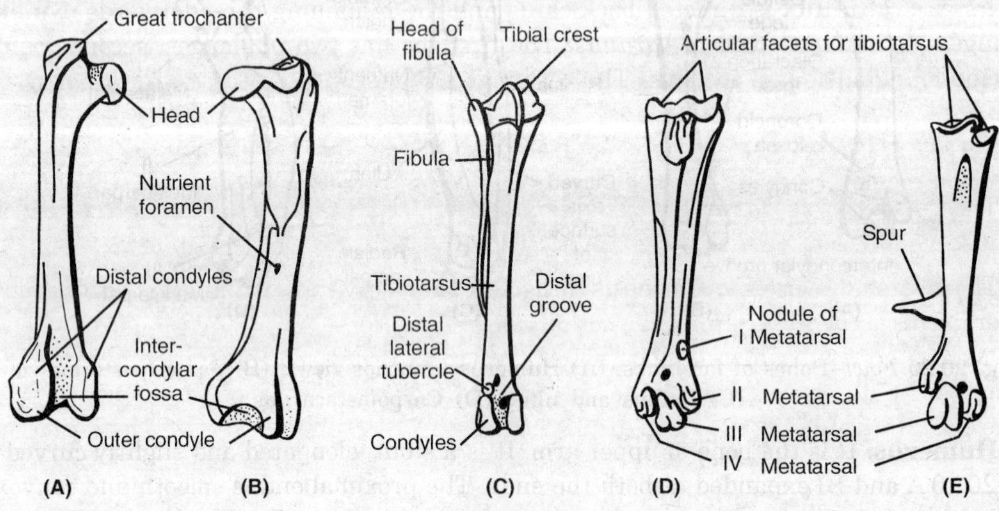

Fig. 20.21 *Fowl*–Bones of hindlimbs. **(A)** Femur (anterior view); **(B)** Same (posterior view); (C) Tibiotarus and fibula (anterior view); **(D)** Trasometatassus of Dmale (posterior view)

Femur: It is the bone of thigh. It is short and powerful bone (Fig. 20.21 A and B) enlarged at both ends. The proximal end is the **head**. It is round and fits into the acetabulum of the pelvic girdle. On the outer side of the head there is an irregular process, the **great trochanter** which articulates with the anti-trochanter of ilium. Distally the femur presents a deeply grooved surface, the **intercondylar fossa** or **groove**, bounded laterally by two prominent **condyles** for articulation with the tibiotarsus. The **patella**, a sesamoid bone forming the knee joint, is found in front of the femur-tibiotarsal joint.

Tibiotarsus and **fibula:** The shank or crus consists of two bones: **tibiotarsus** and **fibula** (Fig. 20.21 C). The tibiotarsus is a long, stout, nearly straight bone formed by the fusion of tibia with proximal row of tarsals. It lies on the inner or preaxial side of the shank. The proximal end is expanded and bears two articular facets for the two condyles of femur. On its anterior face it has a prominent **cnemial crest**. The distal end is pulley-like and bears two articular condyles to articulate with tarsometatarsus. Fibula is reduced to a small slender bone closely applied to the outer surface of tibiotarsus. Its proximal head end articulates with the outer condyle of femur. Its distal end does not reach the ankle and tapers to a sharp point.

Tarsals: The adult bird has no free tarsal bones. The proximal row is fused with tibia and distal row with the metatarsus. An **intertarsal** or **mesotarsal ankle joint** is present between the two rows.

Tarsometatarsus: The foot or pes contains a single stout, straight bone proximally articulating with distal row of tarsals and distally bearing three metatarsals (second, third and fourth) separate from each other and each articulating with a corresponding digit. The first metatarsal is a reduced nodule-like bone attached by ligaments to the inner and posterior surface of tarsometatarsus. In male, the tarsometatarsus bears, on its medial side, a stout conical, slightly curved pointed bony projection, the **fighting spur** (Fig. 20.21 E).

Phalanges: The metatarsus carries four toes made up of small bones called **phalanges**. The first-toe or **hallux** is directed backwards and the remaining three forwards. Number of phalanges is similar to that of *Varanus*. The first toe has two phalanges, second toe three, third toe four and fourth toe five. The terminal phalanx in each toe, bears a horny **claw**.

MUSCULAR SYSTEM

The muscles of birds are greatly specialized to perform different functions. Histogically the muscles are of the usual three kinds: striated, unstriated and cardiac.

There are numerous **cutaneous muscles** that serve for ruffling or fluffing out the feathers. This is necessary for shaking off water from feathers in aquatic birds when they come ashore or for drying the feathers of land birds after a bath. They are also used for fluffing out or depressing the feather in response to temperature changes, erecting crests, pinnae or breeding plumes in sexual displays. These muscles are named in groups corresponding to the feather tracts to which they are distributed. As such the cutaneous muscles of the feathers of humeral tract may be called humeral cutaneous muscles, and so on.

The jaw musculature is modified in response to feeding habits. In birds of prey and seed eaters (which actually crush seeds) the **adductor muscles** which close the jaws are strongly developed, whereas in insectivores or in birds which merely pick up and swallow seeds (pigeon) they are generally reduced. The adductor muscles are known as **adductor mandibulae** and are arranged in four sets. Likwise the gaping muscles or **protractors** also show modifications. They mainly include **depressor mandibular** muscles which are less massive in the seed eaters but strongly developed in those fruit eaters that puncture fruit with a closed bill and then open it to extract pulp or juice.

The muscles of the back are atrophied due to inactivity of the trunk vertebrae. The most important and well developed muscles of birds are the flight muscles, leg muscles for bipedal locomotion and perching muscles.

FLIGHT MUSCLES

The birds fly with the help of the **wings** (modified forelimbs). The muscles which operate the wings during flight are called **flight muscles** (Fig. 20.22). There are many muscles associated with flight but they have been grouped into three categories: **pectoral**, **accessory** and **tensor**.

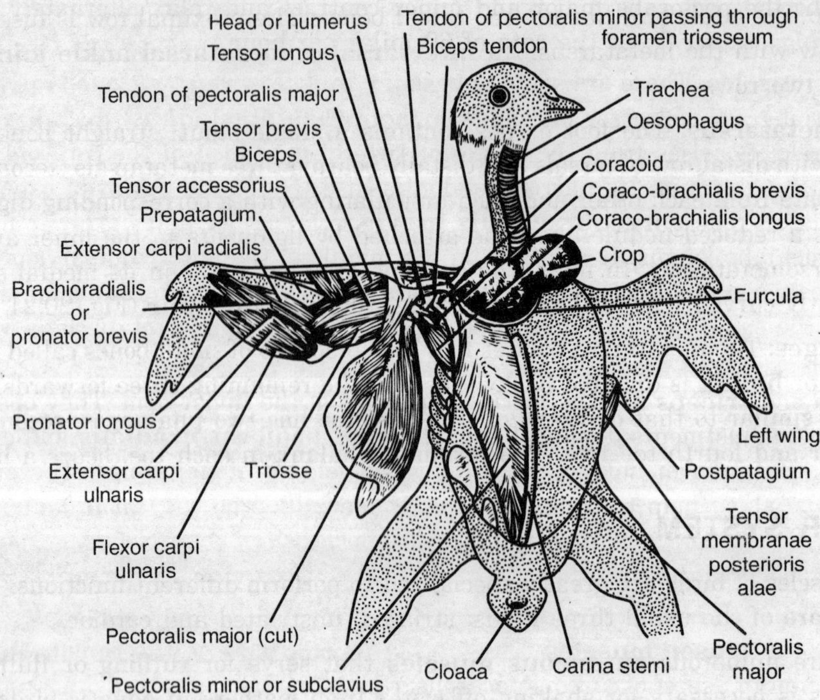

Fig. 20.22 *Columba*–**Flight muscles of breast and wings of right side of the body**

PECTORAL MUSCLES

The largest and most significant flight muscles of bird are pectoral muscles. These muscles are attached to the wings and keel of the sternum and are responsible for up and down movements of wings. In pigeon, they account for nearly one-half of the total body weight. They are of two kinds:

Pectoralis major: It is an enormous triangular muscle arising from the ventral side of the keel of sternum and clavicle, one on each side. It occupies the whole ventral surface and forms the so called breast. It accounts for almost one-fifth of the total body weight. It is dark red in colour due to rich blood supply. Its broad flat tendon is inserted on the ventro-lateral surface of the head of humerus. It is the great depressor muscle pulling the wings to the lower side (downstroke). When the pectoralis major contracts, the wing is pulled downwards and forwards and the body of the bird is lifted up and propelled through the air.

Pectoralis minor: For elevating the wings there is a ventrally placed smaller and elongated muscle the **pectoralis minor** (also called **subclavius** or **deep pectoral** or **supracoracoides**). This muscle also originates on the sternum beneath the pectoralis major and inserts on the dorsomedial aspect of the humerus. Insertion of subclavius on the upper surface of the arm bone is achieved by the passage of its long narrow tendon through the foramen triosseum. It is an elevator muscle causing upstroke of the wing. When it contracts the foramen triosseum acts like a pulley for its tendon pulling the humerus backwards and upwards, thus raising the wing.

During flight the pectoralis major and minor contract and relax alternately in rapid succession. The pigeon can fly at the rate of 60 miles per hour.

Accessory muscles: There are some **accessory muscles** which elevate or depress the wings. A small triangular depressor muscle is **coracobrachialis longus** which lies beneath the pectoral muscle. It arises from the posterior two-third postion of the coracoid and the costal process of the sternum. Its narrow tendon is inserted on the postaxial aspect of the head of the humerus. The **coracobrachialis brevis** is a smaller and narrower muscle in front of coracobrachialis longus. It arises from the middle of the antero-dorsal aspect of the sternum just in between the junction of the two coracoids. The coracobrachialis brevis of both sides arise from the same spot and take opposite courses. It helps to rotate the wing in the glenoid cavity.

The **biceps** and **triceps** are the intrinsic muscles of the upper arm. They operate the elbow and perform adjustments during flight. The **externsor carpiradialis** and **extensor carpiulnaris** of the forearm help in stretching and folding of the wings. There are two **branchioradialis** which help to rotate the radius medially. The **digital muscles** help to move the digits and feathers individually. The **policis muscles** help in moving the bastard wing or alula attached to the first digit.

Tensor muscles: There are some muscles that keep the wings fully spread during flight. These are the **tensor muscles**. The longest among these is the **tensor longus**. It arises from the anterior aspect of carpometacarpus, runs along the outer border of prepatagium and is inserted on the front of the head of humerus. It is a very delicate muscle and thus it often becomes difficult to see it. It keeps the outer border of prepatagium fully stretched. A thin sheet of muscles, the **tensor brevis** spreads between the biceps muscles and the posterior margin of the proximal portion of the tensor longus. Lying across the **prepatagium** there is a comparatively broader muscle the **tensor accessorius**. All these muscles keep the **prepatagium** fully stretched. A thin strip of muscle the **tensor posterius** runs along the outer edge of the postpatagium and keeps it fully stretched during flight.

LEG MUSCLES

The movements of a bird are restricted, there being no movements of abduction and adduction, which a man makes at the hip during walking. The movements are replaced by movements of rotation at the knee. The muscles of legs of a bird are enlarged and strong (Fig. 20.23). The shank and feet have few muscles so they look thin and delicate.

The muscles around the hip joint form a system of couples (or braces) that help in balancing and locomotion. The **lateral** and **medial** (**abductor** and **adductor**) muscles are very poorly developed. The **anterior** muscles (**protractors**) are well developed and include the following:

(i) the **sartorius** (**iliotibialis internus**), a prominent muscle running from the ilium to the tibia,

(ii) the **ilio-tibial**, a large muscle inserted through the patella on a ridge on the front of tibia, thus crossing both hip and knee joint, and

(iii) the **femorotibial** muscles associated with ilio-tibial forming the extensor system of the knee.

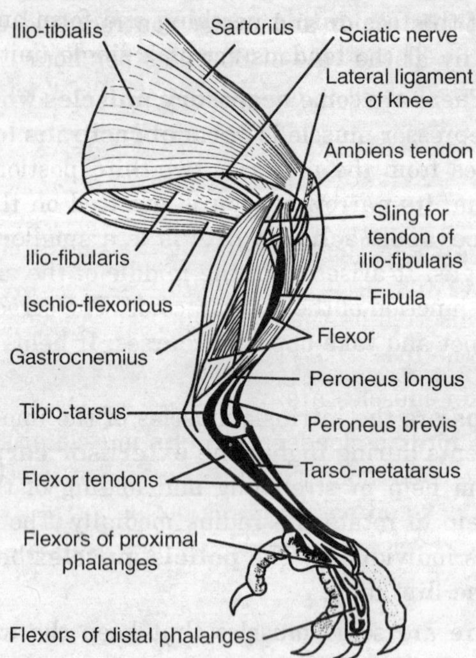

Fig. 20.23 *Columba*–Perching muscles of leg

The lateral side of the hip joint has rather small muscles including the **ilio-trochanteric** muscles that are small braces acting mainly as medial rotators and the **ischio-femoral** and **obturator** muscles that act as lateral rotators. The posterior muscles (**retractors**) of the hip joint are the main locomotor muscles. They lie behind the hip joint and include several muscles such as **posterior ilio-tibial**, **iliofibular**, **caud-ilio-flexorius**, **pub-ischio-femoral**, **ischio-femoral** and **caud-ischio-femoral**. Some of these also act as lateral rotators and obturator muscles.

The **knee joint** comprises three sets of ligaments and a pair of lunate cartilages called **menisci** (singular **meniscus**). The three ligaments are known as the **lateral**, **medial** and **cruciate** according to their positions. The **inter-tarsal joint**, that allows movements of flexion and extension, is like the knee joint and consists of a very strong capsule and lateral and cruciate cartilages, on the lateral side there is a meniscus. The muscles of this joint are the **gastrocnemius** along with the flexor muscles of the toes lying behind the tibia, the **tibialis-anterior** on the front of the tibia and acting across the intertarsal joint and the **extensors** of the toes, also lying on the front of the tibia.

PERCHING MUSCLES AND PERCHING MECHANISM

Some muscles in the upper part of the leg have special arrangement and long tendons for moving the toes. When a bird sits or squats on a perch (a branch, rod or wire) the toes mechanically clinch around the perch. This is brought about by special muscles called, **perching muscles** and the arrangement of these muscles is called **perching mechanism**. The bird can sleep on the perch without any risk of falling. Mainly the **calf muscles** are responsible to bring about flexion of the toes during perching and form an elaborate system of tendons attached to the phalanges. These tendons act as a single unit and there is an interlocking arrangement by which the flexion is maintained by the weight of the body during sleep.

There are two sets of perching muscles: **flexor** and **extensor.**

Flexor

The gripping of the twig by the toes is brought about by eight flexor muscles—six to the anterior toes and two to the hind toe or hallux. These muscles lie on the back of the tibiotarsus and remain inserted on the knee joint. The tendons of the flexor muscles extend up to the phalanges of the toes.

The important flexor muscles are:

Ambiens: In some birds a slender **ambiens** muscle arises from the pectineal process of the ilium and extends down along the inner surface of the thigh. Its tendon runs beneath the patella bone of the knee and passes to the outer side of the tibiotarsus to join the flexor muscle of the second and third toes. Actually ambiens plays a minor role in perching (flexion of two toes only). So it can be cut off without loss of perching ability.

Peroneus medius: It is present on the anterior aspect of the shank attached to the upper part of the tibiotarsus bone. Its tendon divides into three branches which join the three anterior toes.

Gastrocnemius: It is the big calf muscle present on the back of tibiotarsus. Its tendon joins the peroneus and supply the three anterior toes.

Flexor perforans: This muscle is attached to the upper part of the tibiotarsus and its tendon joins the hind toe or hallux. It is also connected to peroneus medius so that a pull on any tendon may flex all the toes.

When a bird sits on a perch the legs are bent at the knee and ankle, the tendons are flexed and the digits automatically grip the perch. All the flexor tendons are interconnected and act as a single unit, that is, a pull on any one of the tendon brings about flexion of all the toes. Flexing of digits is due to a **grip reflex** initiated by a **sensory receptor** present on the lower surface of the foot. It is activated the moment the foot touches the perch.

There is an interlocking mechanism which keep the toes flexed. The under surface of the flexor tendon and the upper surface of the tendon are ridged at the joint between the phalanges and tarsometatarsus which is pressed against the perch by the weight of the body. When the bird perches the ridges interlock so that the toes remain flexed and the bird does not fall. The pulling of the tendon is brought about by the body weight as such to unlock the toes the body has to be raised.

Extensor

A number of extensor muscles are found in front of the tibiotarsus. The tendon of the tibialis anterior muscles passes down in front of the intertarsal joint and divides into three branches to supply the three anterior toes. The contraction of tendons of extensor muscles serve to open the toes, the bird raises its shank and takes off the perch.

COELOM

Beneath the skin are located thin layers of abdominal muscles corresponding to those of other vertebrates and internal to this is the **parietal peritoneum** generally impregnated with streaks of fat. The peritoneum lines a cavity, the **coelom** into which are lodged the various organs or **viscera**. The coelom of bird consists of four chambers (Fig. 20.24), an antero-ventral **pericardial cavity**, two dorso-lateral **pleural cavities** and a posterior **peritoneal cavity**. The whole coelomic space is known as the **pleuroperitoneal cavity**. A membranous partition, the **oblique septum** extends obliquely behind the lungs dividing the pleuroperitoneal cavity into thoracic and abdominal portions. The two pleural cavities and the pericardial cavity lie in the thoracic portion. Each pleural cavity encloses a **lung**. The part of the peritoneum surrounding the lung is called **pleura**. The two pleural cavities are separated from each other by the pericardial cavity lying in the middle between the ventral portions of the pleural cavities. The pericardial cavity encloses the **heart**. The two lobes of the liver are attached to the pericardial sac with the help of **coronory ligament**. The large **abdominal** or **peritoneal cavity** lies posterior to the oblique septum. It contains rest of the visceral organs such as the digestive, excretory and reproductive organs suspended by **mesenteries**. The coelom is filled with a **coelomic fluid**.

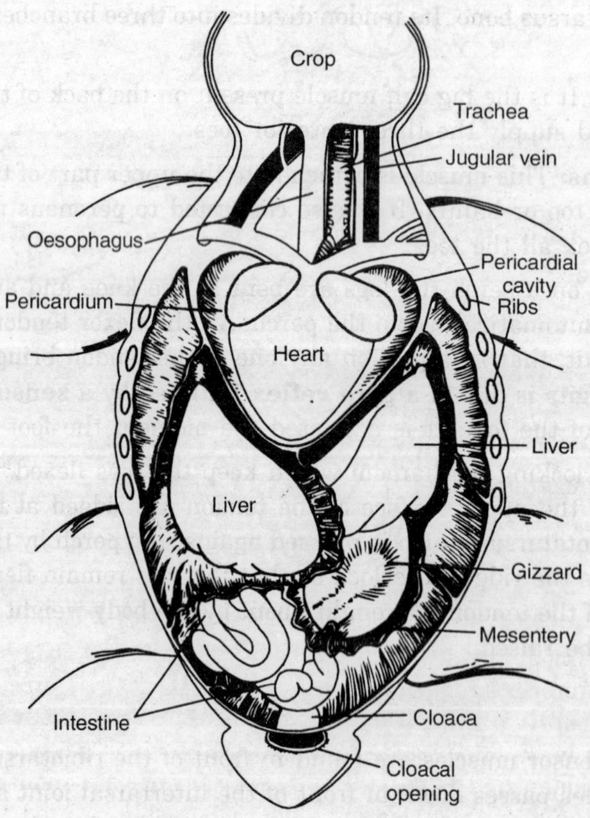

Fig. 20.24 *Columba*–Coelom an viscera

DIGESTIVE SYSTEM

The digestive system of pigeon is well developed and highly modified due to loss of teeth and aerial mode of life. It includes a tubular **alimentary canal** and **digestive glands**.

ALIMENTARY CANAL

The alimentary canal is a long and coiled tube. It comprises **mouth, buccal cavity, pharynx, oesophagus, stomach**, **small intestine** and **large intestine** which opens to the outside by **cloacal aperture**.

Mouth: The mouth opening is a wide aperture bounded by the **upper** and **lower horny beaks** which encase the **jaws** without teeth. The mouth leads into the buccal cavity.

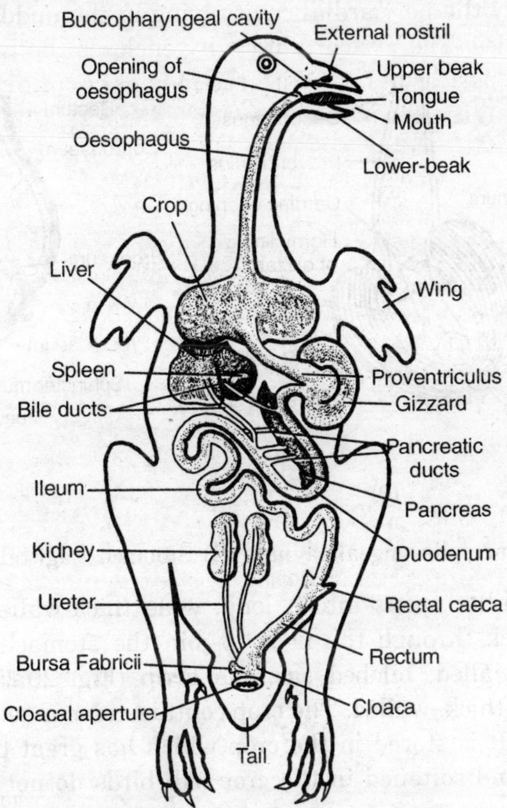

Fig. 20.25 *Columba*–Digestive system

Buccal cavity: The roof of mouth or buccal cavity (Fig. 20.26A) bears a pair of elongated **palatal folds** with free fimbriated margin. It has a deep **palatal fissure** in the median line, therefore the palate is a **split palate**. Dorsal to the palatal folds occur a pair of elongated narrow **posterior nares** lying side by side in the roof of the mouth cavity. The anterior two-third of their length is hidden by the palatal folds. The floor of the mouth cavity is occupied by the **tongue** which is narrow and triangular with a sharply pointed tip. Posteriorly it is produced into two **lobes** fringed with fine horny processes. The tongue is not muscular. The taste buds are few. Numerous mucous glands are present in the oral cavity.

Pharynx: The posteriormost part of the buccal cavity is the pharynx. In the roof of the pharyngeal cavity just behind the posterior nares there is a single median aperture, the opening of paired **auditory** or **eustachian tubes** (the two tubes unite at the point of communication with the pharynx). Each tube extends from this opening to the cavity of the middle ear. The tubes and cavity are derived partly from an evagination from the first visceral pouch. Posterior to this opening the roof bears a pair of folds with fimbriated margins constituting the **soft palate** which hang down like a curtain into the pharyngeal cavity. Posterior to the caudal end of the tongue there is a hardened elevation, the **laryngeal prominence** in the centre of which lies an elongated opening the **glottis** which leads into the trachea. It is supported by **laryngeal cartilages**. Behind the glottis there is a wide opening, the **gullet**.

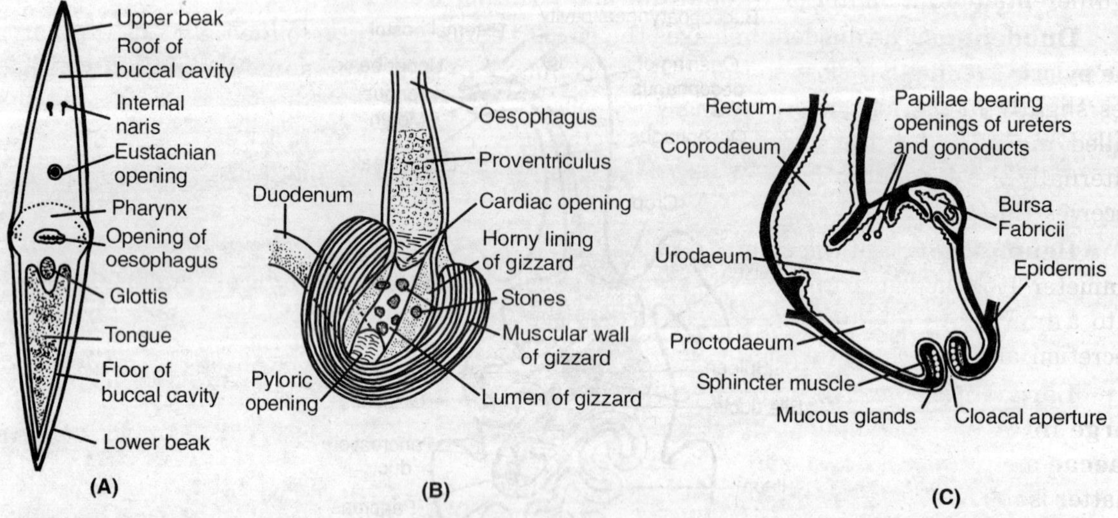

Fig. 20.26 *Columba*–**(A)** Buccopharyngeal caviity; **(B)** Stomach, sagittal section; **(C)** Cloaca L.S.

Oesophagus: The gullet opens into a long, wide thick-walled, distensible tube, the oesophagus, that runs back through the neck to join the stomach. In the neck region, it enlarges to form a thin–walled, bilobed sac, the **crop** (Fig. 20.25) Beyond the crop the oesophagus again becomes thick–walled. The crop contains mucus secreting glands. The birds swallow their food whole. It is stored in the crop which has great power of distension. The stored food is moistened and softened in the crop. All birds do not have crop.

The pigeons, both male and female, have the ability to produce **pigeon milk,** a cheesy and nourishing secretion, especially during the breeding season. Pigeon milk is not produced by any gland but is formed by the disintegration of the epithelial cells lining the crop. Its formation is stimulated and controlled by **prolactin hormone** secreted by the anterior lobe of the pituitary gland. It contains water, proteins, fats and ash and is more nourishing than mammalian milk. It is regurgitated into the mouth of the baby birds till they are old enough to eat by themselves. Pigeon milk is also produced in some parrots.

Stomach: The stomach of pigeon is differentiated into an anterior **proventriculus** and a posterior **gizzard** (Fig. 20.26B).

Proventriculus: It is a small, thick-walled, tubular and gladular structure slightly wider than the oesophagus. Its thick mucus lining secretes gastric juice. The **spleen** is attached by peritoneum to the right side of proventriculus.

Gizzard: The proventriculus opens into the **gizzard** which is a large, hard body, flattened laterally forming a biconvex structure. The opening of proventriculus into the gizzard is called **cardiac opening**. The walls of gizzard are thick and muscular and its inner lining is thick, rough and horny. The cavity of gizzard is very small and contains small **stones** or **grit** called **gastroliths**, which are swallowed by the bird. These pieces help in crushing the food. The gizzard thus becomes a powerful grinding mill. The gizzard opens into the small intestine by **pyloric opening** which is guarded by a sphincter called **pyloric valve** or **pylorus**.

Small intestine: The small intestine or midgut is a narrow tube about 7.5 cm long. It is differentiated into anterior **duodenum** and posterior **ileum**.

Duodenum: The duodenum leaves the gizzard dorsally, a little to the right side, so that the pyloric opening lies close to the cardiac opening of proventriculus into the gizzard which lies slightly to the left side. The duodenum forms a **U-shaped loop.** A part of mesentery called **mesoduodenum** stretches between the two limbs of 'U'. It lodges the pancreas. Internally the duodenum contains villi, crypts of Liberkühn and goblet cells. The duodenum receives three pancreatic ducts and two bile ducts.

Ileum: The rest of the small intestine is ileum. It is a long convoluted tube of uniform diameter. It is the longest part of the alimentary canal. The internal lining of ileum is thrown into a number of minute finger-like processes called **villi**. The villi increases the area of secretion and absorption of the ileum.

Large intestine: The ileum passes, without any change in diameter, into the hind gut or **large intestine**. At the junction of two, a pair of small blind pouches called **rectal** or **colic caecae** are present. They absorb water from food. The large intestine is short because faecal matter is relatively small. It is differentiated into anterior **rectum** and posterior **cloaca.**

Rectum: The rectum is a short narrow tube of the same diameter as the ileum. Its opening into the cloaca is guarded by **anal sphincter**.

Cloaca: It is the terminal part of the alimentary canal. It is a large triparptite chamber (Fig. 20.16 C) divided into three linear compartments: anterior **coprodaeum** which receives the rectum, middle **urodaeum** into which ureters and genital ducts open and posterior **proctodaeum** which opens to the outside by **cloacal aperture** or **vent.** A small glandular thick–walled blind pouch of lymphatic tissue, called **bursa fabricius** lies against the dorsal wall of proctodaeum in young birds but atrophies in adult. It most probably produces lymphocytes. It is also known as **cloacal thymus** as it resembles the thymus gland which is also lymphatic.

DIGESTIVE GLAND

There are a number of digestive glands in pigeon which are associated with different parts of the alimentary canal.

Buccal gland: In the buccal cavity small buccal glands are present which are probably not salivary. They secrete mucus to moisten the food and amylase enzyme. The tongue also has mucus glands.

Salivary gland: In the pharyngeal region paired angular and unpaired sublingual salivary glands are found. They secrete saliva which moistens the food. Some ornithologists believe that it also contains enzymes.

Gastric glands: The epithelial lining of stomach contains several large gastric glands which secrete gastric juice containing peptic enzymes.

Liver: The liver is large, compact, dark red, bilobed structure (Fig. 20.25). The right lobe is larger than the left. The surface of liver is hollowed to receive the heart, duodenum and gizzard. There is no gall bladder in birds. There are two bile ducts, one emerging from each liver lobe. The left bile duct is shorter and stouter and enters the proximal limb of the duodenum about half an inch beyond the gizzard. The right bile duct is slender and passes into the distal limb of the duodenum. The liver secretes bile.

Pancreas: The pancreas is a compact reddish gland lying between the two limbs of the duodenum. Three pancreatic ducts arise from the pancreas and open into the distal limb of the duodenum. The pancreatic secretion contains many enzymes.

Intestinal glands: The epithelial lining of intestine contains many microscopic intestinal glands which secrete many enzymes.

Tubular glands: The lining of gizzard contains numerous tubular glands which produce thick yellowish green secretion.

FOOD AND DIGESTION

The pigeon feeds mainly on cereals, pulses, seeds and sometimes small insects. The food is simply picked up with the beak. The tongue manipulates and lubricates the food in the buccal cavity. As there are no teeth, the food is swallowed. It passes though the oesophagus and stored in the crop.

Inside the crop the food is thoroughly mixed with water, mucus and secretion of buccal glands and softened. The body warmth also helps in moistening and softening the food. The food passes into the proventriculus, where it is chemically acted upon by the peptic enzymes of gastric juice. Gastric secretion is intermittent and under the control of nervous system. The food then passes into the gizzard where it is crushed and ground up by the stones present in the gizzard and muscular contraction of gizzard walls. The gizzard is more acidic than crop but very little digestion takes place here. The partly digested food is called **chyme**. The chyme passes into the duodenum where it is mixed with bile, pancreatic and intestinal juices. The process of protein, fat and carbohydrate digestion in the duodenum is similar to that of other vertebrates. The digested food then passes into the ileum where it is absorbed through the intestinal wall into the blood. The undigested food is very little. It passes into the rectum where reabsorption of water takes place. It then goes into the cloaca and ejected through the cloacal aperture. The faecal matter is almost dry.

The food requirement of bird is quite high. The rate of digestion is rapid due to high rate of metabolism.

CIRCULATORY SYSTEM

The circulatory system of birds is similar to that of other higher vertebrates. It consists of **heart, arteries, veins, blood** and **lymphatic vessels**.

HEART

The heart is completely **four chambered**, consisting of two **auricles** and two **ventricles**. Sinus venosus is absent in birds and is said to be incorporated into the right auricle.

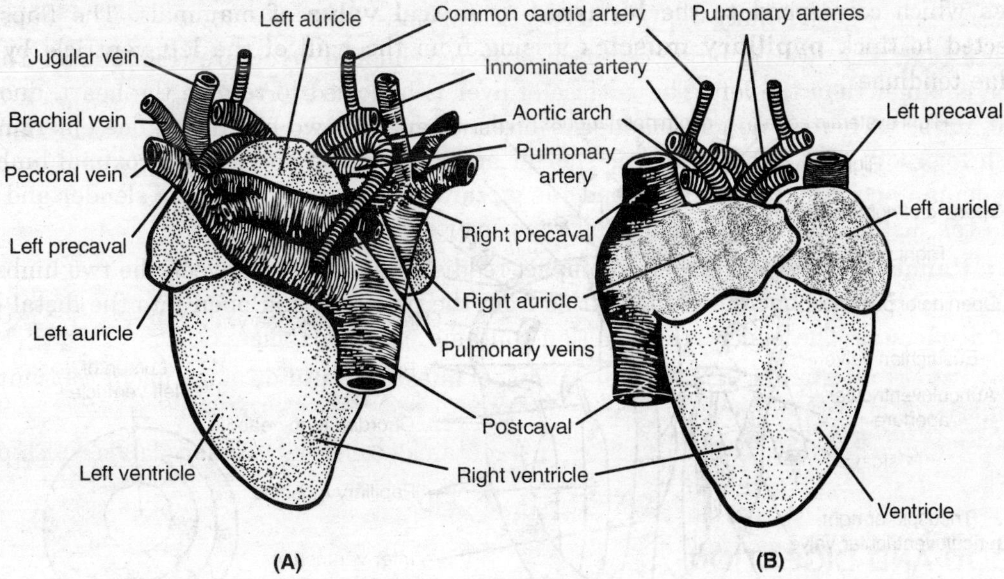

Fig. 20.27 *Columba*–Structure of heart, **(A)** Dorsal view; **(B)** Ventral view

External features

The heart of birds is relatively larger than that of other vertebrates. It lies midventrally in the thoracic cavity partly surrounded by the lobes of the liver. The heart is **conical** in shape with the broad anterior end or **base** directed forwards and the pointed posterior end or **apex** directed backwards (Fig. 20.27). It is surrounded by a thin and membranous sac, the **pericardium** derived from coelomic epithelium. The pericardial wall is made up of outer **parietal layer** and inner **visceral layer**. The two layers enclose a narrow **pericardial cavity** filled with watery **pericardial** or **serous fluid.** The fluid protects the heart from shocks and mechanical injuries and also allows movements during beating. A hollow transverse **coronory sulcus** or the **auriculo-ventricular groove** separates the two auricles and ventricles externally. Similarly, a faint **interauricular septum** dermarcates the two auricles. The auricles are anterior, smaller, dark coloured and thin–walled whereas, the ventricles are posterior, larger, light coloured and thick-walled.

Internal structure

The two auricles are separated from each other by a complete thin muscular partition, the **interauricular septum** (Fig. 20.28) bearing an oval patch, the **fossa ovalis** at the position occupied by the **foramen ovale** in the embryo. The right auricle is larger than the left. The right auricle has three separate openings of the caval veins lying close together. The opening of the postcaval vein is guarded by a muscular fold the **eustachian valve**. The right auricle opens into the right ventricle by a crescentic aperture, the right **auriculo-ventricular**

opening guarded by two muscular flaps, the **auriculo-ventricular valves** connected to the ventricular wall at the base of origin of the pulmonary artery. In the left auricle lies the opening of the pulmonary vein. The left auricle opens into the left ventricle by **left auriculo-ventricular aperture** guarded by two membranous flaps, the **left auriculo-ventricular valves** which correspond to the **bicuspid** or **mitral valve** of mammals. The flaps are connected to thick **papillary muscles** arising from the wall of the left ventricle by two chordae tendinae.

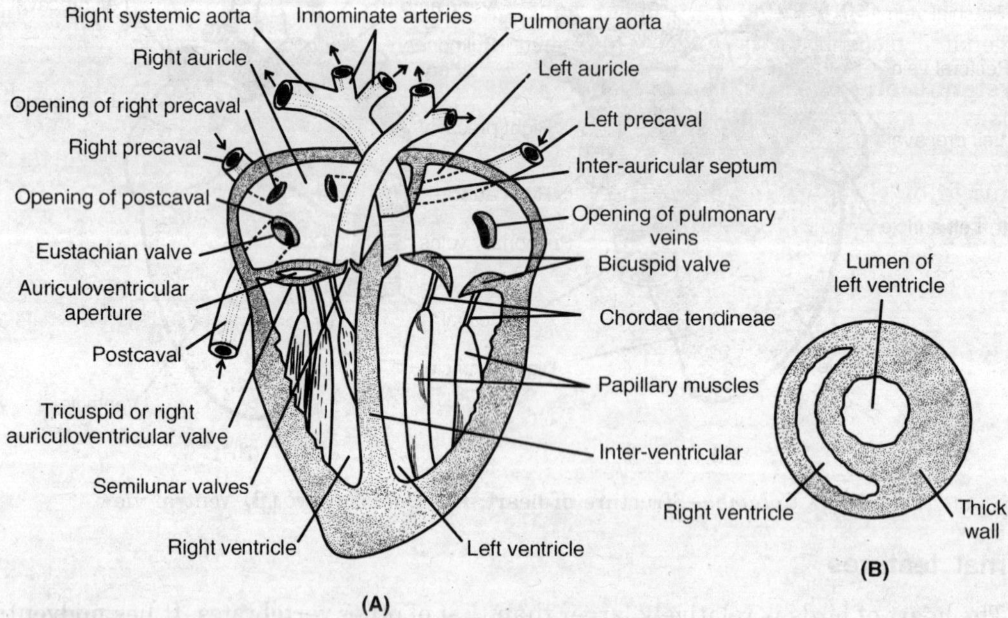

Fig. 20.28 *Columba*–**(A)** Internal structure of heart; **(B)** Cross, section of ventricle

The two ventricles are completely separated from each other by a thick muscular **interventricular septum** which passes obliquely in such a way that the left ventricle becomes larger of the two and includes the whole apex of the heart. The right ventricle is thin-walled and wraps around the left ventricle. In a transverse section the right ventricle appears **crescentic** and the left ventricle appears **circular** (Fig. 20.28 B). The right ventricle gives off a single trunk the **pulmonary aorta** which soon divides into two **pulmonary arteries** each going to a lung. The opening of the pulmonary trunk is guarded by three semilunar valves. The left ventricle gives rise to single **right systemic** or **aortic arch** which is continued into the **dorsal aorta**. The **left aortic arch** is absent in birds. The opening into the aortic arch is guarded by three semilunar valves. Bars of muscles called **trabeculae** or **columnae carnae** traverse the cavities of the ventricles.

Working of heart

The birds are active animals and need a speedy distribution of building materials to the quickly exhausting cells. For this purpose the heart acts as a force pump and drives blood quickly to all parts of the body. The rate of heart beat is rapid to ensure rapid blood circulation, high body temperature (102° F to 112°F) and rapid metabolism for production of energy. Such a rapidly beating heart places a severe strain on the arterial vessels and no wonder that the innominate leading out of the heart is enormously enlarged. The rate of heart beat differs greatly in different species and under different conditions. In birds there is no mixing of **venous** (**deoxygenated**) and **arterial** (**oxygenated**) blood. The right half receives and discharges only venous blood and the left half only arterial blood. Thus, birds like mammals, possess **complete double circulation** which includes **pulmonary** and **systemic circulation.**

Pulmonary circulation: The right ventricle pumps venous blood into the pulmonary aorta from where it goes into the lungs through right and left pulmonary arteries. After purification the oxygenated blood from the lungs is returned to the left auricle by pulmonary veins.

Systemic circulation: The oxygenated blood from the left auricle goes into the left ventricle from where it is driven into the left systemic arch and through the dorsal aorta to all parts of the body. The impure blood from different parts of the body is returned to the right auricle by three caval veins. The blood from the right auricle passes into the right ventricle and then pulmonary circulation starts.

ARTERIAL SYSTEM

In birds all the great arteries spring directly from the ventricle and are situated between the two auricles. The blood leaves the heart through two arterial trunks, the large medially located **systemic** or **aortic arch** and the smaller **pulmonary aorta.**

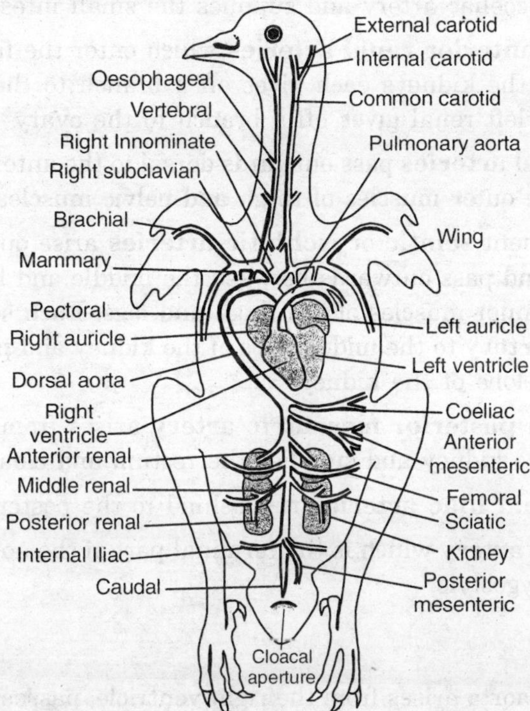

Fig. 20.29 *Columba*–Arterial system

Aortic arch

The **right aortic arch** is the only arch arising arteriorly from the left ventricle. It curves over the right auricle, gives off branches and passes backwards as the **dorsal aorta (Fig. 20.29)**. There is no left systemic arch in birds. Immediately after origin the right systemic arch gives off two small **coronary arteries** to the heart. It then gives off two short and stout right and left **brachiocephalic** or **innominate arteries.** Each of these divides into an anterior **common carotid artery** and a lateral **subclavian artery.** The common carotid runs anteriorly bending towards the mid-ventral side of the vertebral column of the neck. It divides into **internal carotid artery** to the brain and **external carotid artery** to the tongue, muscles of the jaw, other parts of the head and the neck. A branch from common carotid called **comes nervi vagi** runs parallel to the vagus nerve and joins the external carotid of its side. The subclavian artery turns forward, just a little and divides into two branches: the **brahial artery** which supplies the wings and sends a delicate branch to the shoulder girdle, and the **pectoral artery**, a large vessel which supplies blood to the pectoral muscles and sends an **internal mammary artery** to the ventral thoracic wall.

Dorsal aorta: The aortic arch curves over to the right side and runs backwards in the middorsal line forming the dorsal aorta. It gives off a number of branches:

1. A single median vessel, the **coeliac artery** arises just as the aorta emerges below the ventricle and supplies the stomach, gizzard, duodenum, liver and spleen.
2. A single **anterior mesenteric artery** which arises about a quarter of an inch posterior to the coeliac artery and supplies the small intestine.
3. A pair of small **anterior renal arteries** which enter the first lobe of the kidneys. Before entering the kidneys each gives off a branch to the testes in male but in female only the left renal gives off a branch to the ovary.
4. A pair of **femoral arteries** pass outwards dorsal to the anterior lobes of the kidneys and supplies the outer muscles of thigh and pelvic muscles.
5. A pair of prominent **sciatic** or **ischiatic arteries** arise opposite the middle lobes of the kidneys and pass outwards between the middle and last lobe of the kidneys. It supplies the inner muscles of the thigh and legs. Each sciatic artery gives off a **middle renal artery** to the middle lobe of the kidney and **posterior renal artery** to the posterior lobe of the kidney.
6. A single median **posterior mesentric artery** arises from the aorta opposite the hinder end of the kidney and supplies the rectum and cloaca.
7. A pair of **internal iliac arteries** run behind to the posterior part of the pelvis.
8. A single **caudal artery** which is the terminal part of the aorta goes into the caudal vertebrae and pygostyle.

Pulmonary aorta

A single pulmonary aorta arises from the right ventricle, passes ventral to the systemic arch and divides into two **pulmonary arteries**, one going into each lung.

VENOUS SYSTEM

The venous system includes **caval veins, pulmonary veins, hepatic portal system** and **renal portal system.**

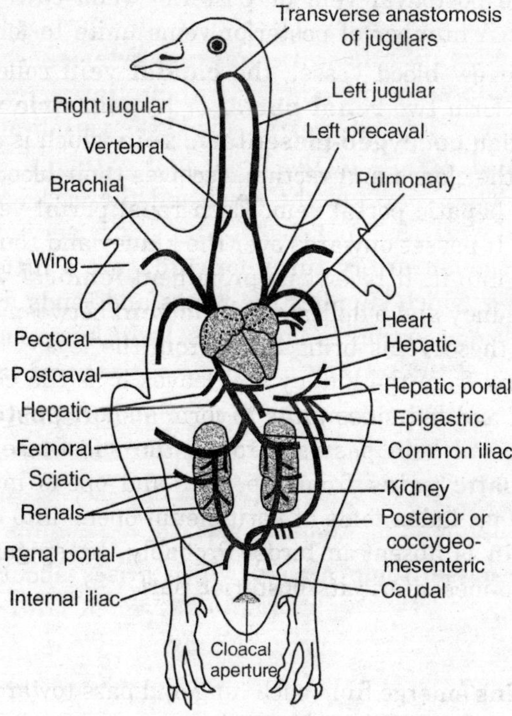

Fig. 20.30 *Columba*–Venous system

Caval veins

There are three caval veins which collect venous blood from various parts of the body and bring it to the right auricle. The two **precaval veins** or **left** and **right anterior vena cavae** collect blood from the anterior part of the body and a single **postcaval vein** or **posterior vena cava** collects blood from the posterior part of the body.

Precaval veins: The left and right anterior vena cavae are the two precaval veins which collect blood from the head, neck, wing and breast through the **jugular** and **subclavian** veins. These veins of the two sides are identical. The **right jugular vein** runs along the right side of the neck close to the vertebral column and almost behind the oesophagus. Near the angle of the jaw it divides into **external** and **internal jugulars**. The external jugular unites with it fellow of the other side beneath the base of the skull to form a transverse loop, the **jugular anastomoses**. This prevents blocking of the flow of blood from the head when one vein is momentarily shut off by twisting the neck. The internal jugular of each side collects blood from the brain. During its course the right jugular receives blood from the spinal cord and neck muscles by a small **vertebral vein** and from the crop by **oesophageal vein.** The **right subclavian vein** is formed by the union of two veins: the **right brachial vein** returning blood from the right wing and the **right pectoral vein** returning blood from the large pectoral muscles. The pectoral vein receives a delicate **internal mammary vein** from the inner surface of the thoracic wall. The left anterior vena cava is corresponding to that of the right side.

Postcaval vein: The postcaval vein or posterior vena cava collects blood from the posterior part of the body. A number of posterior veins unite to form the postcaval vein.

A small unpaired narrow blood vessel, the **caudal vein** collects blood from the tail. Anteriorly it bifurcates to form two **renal portal** or **hypogastric veins.** Just at the point of bifurcation arises a median **coccygeo-mesenteric vein** which is characteristic of birds. It runs forwards parallel to the cloaca and rectum, receives their blood and is continued along the mesentery to join the hepatic portal vein. Each renal portal vein receives an **internal iliac vein** from the pelvis. It passes onwards over the kidney and sends a few small branches, the **afferent renal veins** into it. It receives a prominent **femoral vein** between the middle and anterior lobes of the kidney and a delicate **sciatic vein** between the middle and posterior lobes of the kidney. Both these veins bring blood from the legs. The femoral vein receives small **efferent renal veins** from the kidney and leaves it as the **common iliac vein**. The common iliac vein of right and left sides unite to form a short **postcaval vein** just anterior to the kidneys. The postcaval vein passes through the substance of the right liver lobe, receives two pairs of **hepatic veins** from the liver and opens into the right auricle. An **epigastric vein** from the fat laden folds of peritoneum opens into one of the hepatic veins. **Anterior abdominal vein** is absent in birds. Probably the epigastric vein represents its anterior part and coccygeo-mesenteric its posterior part.

Pulmonary veins

Two **pulmonary veins** emerge from each lung and pass towards the heart immediately posterior to the precaval veins. They unite to open into the left auricle by a common aperture. They carry oxygenated blood from the lungs to the left auricle.

Hepatic portal system

This system presents the same structure as found in other vertebrates. The **hepatic portal vein** is formed by the union of a number of narrow blood vessels coming from various organs of the digestive system. These veins are **gastro-duodenal, anterior mesenteric, posterior mesenteric** and **coccygeo-mesenteric.** On reaching the liver the hepatic portal vein bifurcates into two branches, one going to each liver lobe. The blood from greater part of the alimentary canal is conveyed to the liver from where it goes into the heart.

Renal-Portal system

The renal portal system is greatly reduced in birds. It includes two **renal portal** or **hypogastric veins** which give off a few **afferent renal veins** into the kidney but do not break up into capillaries. Thus, the blood from the posterior region enter the heart directly and not through renal capillaries as in amphians and reptiles.

BLOOD

Blood of pigeon has more red blood cells per ounce than in any other animal so it is the richest blood in the animal kingdom. A space of one square inch may have about 10,000,000 red blood cells. Every part of the body is well nourished by blood. The **red blood cells** are minute and oval. There are no **blood platelets,** but clotting of blood is fast. The **white blood cells** are few in number and include lymphocytes, hetrophils, granulocytes, basophils, eosinophils and monocytes.

LYMPHATIC SYSTEM

The lymphatic system includes a few **lymphatic glands** associated with the blood vascular system. The **lymph vessels** form two **thoracic ducts** which open into the precaval veins. **Lymph hearts** are absent in adult pigeon. The spleen is a small oval reddish gland attached to the right side of the proventriculus by a peritoneal fold. It is also intimately associated with the blood vascular system.

Temperature regulation

The body temperature of birds is variable. In primitive birds like New Zealand kiwis the temperature is lower (about 100°F) and in most passerine birds it may reach to a maximum of 112°F. In most diurnal birds the temperature varies with the time of the day, being highest about mid-day and lowest during night. The temperature varies with activity. The emptiness of stomach may also cause fluctuation of temperature. But in all cases the fluctuating range may be as much as 8 to 10 degrees (from 102°F to 112°F in the passerines).

Birds live in varying environmental conditions to which they adjust by many ways. The birds that live in Arctic zone conserve heat by being densely feathered. The feather coat insulates the body. Sleeping birds commonly tuck their bills in their feathers in cold weather to reduce loss of heat in breathing, and the breathing rate and general metabolism are lowered. Shivering is another method that is employed to adapt to sudden cold, because it converts muscular energy into heat.

In warm weather the air sacs of birds are used as a chief means of dissipating excess heat. The birds have no sweat glands in the skin. The heat passes into the air sacs from where excess heat is removed through the lungs and mouth. In hot weather a bird is usually seen gaping with open mouth. It is not panting for breath, but is speeding up dissipation of internal heat by an increased breathing rate and faster evaporation of moisture. In some species young birds cannot regulate body temperature at hatching and at first need more or less complete brooding. They are, therefore, poikilothermous at birth but as soon as feathers develop homoithermous condition progresses rapidly.

RESPIRATORY SYSTEM

Respiration in birds is pulmonary. The respiratory system is fairly complicated. It includes respiratory tract, respiratory organs and air sacs.

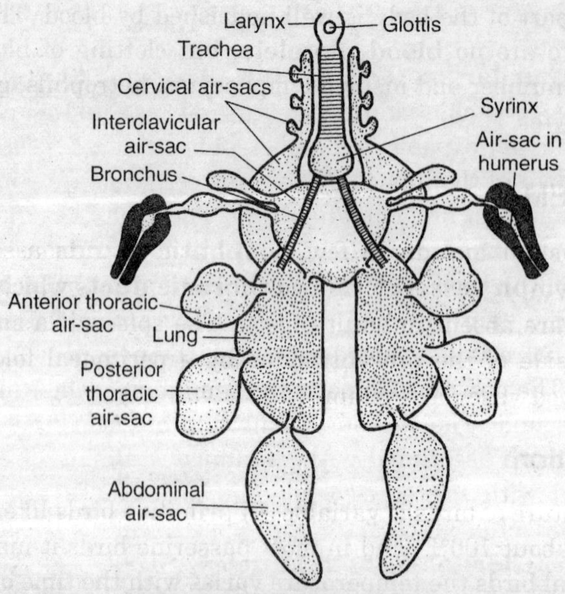

Fig. 20.31 *Columba*–Respiratory system

RESPIRATORY TRACT

The respiratory tract includes nares, trachea and syrinx.

Nares

A pair of slit-like oblique **external nares** or **nostrils** open at the base of the upper beak. It is surrounded by a swollen and sensitive pad of skin called **operculum** or **cere**. The external nares open into the **olfactory** or **nasal sacs** which open directly into the pharynx by **internal nares**.

Trachea

The **glottis** is a median slit-like opening present on the base of the tongue. It leads into the **trachea** or **air tube** (Fig. 20.31). The glottis opens into the **larynx**, an expanded chamber present at the top of the trachea. Morphologically the structure of larynx of bird is similar to that of other vertebrates, but it does not produce sound as vocal cords are absent. Sound is produced by another structure, the **syrinx** situated below the trachea. The wall of larynx is supported by cartilaginous pieces. There is a single large triangular **cricoid cartilage** made up of four pieces, present on the ventral side and a pair of **arytenoid cartilage** one on each side.

The **trachea** or **wind pipe** is a long cylindrical tube which runs back through the neck, ventral to the oesophagus but is displaced to the left side in the middle region by the crop. In the thoracic cavity it divides into two short **bronchi** one entering each lung. The trachea and bronchi are supported by a series of tracheal and bronchial rings respectively. All the tracheal and the first bronchial rings are complete and bony. Rest of the bronchial rings are incomplete in the middle and cartilaginous.

Syrinx

The **syrinx** or **lower larynx** (Fig. 20.32) is the organ of voice of birds. It is concerned with sound production. It is characteristic feature of bird and does not occur in any other vertebrate. The syrinx is an expanded chamber present at the posterior end of the trachea before the commencement of the bronchi. The walls of syrinx are supported by the last three or four tracheal rings and first bronchial half ring. The cavity of syrinx is called **tympanum**. In the dorsal wall of the tympanum a slight vertical fold is present in the median dorsal line extending forward from the level of bifurcation of the trachea. This fold is called **semilunar membrane** (**membrane semilunaris**). Its vibrations are said to produce sound. Running across the tympanum there is a slender bony ridge, the **pessulus** extending from the dorsal to the ventral surface. The pessulus supports the semilunar membrane and is believed to help in the production of sound. The mucous membrane of each bronchus forms an **internal tympaniform membrane** associated with the inner wall and an **external tympaniform membrane** associated with the outer wall. The membranes are regulated by a pair of **intrinsic syringeal muscles** which arise from the sides of the trachea and are inserted into the syrinx and a pair of **sterno-tracheal muscles** which arise from the sternum and are inserted into the trachea. The air expelled from the lungs during expiration passes between the tympaniform membranes and sound is produced by their vibrations. The pitch of the sound produced is altered by changes in the tension of the membranes which is under muscular control. The semilunar membrane only supplements the vibrations of the tympanic membrane.

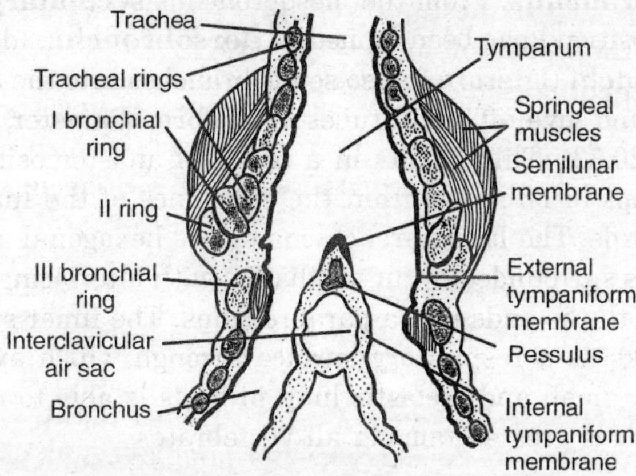

Fig. 20.32 *Columba*–Diagrammatic L.S. of Syrinx

RESPIRATORY ORGANS

The respiratory organs of bird are a pair of **lungs** (Fig. 20.31). The lungs of pigeon are very small in comparison to the size of body. They are bright red coloured, compact and non-elastic bodies closely applied by the dorsal surface to the thoracic vertebrae and ribs. They are covered ventrally by thick sheet of fibrous connective tissue, the **pleura** or **pulmonary aponeurosis,** an anterior continuation of the peritoneum. Dorsally there is no peritoneal covering. Small fan-like **costo-pulmonary muscles** which arise from the junction of the vertebral and sternal ribs, are inserted into the pleura. Their inner edges lie together separated from each other in the median plane by the projecting centra of the thoracic vertebrae and by the dorsal aorta.

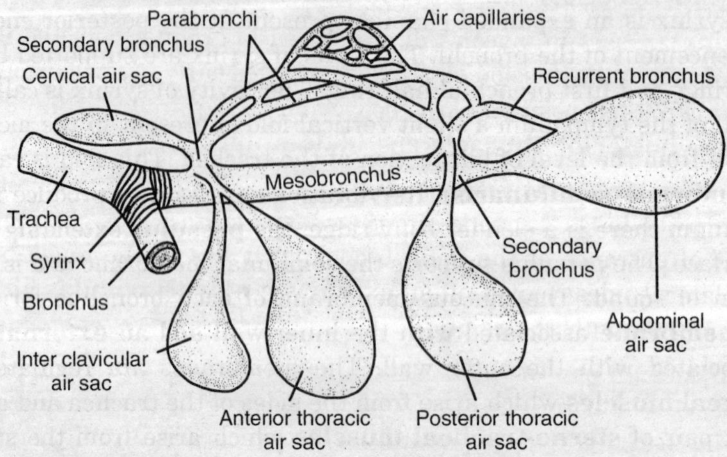

Fig. 20.33 *Columba*–Internal structure of lung and origin of air sacs

The lungs of birds are rather solid and spongy structure as in mammals and not hollow as in other vertebrates. The antero-ventral end of each lung receives a **bronchus**. It is called **primary bronchus**. Inside the lungs it passes into the substance of the lung upto its posterior end as **mesobronchus.** From the mesobronchus **secondary bronchi** arise which according to the their position have been named as **dorsobronchi** (dorsal), **ventrobronchi** (ventral) and **laterobronchi** (lateral). It also sends branches into the air sacs. The secondary bronchi divide further and give off small tubes of uniform diameter, the **tertiary bronchi** or **parabronchi** (Fig. 20.33). This results in a series of anastomosing fine **tubules** or **air capillaries** forming loops or circuits within the substance of the lung. There are no blind pouches or **alveoli** in birds. The lung parenchyma shows hexagonal areas each consisting of a central **parabronchus** surrounded by air capillaries with interlacing blood capillaries. Each air capillary opens at both the ends into a parabronchus. The inner surface of parabronchus and air capillaries works as a respiratory surface through which exchange of gases takes place. Thus, a relatively small and inelastic lung in birds is able to meet the extraordinary demand of oxygen which is greatest among all vertebrates.

AIR SACS

The lungs of all modern birds are characterized by the presence of remarkable system of supplementary **air sacs** (**cellulae aereae**) which facilitate the circulation of air through the lungs, but themselves are not directly respiratory in function, as is evident by the absence of blood capillaries on their surface.

The air sacs are large, thin walled, membranous, non-muscular and non-vascular structures (Fig. 20.31) connected with the ends of the bronchial branches. All arise from the secondary bronchi except the abdominal air sacs which arise from the mesobronchus. They are formed by the dilation of mucous membrane of the bronchus. They occupy spaces between the viscera, beneath the skin (in pelicans), between the muscles supporting the connective tissue, between and around the joints of cervical vertebrae and penetrating the pneumatic cavities of large bones. The opening of bronchi into the air sacs is called **ostia.** All the air sacs except the cervical rejoin the bronchus through **recurrent bronchi** or **saccobronchi**.

There are **nine** air sacs in pigeon. All are paired in origin except the **interclavicular** air sac which is formed by the fussion of two distinct ones early in development. The air sacs are named according to their position in the body cavity. They are **interclavicular, cervical, anterior thoracie, posterior thoracic** and **abdominal**.

Inter clavicular air sac: It is single median air sac somewhat triangular in shape and connected with both the lungs. It opens into the ventral surface of each lung immediately in front of the point of entrance of the main bronchus. It lies in the angle between the two clavicles. On the lateral sides it gives off an **axillary** or **extraclavicular air sac** in the arm pit which communicates with the cavity of the humerus bone.

Cervical air sacs: A pair of small cervical air sacs lie in front of the lungs at the base of the neck, alongside the vertebral column and dorsal to the interclavicular. They give off minute saccules in the neck. They communicate with the anterior angle of the lung.

Anterior thoracic air sacs: The anterior thoracic air sacs cover the ventral surface of anterior two third of the lungs and extend back to overlap the posterior thoracic air sac. They lie at the sides of the thorax and are in extensive contact with the ribs and pericardium. Each communicates with the ventral surface of the lung.

Posterior thoracic air sacs: The posterior thoracic air sacs are a pair of rather small sacs lying in the posterior part of the thoracic cavity just in front of the abdominal air sacs. They lie posterior to the lungs but overlap the ventral surfaces of their hinder end. Each communicate with the outer and posterior angle of the corresponding lung.

Abdominal air sacs: The abdominal air sacs are a pair of large sacs with very thin transparent membranous walls. They lie along the dorsal wall of the abdomen, ventral to the kidneys but dorsal to the intestine. Each communicates with the posterior border of the corresponding lung, near its outer angle.

Functions of air sacs

The air sacs are not respiratory organs but they act as accessory respiratory organs. They lack blood capillaries so no exchange of gases takes place. They simply serve as reservoir of gases. At each expiration they act as bellows forcing their air into the lung, thus leaving no dead space or unrespired air in the lungs. The air sacs are associated with the following functions:

Lightness: The air sacs act as balloons giving lightness and buoyancy in flight and reducing the specific gravity of the body due to contained air. This view is not supported now.

Temperature regulation: The birds have no sweat glands so air sacs help in temperature regulation. They act as a cooling device by losing body heat through **internal evaporation** or **perspiration**, that is, water vapours diffuse from the blood into the cavities of air sacs and pass out through the lungs resulting in loss of body heat.

Cardiac movements: Muller (1980) is of the view that air sacs allow movements of heart which is situated in a rather rigid thorax.

Balancing during flight: The air sacs help in flight and also maintain proper centre of gravity during flight. This is possible because the air sacs are properly arranged on either sides of the body. Since the air sacs are inserted between muscles like pads, they help in reducing mechanical friction during flight.

Resonators: The air sacs help in forceful expulsion of air and controls pitch of sound, that is, they act as resonators.

RESPIRATORY MECHANISM

Respiratory mechanism of birds has many peculiarities. Expiration is an active process. The breathing mechanism in birds is like that of a suction pump. It is different at rest and during flight.

Ventilation at rest: When the bird is at rest the sternum rises and falls alternately by the action of abdominal and intercostal muscles. When the sternum is lowered the air sacs expand and the lungs are compressed. The air is drawn in (**inspiration**) through the nostril, trachea and mesobronchus into the posterior air sacs. At the same time air already present in the lungs moves into the anterior air sacs. When the sternum is raised the air sacs are compressed and the lungs expand. The air rich in oxygen from the posterior air sacs goes into the lungs and the stale air from the anterior air sacs moves into the mesobronchus and trachea and expelled out (**expiration**) through the nostrils.

Ventilation during flight: During flight when the weight is supported by the wings and the sternum is rendered immovable, the same efffect seems to be produced by the elevation and depression of the back synchronized with the strokes of the wing.

The avian lung empties completely with each breath, no residual air being left in. The residual air is confined to the air sacs. The air of lungs is renewed twice at each breath, once during inspiration and again during expiration. The air does not travel through the same passage during inspiration and expiration. The birds that sing a long uninterrupted song (winter wren) are believed to utilize reserve air in the air sacs. When submerged water birds also do the same.

The rate of respiration in birds is normally rapid but varies under different conditions. Hess has studied the breathing rate of pigeon which is 29 times per minute at rest, 180 times while walking and 450 times in flight. Although the demand for oxygen in birds is very high, they can adjust themselves remarkably to different environmental conditions to conserve energy during fasting, sleeping or in cold weather.

NERVOUS SYSTEM

The nervous system of birds is constructed on the reptilian plan, but there are some notable differences. In the brain of bird the olfactory lobes are much smaller in comparison with the hemispheres. The increase in size of the hemispheres indicates an enlargement of instinctive life, and everybody knows that birds with their nest making, their migration and homing habits live a very complex life of instinct.

The nervous system includes the **central, peripheral** and **autonomic nervous systems.**

CENTRAL NERVOUS SYSTEM

The central nervous system includes the **brain** and **spinal cord**.

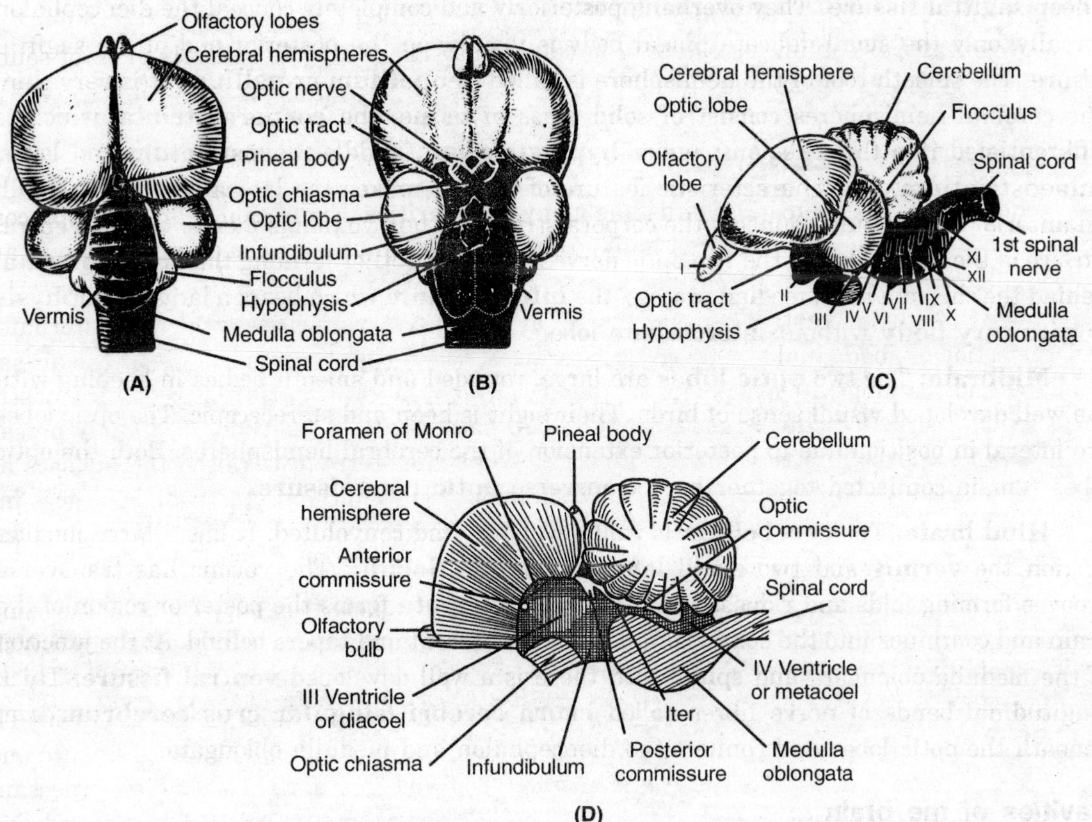

Fig. 20.34 *Columba*–Structure of brain, **(A)** Dorsal view; **(B)** Ventral view; **(C)** Lateral view; **(D)** Sagittal section

BRAIN

The brain of pigeon is short and broad, round and whitish in colour. It completely occupies the cranial cavity of the skull. It is surrounded by two protective membranes or **meninges,** the outer membrane or **duramater** is made up of connective tissue and the inner membrane or **pia-arachnoid** has a rich blood supply. The narrow space between the two membranes is filled with cerebrospinal fluid, connective tissue, fatty tissue and veins. Histologically the brain is made up of two layers of nerve fibres, the **grey** and **white** matter. The grey matter is external in all parts of the brain except optic lobes and medulla oblongata where it is internal.

The brain as usual is divided into three regions : **forebrain, midbrain** and **hindbrain** (Fig. 20.34).

Forebrain: The **olfactory lobes** are very small in correlation with poorly developed sense of smell. The **cerebral hemispheres** are large pyriform, relatively smooth and lacks the deep fissures characteristic of mammalian brain. The two hemispheres are separated by a deep **sagittal fissure**. They overhang posteriorly and completely conceal the diencephalon dorsally, only the small delicate pineal body is visible, on the posterior end of the sagittal fissure. The smooth roof of the hemisphere is called **neopallium** or **pallium**, it is very thin. The cerebral hemispheres consist of solid mass of tissue, the **corpora striata**, which is differentiated into three regions: upper **hyperstriatum,** middle **mesostriatum** and lower **palaeostriatium**. It is characteristic feature of bird. It makes the lateral ventricles small. An anterior commissure connects the corpora striata of the two hemispheres. On the ventral surface of the diencephalon the two optic nerves cross each other forming the **optic chiasma**. Behind the chiasma lies a median process, the **infundibulum** which bears a large **hypophysis** or **pituitary body** without intermediate lobe.

Midbrain: The two **optic lobes** are large, rounded and smooth bodies in keeping with the well developed visual sense of birds. Their sight is keen and stereoscopic. The optic lobes are lateral in position due to posterior extension of the cerebral hemispheres. Both the optic lobes remain connected together by a transverse **optic commissure**.

Hind brain: The **cerebellum** is well developed and convoluted. It has a large median portion the **vermis** and two small lateral lobes the **floculli**. The vermis has transverse grooves forming folds and ridges. The **medulla oblongata** forms the posterior region of the brain and continues into the spinal cord. It is broad in front and tapers behind. At the junction of the medulla oblongata and spinal cord there is a well developed **ventral fissure**. Thick longitudinal bands of nerve fibres called **crura cerebri** (**singular crus cerebrum**) run beneath the optic lobes and connect the diencephalon and medulla oblongata.

Cavities of the brain

The brain is hollow but its **cavities** or **ventricles** are very much reduced (Fig. 20.34 D). The cavities of the two cerebral hemispheres are called **lateral ventricles** or **paracoel.** They are greatly reduced due to great development of corpora striata. The cavity of diencephalon is called **diocoel** or the **third ventricle.** It is a narrow vertical cavity bounded laterally by the optic thalami and communicating in front with the paracoel by a common aperture, the **foramen of Monro**. The roof of the diocoel is called **epithalamus**. It is thin and contains a richly vascular tissue, the **anterior choroid plexus.** The floor of the diacoel is thick and called **hypothalamus**. The cavity of the midbrain is narrow and called **iter**. The diacoel remains separated from it by **posterior commissure**. The iter communicates dorsally with the cavity of the optic lobes, the **optocoel**. The cavity of the medulla oblongata is called **metacoel** or **fourth ventricle**. It is relatively larger with a thin highly vascular roof called the **posterior choroid plexus**. Its sides and floor are thick-walled. The anterior and posterior choroid plexes secrete the **cerebrospinal fluid** which fills the cavities of the brain.

Functions of the brain

The different parts of the brain perform different functions which are as follows:

Olfactory lobe: Smell.

Cerebral hemispheres: Voluntary movements, conditioned reflexes, instinctive behaviour, will, memory, intelligence, emotions, etc.

Diencephalon: Relay impulses to cerebral hemisphere, integrating centre of autonomic system, perception of extreme heat, cold, pain, etc.

Optic lobes: Sight

Crura cerebri: Conduction of sensory impulses

Cerebellum: Voluntary movement, muscular coordination and equilibrium

Medulla oblongata: Involuntary movement

SPINAL CORD

The spinal cord is reduced in length in birds due to the reduction of the vertebral column. It tapers posteriorly but **filum terminale** is absent. In structure it is similar to that of other vertebrates. It has two enlargements, the **brachial** and **lumbar** from which **brachial** and **lumbar plexes** are given off to the fore- and hind-limbs respectively. On the dorsal surface of the lumbar enlargement the two halves of the cord diverge for a short distance enclosing a wide diamond-shaped **rhomboid sinus** containing a gelatinous fluid. The central canal of the spinal cord is continuous with the cavity of the medulla oblongata and is filled with cerebrospinal fluid. Histologically the spinal cord is made up of **outer white** and **inner grey matter**. The grey matter is in the form of 'H'. The spinal cord controls reflex action and conduct impulses to and from the brain.

PERIPHERAL NERVOUS SYSTEM

It includes **cranial** and spinal **nerves**.

Cranial nerves

There are twelve pairs of cranial nerves. They arise from the brain, leave the cranium through special holes and innervate the organs of the body. The different cranial nerves and their nature is as follows:

I. **Olfactory** – sensory

II. **Optic** – sensory

III. **Occulomotor** – mixed

IV. **Trochlear** – motor

V. **Trigeminal** – mixed.

VI. **Abducens** – motor

VII. **Facial** – mixed

VIII. **Auditory** – sensory

IX. **Glossopharyngeal** – mixed.

X. **Vagus** – mixed.

XI. **Spinal accessory** – motor

XII. **Hypoglossal** – motor

The course of these nerves is similar to that of other vertebrates. The spinal accessory passes out with the vagus and is distributed to certain muscles. The hypoglossal is found just posterior to the vagus and is distributed to certain neck muscles. It also sends a branch to the tongue.

SPINAL NERVES

Spinal nerves arise from the spinal cord. Each spinal nerve arises as **dorsal** and **ventral roots**. The dorsal root has a ganglion and carries **somatic** and **visceral sensory fibres**. The ventral root has **somatic** and **visceral motor fibres.** Each spinal nerve is divided into **dorsal ramus**, long **ventral ramus** and **ramus communicans**. The **cervical spinal nerves** arise at segmental intervals and lie between the skin and vertebral column. Their number corresponds with the number of vertebrae in the neck region. The ventral branch of the last cervical nerve together with that of the first of the trunk form the **brachial plexus** to the wing. The next five pass out between the ribs. There is a **lumbosacral plexus** divisible into three parts, the **lumbar, sacral** and **pudendal** plexes. The lumbar plexus is formed by three nerves sending branches to the thigh. The sacral plexus in formed by five nerves which form a large **sciatic nerve** which passes along the dorsal side of the thigh between muscles and proceeds down the legs alongside the femoral artery and vein. The remaining spinal nerves posterior to the sacral plexus form the pudendal plexus from which nerves pass obliquely to the tail and cloacal region.

AUTONOMIC NERVOUS SYSTEM

The autonomic nervous system includes the **sympathetic** and **parasympathetic systems** of nerves and ganglia. The sympathetic system consists of a chain of two cords and segmental ganglia on each side. A sympathetic ganglion lies fused to each spinal nerve in the trunk region as the nerve emerges from the vertebral column. A coeliac plexus of nerves and ganglia is located at about the middle of the rib-bearing region. This extends ventrally from the main sympathetic cord and surrounds the dorsal aorta. Posterior to this region the sympathetic cords are reduced and consists of a single trunk on each side. The sympathetic nerves supply chiefly the alimentary, respiratory, circulatory and urinogenital systems.

SENSE ORGANS

In pigeon, as in other vertebrates all sense organs are present.

TACTILE ORGANS

The tactile sense organs are poorly developed in birds because the body is completely covered with feathers. Some sensory nerve endings are found on the bill which is often used as an exploratory instrument, particularly in ducks, whose beaks have sensory structures for sorting food. **Grandy's corpuscles** present in the covering of the beak of some birds are important tactile organs. On the upper part of the beak dorsal to the nostrils in pigeon, there is a soft swollen **operculum**. It is a sensory structure believed to play stimulatory role during love making. **Rictal bristles** around the bill are sensory structures although they are without nerve endings. They have sensory cells in the feather follicle at their base. **Herbst's corpuscles** present in the dermis are important heat receptors and also perceive vibrations.

GUSTATORY ORGANS

In birds taste discrimination is poorly developed due to scarcity of **taste buds**. A few scattered over the palate or on the dorsal surface of the tongue, however, permit some taste discrimination, but a majority of birds swallow food material quickly without tasting or moistening it. In young birds the food is often placed far back in the throat so there is no use of taste buds. The pigeon being a seed or grain-eater does not have well developed sense of taste, although some others like insect-eating or carrion-feeding birds may have fairly discriminating sense of taste.

OLFACTORY ORGANS

The **organs of smell** or **olfactory organs** are poorly developed in birds because of their aerial mode of life. The **olfactory lobes** of the brain are also reduced. The **external nares** or **nostrils** covered by cere are present at the base of the horny beak. They open into small **nasal chambers** or **olfactory sacs** separated by a median septum, the **mesethmoid** and bounded externally by **ectoethmoid**. From the median septum, three swellings, the **turbinals** or **conchae** project into the nasal cavity. They lie in a row. They increase the olfactory surface. The anterior part of each chamber serves as a **non-sensory respiratory part** or **vestibule** and contains anterior turbinal covered by laminated epithelium. The posterior part of the nasal chamber is the **sensory olfactory part**. It contains middle and posterior turbinals lined by a single layered epithelium, the **Schneidernian membrane** innervated by the olfactory nerve. The two olfactory chambers open into the pharynx by internal nares. The degeneration of the olfactory sense is probably correlated with the increased specialization of other senses such as vision.

AUDITORY ORGANS OR EARS

The sense of hearing is well developed in birds. It is believed that owls depend upon their hearing for hunting and there are some birds that listen for subterranean prey. The singing of birds would have been entirely useless if the power of hearing were not acute. Anatomically the organs of hearing are similar to those of other vertebrates. The **external ear** or **pinna** is lacking in birds, but the **external ear opening** is covered by a special group of feathers called **auriculars**. They are loosely constructed and do not interfere with the reception of sound. In some birds the auriculars are actually lifted slightly while they listen intently.

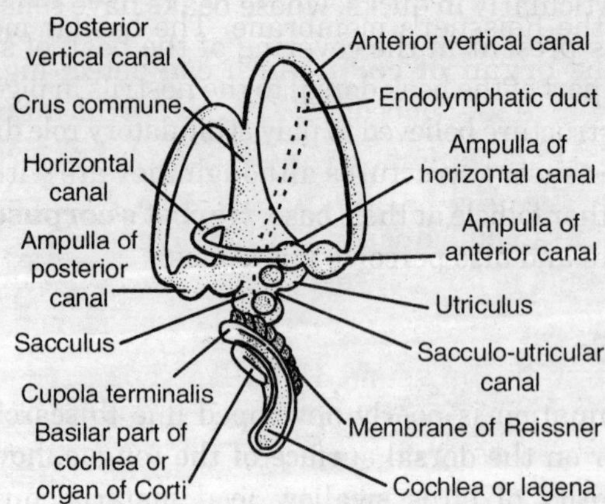

Fig. 20.35 *Columba*– Structure of internal ear

The external aperture is a circular hole on the side of the head, behind and below the posterior border of the orbit. This leads into a narrow passage, the **external auditory meatus** at the base of which the tympanic membrane is situated. The **tympanic membrane** or **tympanum** or **ear drum** is a thin vibratory septum separating the external auditory meatus from the tympanic cavity. The tympanum transmits the vibrations it receives, through the rod-like bone, the **columella** to the inner ear. The columella lies almost horizontally in the tympanic cavity. Its free anterior cartilaginous portion, the **extracolumella** is attached to the inner surface of the eardrum. Its posterior disc-like portion, the **stapes** plugs the fenestra ovalis present on the inner wall of the tympanic cavity. The columella is formed from the uppermost part of the cartilage of the hyoidean arch and probably corresponds to part or the whole of the hyomandibular cartilage of dogfish. Its connection with the ear is purely a secondary character. The **tympanic cavity** is the middle dilated portion of the **tympano-Eustachian passage.** The Eustachian tube is the innermost part of the tympano-Eustachian passage. It is a narrow tube leading from the anterior and lower angle of the tympanic cavity into the pharynx. The Eustachian tubes of the two sides unite and open into the roof of the mouth by a median aperture just behind the posterior border of the narial opening at the back of the palate.

The essential organ of hearing is the **inner ear** or **membranous labyrinth** (Fig. 20.35) embedded in the side wall of the skull. It is surrounded with a fluid called **perilymph** and is filled with a dense fluid called **endolymph**. It consists of three **semicircular canals**, the **anterior vertical, posterior vertical** and **horizontal canals**, lying in planes at right angles to each other. The ends of the canals bear a swelling called **ampulla.** It also consists of relatively smaller **utriculus** and **sacculus** and a well developed **cochlea** or **lagena.** In a transverse section the cochlea shows three chambers: upper scala vestibuli, middle **scala media** and lower **scala tympani.** The scala media forms the **cochlear duct.** It contains endolymph. The other two chambers are filled with perilymph. The junction of the two chambers at the apex of the cochlea is called **helicotrema,** here the cochlear duct terminates. At the posterior end of the cochlea the scala vestibuli is continuous with the fenestra ovalis and scala tympani with the fenestra rotunda. The floor of scala media is formed by the basilar membrane and the roof by the Reissner's membrane. The basilar membrane consists of tall auditory cells which form the **organ of corti** which can detect high frequency sound. The free ends of auditory cells bear hairs embedded in a **tectorial membrane** lying above the organ of corti. From the basal end of auditory cells arise nerve fibres which unite to form **cochlear nerve**, a branch of the auditory nerve. The auditory cells are called **cristae** and **maculae.** The cristae present in ampullae of semicircular canals help in maintaining equilibrium and sense direction. One macula is present each in sacculus, utriculus and lagena. A macula present at the tip of the cochlear duct can perceive low frequency sound.

Birds are known to detect distant sound disturbances but the range in some birds may not be wide. Some workers (Kellogg) have reported that so far as pitch is concerned the birds do not respond to as wide a variety as the human ear is capable of. Expressed in cycle per second the range of hearing in domestic pigeon ranges from about 200 to 7,500, whereas man's hearing ability range from about 20 to 16,000.

VISUAL ORGANS OR EYES

The existence of birds depends upon a keen sense of sight. Having mastered the air, good vision is essential to cover long distances, to avoid obstacles, to find food, to recognize mate and so on. The eyes of birds are relatively larger than in other vertebrates. The structure of eye (Fig. 20.36A) is like that of any other vertebrate with a few distinctive features.

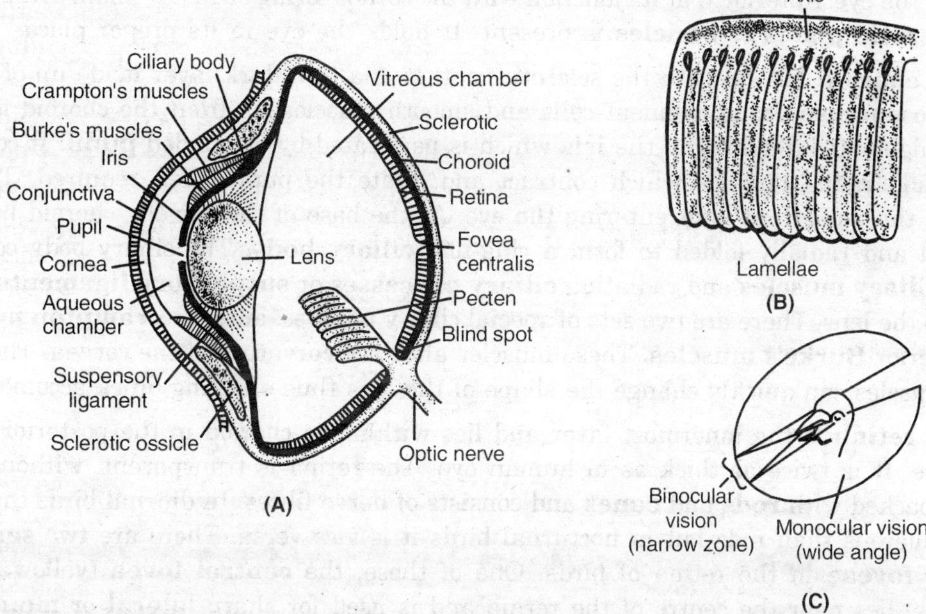

Fig. 20.36 *Columba*–(A) Structure of eye (vs); (B) Pecten; (C) Range of vision

There are two **eyelids** but they are not very conspicuous. The upper eyelid is only slightly movable but the lower one is freely movable and arises upwards to close the eyes. The eyelids do not blink. Eyebrows and eyelashes are absent. A third eyelid, the **nictitating membrane**, is also present. It forms a fold at the anterior angle of the eye. It can be pulled obliquely downward and backward over the front of the eye with great rapidity. The rapid movement of the nictitating membrane is brought about by two muscles, **quadratus** or **bursalis** and **pyramidalis** lying close to the inner or orbital surface of the eyeball. They undercut the eyeball to be inserted into the ventral border of the nictitating membrane. The nictitating membrane mainly functions for cleaning the eyeball. When birds are facing or flying against the wind it acts as a protective covering. In nocturnal birds it cuts down the glare of sunlight during the day and in aquatic birds it shields the eyes under water.

The **Harderian glands** are present on the inner angle of the eye. Its oily secretion lubricates the nictitating membrane. The **lacrymal** or **tear glands** are present below the outer angle of the lower eyelid. Its watery secretion keeps the surface of cornea wet and washes away dust particles.

The eyeball in pigeon is in the form of a biconvex lens and not spherical as in man. It is hollow and its wall is made up of the usual three layers: outer **sclerotic coat,** middle **choroid** and inner **retina**.

The sclerotic forms the outer coat of the eye. In front it bulges out to form a convex transparent horny **cornea** made up of connective tissue. The cornea is the anterior exposed part of the sclerotic. It is covered by a thin, transparent, sensitive and vascular epithelial membrane, the **conjunctiva** which is continuous with the mucous lining of the eyelid. The posterior hidden part of the sclerotic is white, opaque and cartilaginous. It covers the inner surface of the eye. Anteriorly, at its junction with the cornea a ring of 10–12 small, overlapping, bony **sclerotic plates** or **ossicles** is present. It holds the eye in its proper place.

The **choroid** lies beneath the sclerotic coat. It is a thin dark layer made up of highly vascular connective tissue, pigment cells and smooth muscles. In front the choroid forms a circular pigmented diaphragm, the **iris** which is perforated by a rounded **pupil**. It contains circular and radial muscles which contract and dilate the pupil when required. The iris regulates the amount of light entering the eye. At the base of the iris the choroid becomes thickened and radially folded to form a ring-like **ciliary body.** The ciliary body contains smooth **ciliary muscles** and radiating **ciliary processes** or **suspensory ligaments** which run up to the lens. There are two sets of special ciliary muscles: anterior **crampton muscles** and posterior **Burke's muscles.** These muscles alter the curvature of the cornea. The other ciliary muscles can quickly change the shape of the lens thus effecting quick accomodation.

The **retina** is the innermost layer and lies within the choroid in the posterior region of the eye. It is twice as thick as in human eye. The retina is transparent, without blood vessels, packed with **rods** and **cones** and consists of nerve fibres. In diurnal birds cones are more numerous than rods but in nocturnal birds it is vice versa. There are two **sensitive spots** or **foveae** in the retina of birds. One of these, the **central fovea** (yellow spot of mammals) lies near the centre of the retina and is used for sharp **lateral** or **monocular vision.** The other, the **temporal fovea,** lies towards the outer side of the eye and is used for **forward or binocular vision**. The sensitive spots have more cones and provide distinct vision. There is a region on the retina where no rods and cones are present, it is called **blind spot**. The **optic nerve** enters the eyeball at this point. The lower part of the retina contains **yellow droplets** and the upper part contains **red droplets.** These act as filters and increase the perception of colours.

The **lens** is biconvex, transparent, crystalline, soft and colourless. It is suspended by suspensory ligaments behind the iris. The posterior side of the lens is more convex than the anterior side. It is surrounded by a thin fibrous capsule. The lens divides the cavity of the eye into a small anterior **aqueous chamber** filled with a colourless watery fluid, the **aqueous humour,** and a large posterior **vitreous chamber** filled with a thick colourless gelatinous **vitreous humour**. The fluids keep the eyeball in position and also serve to focus the light rays on the retina.

The muscles that support and move the eyeball are essentially the same as those of other vertebrates.

An additional struture called **pecten** (Fig. 20.36B) is present in the eyes of all birds except kiwis. It is a soft, somewhat rectangular, highly vascular and pigmented, comb-like structure which projects into the cavity of the eye from the blind spot. It is more developed in diurnal birds than in nocturnal. Pecten is also present in the eyes of some reptiles. The function of pecten is not certain. It probably nourishes the retina by diffusion through the vitreous humour. It also helps in accommodation by pressing the lens forwards. It regulates the pressure of fluids in the eyes. Its shadow falls on the retina and helps in the perception of movement.

Vision: There is great variation in the actual position of the eye and most birds (except owls) are capable of both **monocular vision** and **binocular vision** (Fig. 20.36 C). The eyes of birds are large so they have keen sight. The eyes are large but the eye muscles are small, so there is very little movement of the eyeball. This is compensated by the long and flexible neck. The eyes of pigeon are laterally directed so they have a wide range of monocular vision, that is, each eye has its own field of vision. In enables the bird to see far-sideways. They have a narrow zone of binocular vision (overlapping of fields of vision of both the eyes) in front.

Several evidences are available which prove good colour vision in birds. The colourful plumages of birds strongly suggest that the birds can see colours in sex recognition. The Australian bowers decorate their bowers with petals of flowers, often they choose colours that occur in the plumage of the female they wish to attract. Cones are sensitive to colours. Their abundance helps colour vision. Besides this the red and yellow oil droplets present in the retina also increase colour vision. The amount and distribution of these droplets differs. For instance in pigeon the yellow droplets are concentrated in the lower part of the retina and it is presumed that these droplets are meant for sky vision and tone down the blue. On the other hand, kingfishers have an abundance of red droplets which may offset the glare of water.

Accommodation: Birds have great power of accommodation. It is very essential because birds change quickly from distant to near vision when they sweep down from height to the ground. The ability to accommodate is effected by change in position of cornea and the shape of the lens, which is brought about by the ciliary muscles. In birds, maximum accommodation is about 20 diopters. In cormorants which need more accommodation to see under water it is 40–50 diopters.

URINOGENITAL SYSTEM

The excretory and reproductive organs are studied together under the **urinogenital system**. They are structurally connected with each other but functionally they are independent. In pigeon the sexes are separate but there is no sexual dimorphism.

URINARY OF EXCRETORY ORGANS

The excretory organs of birds are a pair of kidneys and ureters. The urinary bladder is absent. Its absence probably reduces the body weight, an adaptation to flight.

Kidneys: The **kidneys** (Fig. 20.37 A) are the main excretory organs. They are **metanepheric.** Each kidney is dark brown, flattened, three-lobed organ situated against the dorsal wall in the posterior part of the body cavity. They are embedded in the hollows of the pelvis. Ventrally they are covered by **peritoneum.** Each lobe of the kidney is divided into a number of small **lobules**. A small, elongated, yellowish endocrine gland, the **adrenal body** is attached to the ventral surface of the anterior lobe of each kidney. Each kidney is tightly packed with large number of coiled **uriniferous tubules.** Each tubule has a small **glomerulus** and a short specialized portion, the **loop of Henle.** The loop of Henle reabsorbs water from the glomerular filtrate.

Ureters: The uriniferous tubules unite to form a metanephric duct, the **ureter.** It is a straight, narrow tube arising from the anterior lobe of kidney and runs backwards to open into middle chamber of the cloaca.

Like reptiles the birds are also **uricotelic,** that is, their excretory waste contains urate particles and uric acid. Water is reabsorbed from the excretory wastes in the cloaca and semisolid whitish excretory waste called **guano** is excreted with faeces.

MALE REPRODUCTIVE ORGANS

The male reproductive organs are a pair of **testes** and **vasa deferentia.**

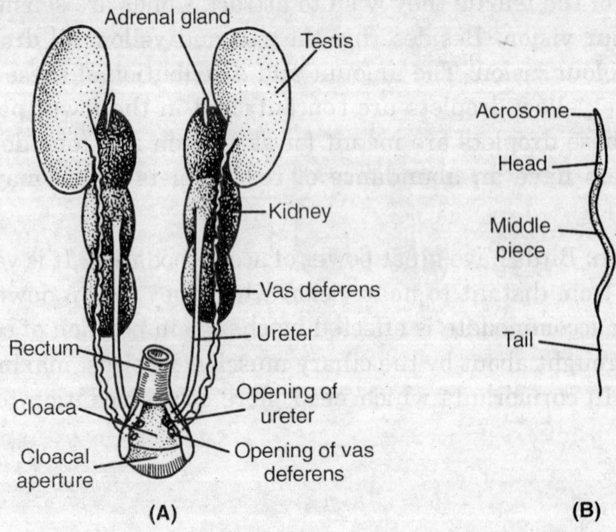

Fig. 20.37 *Columba*–**(A)** Male urinogenital system; **(B)** Structure of sperm

Testes: The **testes** (Fig. 20.37A) are oval whitish bodies lying in contact with the ventral surface of the anterior lobe of the kidney. They are attached ventrally by a fold of peritoneum called **mesorchium.** The right testis is smaller than the left. Their size varies considerably with season. They are greatly enlarged during the breeding season. Both the testes consist of coiled narrow **seminiferous tubules** lined by epithelial cells which produce spermatozoa. In between the tubules are present **interstitial** or **Leydig cells** which produce hormones.

Vasa deferentia: The **vas deferens** or **Wolfian duct** (Fig. 20.37A) is a narrow, convoluted, opaque duct which arises from the inner border, near the posterior end of each testis. It runs posteriorly along the outer side and parallel to the ureter. The anterior end of the vas deferens is called **epididymis.** It is connected with the seminiferous tubules by very fine tubules the **vasa efferentia.** During the breeding season their hinder end, just before opening into the cloaca, is swollen to form the **seminal vesicle** for temporary storage of sperms.

The vas deferens opens dorsally into the urodaeum of the cloaca by a small papilla posterior to the opening of the ureters.

Sperm: A mature **sperm** (Fig. 20.37 B) is a filamentous structure consisting of a **head**, **middle piece** and **tail**. The head contains a nucleus and its anterior end is drawn into an **acrosome.** The acrosome contains Golgi bodies and helps in piercing the vitelline membrane of ovum during fertilization. The middle piece contains mitochondria. The tail is flagellum-like and an axial filament runs through its whole length.

FEMALE REPRODUCTIVE ORGANS

In the adult female there is single ovary and oviduct on the left side. The right ovary and duct are present in the embryo but almost entirely disappear before hatching. A single

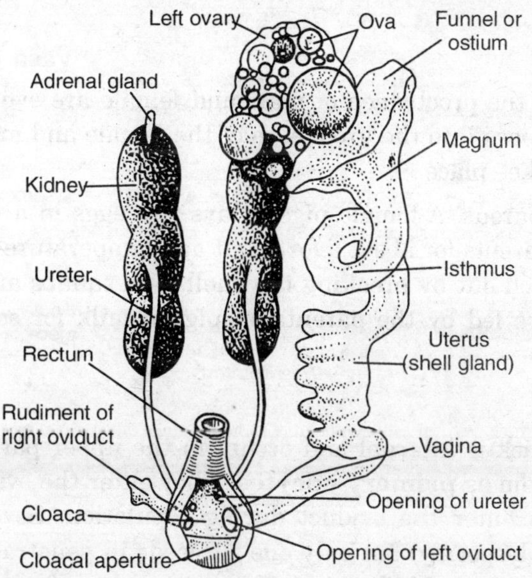

Fig. 20.38 *Columba*–Female urinogenital system

ovary, most probably results in reduction of weight, is an adaptation to flight. During breeding season the ovary and eggs become large occupying almost the whole of the abdominal cavity, so the birds cannot possibly afford to keep two functional ovaries. It also helps in safe manipulation of large eggs with breakable shells.

Ovary: The ovary is a large, irregular structure (Fig. 20.38) situated on the ventral side of the anterior lobe of kidney. It is attached to kidney by a fold of peritoneum called **mesovarium.** The surface of ovary bears numerous **follicles** or **ovisacs** of different sizes, each containing a single **ovum** or **oocyte.** The ova are in different stages of development. Ovulation takes place by the rupture of follicular membrane. The ova are released into the coelom from where they go into the oviduct through its expanded funnel-shaped opening.

Oviduct: The **oviduct** or **Müllerian duct** (Fig. 20.38) is a long and wide convoluted tube running posteriorly to the cloaca. It is attached to the dorsal body wall by a fold of peritoneum called **mesotubarium.** At the anterior end of the oviduct there is a large, expanded, muscular coelomic funnel, the oviducal funnel or **ostium** or **infundibulum**, the margin of which bears long ciliary processes called **fimbriae.** The ostium opens into the coelem near the ovary. The ova enter the oviduct through the ostium. The wall of ostium secretes a thin chalaziferous layer of dense albumen around the fertilized egg. The next part of the oviduct is called **magnum.** It has tubular glands which secrete **albumen** or **egg white**. The magnum is followed by **isthmus** which secretes the parchment-like double shell membranes around the albumen. The next part is the thin-walled **uterus** or **shell gland** which secretes the hard calcareous shell. The last part of the oviduct is the thick-walled, non-glandular muscular **vagina** which secretes pigments and external cuticular layer of the shell. It also secretes mucous so that the eggs may pass out easily. It opens into the urodaeum posterior to the opening of the ureter. A small remnant of the right oviduct is attached to the right side of the cloaca.

COPULATION

During copulation the proctodaea of male and female are everted and pressed against each other, the sperms pass into the urodaeum of the female and move up the oviduct where internal fertilization takes place.

The birds are oviparous. A female pigeon lays two eggs in a rough nest. The eggs are incubated by both the parents for about a fortnight at a temperature of 38° to 40°C. The young ones called **squabs** hatch out by breaking the shell. The squabs are naked and cannot feed by themselves. They are fed by the parents on pigeon milk for some time.

FERTILIZATION

Fertilization in chick is internal and occurs in the upper part of the oviduct. The ova are released in the coelom as **primary oocytes**. They enter the oviduct through the ostium. The sperms from male enter the oviduct during copulation. Several (five or six) sperms surround the oocyte (polyspermy) but only one succeeds in penetrating its head through the egg membrane to bring about fertilization. The mere entry of sperm induces the oocyte to undergo two quick **maturation divisions.** Two **polar bodies** are formed, one at the time of transformation of oocyte to ovum (**first maturation division**) and second just before fertilization. Both the polar bodies degenerate and disappear. The sperm nucleus fuses with the egg. The remaining sperms ultimately die. The fertilized **ovum** or **zygote** then travels down the oviduct where albumen, egg membrane and the shell are secreted. It takes nearly 24 hours to pass down the oviduct.

DEVELOPMENT OF CHICK

The development of chick or the common fowl (*Gallus gallus*) is preferred for study to that of pigeon because of the following reasons:

1. It has been worked out more extensively.
2. The eggs are large in size, easy to procure, easily available throughout the year and can be incubated artificially.
3. The developmental stages can be easily obtained in laboratory conditions for experimental purposes.

The development of all birds is more or less similar. The early stages of development of birds are similar to that of reptiles and even mammals. Development of chick is **direct** as there is no larval stage.

STRUCTURE OF EGG

The hen's egg is cleidoic because it is covered with a waterproof **shell**, **macrolecithal** or **polylecithal** because it contains large quantity of **yolk** and **discoidal** because its cytoplasm remains restricted to a small disc-like area called **blastodisc**, on the top of the yolk.

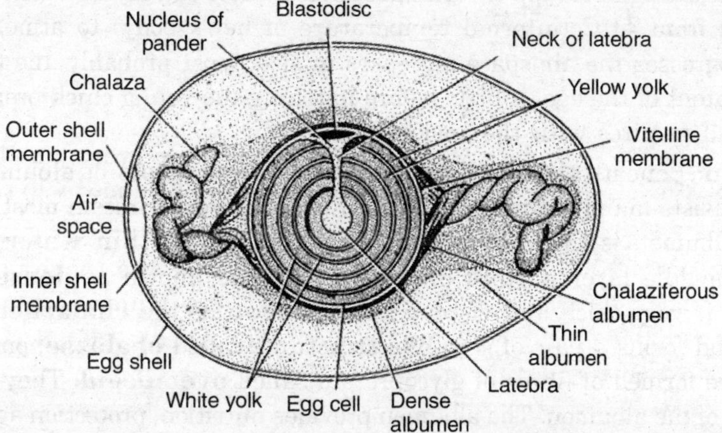

Fig. 20.39 *Fowl*–Structure of egg (diagrammatic L.S.)

The egg is released into the coelom from the ovary in the form of a **primary oocyte,** about 2.5 cm in diameter. It is surround by a delicate egg membrane, the **vitelline membrane,** secreted by the ovary. The oocyte enters the oviduct through its wide mouth, the ostium. The oocyte rotates as it passes downwards in the oviduct, due to peristaltic movements in the wall of the oviduct. Fertilization occurs in the upper part of the oviduct. After fertilization when the egg (Fig. 20.39) passes downwards through the oviduct it receives a thick and viscous layer of **albumen** in the upper part of the oviduct. Since the oocyte rotates, this layer of albumen becomes twisted into a cord-like structure called **chalaza**. This layer of albumen is called **chalaziferous layer**. In the upper middle part of the oviduct, the oocyte again receives a thick coat of dense albumen above the chalaziferous layer. In the lower middle part it receives two closely set **shell membranes** above the albumen layer. The shell membranes are separated at the broad end of the egg to enclose an **air space.** A thin layer of watery albumen is absorbed through the shell membranes in the lower part of the oviduct. The egg finally receives calcareous, porous **shell** in the uterus. The shell is soft and flexible in a freshly laid egg but soon becomes hard and brittle. Now the fertilized egg is laid. It is large, about 5 cm long and 3 cm wide. One end is broader than the other and the broader end comes out of the cloaca first.

The egg consists of the following parts:

1. Egg shell: The egg shell forms the outermost, hard part of the egg. It is usually white in colour, sometimes it may be light brown. It is composed mainly of calcium carbonate (about 94%) and little amount of protein. The shell bears a large number of fine **pores** through which exchange of gases takes place. The total number of pores in an egg is approximately 7000 and the diameter of a pore is 0.04–0.05 mm. In the newly hatched egg the shell remains soft and flexible but soon becomes hard and brittle as it comes in contact with air. The shell provides protection to the inner parts of the egg.

2. Shell membranes: The shell is lined by two white, thin but tough shell membranes which consists of sheets of matted fibres of keratin. The two membranes remain in close contact with each other except at the broader end of the egg where they are separated by an **air space.** According to Kaupp the air space arises after egg laying when the temperature of the eggs falls from 41°C (internal temperature of hen's body) to atmospheric level. As development progresses the air space increases in size, most probably due to evaporation of part of water content of the egg. A little before hatching the young chick breaks the air space with its beak and respires with the enclosed air.

3. Albumen: Beneath the shell membranes lie three layers of albumen or egg white. The albumen consists mainly of water (about 85%) and several protein mostly albumins. The outer layer of albumen is thin and watery and is known as **thin water liquid** or **fluid albumen**. The middle layer is thick and viscous and is known as **dense albumen**. The innermost layer is very thick and viscous and is known as the **chalaziferous layer**. It is spirally coiled and forms a pair of cord-like structures called **chalazae**, one on each end of the egg. They are formed of fibres of glycoprotein called **ovomucoid.** They suspend the egg cell in the centre of the albumen. The albumen provides nutrition, protection against mechanical and chemical injuries and serves as a store of water. The outer and middle layers of albumen are homogenous.

4. The egg cell or ovum proper: The ovum proper is a large rounded cell occupying a central position in the egg. It consists of the following parts.

(a) **Vitelline membrane:** The egg cell is surrounded by a thin, transparent **vitelline** or **fertilization membrane** formed by the fusion of two membranes deposited by the follicle cells of the ovary. It separates the yolky ovum from the surrounding albumen.

(b) **Yolk:** The egg cell consists of a large amount of yolk with a small protoplasmic area called **blastodisc** or **germinal disc** on one side. The yolk is not homogenous but arranged in concentric layers of **white** and **yellow yolk** around a central flask shaped region of white yolk called **latebra.** The layers of yellow yolk are broader than the layers of white yolk. The narrow, elongated neck-like part of latebra is called the **neck of latebra.** It expands below the blastodisc into a broad disc called **nucleus of Pander**.

Most of the yolk is liquid but about 23% is in the form of solid **yolk spheres**. The yolk, as a whole, contains about 49% water, 17% proteins, 33% phospholipids and fats and 1% carbohydrates. The white yolk contains less fat and also less of the fat-soluble pigment carotene due to which is the colour of yellow yolk. Yolk provides nourishment to the developing embryo.

(c) **Blastodisc or germinal disc:** The cytoplasm of the egg is confined in the form of a disc called blastodisc above the nucleus of Pander. The **egg nucleus** or **germinal vesicle** lies in the blastodisc. The blastodisc in a newly laid egg, is divisible into a clear central **area pellucida** and opaque peripheral **area opaca**. Blastodisc undergoes cleavage. The part of the egg cell containing the blastodisc is called the **animal pole.** The end opposite to animal pole is called **vegetal pole**.

CLEAVAGE OR SEGMENTATION

Shortly after fertilization the hen's egg undergoes a series of mitotic divisions called **cleavage** or **segmentation**. The cleavage process occurs only in the **blastodisc**, a small circular area of cytoplasm at the top of yolk. This type of cleavage is called **discoidal** or **meroblastic** in contrast to the holoblastic cleavage of *Branchiostoma* and frog in which the entire ovum divides. The cleavage furrows do not extend into the yolk so it remains permanently

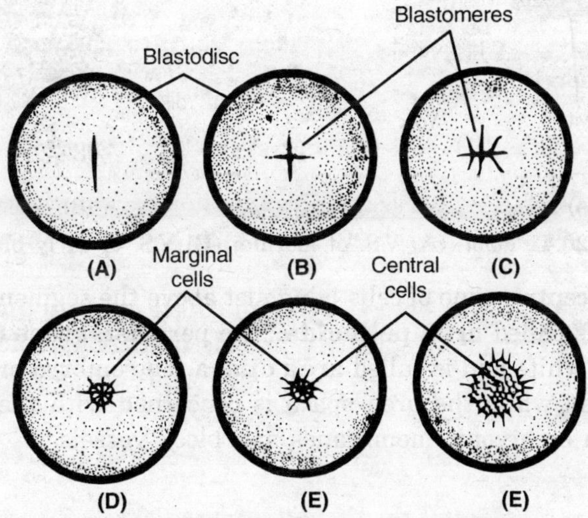

Fig. 20.40 *Fowl*–Surface view of blastodics showing cleavage. **(A)** First division; **(B)** Second; (C) Third; **(D)** Fourth; **(E)** Fifth; **(F)** Early morula

undivided. The **first cleavage** is **vertical** and occurs in the upper part of the oviduct. It incompletely divides the blastodisc into two **blastomeres**. The **second cleavage** is also incomplete, occurs at right angles to the first and produces four blastomeres. It occurs in the middle part of the oviduct. The **third** set of **cleavage furrows** pass parallel to the first and produce eight blastomeres. The **fourth cleavage** begins to cut the apical ends of the eight blastomeres, as a result of which eight **central blastomeres** are separated from eight **peripheral** or **marginal blastomeres** (Fig. 20.40). It occurs in the uterus. After the fourth division there is no definite sequence of division. The whole blastodisc is cut up into a sheet of cells and is called **blastoderm.** In addition to vertical cleavages some horizontal cleavages also occur. The horizontal divisions form the lower border of the blastomeres. The blastoderm now consists of several hundred cells. The diameter of the blastoderm keeps on increasing. Eventually a fluid-filled dorsoventrally flattened narrow space called the **segmentation** or **subgerminal cavity** (Fig. 20.41) appears below the blastoderm. It separates the central cells from the underlying yolk. The peripheral rim of still unsegmented cytoplasm around the blastoderm is called **marginal periblast.** It is slightly darker in colour and remains continuous with the underlying yolk. It unites the blastodisc with the yolk mass forming a **zone of junction.** As development progresses cleavage begins in the periblast region also. The division of nuclei is more rapid than that of cytoplasm, as a result the periblast becomes **syncitial.** The cells of the periblast lie over the yolk, engulf and digest the yolk and form **extra-embryonic endoderm** in future. The activities of the periblast cells supply nourishment to the growing embryo.

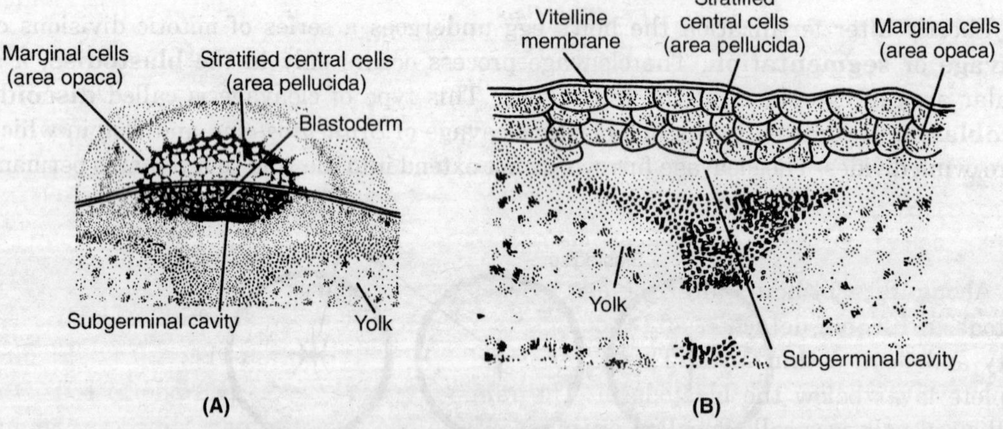

Fig. 20.41 *Fowl*–**(A)** V.S. of morula; **(B)** V.S. of early blastula

At this stage the central zone of cells lying just above the segmentation cavity look clear and transparent and is called **area pellucida.** The peripheral zone of cells lying above the yolk looks opaque and white and is called **area opaca.** The embryo proper is formed entirely from the area pellucida while the area opaca is concerned with the digestion of yolk and formation of the extra embryonic membranes and blood vessels.

BLASTULATION

With the formation of the segmentation cavity the embryo enters the blastula stage. The subgerminal cavity of chick is generally compared with the blastocoel of other vertebrates and this stage of development is called **blastula.** According to Pasteels (1945, 1957), the cavity is not homologous to the blastocoel. This stage in chick is called a **pseudoblastula.** The segmentation cavity is an area where the solid yolk is liquified with the help of certain enzymes secreted by the overlying cells of blastoderm.

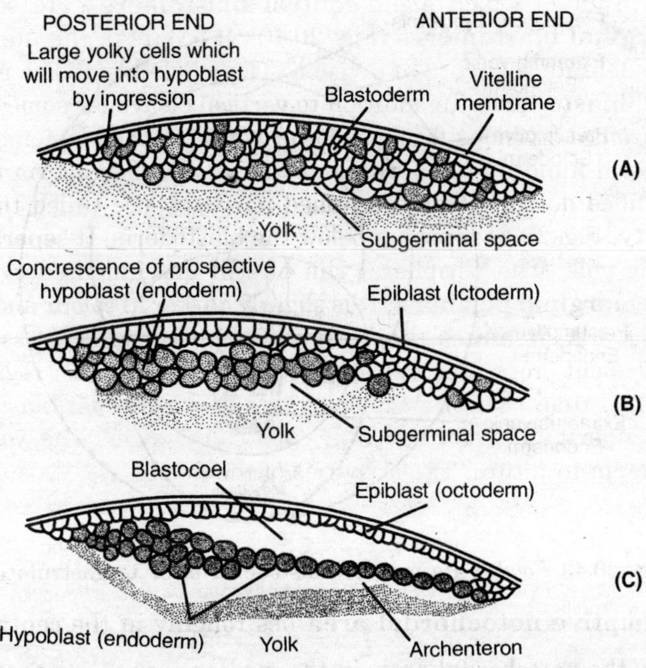

Fig. 20.42 *Fowl*–L.S. of blastoderm showing hypoblast (endoderm) formation, **(A)** Large yolk cells mixed with smaller cells; **(B)** Delamination of endoderm cells; **(C)** Organization of large yolk cells to form hypoblast

About 12 to 14 hours after the egg reaches the uterus some large yolk rich cells of the blastoderm become detached and move down by **polyinvagination** into the subgerminal cavity and form a thin layer. This layer is called the **hypoblast** (Fig. 20.42). It forms a complete layer below the blastoderm. The remaining part of the blastoderm consisting of small sized yolk-less cells is called **epiblast.** These two layers remain joined at the margin of the area opaca. The space between the two layers is called **blastocoel**. The process of separation of hypoblast and epiblast is called **delamination.** The hypoblast contributes cells for the fomation of some extra embryonic membranes but not at all for the formation of the embryo proper. The hypoblast induces the formation of **primitive streak** in the epiblast and determines its **orientation.** The epiblast gives rise to the three germinal layers of the embryo.

Thus, the cleavage converts the germinal disc into a disc-shaped two-layered blastula called **discoblastula** which floats atop the yolk mass.

FATE MAP OF BLASTULA

The fate map of birds constructed by the use of **radioactive tracers (tritiated thymidine** marking) were more superior and showed more details. The fate maps refer only to the epiblast. It shows the following **presumptive areas** (Fig. 20.43):

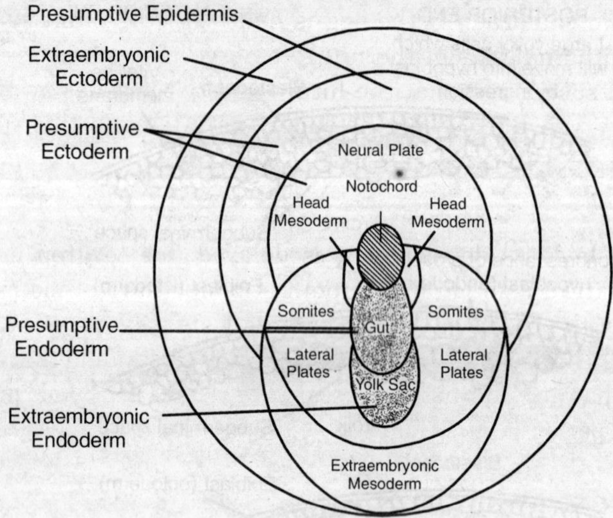

Fig. 20.43 *Fowl*–Fate map of blastoderm prior to gastrulation

1. The **presumptive notochordal area** lies roughly in the centre of area pelhicida.
2. Posterior to the notochordal area, in the median plane, lies an elongated oval area, the **presumptive endoderm**, which later forms the lining of the alimentary canal.
3. Towards the posterior edge of the area pellucida lies the **extraembryonic endoderm** which later forms part of the lining of yolk sac.
4. Various subdivisions of **presumptive mesoderm** lie on both sides of notochordal and endodermal areas and posterior to the endodermal area. They are:
   (i) **Head mesoderm** on the sides of notochordal area.
   (ii) **Somite mesoderm** lateral to the head mesoderm.
   (iii) **Lateral plate mesoderm** posterior to somite areas. The left and right lateral plate mesoderm remain continuous posterior to endodermal area.
   (iv) **Extraembryonic mesoderm** is a large crescentic area along the posterior edge of the area pellucida.
5. **Presumptive ectoderm** lies in the anterior and lateral parts of the area pellucida. It includes the presumptive epidermis and neural plate area.
6. The **extraembryonic ectoderm** forms almost a complete ring around the outer margin of area pellucida.

ORIENTATION

It is not possible to identify the antero-posterior axis of the embryo upto the blastula stage. This is because the first two cleavage planes and the subsequent ones bear no constant relation to the future embryo. It has, however, been observed that there is a relationship between the antero-posterior axis of the embryo and the axis of the entire egg (the line joining the bases of two chalazae). If the egg is held with the blunt surface of the shell to the left, then the longitudinal axis of the future embryo is oriented across the longitudinal axis of the egg shell in such a way that the future anterior end lies away from the observer. This rule was given by Von Baer. There are many variations to this rule.

INCUBATION

When the egg is laid, its development is delayed due to drop in temperature. It is resumed if the temperature is maintained upto the temperaure of mother's body (about 37° to 40°). This is done by the body heat of the parents. One (*Gallus*) or both (*Columba*) of the parent sit upon the egg. This is called **incubation.** In poultry farming the required temperature is maintained in incubators. The hen's eggs are incubated for 21 days till they hatch. In pigeon the incubation period is of 14 days.

GASTRULATION

Gastrulation begins soon after egg laying and ends within 20 hours of incubation. It includes the following processes.

1. Formation of primitive streak.
2. Cell movement within the blastocoel.
3. Regression of primitive streak and extension of embryonic axis.
4. Epiboly of ectoderm.

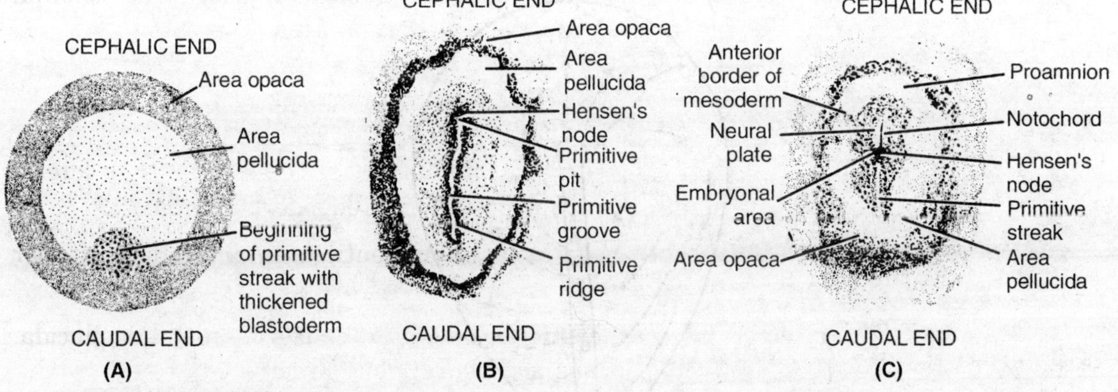

Fig. 20.44 *Fowl*–Formation of primitive streak, **(A)** Short primitive streak (5–6 hours of incubation); **(B)** Definitive primitive streak at 16 hours of incubation; **(C)** 18 hours of incubation showing formation of notochord and mesoderm

Formation of primitive streak

Formation of primitive streak is the main feature of avian gastrulation. The presumptive mesodermal and endodermal cells of the epiblast converge towards the middle line in the posterior part of the area pellucida and forms a short conical thickening called the **initial primitive streak** (Fig. 20.44 A). It appears about six to seven hours after incubation. It elongates by the accumulation of more and more cells from the sides and reaches it maximum length in about 18 to 19 hours of incubation. It is now called the **definitive primitive streak** (Fig. 20.44 B). The early primitive streak is broad and its margins are not well marked. In later stages in contracts in transverse direction and becomes narrow with well marked margins. The fully formed primitive streak extends anteriorly upto about three-fourth the length of area pellucida. The anterior end of the primitive streak appears swollen due to accumulation of cells and is called the **primitive knot** or **Hensen's node.** About 12 to 13 hours of incubation a longitudinal groove, called **primitive groove,** appears in the primitive streak. The lateral edges of the primitive groove are called **primitive ridge**. The primitive groove terminates in the centre of Hensen's node in the form of a funnel-shaped depression called **primitive pit** (Fig. 20.44 B).

As the primitive streak elongates, the adjoining parts of area pellucida also stretches. As a result, the circular area pellucida becomes pear-shaped. The broad end represents the **cephalic end** and narrow end represents the **caudal end** (Fig. 20.44 C).

Cell movement within the blastocoel

The inward migration of cells begins at the stage of short primitive streak. The presumptive mesodermal and endodermal cells of the epiblast move into the blastocoel through the primitive pit and primitive groove. The sheets of cells converge towards the primitive streak and there they become bottleshaped. The sheet breaks up into separate cells which pass singly into the blastocoel. As the cells of the epiblast migrate into the interior the whole blastoderm area disappears from the surface. It is replaced by adjoining area moving towards the primitive streak. The newly arrived cells then migrate into the interior and the process goes on.

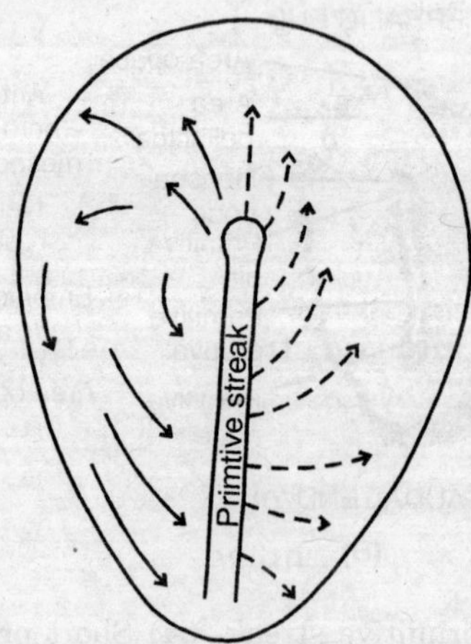

Fig. 20.45 *Fowl*–Movement of cells during gastrulation (continuous lines show movement towards primitive streak and broken lines show movement of cells which have migrated into the interior

Thus the whole primitive streak remains a mass of moving cells (Fig. 20.45). As the cells migrate into the primitive streak, it elongates towards the head region. No true archenteron is formed in the avian gastrula.

The first cells to migrate through the primitive streak are those destined to form the foregut in future. Inside the blasocoel the **presumptive foregut endoderm** cells move anteriorly and eventually displace the cells of anterior part of the hypoblast. Later they form the foregut.

The next cells entering the blastocoel through the Hensen's node also move anteriorly but not as far ventrally as the presumptive endoderm cells. These cells remain between the epiblast and endoderm and form the **head mesoderm** and **chordamesoderm** or **notochordal cells.** The presumptive notochordal cells start moving as a dense mass in the middle line straight forward from the Hensen's node below the surface of the epiblast and forms the **head process** or the **notochordal process.** The cells of the head mesoderm move on both sides of the head process.

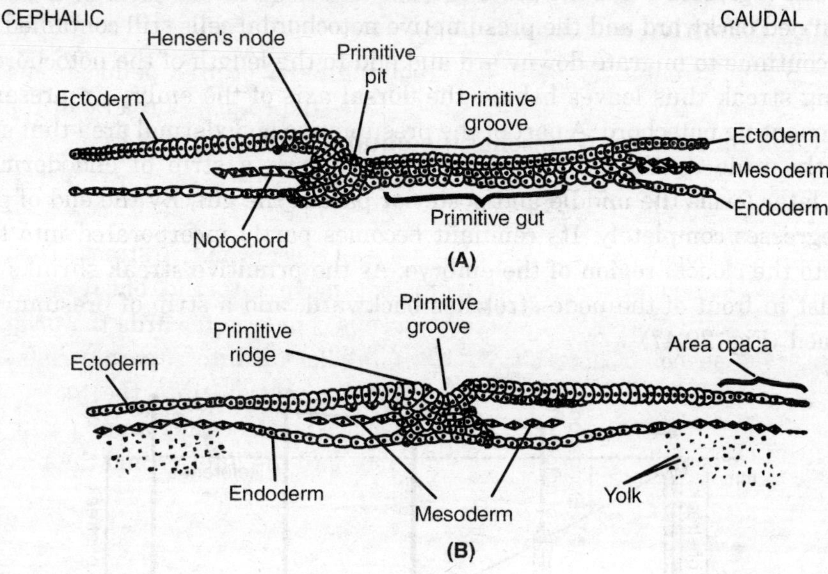

Fig. 20.46 *Fowl*–T.S. 17 hour embryo, **(A)** Showing growing notochord; **(B)** Showing me soderm Formation

Meanwhile the cells continue to migrate inward through the primitive streak. On reaching the blastocoel these cells separate into two streams. One stream moves deeper, these are the **presumptive endodermal cells**. They push the original hypoblast cells to the sides, become continuous with the foregut endoderm and later give rise to all **endodermal organs** of the embryo and form part of the lining of the yolk sac.

The second migrating stream spreads throughout the blastocoel as a loose sheet between the hypoblast and epiblast. This sheet gives rise to **mesodermal portions** of the embryo. The **presumptive somatic mesoderm cells** become arranged as strips on each side of the notochordal process. The **presumptive lateral plate mesoderm cells** become arranged on the side of somatic mesoderm. The cells of **heart** and **kidney** also move along with the lateral plate mesoderm.

The posterior part of the primitive streak contains cells of the **extraembryonic mesoderm** which will form the **extra embryonic membranes.** The displaced hypoblast contributes to the formation of the extraembryonic parts like part of the lining of yolk sac.

By 22 hours of incubation most of the presumptive endoderm cells are in the interior of the blastocoel while presumptive mesodermal cells continue to migrate for a longer time. Now the second phase of gastrulation begins and the primitive streak starts regressing.

The movement of cells within the blastocoel is helped by a **polysaccharide**, the **hyaluronic acid,** secreted by the ectodermal cells of the epiblast. It coats the surface of the incoming cells which keep them separate and allow their migration as individual cells.

Regression of primitive streak and extension of embryonic axis

As the inward migration of cells through the primitive streak ends, it begins to shrink. The anterior end regresses while the posterior end remains stationary. As the streak shortens, the node is carried backward and the presumptive notochordal cells still contained in the node and epiblast continue to migrate downward and add to the length of the notochordal process. The regressing streak thus leaves behind the dorsal axis of the embryo represented by the posteriorly elongating notochord. A part of the presumptive endodermal area that still remains posterior to the node continue moving inward and form a strip of endoderm below the notochord. It later forms the middle and posterior part of the gut. By the end of gastrulation the streak regresses completely. Its remnant becomes partly incorporated into the tail bud and partly into the cloacal region of the embryo. As the primitive streak shrinks the **neural cells** lying just in front of the node stretches backward, and a strip of presumptive **neural plate** is formed (Fig. 20.47).

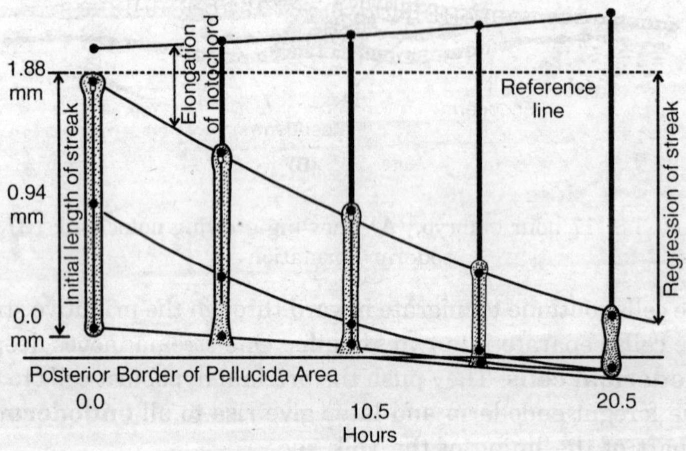

Fig. 20.47 *Fowl*– Graphic representation of regression of primitive streak and growth of notochord

The regression of primitive streak begins about 22 hours after incubation and is completed in the next 20 hours.

Epiboly (expansion) of ectoderm

The areas vacated in the epiblast by the inward migration of presumptive endodermal, mesodermal and notochordal cells, are occupied by expansion of the embryonic ectoderm. Similarly the ectodermal cells of area opaca expand outward as a sheet and completely encircle the yolk.

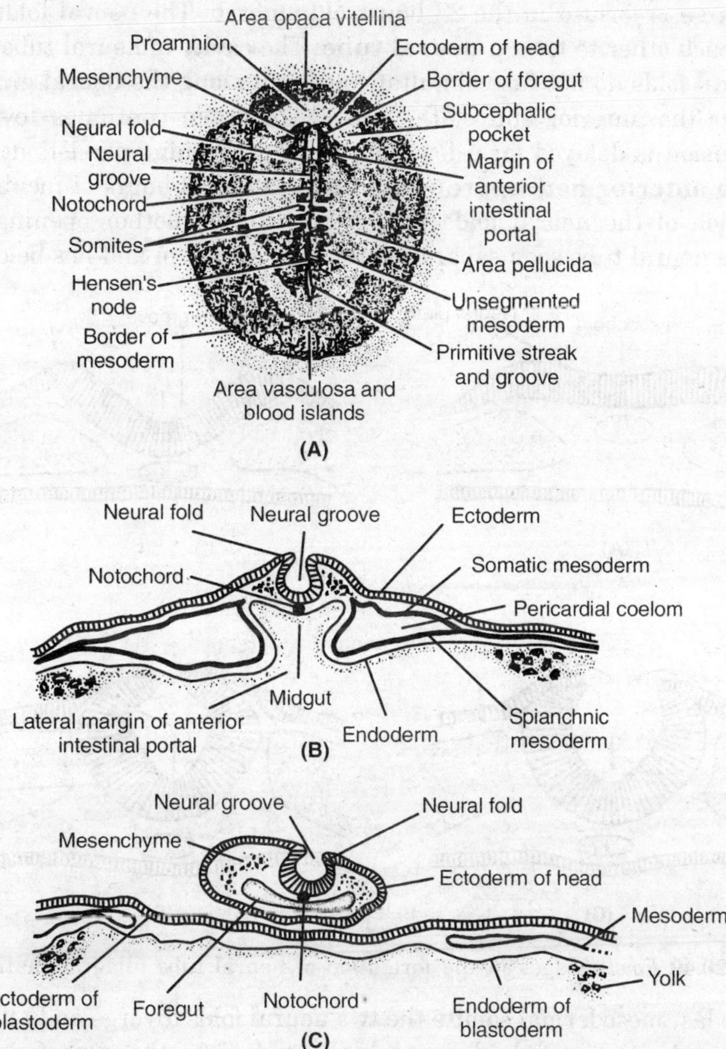

Fig. 20.48 *Fowl*–(A) 24 hour chick embryo (4pairs of somites); (B) and (C) 24 hour embryo

LATER DEVELOPMENT

Formation of notochord (Notogenesis)

In the begining the notochordal process (see gastrulation) is a rod of more or less undifferentiated cells, but it becomes more compact as development proceeds. This is the **notochord**. It soon acquires a sheath of mesenchyme cells derived from the scleretomes.

Formation of neural tube (neurogenesis)

The development of nervous system begins after about 18 hours of incubation. The median strip of endoderm cells above the notochord and in front of the Hensen's node thickens to form the **neural** or **medullary plate.** Soon the neural plate sinks and its margins become elevated forming the **neural** or **medullary folds**. Thus, a **neural** or **medullary groove** is formed in the 22 hours old embryo. The neural folds extend forward and unite with each other to form a **neural tube**. The cavity of neural tube is called **neural canal**. The neural folds do not fuse simultaneously all along the neural groove. Fusion first takes place near the anterior end of the groove and then continues towards both ends. Anteriorly the fusion is delayed for a few hours and on opening persists at the tip for some time. This is the **anterior neuropore** which closes after 33 hours of incubation and leaves a scar. The fusion of the neural folds posteriorly leaves another opening, the **posterior neuropore**. The neural tube soon separates from the ectoderm and lies below it (Fig. 20.49).

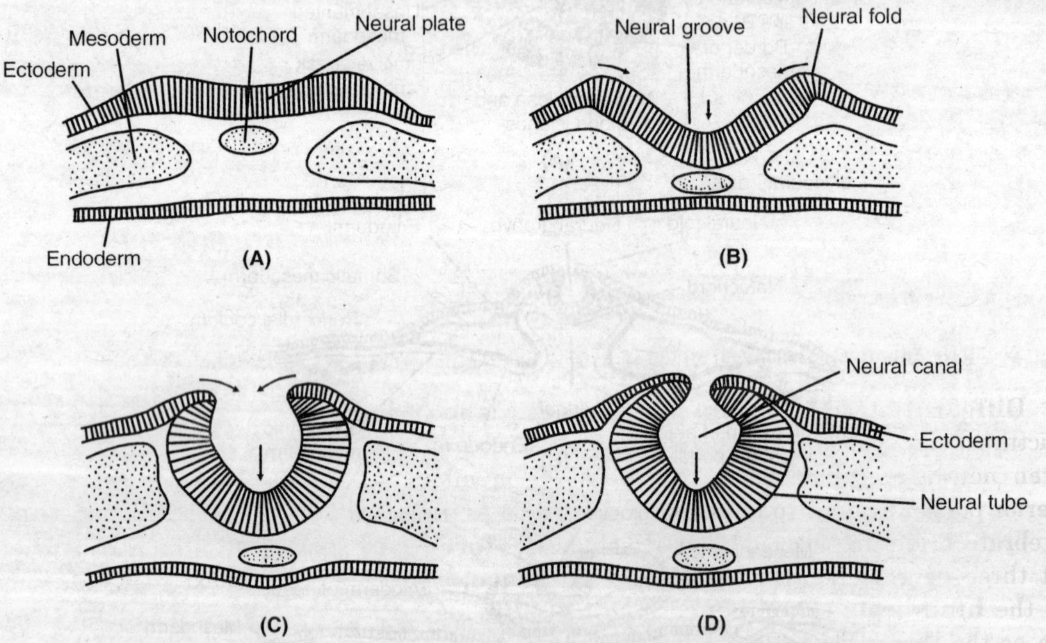

Fig. 20.49 *Fowl*–Stages in the formation of neural tube (diagrammatic T.S.)

Behind the last mesodermal somite the two neural folds diverge and after having passed the diminishing primitive streak they gradually fade into the superficial ectoderm. The posteriorly fading neural folds enclose a shallow area called **sinus rhomboidalis**. This is actually the posterior part of the neural plate which finally forms the complete neural tube.

When the two neural folds fuse dorsally some loose cells appear on either side of the neural tube. These cells form two bands called **neural crests** which later become segmented by the disapperance of cells. They form the dorsal root ganglia of the spinal and cranial nerves as well as ganglia of the autonomic nervous system. They first appear in the anterior region of the embryo and are gradually added posteriorly as the embryo grows.

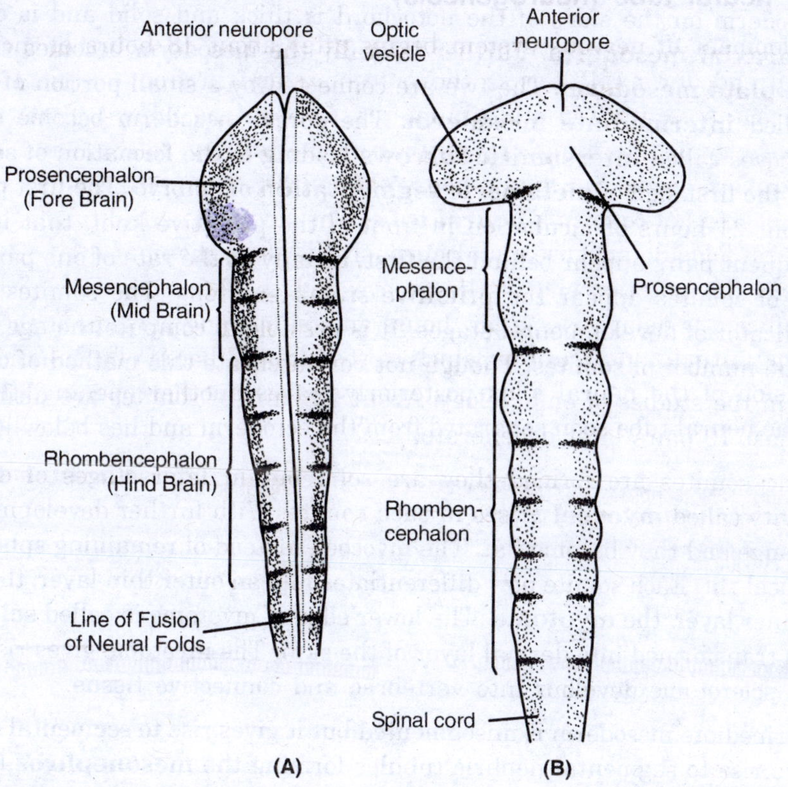

Fig. 20.50 *Fowl*–Dorsal view of brain. (A) 27 hour embryo; (B) 30 hour embryo

Differentiation of brain: The neural folds in the head region present a metameric structure before fusion. The segments are called **neuromeres** but are very vague. About eleven neuromeres have been recognized. Soon after the completion of neural tube, its anterior portion widens to form the brain. Three **primary cerebral vesicles**, typical of the vertebrate brain are formed. These include the **fore-brain** (**prosencephalon**) comprising first three neuromeres, the **mid-brain** (**mesencephalon**) including next two neuromeres and the **hind-brain** (**rhombencephalon**) consisting of the remaining six neuromeres. As soon as the three vesicles are separated the neuromeric constrictions of the first two parts disappear except those separating the vesicles. The neuromeres of the hind brain persist for about 3 days. The forebrain gives rise to sac-like lateral outgrowths called the **optic vesicles.** They begin their appearance after about 29 hours of incubation and in 33 hour chick they are the most conspicuous structure of the brain. The remaining part of the neural tube forms the spinal cord (Fig. 20.50).

Mesoderm and its differentiation

At the early head-fold stage the lateral sheets of mesoderm spread from the sides of the notochord, laterally across the area pellucida and grow upwards to occupy considerable area in the cephalic region of the embryo. It leaves a small mesoderm free area called **proamnion** (a misnomer since it has nothing to do with amnion) in the cephalic region below the head-fold just in front of the notochordal process.

The mesoderm on the sides of the notochord is thick and solid and is called **dorsal, somitic** or **paraxial mesoderm**. Further laterally the mesoderm becomes thinner and is called **lateral plate mesoderm.** The two are connected by a small portion of unsegmented mesoderm called **intermediate mesoderm.** The dorsal mesoderm become segmented by transverse grooves called **intersomitic furrows** leading to the formation of **somites.** Their appearance is the first sign of **metameric segmentation** of embryo. The first pair of somites appear in about 21 hours of incubation in front of the primitive knot, that is, in the neck region. Subsequent pairs appear behind the first, usually at the rate of one pair per hour. As further pairs of somites appear the primitive streak shortens. The somites are the most important indicator of developmental stages. It is possible to compute the age of the embryo by knowing the number of somites. Though not very accurate this method of computation is quite helpful in the studies of embryology. About 52 pairs of somites are ultimately formed of which the last 10 pairs later degenerate.

When the somites are formed they are solid but in later stages of development a transitory cavity called **myocoel** arises in each somite. With further development the shape of somites change and they become flat. The myocoel, instead of remaining spherical becomes a narrow vertical slit. Each somite now differentiates into an outer thin layer, the **dermatome** and a thick inner layer, the **myotome**. The lower edge of myotome is called **sclerotome**. The dermatome is transformed into dermal layer of the skin. The myotome gives rise to voluntary muscles. The sclerotome develops into vertebrae and connective tissue.

The intermediate mesoderm is unsegmented but it gives rise to segmental **nephrotomes** that later give rise to segmental nephric tubules forming the **mesonephros** (adult kidney).

After 26 to 29 hours of incubation certain mesodermal masses of cells called **blood islands** are also established in the posterior half of blastoderm. The cells occyping the middle of each blood island form **blood corpuscles** while the cells on the outside form a **wall** surrounding the corpuscles. Soon the blood islands run together forming a **plexus** of small blood vessels in the central part of area opaca now called **area vasculosa.** The peripheral zone of area opaca into which the mesoderm has not yet penetrated is called **area vitellina**.

Origin of coelom

The **lateral mesoderm** is also unsegmental. In 24 hour old embryo it splits into two layers and encloses a cavity called **coelom** or **splanchnocoel** between the two newly formed layers. The origin of coelom is **schizocoelic**. The upper layer of mesoderm (next to the ectoderm) is the **somatic** or **parietal layer** and the lower layer (next to the endoderm) is the **splanchnic** or **visceral layer**. The ectoderm together with the somatic mesoderm is known as **somatopleure** and the endoderm together with the splanchnic mesoderm is known as **splanchnopleure**. The two sheets of the lateral mesoderm extend from the embryo far out into the extra embryonal area, as such the coelom also extends into that area. The portion of the coelom in the embryonal region is the **intraembryonic coelom** or **splanchnocoel** and that in the extra-embryonal region is the **extraembryonic coelom or exocoel.** There is no sharp demarcation between the two coeloms in early embryos.

By third and fourth day of incubation the intraembryonic coelom becomes marked off by a series of body folds. With the closure of the ventral body wall the intraembryonic coelom separates from the extraembryonic coelom. The folding process also completes the gut ventrally. The right and left coelomic chambers expand ventrally. As a result the gut becomes surrounded by splanchnic mesoderm. The double layered splanchnic mesoderm dorsal to the gut is called **dorsal mesentery** and that ventral to the gut is called **ventral mesentery**. The two mesenteries connect the gut to the body wall and divide the coelom into left and right chambers. Soon the ventral mesentery disappears and the coelomic chambers become continuous ventral to the gut.

Development of Heart

Development of heart begins in 24 hour old chick from mesoderm. The heart is an unpaired structure but it develops from a pair of primordia.

Development of endocardium: A thickening appears in the splanchnic mesoderm near and posterior to the anterior intestinal portal. Within next three hours this thickened mesodermal portion buds off cells into the narrow space between itself and the endoderm. These cells are organised into short tubes called **endocardial tubes** or **primordia**. The walls of the tubes are only one-cell thick. They are destined to form the endocardium or the inner lining of the heart.

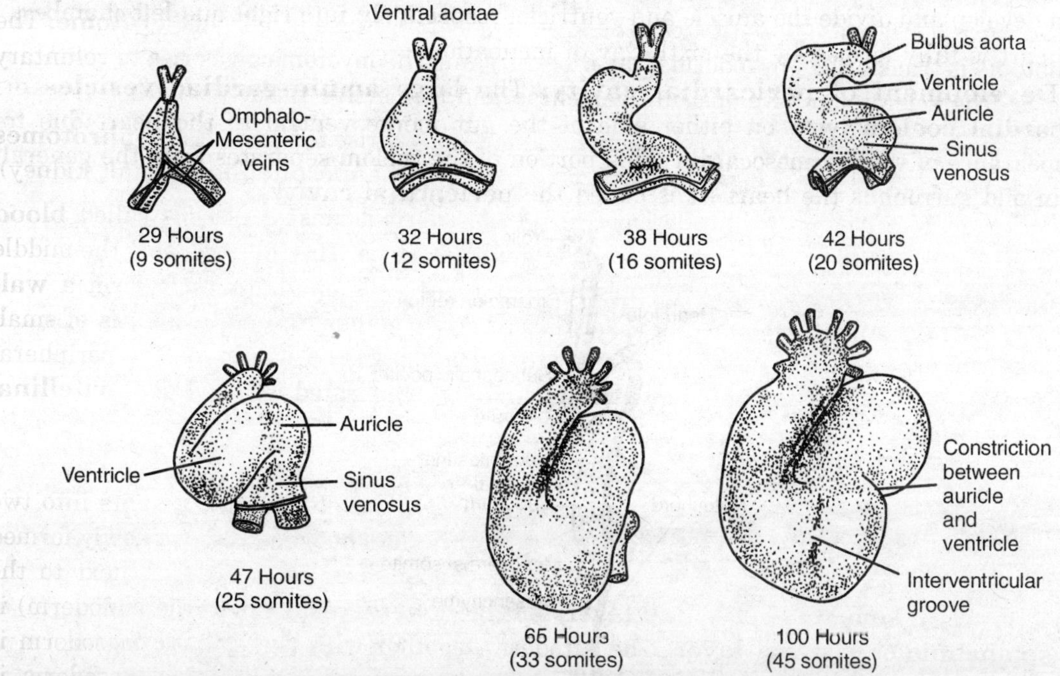

Fig. 20.51 *Fowl*–Ventral view of heart at various stages showing change of shape and regional differentiation

Development of epimyocardium and tubular heart: The two tubes are surrounded by thickened splanchnic mesoderm known as the **epimyocardium**. As the anterior intestinal portal is being pushed posteriorly the endocardial tubes with their epimyocardial envelops approach each other and fuse beneath the newly formed foregut. Consequently a single double walled tube is formed. This is the **embryonic heart**. The outer layer is thick-walled

epimyocardium and the inner layer is thin-walled **endocardium**. The heart is held in place by dorsal and ventral **mesocardium** (meeting points of epimyocardium). The ventral mesocardium disappears soon after it is formed while the dorsal mesocardium persists for a few hours more and eventually disappears except near the sinus venosus where the tubular heart begins to dilate and twist upon itself. The epimyocardium gives rise to a thin outer **epicardium** which later forms membranous covering of the heart and a thick inner **myocardium** which forms the muscles of the heart. In its simplest form the tubular heart is completed within 30 hours of incubation.

Differentiation of tubular heart: In the next 3 or 4 hours the heart tube begins to dilate and bend on itself. As it begins to elongate anteriorly it gives rise to two **ventral aortic roots** and posteriorly it receives the two **omphalomesenteric veins.** As the cardiac tube elongates it bends upon itself, since both its ends are fixed by the blood vessels, and forms a **S–shaped tube** (Fig. 20.51). With further growth it becomes separated into four chamber by transverse constrictions. The **sinus venosus** and **atrium** develop at the junction of the two omphalomesenteric veins. Further flexure of the heart towards the right forms the **ventricle** and at the anterior end where the two ventral aortic roots leave the heart is the bulbus cordis. By the end of fourth day of incubation the **interauricular** and **interventricular septa** develop and divide the auricle and ventricle, respectively, into right and left chambers. The heart is fully formed by the sixth day of incubation.

Development of pericardial cavity: The large **amnio–cardiac vesicles** or **pericardial coelom** lying on either side of the gut merge ventral to the heart due to disappearance of ventral mesocardia. This portion of the coelom separates from the general coelom and surrounds the heart. It is called the **pericardial cavity**.

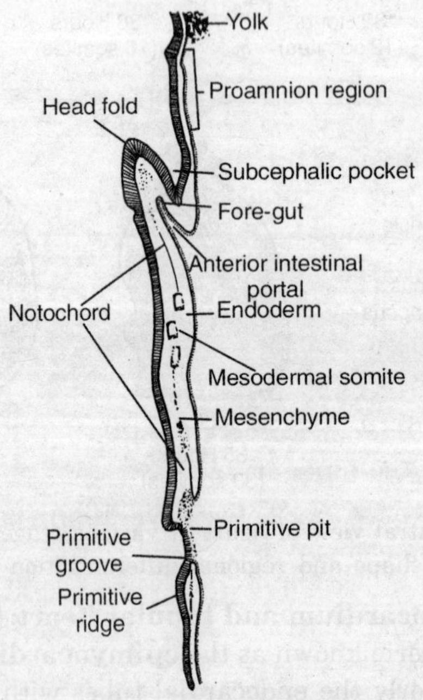

Fig. 20.52 *Fowl*–M.L.S. of 24 hour chick embryo showing head fold and foregut

Development of alimentary canal

On the first day of incubation (at about 18 hours) the **primitive gut** appears at the head end of the embryo. It is lined by endoderm dorsally and yolky mass ventrally. In 23 hours embryo the ectoderm and endoderm in front of the notochord sink downwards forming a **head fold** that later differentiates into head and anterior region of the embryo. A shallow subcephalic pocket is formed beneath the head region of the embryo. The head fold encloses a small cavity with endodermal lining on all sides. This is the **primitive foregut**. It is a blind tube closed at the anterior end and opening posteriorly at the **anterior intestinal portal**. The part of the primary gut posterior to the foregut is called the **midgut.** The floor of the midgut is non-cellular as it is formed of yolk. Such a gut is called **open gut.** The opening of midgut into the foregut is called **anterior intestinal portal**. The foregut continuously increases in length and becomes tubular with the gradual extension of the head fold (about 40 hours of incubation).

The **hindgut** appears quite late, after about 48 hours of incubation. Like the head fold, a tail fold develops at the posterior end of the embryo behind the primitive streak. A subcaudal pocket similar to subcephalic pocket is formed below the tail. The tail fold encloses a sac-like cavity lined by ectoderm. This is the **primitive hindgut.** Anteriorly the hindgut communicates with the midgut through the posterior intestinal portal.

The gut is not completed as a continuous tube until the yolk is totally enclosed in the **yolk sac** which then appears as a bag-like appendage of the gut to which it remains connected by a hollow **yolk sac stalk.** As development proceeds the foregut and hindgut lengthen and the midgut is reduced to a region where the yolk sac stalk opens into the alimentary canal. With further development regional differentiation begins in the gut and the digestive glands are formed.

At about 40 hours of incubation the part of foregut formed is very broad and will give rise to the **pharynx.** On its anterior face the endoderm fuses with the ectoderm to form an **oral plate** which lies in a depression on the under surface of the head. Later a perforation appears on this plate and forms the **stomodaeum** (an ectodermal invagination) which communicates with the pharynx. Along the side walls of the pharynx four evaginations appear on each side. These are the pharyngeal pouches. The lips of these finally fuse with the ectoderm. The first-three break through the surrounding ectoderm and form the visceral clefts. The fourth does not perforate at all.

At the posterior portion of the hindgut the endoderm fuses with the superficial ectoderm forming an **anal plate** which later perforates forming the **proctodaeum** (an ectodermal invagination) which opens into the hindgut.

Digestive glands: The **liver** arises as two **diverticula** of the endoderm of the anterior intestinal portal, one situated immediately above and the other below the posterior end of the ductus venosus. As the anterior intestinal portal retreats backwards the original dorsal liver diverticulum becomes anterior and ventral becomes posterior. These two form the left and right lobes of the liver by 30 somite stage.

The **pancreas** develop from a **dorsal** and a **pair** of **ventral primordia.** The dorsal primordium is an outgrowth of the dorsal wall of the intestine immediately above the posterior liver diverticulum. The ventral primordia arise from the posterior liver diverticulum.

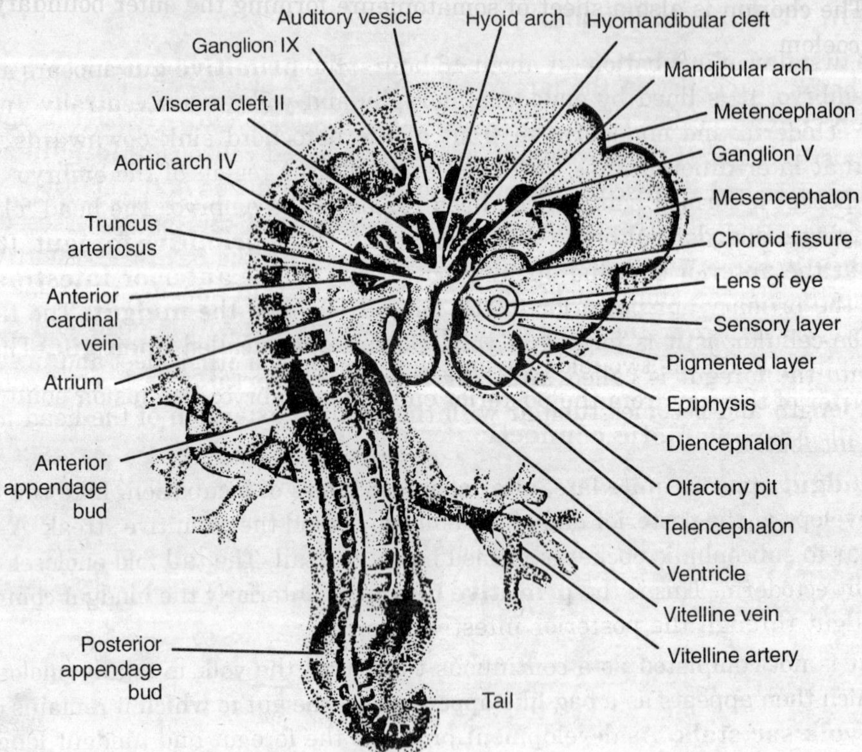

Fig. 20.53 *Fowl*–Whole mount of 72 hour chick embryo

EXTRA-EMBRYONIC MEMBRANES IN CHICK

In addition to the embryo proper the blastoderm gives rise to certain other structures which lie outside the embryo and are therefore called **extra-embryonic.** These structures are the **extraembryonic** or **foetal membranes.** The chick embryo has four different extra-embryonic membranes, namely, the **yolk sac**, the **allantois**, the **amnion** and **chorion** or **serosa**. Of these the first two are present in rudimentary form in frog and other amphibian embryos but amnion and chorion are new structures that develop for the first time in the reptiles. The birds having been derived from reptiles, retained them in their development. At hatching the greater part of these membranes are discarded. The amnion and chorion are completely eliminated and the allontois is also discarded to a great extent. The yolk sac, however, is incorporated into the small intestine. The development of foetal membranes in amniotes is related with the large amount of yolk in the egg and terrestrial life spent within the egg shell.

Amnion and chorion

The amnion and chorion are the first extra-embryonic membranes to develop. Both are derived from extra-embryonic **somatopleure** and are closely associated in origin and development. The **amnion** is a sheet of somatopleure enveloping the embryo and separating it from its immediate contact with the environment. It is like a bag in which the embryo is contained. The chorion is also a sheet of somatopleure forming the outer boundary of extra-embryonic coelom.

At 30 hours of incubation the large and well defined head of the embryo juts forward over the subcephalic pocket. The extra-embryonic somatopleure (ectoderm and somatic mesoderm) just in front of the developing head forms a fold called **amniotic head fold.** In its initial stage this fold is crescentic pointing towards the tail end of the embryo. As it grows towards the posterior side the somatopleure at the sides of the embryo rises up to form **lateral amniotic folds** which soon become continuous with the head fold. Shortly after the second day of incubation similar fold, the **amniotic tail fold** appears posterior to the tail bud. It grows forward over the embryo in a manner similar to the head fold. During the fourth day of incubation the two folds advancing towards each other meet and fuse at a point about a quarter of the way from the posterior end of the embryo. The fusion point is marked by a scar called **seroamniotic connection** or **raphe.**

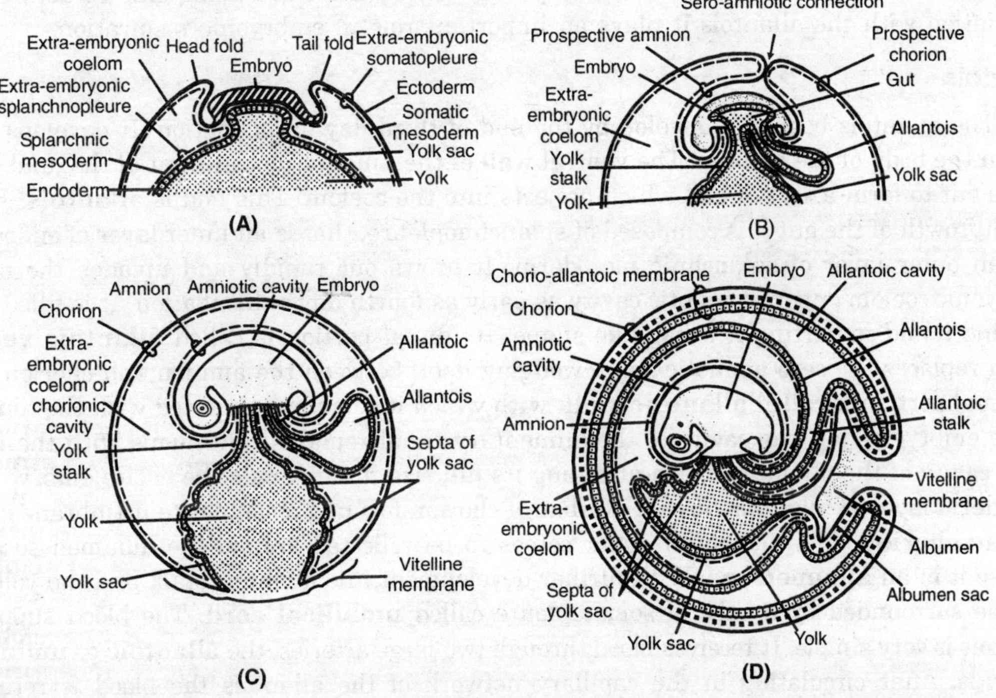

Fig. 20.54 *Fowl*– Schematic diagram showing the development of extraembryonic membranes, **(A)** Formation of head and tail folds; **(B)** Early stage showing allantois; **(C)** Formation of amnion and chorion; **(D)** Fully formed extraembryonic membranes

The amniotic folds are formed as a result of doubling of the somatopleure on itself, so two layers are formed by the union of the amniotic folds. The outer layer is called **chorion** or **serosa** and the inner is called **amnion.** In the amnion the arrangement of layers is reverse since it is formed by the fusion of the inner layers of the folds, the ectoderm is towards the inner side and the mesoderm towards the outer side. The chorion represents the outer half of the amniotic fold so the ectoderm lies on the outer side and the mesoderm on the inner side. Between the two layers there is a cavity, the **chorio-amniotic** or **sero-amniotic cavity** which is completely lined by mesoderm. It represents a part of the extraembryonic coelom. It is in direct communication with the coelom inside the embryo (intraembryonic coelom). The chorion grows rapidly and envelops the yolk sac completely by the end of second week. It expands and adheres to the entire calcareous shell membrane so that the embryo with all its foetal membranes is surrounded by it.

The main function of amnion is protection. The amnion isolates the embryo from the egg shell and thus protects it from adhesion to the shell. Between the embryo and amnion there is an **amniotic cavity** containing a saline **amniotic fluid.** It protects the embryo from the danger of desication. It offers no resistance to the embryo in changing its shape and position. It acts as an efficient **shock absorber** and protects the embryo from external shocks. The amniotic fluid keeps on moving slowly and gently by the muscle fibres developed in the amnion. It prevents the adhesion of the developing organs of the embryo.

The **chorion** also protects the embryo as well as the extra-embryonic structures. In association with the allantois it plays an important role in embryonic respiration.

Allantois

The allantois begins to develop by the end of third day of incubation. It develops from within the body of the embryo. The ventral wall of the hindgut just in front of the anal plate grows out to form a sort of bud which projects into the coelom. This bud is **allantois**. Being an outgrowth of the gut it is composed of splanchnopleure, that is an inner layer of endoderm and an outer layer of splanchnic mesoderm. It grows out rapidly and invades the extra-embryonic coelom or sero-amniotic cavity as early as fourth day of incubation. It is filled with a saline liquid imparting it a bag-like shape. Its distal portion is called **allantoic vesicle** which replaces the sero-amniotic space wedging itself between the amnion and chorion. The proximal portion is called **allantoic stalk** with which it remains connected with the hindgut of the embryo. Thus, the cavity of the alimentary canal remains continuous with the fluid-filled cavity of the allantois. Upon attaining its full size it surrounds the entire embryo. The mesoderm layer of allantois fuses with that of chorion forming a composite membrane called **allanto-chorion**. This membrane later comes to be reflected around the albumen so as to enclose it in an **albumen sac**. With further development, the allantoic stalk and the yolk sac become surrounded by a tube of somatopleure called **umbilical cord.** The blood supply of allantois is very simple. It receives blood through two large arteries, the **allantoic** or **umbilical arteries.** After circulating in the capillary network of the allantois the blood is returned through **allantoic** or **umbilical veins**. In the walls of allantois muscle fibres also develop.

The main functiion of allantois is respiratory. Right from the time of its first appearance the allantois is highly vascularised by the general system of vessels in the splanchnopleure. The formation of allanto-chorion extends the capillary circulation of the allantois over the entire inner portion of the calcareous porous shell. This makes the exchange of respiratory gases through the egg shell rather easy. The allantois helps in the storage of excretory substances which are given off during development. It also helps in the absorption of a large portion of albumen. In the later stages of development the allantoic circulation also absorbs calcium from the egg shell, which is used in the formation of bone, and it also makes the egg shell thin and weak so that it may be easily ruptured at the time of hatching.

Yolk sac

The large amount of yolk present in the egg cannot be contained within the body of the embryo. So the embryo develops a membrane called **yolk sac** around the yolk.

In the embryo of sixteen hours the enteron is represented by a flat, circular cavity lying beneath the primitive streak. Its roof is formed by splanchnopleure and the floor is composed of non-cellular yolk. In 24 hour embryo the foregut differentiates and the hindgut is formed in 48 hour embryo. Between the foregut and hindgut lies the undifferentiated midgut which becomes smaller and smaller as the fore and hindgut increase. The wall of gut gradually extends over the yolk and encloses it in a sac called **yolk sac.** The yolk sac remains connected with the gut by a narrow **yolk stalk**. The wall of yolk sac remains continuous with the wall of gut. As the embryo grows the yolk is consumed and the yolk sac becomes correspondingly smaller, so that two or three days before hatching it has become so diminuitive that it is withdrawn in the tissues of the embryo and appears like a protuberance from the small intestine. As the yolk sac diminishes in size, the splanchnopleure follows the yolk sac into the embryo, where it forms the lining of the midgut. This suggests that the yolk sac is really a part of the gut. Eventually it is incorporated into the small intestine of the chick, where it persists for several days after hatching. The yolk sac is richly supplied with blood vessels.

The yolk provides nourishment to the developing embryo. It is used directly as food by the early embryo. The endoderm lining the yolk sac is specialized both to digest the yolk and to absorb the product of digestion. To increase the surface area the wall of yolk sac becomes folded and the folds are embedded in the yolk forming **yolk sac septa.** The yolk is broken down into diffusible substances, digested by enzymes, picked up by blood capillaries and transported along with blood to the heart from where it is distributed to all parts of the embryo and to extraembryonic tissues and organs.

HATCHING

Hatching takes place after about 21 days of incubation. The chick develops a small calcareous elevation called **egg tooth** or **caruncle** on the tip of the upper beak. A day before or on the day of hatching the chick ruptures the shell membrane by the egg tooth and breathes air from the air space into its lungs. Just before hatching the chick breaks the egg shell with the egg tooth by repeated blows and come out leaving behind the extraembryonic membranes (amnion, chorion and allantois). The newly hatched chick is **precocious** or **nidifugous.** The feathers covering the body are wet. They soon dry up and the chick becomes active.

21

CHAPTER

CLASS AVES:

GENERAL CHARACTERS AND CLASSIFICATION

Of all the animals the birds are best known and most-easily recognized. They have branched off from the reptiles. Both reptiles and birds resemble each other in many respects, so Huxley placed the two classes of vertebrates in a single superclass, **Sauropsida.** Although birds have many reptilian features but they are highly specialized than reptiles. This made Huxley to state that birds are **glorified reptiles**.

The specialized features of birds are:

1. The birds are feather-covered, warm-blooded, air breathing, oviparous, bipedal flying vertebrates.
2. The body is spindle-shaped and consists of four regions; **head, neck, trunk and tail.**
3. The jaws are elongated into a toothless **beak** or **bill**.
4. The birds have two pairs of **limbs**. The **forelimbs** are modified into **wings** for flying. The **hindlimbs** are adapted for walking, running, perching, swimming, etc. They bear four clawed **digits** of which the first one or **hallux** is directed backwards and the remaining three in front.
5. The bird has an epidermal **exoskeleton** in the form of **feathers** covering the body, **scales** on legs, **claws** on toes and **horny sheath** on beak.
6. The skin is dry and without any skin glands, except **oil** or **preen gland** at the root of tail.
7. The endoskeleton is bony. The bones are **pneumatic** or hollow and without bone marrow. This makes the body light. The bones are usually fused.
8. The skull is smooth, with indistinct sutures, diapsid type and **monocondylic,** bearing single occipital condyle.
9. The lower jaw or mandible consists of five or six bones and articulate with the quadrate.
10. The vertebrae have **heterocoelous** centra. Neck long, so cervical vertebrae numerous. A **synsacrum** is present in the vertebral column. It is formed by the fusion of posterior thoracic, lumbar, sacral and anterior caudal vertebrae. The last 3 or 4 caudal vertebrae are fused to form a "ploughshare bone", the **pygostyle**.

11. The sternum is large and bears a **keel** for the attachment of flight muscles.
12. The **ribs** are **double headed** or **bicephalous** and each bears a backwardly directed **uncinate process**.
13. The **pectoral girdle** has a stout **coracoid** and a sabre-shaped **scapula.** The **clavicles** of the two sides fuse to form a V-shaped bone called **furcula** or **wish bone** or **merrythought bone**.
14. The **pelvic girdle** is large, strong and fused with the synsacrum. The acetabulum is perforated.
15. The distal carpals fuse with three metacarpals to form **carpometacarpus**. The proximal carpals are free.
16. The proximal tarsals fuse with tibia to form **tibio-tarsus**. The distal tarsals fuse with II, III and IV metatarsals to form **tarso-metatarsus**. First metatarsal remains free. The ankle joint is inter-tarsal.
17. The heart is four chambered. Sinus venosus and truncus arteriosus are absent. The right systemic arch only persists, the left one disappears. The red blood cells are nucleated.
18. The lungs are compact, spongy and non-distensible and provided with thin walled air sacs.
19. Larynx is without vocal cords. A sound box or syrinx is present.
20. The kidneys are metanephric and three-lobed, urinary bladder is absent. Birds are uricotelic.
21. Brain with greatly developed cerebrum, cerebellum and optic lobes. Cranial nerves twelve pairs.
22. The eyes are surrounded by sclerotic plates and contain vascular pecten.
23. The olfactory organs are poorly developed.
24. The birds have three chambered cloaca.
25. The sexes are separate and they exhibit sexual dimorphism, female has single ovary and oviduct. Male without copulatory organ except in ratitae, ducks, etc.
26. Fertilization is internal.
27. Birds are oviparous. Development is direct.
28. The eggs are cleidoic and amniotic and require external incubation for development.
29. Cleavage is discoidal and meroblastic.
30. The embryo develops four extra-embryonic membranes.
31. Birds exhibit parental care.

CLASSIFICATION

The birds form a more homogenous group than any other class of vertebrates, therefore they fail to present prominent external features for their classification. There are about 9000 living species of birds. The class Aves is divided into two subclasses: **Archaeornithes** and **Neornithes**.

Subclass I: ARCHAEORNITHES (Gr. *archios*, ancient + *ornithos*, bird).

1. It includes ancient and extinct birds that lived about 155 million years ago in the Mesozoic era.
2. Tail long tapering and lizard-like, with more than 30 vertebrae, without pygostyle.
3. Wings primitive, with poor power of flight.
4. Forelimbs with three clawed digits.
5. Jaws provided with teeth.
6. Vertebrae amphicoelous.
7. Sternum without a keel.
8. Ribs without uncinate process, abdominal ribs present
9. Carpals and metacarpals free.
10. Cerebellum small.

Examples: *Achaeopteryx* and *Archaeornis.*

Subclass II: NEORNITHES (Gr. *neos*, modern, *ornithos* bird)

1. Includes modern as well as extinct birds.
2. Tail is short, with 13 or less caudal vertebrae, pygostyle present.
3. Wings usually well developed, with a few exceptions.
4. The digits of forelimbs without claws.
5. The metacarpals fused with distal carpals to form carpometacarpus.
6. The sternum has a keel, ribs bear uncinate process. abdominal ribs absent.
7. Vertebrae heterocoelous
8. Jaws without teeth except in some fossil birds
9. Cerebellum large.

Superorder 1. ODONTOGNATHE (Gr. *odontos*, teeth).

1. Includes extinct Cretaceous birds
2. Teeth present in the beak.

Order 1. Hesperornithiformes

1. Includes marine, flightless, large-sized birds.
2. Teeth pleurodont (present in grooves)
3. Vertebrae amphicoelous.
4. Pectoral girdle reduced.
5. Sternum without a keel.

Examples: *Hesperornis*, *Enaliornis.*

Order 2. Ichthyornithiformes

1. Teeth present or absent is not definite
2. Neck vertebrae amphicoelous.

3. Pectoral girdle well developed.
4. Sternum with a keel.

Examples: *Icthyornis*, *Apatornis*.

Superorder 2. PALAEOGNATHAE or RATITAE

1. Includes **flightless,** terrestrial birds or **ratites.**
2. They are big in size and good runner.
3. The wings are vestigeal or rudimentary.
4. Feathers without interlocking mechanism.
5. Rectrices absent or irregularly arranged.
6. Oil glands absent except in kiwi and tinamous.
7. Palate dromaeognathous or palaeognathous.
8. Teeth absent.
9. Sternum without keel, ribs without uncinate process.
10. Bones are not pneumatic.
11. Syrinx absent.
12. Pectoral muscles poorly developed.
13. Tail vertebrae free, pygostyle small or absent.
14. Male has a copulatory organ.
15. Male incubates the eggs.
16. Young ones are precocious.
17. Distribution restricted. Not represented in India.

Order 1. Struthioniformes

1. Head, neck and legs sparsely feathered.
2. Feathers without after shaft.
3. Head comparatively small, neck long and flexible, beak short and broad.
4. Legs strongly developed, only two toes (third and fourth) present on each foot.
5. Pubic symphysis present.
6. Sternum without keel.
7. Pygostyle absent.

Examples: *Struthio camelus* (ostrich) found in Africa and Arabia.

Order 2. Rheiformes

1. Swift running birds.
2. Head, neck and thigh feathered.
3. Feathers without aftershaft.
4. Legs bear 3 toes with claws.

5. Sternum without keel.
6. Ischial symphysis present.

Example: *Rhea americana* (rhea) of America.

Order 3. Casuariformes

1. Head sparsely feathered, bears a comb-like structure, neck and body densely feathered.
2. Feathers with aftershaft nearly equal to shaft.
3. Wings small or rudimentary.
4. Legs with three clawed toes.
5. Sternum without keel.

Examples: *Casuarius* (cassowaries) of Australia and New Guinea, *Dromaius* (emu) of New Zealand.

Order 4. Apterygiformes

1. Feathers hair-like, without aftershaft.
2. Wings rudimentary or degenerate.
3. Beak long and slender with nostrils at the tip.
4. Neck and legs comparatively short. Legs with four toes.

Example: *Apteryx* (kiwi) of New Zealand.

Order 5. Dinornithiformes

1. Giant, extinct birds, became extinct about 700 years ago
2. Feathers with large aftershaft.
3. Wings almost absent.
4. Beak short.
5. Legs massive bearing four toes.
6. Sternum reduced and without keel.
7. Pectoral girdle completely absent.

Example: *Dinornis* (moas) of New Zealand.

Order 6. Aepyornithiformes

1. Recently extinct birds.
2. Wings vestigeal or tiny.
3. Legs powerful, with four toes.
4. Sternum short, broad and without keel.
5. Coracoid, scapula and wing bones reduced or absent.

Examples: *Aepyornis, Mulleornis* (giant elephant birds) of Africa and Madagascar.

Order 7. Tinamiformes

1. They are great runners (cursorial)

2. Sternum with keel, pygostyle reduced.
3. Egg shell glossy.

Examples: *Tinamus* (tinamous), *Eudromia*.

Superorder 3. IMPENNAE

1. Include aquatic flightless birds.
2. Wings paddle-like
3. Feet webbed.
4. Small scale-like feathers cover the whole body.
5. Come ashore to breed.

Example: *Aptenodytes* (penguins) of Southern Hemisphere.

Superorder 4. NEOGNATHAE or CARINATAE

1. Includes modern flying birds.
2. Wings well developed, feathers with interlocking mechanism.
3. Rectrices arranged regularly.
4. Pterylae regular.
5. Oil gland present.
6. Skull neognathus, that is, vomer is short allowing the palatines to meet. Skull sutures disappear. Quadrate double-headed.
7. Sternum with well developed keel.
8. Ribs with uncinate process.
9. Pygostyle present.
10. Scapula and coracoid meet at a right angle. Clavicles well developed.
11. Ilium and ischium united posteriorly.
12. Copulatory organ absent in male.
13. Young are altricious
14. Distributed throughout the world.

The superorder Neognathae includes several orders. Only a few interesting orders are mentioned below with known examples. For convenience sake these orders have been grouped in six ecological groups on the basis of the habits.

GROUP A. ARBOREAL BIRDS

This group includes those birds which live in and around tree and shrubs. Majority of birds belong to this group.

Order 1. Passeriformes

1. Largest order including half of the known species.
2. Toes three in front and one behind, adapted for perching.
3. Beaks adapted for cutting.

Examples: *Passer domesticus* (common house sparrow), *Corvus splendens* (common house crow), *Corvus macrorhynchos* (Indian jungle crow), *Acridotheres tristis* (common myna), *Saxicoloides fulicata* (Indian robin), *Oriolus oriolus* (golden oriole), *Molpestes* (Bulbul), Swallows, *Sturnus* (starlings), *Alauda* (Larks), *Loxia* (crossbill), Fringilla (finches).

Order 2. Piciformes

1. Includes highly specialized wood-boring birds like woodpeckers, toucans, sap-suckers, etc.
2. Bill is hard and powerful.
3. Tongue is long and protrusible.
4. Insectivorous.

Examples: *Dendrocopos mahrattensis* (Yellow front-pied wood pecker), *Dinopium benaghalensis* (golden-backed woodpecker).

Order 3. Columbiformes

1. Includes doves and pigeons.
2. Beak usually soft and slender.
3. Crop large, producing pigeon milk to feed young ones.
4. Fruit and grain-eating birds.
5. Mostly good fliers.

Examples: *Columba livia* (blue-rock pigeon), *Crocopus* (green pigeon) *Goura crustata* (crowned pigeon), *Streptopelia risoria* (ring dove), *Streptopelia chinensis* (spotted dove).

Order 4. Psittaciformes

1. Includes parrots, parakeets, cocatoos, macaws, etc.
2. Beak stout, narrow, sharp-edged and hooked at the tip.
3. Upper jaw movably articulated with the skull.
4. Feet zygodactylus, that is, two toes in front and two toes behind. Outer toe not reversible.
5. Have considerable memory.
6. Vegetarian
7. Feathers green, blue, yellow or red.

Examples: *Psittacula eupatria* (large Indian parakeet), *Psittacula lameri* (green parrot).

GROUP B. TERRESTRIAL BIRDS

Order 5. Galliformes

1. Includes game birds with short powerful flight.
2. Feet usually massive adapted for scratching and running.
3. Beak short.

4. Feathers with aftershaft.
5. Graminivorous (grain-eating).
6. Palatable.

Examples: *Gallus* (jungle fowl), *Pavo cristatus* (peacock), *Phasianus* (pheasant), *Francolinus* (partridge), *Cotrunix* (quail), *Alectoris grecca* (chukor).

Order 6. Cuculiformes

1. Includes cuckoos.
2. The female usually lays eggs in the nests of other birds, only some species make their own nests.
3. Two toes in front and two behind, outer one reversible.
4. Tail long and beak moderate.

Examples: *Cuculus canorus* (cuckoo), *Eudynamis scolopaceous* (koel).

GROUP C. SWIMMING AND DIVING BIRDS

Order 7. Anseriformes

1. Includes aquatic birds like ducks, geese and swans.
2. Beak broad and flattened, covered with soft cornified epidermis, margins of beak bear many transverse horny ridges or lanellae. Tongue fleshy.
3. Tail usually short, with many feathers.
4. Legs short and feet webbed.

Examples: *Anas* (wild duck or mallard), *Cygnus* (swan), *Anser indica* (bar-headed goose).

Order 8. Coraciiformes

1. Includes kingfishers.
2. Beak strong.
3. Third and fourth toe fused at the base.

Examples: *Halcyon smyrnensis* (white-breasted kingfisher), *Ceryle rudis* (pied kingfisher), *Dichoceros bicornis* (grey hornbill), *Upupa epops* (hoopoe).

Order 9. Gaviiformes

1. Includes loons.
2. Legs short, toes webbed.
3. Tail consists of 18–20 short stiff feathers.

Example: *Gavia,* represented by four species.

Order 10. Podicipediformes or **Colymbiformes**

1. Freshwater divers with legs placed for back on the body, feet lobed.
2. Tail with a tuft of downfeathers.

Example: *Podicipes* (grebes).

Order 11. Procellariformes

1. Includes seabirds like albatrossuc, petrels and shearwaters.
2. Nostrils tubular.
3. Wings long and narrow.
4. Feathers compact and oily in texture.
5. Nest usually on islands.

Examples: *Diomedea* (albatross), *Procellaria* (petrel) and *Puffins* (shearwaters).

Order 12. Pelecaniformes

1. Includes pelicans, darters, gannets and cormorants.
2. Fish-eating and colonial. Nest in colonies on rocks.
3. Beak long, with wide gap for catching and swallowing the prey.
4. Nostril vestigeal or absent.
5. Feet four-toed and webbed.
6. Gular pouch present on the throat.

Examples: *Pelecanus* (pelican), *Phalacrocorax* (little cormorant), *Anhinga* (darter).

GROUP D. SHORE BIRDS AND WADING BIRDS

This group also includes aquatic birds. They live near the shore and rarely swim or dive.

Order 13. Charadriformes

1. The legs are long and slender, toes are webbed.
2. Beak mud-probing.
3. Feathers dense and firm.

Examples: *Lobivanellus indicus* (red-wattled lapwing), *Hydrophasianus chirugus* (pheasant-tailed jacana), *Tringa glariola* (sandpiper), *Capella* (snipe), *Larus* (gull).

Order 14. Ciconiiformes

1. Long-legged marshy wading birds.
2. Neck long and snake-like.
3. Beak javelin or pincer-like for piercing aquatic prey (mainly fishes)
4. Feathers decorative. Strong fliers.
5. Toes without web.
6. Young ones born naked.
7. Some are migratory.

Examples: *Bubulcus ibis* (cattle egret), *Ardea* (great blue heron), *Phonicopterus* (flamingo), *Ciconia* (stork).

Order 15. Gruiformes

1. Small to large-sized wading birds with long legs and partially webbed feet.

2. Beak heavy.
3. Omnivorous diet.

Examples: *Fulica atra* (common coot), *Antegone antegone* (sarus crane), *Choriotis* (bustard).

GROUP E. BIRDS OF PREY

Order 16. Falconiformes

1. Diurnal and strong fliers.
2. Beak short and curved at the tip.
3. Feet with sharp curved claws adapted for grasping and holding the prey.

Examples: *Milvus migrans* (common kite), *Sarcogyps calvus* (king vulture), *Astur badius* (hawk), *Falco peregrinus* (falcon).

Order 17. Strigiformes

1. Nocturnal birds with large rounded head.
2. Eyes large and directed in front. Each located in a disc of radial feathers.
3. Beak short.
4. Feet feathered upto the toes. Claws sharp, adapted for grasping prey.

Examples: *Bubo bubo* (great horned owl), *Ketupa ceylonensis* (brown fish owl).

GROUP F. AERIAL BIRDS

This group include birds which are generally on wings and feet are poorly developed or vestigeal.

Order 18. Micropodiformes or Apodiformes

1. Includes small insectivorous birds.
2. Beak small and weak, with long tubular tongue.
3. Wings long and pointed.

Examples: *Micropodus* (Indian swift), *Trochus* (humming bird).

Order 19. Caprimulgiformes.

1. Shy, nocturnal and insectivorous.
2. Beak small and delicate.
3. Mouth wide, margined with long bristle-like sensory feathers.
4. Feathers soft.

Examples: *Caprimulgus* (goat suckers), *Chordeiles* (night hawk).

SOME COMMON BIRDS

Passer domesticus (sparrow)

Passer domesticus (Fig. 21.1A) is commonly known as **house sparrow** and in Hindi it is called **gauriyya**. It is a small bird, about 10–16 cm in length. It is found in and around human dwellings. It shows distinct sexual dimorphism. The female is of ash-white colour while male is earthy brown with blackish throat and breast and white abdomen. The eyes are small and beak is short, stout and pointed and adapted for eating grains. It also eats insects. The breast and abdomen are bulging. Wing and wing feathers are folded over the body and tail feathers are elongated. The feet are adapted for perching, that is, three toes in front and one behind. It the most familiar bird freely moving and nesting in houses where they live as commensal to man. Sparrows breed throughout the year. They lay 3–8 eggs in nests of dry twigs and straw made in holes in houses. Young ones are naked and blind at the time of hatching.

Upupa epops (hoopoe)

Upupa epops (Fig. 21.1B) is commonly known hoopoe it repeatedly emits a soft musical sound hoo-po-po. In Hindi, it is known as **hudhud.** It is distributed throughout India. It is found in lawns and gardens in and around villages and towns. In size, it is as big as myna. The body is reddish fawn coloured with white and black lines on the back, wings and tail, which look more beautiful when the bird flies. Its beak is about 5 cm long, slender and curved and is used to draw out subterranean insects, larvae and pupae from dry land. It is therefore beneficial to agriculture. It can be easily recognized because it bears a prominent fan-shaped crest on the head. Hoopoe does not show sexual dimorphism.

Psittacula eupatria (parrot)

Psittacula (Fig. 21.1C) is commonly called **Indian parakeet** or **parrot**. In Hindi, it is called **hiraman tota**. It is found in India, Sri Lanka, Bangladesh, Burma and Pakistan. It is of the size of pigeon. The body is slender and covered with briallant green plumage. The female is green all over but the male has a pink and black neck or collar ring. A distinct maroon patch is present on each shoulder. The beak is short, stout, sharp-edged and hooked at the tip, adapted for eating fruits. It also feeds on grains. The tail is long and tapering. The feet are adapted for grasping, holding and climbing. Foot is zygodactylus - I and IV digits directed backwards and II and III digits forward to provide firm grip. Parrots have loud voice. It can be domesticated and trained to speak some words and even sentences. It lives in holes of trees and buildings. It is a serious agricultural pest causing enormous damage to standing crops and ripening fruits.

Eudynamis scolopaceus (koel)

Eudynamis (Fig. 21.1D) is called **koel** or **kokila** in Hindi. It is commonly found in India, Pakistan, China, SriLanka and Australia. It lives on trees in gardens or in grooves. The male has a melodous voice. It sings koo-koo which is often heard during the mango season. The female does not sing but utters sharp repeated Kit-Kit-Kit. Koel is about the size of house crow. It exhibits marked sexual dimorphism. The male is glistening black in colour with yellowish green beak and small blood red eyes with round pupil. The female is brown in colour with white spots. The beak is pointed and curved downwards. Tail feathers are long. Wing feathers folded over the body. Koel is a nest parasite. It has the habit of laying eggs in the nest of crows. The eggs and young ones are looked after by the foster parents.

Fig. 21.1 Common birds. **(A)** *Passer domesticus;* **(B)** *Upupa epops*; **(C)** *Psittaculu eupatria;* **(D)** *Eudynamis scolopaceus*; **(E)** *Acriodotheres tristis;* **(F)** *Corvus splendens*

Acriodotheres tristis (myna)

Acriodotheres (Fig. 21.1E) is commonly known as *myna*. It is found in human dwellings and forests throughout India. It is about 22 cm long. Sexes are similar. The body is dark brown above while the abdomen is pale coloured. The head, throat and breast are black. The beak and legs are bright yellow. There is a patch of yellow skin around each eye. A white patch is present on the underside of the wing which becomes visible in flight. It makes nest in a hole on a tree. It is omnivorous eating fruits, insects and kitchen scraps. Myna can be domesticated and trained to speak words or sentences.

Corvus splendens (crow)

Crow (Fig. 21.1F) is found everywhere in India. It is a bold and intelligent bird. It lives around human habitat on trees in towns, villages and garden. In Hindi, crow is called **kowwa** or **kag**. It is about 32–42 cm long. The body is black with grey neck and breast. Feathers are glossy black. The wings are folded dorsolaterally. Tail feathers are long and feet are adapted for perching. Sexes are similar. The eyes are large and beak is stout and elongated. The crow is omnivorous, feeding on everything, even dead matter. They act as efficient scavengers. They build nests of twigs. They attack eggs and infants of other birds. They roost in large flocks.

Bubo bubo (horned owl)

Bubo (Fig. 21.2A) is commonly known as the **great horned owl** because on its large rounded head it bears two long horn-like feather tufts which look like ears. In Hindi, it is called **ghughu**. It is found throughout the world. It is a large fierce looking bird. The upper surface is reddish yellow, brown and grey and the undersurface is paler with dark vertical lines. The beak is small, sharp and curved and adapted for piercing and tearing. The eyes are large, round and directed forward, each lying in a disc of radial feathers. The wings are folded over the body. The legs are fully feathered, feet with sharp claws adapted for grasping. Sexes are alike. It is a nocturnal bird living in woody places and avoids dense forests. It feeds on small mammals, birds, lizards, insects, fish, crab, etc. It is of great importance to agriculture because it eats rodents.

Another species common in India is the **barn owl** or **screech owl, Tyto alba,** known as **Ulloo,** in Hindi. It is smaller in size than the great horned owl and has no horn-like projections.

Milvus migrans (pariah kite)

Pariah kite (Fig. 21.2B) is called **cheel** in Hindi. It is found in India, Pakistan, Bangladesh, Sri Lanka and Burma. It is a large and heavy bird, measuring about 60 cm in length. The feathers are reddish brown with dark brownstreaks. The sexes are similar. It is solitary or gregarious. The beak is sharp, pointed and hooked. The tail is forked. The feet are raptorial with elongated toes ending in sharp pointed claws. It is active during the day and come into human dwelling. Due to remarkable sight it can view its prey from height. It is a scavenger and feeds on small rodents, birds, reptiles, etc. It first circles round at some height and then suddenly rushes down to pick up the prey.

Another species of kite found in India is *Haliaster indus* commonly known as the brahminy kite. It is found near shores, seaports, harbours, and paddy fields. It is a scavenger feeding on frogs, crabs and fishes. It is smaller than pariah kite. The feathers are bright rusty red in colour. The head, neck and breast are white. The tail is rounded. It is a solitary bird. It makes nests in large trees.

Pavo cristatus (peacock)

Pavo (Fig. 21.2C) is commonly called **peacock** or **peafowl** and in Hindi it is called **mor** and **mayur**. It is the national bird of India. It inhabits dense jungles and forests and is gregarious in habit. It feeds on grains, vegetable shoots, insects, lizards and snakes. It does severe damage to newly sown seeds. *Pavo* shows well marked sexual dimorphism. The male is about the size of vulture, beautifully pigmented with a fan-shaped crest on the head. The head, neck and breast are of brilliant metallic blue colour. The tail is 1.5 m long formed of long trailing bronze green upper tail coverts ending in a shining blue green purple eye. The female is smaller in size, dull coloured and without ornamented tail. Beak is short. The feet are adapted for scratching and running. The peacock dance with its tail coverts spread out like a fan is very famous. It generally dances on cloudy or rainy day. It produces a loud and harsh sound 'may-awe' in morning, evening, moonlight or cloudy days. Peafowl is a shy bird. It is polygamous living in a party or drove of one male and 4 or 5 females.

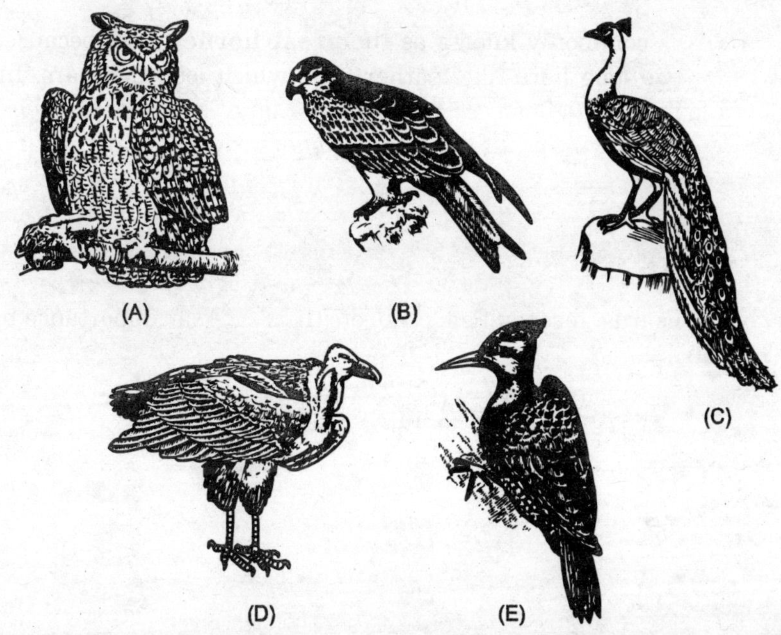

Fig. 21.2 Common birds. (A) *Bubo bubo;* **(B)** *Milvus migrans;* **(C)** *Pavo cristatus;* **(D)** *Sarcogyps calvus;* **(E)** *Brachyternus benghalensis*

Sacrogyps calvus (king vulture)

Sacrogyps (Fig. 21.2 D) is commonly known as **black Pondichery** or **King vulture** and its Hindi name is **raj gidh.** It is a large and heavy, ugly-looking bird, measuring about 7 feet across the wing. It is dirty blackish brown in colour and has two white patches on upper thighs, a white band on the underside of the wings and a white ring at the base of the neck. The head and neck are naked and of red colour. The legs are also red-coloured. The beak is stout and hooked at the tip. The feat bear sharp claws adapted for grasping. Sexes are similar. It is gregarious in habit. It is a carrion feeder. It makes nests on top of banayan or tamarind tree.

The other Indian species is the **white-backed Bengal vulture,** *Pseudogyps bengalensis,* called gidh in Hindi. It is smoky black in colour with white band running along the whole length of the underside of the wing.

Brachypternus benghalensis (woodpecker)

Brachypternus (Fig. 21.2E) is commonly called **woodpecker**. It is found throughout India. It lives on trees. It is an efficient climber. It clings to tree trunk and climbs up by jerks. The upper surface of the body is golden yellow and black in colour and the under surface is white with black spots. The head has a crimson red crest or crown. The beak is long, stout and pointed and adapted for pulling out insects from holes in the tree trunk. The tongue is protrusible. The feet has two toes directed forward and two backward, each ending in a sharp curved claw. It eats insects like beetles, ants, grubs, etc. It lays eggs in holes on trees excavated by both the parents.

22

CHAPTER

AVES:
GENERAL ACCOUNT

ORIGIN OF BIRDS

The birds have branched off from reptiles roughly at the same time in evolutionary history as did the mammals. From anatomical consideration, however, it is clear that these two branches are quite separate and birds of today are more closely related to reptiles than to mammals. Long ago the anatomists predicted that birds emerged out from basal reptilian stock and according to Goodrich, the Sauropsid line. The birds originated about 165 million years ago in the Jurassic period.

The fossil record of birds is very incomplete and fragmentary. The trasitory stages in bird evolution are missing. This led to various speculations regarding the immediate ancestors of bird. Some derived it from Pterodactyles and some from dinosaurs. It now seems likely that the Jurassic bird *Archaeopteryx* is a form intermediate between reptiles and modern birds. It shows avian as well as reptilian features, hence it is a connecting link between the two groups. The newly discovered fossil bird *Protoavis* is nearer to the ancestors of birds than *Archaeopteryx.*

Probable ancestors of birds

Among the various groups of reptiles the **Archosaurian diapsids,** which were the ruling reptiles of the Mesozoic era, showed greater adaptive radiation. They gave rise to the now extinct **Pterosaurs** and **Dinosaurs,** modern **Crocodilia** and **modern birds**. The **Pseudosuchia** or **Thecodontia** were early descendants of the Archosauria. The bipedal Pseu-dosuchians like *Euparkeria* (Fig. 22 1 A) and *Saltopsuchus* (Fig. 22.1 B) are considered as the probable ancestors of birds. They were quite abundant in the Triassic period. They were small, carnivorous bipedal reptiles with hindlimbs much longer than forelimbs. The feet had four toes three directed forward and one backward. The acetabulum formed a cup open below so that the legs were held vertically below the body. The eye orbits were large and had sclerotic rings. The jaws were elongated and provided with teeth which were set in sockets. According to most palaeontologists this form of reptiles gave rise to birds.

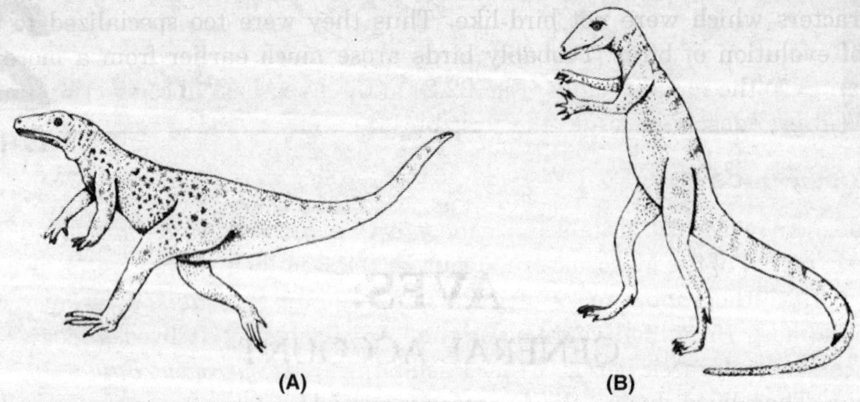

Fig. 22.1 *Probable ancestors*–**(A)** *Euparkeria;* **(B)** *Saltopsuchus*

The pseudosuchians divided early into two orders: **Ornithischia** and **Saurischia** on the basis of the structure of their pelvis. Some early Ornithischians like *Iguanodon* and *Camptosaurus* and Saurischians like *Struthiomimus* and *Ornithomimus* were also bipedal. Among these *Struthiomimus* and *Ornithomimus* were more bird-like in form. The hindlimbs were much longer than the forelimbs. They walked on three toes having three digits and the thumb was opposable and used for grasping. The skull was light. The jaws formed a short beak without teeth. Such wingless forms are considered as the probable ancestors of birds which somehow developed feathers.

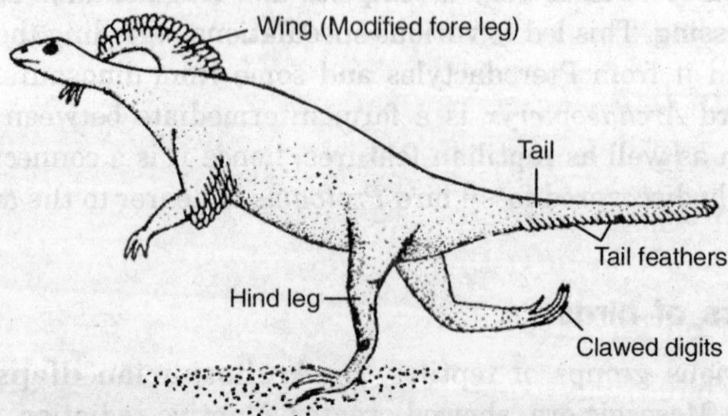

Fig. 22.2 *Hypothetical ancestor*–Proavis

Formerly the pterosaurs or pterodactyls were considered to be the probable ancestors of birds. They were light weight flying reptiles with membrane wings. Now this view is not supported. They are believed to have evolved at a later date from similar ancestors but they never developed feathers. They were simply gliders.

Pro-avis

All the probable avian ancestors did not have a clavicle or wishbone. In *Archaeopteryx* and other primitive birds V-shaped **wish-bone** was present so immediate ancestors of bird could have not been without it. Moreover, all ancestral reptiles were more specialized in several characters which were not bird-like. Thus they were too specialized to be on the direct line of evolution of birds. Probably birds arose much earlier from a more ancestral type. Heilman gave the name *Proavis* (Fig. 22.2) to the hypothetical connecting link between the generalized pseudosuchians and the first birds.

Primitive fossil birds

The oldest known fossil birds are *Archaeopteryx* and *Archeornis*. They lived during the upper Jurassic period of the Mesozoic era, about 140–150 million years ago. Their fossils were collected from the lithographic rocks of Bavaria, Germany. *Protoavis,* a recently discovered fossil bird, is almost 75 million years older that *Archaeopteryx*. It lived almost 225 million years ago in the lower Jurassic period. Other fossil birds are *Hesperornis, Baptornis, Ichthyornis* and *Apatornis*. They lived during the Cretaceous period.

Ancestral bird

Protoavis is regarded as the ancestral bird. This fossil bird was discovered in 1985 by Shankar Chatterjee (an Indian palaeontologists in USA) and his team of the Texas Tech-nical. University and was named by him. The fossils were in the form of some bones like wish bone, keeled sternum and skull with large cranium and orbits, all were bird-like. It also possessed some reptilian features like clawed fingers, long bony tail and teeth in the jaws. It differed from *Archaeopteryx* in having bird-like bones and teeth absent in the back of the jaws. It was more advanced and close to the ancestor of birds than *Archaeopteryx*.

Protoavis was nearly as big as a crow. The bones of forelimbs and hand had a number of nodes to which feathers were attached. The hindlimbs were strong and it could run like dinosaur.

Theories of origin of birds

Two theories have been proposed for the origin of birds:

(*i*) Diphyletic (two lines of descent) origin of birds.

(*ii*) Monophyletic (one line of descent) origin of birds.

Diphyletic origin of birds

Among the fossil birds some like *Archaeopteryx* and *Ichthyornis* are flying and some like *Hesperornis* and *Diatryma* are flightless. Besides flying birds many living birds like ostrich, rhea, kiwi, penguins, etc. are flightless. This led some scientists like **P.R. Lowe** to suggest a diphyletic origin of birds. The flightless and flying birds have descended from different flightless ancestors. According to Lowe the present day flightless birds were never capable of flight and their wings never developed.

Monophyletic origin of birds

Recently the fossil of *Eleuherornis* was discovered from Switzerland. It lived in the Eocene period. It is the probable ancestor of the present day ostrich but has greater affinities with the flying birds. This made many palaeontologists believe that birds have a monophyletic origin. All birds have originated from a single flying ancestor. The Ratitae evolved later from flying ancestors by the loss of flight particulary in areas where food was abundant and there were no enemies. This is the most accepted view and it poses a serious blow to the diphyletic origin of birds.

RATITAE (FLIGHTLESS BIRDS)

The superorder **Palaeognathae** or **Ratitae** is a group of flightless birds. The wings and flight muscles are very much reduced, so they are not capable of flight. They are all alike and ground dwellers. They are good runners because their hind limbs are strong and powerful. The flightless birds are generally large in size and some may reach a height of 4 meters.

The superoder Palaeognathae includes **7 orders** of which only 4 orders have living representatives and 3 are extinct.

Living orders

1. **Struthioniformes**—Ostriches
2. **Rheiformes**—Rheas.
3. **Casuariformes**—Cassowaries and Emus.
4. **Apterygiformes**—Kiwis.

Extinct orders

1. **Diornithiformes**—Moas.
2. **Aepyornithiformes**—Giant elephant birds
3. **Tinamiformes**—*Tinamous*.

Distribution

The flightless birds show **discontinuous distribution.** They are found in widely separated pockets of the world—ostriches in Africa and South-west Asia, rheas in South America, cassowaries and emus in Australia and kiwis in New Zealand. It is quite interesting that all living flightless birds occupy those parts of the world where there are no terrestrial competitors and enemies. Another interesting fact is that they are found in the southern parts of the world and are totally absent in the northern parts. But fossils of some flightless birds have also been found in North America and Europe which proves that they also lived there but due to arrival of carnivorous animals they disappeared.

Characteristic features

1. The bones are solid and without air cavities.
2. The sternum is without keel.
3. The ribs have no uncinate process which normally supports the thorax.
4. The clavicles are small or absent.
5. The ilium and ischium of the pelvic girdle are not fused behind.
6. The palate is palaeognathous type (the palatines do not articulate with the rostrum, the vomer intervening).
7. The wings are reduced or vestigeal.

8. The feathers are soft and curly. They are not arranged in regular feather tracts. The barbs of the feathers are disconnected and the barbules are without hooks.
9. The pectoral muscles are poorly developed.
10. Syrinx is absent.
11. Oil glands are absent.
12. Male has a copulatory organ, the penis.
13. Female has a clittoris.
14. Male incubates the eggs.
15. The newly hatched young one are precocious, *i.e.*, they are not dependent on their parents.

Origin of Ratitae

There are two major opinions regarding the origin of Ratitae.

1. They have evolved from true flying ancestors.
2. They have evolved from ancestors who never developed the power of flight.

Fig. 22.3 Flightless birds. **(A)** *Struthio comelus*; **(B)** *Casuarius casuarius;* **(C)** *Apteryx haiti*; **(D)** *Dromaius novaehollandiae;* **(E)** *Rhea americana*

The first view was proposed by Romer and is most accepted. He was of the view that the Ratitae have evolved from true flying ancestors and only secondarily readapted for a flightless mode of life. This is supported by the fact that the flightless birds possess certain morphological characters showing adaptations to flight. He argues that development of flight in birds is a device to escape enemies on the ground. If there are no enemies then there is no need for flight. So in areas where there was surplus food and no enemies the Ratitae stopped flying and gradually became flightless. This can be very well explained if the distribution of present day Ratitae is studied. This view can further be supported by studying the ancestry of some Ratitae. In the early Coenozoic era a number of neognathous birds were ground dwelling. The mammals also evolved at the same time, so there might have been strong competition between the ground-dwelling birds and mammals to conquer the ground vacated by the large reptiles which had become extinct. The mammals being carnivorous succeeded and so only a few birds. These probably gave rise to the modern flightless birds.

According to Young, the ancestors of Ratitae were not flying birds. The Ratitae never passed through a flying stage, they diverged early from the main stock of flying birds.

COMMON FLIGHTLESS BIRDS

Struthio camelus (African ostrich)

Ostrich (Fig. 22.3 A) is the largest living bird. It is found in the arid and semiarid regions of Africa and Arabia. It is herbivorous, feeding mainly on fruits and seeds. Ostrich may grow up to a height of 2.5 metres and weighs about 150 kg. The head, neck and legs are not covered with feather. The head is small, with large eyes and is mounted over a long pink-skinned neck. The legs are strong and powerful. It takes steps of 5 metres and can run at a speed of 80 km per hour. It runs on its toes which have fatty pads below. Each foot has two clawed toes (third and fourth) which serve as weapons of defence. The wings are small and usually folded. The feathers are soft and without after shaft. The male and female differ in the colour of their plumage. In female the feathers are brownish and in male the tail and wing feathers are white and rest black. Ostrich generally travels in groups. The male is polygamous. One male has several females. The male has a copulatory organ. Copulation is preceded by some courtship display by the male in which it drops to its heels before a female. The female lays eggs in nests made in sand by the male. The eggs are very large, about 6 inches long and 5 inches in diameter, weighs about 1.5 kg and takes about 50 minutes to boil. The eggs hatch in 42–48 days. The young ones are precocious. The ostriches are domesticated for their skin and feathers.

Casuarius casuarius (Cassowary)

Cassowary is found in densely woody parts of Australia and New Guinea. It is a shy, nocturnal bird. The head is small with compressed beak and a helmet-shaped bony knot on top (Fig. 22.3B). The young ones are brown while adults are black in colour. The skin of head and neck is brightly tinted. The quill feathers of the wing are like stiff long rods and hang down freely. The aftershaft is as long as the main shaft. The rectrices or tail feathers are absent. There are three toes on each foot. The bird is quarrelsome and quite vicious, if disturbed, the male may even attack man. The native people sometimes domesticate young birds. The flesh of cassowary is palatable.

Apteryx (Kiwi)

Kiwi is found in New Zealand. It is the smallest living flightless bird, almost as big as the domestic hen (Fig. 22.3C). The feathers are reddish or greyish-brown in colour, without aftershaft, long hair-like and hang down loosely. The wings are vestigeal and tail is absent. The head and eyes are remarkably small. The beak is long, slender, strong and slightly curved downwards. The nostrils are situated at the tip of the beak. The legs are scaly, short, powerful and backwardly situated. The feet is 4-toed, 3-toes in front and hallux behind. Kiwi is nocturnal, insectivorous and worm eating. It is burrowing in habit and lives in burrows during the day. The female is larger than the male. As compared to its size the eggs are quite large in size and weighs almost one-fourth the weight of the parent. Its flesh is relished by the local people and feathers are used for making ornaments. Kiwi is an endangered species so New Zealand Government has set-up some sanctuaries to protect it.

Dromaius novaehollandiae (Emu)

Emu (Fig. 22.3D) is the second largest living bird. It is found in the deserts of Australia. It has more slender built than ostrich. The neck is short and covered with feathers. The feathers have aftershaft as long as the main shaft. Emu has a tracheal pouch connected by a slit with the wind-pipe and fully developed in adults. It feeds on herbs, roots and fruits. It is fond of bathing and swimming. Emu is monogamous. Male makes the nest and hatches the first brood of eggs and the female hatches the second brood. Incubation is completed in about 60 days. The young ones are longitudinally striped. The stripes disappear at maturity. The feathered skin in used for making rags and mats. Its body fat is used as a lubricant.

Rhea americana (American ostrich)

Rhea (Fig. 22.3E) is found in the plains of South America. It is smaller (about 1.5 metres high) and less specialized in running than the African Ostrich. It loves bathing in water and is capable of swimming. It is omnivorous. The wings are small, the feathers are grey and without aftershaft. The tail feathers are absent. The head and neck are covered with short feathers. The legs are long and stout. The feet are covered with large scales. Each foot bears three forwardly-directed clawed toes (second, third and fourth). The male is polygamous. The females lay eggs in nests made by male. The male incubates the eggs. Incubation is completed in 40 days. The plumage is inferior to that of the African ostrich, so it used to make feather dusters.

BIRDS ARE GLORIFIED REPTILES

Birds are a specialized group of vertebrates. Young (1958) stated that birds are "the masters of air". Birds have originated from reptiles in the Mesozoic era. Externally reptiles and birds are quite different but closer examination reveals that both of them are basically similar. De Beer stated that nothing is more certain than the fact that "birds have descended from reptiles." As reptiles and birds are very similar Huxley placed these two classes of vertebrates in a single superclass **Sauropsida.** Birds are better evolved in their organisation but have many reptilian features so Huxley stated that "birds are glorified reptiles."

The evidence of reptilian origin of birds can be furnished by their comparative anatomy, embryology and paleontology.

ANATOMICAL EVIDENCE

Exoskeleton

1. **Scales:** The body of reptiles is covered by epidermal scales which persists in birds on the legs and feet.
2. **Feathers:** Birds have feathers which are homologous to the scales of reptiles. Both are epidermal derivatives.
3. **Beak:** The beak of turtles and tortoises is like the beak of birds. The horny covering (rhamphotheca) of the beak of bird is comparable to the reptilian scales. In some birds like puffin the outer covering of beak is shed off every year. This process is similar to the moulting of scales in reptiles. Moulting is also seen in many birds which cast off and renew their feathers periodically.
4. **Claws:** The horny claws present on the toes have the same basic structure in reptiles, birds and mammals.

Endoskeleton

1. **Skull:** The skull in both is well ossified, diapsid type and with single occipital condyle (monocondylic). The jaw suspensorium is autostylic. The basisphenoid is well developed and parasphenoid is poorly developed. An interorbital septum is present (absent in snakes and legless lizards).
2. **Lower jaw:** The lower jaw is complex and is made up of five or six bones in each half. It articulates with the skull by the quadrate bone of upper jaw.
3. **Vertebrae:** The vertebrae are without epiphyses (present in mammals). The neck vertebrae variable in number.
4. **Ribs:** Both cervical and thoracic ribs present. In young birds the cervical ribs are free but in adult they fuse with the centra of cervical vertebrae. The thoracic ribs form a complete basket and meet the sternum in the mid-ventral line. Ribs are provided with uncinate processes in birds, crocodiles and *Sphenodon*.
5. **Girdles:** The pectoral and pelvic girdles are similar in structure in both the classes. There is a well developed single coracoid in birds but in reptiles it is divided into precoracoid and coracoid proper.
6. **Limbs:** The skeletal frame of the hindlimbs is strikingly similar. The number of phalanges (2, 3, 4, 5) is the same.

Digestive system

In birds the alimentary canal has a gizzard and among reptiles crocodiles have a gizzard, they habitually swallow stones to help in digestion. In both reptiles and birds the alimentary canal ends in a tripartite cloaca (the three chambers are coprodaeum, urodaeum, and proctodaeum). The division in birds is not very distinct. A caecum is present.

Respiratory system

The lungs are the organs of respiration in both reptiles and birds. The air sacs characteristic of birds occur only in *Chamaeleon* among lizards. The trachea has complete tracheal rings.

Circulatory system

The heart is completely four chambered in birds and in crocodiles among reptiles. The blood in both groups is similar and the red blood cells are nucleated. In birds the right aortic arch forms the systemic aorta and the left aortic arch disappears. In crocodiles the left aortic arch is much smaller than that of the right side.

Nervous system

In both the groups the neopallium is smooth, the cerebellum is divided into a median vermis and lateral floculi and the mid-brain contains two optic lobes called corpora bigemina. Both have 12 pairs of cranial nerves except snakes.

Sense srgans

In birds, crocodiles and dinosaurs, the front part of the eye has a ring of minute bony plates, the sclerotic ossicles. The eyes of birds and crocodiles contain a comb-like pecten which projects into the vitreous chamber from the region of the blind spot.

The external ear or pinna is absent in both the groups. In many birds and crocodiles a valve-like flap projects into the auditory meatus from its outer margin. The middle ear contains a single rod-like ear ossicle called columella which connects it with the inner ear. The cochlea is not coiled.

Urinogenital system

The kidneys are metanephric. Both are uricotelic, i.e., they excrete semisolid urine in the form of uric acid. The cloaca reabsorbs water. The rectum and genital ducts open into the cloaca. Urinary bladder absent in birds and snakes. The mesonephric duct functions as vasa efferentia. The oviduct is differentiated into different glandular areas. The intromittent organ is present in males in all reptiles except *Sphenodon* and among birds in Ratitae and ducks.

EMBRYOLOGICAL EVIDENCE

Both reptiles and birds show following similarity in embryology.

1. Copulation occurs in both.
2. Fertilization is internal.
3. Both are oviparous. The eggs are large and telolecithal and laid on land.
4. The eggs are cleidoic, i.e., surrounded by a thick hard calcareous shell.
5. The ovum is surrounded by albumen, an egg membrane and a shell, all of which are secreted by special glands present in the wall of the oviduct.
6. Sperms are similar in size and structure.

7. Development requires incubation.
8. Cleavage is meroblastic and discoidal.
9. Extraembryonic membranes—amnion, chorion, allantois and yolk sac are present to protect the embryo.
10. The mechanism of gastrulation and mesoderm formation is similar.
11. The blastoderm has two regions: area pellucida and area opaca.
12. Primitive streak appears during gastrulation.
13. Both groups develop gill slits in the embryonic stages which disappear in the adults.
14. In reptiles an interclavicle is present in the pectoral girdle. A similar interclavicle develops in the embryo of birds.
15. Jacobson's organ is present in reptiles and in birds it is present in the embryonic stage.

PALAEONTOLOGICAL EVIDENCE

The palaeontological studies of fossil birds and reptiles give strongest support to the reptilian ancestry of birds. It is, however, unfortunate that fossil birds are not well represented because of arboreal life and soft skeleton. One of the earliest-known fossil bird is *Archaeopteryx lithographica*. It is considered a connecting link between reptiles and birds.

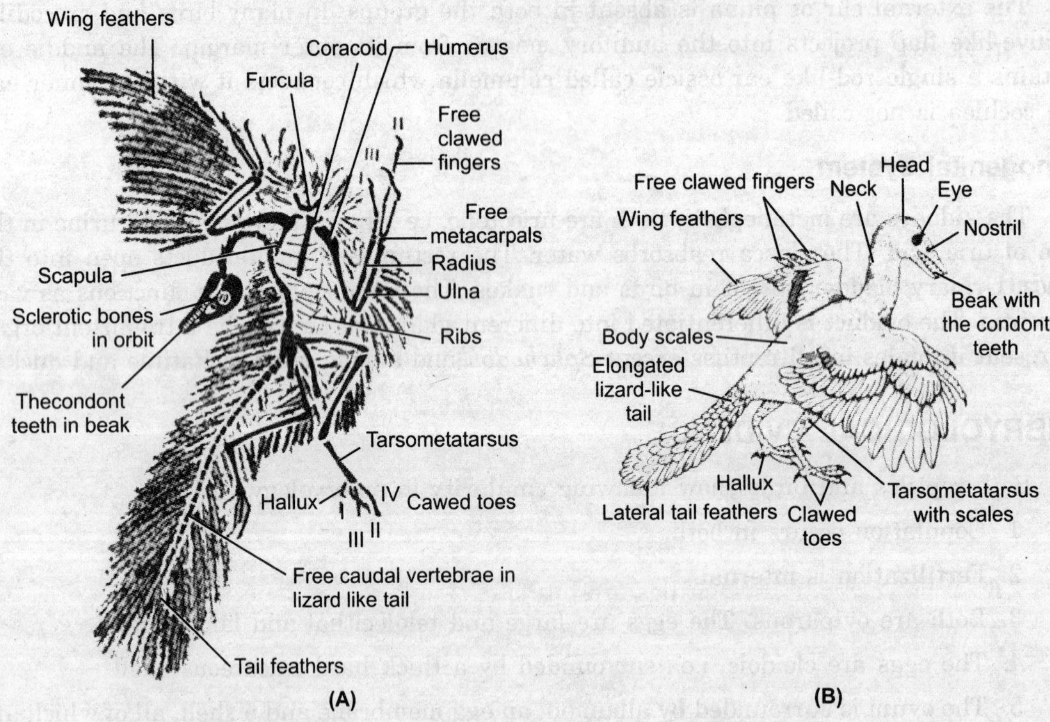

Fig. 22.4 *Archaeopteryx lithographica.* **(A)** Fossil showing skeleton and feathers; **(B)** Detailed structure

Archaeopteryx lived in the Jurassic period about 140 million years ago. It was discovered in 1861 from Bavaria, Germany by Andres Wagner. A second skeleton (Fig. 22.4 A) was discovered in 1877 and a third in 1956 from the same locality. *Archaeopteryx* possesses both reptilian and avian features.

Reptilian characters

1. Bones are not pneumatic.
2. Scales are present on body and limbs.
3. Jaws are provided with homodont teeth lodged in sockets.
4. Vertebrae are amphicoelous as in *Sphenodon*.
5. Cervical vertebrae few (9 or 10).
6. Free cervical and abdominal ribs are present in addition to thoracic ribs as in *Sphenodon* and crocodiles. Ribs single-headed and without uncinate process.
7. The tail is long, lizard-like and with about 20 free caudal vertebrae. Pygostyle absent.
8. Sternum without keel, poorly developed or absent.
9. Eyes provided with sclerotic ossicles.
10. Forelimbs provided with three free clawed fingers. The phalanges are 2, 3 and 4 in first, second and third fingers respectively.
11. Carpals and metacarpals are free, there is no carpometacarpus.
12. Pelvic girdle has an elongated ilium and a backwardly directed pubis.
13. Brain simple with cylindrical cerebral hemispheres and unexpanded cerebellum.

Avian characters

1. Presence of feathers (contour) over the body. If the feathers of *Archaeopteryx* were not preserved, it would have been classified as a reptile.
2. Forelimbs modified as wings bearing flight feathers (remiges).
3. Tail bears two lateral rows of elongated tail feathers (rectrices).
4. Skull large, bones intimately fused, brain case large and rounded and orbits large.
5. Jaws elongated into a beak.
6. Bones of limbs and girdles are bird like.
7. A V-shaped furcula bone present.
8. Tibia and fibula are separate.
9. Each foot bears 4 clawed toes with 2, 3, 4 and 5 phalanges respectively. The first toe or hallux is directed backward and is opposable.
10. Eyes provided with sclerotic ossicles.

Conclusion

A superficial examination reveals that a bird appears much different from a reptile. But a close examination reveals that a bird is nothing but a reptile clothed with feathers (Fig. 22.4 B). Because of feathers the birds have conquered the air. Hence birds are described as glorified reptiles.

AERIAL OR FLIGHT ADAPTATIONS

Young (1958) described birds as the **masters of air.** The entire body of birds have been modified for flight. Birds are good fliers and the fastest of all animals. In fact, a bird is like a flying machine. It flies on the principle of an aeroplane, a heavier than air machine. The morphology, anatomy, physiology and embryology of birds are highly modified for flight. The birds shows following flight adaptations:

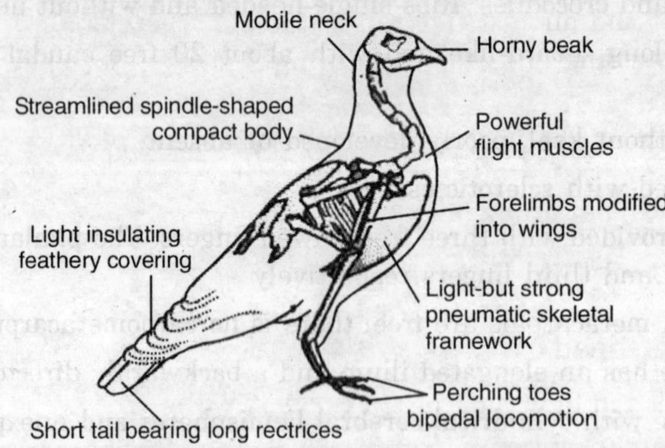

Fig. 22.5 Diagrammatic representation of flight adaptations

1. Body shape and structure

Birds have a **stream-lined, boat-shaped** or **spindle-shaped** body which offers minimum resistance to the air while flying. The body is quite compact. It is light but strong dorsally and heavy ventrally, it helps in maintaining balance in the air.

The internal structures are well centralized. The attachment of wings high up on the thorax, the high position of light organs like lungs and air sacs and low and central position of heavy structures like muscles, sternum and digestive organs beneath the line of attachment of the wings lowers the centre of gravity.

2. Feathers

The feathers are characteristic features of birds. They fully cover the body. They perform following functions:

(a) The feathers covering the body reduce friction.

(b) The feathers are light so they reduce weight.

(c) the feathers **insulate** the body and prevent loss of body heat. This enables the bird to tolerate extreme cold at height and maintain a constant body temperature.

(d) The body feathers hold a blanket of air around the body and this helps in **buoyancy.**

(e) The wing feathers help in striking the air and tail feathers are used for lifting and steering.

3. Wings

The forelimbs are modified into **wings** which are used for propulsion through the air. The convex upper surface and concave lower surface cause decrease in air pressure above and increase below and this helps in driving the bird forward and upward during flight.

4. Tail

In birds the tail is reduced to a degenerate skeletal stub, thus centralizing the weight. In its place long, light and air resisting tail feathers or **rectrices** are present. They act as a rudder for lifting, steering and stopping during flight. It also helps in balancing during perching.

5. Beak

The jaws are drawn out into a **horny beak.** The beak compensates for the modification of forelimbs into wings. It is used for picking up food, in nest building, for preening and for offence and defense.

6. Flexible neck and head

The head is small and light and the neck is long and flexible. It can be retracted during flight, thus shortening the body-axis. The flexible neck allows free movement of head which helps the beak in performing its functions easily.

7. Skin

The skin is quite loose. It allows free movement of flight muscles.

8. Flight muscles

The wing is operated by a set of muscles called **flight muscles.** They are highly developed in flying birds and constitute nearly one-sixth of the total body weight. There are four set of flight muscles. They are:

(a) **Pectoralis major**

(b) **Pectoralis minor**

(c) **Coracobrachialis**

(d) **Tensor muscles**

The pectoralis major and coracobrachialis lower or depress the wings (**down stroke**) and the wing is raised or elevated (**up stroke**) by pectoralis minor. The tensor muscles keep the wing stretched while flying. The flight muscles are well aerated and provide the sustained power to maintain flight.

9. Loss of weight

For flight less weight is very essential. This is achieved by the following features:

(a) Bones are **pneumatic,** i.e., they are provided with air sacs. The skeletal framework is light but strong.

(b) Skull bones are thin and light and firmly fused together.

(c) Thoracic ribs bear **uncinate processes** which provide compactness by concentrating the mass.

(d) Absence of teeth.

(e) Absence of ovary and oviduct of right side in female.

(f) Absence of urinary bladder.

(g) Production of solid excretory waste.

(h) Presence of feathers.

10. Air sacs

The air sacs are thin walled, non-muscular sacs connected to the lungs. There are four paired and one unpaired air-sacs. The paired air sacs are **cervical, anterior thoracic, posterior thoracic** and **abdominal** and unpaired **interclavicular air sac.** They occupy the space between the internal organs and extend into the cavities of bones. They serve as reservoirs of air. In other vertebrates air is supplied to the lungs only once and that is during inspiration. But in birds during expiration also air is supplied to the lungs from the air sacs. Thus in birds blood is oxygenated twice during a single breathing. This is called **double oxygenation** or **double ventilation.** Thus the respiratory system of birds is highly adapted to supply more oxygen to flight muscles. The air sacs also help in internal perspiration which help in regulation of body temperature.

11. Warm bloodedness

Birds are warm blooded animals. This is achieved by perfect aeration of blood and a great output of energy over a long period. The high body temperature (40–46°C) helps the birds to take flight at high altitudes and in all seasons.

12. Efficient circulatory system

Active flight and warm bloodedness require high output of energy which requires more oxygen. The circulatory system of birds is highly adapted for carrying more oxygen:

(a) The heart is relatively large and completely divided into four chambers.

(b) Efficient double circulation of blood.

(c) The blood contains high proportion of haemoglobin.

13. Digestive system

Since metabolic rate is high birds require large amount of food and rapid digestion. Digestive system is compact but effective. Food eaten by birds is generally of high caloric value, easily digested and with minimum undigested waste. The rectum is short because of small amount of waste material which is voided at once.

14. Excretory system

The excretory system of birds is highly modified to reduce weight.

(a) Urinary bladder is absent.

(b) The excretory liquid is reabsorbed in the kidney tubules and cloaca such that excretory waste is in the form of semisolid **uric acid** and **urate particles (uricotelic excretion)** which are voided at once.

15. Vision

The birds depend on their sight rather than smell in contrast to reptiles and most mammals. The eyes are large and their ability to accommodate rapidly is also well developed. This enables them to see objects from very high altitudes. A comb-like structure called **pecten** is present in the eye of birds. It probably helps in accommodation.

16. Endoskeleton

(a) The **heterocoelous** cervical vertebrae offer great flexibility to the neck.

(b) In the vertebral column about fourteen vertebrae are fused together to form a plate-like structure called **synsacrum.** The vertebrae are **one thoracic** (last), **six lumbar, two sacral** and **five caudal.** It acts as a **girder** to support the entire weight of the body and also provides a firm **fulcrum** for the action of wings.

(c) The shortening of caudal vertebrae and formation of **pygostyle** (fusion of last four caudals) provides stability in the air.

(d) The fusion of pelvis with the synsacrum provides firm attachment to the legs, supports body weight during locomotion on the ground and counteracts the effect of shocks when the bird alights.

(e) The sternum bears a mid-ventral ridge called **keel** or **carina** to which the flight muscles are firmly attached.

(f) The clavicles of the pectoral girdle are fused ventrally to form a V-shaped **furcula** or **merry thought bone** or **wish bone**. It works like a spring between the two wings.

(g) The mid-ventral symphysis between the pubes and ischia is absent. This results in a more posterior displacement of the viscera, shifting the centre of gravity towards the legs.

17. Perching

The hindlimbs and their muscles are well developed and help in perching. Thus a resting or sleeping bird is automatically clamped to the perch and does not fall down.

MODES OF FLIGHT

The birds use many methods to fly. There are four modes of flight. The same bird may use different modes at different times. The tail in all cases is used as a rudder to support and balance the body.

Gliding: It is the most primitive mode of flight and is exhibited by some shore birds like duck, gulls and herons and other birds like pigeon and falcon. The bird keeps the wings stretched and glides for short distance without flapping them. Gliding movement can be exhibited for a short time only as it depends on the speed acquired by previous wing strokes, height from where the bird is descending or air currents.

Flapping: This is the most common mode of flight and is exhibited by majority of the birds. Each flapping movement includes an **effective** or **downstroke** and a **recovery** or **upstroke** of the wing. Initially the wings are held vertically upward and fully spread. In the effective stroke the wings move obliquely forward, downward and backward and their distal ends are turned upwards. Thus, there is both lift and thrust. In the recovery stroke the wings are partly folded and their primary feathers spread out to allow air to slip through, thus making it easier to lift them. They move up and backwards. As a result of flapping the bird is propelled forward and sustained in the air.

Hovering: It is a peculiar mode of flight and is a slight deviation from normal flapping. The body is kept vertical in the air and the tips of the wings describe a figure of '8'. The bird remains poised over a flower or some object on the ground. The most common example is humming bird.

Soaring or sailing: It is a highly specialized mode of flight characteristic of birds with a large wing span, such as, albatross, vultures, falcon, stork, etc. The bird describes great circles at a high level without any wing movement.

ORIGIN OF FLIGHT IN BIRDS

Birds are described as feathered reptiles. Reptiles were non-fliers but birds are fliers. So the major trend in the origin of birds from reptiles is the origin of flight. Sufficient transitional forms to explain the origin of flight are lacking. However, several theories have been postulated to explain the origin of flight.

1. Diving origin of flight

This theory was postulated by Newmann. According to this theory the aquatic fish-eating reptiles gave rise to birds, thus the pro-avis was aquatic. Flight started in the form of soaring over water before diving for fishes. If this theory is accepted then all land birds originated from some aquatic form. But the primitive bird *Archaeopteryx* was a land bird. So this theory is not accepted.

2. Cursorial origin of flight

The theory was postulated by Nopsca, a Hungarian palaeontologist. He is of the view that the first bird evolved from some long-tailed cursorial reptile. The hypothetical pro-avis ran over the earth on its strong hind legs, stretched out its forelimbs for maintaining balance and probably flapped them to aid the speed of running. These flapping forelimbs or **prowings** developed more surface, probably by the **scales** and these specialized flight scales soon evolved into **feathers.** Soon the body scales of the pro-avis became modified into feathers. According to this theory feathers are regarded as specialized scales.

The first step in flying would be prolonged leaps aided by the flapping prowings, then short soaring flights followed by longer flights by flapping of the wings. Ultimately the forelimbs became organs of flight or wings rather than aid to rapid running.

3. Arboreal origin of flight

According to this theory the pro-avis was an arboreal animal. It glided from tree top to the ground or from one tree to another. Based on whether pro-avis had a pair of wings or four wings, there are two theories to explain the arboreal origin of flight:

Pair-wing theory: This theory was proposed by Steiner and is based on the study of *Archaeopteryx*. It is an arboreal bird, not a true flier but a glider. It glided from tree to tree or branch to branch with the help of its wings. The hindlimbs of ancestral bird were modified for springing and fore-limbs for gliding and balancing the body in the air.

Four-wing theory: This theory was proposed by Beebe. According to this theory the pro-avis had feathers on both forelimbs and hindlimbs. This is proved by the fact that vestigeal flight feathers occur on the thigh of modern birds. In the course of evolution the forelimbs enlarged and converted into wings which supported the animal in flight, the feathers of hindlimbs degenerated and they were used for perching and the tail shortened. Thus, flight evolved from climbing and gliding.

4. Compromise theory

This theory was proposed by Gregory and is a compromise between cursorial and arboreal theory. According to this theory the pro-avis had adaptations for both cursorial and arboreal life. It was a good runner, a good climber and also had perching ability. It could run by constantly flapping the forelimbs which in the course of evolution developed into wings. This is the most accepted theory.

MECHANISM OF FLIGHT

The wings are the means of flight in birds. To carry the weight of the body the wings strike vertically and to carry the bird forward they strike obliquely. The vigorous down-strokes help to raise the body in the air. The tail acts as a rudder for steering during flight and to suddenly stop flight.

A bird flies on the principle of indirect movement. It moves the air which by its displacement moves the bird. The wings of a bird acts as an aerofoil (sheet of air) and as a propeller for forward motion. The forward motion of the bird creates a flow of air over the wings and forces are developed in several opposing planes as a result of which the bird does not fall down but glides effortlessly. The wings are slightly convex above and flat or slightly concave below. When air flows over the wings its speed is faster above than on the lower surface. A force is developed due to differential speed. The high pressure area on the lower surface of the wing causes an upward thrust below the wing which tends to **lift** the bird and keep it aloft and moving forwards and upwards. The air also pushes the wing horizontally backwards and tends to **drag** or slow the bird down. The resultant of these two forces (lift and drag) is the **total aerodynamic force** which is equal and opposite to the weight of the bird. The lift force is perpendicular to the air stream and the drag force is parallel to the air stream. In active flight the total aerodynamic force results by flapping the wing.

The weight of the body of bird acting downwards is opposed by two forces working at right angles to each other, a sinking force and a driving force, pushing the body obliquely downwards. These two forces are opposed by two equal forces acting in opposite direction. The sinking force is opposed by the lift force and the driving force by the drag force.

BIRD MIGRATION

Migration of birds is one of the most spectacular event in the animal world. It has intrigued mankind for centuries and even now after more than 200 thousand years of observations climaxed by a century or more of experimental study many aspects of migration are not clearly understood.

Migration may be defined as seasonal movement of a number of birds as a habit to obtain the advantage of favourable conditions in other regions (Cahn). Migration is different from unidirectional movement and those in which animals are helplessly carried by some other agency. In fact, bird migration is a two way journey between their summer and winter homes. All birds do not migrate. Those bird which remain in one country throughout the year are called **resident birds** and those which migrate are called **migratory birds.**

Kinds of migration

Birds migrate in a variety of ways. Following kinds of migration has been observed in birds:

1. Latitudinal migration: This is the most common type of bird migration. The bird moves from north to south and vice versa. Most birds move into the wide land masses of the north temperate and subarctic region where there are facilities for feeding and nesting during warmer months and then retire to South during winter when the north is covered with ice and snow. Many birds of North America and Eurasia cross the equator and spend winter in warmer parts of South America and Africa. Some Siberian birds spend their winter in the plains of Himalayas in India. The American golden plover passes the nine months of winter, 8000 miles south in the plains of Argentina thus enjoying two summers in a year. In the Southern hemisphere, an opposite but lesser migration occurs when the seasons change.

2. Altitudinal or vertical migration: Some birds migrate up and down the mountain slopes as whether changes. In India many species of birds migrate from the lowlands to the slopes of Himalayas several thousand feet above sea level during summer and return to valleys as winter begins. Vertical migration also occurs in grebes and coots of Andes in Argentina and violet green swallows of Great Britain.

3. Longitudinal migration: Some birds move from east to west and vice versa to avoid winters. For example, sterlings move from East Europe or Arid (breeding place) towards Atlantic coast (feeding place).

4. Partial migration: Many species of birds of temperate region show partial migration. Some new individuals are added to the resident population of the same species for a short period. For example, barn owls, blue birds and blue jays of Canada and northern United States move southwards and mix with the resident population of that region.

5. Irregular or erratic or vagrant migration: Some birds such as heron, thrushes and warblers, after breeding move away from their home with the young ones and disperse in all directions over a few hundred miles in search of food and safety from enemies.

6. Seasonal migration: Some birds migrate regularly at definite seasons year after year. This is called seasonal migration. For example, swifts, swallows, cuckoos and nightingales migrate from south to north during summer. These birds are called **summer visitors.** Certain birds like fieldfare, snow bunting and redwig migrate from north to south during winter. They are called **winter visitors.** Some birds like snipes and sandpipers are called **birds of passage** as they are seen for a short time twice in a year on their way to north or south.

7. Daily migration: Some birds fly from their nest early in the morning and return to the same nest in the evening or vice versa. Thus daily migration is of two types:

Diurnal migration: Some birds like crows, swallows, robins, hawks, blue birds, jays, cranes, ducks, geese, swan, pelicans, etc. fly during the day. They are called **diurnal migrants.** They usually travel in flocks which may be well organized (ducks, geese, swans) or loose (swallows).

Nocturnal migration: Some small sized birds like sparrows, warblers, thrashes, etc. are **nocturnal migrants** because they fly at night and rest during the day. They thus escape their enemies.

Purpose of migration

Birds migrate to obtain better surroundings for food, nesting and breeding and for getting favourable climatic conditions.

Causes of migration

Bird migration is an **inborn custom** or an **instinct.** There are several factors which act as stimulus in initiating migratory instinct. These are scarcity of food, shortening of day length, fall in temperature and ripening of gonads.

Speed of migration

The speed or velocity of flight in migration varies from species to species. The birds usually fly with tremendous speed ranging from 30 miles per hour in case of small birds like golden plover to 200 miles per hour for large birds like swift, etc. They can travel non-stop at this speed in a day or a night, usually they travel for 5-6 hours and then stop for food or drink. The body fat serves as an efficient fuel during flight. The longest non-stop flight has been recorded for golden plover which covers a distance of 2,400 miles from Hudson Bay and Alaska to South America.

Altitude of flight

Some birds migrate close to the earth while others migrate up to 3,000–5,000 feet but rarely higher. They generally prefer to fly below the level of clouds. Some species have been reported to cross the Andes and the Himalayas at 20,000 feet or more.

Range of migration

Birds travel thousands of miles to reach their destination. The distance covered depends on species and environmental conditions. The bird which covers the longest distance is the arctic tern which travels 11,000 miles from ice free coasts of Labrador inside the Arctic circle to Antartica in winters and returns again through the same route in summers. White storks cover a distance of 8,000 miles and golden plover 2,400 miles. Himalayan snow partridges come down only a few feet while the chicadas may descend nearly 8,000 feet.

Regularity of migration

Bird show striking regularity and accuracy in migration. They are very punctual in their time of arrival. The purple martins are known to return on the same day every season.

Segregation during migration

Some birds such as kingfishers, swifts and night hawks travel in separate companies but some birds like turkeys, blue birds, swallows, vultures travel in mixed companies due to similarity in their size and food habit. In some species the males and females travel separately. Males arrive first to build nests and the young ones accompany their mothers. The birds usually follow a strict order during flight. Normally the adults travel first followed by the young ones. The ripening of gonads usually stimulate the adults to migrate first. In the return journey the order is reversed—the young birds start first and follow the same path which their adults had followed in the onward journey.

Routes of migration

Another remarkable and astonishing feature of migration is that the birds come and go by the same route year after year. For example, when the swallows return from their winter abode in South Africa to Europe they visit the same village and also the same tree or building for nesting.

Bird navigation

How do birds navigate or find their way has been the greatest mystery among the ornithologists. There are a few explanations or theories.

1. **Visual landmarks:** Some migratory birds find their way by recognizing certain **landmarks** or **topographical features** such as rivers, river valleys, coastal lines, oceans, seas, oceanic islands, deserts, mountain ranges, etc. These landmarks enable them to recognize and remember the routes and entrances to the country. The birds migrating at night, however, cannot easily make use of these landmarks.

2. **Magnetic fields:** Some naturalists like Van Middendroff and H.L. Yeaglev are of the view that birds are guided by the earth's magnetic field. Their inner ear reacts with the mechanical **coriolis effect** produced by the rotation of the earth. But there is no evidence to support it.

3. **Celestial bodies:** Saüer (1957) demonstrated that stars serve as a guiding force for nocturnal birds. This is true for warblers. Gustava Kramer (1949) and Mathews emphasized that the diurnal migratory birds are guided by the position of the sun. Experimenting with sterlings in cages, they showed that birds possess an **internal time clock** or **biological clock** with the help of which they make necessary adjustments in their course according to the position of the sun.

4. **Experience:** It is believed that birds learn to find their way by experience. This is possible when old members become leaders and guide the younger generation. However, most birds do not learn from their elders as they do not fly in flocks and in many cases, young birds make their first journey independently. Individual memory may play some part in migration.

5. **Air currents:** Some ornithologists are of the view that soaring birds like vultures migrate with the help of air currents.

6. **Homing instinct:** Some ornithologists believe that homing instinct plays a major role in navigation. But experiments with carrier pigeons have proved that vision helps in navigation.

In short, how birds find their way is still unknown.

Origin of migration

Origin of migration has always remained a problem for biologists. There are two theories to explain the origin of migration. They are:

1. Northern ancestral home theory: According to this theory all the birds lived in the northern hemisphere when these regions were warmer than they are now. When the season changed and these regions became colder the birds were driven south wards to find food and escape cold weather. When the climate was normal in the north, they returned to their original homelands. By constant repetition of this behaviour over many generations, birds have developed the habit of migrating.

2. Southern ancestral home theory: According to this theory the southern hemisphere was the original home of the birds. Owing to over population they were driven to the northern hemisphere. They returned to south for nesting and breeding. As this process was repeated for several years, the birds developed the habit of migrating.

The most convincing view that explains the process of migration is evolution. In the past the tropical birds spread out in the cold northern regions where food was abundant but when winter came they were forced to move to the south. This later became an inborn custom.

Advantages of migration

1. The birds avoid harsh environmental condition by migrating in winter.
2. They get long day light hours for searching food when they migrate in winter and thus avoid scarcity of food due to snowfall and freezing.
3. They get suitable and uncongested dwelling places.
4. In summer when they return to the north they get suitable nesting grounds and abundant food.

PALATE IN BIRDS

Palate is an assemblage of bones situated in the roof of buccal cavity of vertebrates. It separates the oral cavity from nasal passage. It helps the nasal passage to open far behind into the buccal cavity. The palate helps the animal to chew and breathe at the same time. Palate first appeared in some reptiles and is well developed in birds and mammals.

There are four types of palate in birds.

1. Dromaeognathous or palaeognathous
2. Schizognathous
3. Desmognathous
4. Aegithognathous.

Dromaeognathous or Palaeognathous

Dromaeognathous palate (Fig. 22.6 A) is found in flightless birds or Ratitae:

1. Vomer is large and broad and is connected posteriorly with the palatines.
2. Palatines do not articulate with the parasphenoidal rostrum because the vomer intervenes between the two.
3. Basipterygoid processes of the basisphenoid are large and articulate with the posterior part of pterygoid.
4. Maxillo-palatine processes are short and do not unite with one another or with the vomer.
5. Pterygoid is unmovably fixed to vomer.

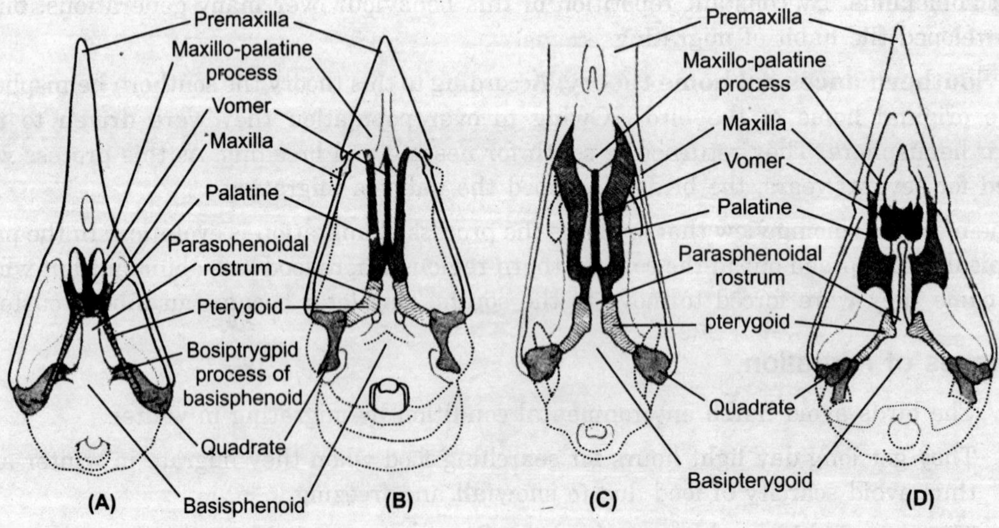

Fig. 22.6 Types of palate in birds. **(A)** *Dromaeognathous;* **(B)** *Schizognathous*; **(C)** *Desmognathous*; **(D)** *Aegithognathous*

Schizognathous

Schizognathous palate (Fig. 22.6B) is found in birds like pigeon, fowl, gull, crane, woodpecker, penguin, etc.

1. Vomer is small and pointed or absent.
2. Palatines and pterygoids articulate with the parasphenoidal rostrum where they meet each other.
3. Maxillo-palatine processes are small and do not unite with each other or with the vomer.
4. Basipterygoid processes are small or absent.
5. Pterygoids movably articulated.

Desmognathous

Desmognathous palate (Fig. 22.6C) is found in duck, goose, stork, parrot, heron, birds of prey, etc.

1. Vomer, when present, is slender and tapering anteriorly or it may be absent.
2. Palatines and pterygoids articulate with the parasphenoidal rostrum.
3. Maxillo-palatine processes are large and spongy and unite with each other in the middle to form a flat spongy palate.
4. Basipterygoid processes absent.

Aegithognathous

Aegithognathous palate (Fig. 22.6D) occurs in crow, swift, bulbul, etc. It is similar to schizognathous type of palate. Vomer is short and broad and truncated in front instead of being pointed. Porteriorly it is deeply cleft embracing the parasphenoidal rostrum.

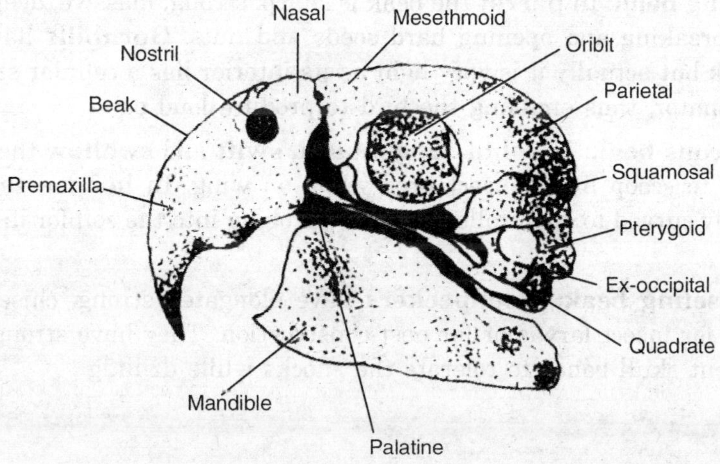

Fig. 22.7 Beak and skull articulation in parrot

In parrot specialized form of desmognathous palate called **sliding palate** (Fig. 22.7) is found. The nasals and the ascending process of the premaxilla are very thin and elastic where they join the skull, thus there is large amount of movement in the vertical plane. A cranio-facial joint or hinge occurs between the upper beak and the skull.

Importance of palate

The palate of birds plays an important role in the classification of birds. Modern birds are classified into two super-orders—**Palaeognathae** and **Neognathae.**

Palaeognathae includes all flightless birds (Ratitae). They have **palaeognathous palate** in which the vomer is large and the two palatines do not meet each other and with the rostrum. Neognathae includes all flying birds. They have **neognathous palate** which is divided into three subtypes—**schizognathous, desmognathous** and **aegithognathous.** In these types the vomer is small or absent and the palatines meet the rostrum.

TYPES OF BEAKS IN BIRDS

In birds teeth are absent. The bones of the upper and lower jaws become elongated to form a **beak** or **bill** covered by a horny sheath called **rhamphotheca.** The beak serves both as mouth and hand. According to the type of food and feeding habit beak of birds show many variations in shape, size and structure. A few common types of beaks in birds are being described here.

1. Seed-eating beak: Some small birds like **sparrow, finches** and **cardinals** have short, peg-like, conical beaks which are used to eat grains or seeds. In some seed eating birds the beak is weak and is used to pierce small seeds, while some have strong beak which is used to crush hard-shelled seeds and fruit stones.

2. Cutting beak: Crows have long and slender beak with cutting edges. It is used to cut plants.

3. Fruit-eating beak: In **parrot** the beak is sharp, strong, massive, deeply hooked and well adapted for breaking and opening hard seeds and nuts. **Hornbills** have a large and heavy-looking beak but actually it is very light as its interior has a cellular structure. These cells act as a resonator, thus enabling the bird to produce loud cry.

4. Insectivorous beak: In **robin, frog-mouth, swift** and **swallow** the beak is small, delicate and wide to scoop living insect prey while on wing. In **hoopoe** the beak is long slender and slightly curved to open rolled leaves or probing into the soil for insect larvae and pupae.

5. Wood-chiseling beak: Woodpeckers have elongated strong, chisel-like beak for drilling into wood for insect larvae or for nest construction. They have strong neck muscles and shock-absorbent skull bones to tolerate the shocks while drilling.

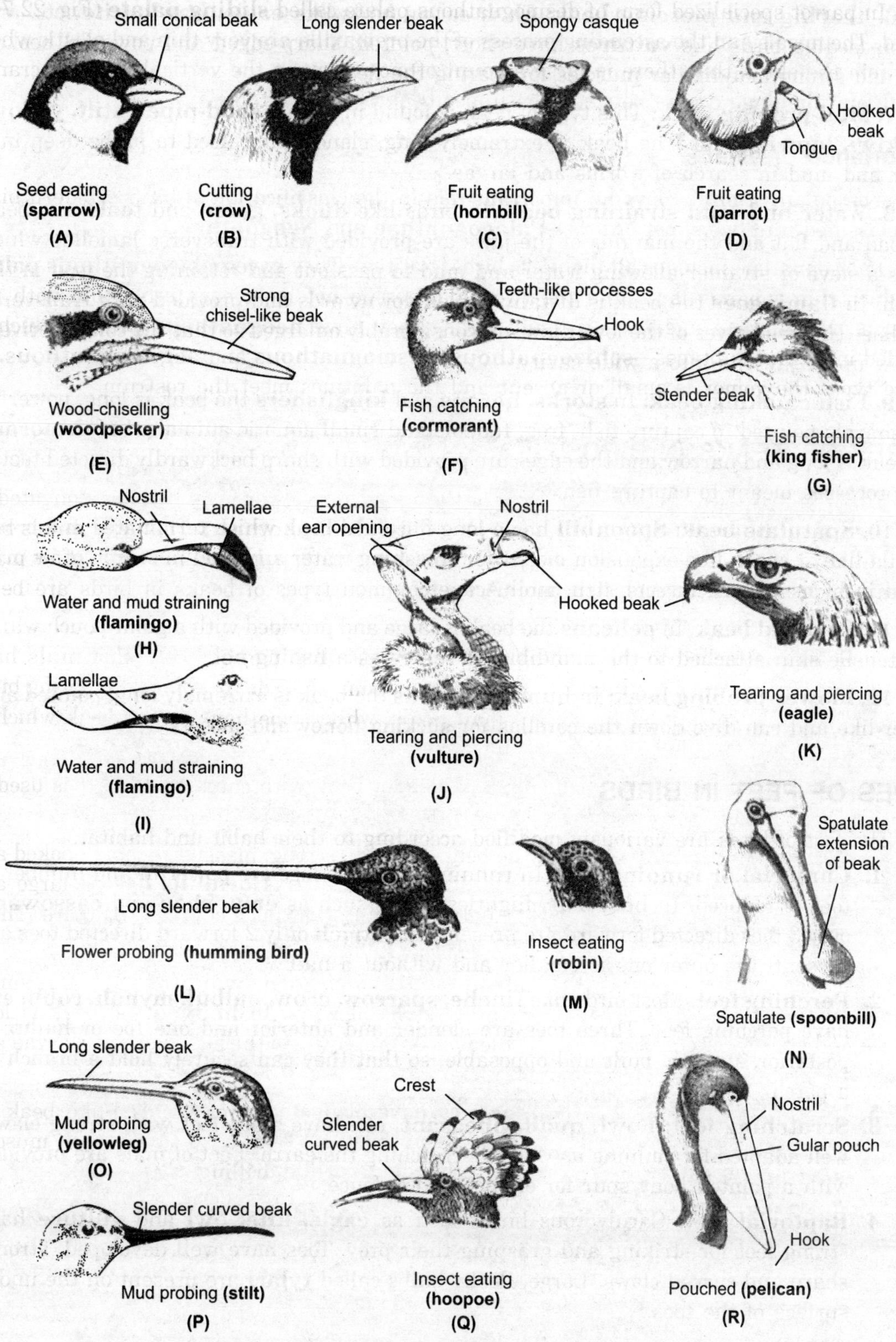

Fig. 22.8 Type of beaks in birds

6. Tearing and piercing beak: Flesh eating birds like **eagle, hawk, kite,** etc. and carrion feeding birds like **vultures** have short, pointed, sharp-edged and powerful hooked beak and strong mandibular muscles for tearing the flesh.

7. Mud-probing beak: This type of beak is found in **snipe, sand-piper, stilt, yellow-leg, kiwi, lapwing,** etc. The beak is extremely long, slender and used to probe deep into water and mud in search of worms and larvae.

8. Water and mud straining beak: In birds like **ducks, geese** and **teals,** the beak is broad and flat and the margins of the jaws are provided with transverse lamellae which act as a sieve or strainer allowing water and mud to pass out and retaining the food in the mouth. In **flamingoes** the beak is distally curved downwards and provided with transverse lamellae. The two halves of the lower jaw are considerably enlarged so that the comparatively narrow upper jaw fits into a wide cavity.

9. Fish-catching beak: In **storks, herons** and **kingfishers** the beak is long, powerful and sharply pointed to capture fish, frog, tadpole and small aquatic animals. In **cormorant** the beak is long and narrow and the edges are provided with sharp backwardly directed teeth-like processes meant to capture fish.

10. Spatulate beak: Spoonbill has a long flattened beak which terminates in a broad spatula-like or spoon-like expansion meant for splashing water and mud in search of its prey which includes worms, insects, fish, molluscs, etc.

11. Pouched beak: In **pelicans** the beak is large and provided with a gular pouch which is extensile skin attached to the mandible. It serves as a fishing net.

12. Flower-probing beak: In **humming birds** the beak is extremely long, pointed and rapier-like and can dive down the corollas for sucking honey and insects.

TYPES OF FEET IN BIRDS

The feet of birds are variously modified according to their habit and habitat.

1. **Cursorial or running feet:** In running birds the legs are powerful and number of toes is reduced. In **bustards,** flightless birds such as **emu, rhea** and **cassowary,** only 3 toes directed forward are present. In **ostrich** only 2 forward directed toes are present, the outer one is smaller and without a nail.
2. **Perching feet:** Most birds like **finche, sparrow, crow, bulbul, mynah, robin** etc. have perching feet. Three toes are slender and anterior and one toe or hallux is posterior, strongly built and opposable, so that they can securely hold a branch or a perch.
3. **Scratching feet: Fowl, quail, pheasant,** etc. have stout feet with strong claws, well adapted for running as well as scratching the earth. Feet of male are provided with a pointed bony spur for offence and defence.
4. **Raptorial feet:** Carnivorous birds such as **eagle, kite, owl** and **vulture** have strong feet for striking and grasping their prey. Toes have well developed, strong, sharp and curved claws. Large, fleshy bulbs called **tylari** are present on the under surface of the toes.

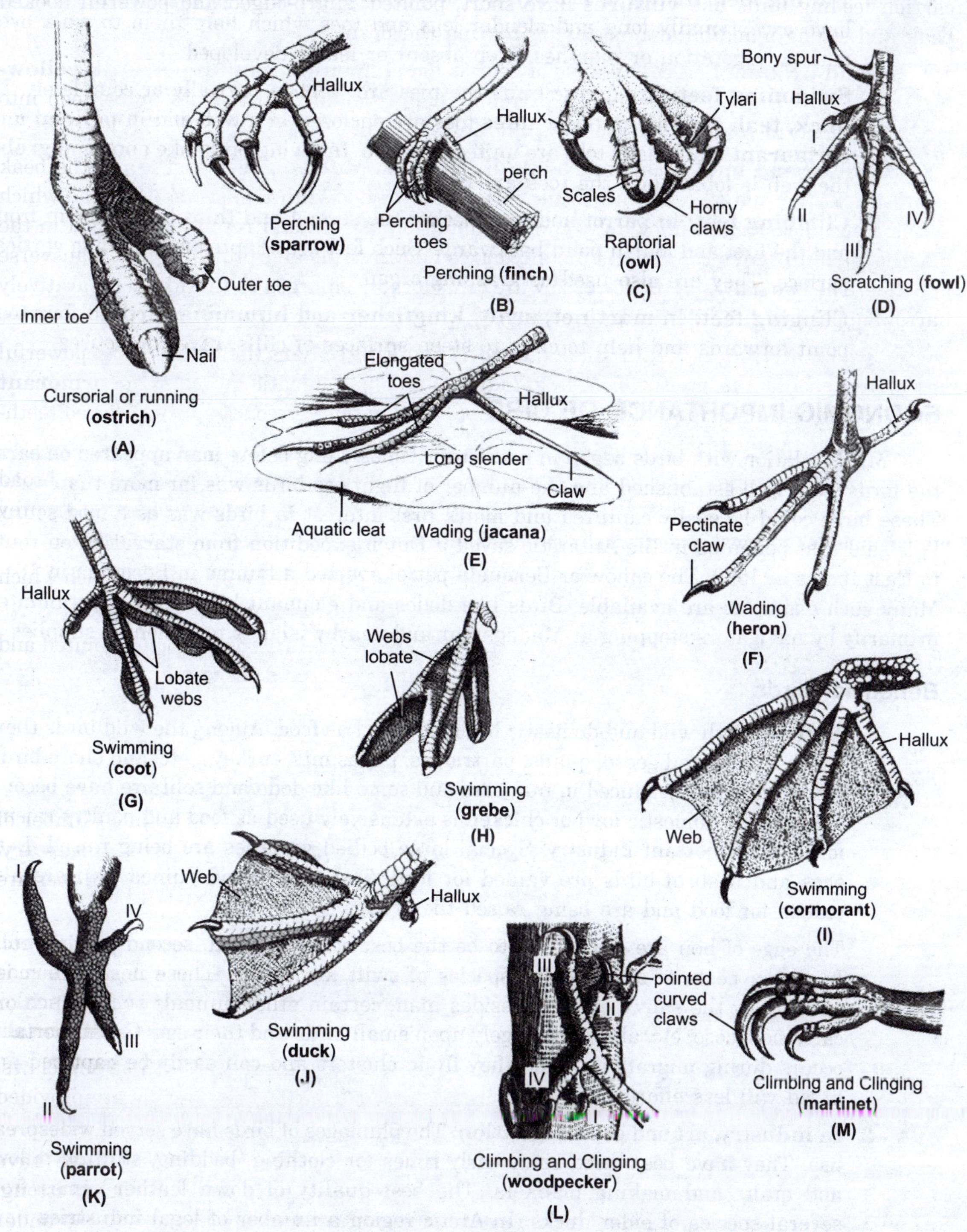

Fig. 22.9 Type of feet in birds

5. **Wading feet:** Wading or marshy bird such as **heron, jacana, snipe, lapwing,** etc. have exceptionally long and slender legs and toes which help them to walk over aquatic vegetation or marshes. Web absent or feebly developed.
6. **Swimming feet:** In aquatic birds the toes are webbed partially or completely. In **duck, teal,** etc. only anterior three toes are enclosed in a web and in **pelican** and **cormorant** all the four toes are united in a web. In diving birds like **coot** and **grebe** the web is lobate and the toes are free.
7. **Climbing feet:** In parrot and woodpecker the second and third toes point in front and the first and fourth point backwards. Such feet are adapted for climbing vertical surface. They are also used as grasping organ.
8. **Clinging feet:** In **martinet, swift, kingfisher** and **humming bird** all four toes point forwards and help to cling to steep surfaces of cliffs, caves or houses.

ECONOMIC IMPORTANCE OF BIRD

Man's relation with birds began in prehistoric times. Long before man appeared on earth the birds were well established and the number of flightless birds was far more than today. These birds could be easily captured and man's first interest in birds was as a food source. It is said that penguins in the Antarctic saved a Dutch expedition from starvation en route to East Indies in 1599. The cahow or Bermuda petrel averted a famine in Bermuda in 1614. Many such examples are available. Birds like dodos and elephant birds were exterminated primarily by navigators stopping at Madagascar and nearby islands to replenish supplies.

Beneficial birds

1. **As food:** Both wild and domestic birds are used as food. Among the wild birds there are the ducks and geese, quails, partridges, pheasants, turkeys, etc. and shore birds. Many birds have reduced in numbers and some like dodo and solitaire have become extinct. The domestic fowl or chicken is extensively used as food and poultry raising forms an important industry. Special large bodied varieties are being raised. Both eggs and flesh of birds are valued for food. The turkey and guinea fowl are also valued for food and are being raised these days.

   The eggs of hen are considered to be the best standard food, second only to milk. In China nests of a particular species of swift are edible. These nests are made entirely by the saliva of bird. Besides man, certain other animals such as snakes, cats, mongoose, etc. also prey largely upon small birds and their egg. Great mortality occurs during migrations when they fly in clusters and can easily be captured and killed will less effort.

2. **In industry, art** and **ornamentation:** The plumages of birds have served widespread use. They have been used since early times for clothing, bedding, stuffing pillows and quilts and making blankets. The best quality of down feather comes from several species of eider ducks. In Arctic region a number of legal industries have been set up to harvest down feathers of eider on a large scale. From time immemorial feathers have been used for arrow-making, headgear and ornamental wear. The Red

Indians of America use wing and tail feathers of golden eagles for ceremonial head dresses and feathers of flicker and valley quails for decoration. The feathers of wild birds were used to decorate women's hats. The crested cranes were once destroyed in large number because of their feathers demanded by fashion. Tail feathers of *Rhea* are used as dusters in Brazil and Argentina. Shuttlecocks are manufactured from feathers.

3. In agriculture:

   (a) As fertilizer: The faecal matter of birds called guano has been extensively used as fertilizer as it contains nitrogen, phosphorus, calcium and iron, etc. On islands off the coast of Chile, where little rain falls and which are breeding places for a large number of migratory birds and sea birds, the guano accumulates in enormous quantities and is exported to other countries. The guanay cormorant, *Phalacrocorax bougainvillii* has been called the most valuable bird of the world as it provides a very good type of guano.

   (b) As pollinators: Birds serve as very good agents of pollination. The humming birds living in meadows on flowering plants are good pollinators. Some fruit eating birds help in the dispersal of fruits and seeds.

   (c) In biological control: Birds are good friends of farmers as they help in controlling injurious crop pests. There are a wide variety of insect eating birds that keep the insect enemies of crops and fruits under control. Birds like finches and sparrows help to keep weeds in control. Sparrows generally eat all seeds of noxious weeds.

   Birds of prey such as hawks, eagles and owls kill rodents like field mice, hares, rabbits, ground squirrels, etc.

4. **As predators:** Peacocks, eagles, kites and other carnivorous birds destroy several poisonous creatures like scorpions and snakes, etc.

5. **As scavengers:** Carrion eating birds like vultures, eagles, hawks and crows are of great sanitary value as they feed upon dead bodies and decaying organic substances. They check the spread of diseases and thus make the environment healthy. Some birds such as ox-peckers, Rhinoceros bird, African starlings, and crocodile birds of India eat the ectoparasites of these animals and thus render a valuable service to them.

6. **In medicine:** Flesh and feathers of birds have been used in many ayurvedic and Unani medicines. Their bodies are warm and so in Unani system it is prescribed to keep them in contact with the chest in lung diseases. The egg of fowl is used in tonics and various medicines. Flesh of pigeon is said to be good for the patients of paralysis. The egg of fowl has great experimental value being used as a medium to grow cultures of viruses, bacteria, fungi, nematode larvae, etc.

7. **As messengers:** Pigeons with their remarkably developed intelligence, sense of direction and power of undertaking long non-stop flights at high altitudes have been used as carriers of messages.

8. **As signals:** Many birds indicate changes in seasons. In India appearance of cuckoo suggests the onset of spring. The peculiar voice of peacocks in cloudy weather indicates rainfall. Some birds help the hunters by giving danger signals, some sparrows cry before carnivorous animals such as tigers hidden in bushes. The typical path sparrows cry along the path of king cobra.

9. **For amusement:** Duck, quails, herons, etc. have been sought as game animals in Europe for centuries. Many birds have been domesticated for their ability as singers. For centuries man has kept parrots, mynahs, parakeets and several other birds as pets in cages and aviaries for their plumage and their ability to repeat human words. Certain birds like parrots, bulbul, etc. were trained to perform strange feats in public and people earn money by them.

   The birds are of great value as sports. There are several varieties of game birds such as grouse, flocks of wild pigeon and wild turkey that have attracted mankind in all parts of the world. Water fowl is another bird of great interest to bird hunters. Hunting ducks on wing or on water is a pleasure to accomplish and requires good knowledge and experience.

10. **Aesthetic value:** Birds are of great value to man not only because of their bright colours and interesting ways but because of their cheerful songs. Bird study with field glasses and cameras has provided a never ending healthful outdoor recreation for millions of bird watchers all over the world. Their melodious songs provide a temporary relief to the stressed people of our modern civilization.

11. **Bird oil:** Birds are sometimes used for oil, though this is hardly practical for commercial purposes. In northern South America in caves along the sea coast the young of oil bird (*Steatornis caripensis*) are fed on rich fruit of palm trees till they become helpless globe of fat. They are collected by natives and fat is melted in earthen pots and used as butter. The oily bodies of small petrels are lighted to serve as torches. Oil extracted from penguins is used for candles in Antarctica.

Injurious birds

Many birds are known to be definite enemies of man.

1. **Menace to agriculture:** The insect eating birds may destroy beneficial insects along with injurious ones. Some birds eat newly planted seeds, young plants or mature seeds and fruits and thus damage crop. It sometimes becomes necessary to control such damages. In China sparrows have been exterminated as a result of countrywide campaign against them.

2. **Destroyer of game birds and other animals:** Many birds like hawks, owls, etc. attack and kill game birds, song, birds and chick in poultry farms. Herons, which feed on fishes cause great damage to fish industry.

3. **Pests of fruits and stored grains:** Fruit eating birds like parrot damage fruits in orchards and gardens. They eat ripe fruits and throw unripe fruits on the ground. Some birds damage garden vegetables. Sparrows are a serious pest of stored grain in granaries.

4. **Spread of diseases:** Many birds serve as secondary host of disease causing germs and help in spreading the disease. Many birds help in the distribution of harmful larvae which, otherwise could not reach due to barriers and slow speed to suitable hosts, plants as well as animals.

5. **Pests of honeybees:** Little birds called honey guides, found in Africa, feed on grubs of bees and wasps, but they are not strong enough to attack a nest. They do this job in partnership with badger. The badger is fond of honey but its legs are too short for making long treks in search of nest of bees. When the bird finds a hive it guides the badger to it. The badger in its sting-proof hide of coarse hair breaks the hive and eats honey while the birds feeds on grubs. Sometimes man also joins them to get honey.

6. **As hazards to aeroplane:** Birds are a great nuissance to aeroplanes. They are sometimes sucked into the rear part of the plane and cause accidents. Efforts are being made to keep airports and adjoining areas free of bird population.

23
CHAPTER

CLASS MAMMALIA

TYPE ORYCTOLAGUS

The type for study of the class **Mammalia** is the **rabbit,** *Oryctolagus cuniculus.* It is dissected in the laboratory in most Universities because it is a harmless animal, easily available and easy to handle and can be easily reared in laboratory. In the United States and Canada the name rabbit is given to any hare. However, hare should be restricted to the European *Lepus cuniculus* which differs from its relatives in many characters. The common species of rabbit is *Oryctolagus cuniculus* which has a worldwide distribution. Both rabbit and hare belong to the order *Lagomorpha* (**Gr.,** *lagos,* hare). The North Indian species of hare is *Lepus ruficaudatus* and the South Indian species is *Lepus nigricollis*. The major differences between the two animals are:

Rabbit	Hare
1. Usually smaller, 30–45 cm long.	Larger, 50–70 cm long.
2. Lives in burrows with branching tunnels and several exits.	Lives in temporary excavations on ground or grass or bush.
3. Live in groups (gregarious), crepuscular, *i.e.*, feed in the morning and evening.	Solitary, nomadic and nocturnal.
4. Fast runner.	Very fast runner.
5. Can be domesticated.	Wild, cannot be domesticated.
6. Forelimbs shorter than hindlimbs. Claws strong for digging.	Both are equal. Claws weak.
7. External ears shorter.	External ears longer, with black tip.
8. Young ones blind, naked and weak.	Youngs ones covered with hairs, eyes open, and strong.

DISTRIBUTION

The European rabbit *O. cuniculus* has a cosmopolitan distribution. Its original home is believed to have been Western Europe and Africa, but now it is found everywhere in the world.

Habitat

Rabbits are widespread and highly adaptable. They occur in open fields, grasslands and woodlands.

Habit

The rabbits are fossorial in habit. They dig out extensive burrows called **warrens** in soft soil with the help of their forelimbs which are specialized for this work. The burrows are spacious, extensively branched and with many exits. The entire family can live inside the burrow. They generally do not move far away from the burrow and take hasty refuge when alarmed.

Rabbits are entirely terrestrial and gregarious animal. They move in groups. They are crepuscular, their favourite feeding time being early morning and evening. They are exclusively herbivorous, sometimes, they become coprophagous, *i.e.*, they eat their soft night droppings without mastication, thus the same food passes through the gut twice, to provide maximum nourishment.

Rabbits move by walking, running and leaping. They are fast runners, average speed being 30–40 km per hour. They sometimes do great damage to plantation by nibbling the bark of trees. They are excessively destructive when they obtain access to gardens.

Rabbits are docile animal. They have been domesticated for a long period and a large number of varieties are known. Rabbits are **polygamous,** *i.e.,* one male lives with many females. The length of life of a rabbit is seven to eight years and it breeds four times a year. Each litter many contain five to eight young ones which become sexually mature in four to six months. Parental care is well developed.

The rabbits are defenceless animals and a large number of them are destroyed by their enemies. The gregarious habit, to some extent, is defensive. They thump the hindlegs to give a warning of approaching danger. Natural enemies are birds of prey, large snakes, weasels, polecats, badgers, etc.

EXTERNAL FEATURES

Shape, size and colour

The rabbit is a small, bilaterally symmetrical animal of the size of cat. It is about 16 inches long and weights 2–4 pounds. The whole body is covered with a soft and uniform hairy coat or fur which conserves body heat and maintains constant body temperature of 38.8°C. The colour is naturally reddish brown, the under surface and the lower part of the tail is white. The domesticated rabbits show a wide variety of colours.

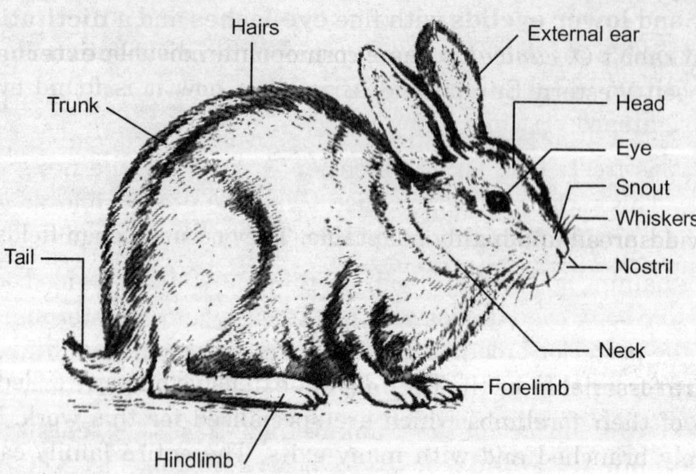

Fig. 23.1 *Rabbit.* External features

Form of body

The rabbits show the typical mammalian form of the body which is divided into head, neck, trunk and tail (Fig. 23.1).

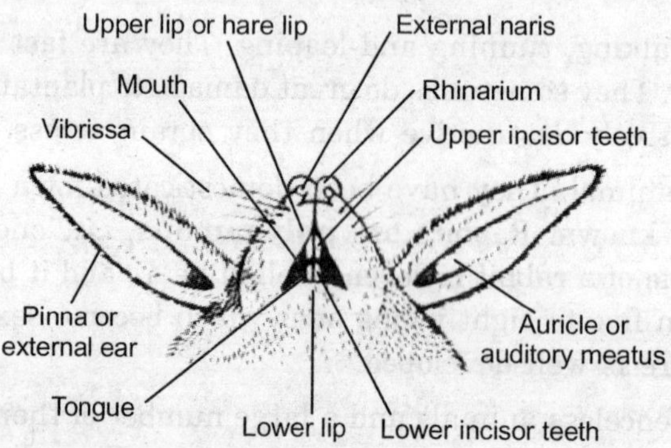

Fig. 23.2 *Rabbit:* Ventral view of head showing hare lip

Head: The anterior end of the body is the head. It is large and pear shaped and the anterior end is produced into a blunt snout or **muzzle.** It bears a terminal mouth, external nares, vibrissae, eyes and external ears.

The **mouth** is a small slit-like opening bounded by soft, fleshy and movable **upper** and **lower lips.** The upper lip is called **hare-lip** (Fig. 23.2). It is divided by a vertical median cleft in front exposing the large upper incisor teeth which remain outside the mouth cavity even when the mouth is closed. From the side of the upper lip project the sensitive tactile **vibrissae** or **whiskers.** The **external nares** or **nostrils** are a pair of oval slits present at the tip of the snout immediately above the mouth. Each is surrounded by moist skin called

rhinarium. The **eyes** are large oval structures present at the sides of the head. Each eye has movable **upper** and **lower eyelids** with fine **eye lashes** and a **nictitating membrane.** The outer ear in mammals consists of a large, prominent, movable **external ear** or **pinna** and an **external auditory meatus.** The pinnae are held erect when the animal is alert and is laid back when frightened or while running.

Neck: The head is freely held on an extension of the body, the neck, present between the head and trunk. The neck is short but distinct and quite flexible. It gives freedom of movement to the head and enables the animal to look in all directions.

Trunk: The remaining part of the body is the trunk. It is divided into two distinct regions: (*i*) the anterior bony cage having ribs at the sides and the sternum ventrally is the chest or thorax. (*ii*) the posterior broad, soft-bellied part is the abdomen. Internally the cavity of thorax is separated from that of the abdomen by a muscular partition, the **diaphragm.**

In females the ventral surface of the trunk bears 4–5 pairs of **teats** or **nipples,** which are openings of the **mammary glands.** Teats are poorly developed or rudimentary in males. The ventral wall of abdomen is soft and distensible. There is no **cloaca.** A small rounded aperture, the **anus** is present at the posterior end of the trunk just beneath the base of the tail. It is the outer opening of the rectum. At the sides of the anus are a pair of hairless depressions of the skin called **perineal pouches** into which open the ducts of **perineal scent glands,** the secretion of which has a strong smell characteristic of rabbits. In front of the anus lies the **urinogenital aperture** which in male opens at the tip of small cylindrical muscular copulatory organ or **penis.** In female, the urinogenital aperture or **vulva** lies at the base of the **clitoris.** In male the testis are enclosed in small bag-like **scrotal sacs** which lie on either sides of the penis.

Tail: The tail is small and bushy. The ventral surface of the tail is white and is used to give a warning signal when in danger. It serves as a balancing organ during movement.

Limbs: The trunk bears two pairs of **pentadactyl limbs** or **appendages.** The anterior pair or the **forelimbs** are short and strong and adapted to the burrowing habit. Each forelimb consists of **upper arm** or **brachium,** middle **forearm** or **antebrachium** and distal **hand** or **manus.** Each hand includes **wrist** or **carpus, palm** or **metacarpus** and five **fingers** ending in sharp horny **claws.** They are used as levers to take the shock of alighting at the end of a leap.

The posterior pair or the **hindlimbs** are much longer enabling to progress rapidly by short leaps. Each hindlimb consists of proximal **thigh** or **femur,** middle **shank** or **crus** and distal **foot** or **pes.** The foot incudes **ankle** or **tarsus, sole** or **metatarsus** and four clawed **toes.** The **first toe** or **hallux** is absent. When at rest the thigh is directed forwards, the shank backwards and the ankle and foot forwards. Thus, by straightening the limb a powerful forward thrust results.

In the primitive tetrapods the limbs extend laterally, while in mammals they project ventrally and being elongated they carry the body high above the ground. This change has involved a rotation of 90° in each limb. The hindlimb is rotated forward so that the original dorsal surface is anterior and the toes point forward. The knee or joint between the thigh and shank is likewise directed anteriorly. On the other hand, the forelimb has rotated posteriorly so that original dorsal surface faces posteriorly and the pre-axial surface faces laterally. The result of this is that the elbow or joint between the upper and fore arm is directed posteriorly. As a result of rotation of the upper arm the fingers would point posteriorly but by an additional torsion they have been brought around to the front again. This position involves a crossing of the two bones of the forearm, the distal end of the pre-axial bone being brought internal to the distal end of the post-axial bone. This position of the forearm is known as the **prone position** and is imitated in the human arm when the arm hangs by the side with the back of the hand directed forward. If now the arm is raised sidewise, shoulder high with the palm facing forward, the two bones return to the primitive parallel position, the **supine position.** Unlike man in most other mammals the forearm is fixed in prone position.

The position of parts of the foot in walking is different in different mammals. The rabbit walks on the digits with the remainder of the hand and foot elevated. This mode of walking is called **digitigrade.** Man walks on the whole sole of the foot this is the primitive method and is called **plantigrade.** Horses, cattle and other ungulates walk on their nails which becomed broadened into hoofs, this is called **unguligrade.**

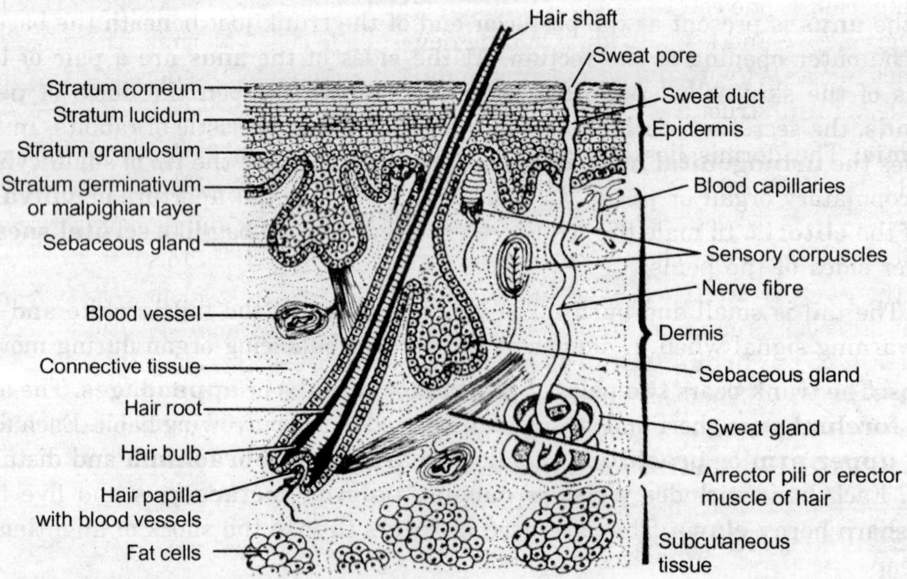

Fig. 23.3 *Rabbit:* Vertical section of skin

SKIN

The body of all vertebrates is invested by an outer covering called **integument** or **skin.** The skin of mammals is thick. It consists of two distinct layers (*i*) Outer **epidermis** (ectodermal in origin) and (*ii*) inner **dermis** (mesodermal in origin). Below the dermis lies the **musculature** with which it is connected by **areolar connective tissue.** Each layer has a composite structure (Fig. 23.3). The skin is not used as a respiratory surface as such the necessity of mucous glands no longer arises.

Epidermis: The epidermis is the upper, thick layer formed of stratified epithelium arranged in several layers:

1. **Stratum corneum:** This is the outermost layer of hard, scale-like, dead, fully keratinized and flattened cells. The keratin present in this layer prevents the passage of water and solutes. This layer is constantly worn out and cast off periodically. New cells are added from below. The cornified layer is especially thick on palm and soles.
2. **Stratum lucidum:** The stratum corneum is followed by a thin layer of transparent, hyaline, shiny and refractive cells, the stratum lucidum. It is conspicuous in the palm and sole. A chemical substance called **eleiden** is present in this layer.
3. **Stratum granulosum** or **granular layer:** This layer lies below stratum lucidum. It is a thick layer made up of granular cells containing keratohyaline granules.
4. **Stratum spinosum** or **transitional layer:** It is made up of several layers of polyhedral cells. The upper layer of cells become flattened and keratinized.
5. **Stratum germinativum** or **stratum Malpighi** or **germinal layer:** It is the innermost, one cell thick layer made up of columnar cells. The cells of this layer are very active and keep dividing adding new cells to the epidermis. The germinal layer lies in contact with the basement membrane secreted by the dermis. The glands and all keratin structures are derived from this layer.

Dermis: The dermis lies beneath the epidermis. It is made up of dense connective tissue and contains blood vessels, lymph vessels, nerve fibres, muscle fibres, elastic fibres and pigment cells. The blood vessels play an important role in nutrition and temperature regulation. The elastic fibres give elasticity to the skin and brings the skin back to shape after distortion or when the skin is stretched. The nerve endings penetrate the epidermis. Nerves and microscopic sensory organs are numerous in the dermis. They give sensitivity to the skin. The pigment cells or **chromatophores** or **melanophores** are branched with long processes. They contain melanin. They are differentiated from melanoblasts which arise from the neural crest cords of the embryo. The deeper parts of the dermis is formed of fatty or adipose tissue. This fatty layer is called **stratum adiposus.** It stores fat which serves as reserve food.

SKIN DERIVATIVES

Hair

The characteristic feature of mammalian skin is the presence of **hairs** which form an insulating layer by entangling air between them in order to maintain the body temperature. Hairs are elongated, thread-like, cylindrical outgrowths of the epidermis. Each hair consists of a **root** buried in the skin and a **shaft** projecting from it (Fig. 23.4A). The root is lodged in a pit-like invagination of the epidermis called **hair follicle.** It is swollen at the base forming a **bulb,** below which lies a concentration of vascular dermal tissue, the **dermal papilla** which feeds the hair. The root is surrounded by epithelial and connective tissue sheaths, the **root sheaths** which can be distinguished into an **external root sheath** and an **internal root sheath.** The shaft is made up of three layers: outer **cuticle,** middle **cortex** and inner **medulla** (Fig. 23.4 B). The cuticle is a cornified layer made up of overlapping cells.

The cells of the cortex are elongated and form greater part of the hair. The pigments present in the intercellular spaces of the cortex give colour to the hair. The medulla is made up of cuboidal cells arranged in double row. It forms the central core of the hair and contains air spaces in stouter hairs. The medulla may be absent in finer hairs.

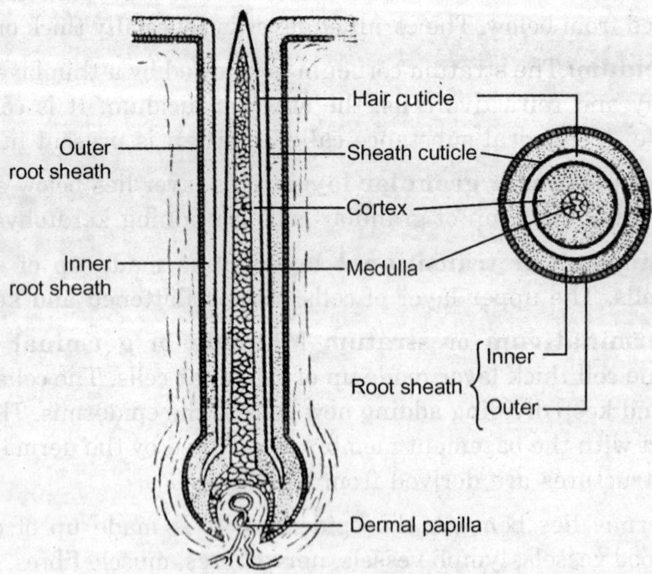

Fig. 23.4 *Rabbit*– Structure of hair. **(A)** Longitudinal section; **(B)** Transverse section

Only the base of the hair is living and the part above the skin surface is a dead structure. A sebaceous gland opens into the hair follicle. It produces an oily secretion which prevents the hair from becoming brittle. A small smooth muscle called **erector pili** is associated with the hair follicle, it keeps the hair erect.

Development of hair: The cells of the germinative layer of the epidermis starts multiplying at any one place, where a hair has to appear, forming a concentration of epidermal cells. This concentration increases in size due to rapid growth, sinks into the corneum and forms the **hair germ.** The tip of the hair germ enlarges forming a **bulb** at the base of which some dermal cells accumulate. These dermal cells form the **dermal papilla** that provides nourishment to the developing hair. Now the cells of the germinative layer just in contact with the dermal papilla produce, by rapid multiplication, a conical **shaft** of the hair. The cone grows for sometime and then separates from the surrounding cells as they split off. The basal portion of the shaft thus separated is the **root** of the hair. As development proceeds, the cells of the root become horny. The lining of the follicle is also somewhat cornified and is known as the **outer root sheath.** This clings to the hair as a white coat when the hair is pulled out. The sheath later on differentiates further forming a thin **inner root sheath.** The shaft grows more and more and finally emerges out after piercing through the epidermis. The hairs emerge at an acute angle having a slant in some special direction. This is responsible for the formation of **streams** of hairs on various parts of the body. When such streams meet either

rhomboids or **vortices** are produced. The vortices are usually called **cowlicks.** Such rhomboids and vortices also appear on human body and are evident in regions where hair is short.

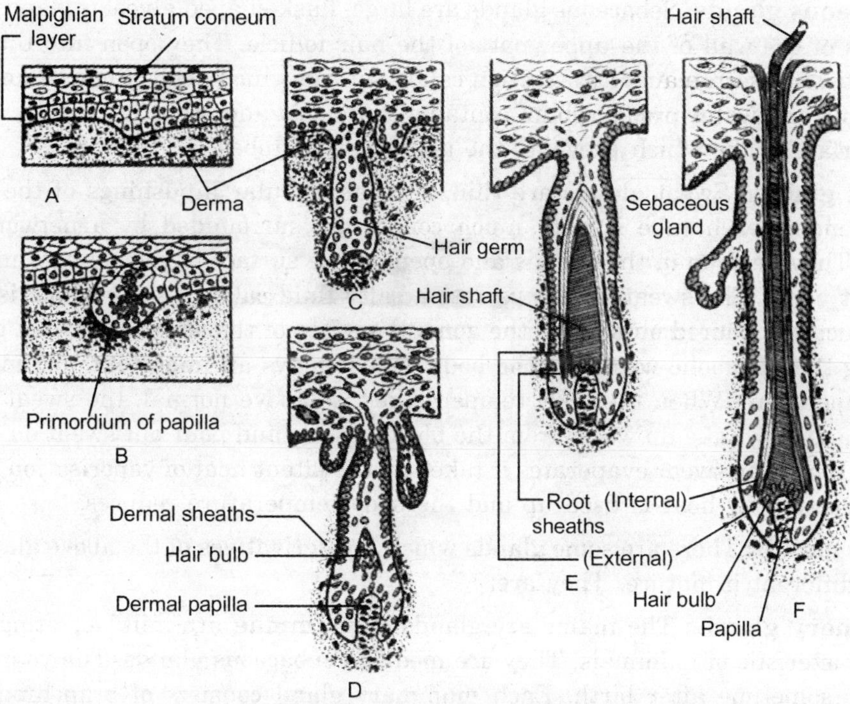

Fig. 23.5 *Rabbit*– Development of hair

Functions of hairs

In mammals hairs perform a variety of functions.

1. It forms an insulating layer by trapping air which prevents loss of heat from the body.
2. It gives colour to the animal.
3. In furred mammals it serves as a blanket in cold weather.
4. Vibrissae have nerve endings so they serve as sensory structures.
5. Eyelashes protect the eyes from dust.
6. The hairs in the nose trap dust particles.
7. The hairs at the tip of the tail in some mammals like horses and cows help in driving away flies and other harmful insects.

Epidermal glands

There are two kinds of epidermal glands in mammals. They are: (*i*) **Sebaceous glands** and (*ii*) **Sweat** or **sudorific glands.**

Sebaceous glands: Sebaceous glands are large, flask-shaped alveolar glands formed as outpushings of the wall of the upper part of the hair follicle. They open into the follicle by a small duct. They secrete an oily secretion called **sebum** which keeps the hair and skin soft, greasy, supple and water proof. Sebum contains lactic acid and other fatty acids so it makes the skin surface acidic which prevents the growth of microbes.

Sweat glands: Sweat glands are thin, sinuous, tubular inpushings of the epidermis. The lower end of each tube is very much coiled and surrounded by a network of blood capillaries. They lie deep in the dermis and open on the surface of the skin by minute pores called sweat pores. The sweat glands secrete a salty fluid called **sweat** which is conducted along the duct and poured outside on the general surface of the skin. The sweat glands help in removing the metabolic wastes of the body. It also plays an important role in regulation of body temperature. When the body temperature rise above normal, the sweat glands are stimulated and they take up water from the blood vessels and pour out sweat on the surface of the skin. When the sweat evaporates it takes up the latent heat of vaporisation of the skin. Thus the excess body heat is used up and the body temperature reduces.

Other glands: There are some glands which are derivatives of the above glands but are altogether different in nature. They are:

Mammary glands: The mammary glands or **mammae** are milk secreting epidermal glands characteristic of mammals. They are modified sebaceous glands. The young ones live on milk for sometime after birth. Each mammary gland consists of branching system of tubules and alveoli and their ducts open on small projections of the skin called **nipples** or **teats.** The mammary glands develop in both the sexes from the ventro-lateral **milk ridges** in the foetus. During development milk line or ridge appears on either side on the ventral surface of the body. Soon this breaks up into **milk points,** the intermediate parts of the milk line between the milk points disappear. Each milk point may develop into a definite milk gland or mamma, but all of them do not develop fully. For this reason, the number in adult is less than that in the embryo. Mammary glands remain undeveloped and functionless in males but are well developed in sexually mature females especially when nursing young ones.

The number and distribution of mammae (hence of nipples) differs with species. In apes and man single pair of thoracic teats occur. The tree dwellers, among these, are required to carry the young in the arms. Similar thoracic teats occur in bats also. Flying lemurs have a single pair of axillary teats situated in the arm-pit, while in the Cetacea teats occur in the groin and grow as inguinal teats. In this case the baby whale or propoise holds on to the teats like ectoparasite (baby trailer) as the mother swims in water. Insectivores have two pairs of teats, one thoracic and one inguinal. Mammals such as pigs, dogs, edentates, etc., have a series of teats scattered all along milk-line including axillary, thoracic, abdominal and inguinal. Generally speaking there is a provision for the number of young in a litter. Besides these, additional supernumerary teats may develop in any mammal including man. It is interesting to note that the mammary apparatus develops equally in both the sexes up to the time of puberty, when it degenerates in the male and becomes potentially functional in the female.

The teats or nipples may be of two types, **true** and **false** nipples. In true nipple the gland bearing area becomes elevated and the ducts open at the tip of the nipple. It false nipple the ducts open in a depression and the skin around this depression becomes elevated forming what is called a milk canal. The glands empty at the bottom of the milk canal which conducts it to the outside.

Meibomian glands: These glands are present on the margin of the eyelids just inside the eyelashes. They are modified sebaceous glands with long ducts into which several alveoli open. They secrete an oily secretion for the lubrication of eyelashes.

Lacrymal or tear glands: These glands are found in association with the eyes and secrete tears which keep the eyes and eyelids moist and clean.

Scent glands: The scent glands are present at different positions in different mammals. The secretion of these glands serve to attract members of the opposite sex and to drive away enemies. In rabbits **perineal scent glands** are present under the skin near the anus and open into hairless depressions the **perineal pouches.** Their secretion gives characteristic odour to the rabbit.

Wax glands: Wax glands are present in the auditory canal. They secrete a fatty secretion called **ear wax** or **cerumen.** It lubricates and protects the tympanic membrane.

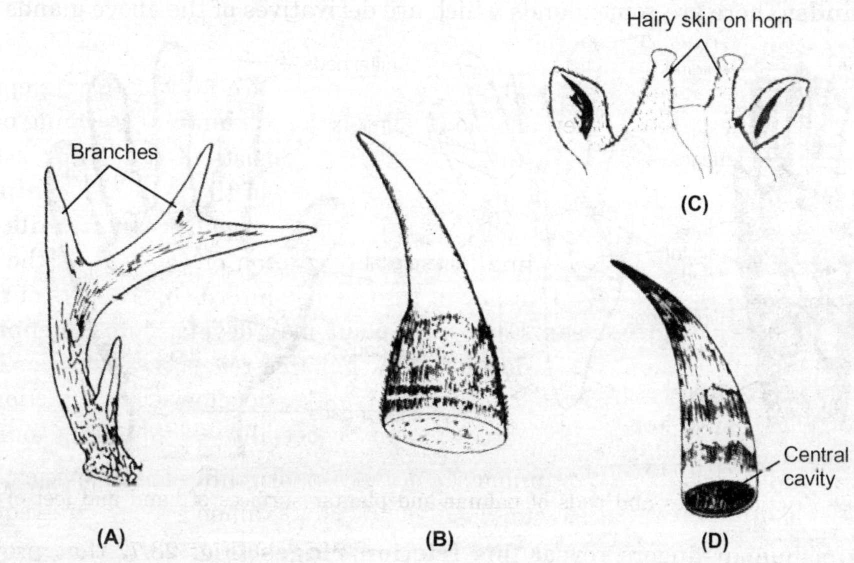

Fig. 23.6 *Rabbit*– Different types of horns. **(A)** *Antlers of male black tailed deer*; **(B)** *Fibre horn of Rhinoceros;* **(C)** *Horn of male giraffe;* **(D)** *Horn of cow*

Other epidermal structures

The **hair** are variously modified to form various structures such as the **horns** of *Rhinoceros*, **scales** of scaly anteater, **spines** in spiny anteater, **quills** of porcupines or **bristles**, etc.

In the amniotes the stratum corneum gives rise to epidermal scales, claws, nails, horns, spikes, beak covering (turtles, birds, duck-bill), spur covering, feathers and hair. The epidermal scales are characteristic of reptiles and in birds and mammals, they may be present in restricted areas. The claws are characteristic of most of the reptiles, birds and mammals. In mammals the claws may be modified as nails and hoofs. The horns characteristic of ungulate mammals are also found in some reptiles. The horns of mammals are of four types: (*i*) **hair horns,** (*ii*) **antlers,** (*iii*) **true horns** and (*iv*) **giraffe-horns.**

The **hair horns** (Fig. 23.6 B) are modified agglutinated hairs and occur in *Rhinoceros*. The **antlers** (Fig. 23. 6 A) are branched bony horns characteristic of deer family and are shed annually. During development stages the antlers are covered with soft skin or velvet which soon wear off leaving naked bone. Anthers occur in male except in reindeer. Antlers are not integumentary structures. **True horns** (Fig. 23.6 D) are present in the bovine animals such as cattle, sheep and antelopes. They have a bony core covered with a horny epidermal cap that is not shed except in pronghorn antelope in which the horny cap is shed annually. The true horns are never branched, never shed and usually present in both sexes. The epidermal cap is a modification of the stratum corneum. The **giraffe horns** (Fig. 23.6 C) are stunted antlers always covered with unmodified epidermis.

Other modifications

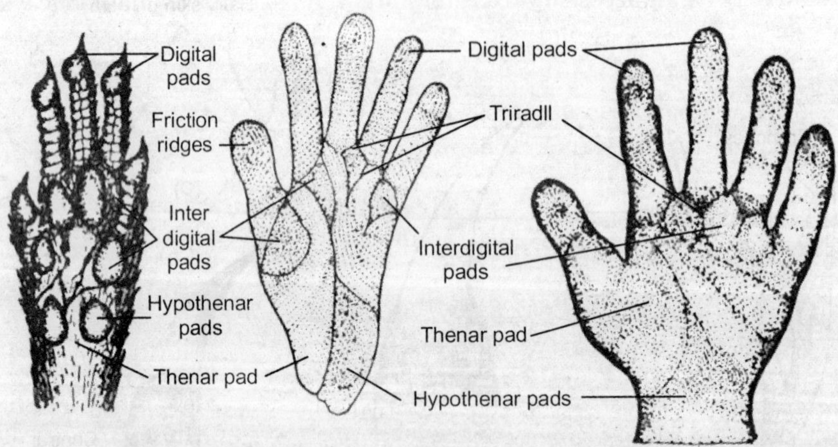

Fig. 23.7 *Rabbit*–Friction ridges and pads of palmar and plantar surfaces of hand and feet of primates

The tips of human fingers reveal fine **friction ridges** (Fig. 23.7) that prevent the fingers from slipping when brought into contact with objects. The arrangement of ridges shows different patterns but in most they are arranged at right angles to the direction in which there is greatest tendency to slip. They are mostly developed in the **palmar** and **plantar** surfaces (surfaces that come habitually into contact with other objects) of the hand and feet of man and other primates.

Friction areas are associated with pad-like epidermal cushions or **tori** that appear on palm and sole. Typically ten tori occur on each hand or foot, five **digital** at the tips of fingers or toes, three **interdigital** near the base of the digit, one **thenar** at the posterior part of palm or foot on the side of the thumb and one **hypothenar** placed on the side of the little finger. **Corns** and **calluses** in man are abnormal thickenings of the stratum corneum developing at places where the integument is subject to constant friction. The tori appear in human embryo but disappear in adult, but in other mammals they are present as distinct elevations throughout life.

Functions of the skin

1. The skin helps in giving characteristic shape to the body.
2. Skin is the main barrier between the organism and its surrounding. It protects the body from mechanical injury, undue loss of water and ultraviolet radiation. It prevents the entry of bacteria and fungi.
3. The skin also helps in maintaining constant body temperature.
4. The skin derivatives like claws, horns, nails, hoofs, etc., serve as organs of defence and as a tool for digging.
5. It serves as an organ for storage of reserve food. Fats are stored as subcutaneous fat. In whales the fatty layer is called blubber.
6. The sweat glands help in regulating the body temperature by evaporating sweat.
7. The skin is a very important sensory organ.
8. The skin acts as an accessory excretory organ as it removes metabolic wastes of the body.
9. The skin helps in the synthesis of vitamin D in the presence of sunlight.
10. The skin serves as an organ of secretion. The oily secretion sebum is responsible for water proofing the skin. The mammary glands secrete milk for feeding the young ones.
11. Special gland secretions help in sexual selection. They attract the opposite sex.
12. In flying mammals the skin forms the patagium which helps in flying.
13. The skin serve as an organ of absorption as it absorbs oils and ointments.

Muscles

Immediately beneath the skin the whole ventral region of the trunk and neck is covered by thin **cutaneous muscle** that enables the rabbit to twitch its skin. The mammary glands of female lie internally to this muscle. Along the mid-ventral line of the abdomen passes a whitish band of connective tissue extending from xiphisternum to pubis. This separates musculature into two longitudinal bands, the **recti adbominis.** Laterally the abdominal wall consists of three thin layers of muscles with their fibres running in different directions. These are the **external oblique, internal oblique** and **transversalis.** A fibrous cord, **Poupart's ligament,** extends upwards and forwards from each pubis to the corresponding ilium. Beneath this ligament pass blood vessels and nerves outwards to the leg. The muscles in the region of the thorax are broken up into a series of **intercostal muscles** by the ribs. Like the muscles of abdomen the intercostal muscles are also arranged in two layers, **external** and **internal** and aid respiration. Extending from the thorax to the forelimb of either side are large **pectoral muscles.** There are many more muscles in the neck in the ventral middle line of which lies the wind pipe, covered by cutaneous epithelium. The hyoid bone between the jaws is also burried into a mass of muscles. Before the muscles lie a thin shining membrane of flat mesodermal cells, the **peritoneum** or **coelomic epithelium.**

Endoskeleton

The skeleton of adult rabbit consists of bones with a small amount of cartilage. The bones are of two kinds, **cartilage** or **replacing** and **membrane** bones. A third type of bone called **sesamoid bone** is also found in rabbit. These are small nodules of bones developed within the tendons of certain of the link muscles close to the joints. They modify the action of these muscles by altering the direction in which they pull. The knee cap or patella of the knee joint is such a bone. At the ends of the successive vertebral centra as well as the ends of long bones of limbs **epiphyses** are present. The successive vertebrae are connected by intervertebral ligaments. The rabbit's skeleton is also divided into **axial** and **appendicular** portions.

Axial Skeleton

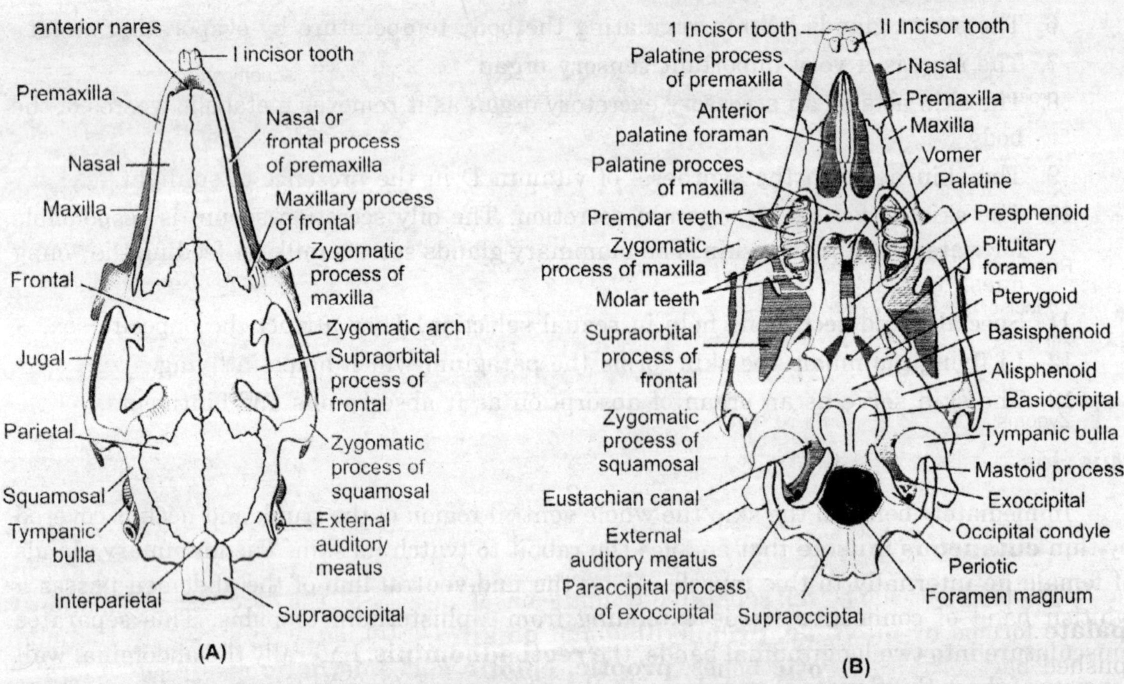

Fig. 23.8 *Rabbit*–Skull **(A)** Dorsal view; **(B)** Ventral view

The axial skeleton includes the skull and vertebral column.

SKULL

The mammalian skull is completely ossified with the exception of a small part of the ethmoid region. The number of bones is, however, considerably less than that found in reptilian skull. This is due, in part, to a loss of bones and in part to a fusion of persisting bones. This fusion is most marked in human skull. The most primitive reptiles had about seventy two separate bones in the skull while the human skull has only twenty eight inclusive of the six ear bones.

The skull is a hard, bony case, in which the limits of separate bones are marked by wavy or jagged lines, the **sutures** (Fig. 23.8). Along these lines the bones dovetail into each other forming firm immovable joints. The skull is **dicondylic,** *i.e.,* each exoccipital bears an occipital condyle to articulate with the atlas vertebra. The skull is **tropibasic,** *i.e.,* a vertical **interorbital** septum lies between the two orbits so that the cranium does not extend forward into orbital region. The orbits are quite large. Since there is a general tendency to increase in the size of the brain the skull has a posterior **cranial part** enclosing the brain and an anterior **facial part** supporting the nose, eyes and jaws. The facial part of the skull is not in line with the cranium but is deflected downwards at an angle of about 60°.

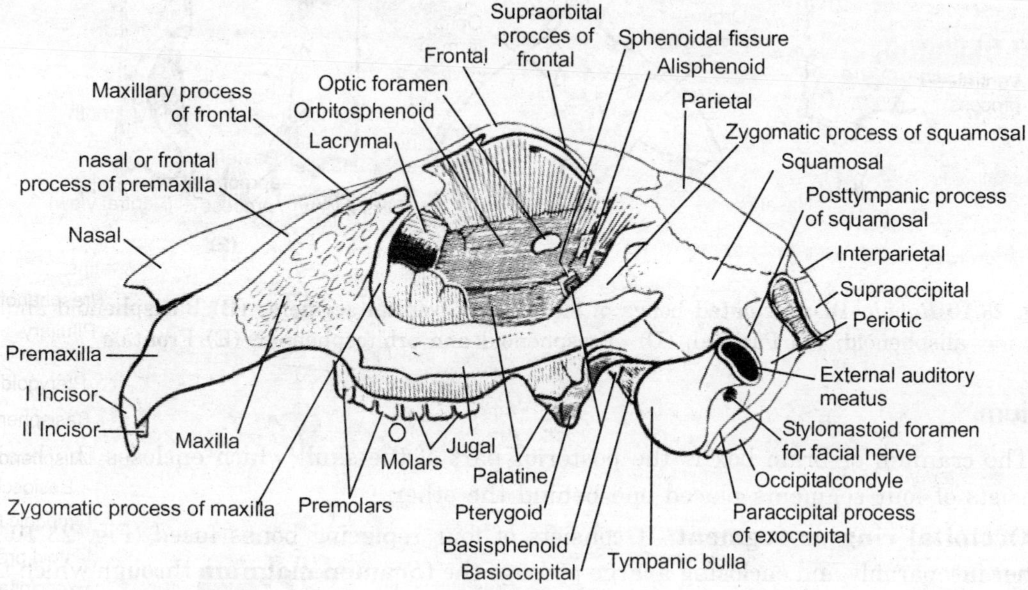

Fig. 23.9 *Rabbit–* Skull lateral view

The food passage is well separated from the nasal passage by the development of a **bony palate** formed by **maxillae, premaxillae** and **palatines,** the internal nares are therefore pushed back. The three **otic** bones, **prootic, epiotic** and **opisthotic** are fused to form a single composite bone, the **periotic** on either side. The tympanic bone is closely applied with it and forms a swollen **tympanic bulla.** The middle ear cavity contains three ear ossicles, **malleus, incus** and **stapes** representing the **columella, quadrate** and **articular** bones respectively. A single dentary bone forms the lower jaw on each side. The jaw suspensorium is **craniostylic,** *i.e.,* mandible articulates with squamosal bone of skull. Hyoid bone is reduced and consists of a transverse bar with two pairs of cornua.

Parts of skull

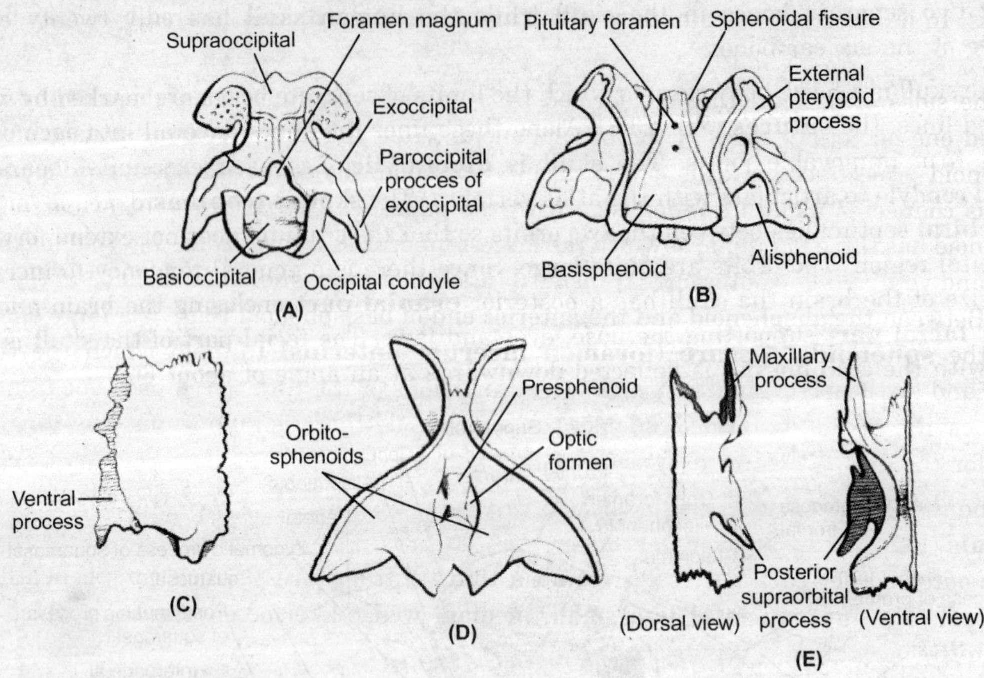

Fig. 23.10 *Rabbit*–Disarticulated bones of skull. **(A)** Occipital segment; **(B)** Basisphenoid and alisphenoid; **(C)** Parietal; **(D)** Presphenoid and orbitosphenoid; **(E)** Frontals

Cranium

The cranium or brain box is the posterior part of the skull which encloses the brain. It consists of four segments placed one behind the other.

Occipital ring or segment: It consists of four replacing bones fused (Fig. 23.10 A) together inseparably and enclosing a large passage the **foramen magnum** through which the spinal cord comes out. Of these bones the **basioccipital** is a flat median bone forming the floor. It articulates in front with the basisphenoid and laterally with the periotic. The paired **exoccipitals** form the side and also greater part of the occipital condyles. Each exoccipital is produced at its outside into a long downwardly directed **paroccipital (jugular) process,** which is closely applied to the postero-ventral surface of the tympanic bulla. A large median **supraoccipital** completes the occipital ring above. Its dorsal surface is raised into a prominent shield-shaped elevation, giving attachment to the muscles raising the head.

Parietal segment or ring: It lies just in front of the occipital ring and consists of both cartilage and membrane bones. It includes ventral **basisphenoid,** lateral **alisphenoid,** dorsal **parietals** and **interparietal** (Fig. 23.10 B).

The base of the parietal ring is formed by the **basisphenoid.** It is a flat and triangular bone with the apex directed upwards to join the presphenoid and the base articulates with the basioccipital by a thin cartilage. On its dorsal surface there is a depression, the **sella turcica,** lodging the pituitary body. This is bounded by large overhanging posterior **clinoid process.** In its floor there is a minute **pituitary foramen** which is covered by a membrane in life.

The sides of parietal ring are formed by a pair of wing-like bones, the **alisphenoid,** attached one on either sides of the basisphenoid. From the antero-ventral surface of each alisphenoid projects downwards a vertical plate of bone, the **external pterygoid process** which is connected along its inner edge with the palatine. The postero-ventral part of the alisphenoid has three foramina, of which the largest innermost transmits the internal maxillary artery and vein and two outer ones transmit branches of the mandibular division of the fifth nerve. Between the alisphenoid and the anterior end of basisphenoid there is a large slit-like notch the **sphenoidal fissure (foramen lacerum anterius)** through which the third, fourth, and sixth nerves and first and second divisions of the fifth nerve pass out from the cranial cavity. At the posterior border of the alisphenoid there is an opening, the **foramen ovale** for the exit of the third division of the fifth nerve.

The roof of the parietal ring is formed by a pair of thin flattened membrane bones, the **parietals** (Fig. 23.10 C) which meet in the mid-dorsal line by a prominent suture. From its postero-outer border each parietal gives off a thin ventral process which runs beneath the squamosal. The **interparietal** is a small, median, wedge-like membrane bone which lies between the parietal and supraoccipital.

Frontal ring or segment: It consists of a **presphenoid** below, paired **orbitosphenoid** forming the sides and the **frontals** forming the roof (Fig. 23.10 D and E).

The **presphenoid** is a small, median, laterally compressed cartilage bone forming the floor of the frontal ring. It is connected to the basisphenoid by cartilage. It forms the lower margin of the **optic foramen,** a large opening for the exit of the optic nerve. The upper and posterior border of presphenoid is produced backwards into a small **anterior clinoid process,** binding sella turcica in front.

The **orbitosphenoids** are a pair of relatively large lamellar bones forming the sides of the frontal ring. They surround the optic foramen. In front of the optic foramen the orbito-sphenoids are continued forward as a pair of vertical laminae forming parts of the inner walls of orbit.

The **frontals** are large membrane bones forming the roof and sides of the frontal segment. They unite suturally in the mid-dorsal line and with parietals behind. The outer surface of each frontal forms the upper part of the wall of orbit and bears a prominent crescentic supraorbital process. Anteriorly it articulates with the nasal bone. A slender maxillary process of the frontal runs in front between the maxilla and premaxilla. Ventrally it articulates with the orbitosphenoid.

Ethmoid ring or segment: The ethmoid segment lies entirely in front of the cranial cavity and is closely associated with the olfactory capsules. It encloses the nasal cavities and separates them from the cranial cavity. It includes a single, median, vertical, plate-like cartilage bone the **mesethmoid,** which extends downwards in front of presphenoid. It forms the **nasal septum** which separates the two nasal chambers. The posterior part of mesethmoid is transformed into a laterally expanded **cribriform plate** which carries a number of perforations for olfactory nerves.

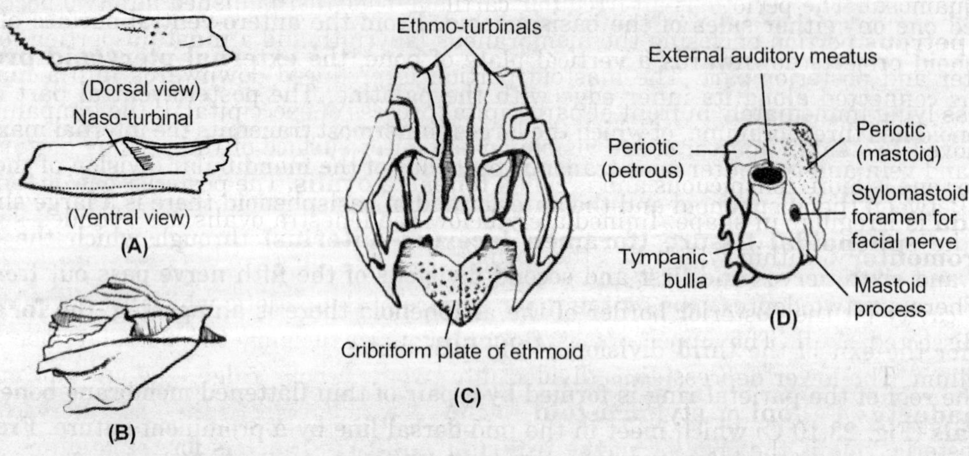

Fig. 23.11 *Rabbit*–Bones of olfactory capsule. **(A)** Nasal; **(B)** Maxilloturbinal; **(C)** Ethmo-turbinals; **(D)** Bones of auditory capsule-periotic and tympanic bulla

Sense capsules

There are three pairs of sense capsules: **olfactory capsules** enclosing the organs of smell, **auditory capsules** containing organs of hearing and **orbits** lodging the eyes.

Olfactory capsules: The olfactory capsules enclose the organs of smell. The roof of olfactory or nasal chambers consists of a pair of long, narrow membrane bones, the **nasals** (Fig. 23.11 A) and a small part of the frontals. The two nasals unite with each other in the median plane. The anterior end of each nasal forms the upper boundary of nostrils. On the outer side each nasal articulates with premaxilla.

Ventrally the nasal chambers consists of a pair of slender elongated bones, the **vomers** which are fused together and attached to the nasal septum.

The **turbinals** are peculiar, delicate, grooved bones which occupy the lateral walls of nasal cavities and fill most of the interior. The most posterior of these is **ethmoturbinal** (Fig. 23.11 C) or **ethmoid labyrinth** situated just in front of the cribriform plate. It is a greatly folded structure, the folds enclosing spaces known as **ethmoid cell.** In front of the ethmoturbinal there is another small bone, the **maxilloturbinal** (Fig. 23.11 B) borne on the inner surface of maxilla. Each nasal bears on its inner surface a single elongated ledge of bone, the **nasoturbinal.** It lies above the uppermost scroll of the ethmoturbinal. The function of turbinals is to increase the respiratory and olfactory surface of the nose.

Auditory capsules: The auditory capsules enclose the organs of hearing. They are present in the postero-lateral region of the cranium and are closely fused with each other. In early embryo the auditory capsule is cartilaginous and consists of three bones: **prootic, epiotic** and **opisthotic** but in adults they fuse to form a composite bone the **periotic** (Fig. 23.11 D) which fits loosely into the side walls of the cranium between occipital segment and squamosal. The periotic is an irregular cartilage bone distinguished into two portions: a dense **petrous** portion enclosing the membranous labyrinth and a **mastoid** portion forming its outer and posterior part. The mastoid portion is produced downwards into a **mastoid process** lying immediately in front of paroccipital process of exoccipital. If the tympanic bone is removed two small apertures are visible on the outer surface of the petrous portion. The anterior one is more conspicuous and is called **fenestra ovalis.** The posterior one or **fenestra rotunda** is irregular in shape. Immediately below the fenestra ovalis is a rounded swelling the **promontory,** within which lies the cochlea.

There are two depressions on the inner surface of the periotic which are revealed only in a dissected skull. The upper one is **floccular fossa** lodging the floccular lobe of the cerebellum. The lower depression is divided into two by a bony ridge. The anterior of these is **aquaductus Fallopi** or **stylomastoid foramen** through which the facial nerve passes. The posterior one is the opening of the **internal auditory meatus** for the auditory nerve.

A pair of somewhat flask-shaped membrane bones, the **tympanics** lie closely applied to periotics. The expanded body of the flask, the **tympanic bulla** encloses the **tympanic cavity** which contain three ear ossicles, **malleus, incus** and **stapes.** The neck of the flask encloses the **external auditory meatus** which is directed upwards and supports the cartilage of large external ear. Within the bulla at the base of meatus an incomplete bony ring is present. The **tympanic membrane** is stretched around this ring in life. The stylomastoid foramen for the exit of the facial nerve from the skull opens on the posterior side of tympanic.

Orbits: The orbits are situated on the sides of frontal segment in the cranial region. A median bony interorbital septum separates the two orbits. The orbits lodge the eyes. Each orbit is formed by several bones like the **frontal** which project above the supra-orbital ridge, **orbitosphenoid**, **alisphenoid** and **ethmoid**. In the anterior wall of the orbit is the **lacrimal** bone wedged in between the frontal and maxilla. At the anterior end of the lacrimal bone is an opening of the posterior end of the nasolacrimal duct, by means of which the tears drain into nasal cavity. The sides of the orbit are protected by zygomatic arch. The lower side of the orbit is unossified.

Visceral skeleton

Visceral skeleton includes upper and lower jaws and hyoid apparatus.

Upper jaw: The upper jaw forms the roof of the snout. It is made up of a series of bones closely associated with the cranium and olfactory capsules. The bones are premaxillae, maxillae, jugals, pteryoid and palatines (Fig. 23.12).

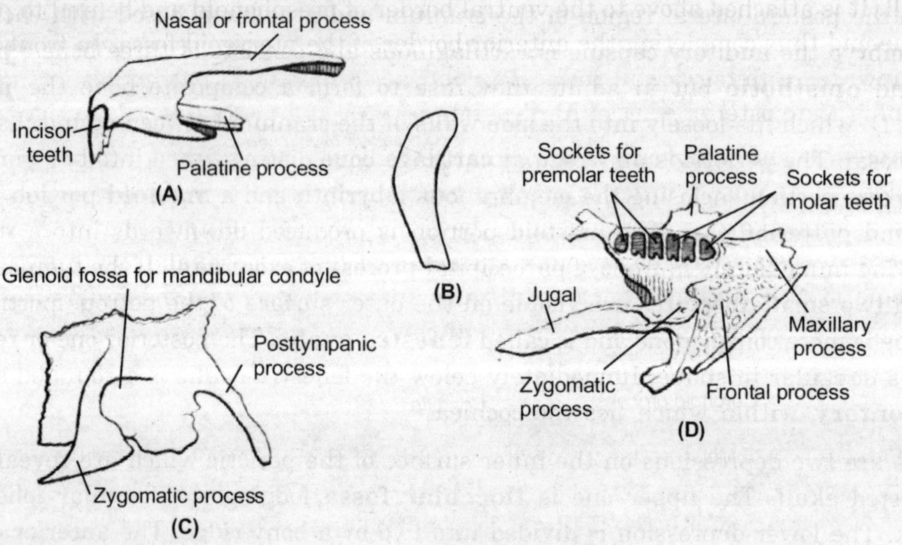

Fig. 23.12 Bones of upper jaw. **(A)** Premaxilla; **(B)** Pterygoid; **(C)** Squamosal; **(D)** Maxilla and jugal

The two **premaxillae** are relatively longer in rabbit. They form the anteriormost part of the upper jaw and are fused with each other in the median plane. The anterior end of each premaxilla bears two sockets for two incisor teeth. A long **nasal process** is given out from each premaxilla which extends backwards between the nasal and maxilla up to the frontal. A **palatine process** arises from each premaxilla, both meet each other along the midventral line and reach up to the palatine processes of maxillae, thus dividing the anterior palatine foramen into two halves. The premaxillae form the anterior part of the hard palate by means of the palatine processes.

The **maxilla** is the main bone of the upper jaw forming its greater part. In rabbit it is much fenestrated. Each maxilla is a large irregular bone bearing sockets for cheek teeth. It gives off a number of processes. An **orbital process** which forms a part of the anterior wall of the orbit ventral to the lacrimal bone. Anteriorly the maxilla joins the premaxilla by a **maxillary process.** It extends up to frontal bone above by a **frontal process.** A median **palatine process** which meets its fellow of the opposite side in the mid-ventral line and forms the anterior part of the hard palate. A stout **zygomatic process** from its outer surface constitutes the beginning of the **zygomatic arch** which bounds the orbit below. The premaxilla and maxilla contribute to the anterior and posterior half, respectively, of **diastema** or **toothless gap** between the incisors and premolars.

The **molar** or **jugal** is a laterally compressed bone which extends between the zygomatic processes of the maxilla and squamosal and form greater part of the zygomatic arch.

The **pterygoids** are a pair of narrow vertical plates of bone, attached to the base of the skull at the junction of basisphenoid and alisphenoid. The posterior border of each pteryoid is free and ends below in a curved **hamular process.** The anterior border unites with palatines and alisphenoids. On its outer side is a triangular **pterygoid fossa**.

The **palatines** are a pair of irregular-shaped membrane bones on the mid-ventral aspect of the skull. It is attached above to the ventral border of presphenoid and behind to pterygoid and alisphenoid thus completing the anterior border of the pterygoid fossa. In front they are produced inwards into horizontal **palatal processes** which meet each other in the median plane forming the posterior part of hard palate.

Squamosal: The squamosal is somewhat rectangular bone lying ventral to the parietal on either side. A long **post-tympanic process** arises from the posterior side of each squamosal and is closely applied over the periotic just above the external auditory meatus. A prominent **zygomatic process** arises from the outer surface of each squamosal and forms posterior part of the zygomatic arch. A **glenoid** or **mandibular fossa** is present on the undersurface of the zygomatic process for articulation of the lower jaw.

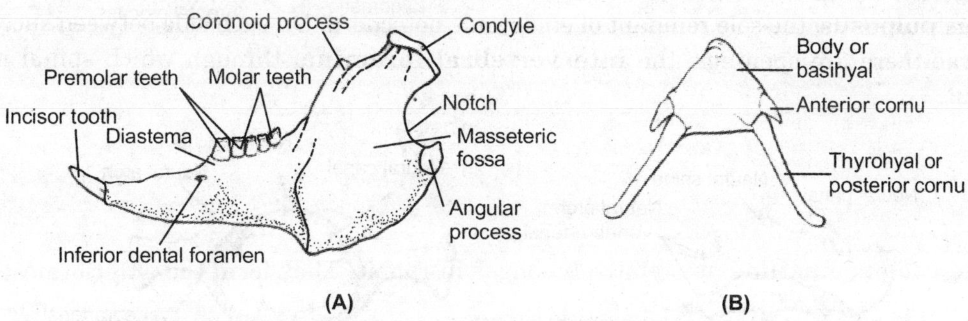

Fig. 23.13 *Rabbit*–**(A)** Lower jaw, left half; **(B)** Hyoid apparatus

Lower jaw: The lower jaw or **mandible** (Fig. 23.13 A) consists of two halves or **rami** united in front by **mandibular symphysis.** Each rami consists of a single membrane bone, the **dentary**. All other bones seen in lower jaw of lizard are absent in rabbit. Each dentary is somewhat triangular bone bearing sockets for teeth along its upper margin. A diastema or tooth gap between the incisor and premolar corresponds with that of the upper jaw. On the postero-lateral surface there is a depression, the **masseteric fossa** for attachment of **masseter muscles** (jaw muscles). While grinding the food, rabbits move their mandibles fowards and backwards and not crosswise. Each dentary has a ventral **angular process** which is rounded and inflected and a thick, elongated dorsal **condyle** set much above the level of teeth. On the posterior side there is a deep notch which separates angular process from condyle. The condyle articulates with the glenoid fossa of squamosal. In front of the condyle there is a **coronoid process** for the attachment of temporal muscles. Near the anterior tip of mandible on the outer surface there is a prominent **dental** or **inferior dental foramen** through which the nerve of lower jaw exits. On the inner surface there is a **mandibular foramen** through which the nerve enters. These also allow passage of blood vessels.

Hyoid apparatus: The hyoid apparatus (Fig. 23.13 B) lies in the floor of the mouth between the rami of the mandible and supports the tongue and larynx. It consists of a strong median **body** or **basihyal** and two pairs of **horns** or **cornua.** The anterior cornua is situated below the tongue and is made up of a series of small bones: **ceratohyal, epihyal, stylohyal** and **tympanohyal.** The tympanohyals are attached to the otic region of the skull. The posterior cornua consists of a single bone, **thyrohyal.** The anterior cornua are the ventral end of the hyoidean arches, the dorsal end of each of which has aborted. The posterior cornua are the lower ends and only persistent part of first pair of branchial arches.

Vertebral column

The vertebral column of mammals is differentiated into the usual five regions: **cervical, thoracic, lumbar, sacral** and **caudal.** The total number of vertebrae in rabbit is about 45–47. The vertebral formula is C_7, T_{12-13}, L_{6-7}, S_5, Cd_{16}. The vertebrae of these regions are markedly distinct from each other. A bony plate called **epiphysis** is present at the end of each vertebra, so the articular surfaces of centra are more or less flat (**amphiplatyan** or **acoelous**). However, in the embryonic stage a thin **epiphysial cartilage** separates the centrum and epiphysis which later fuses with the centrum. Between the ends of the vertebrae are found cartilaginous discs, the **intervertebral discs.** In the centre of the disc is a soft area, the **nucleus pulposus,** the sole remnant of embryonic notochord. On each side between successive vertebrae there are openings, the **intervertebral foramina,** through which spinal nerves pass out.

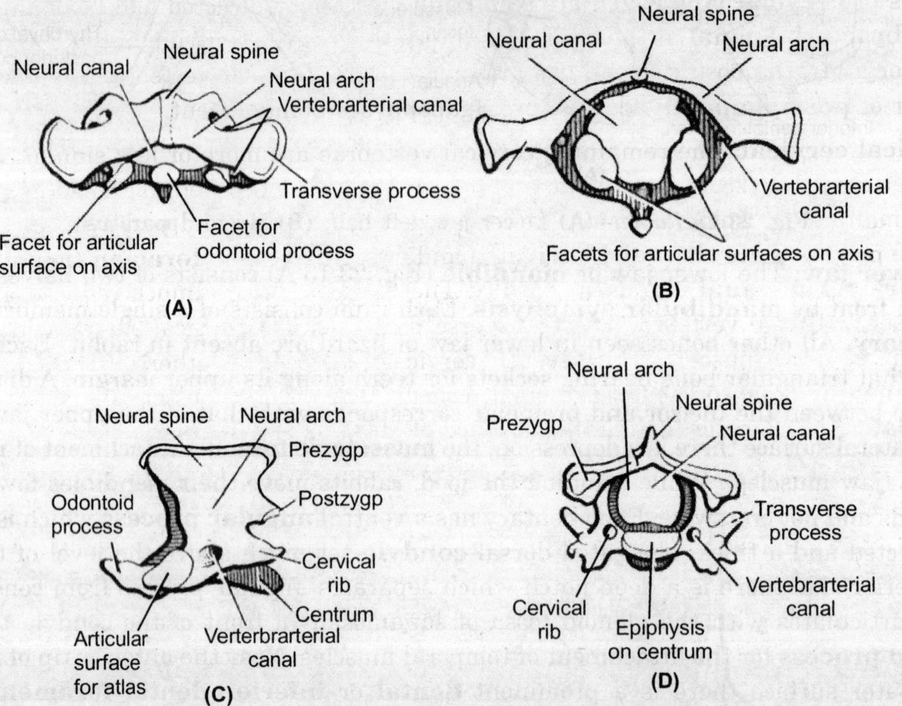

Fig. 23.14 *Rabbit*–Vertebrae. **(A)** Atlas (dorsal view); **(B)** Same, posterior view; **(C)** Axis (lateral view); **(D)** Typical cervical (anterior view)

Cervical vertebrae

There are **seven** cervical vertebrae. The first two are differentiated as **atlas** and **axis** and remaining 3–7 are more or less alike and are called **typical cervical.**

Atlas: It is very different from atlas of others. It is ring-shaped (Fig. 23.14 A) and consists of a large **neural arch,** reduced **neural spine** and is without **centrum** and **zygapophysis.** The centrum is present in the embryonic stage but later fuses with axis vertebra. On the sides are present enlarged and flattened wing-like **cervical ribs** which are often called transverse processes. They are perforated by a pair of openings, the **vertebrarterial canals.** The **neural canal** is large and is divided by a **ligament** into two parts, dorsal part for the passage of spinal cord and ventral part for odontoid process of the axis vertebra. The anterior end of the atlas carries articulating surfaces, the **occipital facets** for articulation with the occipital condyles of the skull. Posteriorly it bears two stout **lateral facets** which articulate with axis and a midventral facet for odontoid process.

Axis: The second cervical vertebra, **axis** (Fig. 13.14 C) or **epistropheus,** has a very long and broad **neural arch,** provided with an elongated **neural spine** which projects forward over the atlas. At its anterior end the centrum bears a pointed projection, the **odontoid process** which fits into the atlas allowing the turning of head. The odontoid process is in reality the centrum of atlas. The axis bears laterally the so called **transverse processes** or **cervical ribs** which are small and posteriorly directed and bear an opening, the **vertebrarterial canal** at the base. The anterior end of the centrum has two articulating surfaces for atlas, the posterior end has articular surface for next vertebra. The neural arch has a pair of **postzygapophyses,** the prezygapophyses being absent.

Typical cervical: The remaining cervical vertebrae are more or less similar. A typical cervical vertebra (Fig. 23.14 D) is broad and have a well developed neural arch, a large neural canal, a small neural spine, a small flattened centrum and pre- and post-zyapophyses. The transverse processes are short and bifurcated and a foramen called **foramen transversaria** or **vertebrarterial canal** is present basally. The cervical ribs are much reduced and more or less fused with the vertebra. The transverse processes and cervical ribs provide site for attachment of neck muscles. The seventh cervical differs from others in having a more elongated neural spine, presence of a small concave facet at the posterior edge of centrum for articulation of thoracic ribs and in absence of foramen transversaria.

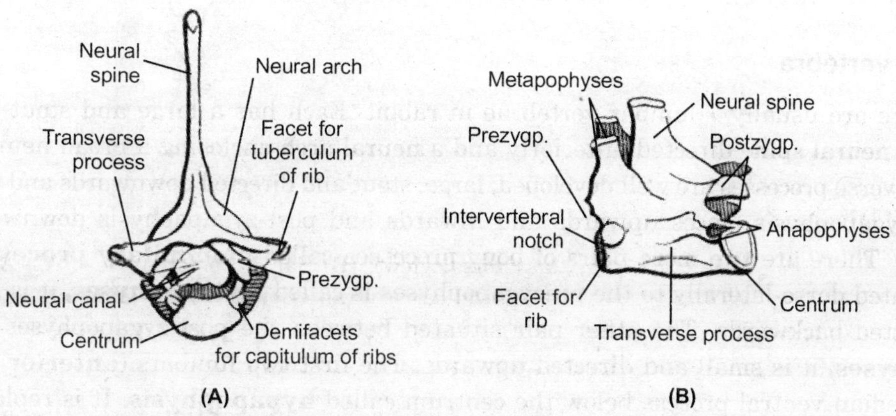

Fig. 23.15 *Rabbit*–Vertebrae. **(A)** Anterior thoracic (anterior view); (B) Posterior thoracic (laterial view)

Thoracic vertebra

There are twelve thoracic vertebra in rabbit, the number varies in other mammals (man has twelve).

Anterior thoracic: The first seven or eight thoracic vertebra have tall neural spine directed backwards and short and thick centrum which bears at its anterior and posterior ends small half-facet for articulation with the capitulum of the rib. The neural arches bear small prezygapophyses directed outwards and upwards and small post-zygapophyses facing inwards and downwards. The short and stout transverse processes ventrally bear at their ends a facet for articulation with the tuberculum of rib (Fig. 23.15 A).

Posterior thoracic: The posterior four or five thoracic vertebrae are some what different from the anterior ones in having longer and stouter centra, prominent zygapophyses, short neural spines and reduced transverse processes. Tubercular facets are absent but complete capitular facet for a rib is present, one on either side, near the anterior end of centrum (Fig. 23.15 B).

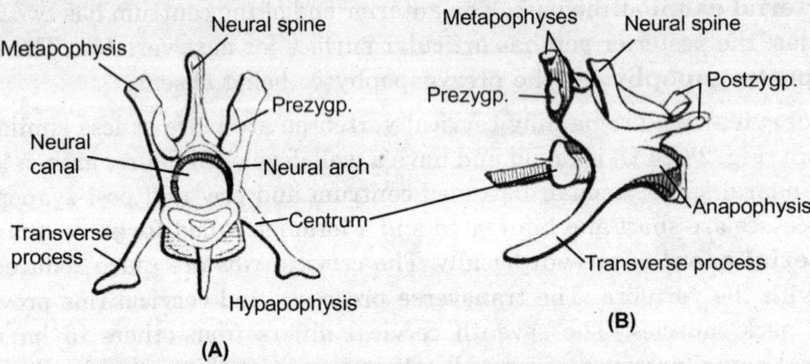

Fig. 23.16 *Rabbit*–Vertebrae. **(A)** Anterior lumbar (anterior view); **(B)** Posterior lumbar (lateral view)

Lumbar vertebra

There are usually 7 lumbar vertebrae in rabbit. Each has a large and stout centrum, crest-like neural spine directed anteriorly and a neural arch enclosing a broad neural canal. The transverse processes are well developed, large, stout and directed downwards and forwards. The prezygapophysis faces upwards and inwards and post-zygapophysis downwards and outwards. There are two more pairs of bony processes called **mammilary processes.** One pair situated dorso-laterally to the prezygapophyses is called **metapophyses,** it is large flat and directed backwards. The other pair situated between the post-zygapophyses is called **anapophyses,** it is small and directed upwards. The first two lumbars (**anterior lumbar**) bear a median ventral process below the centrum called **hypapophysis.** It is replaced by a short ridge in posterior five lumbars (Fig. 23.16).

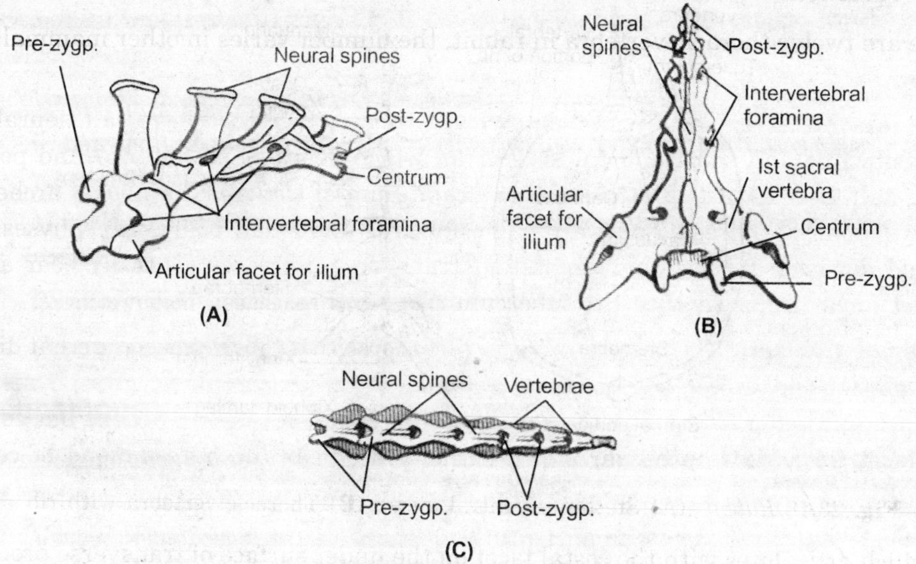

Fig. 23.17 *Rabbit*–Vertebrae. **(A)** Sacrum (lateral view); **(B)** Sacrum (dorsal view); **(C)** Caudal (dorsal view)

Sacrum

The sacrum (Fig. 23.17 A and B) is a single compound bone composed of a variable number of vertebrae fused together to support the pelvis. There are four sacral vertebrae in rabbit (five in man) of which only the first two really contribute to attachment. The first sacral is the largest and has strong transverse processes and fused sacral ribs which are attached to ilia bone of pelvic girdle. The neural spine is upright, metapophyses reduced and anapophyses absent. The articular surfaces of ilium and sacral vertebra are covered by cartilage to which strong ligament, **interosseus sacroiliac ligament**, is attached. The boundaries between the fused sacrals are readily made out by means of their neural spines. The remaining sacral vertebrae become smaller behind. The neural spines are directed backwards. They bear lateral tubercles which represent fused pre- and post-zygapophyses, other processes are reduced.

Caudal vertebrae

The caudal vertebrae (Fig. 23.17 C) are variable in number in mammals. In rabbit there are sixteen caudal vertebrae. They gradually decrease in size posteriorly. The neural arches, transverse processes and zygapophyses diminish caudally, the last vertebra consisting of rounded cylindrical centra only.

Ribs

The ribs are curved, rod-like bone consisting of a long dorsal bony part, the **vertebral rib** and a small ventral cartilaginous portion, the **sternal rib** or **costal cartilage.** The vertebral rib is articulated to the vertebra by two heads **(bicephalous)**, an upper **tuber-**

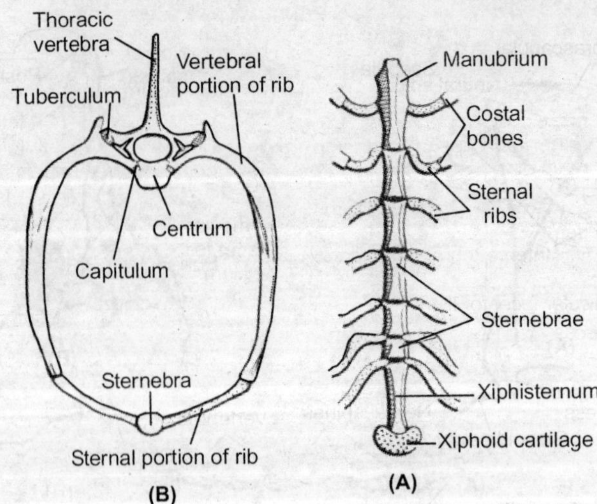

Fig. 23.18 *Rabbit*–**(A)** Sternum (ventral view); **(B)** Thoracic vertebra with rib

culum which articulates with the costal facet on the under surface of transverse process, and lower **capitulum** attached between two centra to demifacets, *i.e.,* half a facet on the posterior end of one centrum and half a facet on the next posterior centrum, and absent on last three centra, the ribs of which are provided with capitular head only. The narrow part of rib between the two heads is called **neck,** the point of greatest curvature, just a little beyond the tuberculum, the **angle** and remainder of rib, the **body.** The sternal ribs reach the median ventral line and are attached to the sternum. On the basis of their attachment the ribs are of two types: those which meet the sternum below by their sternal parts are called **true ribs,** those which have no sternal part and are not attached to the sternum are called **floating ribs.** The rabbit has seven true ribs (Fig. 23.18 B) and five floating ribs.

Sternum

The sternum of rabbit (Fig. 23.18 A) is an elongated, laterally compressed structure lying embedded mid-ventrally in the thoracic wall. It is a segmented structure made up of seven elongated bony pieces called **sternebrae.** The first sternebra is called **manubrium** or **presternum.** Ventrally it bears a **keel** to which the first pair of rib is attached. The sixth sternebra is smallest. The seventh is long and called **xiphisternum.** It terminates in an expanded **xiphoid** or **xiphisternal cartilage.** The rest of the true ribs are attached at inter-sternebra junctions.

Appendicular skeleton

The appendicular skeleton includes the skeleton of two pairs of limbs and girdles that support them.

Pectoral girdle

The pectoral girdle consists of two separate halves (Fig. 23.19 A) and is placed dorsal to the anterior thoracic ribs in between the forelimbs. In rabbit each half consists of two bones—a small membrane bone, the **clavicle** and a large replacing bone the **coracoscapula** or

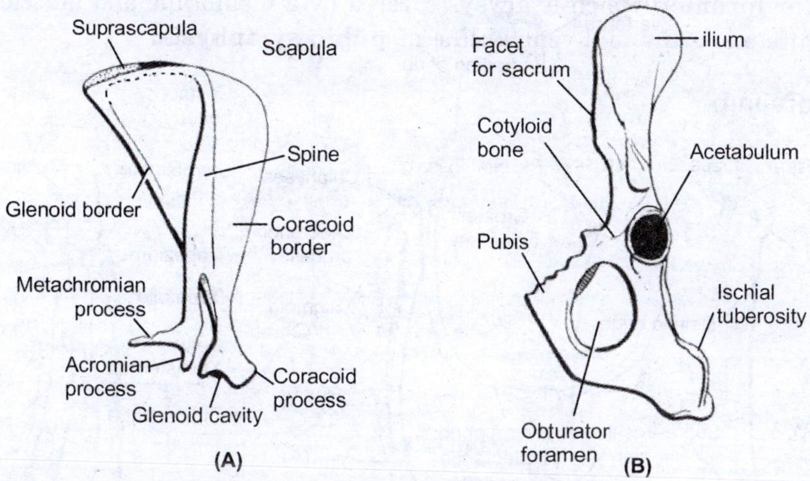

Fig. 23.19 *Rabbit*–**(A)** Pectoral girdle (right half); **(B)** Pelvic girdle (left half)

shoulder blade. The clavicle is a thin, slender, rod-like bone extending between the acromion process of scapula and the manubrium of the sternum. It is imbedded in the muscles of the front of shoulder and is not articulated to the rest of the girdle, so it falls off in prepared skeleton. The coracoscapula is mainly formed of **scapula** which is a large, flat, triangular bone, the apex of which is directed downwards and bears a concave **glenoid cavity** which articulates with the head of the humerus. From the anterior side of the margin of the glenoid cavity a small beak-like **coracoid process** projects medially, this is the vestige of coracoid bone. The outer surface of scapula bears a prominent vertical ridge, the **spine** which divides it into antero-dorsal and postero-ventral portions. The ventral end of spine terminates into a pointed acromion process, just above this and projecting laterally and posteriorly is the long **metacromian process.** The dorsal edge of scapula is made up of a thin strip of cartilage, the **suprascapula.**

Pelvic girdle

The pelvic girdle of rabbit is formed of two equal halves (Fig. 23.19 B), the **os innominate** joined together midventrally by **pubic symphysis.** It is very similar to that of reptiles in shape and to that of birds in that the three pelvic bones, **ilium, ischium** and **pubis** are fused into a single **innominate** or **hip bone.** A deep cup-like cavity, the **acetabulum** is present on the outer middle side of innominate. It articulates with the head of femur. Ilium and ischium take part in the formation of acetabulum but the pubis does not meet it because a small **cotyloid bone** is present between the two.

The **ilium** is the most anterior, largest and blade-like bone which is joined to the sacrum near its middle by a prominent crescentic facet. Posteriorly it extends up to the acetabulum. The postero-dorsal part of innominate is formed by the **ischium.** Its posterior end forms prominent projection called **ischial tuberosity.** The ventromedial part of innominate is formed by the **pubis.** Each pubis is a small, slender bone separated from ischium by a large oval **obturator foramen** which is always covered by a membrane and muscles in life. The two pubes unite along the mid-ventral line at **pubic symphysis.**

Bones of forelimb

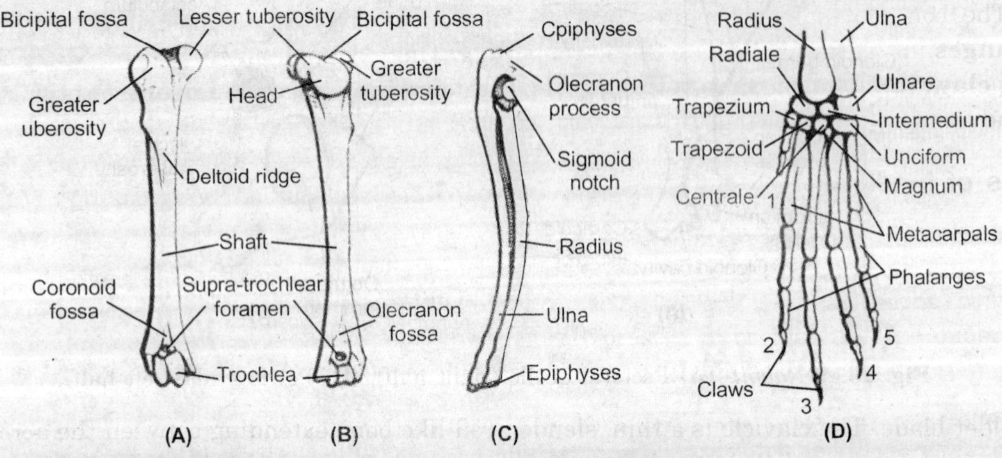

Fig. 23.20 *Rabbit*–Bones of forelimb. **(A)** Humerus (anterior view); **(B)** Same (posterior view); **(C)** Radio-ulna; **(D)** Bones of hand

Humerus: Humerus (Fig. 23.20A and B) is the bone of upper arm. It is a relatively short but stout bone having a rounded **head** fitting into the glenoid cavity of scapula. Close to the head there is a **greater tuberosity** on the outer and **lesser tuberosity** on the inner side. Between the two tuberosities there is a **bicipital groove** for attachment of muscles. The anterior surface of humerus below the head bears a **deltoid ridge** for attachment of muscles. The distal end of humerus is expanded into epicondyles and bears a pulley-like trochlea for articulation with ulna. It allows only forward and backward movements of the forearm. Immediately above the trochlea, in front and behind, are two depressions, the **supra-trochlear fossae,** which communicate with each other through **supra-trochlear foramen.** The posterior or **olecranon fossa** is larger and lodges the olecranon process of ulna when the arm is extended. The anterior or **coracoid fossa** is smaller.

Radius and ulna: The forearm consists of two separate bones, radius and ulna (Fig. 23.20 C), of which the latter is larger. The two ends of radius and ulna are tightly bound so that they cannot move over each other. The proximal end of ulna forms a prominent projection, the **olecranon process** or **elbow** which articulates with olecranon fossa of humerus. Below the process there is a deep semicircular concavity, the **semilunar or sigmoid notch** which articulates with the trochlea of the humerus. The distal border of the notch forms another projection, the **coronoid process.** The radius is situated on the inner side. Distally the raduis and ulna are provided with **epiphysis** and articulate with the bones of wrist.

Bones of hand (Fig. 25.20 D): The **wrist** or **carpus** is made up of several small bones, the **carpals,** arranged in two transverse rows. The proximal row consists of three pieces, **radiale, intermedium** and **ulnare.** The distal row consists of five bones—**centrale, trapezium, trapezoid, magnum** and **unciform**. The unciform is actually two fused carpals.

A sesamoid bone **pisiform** is present on the ventral side of carpus. The radiale is situated below the radius, ulnare below the ulna and intermedium between them. Similary, trapezium and trapezoid are situated below radiale, centrale and magnum below intermedium and uniciform below ulnare.

The bones of palm or manus are five long **metacarpals** and five **digits** with two **phalanges** in the first and three in the remaining. The terminal phalanx of each digit bears a long **claw.** In rabbit the limb is fixed with palm downwards, that is in **prone** position. This is of advantage in burrowing.

Bones of hindlimb

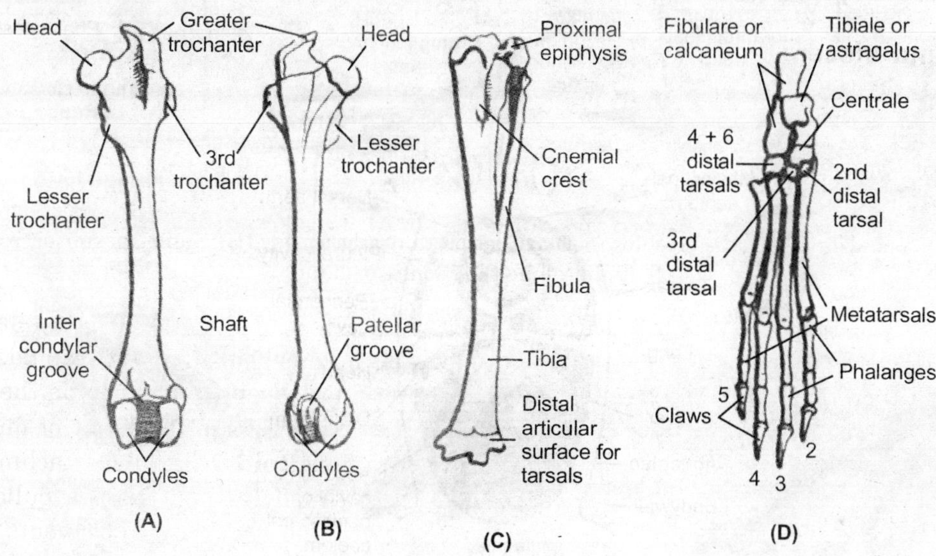

Fig. 23.21 *Rabbit*–Bones of hindlimb. **(A)** Femur (back view); **(B)** Same (front view); **(C)** Tibiofibula; **(D)** Bones of foot

Femur: Femur (Fig. 23.21 A and B) is the bone of thigh. It is the longest bone with a long cylindrical shaft and expanded extremities. The proximal end has a rounded **head** and three prominences, a **greater trochanter** lateral to the head, **third trochanter** below the greater trochanter and a **lesser trochanter** below the head. These serve for muscle attachment. The deep groove between the head and greater trochanter is called **digital fossa.** At distal end the femur bears a pair of expanded **condyles** (medial and lateral) with an **inter-condylar groove** between them. The condyles provide articular facets for articulating with tibia. At the knee joint a **patella** is present.

Tibio-fibula: The shank is composed of two long bones: **tibia** and **fibula** (Fig. 23.21 C) which are fused in the distal half. The tibia is present on the inner side and is stouter than fibula. The proximal end bears two **proximal epiphysis** which articulate with the condyles of femur. The dorsal end of tibia has a **cnemial crest** for attachment of muscles. Distally tibio-fibula articulates with the ankle.

Bones of foot (Fig. 23.21 D): The ankle or tarsus is composed of six **tarsal bones** arranged in two rows. The proximal row consists of two **bones, tibiale** or **astragulus** on preaxial side and **fibulare** or **calcaneum** on postaxial side. Actually astragulus is formed by fusion of two bones **tibiale** and **intermedium.** The astragulus bears a pulley like surface for articulation with tibia. The calcaneum is produced into a prominent **calcaneal process** or **spur** just behind its articulation with tibio-fibula. The middle row consists of a single bone the **centrale** which is a curved bone reaching the medial side of the foot. The distal row has three trarsals—**mesocuneiform, ectocuneiform** and **cuboid.**

The **sole** or **pes** consists of four long and slender **metatarsals,** the first one is rudimentary or absent. There are only four **toes** (first toe or **hallux** is absent). Each toe consists of three **phalanges.** The terminal phalanges bear a **claw.**

Coelom and viscera

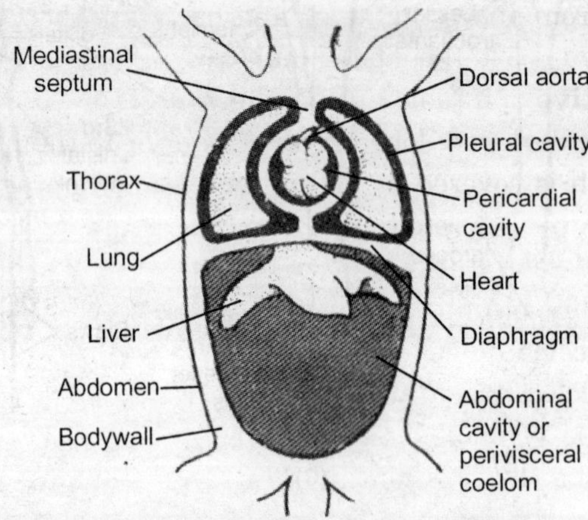

Fig. 23.22 *Rabbit–*Diagramatic representation of body cavities

Coelom: The trunk of mammals contains a large mesodermal **body cavity** or **coelom** which encloses most of the **internal organs** or **viscera.** The coelom is filled with **coelomic fluid** which allows free movement of visceral organs. The trunk is divided into an anterior thoracic and posterior abdominal region. Each of these regions contain cavities which are partitions of the coelom (Fig. 23.22). The **thoracic coelom** is divided into three portions, two laterally situated **pleural cavities** and a median **pericardial cavity** located between them. The pleural cavities lodge the soft, spongy lungs whereas the pericardial cavity accommodates the heart. A delicate partition called **mediastinal septum** stretches from the heart to the median ventral line. It is made up of two medial walls of the right and left pleural cavities in contact with each other. The space between the two walls is called **mediastinum.** A moist-membrane, the **pleura,** lines the pleural cavity. It is a part of the peritoneum. The pleura is divided into **parietal** and **visceral** parts. The parietal pleura covers the inner side of the pleural cavity, the anterior face of diaphragm and forms mediastinal septum in the midventral line. The visceral pleura covers the lungs to which it is indistinguishably fused.

The pericardial coelom is enclosed by thin pericardium, an extension of the peritoneum which is not attached to the heart except at the anterior end where the great vessels enter and leave the heart. The surface of heart is invested by the **visceral pericardium** that is inseparably fused to the heart wall. The **parietal pericarduim** is separated from the heart by the pericardial cavity. As the heart lies in the mediastinum the parietal pleura of the mediastinal septum forms the third layer of coelomic epithelium surrounding the heart. The thymus is also situated in the mediastinum ventral to anterior part of heart.

The thoracic coelom is separated from the abdominal coelom by a curved muscular sheet the **diaphragm.** The **abdominal** or **peritoneal cavity** is lined by a membrane the **peritoneum.** The portion of peritoneum inside the body wall is called **parietal peritoneum** and that extending over the visceral organs is the **visceral peritoneum** or **serosa.** In passing to and from the body wall to the visceral organs the peritoneum forms double walled **mesenteries** or **ligaments.** The **dorsal mesentary** is intact in mammals and **ventral mesentary** persists in the region of liver and urinary blader.

Viscera: As is evident from above the lungs and heart are the major organs occupying the thoracic coelom. The organs of, the abdominal cavity include the **liver, alimentary canal, excretory, reproductive** and some **endocrine organs,** etc. Fitting into the concavity of the diaphragm is the large, dark-red, lobed **liver.** Posterior to the liver the cavity is filled by the coils of intestine which is covered by a thin membrane impregnated with streaks of fat. This is the **great omentum** more conspicuous in other mammals than in rabbit. At the posterior end of the peritoneal cavity lies the pear-shaped urinary bladder. The stomach lies dorsal to the left liver lobe. The spleen is attached to its left border. At about the level of the posterior ends of liver lobes occur the bean-shaped **kidneys** on the dorsal wall of abdominal cavity. A small, rounded, yellow adrenal gland is found in front of each kidney. In female rabbit oviducts and uterus are visible. The testes in male lies in **scrotal sacs** outside the peritoneal cavity. The abdominal cavity is continued into the cavity of scrotal sacs by **inguinal canals.**

DIGESTIVE SYSTEM

The digestive system is concerned with digestion of food and egestion of undigested wastes. It includes the alimentary canal and digestive glands.

Alimentary canal

The alimentary canal or digestive tract is a long coiled tube of varying diameter extending from mouth to anus. It includes following parts:

Mouth: The mouth is a transverse slit-like aperture situated at the tip of snout. It is bounded by two muscular, soft, fleshy and movable **lips** covered with hairy skin on outside and lined with mucous epithelium on the inner side. The upper lip bears a cleft which extends up to the nostrils, as a result the upper incisor remain exposed even when the mouth is closed. Each lip is provided with **labial glands** on the inner surface. It secretes mucous which keeps the inner surface of the lips moist.

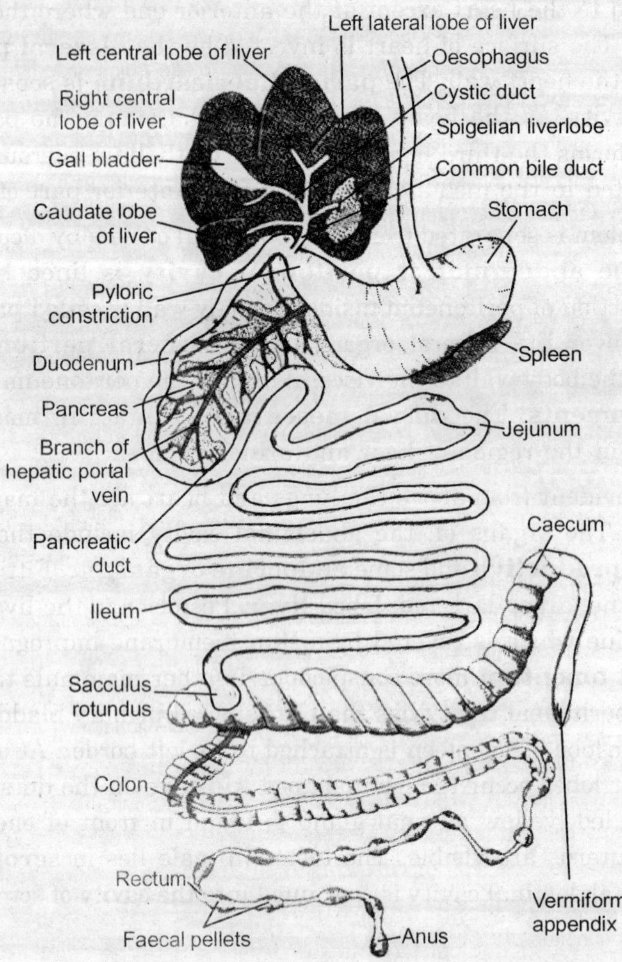

Fig. 23.23 *Rabbit*–Alimentary canal and digestive glands

Vestibule: The mouth leads into a narrow space called vestibule lying between the lips, cheeks and teeth. It is lined by mucous epitheluim.

Buccal cavity: The vestibule opens into a large spacious buccal or oral cavity (Fig. 23.24 A) behind the jaws. It is lined with mucous epithelium and contains the following parts:

***(i)* Palate:** The roof of buccal cavity is formed by the palate. Its anterior part is called **hard palate** and is supported by maxilla, premaxilla and palatine bones. The mucous mebrane lining the hard palate is thrown into numerous roughened transverse ridges known as **palatine rugae** which prevent the escape of food from mouth cavity. At the anterior end of hard palate just behind the incisor teeth occur a pair of openings of the nasopalatine ducts which connect the mouth and nasal cavity. In each canal a sensory **Jacobson's organ** is

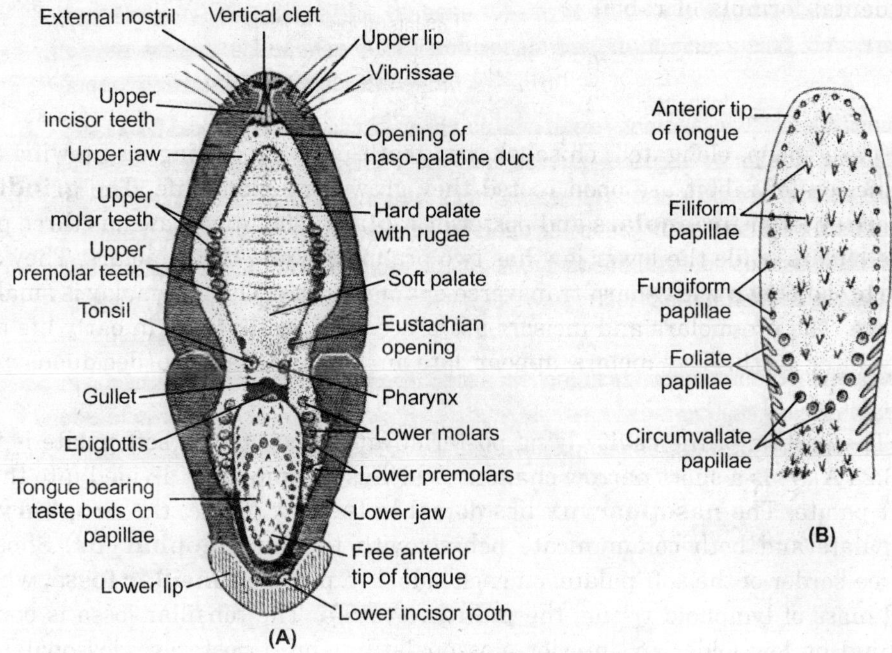

Fig. 23.24 *Rabbit*–**(A)** Buccopharyngeal cavity; **(B)** Tongue showing different types of papillae

present which helps to recognize different kinds of food. The posterior part of the palate is formed by soft, smooth and fleshly **soft palate** made up of connective tissue and is without bony support. The nasal or respiratory passage lies above the palate. It is divided by a nasal septum into two parts above the hard palate but remains undivided above the soft palate.

***(ii)* Tongue:** The floor of the oral cavity is occupied by the **tongue,** a fleshy, muscular, movable organ. The greater part of the undersurface of tongue remain attached to the buccal floor and the free tip is rounded. The anterior margin of attachment of the tongue to the floor is in the form of a vertical fold, the **frenulum.** The dorsal surface of the tongue bears a median groove and numerous **papillae** containing **taste buds.** The papillae (Fig. 23.24 B) are of four types (*i*) **fungiform,** mushroom-shaped papillae situated at the margin of the tongue, (*ii*) **filliform,** filamentous papillae on the upper surface of the tongue, (*iii*) **circumvallate,** large-sized papillae at the base of the tongue, and (*iv*) **foliate,** broad leaf-shaped papillae at the sides of the base of the tongue. Halfway between the lower incisors and frenulum in rabbit are located a pair of slit-like openings of the ducts of submaxillary glands. The main function of tongue is mixing of food with saliva in the mouth cavity.

***(iii)* Teeth:** The teeth are present on both jaws. The teeth of rabbit are **thecodont,** *i.e.,* they are situated in cup-like sockets of the jaw bone. They are **diphyodont,** *i.e.,* two sets of teeth are present during life time. The first set is known **milk or deciduous teeth** which are replaced by *permanent teeth* in the adults. The teeth of mammals are **heterodont,** *i.e.,* they are of different types. They are incisors, canines, premolars and molars. In rabbit canines are absent so a great gap called **diastema** is present in between the incisors and premolars. The dental formula of rabbit is

$$i\ \frac{2}{2},\ c\ \frac{0}{0},\ pm\ \frac{3}{2},\ m\ \frac{3}{3} = 28$$

The incisors are sharp, elongated, chisel-shaped teeth used for cutting or gnawing the food. Since the incisors of rabbit are open rooted they grow throughout life. The **grinding** or **cheek teeth** are anterior **pre-molars** and posterior **molars.** The upper jaw has three pre-molars and three molars while the lower jaw has two pre-molars and three molars. They are similar and have broad crown with sharp transverse enamel ridges. The last molar is smaller and without groove. The pre-molars and incisors are preceded by milk teeth in early life and are shed shortly after birth. The molars appear late in life and have no deciduous pre-decessors.

Pharynx: The portion of the cavity lying posterior and dorsal to the soft palate is the pharynx (Fig. 23.24 A). It is a short narrow chamber which is incompletely divided into three parts by the soft palate. The **nasopharynx** lies dorsal to the soft palate, the **oropharynx** below the soft palate and both communicate behind with the **laryngopharynx.** Shortly anterior to the free border of the soft palate, on each side, is a pit, the **tonsillar fossa,** which contains a small mass of lymphoid tissue, the **palatine tonsil.** The tonsillar fossa is bound in front and behind by low folds, an anterior glossopalatine and a posterior glossopalatine arch. The external nares open into the nasal passage and the internal nares or **choanae** have been pushed far back into the pharynx both by the palate and nasal passage. Posterior to the internal nares on the lateral wall of nasopharynx are located a pair of oblique slits, the openings of the **auditory** or **Eustachian tubes,** the canals which connect the pharynx with the cavity of middle ear. On the floor of laryngopharynx just, behind the tongue, there

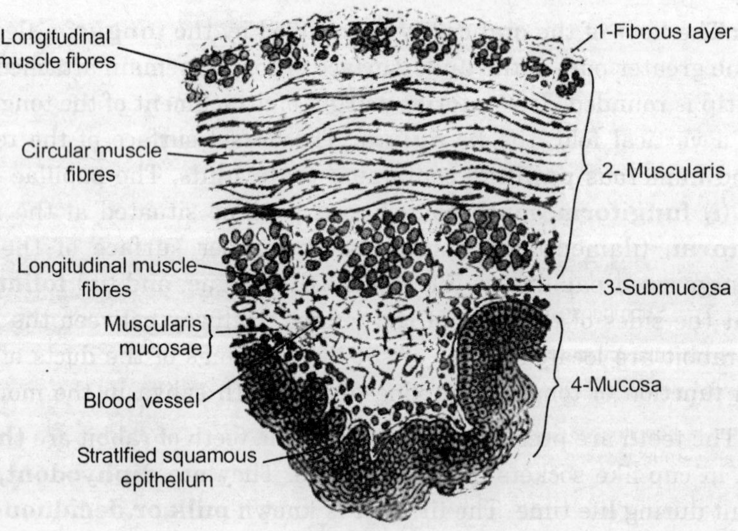

Fig. 23.25 *Rabbit*–T.S. of oesophagus

is a vertical slit, the **glottis** leading into the larynx. It is guarded by a thin, cartilaginous, bilobed process the **epiglottis.** In the pharynx the paths for food and air are crossed (as in all air-breathing vertebrates), but due to formation of palate and the posterior migration of internal nares, the air no longer enters the oral cavity as in Amphibia and most reptiles, but proceeds directly into the pharynx. Posteriorly the laryngopharynx leads into the oesophagus through a wide opening, the **gullet.**

Oesophagus: It is a long, narrow, elastic and muscular tube running, down through the neck dorsal and parallel to the trachea. It traverses the thorax dorsal to the heart and lung and after piercing the diaphragm enters the abdomen and opens into the stomach. Histologically the oesophagus (Fig. 23.25) is made up of outer thin **fibrous layer** followed by thick **muscle layer.** The muscle layer consists of two **longitudinal muscle layers,** one above and one below a thick layer of **circular muscles.** The muscle layer is followed by a thin **submucosa** layer and inner **mucosa** layer made up of stratified epithelium. The lumen of the oesophagus is considerably reduced due to folding of this layer. The longitudinal muscles of oesophagus help in forward movement of food to the stomach. No digestion occurs in the oesophagus.

Stomach: The stomach (Fig. 23.23) is the most dilated portion of the alimentary canal. It is a bean-shaped sac which lies transversely on the left side of the anterior part of the abdomen. The entrance of oesophagus into the stomach has a valve called **cardia** or **cardiac sphincter valve** made up of circular muscles. It prevents regurgitation of food. The large anterior part of stomach is called **cardiac stomach** and the small posterior part is called **pyloric stomach.** The pyloric end terminates at the **pyloric constriction.** The shorter,

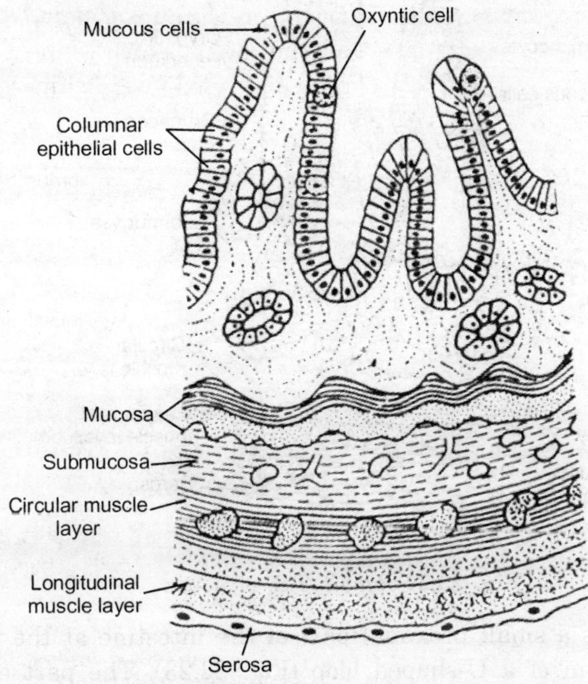

Fig. 23.26 *Rabbit*–T.S. of stomach

slightly concave inner surface of stomach is called **lesser curvature** and the large, convex outer surface is the **greater curvature.** The sac-like bulge between cardiac and pyloric regions is called **fundus.** The pyloric stomach opens into the duodenum and the opening is guarded by **pyloric sphincter valve.** Dorsally the stomach is suspended by the **mesogaster.** A small portion of mesogaster is attached to the stomach and its greater part forms the **greater omentum.** It is formed by reduplication of the peritoneum, as such is double walled. A red coloured spleen is attached to the outer curvature of the stomach. The portion of greater omentum between spleen and stomach is the **gastrosplenic ligament.**

The wall of stomach (Fig. 23.26) presents a typical structure. The outermost layer is serosa which is made up of simple squamous epithelium. It is followed by muscle layer made up of outer layer of longitudinal fibres and inner layer of circular fibres. The muscle layer is thickened at the pyloric end. Below the muscle layer there is a connective tissue layer, the submucosa. Finally there is the inner layer of mucous membrane which is raised into numerous irregular longitudinal folds which form deep gastric pits.

Small intestine: Following the stomach is the small intestine. It is the longest part of the alimentary canal. It is long, elongated and much convoluted tube differentiated into the following regions:

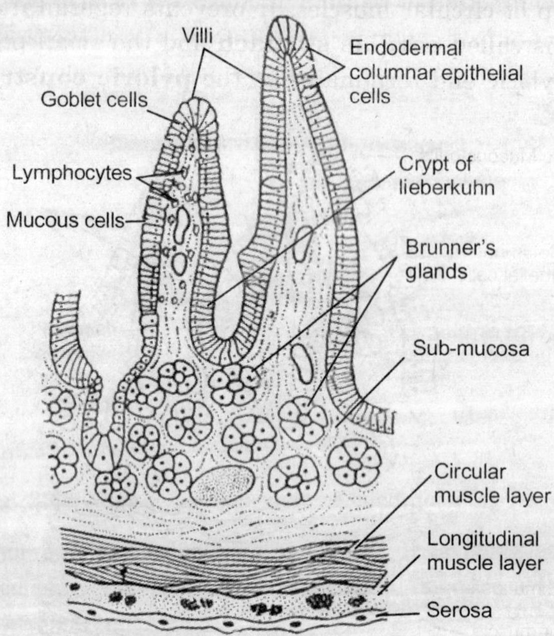

Fig. 23.27 *Rabbit*–T.S. of ileum

***(i)* Duodenum:** It is a small proximal part of the intestine at the pyloric end of the stomach. It takes the form of a U-shaped loop (Fig. 23.23). The part of this loop which descends posteriorly is the **descending limb,** the short portion at the turn is the **transverse limb** and the part that ascends anteriorly is the **ascending limb.** The duodenum is supported by a part of dorsal mesentery called **mesoduodenum** in which the **pancreas** is located. The pancreatic duct opens at the begining of the ascending limb while the bile ducts opens at the begining of the descending limb. The location of pancreatic duct in rabbit is unusual in mammals. In other mammals (for example, cat) there are two pancreatic ducts, the **principal one** joins the bile duct at an enlarged **ampulla** of **Vater** and opens into the duodenum. The second or **accessory pancreatic duct** enters the duodenum behind the opening of the first.

***(ii)* Jejunum and ileum:** Following the duodenum is the jejunum (Fig. 23.23) followed by ileum, but there is no definite boundary between the two. Both are long and greatly coiled and are held to the abdominal wall by folds of mesenteries. The great length provides extensive area for absorption. The absorptive area is further increased by development of folds and number of tiny finger-like processes called **villi.** In the centre of each villus is a lacteal vessel containing lymph (being a part of the lymphatic system) and is concerned with absorption of fat. In other histological details the small intestine of rabbit shows little differentiation from that of frog. The serosa, musculature, submucosa and mucosa, all show the same relationship (Fig. 23.27). The distal end of ileum is expanded to form a rounded sac-like structure, the **sacculus rotundus,** which opens into the caecum. An **ilio-caecal valve** is present at the opening by which the contents of ileum pass into the caecum before passing into colon.

***(iii)* Caecum:** At the junction of ileum and colon occurs a long, wide, thin walled tube, the **caecum** which terminates in a narrower thick-walled, blind tube, the **vermiform appendix.** The caecum varies in length in different mammals and in rabbit, whose food is exclusively vegetable, the caecum is about one and a half foot long consisting of a series of pouch-like dilations. The vermiform appendix is about five or six inches long. The inner surface of caecum is raised along the line of external constriction into a prominent spiral valve, a quarter of an inch or more in width. The plant cellulose is digested in the caecum.

On the wall of the intestine are abundantly present small portions of lymphatic tissue called **lymph nodules.** Aggregations of lymph nodules are called **Peyer's patches.** They occur along the entire small intestine on the side opposite to that attached to the mesentery, in the form of thickened oval spots on the surface. There is a larger patch at the junction of sacculus rotundus and caecum.

Large intestine: The large intestine is one meter or more in length. It consists of two regions.

***(i)* Colon:** The large intestine beyond the caecum is called **colon** (Fig. 23.23). It is wide, about 50 cm long and supported by a dorsal mesentery, the **mesocolon.** The first part of colon is called **ascending colon** which is somewhat long and follows a rather winding course. It is followed by a short transversely running portion, the **transverse colon** and on the left it turns abruptly as the **descending colon.** The wall of colon bears three longitudinal muscular bands or **taeniae** of colon and shows regular sacculations, the **haustra.**

***(ii)* Rectum:** The terminal portion of the descending colon is called rectum. It is a narrow tube about 70 cm long and has a beaded appearance due to faecal pellets present inside.

The large intestine absorbs water from food and helps in the formation of faecal pellets.

Anus: The rectum opens to the outside through the anus. It is a sphinctered opening present at the base of tail.

Digestive gland

There are a number of glands associated with the digestive system.

Mucous gland: Mucous glands are present in the mucous lining of the whole digestive tract but are more abundant in vestibule. They secrete mucous, a thick, slimy secretion which moistens and lubricates the food and help in swallowing. The mucous lining of the alimentary canal protects it from the action of enzymes.

Salivary glands: The salivary glands are masses of glandular tissue that pour salivary secretions in the mouth cavity. There are four pairs of salivary glands in rabbit and are named according to their position. The **parotid gland** is the largest of all and is located in front of the base of the pinna just under the skin. The duct of parotid, the **stenonian duct,** leaves the gland at its anterior border and runs forward just beneath the skin and opens on the inner side of the cheek opposite to the second upper pre-molar tooth. The **infraorbital gland** is an irregular lobulated mass lying below the orbit. Its duct runs downwards to open close to the opening of the parotid gland. The **submaxillary gland** is a compact roundish mass of reddish colour lying close to its fellow between the angles of the mandibles and in front of the larynx. Its duct is called **Wharton's duct** that leaves the outer side of the gland and runs forwards along the inner side of the jaw opening just behind the lower incisors. The **sulingual** gland is a small elongated, flattened reddish body present below the tongue and open by several short ducts on the floor of the mouth.

The secretion of salivary glands is a watery secretion called **saliva.** It mixes with the food in the buccal cavity. It contains ptyalin enzyme which converts starch into maltose.

Gastric glands: The mucous lining of stomach contains numerous long tubular gastric glands. The gastric glands of cardiac region are called **cardiac gastric gland** and of the pyloric region are called **pyloric gastric gland.** Both produce mucous. The gastric gland of fundic region are called **fundic gastric gland.** They contain mucous secreting cells, **zymogen cells** or **propepsin, prorenin** and **oxyntic cells** or hydrochloric acid producing cells.

Liver: The liver is a very large gland made up of **five lobes** (Fig. 23.23). Three lobes on the left side are **left central, left lateral** and a small **spigelian lobe** and the two lobes on the right side are **right central** or **cystic** and **caudate.** It has a convex anterior surface which fits into the posterior concave surface of the diaphragm attached by a fold of peritoneum called **falciparum ligament.** The posterior concave surface fits over the stomach and first part of the small intestine. The **gall bladder** is large thin walled and green coloured and lies in a groove between the right and left central lobes. The gall bladder stores **bile** which is a greenish fluid secreted by the liver cells. A **cystic duct** arises from the gall bladder and a **hepatic duct** arises from each liver lobe. All unite to form the **common bile duct** which opens into the proximal limb of the duodenum.

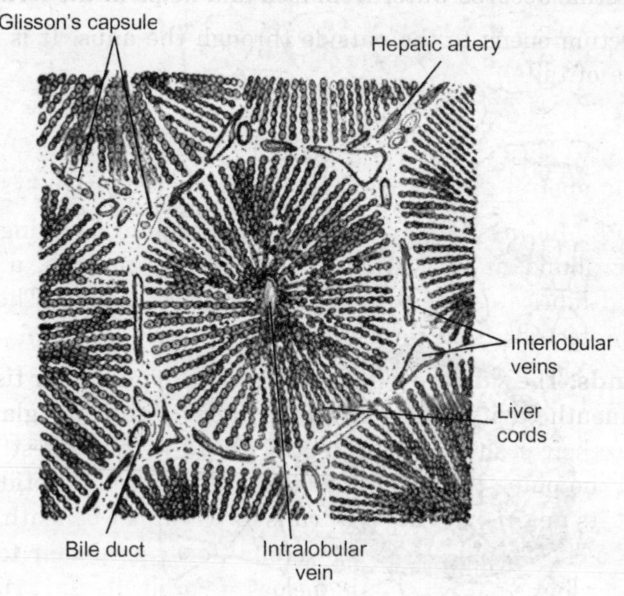

Fig. 23.28 *Rabbit*–T.S. of liver

The liver is a solid glandular organ made up of **hepatic lobules** (Fig. 23.28) separated from one another by loose connective tissue capsules enclosing branches of portal vein, hepatic artery and bile duct. The capsules are known as **Glisson's capsule.** Each lobule consists of a mass of polyhedral **hepatic cells** which are granular in appearance containing spherical nuclei. A specialized **Kuffer cells** are also found in liver which destroy germs. A network of **blood vessels** pierces the lobules there being thus **interlobular blood** vessels lying in between lobules and **intralobular vessels** lying in the centre of lobules. The hepatic cells are arranged in longitudinal cords. In between the hepatic cells are intercellular spaces which receive the secretions of hepatic cells. These are the commencements of the **bile channels,** which merge to form large **interlobular bile ducts,** ultimately uniting to form the bile duct. The bile ducts are lined by columnar epithelium outside which is a basement membrane. In larger ducts some fibrous and plain muscular tissue may also be present.

Pancreas: Pancreas is an irregular, pinkish gland lying in the mesentery between the two limbs of the duodenum (Fig. 23.23). It has a single pancreatic duct which opens into the distal limb of duodenum. Histologically the pancreas (Fig. 23.29) is made up of a large number of branching tubules called **acini** embedded in connective tissue containing blood and lymph vessels, nerves and pancreatic ductules. The cells of acini secrete pancreatic juice which contains enzymes such as lipase, amylase and trypsinogen. Clusters of cells called **islets of Langerhans** are present between the acini. These are endocrine glands which secrete the hormone insulin which plays an important in carbohydrate metabolism. Its deficiency causes diabetes.

Intestinal gland: In the mucous lining of the intestine there are numerous microscopic tubular glands which secrete the intestinal juice, **succus entericus.** In the wall of duodenum

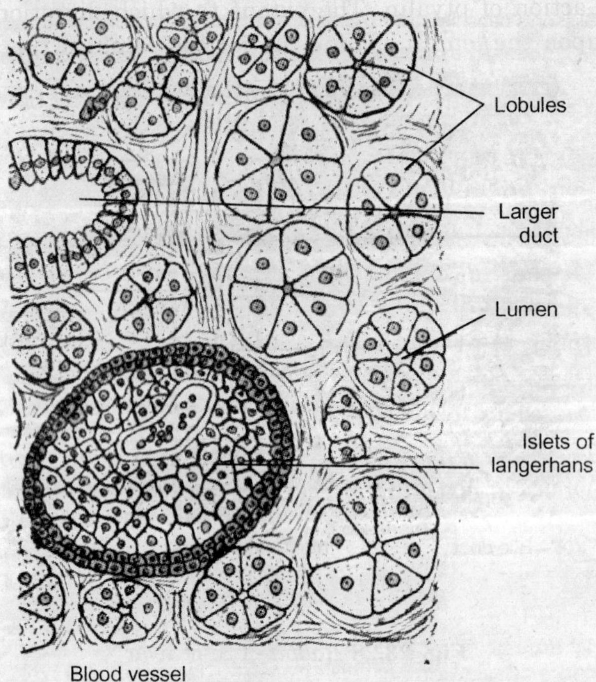

Fig. 23.29 *Rabbit*–T.S. of pancreas

special glands called **Brunner's glands** are present at the base of villi. They open into **crypts of Lieberkuhn** present between the villi and surrounded by columnar and goblet cells. Both Brunner's glands and crypts of Lieberkuhn secrete intestinal juice. The succus entericus is a watery alkaline fluid containing **mucin** and other important enzymes.

Food and feeding

The rabbit is herbivorous, *i.e.,* it feeds on green vegetables, leaves, grasses, etc. It gnaws or cuts its food into small pieces with the help of chisel-shaped incisors. The small pieces of food are picked up by the movable lips and passed into the buccal cavity. Here the food is masticated by the grinding cheek teeth (pre-molars and molars) and at the same time mixed with saliva poured by the salivary glands. During mastication the food forms a pulpy mass and turned round with the help of tongue to form a **food bolus.** The bolus is swallowed after mastication is completed. The tongue pushes the bolus behind the pharyngeal muscles forcing it into the oesophagus. This is known as **deglution**. From the oesophagus it is forced into the stomach by peristaltic contractions. The soft palate prevents the bolus from entering into the nasal passage and epiglottis prevents it from entering the larynx.

Physiology of digestion

Buccal cavity: Digestion of food starts in the buccal cavity where the food mixes with **saliva.** The saliva contains **mucin** and digestive enzyme **ptyalin.** This enzyme acts on **starch** and converts it into **maltose** and **dextrin.** The pH of saliva is approximately 6 or

7 which is suitable for action of ptyalin. The extent to which digestion takes place in the buccal cavity depends upon the length of time the food is retained there and thoroughness of mastication.

Stomach: In the stomach the food is subjected to the action of **gastric juice.** A hormone **gastrin** secreted by the stomach stimulates the gastric glands to produce gastric juice. The gastric juice contains mucus, HCl, enzymes renin, pepsin and gastric lipase.

The enzyme renin acts on soluble milk protein **caseinogen** and converts it into insoluble **casein** in the presence of **calcium ions.** Pepsin acts on proteins and break them down into peptones. Gastric lipase acts on fats.

The pyloric sphincter helps to retain the food for sometime in the stomach, During this period the food is churned and mixed thoroughly with gastric juice until it takes the form of a thick creamy acid fluid called **chyme.**

Small intestine: The chyme passes into the small intestine. The hormone secretin is produced by duodenum which induces the pancreas to become more active, the liver to pour bile and the pyloric to relax. The chyme mixes with bile from the liver, pancreatic juice from the pancreas and succus entericus from the intestine.

The bile juice is an alkaline dark green coloured fluid containing water, bile salts and bile pigments. It has no digestive enzymes, so it does not take part in digestion but it plays a very important role in the process of digestion. The salts of bile are sodium bicarbonate, glycocholate and taurocholate, the two latter ones activate pancreatic lipase and bring about emulsification of fat (break down fat into smaller globules), so that they can be easily acted upon by pancreatic lipase. Sodium bicarbonate neutralizes the acidity of chyme and makes it alkaline (pH 8) because the enzymes of small intestine require alkaline medium for their action. Bile salts prevent purification of chyme. The bile pigments **bilirubin** and **biliverdin** are excretory wastes produced in the liver by breakdown of haemoglobin of worn out red blood cells. They only give green colour to bile and are passed out as such with faeces.

The hormone **secretin** secreted by the duodenum is produced under the influence acidic chyme. It goes into the pancreas and stimulates the production of **pancreatic juice**. The pancreatic juice is alkaline (pH 8.8) and it neutralizes chyme. It contains **proteolytic trypsin, diastatic amylase** and **lipolytic lipase.** Trypsin occurs in the form of inactive proenzyme **trypsinogen** which is activated by intestinal **enterokinase.** It converts peptones into amino acids. The lipase **steapsin** converts fatty emulsion into fatty acids and glycerol and amylase converts starch to maltose.

The succus entericus is produced by the intestinal glands. It is an alkaline watery fluid containing mucin and many enzymes. Erepsin, a proteolytic enzyme converts peptones into amino acids, enterokinase an activator enzyme, activates trypsinogen, lipase, a lipolytic enzyme converting fat into fatty acid and glycerol and carbohydrate digesting enzymes maltase (maltose to glucose), invertase (sucrose to glucose and fructose) and lactase (lactose to glucose and galactose).

Digestion of cellulose takes place in caecum where cellulose decomposing bacteria and protozoans are present. They break cellulose into soluble sugars and in return get nitrogenous food from the animal (symbiotic relationship). The process of digestion and absorption of cellulose is a slow process, therefore food remains for a long time in the caecum.

As the food passes down the duodenum it receives more water and forms a fluid **chyle** which is slowly forced by peristalsis through the long small intestine. The ends products of digestion are amino acids, glucose or similar monosaccharide sugars, fatty acids and glycerol.

Absorption: Ordinarily food is not absorbed in the buccal cavity or the stomach (only drugs and alcohol may be absorbed). Major absorption takes place in the jejunum and ileum. The digested food passes from the gut into capillaries of villi and conveyed to the liver by hepatic portal vein. The liver acts as a regulating mechanism. Only required amount of sugar is allowed to enter the blood leaving the liver by hepatic vein. Excess of sugar is transformed into glycogen and stored in the liver. Proteins cannot be stored in the liver. The amino acids pass into the blood stream in suitable quantities but excess is subject to the process of **deamination** in the liver forming relatively harmless substance **urea** which is excreted by the kidneys and into other products which can be transformed into glucose. The fatty acids and glycerol pass into the lymph vessels or lacteals directly as minute fat globules. The lymph vessels therefore contain a whitish fat emulsion. This enters the blood stream through ducts connecting lymph vessels with the blood vessels. The absorbed carbohydrates and fats are used as sources of energy for vital activities and proteins are used for building up new protoplasm. By the time digested products are absorbed, the fluid content of the alimentary canal passes into the colon. Here water is absorbed and this helps in water conservation.

Egestion or defecation: After absorption the food residuum reaches the rectum as semisolid material and forms **faeces.** In rabbit it forms dehydrated **pellets** and is egested or extruded at intervals through the anus.

Rabbits are coprophagous, *i.e.,* they feed on their own excreta. They feed on night excreta which is soft and moist due to incompletely digested cellulose. The day excreta is dry and hard so they do not eat it.

CIRCULATORY SYSTEM

The circulatory or blood vascular system of mammals is of closed type. It includes a pumping organ, the **heart,** a circulating fluid, the **blood** and **blood vessels (arteries, veins** and **capillaries**) for transporting blood. It is concerned with the distribution of materials to all parts of the body.

Heart

The heart is a pear-shaped, muscular, four chambered organ of red colour lying mid-ventrally between the two lungs in the anterior part of the thorax. The broad **base** of the heart is directed towards the anterior side and the pointed **apex** towards the posterior side. The heart is completely enclosed by a transparent, thin walled, double layered sac, the **pericardium.** The space between the two layers is **pericardial cavity** and it is filled with a watery fluid, the **pericardial fluid.** The fluid allows free movement of the heart and protects it from mechanical injury. The pericardium is connected to the ventral thoracic wall and the posterior diaphragm to keep the heart in position.

External structure

The heart is four chambered consisting of two auricles and two ventricles (Fig. 23.30). The **sinus venosus** and **truncus (conus) arteriosus** is absent. Externally a transverse groove called **coronary sulcus or auriculoventricular groove** divides the heart into a small, dark red anterior **auricular region** and a larger, paler posterior **ventricular region.**

Auricles: The auricular region consists of two **auricles.** The right auricle is slightly larger than the left. The lower end of each auricle is produced into a swollen flap, the **auricular appendix,** which slightly project over the ventricle of its side.

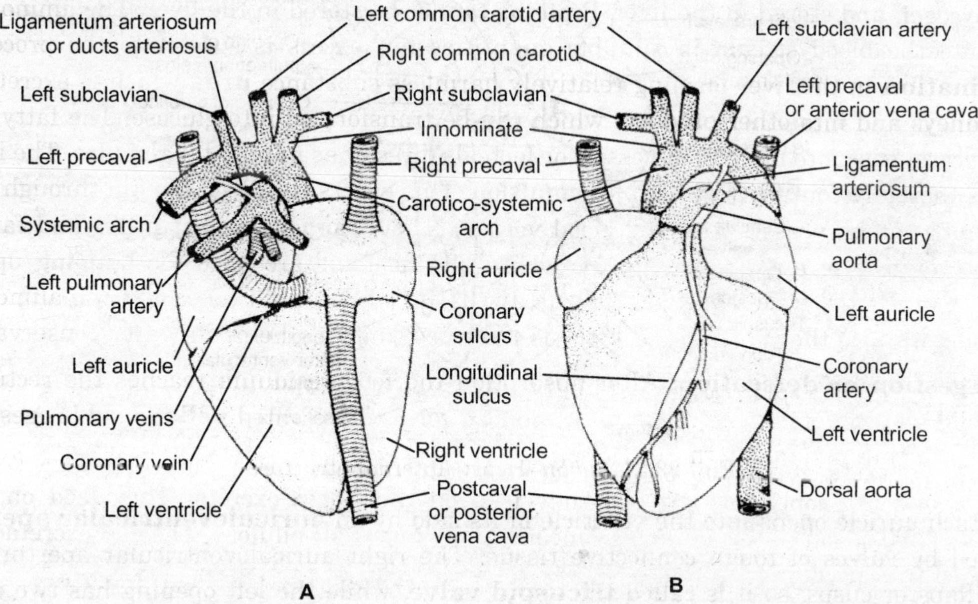

Fig. 23.30 *Rabbit*–Heart. **(A)** Dorsal view; **(B)** Ventral view

Ventricles: The ventricular region consists of two **ventricles.** It is divided into left and right parts by an **interventricular groove** which runs from the left side of the base obliquely to the right and terminating to the right of the apex. Thus the right ventricle remains smaller and the left one becomes larger and includes the rounded apex.

Internal structure

Internal structure of heart (Fig. 23.31) can be easily seen by dissecting the heart longitudinally. Internally it consists of two completely divided auricles and ventricles.

Auricles: The two **auricles** or **atria** are thin-walled chambers separated from each other by a thin, muscular, vertical **interauricular septum.** In the embryo a small oval opening, the **foramen ovale** is present in the septum and the two auricles communicate through it. In the adult it closes and is represented by a small depression called **fossa ovalis.** Internally the auricular wall forms low muscular ridges called **musculi pectinati.** The wall of left auricle is thicker than that of the right.

The **caval veins** (two **precavals** and one **postcaval**) collect blood from different parts of the body and open into the right auricle by three separate openings. A membranous

eustachian valve guards the opening of the post caval vein. Near it lies the **sinuauricular node,** a remnant of the embryonic sinus venosus. The two **pulmonary veins** bringing oxygenated blood from the lungs open into the left auricle by a single pulmonary opening without any valve.

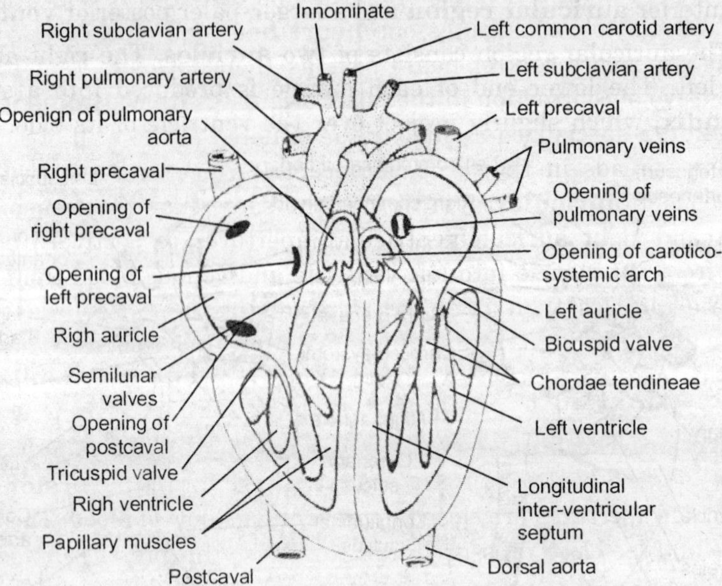

Fig. 23.31 *Rabbit*–Heart. Internal structure

Each auricle opens into the ventricle of its side by an **auriculoventricular aperture** guarded by valves of tough connective tissue. The right auriculoventricular aperture has three flaps or cusps, so it is called **tricuspid valve,** while the left opening has two cusps, so it is called **bicuspid valve.** The bicuspid valve is also called **mitral valve** because it resembles the hat called **miter** worn by the bishops. The cusps are connected to finger-like processes called **columnae carnae** present in the wall of the ventricle, by **chordae tendinae.** These valves prevent the backflow of blood from the ventricles to the auricles.

Ventricle: The ventricles are thick walled muscular chambers. The two ventricles are completely separated by a thick, oblique and vertical **interventricular septum.** The inner surface of ventricular wall bears numerous irregular muscular ridges called **columnae carnae** and a few large conical elevations called **papillary muscles.** The free ends of bicuspid and tricuspid valves are attached to columnae carnae by tough connective tissue strands called **chordae tendinae.** The chords prevent the flaps from being inverted into auricular cavities when the ventricles contract. The two main arteries arise from the ventricles. The **pulmonary arch** or **aorta** arises from the right ventricle and carries venous blood to the lungs for oxygenation. The **systemic arch** or **aorta** from the left ventricle and carries oxygenated blood to different parts of the body. The two aorta leave the ventricle, cross each other and communicate via a muscular strand, the **ligamentum arteriosum.** In the embryonic condition the two aorta are connected by ductus arteriosus or **Boteli**. At the base of each aorta three **semilunar valves** are present which prevent the backflow of blood from the aorta to the ventricle.

Working of heart and course of circulation

The heart acts are a force pump and keeps the blood in circulation through blood vessels throughout life. The action of heart is the result of two alternately but simultaneously occurring processes: **systole** or contraction of heart and **diastole** or relaxation of heart. One complete systole and diastole constitutes one **heart beat** or **heart** or **cardiac cycle.** The cardiac muscles of the heart control heart beat as they have inherent tendency to beat rhythmically. The wave of contraction in the heart originates from **sinuauricular node (SA node)** or **pacemaker,** a specialized cardiac tissue situated in the wall of right auricle. This wave of contraction spreads in the form of concentric rings over the two auricles which contract simultaneously (**auricular systole**). At the same time the bicuspid and tricuspid valves of the left and right auriculo-ventricular apertures respectively, open forcing the oxygenated blood from left auricle into left ventricle and venous blood from right auricle to right ventricle. The auricles then relax (**auricular diastole**). Now wave of contraction originates from the **auriculo-ventricular node** (**AV node**), a special cardiac muscle situated in the wall between the two ventricles. From AV node arises the **bundle of His** which bifurcate into two branches, the **auriculo-ventricular bundles (AV bundles),** in the interventricular septum. The fibres of AV bundles called **Purkinje** fibres ramify into the ventricular wall. When the ventricles begin to contract (**ventricular systole**) the right and left auriculoventricular apertures are closed to prevent backflow of blood. The blood is blocked inside the ventricle and the blood pressure increases. When this pressure exceeds the pressure in the arteries the semilunar valves of the aorta open and blood is forced into them. Thus the oxygenated or arterial blood from the left ventricle goes into the systemic aorta and distributed to all parts of the body. The venous blood from the right ventricle goes into the pulmonary arch and carried to the lungs for purification. When the ventricles relax

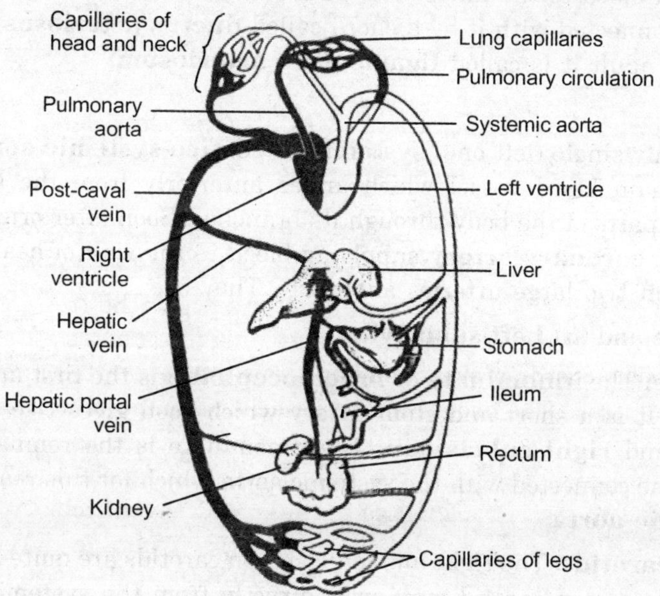

Fig. 23.32 *Rabbit*–Diagram showing double circulation

(**ventricular diastole**) the pressure of blood inside the aorta increases and the semilunar valves of the aorta close. The auricles are filled with blood and waves of contraction start again after a short pause.

Double circulation: In mammalian heart there remains a complete separation of venous and oxygenated blood. The right half of the heart has venous blood which is sent to the lungs for purification through pulmonary aorta. The left side of the heart contains oxygenated blood which is sent to different parts of the body through systemic circulation. Thus in one complete circuit the blood passes through the heart twice: once through the right side and second time through the left side. This type of circulation is called **double circulation** (Fig. 23.32). Due to complete separation of the ventricles there is no mixing of venous and arterial blood in the heart.

Arterial system

The blood vessels which carry blood away from the heart to different parts of the body are called **arteries** and they together form the **arterial system.** All arteries except **pulmonary artery** carry oxygenated blood. In mammals there are two large blood vessels called **aorta** or **aortic arches** which originate from the ventricle and then send branches to all parts of the body. They are:

(1) **Pulmonary aorta** and (2) **Systemic aorta.**

1. Pulmonary aorta

The pulmonary aorta (Fig. 23.33) originates anteriorly from the right ventricle, runs forward between the auricles, curves over the left auricle and finally divides into right and left **pulmonary arteries,** each of which runs to the lung of its side. The pulmonary artery carries deoxygenated blood to the lungs. The pulmonary aorta passes beneath the systemic aorta and remains connected with it by a chord called **ductus arteriosus** in the embryonic condition but in the adult it is called **ligamentum arteriosum.**

2. Systemic aorta

In mammals only single (left one) **systemic** or **cortico-systemic aorta** (Fig. 23.33) is present. It is the main blood vessel which arises anteriorly from the left ventricle and supplies blood to all parts of the body through its branches. Soon after origin it gives off two small **right** and **left coronary artery** supplying blood to the wall of heart. It turns to the left side and gives off two large arteries anteriorly. They are:

(*i*) **Innominate** and (*ii*) **Left subclavian.**

(i) **Innominate:** The **innominate** or **brachiocephalic** is the first artery to arise from the systemic aorta. It is a short and stout artery which soon gives rise to **right** and **left common carotid** and **right subclavian.** The innominate is the remnant of the ventral aorta that has become connected with the systemic aorta which for this reason is often called the **cortico-systemic aorta.**

(a) Common carotids: The right and left common carotids are quite long blood vessels. Sometimes the left common carotid may arise directly from the systemic aorta. Both the

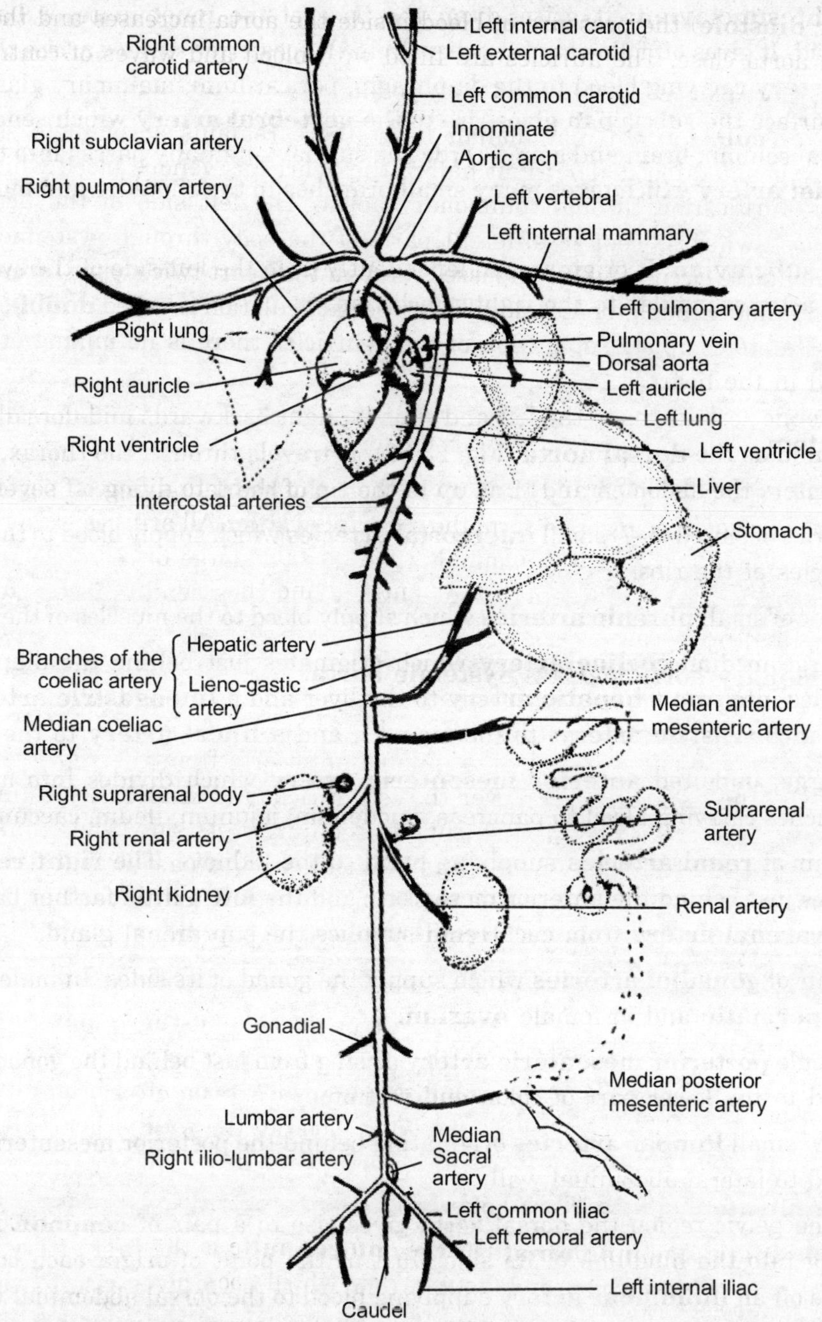

Fig. 23.33 *Rabbit*–Arterial system

common carotids run anteriorly through the neck almost parallel to the trachea and divide into **external** and **internal carotids.** The external carotid supplies blood to the back of head, tongue, lips, pinna and muscles of the jaws and the internal carotid supplies blood to different parts of the brain.

(b) Right subclavian: It arises from the base of the innominate or base of right common carotid. It gives off following branches: from the posterior surface arises the **internal mammary artery** carrying blood to the diaphragm, pericardium, mammary glands, etc. On its anterior surface the subclavian gives rise to the **vertebral artery** which sends branches to the vertebral column, brain and spinal cord. The subclavian finally passes into the forelimb as the **brachial artery** which gives many small branches to the shoulder and muscles of the region.

(ii) **Left subclavian:** It originates independently from the left side of the systemic arch and gives off arteries similar to the right subclavian.

Dorsal aorta

The systemic aorta turns to the left and runs straight backwards mid-dorsally below the vertebral column as the **dorsal aorta** (Fig. 23.33). It travels through the thorax, pierces the diaphragm, enters the abdomen and runs up to the tip of the tail giving off several arteries:

1. A series of paired (5–7) small **intercostal arteries** which supply blood to the intercostal muscles of the ribs.
2. A pair of small **phrenic arteries** which supply blood to the muscles of the diaphragm.
3. A large median **coeliac artery** which originates just behind the diaphragm and divides into two: a **hepatic artery** to the liver and a **lineogastric artery** sending a branch (**gastric artery**) to the stomach and a **lineal artery** to the spleen.
4. A large, unpaired **anterior mesenteric artery** which divides into a number of branches carrying blood to pancreas, duodenum, jejunum, ileum, caecum and colon.
5. A pair of **renal arteries** supplying blood to the kidneys. The **right renal artery** arises just behind the anterior mesenteric and the **left** a little farther back. A small **suprarenal artery** from each renal supplies the suprarenal gland.
6. A pair of **gonadial arteries** which supply the gonad of its sides. In male it is known as **spermatic** and in female **ovarian.**
7. A single **posterior mesenteric artery** arising from just behind the gonadial supplies blood to the lower part of colon and rectum.
8. A few small **lumbar arteries** originating behind the posterior mesenteric supplying blood to lateral abdominal wall.
9. In the pelvic region the dorsal aorta gives rise to a pair of **common iliacs,** each going into the hindlimb of its side. Just at the point of origin each common iliac gives off an **iliolumbar artery** supplying blood to the dorsal abdominal wall. It then divides into **internal iliac** or **hypogastric** supplying the inner side of the leg, a small **vesicular** to the urinary bladder, ureter, uterus in female and seminal vesicle in male and an **external iliac** or **femoral** to the outer region of leg.
10. The dorsal aorta terminates posteriorly as a single **caudal artery** which supply blood to the tail region.

Venous system

The blood vessels which collect blood from different parts of the body and bring it to the heart are called **veins** and they together form the **venous system.** All the veins except pulmonary vein carry impure blood to the heart. The venous system includes the following sets of veins: 1. Pulmonary veins, 2. Caval veins and 3. Hepatic portal system.

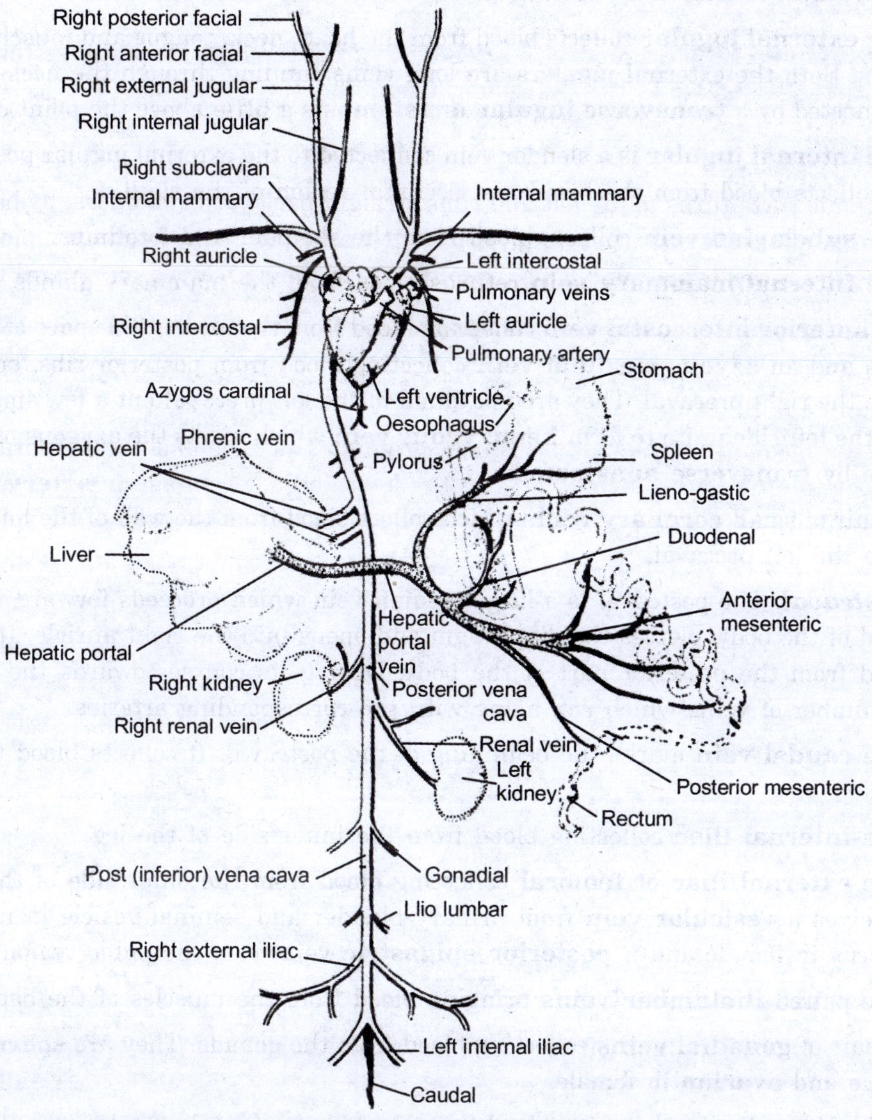

Fig. 23.34 *Rabbit*–Venous system

1. Pulmonary veins: The left and right pulmonary veins (Fig. 23.34) collect oxygenated blood from the lungs and unite to form a common pulmonary vein which opens into the left auricle by a single opening.

2. Caval veins: There are three caval veins—the **left** and **right anterior vena cavae** or **precavals** and a **posterior vena cava** or **postcaval** (Fig. 23.34). They collect venous blood from all parts of the body except the lungs and open into the right auricle by separate openings.

(a) *Precavals*: The left and right precavals collect blood from the anterior part of the body. Each precaval is formed by the union of the following veins:

(i) The **external jugular** collects blood from the head, neck, tongue and muscles of the jaws. Both the external jugulars are long veins running through the neck and are connected by a **transverse jugular anastomosis** a little above the point of origin.

(ii) The **internal jugular** is a slender vein connected to the external jugular posteriorly. It collects blood from the brain and occipital region of the skull.

(iii) The **subclavian vein** collects blood from the shoulder and forelimbs.

(iv) The **internal mammary vein** returns blood from the mammary glands.

(v) An **anterior intercostal vein** collecting blood from the intercostal space of anterior ribs and an **azyous cardinal vein** collecting blood from posterior ribs, both open into the right precaval. They are not found in the left precaval but a few small veins on the left side unite to form **hemizygous vein** which meets the azygous of the left side by **transverse anastomosis.**

(vi) A pair of small **coronary veins** which collect blood from the wall of the heart open into the left precaval.

(b) *Postcaval*: The postcaval is a large median vein which proceeds forward from the posterior end of the body, pierces the diaphragm and opens into the right auricle. It collects venous blood from the posterior part of the body. During its course towards the heart it receives a number of veins which run along with the corresponding arteries.

(i) The **caudal vein** marks the beginning of the postcaval. It collects blood from the tail.

(ii) The **internal iliac** collecting blood from the inner side of the leg.

(iii) The **external iliac** or **femoral** collecting blood from the outer side of the leg. It receives a **vesicular vein** from urinary bladder and seminal vesicle in male and uterus in female and a **posterior epigastric** vein from the pubic region.

(iv) The paired **iliolumbar veins** bringing blood from the muscles of the back.

(v) A pair of **gonadial veins** collecting blood from the gonads. They are **spermatic** in male and **ovarian** in female.

(vi) Paired **renal veins** bringing blood from the kidneys (renal portal system is absent in rabbit). Each receives a **suprarenal vein** from the suprarenal gland.

(*vii*) The postcaval receives two pairs of **hepatic veins** from the liver.

(viii) A pair of **phrenic veins** collecting blood from the diaphragm also join the post caval.

3. Hepatic portal system: The blood from different parts of the alimentary canal is not carried directly to the heart but is carried to the liver by hepatic portal vein. The hepatic portal vein (Fig. 23.34) is formed by the union of four veins:

(i) a **lineogastric** from the stomach and spleen

(ii) **duodenal** from the duodenum

(iii) **anterior mesenteric** from the ileum, caecum and colon and

(iv) **posterior mesenteric** from rectum.

The blood from liver is collected by hepatic vein which joins the postcaval.

Blood

The **blood** of mammals, as in other vertebrates, is a liquid connective tissue which flows in the blood vessels. It is a circulatory media which plays an important role in the transport of food, gases, hormones, excretory wastes, etc. The blood consists of a liquid **matrix,** the **plasma,** in which are suspended some **elements,** such as, **red blood cells** or **corpuscles (RBC)** or **erythrocytes, white blood cells or corpuscles (WBC)** or **leucocytes** and **platelets** or **thrombocytes.** The red colour of blood is due to presence of iron-containing respiratory pigment, the **haemoglobin,** in the red blood cells. It is responsible for the transport of oxygen.

The blood of rabbit resembles that of other vertebrates in all structural and functional details but exhibit some minor differences:

1. The erythrocytes in mammals are small (about 7.5 μ in diameter), circular, biconcave and non-nucleated but in frog they are large (about 25 μ in diameter), oval, biconvex and nucleated.
2. The number of erythrocytes is more in rabbit than in frog.
3. The blood platelets of rabbit are rounded, non-nucleated and more in number while in frog they are spindle-shaped, nucleated and less in number.

Lymphatic system

Beside the blood vascular system, a **lymphatic system** is also present in all vertebrates. It includes the circulating fluid **lymph, lymph capillaries, lymph vessels** and **lymph nodes.**

Lymph: Blood never leaves the blood vessels (arteries and veins) due to their thick walls but the wall of blood capillaries are extremely thin, so blood plasma readily seeps out through them into the intercellular spaces of the tissues. This colourless part of blood is called **lymph** or **tissue fluid.** It is similar to plasma in constitution but lacks erythrocytes and some plasma proteins and leucocytes are more than in blood. There is four or fives times more lymph than blood in the body. Like blood lymph is also a circulatory fluid which carries food and oxygen from the blood to the tissues and receives substances from the tissues to re-enter blood.

Lymph capillaries: The lymph is drained into a network of extremely permeable blindly ending **lymph capillaries** located in the tissue spaces. The fluid pressure inside the lymph capillaries is almost zero.

Lymph vessels: The **lymph vessels** are formed by the union of lymph capillaries. They are thick walled and provided with **valves.** In mammals the **lymph hearts**, found in lower vertebrates, are absent. The lymph flows due to squeezing pressure of the surrounding muscles. The valves keep the lymph flowing in one direction. There are two main lymphatic ducts—the **thoracic duct** and the **right lymphatic duct.** All the lymph vessels of the posterior side of the body and from the left side of the head, neck and thorax communicate with the thoracic duct which finally opens at the point where left subclavian meets the left external jugular vein. The lymph vessels from the right side of the head, neck, thorax and right arm open into the right lymphatic duct which opens at the point where right subclavian meets the right external jugular vein. The lymphatic system is a closed system but it does not form a complete circuit in the body.

Lymph nodes: In mammals the lymph vessels are provided with **lymph nodes** at certain places. Lymph nodes are not found in other vertebrates. Each lymph node consists of a mesh work of connective tissue fibres enclosed in a capsule. In rabbit and other mammals lymph nodes are found in the head, neck, arm pits, tonsils, groins, near large blood vessels and Peyer's patches in the intestine. The lymph nodes produce lymphocytes of the body's defense system.

The lymphatic system helps in maintaining a constant chemical environment around the tissues as it transports food, oxygen, hormones and other chemicals from the blood to the cells and waste products, carbon dioxide, etc. from the cells to the blood. It is a very important part of the immune system. During some infection the lymph nodes may swell up and become painful.

RESPIRATORY SYSTEM

The respiratory mechanism of mammals is more complicated. The **lungs** are the sole organs of respiration and the respiratory surface they enclose has a complex structure. The thoracic cavity in mammals is completely separated from the abdominal cavity by a muscular partition, the **diaphragm.** Another characteristic feature is the complete separation of the **respiratory channel** from the **food canal** by the development of a **secondary palate.** The result of this is that the internal nares are pushed in and lie in close vicinity of the glottis. The respiratory system of rabbit includes two parts: (*i*) respiratory tract or passage (*ii*) respiratory organ or lungs.

(*i*) Respiratory passage: The respiratory passage is the path along which the air moves. The fresh air enters the lung through it and the foul air after gaseous exchange passes out through the same passage. The respiratory tract is lined by richly vascularized, ciliated mucous secreting epithelium. It includes the following parts:

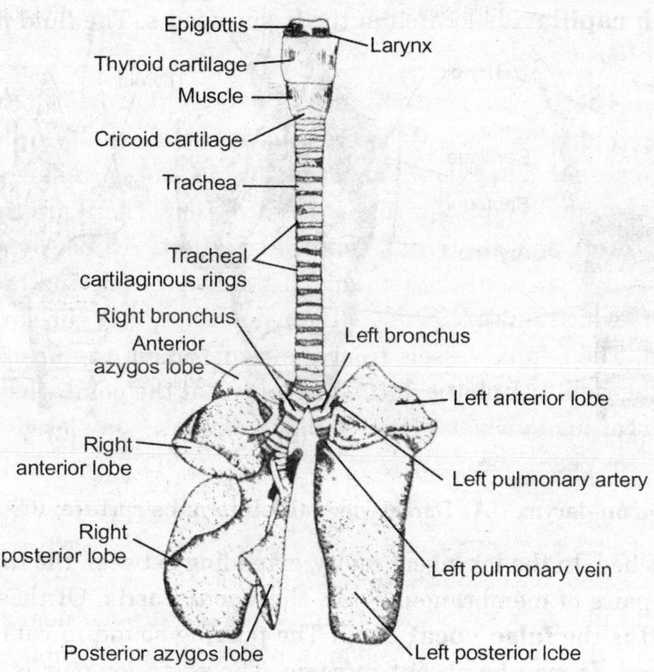

Fig. 23.35 *Rabbit*–Respiratory system

External nares: External nares are a pair of openings present near the tip of snout. The air is drawn in through it.

Nasal chambers: The external nares open into nasal or olfactory chambers which are separated from each other by vertical **nasal septum** and from the buccal cavity by the **hard palate**. Each nasal chamber is lined by vascularized mucous epithelium. The inner surface is increased by thin scroll-like **turbinal bones.** The nasal chambers serve two functions: first they act as olfactory organs and secondly they help in respiration. The air passing through the nasal chambers becomes warm and moist and is freed of dust particles which adhere to the moist mucous lining and hairs present in the nasal chambers.

Internal nares and **glottis:** The two nasal chambers communicate behind with the **nasopharynx,** present above the soft palate, through the **internal nares.** Posteriorly the nasopharynx communicate with the **laryngopharynx,** by a large vertical aperture the **glottis** which is covered by a flap-like bilobed, cartilaginous epiglottis. It serves to prevent food from entering the glottis during swallowing.

Larynx: Tho **larynx** or **voice box** is a small expanded chamber present at the anterior end of the trachea (Fig. 23.36). Its wall is supported by four cartilages. The ventral wall is supported by a large shield-shaped **thyroid cartilage.** A little posterior to the thyroid cartilage is the **cricoid cartilage** which forms a ring around the larynx. It is broad dosally and narrow ventrally. The dorsal side of larynx is supported by a pair of small and curved **arytenoid cartilage.** At the apex of each arytenoid cartilage a small nodule-like **cartilage**

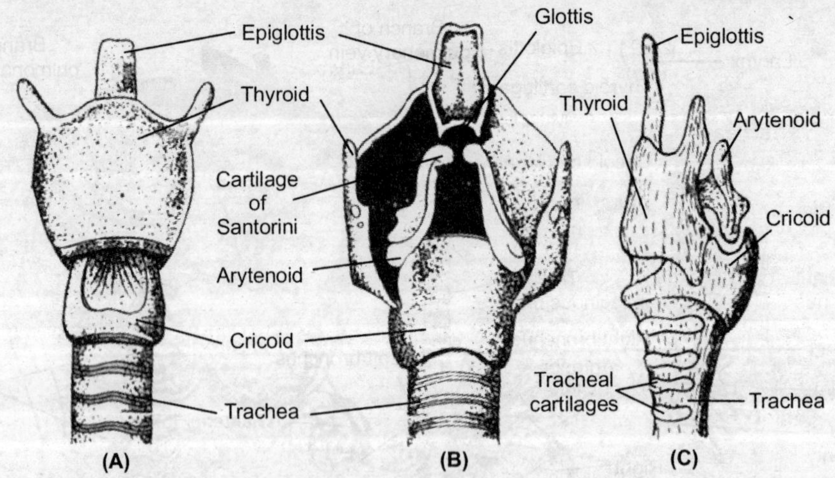

Fig. 23.36 *Rabbit*–larynx. **(A)** Dorsal view; **(B)** Internal structure; **(C)** Lateral view

of **Santorini** is attached. In the laryngeal cavity, extending between the thyroid and arytenoid cartilages, are two pairs of membranous folds, the **vocal cords.** Of these the anterior pair is more delicate and is the **false vocal cord.** The purring sound in cats is due to vibration of the false vocal cord. It may be absent in some. The posterior pair is known as the **true vocal cord.** In some mammals, such as the *Hippopotamus,* the true vocal cords may also be absent. The vocal cords arise as folds from the lateral walls of the larynx and consists of elastic connective tissue covered by mucous membrane. They close the glottis except for a slit between their free edges and it is through this gap that air enters during the respiratory movements. In resting or normal condition the vocal cords lie apart, at an acute angle to one another but for sound production they come close, almost parallel to each other. They vibrate as the air under pressure from the lungs rushes out through the narrowed glottis resulting in **sound production** or **phonation.** The movement of vocal cords is maintained by the action of laryngeal muscles. Loud sound is not produced in rabbit. In human beings speech is the result of combined action of vocal cords, tongue, pharynx, mouth and lips. The protrusion of larynx in humans is visible as Adam's apple in the neck.

Trachea: From the larynx the **trachea** or **wind pipe** (Fig. 23.35) proceeds posteriorly. It is a long tube running through the length of the neck beneath the oesophagus. On emerging in the thoracic cavity it divides into right and left primary **bronchi** or **bronchial tubes.** Each bronchus passes into the lung of its side. The wall of trachea and primary bronchi are supported by cartilaginous **tracheal rings** which are incomplete dorsally. The function of these rings is to prevent the trachea from collapsing and to allow free passage of air from the pharynx to the lungs. Internally the trachea and bronchi are lined by ciliated epithelium with numerous mucous glands. The mucous keeps the inner surface moist and slimy and holds dust particles and bacteria coming inside with air, thus protects the respiratory tract from infection.

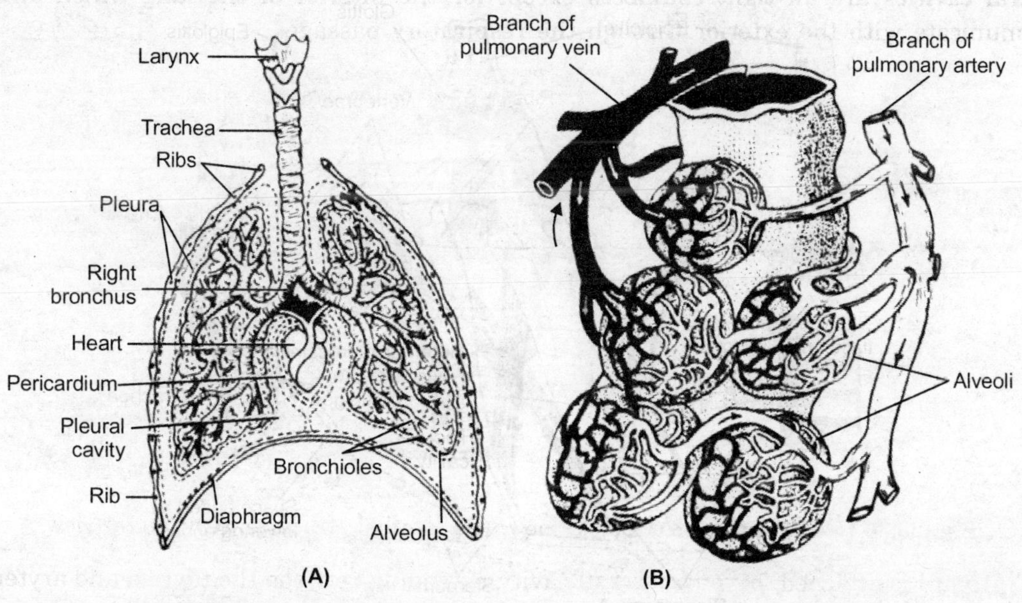

Fig. 23.37 *Rabbit*–**(A)** Median longitudinal section through thorax showing internal structure of lungs; **(B)** Figure showing arrangement of blood vessels in alveoli

(ii) **Respiratory organ:** A pair of **lungs** (Fig. 23.35) are the only respiratory organs of mammals. They are pinkish, spongy, elastic and highly vascular structures lying in the pleural cavities of the thorax. The right lung is somewhat larger than the left. It is divided into four lobes: **anterior azygous, right anterior, right posterior** and **posterior azygous.** The left lung is divided into two lobes: **left anterior** and **left posterior.** The lungs are freely suspended within the pleural cavity, attached only by their anterior end through which the bronchus, blood vessels and nerves enter. The pleural cavity is filled with a watery **pleural fluid.**

Inside the lungs the bronchi divide and subdivide into finer branches called **bronchioles** (Fig. 23.37). Each bronchiole divides into still finer branches the **alveolar ducts** which terminate in small dilated **air sacs** or **infundibula.** The walls of each air sac form tiny hollow **air cells** or **alveoli.** The walls of alveoli are thin, made up of elastic connective tissue and richly supplied with capillaries of pulmonary artery and vein (Fig. 23.37 B). The alveoli are the site of gaseous exchange.

Mechanism of breathing

The mechanism of breathing involves taking fresh air into the lungs (**inspiration**) and giving out respired air from the lungs **(expiration)**. These breathing movements bring about the **ventilation** of lungs. The movement of the diaphragm and the ribs plays a very important role during breathing. The muscular diaphragm in the position of rest is dome-shaped, the curvature of the dome is directed towards the thoracic cavity. The ribs that are embedded in the wall of the thorax are movable because of the presence of the **inter costal muscles.** The pleural cavities are air-tight chambers except for the interior of the lung which directly communicate with the exterior through the respiratory passage.

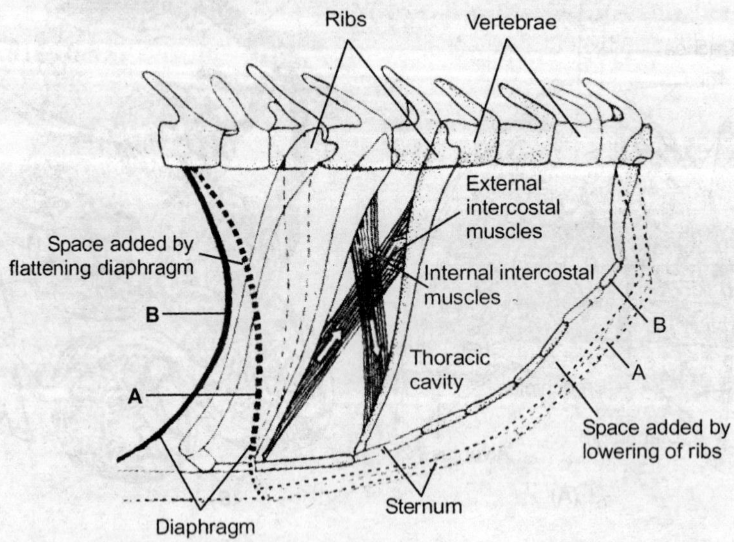

Fig. 23.38 *Rabbit*–Mechanism of breathing. **(A)** Position during inspiration; **(B)** During expiration

Inspiration: During inspiration the muscles of the diaphragm contract thereby **flattening** or **lowering** it (Fig. 23.38 A). At the same time the external intercostal muscles cause forward movement of the ribs. As a result the internal capacity of the thorax is increased. This lowers the pressure within the thoracic cavity causing the lungs to expand. As a result air rushes in through the respiratory passage due to higher pressure outside. Air rushes into the lungs till the air pressure inside becomes equal to that of the air outside. Inside the lungs exchange of gases takes place in the alveoli by simple diffurion, the oxygen from air diffuses into the blood and carbon dioxide from the blood diffuses into the air, at the same time.

Expiration: During expiration the muscles of the diaphragm relax (Fig. 23.38 B) and it returns to the normal dome-shaped position. The external intercostal muscles relax and the internal intercostal muscles contract thereby the ribs return to their original position. The volume of the thoracic cavity decreases expelling the air through the same route.

The lungs, however, are never emptied completely, some residual air (residuum) is always left in the lungs. This residuum mixes with fresh air, as a result alveolar air in a mammal always has less oxygen content and more carbon dioxide content than the atmospheric air. As compared to bird a mammalian lung is less efficient.

In the alveoli exchange of gases, intake of oxygen and output of carbon dioxide, takes place. This is **external respiration.** The true respiratory activities take place in the tissues releasing energy needed by the body. This is called **tissue respiration.** A chain of complex chemical reactions catalysed by special respiratory enzymes are responsible for the release of energy.

NERVOUS SYSTEM

The nervous system of rabbit, as in other vertebrates, consists of:

1. **Central nervous system** – **brain** and **spinal cord.**
2. **Peripheral nervous system** – **cranial** and **spinal nerves.**
3. **Autonomic nervous system** – **sympathetic** and **parasympathetic nerves** and **ganglia.**

Central nervous system

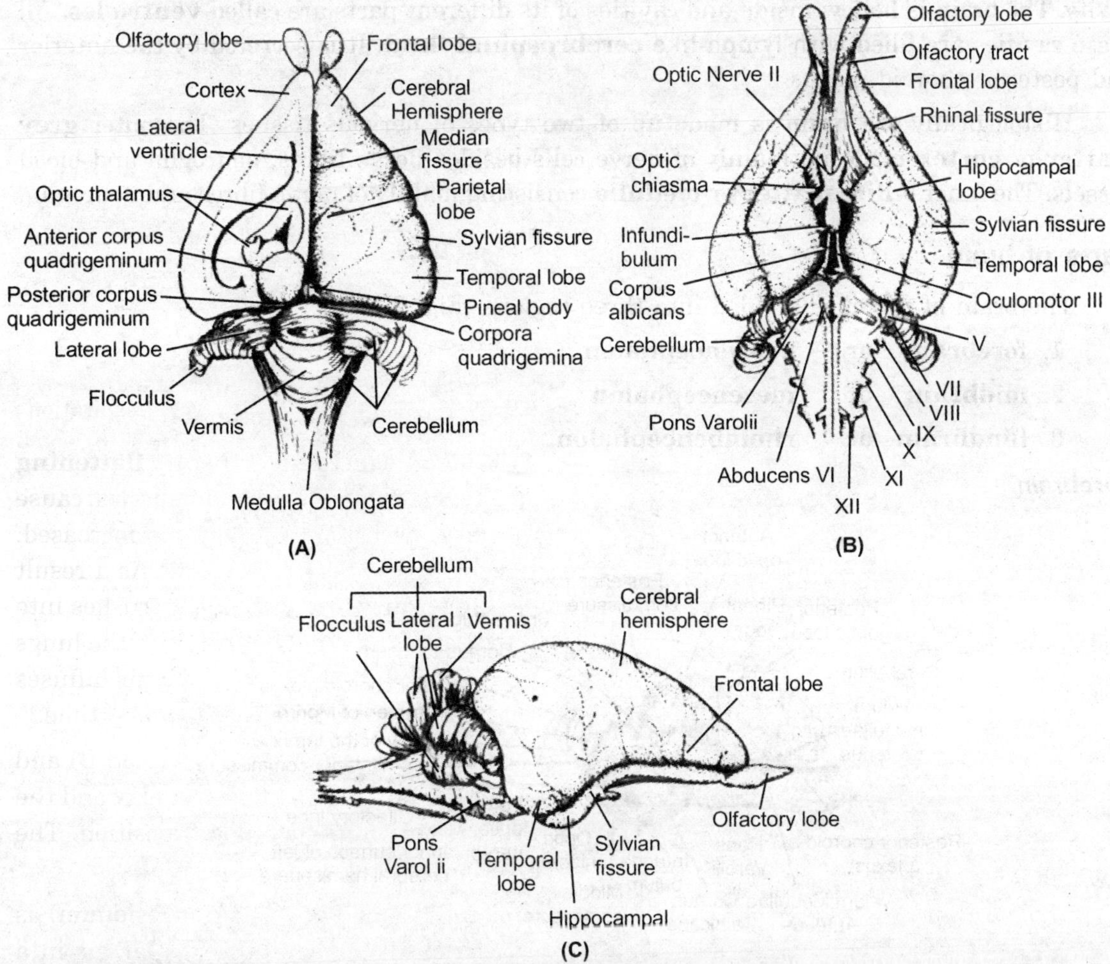

Fig. 23.39 *Rabbit*–Structure of brain. **(A)** Dorsal view; **(B)** Ventral view; **(C)** Lateral view

Brain

The brain of rabbit presents the same typical subdivisions as found in other vertebrates. One major difference is the reduction of the smell brain or the olfactory lobes. As in the birds, the loss of smell in mammals has been compensated by the elaboration of senses of sight and hearing.

The brain is a soft, white structure enclosed in the cranial cavity of the skull. It is almost twice as long as broad. The brain is surrounded by some membranes known as **meninges** (singular, menix). In rabbit there are three membranes: **duramater, piamater** and **arachnoid.** The duramater is a thick, tough, connective tissue membrane which lines the cranial cavity. The piamater is a thin and highly vascular layer which invests the brain closely. It is fused with the roof of the brain at two places forming the **anterior** and **posterior choroid plexus.** In between these two membranes there is a thin, delicate and highly vascularized **arachnoid layer.** The space between arachnoid and duramater is called **subdural cavity** and the space between arachnoid and piamater is called **sub-arachnoid** cavity. The brain is hollow inside and cavities of its different parts are called **ventricles.** All these cavities are filled with lymph-like **cerebrospinal fluid.** It is secreted by the anterior and posterior choroid plexus.

Histologically the brain is made up of two types of nervous tissues. The outer **grey matter** or **cortex** made up mainly of nerve cells besides nerve fibres, neuroglia and blood vessels. The inner **white matter** or **medulla** consisting mainly of nerve fibres and neuroglia.

Parts of brain

The brain of rabbit is divided into three regions (Fig. 23.39):

1. **forebrain** or **prosencephalon**
2. **midbrain** or **mesencephalon**
3. **hindbrain** or **rhombencephalon.**

Forebrain

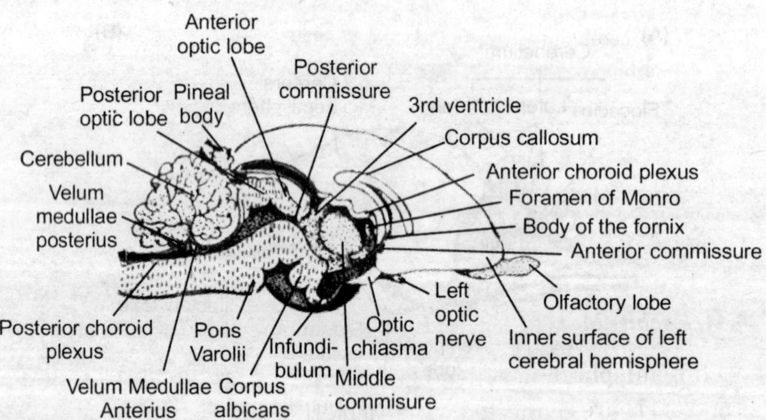

Fig. 23.40 *Rabbit*–Brain, sagittal section

The forebrain consists of a pair of **cerebral hemispheres** and **olfactory lobes** and a **diencephalon.**

Olfactory lobes: The olfactory lobes are the anteriormost part of the forebrain. They are small, club-shaped structure, clearly visible from the ventral side due to overhanging cerebral hemispheres. Each lobe consists of an **olfactory bulb** and an **olfactory duct.** The cavity of the olfactory lobe is narrow and is called **I ventricle** or **rhinocoel.** The olfactory lobes are concerned with the sense of smell.

Cerebral hemispheres: The cerebral hemispheres are large and form anterior two-third of the brain. They are broad, smooth-surfaced, triangular bodies, narrow in front and broad behind. They are closely apposed to each other along the median plane but separated from one another by a deep longitudinal groove, the **median fissure.** The two cerebral hemispheres are so large that they completely overshadow the olfactory lobes and project backwards to such an extent that they hide from view the diencephalon and midbrain and almost touch the cerebellum behind. The whole surface of the hemispheres is divided into many regions by means of fissures. A shallow oblique **Sylvian fissure** at the sides, divides each hemisphere into a large antero median **frontral lobe** and a smaller posterolateral **temporal lobe.** A longitudinal **rhinal fissure** on the ventral side demarcates the olfactory duct. A longitudinally running **hippocampal fissure** or **sulcus** demarcates the **hippocampal lobe.**

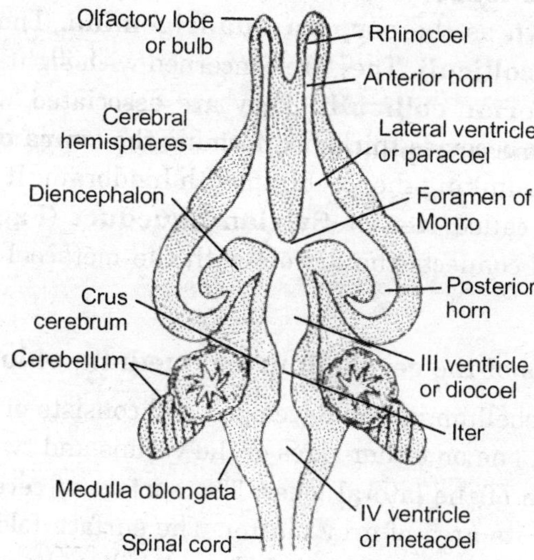

Fig. 23.41 *Rabbit*–Median longitudinal section showing cavities of brain

The cavity of each hemisphere is called **II ventricle** or **lateral ventricle** or **paracoel.** The two paracoels (Fig. 23.41) are connected with each other and with the cavity of diencephalon by a passage called **foramen of Monro.** They communicate with the rhinocoel in front. In the floor of the median fissure above the roof of the lateral ventricle there is a transversely running sheet of fibres forming the **corpus callosum,** which connects the two hemispheres. The floor and lateral walls of the paracoel are formed by **corpus striatum**. The cerebral hemispheres are the centre of intelligence, thought, reasoning, memory, etc.

Diencephalon: The diencephalon or thalamencephalon is small and narrow and lies covered by the cerebral hemispheres. Dorsally the roof of diencephalon bears the **pineal stalk** carrying a small rounded **pineal body** (an endocrine gland). Remaining part of the roof is non-nervous but highly vascular and is called the **anterior choroid plexus.** Ventrally the floor of diencephalon bears a median rounded **infundibulum** the lower end of which carries an oval pouch the **Rathke's pouch** or **hypophysis.** The infundibulum and hypophysis together form the **pituitary body** (an endocrine gland). The pituitary body has a rounded swelling called **corpus albicans** on the posterior side. An **optic chiasma,** formed by crossing of two optic nerves, lies in front of the infundibulum. The cavity of diencephalon is called **diocoel** or **III ventricle** (Fig. 23.41). Its roof is called **epithalamus,** floor **hypothalamus** and sides **optic thalami.** The two optic thalami are connected by a large transverse **middle commissure.**

The diencephalon controls different chemical activities, metabolism, temperature, reproductive activities and the autonomous nervous system.

Midbrain

The midbrain is the small middle part of the brain. It is also covered by the backward extension of the cerebral hemisphere. The dorsolateral surface of the midbrain becomes thickened to form the **optic lobes** which in rabbit are represented by four small rounded eminences collectively known as the **corpora quadrigemina.** The two anterior lobes are larger and called **superior colliculi.** They are concerned with sight. The two posterior lobes are smaller and called **inferior colliculli.** They are associated with acute hearing. The ventral wall of the midbrain becomes thickened forming the **crura cerebri** which are in the form of bands of nerve fibres linking the forebrain with hindbrain. It controls muscle activity. The cavity of midbrain is called **iter** or **Sylvian aqueduct** (Fig. 23.41). It is a narrow longitudinal passage which connects the diocoel with the metacoel or IV ventricle.

Hindbrain

The hindbrain consists of the **cerebellum** and **medulla oblongata.**

Cerebellum: The cerebellum is well developed and consists of a large median lobe, the **vermis,** two **lateral lobes,** one on either sides of the vermis and two **floccular lobes** which are ventro-lateral extension of the lateral lobes. The surface of cerebellum is thrown into a number of folds increasing its area of grey matter. The surface foldings result in a number of elevations called **gyri** and grooves called **sulci.** The cerebellum helps to maintain equilibrium and coordinate voluntary muscular movements. The cerebellum has no cavity.

Medulla oblongata: The medulla oblongata forms the posteriormost part of the brain. It is triangular in shape, being broad anteriorly and posteriorly tapers into the spinal cord. The roof of medulla is non-nervous but highly vascular and forms the **posterior choroid plexus.** Extending across the ventral side of the medulla and connecting the two halves of the cerebellum is a stout transverse band of nerve fibres called **pons Varolii.** The cavity of medulla is called **IV ventricle** or **metacoel** which is continuous with the central canal of the spinal cord. The medulla controls involuntary activities of the body such as respiration, digestion, excretion, etc.

Spinal cord

The medulla oblongata leaves the cranium through the foramen magnum and continues posteriorly as the **spinal cord.** The spinal cord is a thick, white cord of nervous tissue lying within the neural canal of the vertebrae and terminates in the lumbar region as a non-nervous thread-like **filum terminale** which extends into the tail. Like brain, the spinal cord is also enclosed in three membranes **duramater, piamater** and **arachnoid.** The space between the membranes is filled with cerebrospinal fluid. The histological structure of spinal cord is similar to that of other vertebrates. A **central canal** runs through the whole length of the spinal cord, it is filled with cerebrospinal fluid. There is a **median dorsal fissure** or sulcus along the somewhat flattened middorsal line and a **median ventral fissure** along the midventral line. The midventral fissure is deeper and broader than the mid-dorsal fissure and also contains blood vessels.

A T.S. of spinal cord is somewhat spherical in shape (Fig. 23.42). In the centre grey matter is present. It is made up of nerve cells. It is butterfly-shaped or H-shaped with a small central canal in the centre. The grey matter is produced dorsolaterally and ventrolaterally into two **dorsal** and two **ventral horns** respectively. The roots of spinal nerves are attached to these horns. The grey matter is surrounded by white matter which is made up of nerve fibres. A thin vertical dorsal septum extends from the dorsal fissure up to the central canal. The white matter is thus divided into two compartments by the dorsal septum and ventral fissure. Each compartment is further divided into **dorsal funiculus, lateral funiculus** and **ventral funiculus** by the two horns of the grey matter. The spinal cord plays important role in reflex activities. It also transmits impulses to and from the brain.

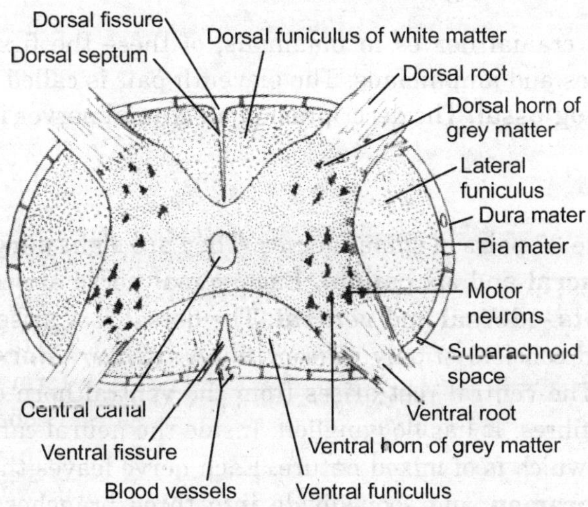

Fig. 23.42 *Rabbit*–T.S. of spinal cord

Peripheral nervous system

Cranial nerves

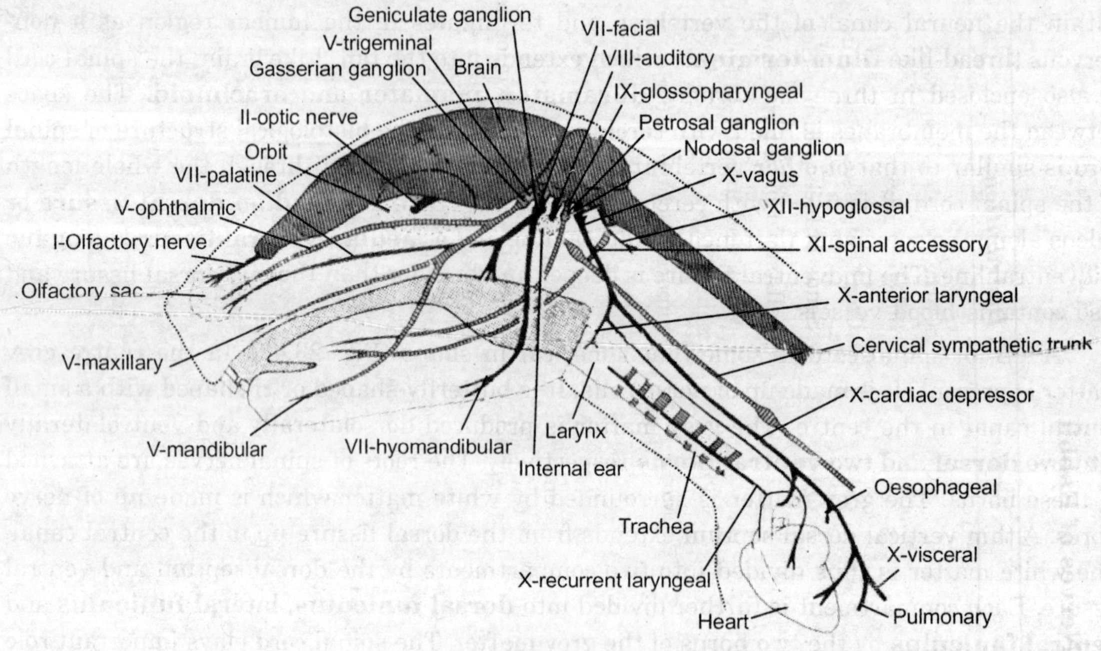

Fig. 23.43 *Rabbit*–Cranial nerves

There are twelve cranial nerves in mammals, of these the first ten pairs are almost similar to those of fishes and amphibians. The eleventh pair is called **spinal accessory** and the twelfth pair is **hypoglossal.** The description of all cranial nerves is given in the following table.

Spinal nerves

In rabbit there are 37 pairs of spinal nerves. They are eight **cervical,** twelve **thoracic,** seven **lumbar,** four **sacral** and six **caudal.** Each spinal nerve arises from the sides of the spinal cord by two **roots—dorsal** and **ventral.** The dorsal root arises from the dorsal horn of the grey matter and consists of only sensory fibres. It bears **dorsal ganglion** made up of only sensory cells. The ventral root arises from the ventral horn of the grey matter and consists of only motor fibres. It has no ganglion. Inside the neural canal both the roots unite to form a single nerve which is of mixed nature. Each nerve leaves the neural canal through the **intervertebral foramen** and soon divide into three branches: **dorsal, ventral** and **visceral.** The **dorsal branch** or **ramus** supplies the muscles and skin of middorsal region. The **ventral branch** or **ramus** innervates the lateral and ventral parts of the body and most of the skin. The **visceral branch** or **ramus communicans** together with the sympathetic trunk innervate the visceral organs.

Table 23.1. Cranial Nerves of Rabbit

Number	*Name*	*Branch if any*	*Origin from brain*	*Nature*	*Distribution*
I	Olfactory	—	Antero-ventral surface of olfactory lobe	Sensory	Carry impulses from nasal sacs to olfactory lobe.
II	Optic	—	Optic chiasma	Sensory	Carry impulses form retina of eye.
III	Occulomotor	—	Crura cerebri of mid brain	Motor	Sends branches to muscles of eyeball (superior, inferior and internal rectus and inferior oblique muscles, celiary body.
IV	Trochlear (Pathetic)	—	Junction of mid-and hind-brain	Motor	Superior oblique muscles of eyeball
V	Trigeminal		Pons Varolii	Mixed	—
		V_1 Ophthalmic	travels above the orbit	Sensory	Skin of snout, eyelids
		V_2 Maxillary	travels below the orbit	Sensory	Upper jaw and lip, skin of nose, lower eyelid, vibrissae.
		V_3 Mandibular	along the margin of lower jaw	Motor	Lower jaw and lip, tongue external ear.
VI	Abducens	—	Ventral side of medulla	Motor	Posterior rectus muscles of eyeball.
VII	Facial		Pons Varolii	Mixed	—
		VII_1 Palatine	—	Sensory	Roof of buccal cavity and nasal chambers.
		VII_2 Chorda tympani	—	Motor	Tongue, salivary glands and taste buds.
		VII_3 hyomandi-	—	Motor	Muscles of lower jaw, neck and pinna.

		bular			
VIII	Auditory		Sides of medulla	Sensory	
		VIII_1 Vestibular	—	sensory	Internal ear
		VIII_2 Cochlear	—	sensory	Cochlea
IX	Glossopharyngeal	—	Sides of medulla	Mixed	Tongue and muscls of pharynx.
X	Vagus		Sides of medulla	Mixed	—
		X_1 Anterior laryngeal	—	Motor	Muscles of laryngeal region
		X_2 Recurrent laryngeal	—	Motor	Muscles of larynx
		X_2 Cardiac depressor	—	Motor	Muscles of heart
		X_4 Pneumogastric	—	Sensory and Motor	Lungs, oesophagus, stomach and ileum.
		X_5 Deprsessor	—	Motor	Muscles of diaphragm.
XI	Spinal accessory	—	Sides of medulla	Motor	Muscles of pharyngeal, laryngeal and cervical region.
XII	Hypoglossal		Ventral surface of medulla	Motor	Muscles of neck and tongue.

Plexus: The ventral branches of spinal nerves traverse the intercostal muscles and form **brachial plexus** anteriorly and **pelvic** or **lumbosacral plexus** behind.

Brachial plexus: The ventral branches of fourth to eighth cervical nerves and the ventral ramus of the first thoracic nerve are united by inter communicating branches (**ansae**) in the region of forelimbs to form the **brachial plexus,** which innervates the muscles of the shoulder, breast and forelimbs. The fourth cervical sends a **phrenic nerve** to the diaphragm.

Lumbosacral plexus: The ventral branches of the last four lumbar nerves form a **lumbar plexus** and those of the sacral nerves form a **sacral plexus.** The two plexi are united with each other to form the **lumbosacral plexus.** A number of nerves arise from this plexus and a few important ones are named here. The fourth lumbar nerve gives rise to the **lateral cutaneous,** a stout nerve to the thigh, a slender **genito-femoral** which innervates the abdominal wall of the inguinal region. The fifth lumbar sends a branch, the **lumbar cutaneous** and with the sixth forms a stout **femoral nerve** which enters the medial surface of the thigh along with the femoral artery and vein. The connecting band between sixth and seventh gives off a slender **obturator nerve.** The seventh lumbar together with the first sacral forms a stout **sciatic nerve.** The **pudendal** and **posterior cutaneous** arising from the sacral plexus innervate the rectum and its muscles respectively. The fourth sacral passes laterally and then turns posteriorly to the rectum.

Autonomic nervous system

The autonomic nervous system mainly controls involuntary activities of the body. It includes (*i*) **sympathetic nervous system** and (*ii*) **parasympathetic nervous system.**

Sympathetic nervous system: The sympathetic system in rabbits consists of two ganglionated sympathetic nerves or cords which leave the brain through the carotid canals in the cranium and extend from the neck to the end of the abdomen in the middorsal region on either side of the vertebral column. They enter the tail in the form of fine nerve fibres. Each sympathetic nerve has 24 ganglia of which three are **cervical,** ten **thoracic, six lumbar,** four **sacral,** and one **caudal.** Communicating branches called **rami communicantes** (singular, **ramus communicans**) arise from each sympathetic ganglion and meet the corresponding spinal nerve. Nerve fibres from the sympathetic cords innervate the eyes, heart, lungs, visceral organs and glands.

Parasympathetic nervous system: The parasympathetic system is formed by fibres of III, VII, IX and X cranial nerves and II, III and IV sacral nerves. The parasympathetic ganglia lie in the head, neck and sacral regions. The parasympathetic nerve from the III cranial nerve supply the muscles of iris and from VII and X supply the lacrimal gland, salivary gland, lungs, heart, liver and different parts of the alimentary canal. The parasympathetic nerve of the sacral region form a pelvic ganglion on each side and send nerve fibres to the kidney, urinary bladder, rectum and gonads.

The sympathetic and parasympathetic nerves work independently and their effects are antagonistic (opposite) to one another. The sympathetic nerves stimulates the organ for action while parasympathetic inhibits or stops the function. Some common effects of sympathetic stimulation are increased heart beat, increase in blood pressure, dilation of pupil of eye, inhibition of activity of digestive tract, etc. Parasympathetic stimulation, on the other hand decreases heart beat and blood pressure, constricts pupil of eye, speeds up activity of the alimentary canal, etc.

The autonomous system also controls the secretion of exocrine and endocrine glands.

Sense organs

There are five specialised sense or receptor organs in rabbit. They are:

1. **Organs of touch** or **tangoreceptors (skin)**
2. **Organs of taste** or **gustatoreceptors (tongue)**
3. **Organs of smell** or **olfactoreceptors (olfactory sacs)**
4. **Organs of sight** or **photoreceptors (eyes)**
5. **Organs of hearing** and **equilibrium** or **statoacoustic receptors (ears)**

Organs of touch

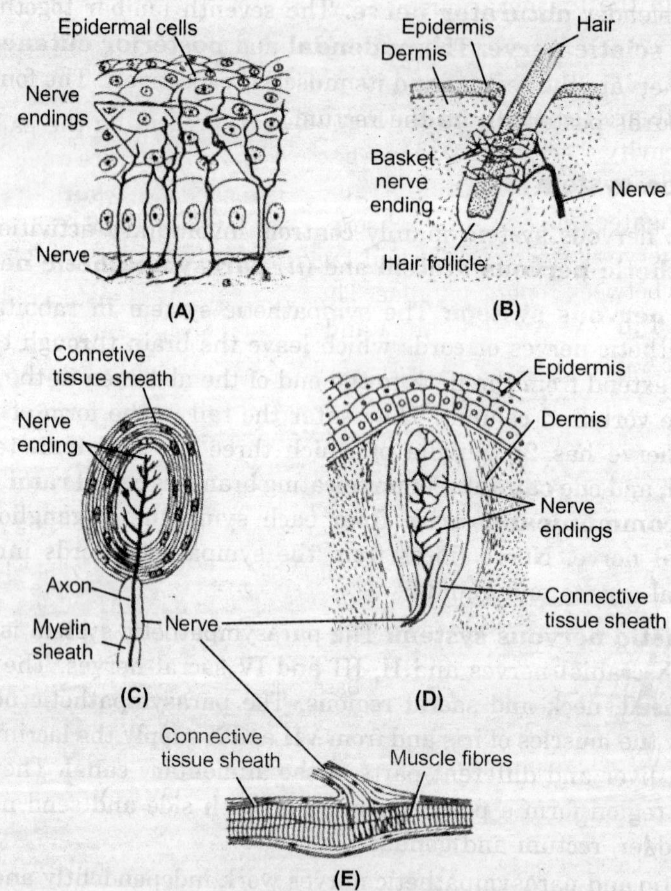

Fig. 23.44 *Rabbit*–Cutaneous receptors. **(A)** Free nerve ending; **(B)** Basket nerve ending; **(C)** Pacinian corpuscle; **(D)** Meissner's corpuscle; **(E)** Neuro-muscular bundle

The skin is the organ of touch. It is highly sensitive and provided with various types of receptors. These receptors are microscopic and are scattered beneath the epidermis. They may be in the form of:

(*a*) Free nerve endings: These are fine branching fibres (Fig. 23.44 A) of sensory nerves which lie below or may extend into the epidermis. They are sensitive to touch, pressure, cold, warmth, pain, etc. Sometimes nerve endings form a fine network around a hair follicle (**basket nerve ending**) (Fig. 23.44 B).

(*b*) Encapsulated nerve endings: These are found in those parts of the skin where hairs are absent, for example **Meissner's** (Fig. 23.44 D) and **Pacinian capsules,** or **corpuscles.** These corpuscles are oval in shape. Each corpuscle is formed (Fig. 23.44 C) of a sheath of connective tissue arranged in concentric rings around a core of nerve fibres. The Meissner's capsules are present beneath the epidermis and are sensitive to touch. The Pacinian capsules occur in the dermis and are sensitive to pressure.

(*c*) Neuromuscular bundles: Specialized nerve endings (Fig. 23.44 E) found in skeletal muscles, tendons and joints are called neuromuscular bundles or neuromuscular spindles. They detect changes in the tension of muscles.

Organ of taste

The organ of taste are the **taste buds** (Fig. 23.45 A) present on the papillae of tongue and roof of the buccal cavity. Each taste bud is a barrel-shaped structure made up of two types of cells, the **neurosecretory** or **receptor** or **gustatory cells** and **supporting cells.** The gustatory cells are elongated and bear fine sensory hairs on their inner free ends. Their outer ends are supplied by nerve fibres from VII and IX cranial nerves. Taste buds open by **taste pores** into depressions between tongue papillae. The receptor cells of taste buds are stimulated by substances dissolved in mucous and saliva and finally tastes such as sweet, bitter, salty or sour are felt.

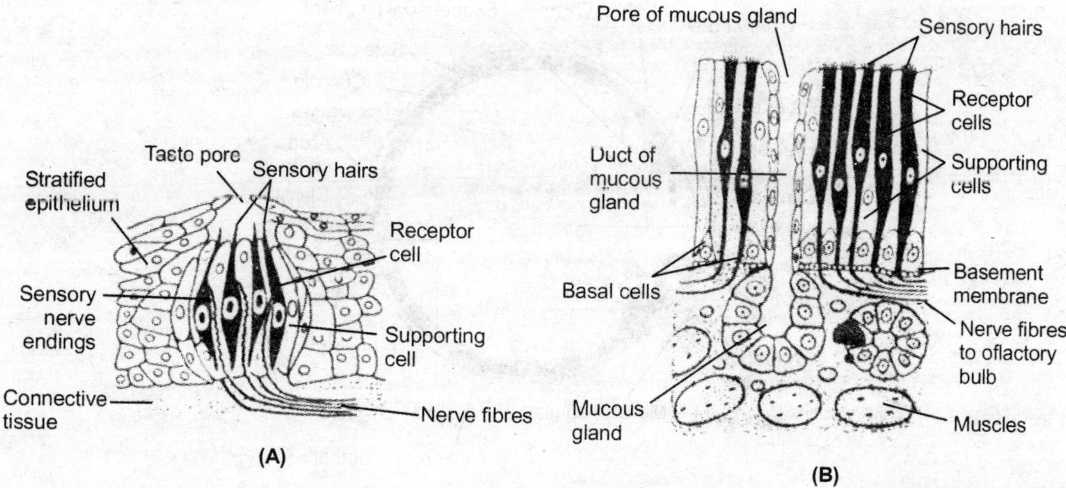

Fig. 23.45 *Rabbit*–**(A)** Vertical section of a taste bud; **(B)** Olfactory epithelium with mucous gland

Organs of smell

The nasal cavity is long and is divided into right and left **cavities** or **fossae** by a perpendicular **internasal septum.** Each nasal fossa contains spindle-shaped **olfactory cells** (Fig. 23.45 B), **mucous cells** and columnar **supporting cells.** The lateral and posterior walls of the nasal fossa are occupied by delicate scrolled and folded bones, the **turbinals** or **conchae.** These are divisible into three parts: (*i*) the **inferior concha (maxilloturbinal)**, much folded and located on a separate small bone of the skull, (*ii*) the **middle concha (nasoturbinal)**, a long single fold dorsal to the preceding one, and (*iii*) **superior concha (ethmoturbinal),** part of ethmoid bone. The spaces enclosed by bony folds are known as **ethmoid cells.** The folds are covered with **olfactory epithelium** or **Schnederian epithelium.** The olfactory cells bear several delicate olfactory hairs connected with the olfactory lobes of the brain. The mucous cells produce mucous which holds dust particles and also dissolves odoriferous substances from air. The epithelium of the ethmoturbinal is the main olfactory region as it is sensitive to smell. The epithelium of maxilloturbinal moistens and warms the air. The nasoturbinal epithelium has no olfactory cells.

Each nasal passage also has an oval sensory pad with minute papillae and ridges. They are tactile and act as distance receptors. In rabbit Jacobson's organ is also present in the roof of the buccal cavity. It is lined by olfactory epithelium and smells the contents of the mouth. The olfactory sensors in mammals are distance receptors as they are stimulated by very minute concentration of odours of chemicals present in the air. Quantitative estimates indicate that the olfactory system is more sensitive that the gustatory system.

Organs of sight

The organs of sight are a pair of **eyes** which lie, one on either side of the head in sockets called **orbits** in the skull. The structure of eyes is similar to that of other vertebrates.

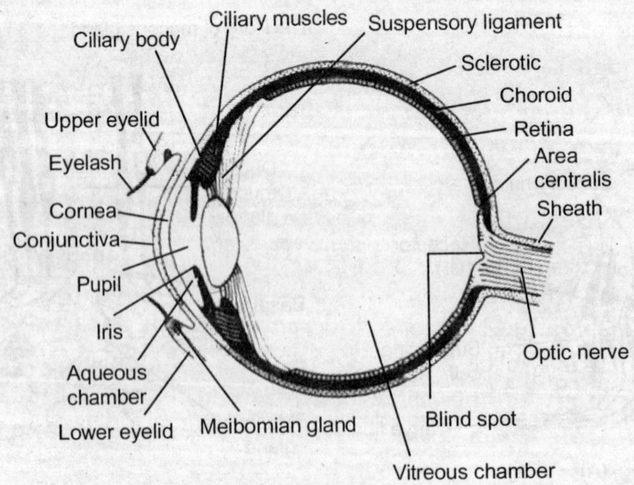

Fig. 23.46 *Rabbit*–Eyes, vertical section

The eye or eyeball is a spherical structure (Fig. 23.46) and about four-fifth part remains embedded in the orbit. The movement of eyeball is brought about by six sets of muscles, four **rectus** and two **oblique.** One end of the muscle is attached to the eyeball and the other end to the orbit.

The exposed part of the eyeball is known as **cornea.** It is protected by two mobile eyelids, the upper and lower. The free edge of both the eyelids is provided with stiff hairs or eyelashes which protect the eyes from dust particles. The eyelids can close when required. The movement of upper eyelid is brought about by a thin sheet of muscle, the **levator palpebrae superioris** and of lower eyelid by **depressor palpebrae inferioris.** Within the eyelid lies the third eyelid, or **nictitating membrane,** which is a hairless fold of opaque white skin and can be pulled across the eyeball from the anterior angle of the eye called **canthus.** In human eye the nictitating membrane is vestigeal and is represented by a pikish mass at the anterior corner of the eye. There are many minute glands on the eyelids. On the margins of upper and lower eyelids are **Meibomian glands,** located on each side. They secrete an oily secretion which lubricates the eyelids. The **Harderian glands** are situated below the lower eyelid. The secretion of these glands lubricates the conjunctiva. The **lacrimal glands** are situated below the outer corner of the upper eyelid. They secrete a slightly saline, watery secretion, the tears. It keeps the surface of the eye moist, clean and free from bacteria and also provides nourishment to the cornea. The tears are drained into nasal chambers by means of **naso-lacrimal duct.**

A vertical section of the eyeball reveals its internal structure. The eyeball is hollow inside and is made up of three layers: **sclerotic, choroid** and **retina.**

The **sclerotic** is the outer tough layer made up of dense connective tissue fibres. The hidden part of sclerotic is called **sclera** and is opague white. A small exposed part of sclerotic in front is transparent and is called **cornea.** It bulges out slightly and is covered by a thin, delicate, transparent epithelial layer, the **conjunctiva.** The conjunctiva is continuous with the epidermal lining of the eyelids. It is supplied with blood vessels and nerve fibres. Light enters the eye through cornea.

The **choroid** is the middle layer. It is made up of thin and richly vascular connective tissue with dark brown pigments. The pigments absorb light rays, thus reducing internal reflection of light. At the junction of sclera and cornea the choroid swells up to form the **ciliary body** which contains **ciliary muscles.** In front of the ciliary body the choroid separates from the cornea forming the **iris**. The iris is an opaque pigmented disc which prevents the light from entering the eye except through a round aperture, the **pupil** between the two iris. The iris contains circular and radial smooth muscles. These muscles control contraction and relaxation of the pupil, thus regulating the amount of light entering the eye. Just behind the iris a crystalline, solid, biconvex **lens** is present. It is enclosed in a thin, transparent **lens capsule** which is held in position by **sensory ligaments** attached with the ciliary body.

The **retina** forms the innermost layer of the eyeball. It is transparent and sensitive to light. The outer layer of retina is pigmented, non-nervous and closely applied with the choroid. The inner nervous layer is the **retina proper.** It is made up of two types of sensory cells, **rods** and **cones.** The rods are sensitive to light of low intensity. So more suitable for night vision but the image formed is diffuse and not very clear. The cones are sensitive to light of high intensity so suitable for day vision. The image formed is sharp and clear. The rods contain a photosensitive pigment, **rhodopsin** or **visual purple** which can distinguish light and darkness. The cones are responsible for colour vision in primates. The light sensitive pigment of cones is **iodopsin.** The rods and cones are not uniformly distributed. The cones are more numerous in the centre and rods towards the periphery. All the nerve fibres of the retina converge to form the **optic nerve** and the point where the optic nerve leaves the eye is called **blind spot** as it has no rods and cones. Near the blind spot in the area centralis there is an oval depression called **yellow spot** or **fovea** where brightest image is formed. In this area only cones are present.

The lens and iris divides the cavity of the eyeball into two unequal chambers. The small anterior chamber between the cornea and lens is called **aqueous chamber** and the bigger posterior chamber between the lens and retina is called **vitreous chamber.** The aqueous chamber is filled with a watery liquid called **aqueous humour.** It is secreted by the ciliary body and drained into the aqueous chamber through **canal of Schlemm.** The aqueous humour nourishes the cornea and lens and maintains **intraocular pressure.** Any disturbance in pressure causes **glaucoma,** a disease in which retina is injured. The vitreous chamber is filled with a gelatinous secretion called **vitreous humour.**

Working of the eye

The working of eye is similar to that of a photographic camera. The light rays reflected from an object are refracted by the cornea and then focussed by the lens on the retina and a sharp inverted image is formed. The retinal image is interpreted by the brain and the animal sees the object in an upright position.

Vision: In lower mammals like rabbit, rat, etc. the eyes are situated on each side of the head so each eye covers a different field of vision and there is no overlapping. This type of vision is called **monocular vision.** In higher mammals like man and other primates the eyes face forward and the field of vision of the two eyes overlap each other. Thus both the eyes can see a single object. This type of vision is called **binocular** or **stereoscopic vision.** The animal can also estimate the distance of the object.

Accommodation: The ability to clearly focus the objects situated at various distances is called **accommodation** and the power of accommodation is well exhibited by the eyes of rabbit. It is brought about by changing the curvature of two lens. In normal condition when the eyes are at rest, they are adjusted to see distant objects. The ciliary muscles are fully relaxed and the diameter of the ciliary body increases due to outward pressure of the fluid in the eyeball. This increases the suspensory ligaments which pull the lens capsule so that the lens becomes flattened and the eye is adjusted for seeing distant objects. For seeing near objects the process is reversed. The ciliary muscles contract and the diameter of the ciliary body decreases. The suspensory ligaments relax and the lens becomes more convex. The focal length of the lens is reduced and the eye is adjusted to see near objects.

Chemistry of vision (Photochemistry): The photosensitive pigment rhodopsin present in the rods plays an important role in vision. In the presence of light it breaks down into **retinene,** a derivative of vitamin A, and a protein **opsin** and energy is released. This energy stimulates a neuron and a nerve impulse is sent to the brain through the optic nerve. An image is formed on the retina and the animal sees the object. In the presence of ATP rhodopsin is resynthesized from retinene and opsin. The breakdown and resynthesis of rhodopsin goes on continuously. In dim light large amount of rhodopsin is present. When one comes out of dark room in bright light the vision is blurred and it becomes clear when the level of rhodopsin in the rods decreases. Similarly when one goes into a dark room from bright light, again vision is not clear until the level of rhodopsin in the rods increases. Rhodopsin can be synthesized from vitamin A in the presence of enzyme **retinal isomerase.** Deficiency of vitamin A may cause night blindness.

Organs of hearing (Statoacoustic organs)

The ears are the organs of hearing and equilibrium. The ear of mammals consists of three parts: **external, middle** and **internal ear.** All the three parts of the ear are concerned with hearing but equilibrium is maintained only by the internal ear.

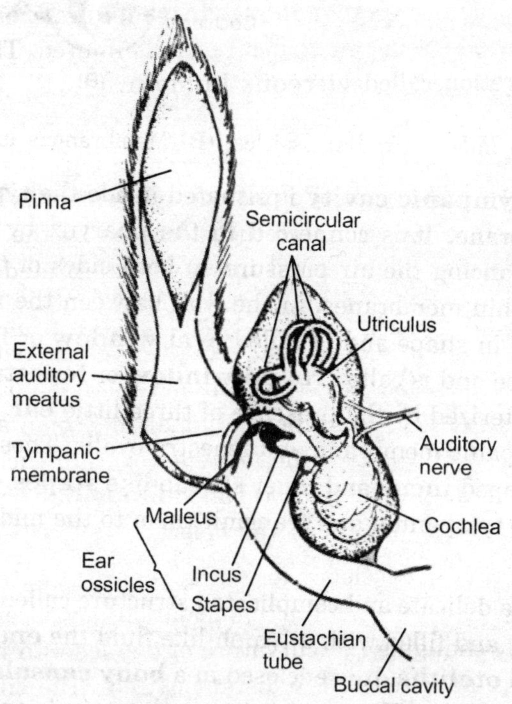

Fig. 23.47 *Rabbit*–Ear region in vertical section

The **external ear** includes the **pinna** or **auricle** or **concha** and **external auditory meatus** (Fig. 23.47). The pinna is large, oval, visible, skin coloured, freely movable structure projecting from the side of the head. Each is folded longitudinally and is supported by cartilage and surrounds the opening of the external auditory meatus. The external auditory meatus is a tubular passage which leads into the skull and terminates at a cone-shaped membrane, the **tympanum** or **eardrum.** It is lined by skin containing hairs, oil and wax glands which trap dust particles and small creatures and prevent them from entering and damaging the tympanum. The function of external ear is to catch and concentrate sound vibrations and to detect the direction of sound vibrations. In man the pinna is not very efficient because it is immovable.

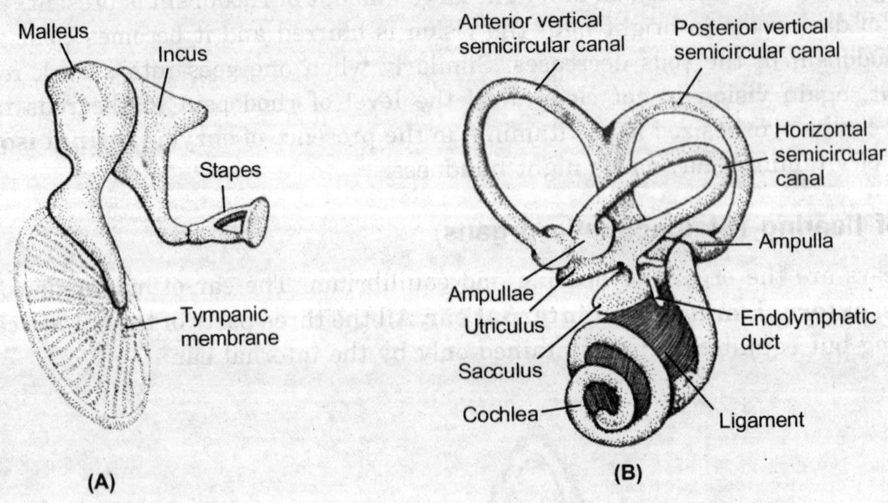

Fig. 23.48 *Rabbit*–**(A)** Ear ossicles; **(B)** Membranous labyrinth

The **middle ear** or **tympanic cavity** is situated inside the **tympanic bulla.** It starts from the tympanic membrane. It is connected to the pharynx by a tubular passage, the **eustachian tube,** for balancing the air pressure on both sides of the ear drum. There are two openings covered by thin membranes, in the wall between the middle and internal ear. The upper opening is oval in shape and is called **oval window** or **fenestra ovalis** and the lower one is round in shape and is called **round window** or **fenestra rotunda.** The middle ear of mammals is characterized by the presence of three little **ear bones** or **ossicles** (Fig. 23.48 A) between the tympanic membrane and fenestra ovalis. These are outer hammer-like **malleus,** middle anvil-shaped **incus** and inner stirrup-like **stapes.** The ear ossicles take up sound vibrations from the tympanum and transmit them to the middle ear through fenestra ovalis.

The **internal ear** is a delicate and complicated structure called **membranous labyrith** (Fig. 23.48 B). It is hollow and filled with a lymph-like fluid the **endolymph** which contains calcareous particles called **otoliths.** It is enclosed in a **bony capsule** or **labyrinth** of similar shape formed by the periotic bone. The space between the capsule and membranous labyrinth is filled with a liquid called **perilymph**. The two liquids (endolymph and perilymph) are not continuous at any point.

The membranous labyrinth includes the following parts: the **body proper (utriculus** and **sacculus)**, **semicircular canals** and **cochlea.**

The utriculus is the large dorsal or upper chamber and the sacculus is the small lower chamber. The two are connected by a small duct called **sacculo-utricular canal.** A tubular **endolymphatic duct** arises from the sacculus. It ends blindly into an endolymphatic sac in the cranium. Each chamber has special sensory cells called **macula** (**macula utriculi** in utriculus and **macula sacculi** in sacculus), with fine projecting hairs. They help in maintaining body posture. The bony labyrinth around the two chambers is called vestibule.

Three narrow **semicircular canals** arise from the utriculus. They are **anterior vertical, posterior vertical** and **horizontal** canals at right angles to each other. They are situated inside bony semicircular channels. The anterior and posterior vertical semicircular canals from a common canal called **crus commune**, for some distance after their origin from the utriculus. Each semicircular canal has a swollen **ampulla** at its lower end. Each ampulla has sensory hair cells called **cristae**. The cristae differ from maculae in having longer sensory hair and lacking otoliths. They convey information to the brain about rotational and accelerating movements.

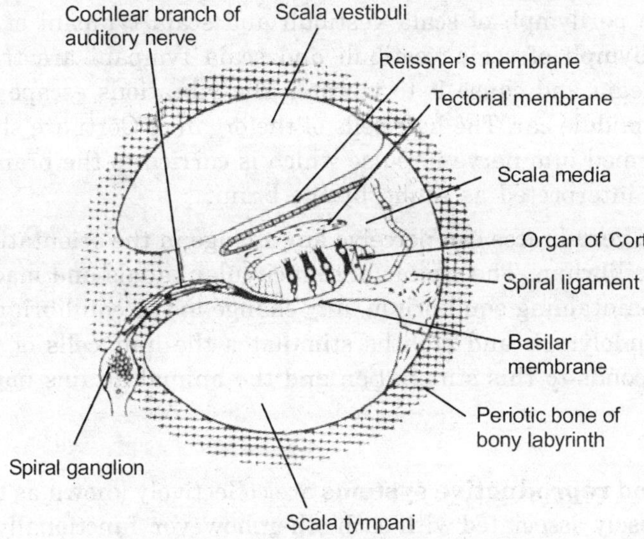

Fig. 23.49 *Rabbit*–T.S. cochlea

From the sacculus arises a spirally coiled tube called the **cochlear duct** or **lagena** which is enclosed in a similarly coiled bony **cochlear canal,** a part of the bony labyrinth. The two together form the **cochlea,** the organ of hearing. The cavity of cochlea (Fig. 23.49) is divided by two membranes, the **Reissner's membrane** and **basilar membrane,** into three chambers, the upper **scala vestibuli,** middle **scala media** and lower **scala tympani.** The scala media is filled with endolymph and ends blindly at the tip of the spiral ganglion. The upper and lower chambers, being part of the bony labyrinth, are filled with perilymph. They communicate around the tip of the spiral ganglion by a small opening called **helicotrema.** At the basal end scala vestibuli communicates with feneratra ovalis and scala tympani with fenestra rotunda.

The scala media is lined by simple epithelium. The epithelium of basilar membrane becomes modified to form the **organ of Corti,** the organ of hearing. It consists of sensory hair cells and tall columnar supporting cells. The hair cells send out nerve fibres which form the cochlear branch of the auditory nerve. The organ of Corti is covered by a thin gelatinous ribbon-like sheet of connective tissue called **tectorial membrane.**

The auditory nerve or the VIII cranial nerve divides into two branches: **vestibular branch** supplying the semicircular canals, utriculus and sacculus and a **cochlear branch** to the cochlea.

Working of ear: The ear of mammals performs two functions: **hearing** and **equilibrium.** The sacculus and lagena are concerned with hearing and utriculus and semicircular canals help in maintaining equilibrium.

Hearing: The pinna collects the sound waves which travel down the external auditory meatus and cause the ear drum to vibrate. These vibrations are transmitted across the ear ossicles of the middle ear and through the fenestra ovalis into the perilymph of the internal ear. The vibrations of the membrane of fenestra ovalis cause alternate increase and decrease in the pressure of the perilymph of scala vestibuli and scala tympani of the cochlea. The vibrations in the perilymph of scala vestibuli and scala tympani are transmitted to the endolymph of scala media and cause it to vibrate. The vibrations escape through fenestra rotunda back into the middle ear. The hair cells of the organ of Corti are stimulated and the vibrations are transformed into nerve impulse which is carried to the brain by the auditory nerve. The impulse is interpreted as sound by the brain.

Equilibrium: All vertebrates can perceive any change in the orientation of their bodies and then maintain equilibrium. The cristae of semicircular canals and maculae of utriculus and saccules help in maintaining equilibrium. Any change in the equilibrium of the body due to movement of the endolymph and otoliths stimulates the hair cells of the maculae and cristae. The brain responds to this stimulation and the animal attains normal position.

Urinogenital system

The **excretory** and **reproductive systems** are collectively known as the **urinogenital system** as they are closely associated with each other, however, functionally the two systems have nothing in common.

Excretory system

The excretory system of rabbit includes a pair of **kidneys,** a pair **ureters** and a **urinary bladder.**

Kidneys

The **kidneys** of mammals are **metanephric** (Fig. 23.50). They are paired, small, bean-shaped structures of reddish brown colour, situated on the dorsal body wall of the abdominal cavity, one on either side of the vertebral column. The right kidney is considerably anterior to the left one. Each kidney is covered by a fold of **peritoneum** on its ventral surface only and is embedded in a mass of fat. The outer surface of kidney is convex and the inner side is deeply concave having a notch called **hilum** or **hilus.** The blood vessels enter (**renal artery**) and leave (**renal vein**) the kidney, through the hilus. A white tube, the **ureter** also passes out of the kidney from the hilus. At the point of origin from the kidney the ureter is slightly flattened to form a funnel-shaped space called **pelvis.**

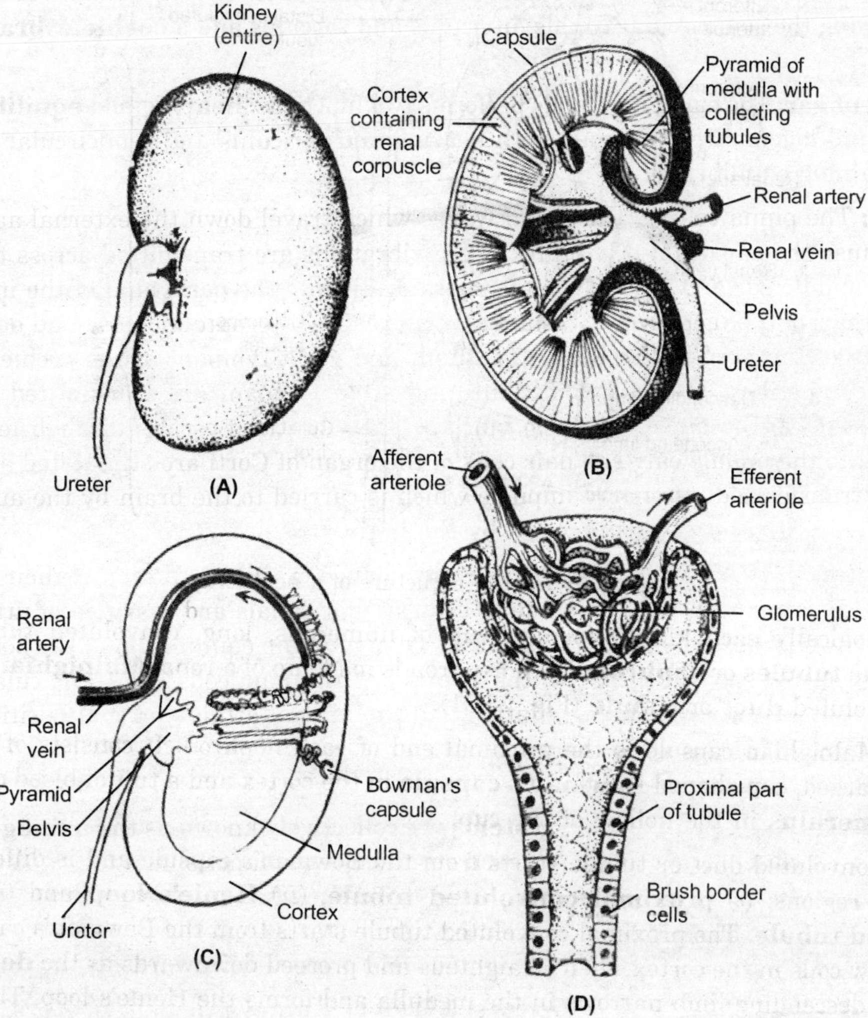

Fig. 23.50 *Rabbit*–Kidney. **(A)** Entire; **(B)** Longitudinal section; **(C)** Diagrammatic representation of the main region; **(D)** Sagittal section of malpighian body

Internal structure of kidney

In a longitudinal section the kidney is seen enclosed in a thin **capsule** of tough, fibrous connective tissue (**tunica fibrosa**). The substance of the kidney is divided into two areas the peripheral homogenous, light coloured region, the **cortex** and central, striated and darker region, the **medulla** (Fig. 23.50). The cortex contains the renal capsule (Malpighian bodies) and the convoluted and looped portions of the kidney tubules. The medulla is formed of lobes called **pyramids.** The portion of cortex between the pyramids is called **renal column of Bertini.** The pyramids are connected with the pelvis.

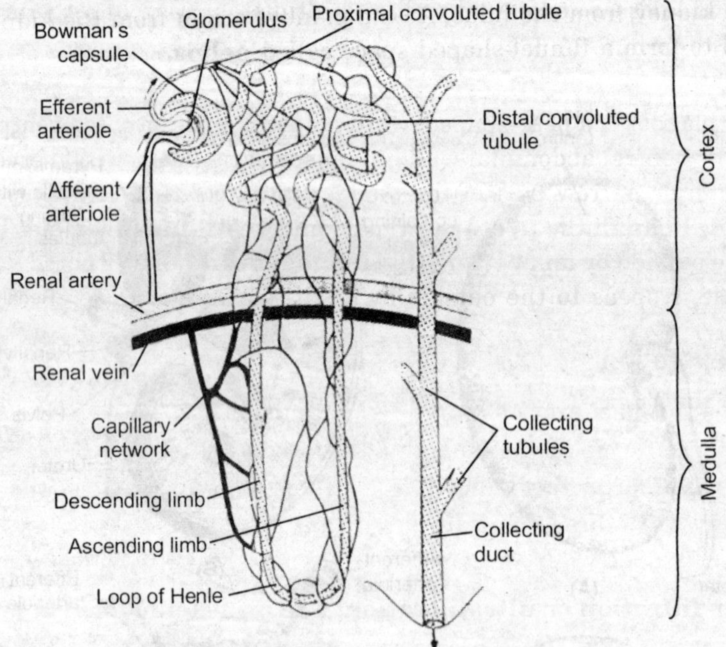

Fig. 23.51 *Rabbit*–Structure of a nephron

Histologically each kidney is made up of numerous, long, convoluted tubes called **uriniferous tubules** or **nephron.** Each nephron is made up of a renal **Malpighian capsule** and a convoluted **duct** or **tubule** (Fig. 23.51).

The Malpighian capsule is the proximal end of each nephron. It consists of a double-walled, enlarged, cup-shaped **Bowman's capsule** in the cortex and a tuft of blood capillaries called **glomerulus** in the hollow of the cup.

The convoluted duct or tubule starts from the Bowman's capsule and is differentiated into three regions: (*i*) **proximal convoluted tubule,** (*ii*) **Henle's loop** and (*iii*) **distal convoluted tubule.** The proximal convoluted tubule starts from the Bowman's capsule and makes a few coils in the cortex, then straightens and proceed downwards as the **descending limb.** The descending limb narrows in the medulla and forms the Henle's loop. The Henle's loop again proceeds towards the cortex as **ascending limb** which continues as the distal convoluted tubule. The distal convoluted tubule joins a large **collecting tubule** which finally opens into the pelvis.

Blood supply of kidney: The **renal artery** enters the kidney through the hilus and divides into several small branches called **arterioles.** The **afferent arteriole** enters the glomerulus and is twice in diameter than the **efferent arteriole** which leaves the glomerulus. The efferent arteriole then forms fine capillaries around the convoluted tubule. These capillaries then unite to form a **venule.** All the venule finally unite to form the **renal vein** which leaves the kidney through the hilum.

Ureters

The ureters are thick walled, cylindrical tubes which passes out of the hilum and runs backwards along the dorsal abdominal wall and ultimately open into the urinary bladder.

Urinary bladder

The urinary bladder is a pear-shaped, white, muscular sac situated ventral to the rectum and attached to the ventral abdominal wall by a suspensory ligament. It serves as a reservoir of urine. The posterior narrow part of the bladder is called neck. It is muscular and opens into a **urinogenital canal,** the **urethra** in male and **vestibule** in female. The urinogenital canal opens to the outside by an aperture, the **urinogenital pore** situated at the tip of penis in male. In female it opens to the outside by a slit-like **vulva.**

Physiology of excretion

The kidney is the chief excretory organ and its main function is **urine formation.** The chief nitrogenous waste product in mammals is **urea.** It is formed in liver and carried through blood circulation to the kidneys from where it is excreted in the form or **urine.** Urine formation involves three steps: (*i*) glomerular filtration, (*ii*) tubular reabsorption and (*iii*) secretion.

Glomerular filtration or **ultrafiltration:** The diameter of afferent arteriole entering the glomerulus is more than that of efferent arteriole leaving it, so more blood enters the glomerulus than that can leave. As a result the blood in glomerulus is under high pressure. Ultrafiltration occurs in the glomerulus, whereby substances like urea, glucose, water, amino acids, sodium and potassium ions, uric acid, etc., dissolved in blood are filtered into the lumen of the Bowman's capsule and some substances like proteins, fats, cells, etc. pass on to the efferent arteriole. This is purely a physical process. The filtered fluid is called **glomerular** or **capsular filtrate.**

Tubular reabsorption: The glomerular filtrate contains many substances such as, water, salts, glucose, amino acids, etc. which are useful to the body. As the filtrate passes down the tubule, these useful substances are reabsorbed through the capillary network which surrounds the tubule. This is known as **tubular reabsorption** or **selective absorption.** Some substances like glucose, amino acid and some urea are reabsorbed in the proximal convoluted tubule, salts in the proximal and distal convoluted tubules and most of the water in loop of Henle.

Tubular reabsorption may take place either by **passive diffusion** (urea and water) or by **active transport** (amino acids, sugar, salts, etc.). In active transport the substances are actively taken up by the cells of tubules and secreted into the capillary blood against a concentration gradient. This requires energy.

Tubular secretion: At the same time some substances like creatinine, ammonia, hydrogen and potassium ions, various drugs, etc. are added from the capillary blood to the filtrate by tubular secretion.

After the above processes the highly concentrated urine is left in the tubule. It is rich in water, urea, uric acid, ammonia, creatinine, etc. The urine finally passes into the collecting tubule, reaches the pelvis, carried by the ureter to the urinary bladder where it is stored till discharge.

Besides excretion, the kidneys also help in maintaining normal internal environment by regulating body fluids, *i.e.,* it acts as an homeostatic organ. For example, it regulates the amount of water in the body (**osmoregulation**). If the animal drinks plenty of water the urine becomes dilute and excess of water is removed with urine. On the other hand, if there is scarcity of water in the body, the urine becomes concentrated. The volume of urine is regulated by antidiuretic hormone produced by the posterior lobe of the pituitary gland.

REPRODUCTIVE SYSTEM

In rabbit there is well marked sexual dimorphism.

Male reproductive system

The male reproductive system includes a pair of **testes,** a pair of **epididymis**, a pair of **vasa deferentia, urethra, penis** and a number of **accessory glands.**

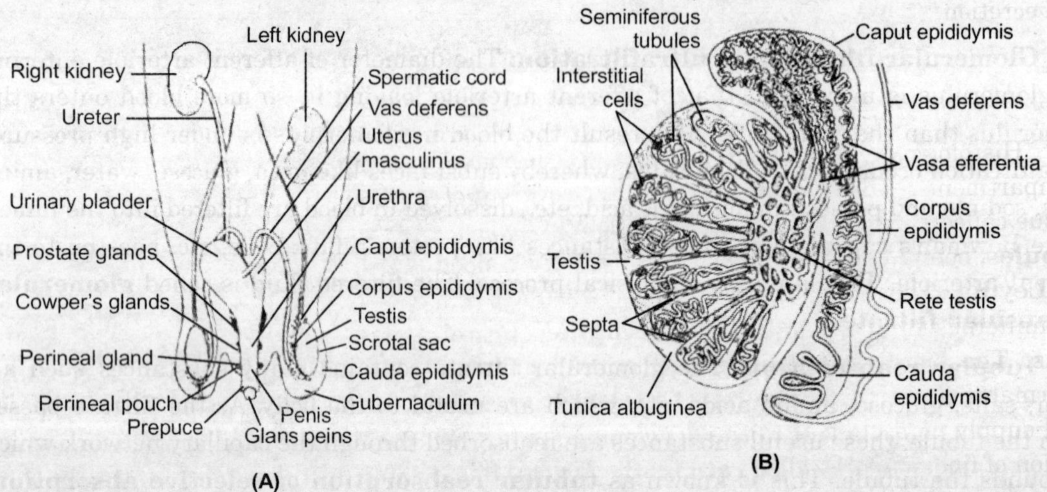

Fig. 23.52 *Rabbit*–**(A)** Male reproductive organs; **(B)** Longitudinal section of testis and epididymis

The **testes** are a pair of elongated ovoid bodies varying greatly in size at different times and about an inch and a half when fully developed. Each testis is lodged in a thin walled sac of hairy skin called **scrotal sacs.** The scrotal sacs lie outside the abdominal cavity, located ventrally in the pubic region, on either side of the penis. The scrotal sac is in communication with the abdominal cavity through the **inguinal canal.** In young animal the testis lie within the abdominal cavity close to the kidneys but at puberty they descend down the inguinal canal into the scrotal sacs. The development of sperms in mammals takes place within a limited range of temperature. The temperature in the abdominal cavity is high so the sperms develop in scrotal sacs where the temperature is somewhat lower.

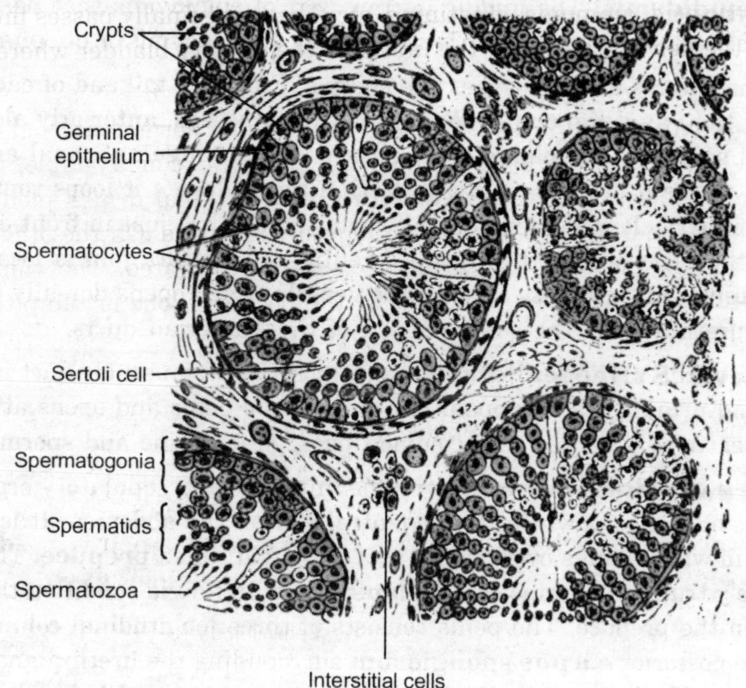

Fig. 23.53 *Rabbit*–T.S. of testis

Histologically (Fig. 23.53) each testis is made up of a large number of cone-shaped compartments called **lobules.** The outer covering of the testis is a tough coat of connective tissue called **tunica albuginea.** Each lobule consists of a mass of long, coiled **seminiferous tubules,** bound by connective tissue, containing blood vessels and **interstitial cells** or **cells of Leydig.** These cells secrete hormone which control secondary sexual characters. The seminiferous tubules are lined by **germinal epithelium** which is made up of two kinds of cells. The small **spermatogenic cells** which undergo spermatogenesis and produce spermatozoa. Certain large cells called **Sertoli cells** occur in between spermatogenic cells and supply nourishment to the developing sperms. All the seminiferous tubules open into a region of fine network called **rete testis**. Several fine ciliated ductules called **vasa efferentia** arise from this region and open into the epididymis.

The **epididymis** (Fig. 23.52) is a long, highly coiled structure located along the inner surface of the testis. It is made up of three distinct parts:

(a) **Caput epididymis:** The epididymis begins at the anterior end of the testis where it forms a cap-like structure of coiled mass called **head** of the epididymis or **caput** epididymis. It lies burried in fat body. It is connected with testis by fine ductules called **vasa efferentia** and to the dorsal abdominal wall through a **spermatic cord** which consists of blood vessels (spermatic artery and vein), nerves and connective tissue.

(b) **Cauda epididymis:** At the posterior end of the testis the epididymis forms another coiled mass called the **tail** of epididymis or **cauda epididymis**. It is connected with the scrotal sac by a short and stout ligament called **gubernaculum.**

(c) **Corpus epididymis:** The middle narrow part of epididymis that passes down the dorsal surface of the testis as a coiled tube is called **body** of epididymis or **corpus epididymis.**

The epididymis stores and nourishes the sperms. From the tail end of each epididymis the duct of testis, the **vas deferens** or **Wolffian duct** proceeds, anteriorly along the inner side of the scrotal sac, as a convoluted tube, passes into the inguinal canal and enters the abdominal cavity, where it becomes straight. Within the abdomen it loops round the ureter from the outer to inner side and opens dorsally into the urethra just in front of the opening of the ureter. At the place where the vasa deferentia and urethra join is located a blindly ending sac, the **uterus masculinus** or **seminal vesicle** which opens dorsally by a cresentic opening into the urethra. It is the sole remnant of the Mullerian ducts.

The **urethra** or the **urinogenital duct** is a thick-walled muscular duct into which the vasa deferentia and urinary bladder open. It traverses the penis and opens at its tip as the **urinogenital aperture.** It serves as a common passage for urine and spermatic fluid.

The **penis** is a cylindrical, muscular, erectile, intromittent, copulatory organ located in front of the anus and hangs down from the ventral wall of the abdomen. It is covered by a loose sheath of skin which hangs over the tip as a loose fold called **prepuce.** The tip of penis is soft and obliquely truncated and is called **glans penis.** In flaccid condition the glans penis is retracted within the prepuce. The penis consists of three longitudinal columns of spongy erectile tissue, the posterior **corpus spongiosum** surrounding the urethra and two anterior **corpora cavernosa.** During sexual activity these columns become filled with blood causing erection of penis.

Accessory glands

A number of accessory glands are associated with the male reproductive system. They open into the urethra. The secretion of these glands and that of epididymis and uterus masculinus is called **seminal fluid** or **semen.**

Prostrate gland: A large prostrate gland is present at the base of uterus masculinus. It opens into the urethra by number of small ducts. Its secretion nourishes and activates the spermatozoa.

Cowper's glands: A pair of **bulburethral** or Cowper's gland are situated at the base of penis posterior to the prostrate gland. Its secretion neutralises the acidity of the urethral passage.

Perineal glands: A pair of scent or perineal glands are situated at the base of the penis behind the Cowper's gland. These glands open to the outside by ducts into a hairless **perineal depression** on either side of anus. The secretion of these glands provides a characteristic odour to the rabbit.

Rectal glands: A pair of large rectal glands of unknown function, are situated at the sides of the rectum.

Female reproductive system

The female reproductive system includes a pair of **ovaries, oviducts** and **uteri** and **vagina, vestibule** and **clitoris** and some **accessory glands.**

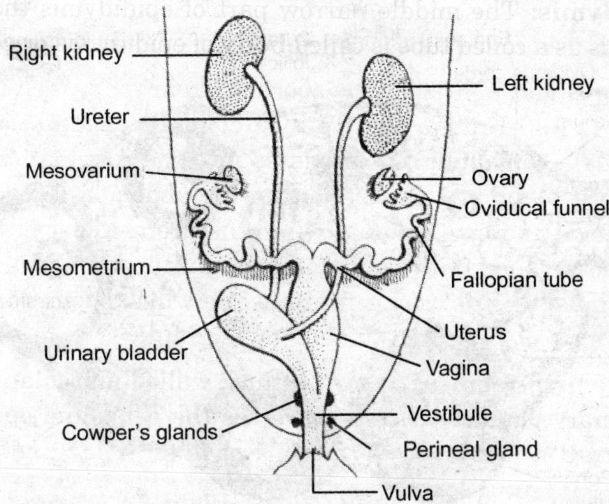

Fig. 23.54 *Rabbit*–Female reproductive organs

The **ovaries** are a pair of small ovoid bodies of pale yellowish colour. They lie behind the kidney and are attached to the dorsal wall of the abdomen by a fold of peritoneum called **mesovarium.** The surface of ovary bears several blister-like, small, rounded projections called **graafian follicles,** each containing a developing ovum.

The mammalian ovary is a compact body the bulk of which is made up of mass of fibrous connective tissue surrounded by a layer of spindle-shaped **germinal epithelial** cells, the two together forming the **stroma.** The stroma contains actively dividing germinal cells called **follicles** in different stages of development. In each follicle a single cell enlarges to form the oocyte or ovum while other cells surround and nourish it. As the follicle enlarges and sinks deeper and deeper into the stroma it requires a fluid filled space, the **antrum** or **follicular cavity**, the fluid being called the **liquor folliculi.** This crescentic space gradually becomes larger and separates the oocyte from rest of the follicle cells except at one point the **discus proligerus** or **cumulus oophorus.** The ripe follicles are called **Graafian follicles** which appear on the surface of ovary as small, semitransparent projections. The fully mature oocyte is surrounded by a transparent membrane, the **zona pellucida** containing yolk and fat droplets. It is surrounded by a layer of columnar cells called **corona radiata.** The follicular cavity is lined by **membrana granulosa.** In the stroma groups of **interstitial cells** are present which produce sex hormone (estrogen).

The membrana granulosa of the Graafian follicle (Fig. 23.55) is surrounded by a vascular connective tissue layer the **theca interna** containing blood vessels and nerves. This in turn is surrounded by another connective tissue layer the **theca externa.** The Graafian follicle ruptures releasing the oocyte into the body cavity (ovulalion). The remaining cells of the follicle proliferate, fill up the empty space of the discharged follicle to form a yellow body, the **corpus luteum** which remains active throughout embryonic development. The cells of corpus luteum secrete the hormone **progesterone** which causes enlargement of uterus and stimulates lactation. If however, fertilization does not occur it degenerates leaving a scar called corpus albicans.

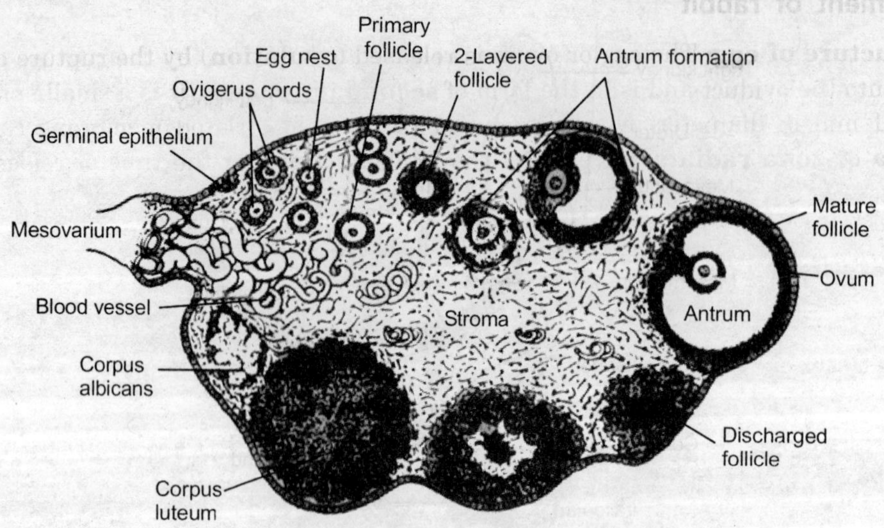

Fig. 23.55 *Rabbit*–T.S. of ovary

The **oviducts** are derived from the Müllerian ducts of the embryo. They are long covoluted tubes consisting of two parts. The anterior ends are narrow and called **fallopian tubes** while the posterior wider ends are called **uteri.** The anterior end of fallopian tubes are funnel-shaped (**oviducal funnel)** and its opening is called **ostium.** Its margin is frimbriated. The ostium is closely applied to the outer border of the ovary so that the eggs may pass into it soon after they are released. A pair of folds of peritoneum, the **broad ligament** suspend them from the dorsal wall of the abdomen. The eggs pass down the fallopian tube and fertilization occurs here. The **uterus** is thick walled, muscular, highly distensible tube, attached to the dorsal wall of the abdomen by a mesentery. It is richly supplied with blood vessels. The embryo or foetus develops in the uterus.

The two uteri fuse with each other posteriorly to form a common median tube, the **vagina,** lying dorsally, above the urinary bladder. It is a wide tube that runs straight back and in the pelvic cavity meets the neck of the urinary bladder to form a common urinogenital canal called the **vestibule** which runs ventral to the rectum and opens to the exterior by a slit-like **vulva.** The walls of the vestibule are very vascular and correspond to the corpus spongiosum in the male. The vagina receives the penis of the male during copulation.

At the entrance of the vestibule is a small root-like body projecting into it. This is the **clitoris** corresponding to the penis of male. It ends in a soft conical body the **glans clitoris** and is connected with ischia by two **corpora cavernosa** similar to those of male but smaller. The clitoris is erectile but the urethra does not pass through it.

The **Cowper's glands** or **Bartholin's glands** are a pair of small glands embedded in the dorsal wall of the vestibule. They are reduced or absent in female rabbit. Their secretion lubricates the vaginal passage. The **perineal** and **rectal glands** are also present and are similar in structure and disposition as in male. **Prostrate glands** are absent in female rabbit.

Development of rabbit

Structure of egg: The egg or ovum is released (**ovulation**) by the rupture of the wall of ovary into the oviduct and is in the form of **secondary oocyte.** It is a small spherical cell (about 0.1 mm in diameter in rabbit) enclosed by a thin striated membrane called **zona pellucida** or **zona radiata** secreted by the follicular cells. At the time of release the egg

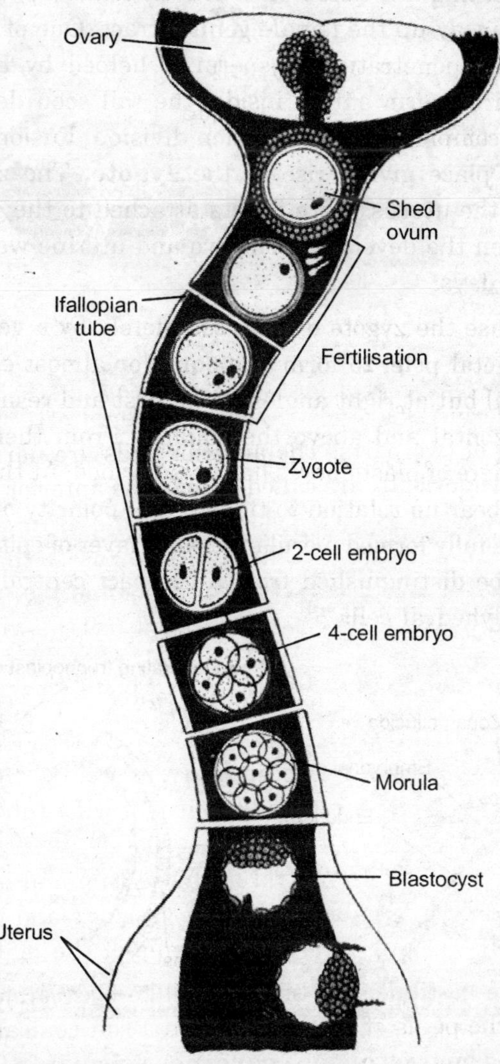

Fig. 23.56 *Rabbit*–Early stages of development from release of egg upto implantation

is surrounded by a number of large follicle cells called **corona radiata.** Soon after fertilization these cells disappear and in their place a dense coat of **albumen** is secreted by the walls of the oviduct. The amount of yolk in the mammalian egg is reduced (**microlecithal**) but is distributed evenly in the cytoplasm. It is not surrounded by the primary egg membrane. In the hemisphere opposite to that from which the polar bodies are extruded the yolk is slightly more abundant. This exhibits a distinct polarity. The nucleus is eccentric, it lies above the equator in an area quite free from yolk, this also shows polarity. The animal pole carries one or two extruded polar bodies depending upon the stage of maturation.

Fertilization: When mature, the ovum is released by a rupture of the ovary and passes into the upper end of Fallopian tube. The ovum begins to pass down the fallopian tube towards the uterus and if mating has taken place it soon comes in contact with spermatozoa which have travelled all the way up the female genital tract. One of the spermatozoa passes into the egg cytoplasm. The penetration of sperm is helped by **hyaluronidase** enzyme present in semen. The entire sperm enters inside, the tail soon degerates. Soon after the entry of the sperm the egg completes its maturation division. Fusion of the egg nucleus and sperm nucleus then takes place giving rise to the **zygote.** The zygote passes down the fallopian tube and reaches the uterus. Here it gets attached to the wall of the uterus. Soon a placenta is formed between the developing embryo and uterine wall. The gestation period in rabbit is of about 22–30 days.

Cleavage: After a pause the zygote divides completely by a vertical furrow, extending from the animal to the vegetal pole, to form two equal or almost equal **blastomeres.** The next cleavage is also vertical but at right angles to the first and results in a 4-celled embryo. The third cleavage is horizontal and above the equator. From then on, cleavages proceed irregularly until a small sphere of blastomeres has been formed. At this stage the blastomeres are all alike and appear to bear no relation to the original polarity of the egg. This solid ball of cells is called **morula.** In a fully formed morula the outer layer of epithelial cells (**trophoblast** or **trophoectoderm**) can be distinguished from a compact central mass (the **inner cells mass**) of slightly larger polyhedral cells.

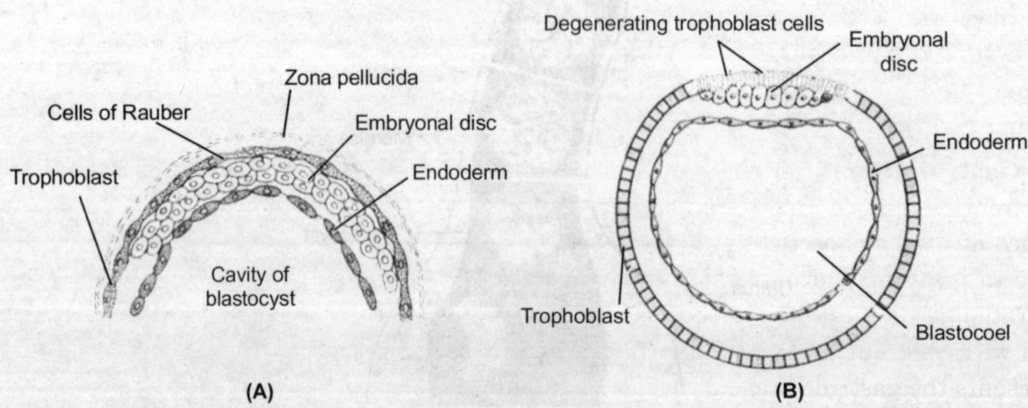

Fig. 23.57 *Rabbit*–**(A)** V.S. of blastocyst showing formation of endoderm; **(B)** V.S. of fully formed blastodisc

Blastocyst: The morula is not much larger than the fertilized egg although it consists of many cells. The cells of the embryo continue to divide and further growth is very rapid. Now certain other changes occur. Fluid begins to accumulate between the cells and this results in the formation of a **blastocyst** or **blastodermic vesicle** (Fig. 23.57 A)**.** By this time the embryo has reached the uterus and begins to absorb fluid secreted by the mucous membrane. The embryo swells rapidly as the amount of fluid in the interior increases. The fluid-filled space is called the blastocoel, which increases restricting the inner cell mass to one side, which will develop into the body of the organism. The surrounding cells constitute the **trophoblast.** The trophoblast cells which lie immediately above the inner cell mass are called **cells of Rauber.** With further development the blastocoel increases so much so that the inner cell mass appears as a knob-like thickening, called the **embryonal knob,** (Fig. 23.57 A) at one pole.

By this time the embryo becomes apposed and fixed to a small part of the uterine wall by means of the trophoblast. The joint reaction and development of the uterine mucosa and trophoblast is called **implantation** (Fig. 23.56) and leads to the development of the placenta through which the embryo exchanges nutrients and excretory products with its mother. This stage onwards the embryo remains attached to the uterine wall, from where it does not move until born.

The development of a mammal takes place in two phases: the first begins with fertilization and ends with the apposition of the blastocyst to the uterine wall and the second begins when the individual becomes intimately attached to its mother via the placenta and ends at birth. There is a very important difference between these two phases. The individual is small in the first or **free-living phase** and does not form specialized sets of cells, but from the beginning of the second or **fixed phase,** organ formation commences and this is accompanied by a rapid increase in the size of foetus. In the free living phase the uptake of nutrients and the elimination of waste products takes place by direct exchange with the local environment, in this case the secretion of the fallopian tube. In the fixed phase the embryo develops a system concerned with exchange of nutrients and excretory products between all its cells and mother. This system comprises the foetal blood, heart and circulation. The mother's own capabilities become sufficiently great during pregnancy to deal with the extra-metabolic activities of the growing embryo.

Gastrulation: After the growth of the blastocyst for sometime the inner (unattached) cells from side of the inner cells mass (roof of blastocoel) begin to proliferate along the inner surface of the trophoectoderm to form the yolk sac. These cells have more darkly staining properties and extend over the inner cell mass forming a layer called the **endoderm.** Its separation is the first step in the conversion of the blastocoel into the **yolk sac** which is not filled with yolk but merely with a fluid. There is no doubt that at this stage the blastocyst represents the gastrula stage although the amniotic cavity has already made its appearance. The embryonal knob flattens soon after the separation of the endoderm layer and gradually increases in size. The cells of the embryonal knob become rearranged forming a fairly regular epithelium. This is the area that can be compared with the blastodisc and it gives rise to the future embryo. It may now be called the **embryonal disc.** As the growth proceeds the cells of Rauber covering the embryonal disc disintegrate and are then sloughed off or absorbed by the trophoblast, exposing the embryonal disc. By this time the zona pellucida of the original ovum and albuminous layers surrounding it also break and are absorbed.

Further development of the embryonal disc resembles, in broad outline, that of reptiles and birds. A **primitive streak,** with all its components (**primitive knot, primitive pit,** etc.) develops on the embryonal disc. It initiates the formative movements of cells, exactly as in chick, ultimately separating the mesoderm and notochordal process. For sometime, however, the notochordal process remains a hollow structure traversed by a **chorda canal** (also formed in reptiles). Formation of **neural plate** follows the primitive streak. This finally develops the **neural groove** flanked by the **neural folds** which meet enclosing the **neural tube,** the forerunner of the brain and spinal cord. The differentiation and folding of the embryo follows the same steps as seen in the chick. At this stage, it becomes difficult to differentiate the mammalian embryo from the chick embryo.

Embryonic membranes

In higher vertebrates during development some membranous structures are formed which do not take part in the formation of the embryo but serve some important functions of the embryo like nutrition, respiration, excretion and protection. They are called **foetal** or **embryonic** or **extra-embryonic membranes.** They are **amnion, chorion, allantois** and **yolk sac.** Their formation is similar to that of chick. The formation of amnion and chorion differs in different mammals but when formed they bear the same relationship with the embryo.

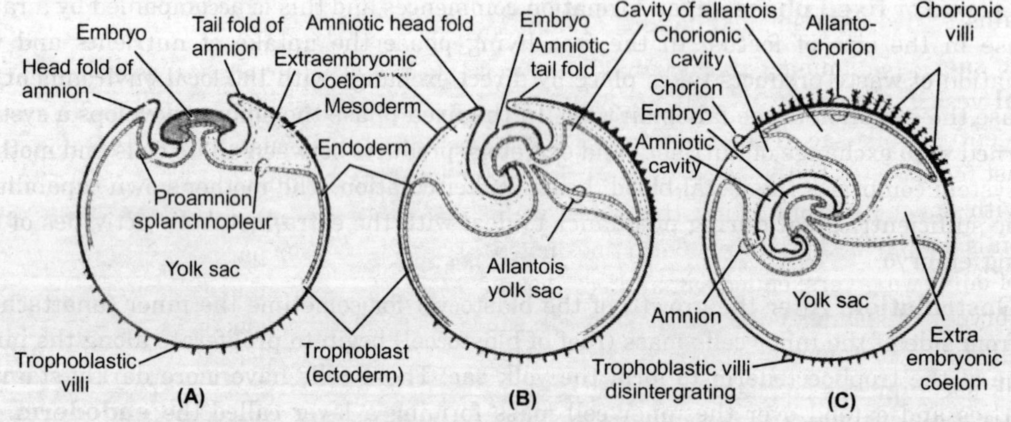

Fig. 23.58 *Rabbit*–Formation of foetal membranes. **(A)** Head and tail folds of amnion growing towards each other; **(B)** Allantois formed; **(C)** Formation of foetal membranes and placenta completed

Amnion and chorion: The amnion and chorion arise as **amniotic folds** around the margin of the embryonal disc. The **head fold** is formed on the anterior side and the **tail fold** posteriorly (Fig 23.58 A). Each fold encloses a cavity, the **extra-embryonic coelom.** The tail fold is larger and grows more rapidly than the head fold. The outer layer of each fold has an outer layer of ectoderm and an inner layer of somatic mesoderm. In the inner layer this arrangement is reversed, the outer layer is of somatic mesoderm and inner layer of ectoderm. The head and tail folds advance towards each other and ultimately fuse with each other, the point of fusion or **seroamniotic connection** lies above the anterior region of the embryo. As a result two membranes are formed, the outer layer is **chorion** and the inner layer is **amnion**. The amnion is made up of outer layer of mesoderm and inner layer of ectoderm. It encloses a fluid filled amniotic cavity lined by ectoderm. It protects the embryo from jerks and dessication. The chorion is made up of outer ectoderm and inner mesoderm. The cavity, between the amnion and chorion is **chorionic cavity.** The ectoderm of chorion develops villi, the **chorionic villi,** like those of the lower half of the blastocyst, which penetrate into the walls of the uterus.

Allantois: The endoderm beneath the hind end of the embryo grows out as a sac or vesicle and carries with it the overlying layer of splanchnic mesoderm. This vesicle is **allantois** (Fig. 23.58 B). Its formation begins even before the amniotic cavity has become closed. In the chick embryo also it is formed in the same manner. It enlarges rapidly and extends into the extraembryonic coelom between the amnion and chorion. The allantois has a rich vascular supply. With further development some of these vessels become larger forming **allatoic (umbilical) arteries** and **veins**. The outer splanchnic mesoderm layer of allantois comes in contact with the mesodermal layer of chorion and both fuse together to form allanto-chorion or chorio-allantois. The outer layer of extraembryonic wall becomes three-layered, outer ectoderm, middle double mesoderm and inner endoderm. The allanto-chorion plays important role in excretory and respiratory processes as in birds and in mammals it also transfers nutrients being a major part of the plancenta. In man and apes the allantois is rudimentary and remains buried as a small tube in the umbilical cord.

Yolk sac: In mammals the original blastocoel becomes converted into **yolk sac** or **umbilical** vesicle (the name usually applied in mammals). The growing layer of endoderm lines the entire yolk sac. The wall of yolk sac, at this stage, consists of endoderm and trophoblast (ectoderm). In the upper part of the yolk sac the splanchnic mesoderm comes in contact with endoderm completing the upper wall (splanchnopleurae) in which a vitelline circulation is established. The mesoderm does not penetrate the lower part of yolk sac which consists of only two layers, endoderm and ectoderm. The yolk sac is connected to the enteron of the embryo by a slender **yolk stalk.** The yolk sac does not contain yolk as in reptiles and birds but it helps in conveying nourishment to the embryo. After sometimc the lower wall of the yolk sac ruptures so that the cavity of the yolk sac communicates with that of the uterus. The uterine glands produce certain nutritive secretions which fill the ruptured yolk sac and from there they are absorbed into the vitelline circulation and passed on to the enteron for nourishment. The nutritive uterine secretion in some mammals (carnivores and ungulates) is produced in large quantity and is known as **uterine milk.**

PLACENTA IN MAMMALS

In the evolution of reptiles to placental mammals the egg has lost its yolk and has returned to an isolecithal condition with less yolk than the egg of Amphioxus. This necessitates the early development of structures to draw nourishment for the embryo from the uterine tissues of the mother. To carry out such functions the embryonic membranes have become modified and have developed structures that enable the embryo to obtain nourishment from the uterus. This embryonic organ is called **placenta** and it is used for the supply of oxygen and food and for the elimination of carbon dioxide and other waste from the embryo. The process of formation of placenta is called **placentation.**

Placenta may be defined as a special kind of tissue connection between the mother and the foetal membranes for physiological exchange of materials. It is formed by the interlocking of both foetal and maternal tissues. The foetal part is called **foetal placenta** and the part derived from the uterine wall is called **maternal placenta.**

Placenta is present in metatherian and eutherian mammals which are **viviparous,** *i.e.*, the young ones develop inside the uterus of the mother. The protetherian mammals are **oviparous,** they lay large yolk-filled, shelled eggs, so placenta is absent.

Types of placenta

The mode of origin, shape and structure of placenta varies is different mammalian orders. The main factors involved in the classification of placenta are:

1. The foetal membranes involved in its formation.
2. Distribution of villi and shape of placenta.
3. Nature of contact between the tissues involved in its formation.

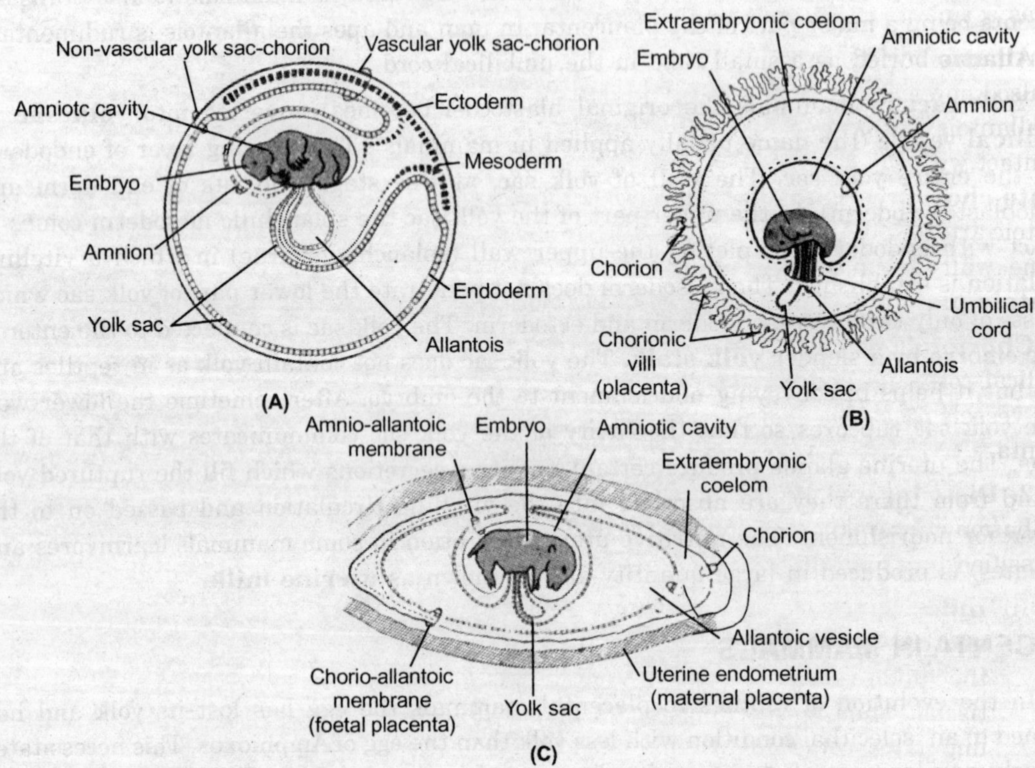

Fig. 23.59 *Rabbit*–Types of placenta. **(A)** Yolk sac placenta (opossum); **(B)** Allantoic placenta (pig); **(C)** Chorionic placenta (man)

1. The foetal membranes involved: Of the four foetal membranes, chorion, yolk sac and allantois are involved in the formation of placenta and the amnion does not take part. Based on the foetal membranes involved in its formation placenta is of three kinds: yolk sac, allantoic and chorionic.

Yolk sac placenta: It is a primitive type of placenta (Fig. 23.59 A) developed in a few primitive Metatherians like *Didelphis virginiana* (opposum, a marsupial) and *Dasyurus* (marsupial cat). The placenta is formed by yolk sac and chorion. It is also called **chorio-vitelline placenta.** The yolk sac developed from the lower part of the blastocyst is very large and ecloses the entire embryo. It comes in direct contact with the chorion which develops wrinkles and corrugations on its outer surface where it comes in contact with the uterine wall. These wrinkles hold the blastocyst to the uterine wall and thus reduce the possibility of its premature exit. The yolk sac wall develops vitelline blood vessels. The uterine secretion secreted by the uterine glands is absorbed by the chorion and transported by vitelline circulation to the embryo. The allantois remains poorly developed.

The yolk sac placenta is not very well developed, so embryonic growth remains limited and immature young is born. It is kept in **marsupium**, a pouch in the abdomen of female, and fed on milk until fully formed.

In Eutheria yolk sac placenta is usually not formed but in some mammals like hedgehog and rabbit it may develop in early stages. In man yolk sac is small and lies in the umbilical cord as a small tube.

Allantoic placenta: This placenta is formed by chorion and allantois (Fig. 23.59 B), so it is also known as **chorio-allantoic placenta**. It is found in majority of eutherian mammals. The allantois is a large sac-like structure arising from the hindgut of the embryo. It comes in contact with the chorion. The mesoderm of both fuse together to form a membrane called **allanto-chorion** or **chorio-allantoic** membrane. This membrane is well vascularized with allantoic arteries and veins. Its outer surface develops finger-like outgrowths called **villi.** The uterine wall develops depressions called **crypts.** The villi penetrate the crypts forming **allantoic placenta.** This facilitates absorption of nutrients from the maternal side.

Chorionic placenta: In man and apes the allantois is small and buried into the umbilical cord and does not reach the chorion (Fig. 23.59 C). However, its mesoderm and blood vessels reach the chorion. The villi of chorion reach the uterine wall forming chorionic placenta.

2. Distribution of villi and shape of placenta: Based on the shape of placenta, distribution of chorionic villi and nature of contact will the uterine wall, the allantoic placenta is classified into following types:

(i) **Indeciduate** or **non-deciduous placenta:** The chorionic villi are simple projections and have loose connection with the crypts in the uterine wall. At the time of birth the villi are easily withdrawn from the uterine epithelium causing no loss of maternal tissue. Thus no bleeding occurs at parturition. This type of placenta is found in ungulates, cetaceans, sirenians and lemurs. The indeciduate placenta may be of the following forms:

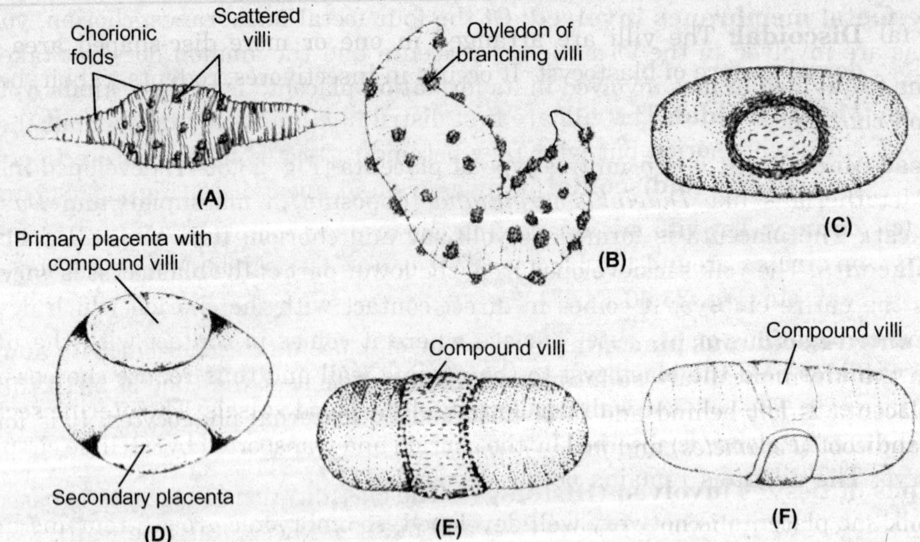

Fig. 23.60 *Rabbit*–Types of placenta. **(A)** Diffuse (pig); **(B)** Cotyledomary (cow); **(C)** Monodiscoidal (man); **(D)** Bidiscoidal (monkey); **(E)** Complete zonary (dog); **(F)** Incomplete zonary (racoon)

(a) **Diffuse:** The villi are uniformly distributed throughout the surface of alantochorion except at the extreme ends. It is found in horse, pigs and other ungulates (Fig. 23.60 A).

(b) **Cotyledonary:** In this type of placenta the villi are not uniformly distributed but occur in tufts or bunches called **cotyledons** (Fig. 23.60B). The uterine wall is provided with thickened sockets called **caruncles** into which the cotyledons fit. The villi of cotyledons fit into corresponding crypts in the caruncles. It is found in cow, goat, sheep and deer.

(c) **Intermediate:** The cotyledons are present and between them occur the villi. It is a rare type present in giraffe and camel.

(ii) **Deciduous** or **deciduate placenta:** In this type of placenta the union between the chorion and uterine wall (endometrium) is much more intimate. The foetus along with the foetal membranes is enclosed by the uterine epithelium. The chorionic villi penetrate the uterine mucosa and secrete certain enzymatic material that erodes the maternal tissue establishing intimate relation between the villi and maternal tissue. At the time of birth a large part of maternal tissue is lost along with the foetal membrane causing bleeding.

The uterine wall participating in the formation of placenta is called **decidua**. It has three important parts: (*i*) **decidua basilaris,** the upper part to which the embryo becomes attached, (*ii*) **decidua capsularis,** the part which surrounds the blastocyst and separates it form the cavity of the uterus and (*iii*) **decidua parietalis,** the part which forms the inner lining the uterine wall.

Deciduate placenta is found is primates, Chiroptera, Rodentia, Insectivora, etc. The deciduous placenta is of the following types:

(a) **Discoidal:** The villi are arranged in one or more disc-shaped area on the dorsal surface of blastocyst. It occurs in insectivores, rodents, rabbit, bear, etc.

(b) **Metadiscoidal:** The villi are first distributed uniformly throughout the entire surface of chorion but later they become restricted to one (**monodiscoidal** in man) or two (**bidiscoidal** in monkeys and apes) discs (Fig. 23.60C and D).

(c) **Zonary:** The villi are arranged in one or more complete or incomplete **girdles** or **circles** around the blastocyst. It occurs in carnivores (cat, dog), elephant, seal, etc (Fig. 23.60 E).

(iii) **Contra-deciduous placenta:** In this type of placenta the chorionic villi and crypts in the uterine wall are so intimately connected that at the time of birth the foetal placenta is left behind and later absorbed by maternal phagocytes. It is found in bandicoot (*Perameles*) and mole (*Talpa*).

3. Types of tissues involved (Histology): The classification of placenta is based upon the relationship of the embryonic tissue (villi) and maternal tissue (uterine wall). There are 6 tissues which are involved in the formation of placenta:

1. Maternal endothelium.
2. Maternal connective tissue (syndesmos).
3. Maternal epithelium.
4. Chorionic epithelium.
5. Foetal connective tissue.
6. Foetal endothelium.

Histologically there are 5 types of placenta.

1. **Epithelio-chorial placenta:** It is the simplest type of placenta in which all the six tissue layer of placenta are retained (indeciduate, allantoic placenta). It is found in pigs, lemurs, horse, etc.
2. **Syndesmochorial placenta:** The uterine epithelium is eroded and the chorionic epithelium comes in direct contact with the maternal connective tissue (cotyledonary indeciduate). It is found in cow and sheep.
3. **Endothelio-chorial placenta:** In this type, the uterine epithelium as well as connective tissue are eroded and thus the chorion makes direct contact with maternal endothelium (zonary deciduate). It is found in cat, dogs, etc.
4. **Haemochorial placenta:** In this type, in addition to maternal epithelium and connective tissue, the endothelium of blood vessels is also eroded. The chorionic villi directly dip into maternal blood sinuses (discoidal deciduate). It is found in man, apes, some rodents and insectivores.
5. **Haemoendothelial placenta:** In this type, the chorion is also completely eroded so that the foetal blood vessels bathe directly into maternal blood pool. It is found in rat, guinea pig, rabbit, etc.

Functions of placenta

There is never any direct communication between the foetal and maternal blood streams. Several reasons have been put forward for this. Firstly, the foetal blood vessels are delicate and would not be able to withstand the pressure of blood in material arteries. Then in the maternal blood certain proteins specific to individuals are circulating. These, if transferred to the embryo, might prove harmful. The placenta, therefore, acts as a **physiological barrier** and selects only such materials that suit the embryo. The chief functions of placenta are:

1. If forms a **semipermeable membrane** between the mother and foetus and allows only smaller molecules to pass through it. The molecules which pass through the placenta are O_2, CO_2, water, chlorides and phosphates of sodium, potassium and magnesium and organic substances like monosaccharides, amino acids, vitamins, hormones and waste products like urea. It prevents direct mixing of maternal and foetal blood.
2. Blood proteins and fats, being large molecules cannot pass through the placenta. The proteins are broken down into amino acids and fats into fatty acids and glycerol and then transported to the embryo. The foetus rebuilds complex proteins and fats. The placenta manufactures fructose from glucose.
3. The placenta prevents direct mixing of maternal and foetal blood.
4. Ascorbic acid is selectively by absorbed by the placenta. At the time of birth the concentration of ascorbic acid in placental blood is 2 to 4 times higher than in maternal blood.
5. Antibodies of diphtheria, smallpox and measles, tetanus toxins and Rh antibodies readily pass through the placenta to the foetus which also acquires immunity against such diseases.
6. Many drugs consumed by the mother pass through the placenta, some may cause harm to the foetus, for example, **thalidomide** taken as a sedative by the mother in early pregnacy may effect limbs, heart and digestive tract of foetus.
7. Certain pathogenic organisms can pass through the placenta and attack the foetus, for example, syphilis, small pox, chicken pox, measles etc. Infection of placenta by pathogenic organsms is called **placentitis.**
8. The placenta functions as an **endocrine gland.** It secretes **hormones** like oestrogen, progesteron, chorionic gonadotrophin, lactogen, etc. In rabbit it secretes relaxin.
9. The placenta, foetal and maternal tissue expelled out after birth often serves as a source of nourishment in many mammals because they are eaten by the mother.

It has been pointed out that there is no direct communication between foetal and maternal blood. But if accidently the placental capillaries break then some red blood cells of the foetus may enter maternal circulation. If by chance the mother is Rh negative and foetus Rh positive, the positive red blood cells will induce production of Rh antibody in the mother, which will easily enter foetal circulation and damage red blood cells causing jaundice or even death of foetus.

Parturition (birth)

The mother carries the developing young in the womb for a fixed perior known as **gestation,** which varies from a few weeks in smaller mammals to nearly two years in elephant. Towards the end of this period the foetus starts looking like miniature adult. Then the wall of uterus undergoes rhythmical contractions, the connection between the foetal and maternal layers in the placenta is broken and the foetus is gradually expelled through the vagina which becomes dilated for this purpose. Although the connection between the foetus and the placenta is cut-off in the earlier stages of birth, yet a portion of the umbilical stalk remains attached to the foetus for a time. Finally it shrivels and is absorbed, its position being indicated by the **umbilicus** or **navel.** Sometimes after the birth of the young animal, in most mammals, the foetal contribution to the placenta and other foetal membranes are expelled.

24 CHAPTER

CLASS MAMMALIA:

GENERAL CHARACTERS AND CLASSIFICATION

The mammals are the final product of evolution. They are the highest group in the animal kingdom. The class Mammalia includes man, the master of the present age. The salient features of this class are:

1. The mammals (L., *mamma,* breast) are warm blooded (homeothermous) animals that have mammary glands by means of which the females suckle their young ones. In males, the mammary glands are undeveloped.
2. The body is usually covered with hairs which are moulted periodically.
3. The body is divided into head, neck, trunk and tail. The limbs are pentadactyl and variously modified for walking, running, climbing, burrowing, swimming or flying. Hindlimbs absent in Cetacea and Sirenia.
4. The skin is provided with sweat, sebaceous, and sometimes scent glands.
5. The exoskeletal structures are lifeless horny epidermal hairs, scales, claws, nails, hoofs, horns, etc.
6. Skull is provided with two occipital condyles. The cranial cavity is very large and the temporal region has a zygomatic arch. Pterygoids small and scale-like.
7. Nasal region is usually elongated. A secondary palate separates the nasal passage from the food passage.
8. The auditory capsule projects outwards as a swollen structure, the tympanic bulla.
9. The bones absent in the skull are prefrontal, post-frontal, quadratojugal, supra-orbital and basipterygoids.
10. Jaw suspensorium craniostylic.
11. Lower jaw is formed of a large dentary bone which articulates with the squamosal of the skull.
12. Vertebrae acoelous (flat centrum) and with terminal epiphysis. Cervical vertebrae seven.

13. Coracoid and precoracoid of pectoral girdle are reduced or vestigeal.
14. Ribs bicephalous.
15. Teeth thecodont (present in sockets), heterodont, (different types like incisors, canines, premolars and molars) and diphyodont (two sets– milk teeth and permanent teeth).
16. A muscular diaphragm separates the thoracic and abdominal cavities.
17. Alimentary canal terminates by anus, cloaca absent.
18. Respiration by lungs contained in pleural cavity. Larynx with vocal cords.
19. Heart four-chambered, only left aortic arch present and renal portal system absent.
20. Red blood cells biconcave and non-nucleated.
21. Kidneys metanephric and ureters open into a urinary bladder. Excretory waste is urea (ureotetic) excreted as urine.
22. Brain highly developed. Mammals show greatest intelligence. Corpus callossum present and it connects the two cerebral hemispheres. Optic lobes four in number called corpora quadrigemina.
23. Fleshy and cartilaginous external ear or pinna present to collect sound waves and protect external ear opening. Three ear ossicles—malleus, incus and stapes present in the middle ear. Cochlea of inner ear highly coiled.
24. Sexes separate, usually with well marked sexual dimorphism. Males with a copulatory organ, the penis. The testes usually in scrotum, external to abdomen.
25. Fertilization internal. Eggs usually minute, without protective shell and are retained in the uterus for development.
26. Developing foetus is attached to the uterine wall by placenta through which it derives nutrition, respires and removes wastes.
27. All mammals except Prototheria, are viviparous, *i.e.*, they give birth to young ones.
28. Mammals show parental care of high level.
29. Mammals live in all sorts of habitats from polar regions to the tropics and from the sea to the densest forests and driest deserts. Many mammals are nocturnal.

CLASSIFICATION

Class Mammalia includes 5,000 living species, about 15,000 subspecies and many extinct forms. According to G.G.Simpson there are 18 living and 14 extinct orders of mammals.

Class Mammalia is broadly divided into three subclasses: **Prototheria, Allotheria** and **Theria.**

Subclass I. Prototheria

(**Gr;** *Protos,* first + *therios,* beast).

1. Primitive, reptile-like, egg laying mammals.
2. Mammary glands unspecialised and without teats.

Order 1. Monotremata (Gr; *monos,* single + *trema,* opening).

1. Pinna absent.
2. Cloaca present.
3. Teeth in young, adults with horny beak.
4. Confined to Australian region.

Examples. *Ornithorhynchus* (duckbill platypus), *Tachyglossus* = *Echidna* (spiny anteater).

Subclass II. Allotheria

Extinct mammals chiefly known through the fossils of mandibles and teeth. Nothing is known about the soft parts but the skeleton appears to agree to some extent with that of the Monotremata. Occur from Jurassic to Eocene.

Order 1. Multituberculata.

Examples. *Plagiaulax, Microlestes*

Order 2. Triconodonta

Example. *Triconodon*

Subclass III. Theria

1. Marsupials and placental mammals.
2. External ear or pinna present.
3. Mammary glands provided with teats.
4. Female give birth to young ones (viviparous).

Infraclass A. Metatheria

1. Intermediate between monotremes and placental mammals.
2. Females with a pouch (marsupium) in the abdomen.
3. With or without rudimentary yolk sac placenta.
4. Occur mostly in Australia and South America.

Order 1. Marsupialia

1. Young born in a very undeveloped state and sheltered in the marsupium which contain teats of mammary gland.
2. Uterus and vagina double.
3. Usually 3 premolars and 4 molars on each side in each jaw.

Examples. *Didelphis* (opossum), *Macropus* (kangaroo), *Phascolarctos* (koala).

Infraclass B. Eutheria (Placentalia)

1. Higher placental mammals.
2. Marsupium absent.
3. Vagina single.
4. Young born in advanced stage.

There are several orders of extinct and living eutherians. Only living orders are being described.

Order 1. Insectivora

1. Small primitive mammals with a long pointed snout.
2. Body covered with fur or spines.
3. Feet plantigrade or semiplantigrade, generally pentadactyl with clawed digits.
4. Molars rooted, with pointed, peg-like cusps.
5. Widely distributed, terrestrial and noctural.
6. Testes never descend into scrotum.

Examples. *Talpa* (mole), *Sorex* (shrew), *Erinaceus* (hedgehog).

Order 2. Dermoptera

1. Includes flying lemurs which resemble flying squirrels.
2. Four equal sized limbs.
3. A hairy, muscular, parachute-like pataguim (fold of skin) extending from neck to the wrist and down the sides of the body to the ankle including the tail.
4. Nocturnal and feed on fruit and leaves.
5. Can glide with the help of pataguim.

Example. *Cynocephalus* (= *Galeopithecus*). Only one living genus with two species.

Order 3. Chiroptera

1. Nocturnal flying mammals (bats).
2. Forelimbs modified as wings (patagium), second to fifth digits greatly elongated to support a broad web of skin extending back to the short hindlimb.
3. Tail if present included in the pataguim.
4. Teeth small, sharp and heterodont.
5. Sternum with keel.
6. Eye small, vision weak.
7. Ear pinna large.
8. Capable of true flight.

Suborder *i.* Megachiroptera

1. Frugivorous, large animals which sleep by day on tree hanging head downwards, with wings folded, cloak-like around the body, in deserted areas.
2. Live in flocks.
3. Eyes large, snout elongated and without foliaceous appendages, ear pinna large and without appendages, first and second digits clawed.
4. Tail absent.
5. Molars with longitudinal grooves.

6. Found in tropical and subtropical regions of Eastern hemisphere. Commonly known as flying foxes or fruit bats.

Examples. *Pteropus, Cynopterus*

Suborder *ii*. Microchiroptera

1. Insectivorous, small animals which hang by hind claws head downward during day in crevices of rocks or trees or in caves or abandoned buildings.
2. Found in temperate and tropical regions of both hemispheres.
3. Some gregarious, others solitary.
4. Eyes small, snout short, with foliaceous appendages. Ear pinna large with leaf-like appendages (targus). Only first digit clawed.
5. Tail long, enclosed in inter-femoral membrane.
6. Molars with sharp cusps.

Examples. *Myotis* (small brown bat), *Vespertilio, Tadarida mexicana* (free tailed bat), *Desmodus* (blood-sucking vampire bat).

Order 4. Edentata

1. Teeth absent or reduced to molars in anterior part of the mouth, enamel absent. Dentition monophyodont.
2. Toes with strong, curved claws for scratching.
3. Testes abdominal.
4. Some species may be armoured.
5. Solitary, nocturnal and arboreal.

Examples. *Myrmecophaga* (giant anteater), *Dasypus* (armadillo), *Bradypus* (3-toed sloth).

Order 5. Pholidota

1. Body covered with overlapping epidermal scales which can be erected. Hair is present between the scales.
2. Teeth absent. Tongue long, sticky and protrusible and used for catching insects.
3. Limbs short and have five digits.
4. Terrestrial and burrowing animals. Can roll themselves into a ball for protection.

Example. Only one genus *Manis* (pangolin or scaly anteater).

Order 6. Tubulidentata

1. Pig-like thick-skinned animals with long snout.
2. Teeth 4 or 5 in each jaw, without enamel, incisors and canines absent. Teeth perforated by tubules of pulp.
3. Tongue long, sticky and protrusible.
4. Forelimbs short and stout with four toes, hindlimbs with five toes.
5. Ear long, erect and pointed.

Example. Only one genus *Orycteropus* (Cape anteater or aardvark) of South Africa.

Order 7. Rodentia

1. Small terrestrial herbivorous, plantigrade, gnawing mammals.
2. Incisors chisel-shaped, canines absent leaving a small gap called diastema between incisors and check teeth.
3. Body covered with fine fur which may be modified into spines.
4. Nails usually claw-like.

Examples. *Rattus* (rat), *Mus* (mouse), *Hystrix* (porcupine), *Cavia* (guinea pig), *Funambulus* (squirrel), *Castor* (beaver), *Bandicota* (bandicoot).

Order 8. Lagomorpha

1. Small to moderate in size, with small stubby tail.
2. Incisors chisel-like, grow continually, upper incisors two pairs. Canines absent leaving a large gap diastema.
3. Gnawing animals, eat leaves, stem and bark.

Example. *Lepus* (hare), *Oryctolagus* (rabbit), *Ochotona* (pika).

Order 9. Carnivora

1. Includes beasts of prey like dog, cat, tiger, lion and bear.
2. Range in size from tiny weasel to large sized bear.
3. Many of them are known for their speed particularly the cat family.
4. Flesh-eating animals, show adaptations for hunting. The canines are large. The last premolar in the upper jaw and the first molar in the lower jaw work against each other forming the **carnasial teeth.**
5. Skull has lamboidal crests for attachment of powerful jaw and neck muscles.
6. Mammae are abdominal.

Suborder *i.* Fissipedia

Terrestrial Carnivora with large and strong projecting canines, poorly developed incisors, six in each jaw, cutting premolars, molars tuberculate, carnasial teeth present. Limbs ungulate and provided with strong claws. Most of them are flesh eaters. Pelts of many form valuable fur.

Examples. *Canis familiaris* (domestic dog), *C. lepus* (wolf), *Urus* (bear), *Herpestes mungo* (mongoose), *Panthera leo* (lion), *Felis domesticus* (domestic cat), *Panthera tigris* (tiger), *Acinonyx* (cheetah), *Hyaena* (hyaena), *Ailuropoda* (panda).

Suborder *ii.* Pinnipedia

Marine carnivores with short pentadactyle limbs modified into flippers. Body streamlined, powerful swimmers, slow and clumsy on land where they come to rest and raise their young ones. They feed on fish and other marine animals.

Examples. *Phoca* (seal), *Mirounga* (sea elephant), *Odobenus* (walrus), *Callorhinus* (fur seal).

Order 10. Cetacea

1. Whales and dolphins.
2. Body fish-like, naked but with a thick layer of subcutaneous blubber.
3. Nasal apertures on top of the head.
4. Caudal fin-expanded horizontally and divided into two fleshy flukes which help in propulsion.
5. Forelimbs modified into broad paddle-like flippers, hindlimbs absent.
6. Minute ear opening devoid of external ear or pinna.
7. Eyes small and at the angle of the mouth.
8. Live in schools (herds) and swim with great speed.
9. Carnivorous and predaceous, found in all seas.
10. Largest animal that have ever lived.

Suborder *i*. Odontoceti

It includes toothed whales, relatively small in size. The two nostrils fuse and open into a single blow hole on top of the head. Have a long beak with homodont teeth. Feed mainly on fish. Upper surface of skull asymmetrical. Nasal passages directed up and posteriorly. The nipples in females lie close to the anus. Sternum many jointed, body oil used in cosmetics, candles, lubricants, etc.

Examples. *Platanista gangetica* (Ganges dolphin), *Physetes catodon* (sperm whale), *Delphinus* (dolphin), *Phocaena* (porpoise), *Orcinus* (killer whale), *Monodon* (nar whale).

Suborder *ii*. Mysticeti

It includes whalebone whales or baleen or toothless whales. They are relatively large in size. The two nostrils are separate. Teeth completely absent. Many horny triangular whalebone or baleen plates hang down from the upper jaw. These plates help in straining planktons on which they feed. Upper surface of skull symmetrical. Nasal passages directed up and anteriorly. Sternum consists of a single piece. Body oil edible.

Examples. *Balenoptera musculus* (blue whale), *Balaena* (right whale).

Order 11. Sirenia

1. Includes sluggish aquatic mammals commonly known as sea-cows.
2. They are short-necked, thick-skinned, naked or with very sparse hairy covering, herbivorous mammals browsing on sea weeds and other aquatic plants.
3. Nostrils anteriorly directed, pinna absent.
4. Forelimbs paddle-like and very mobile, hindlimbs absent.
5. Body streamlined, with rounded muzzle and a prominent cleft in the lip.
6. Tail horizontally expanded into a caudal fin.
7. Mammae two, pectoral. Testes abdominal.

Examples. *Halicore = Dugong* (dugong), *Trichechus* (manatee).

Order 12. Hyracoidea

1. Includes small rodent-like herbivorous mammals, distantly related to elephants.
2. They have close set fur, small ears, split snout and reduced tail.
3. Incisors $\frac{1}{2}$. Upper incisors are curved and chisel-shaped, lower incisors comb-like. Canines absent. Check teeth lophodont. Diastema present.
4. Testes abdominal mammae six pairs, four pairs inguinal and two pairs axillary.

Example. *Hyrax* = *Procavia* (conies) found in South Africa.

Order 13. Proboscida

1. Includes elephants having a long proboscis.
2. Largest living land animal with an enormous head.
3. Proboscis is a long, mobile, muscular expansion of the nose bearing nostrils at the tip. It is prehensile and capable of movement in all directions.
4. Limbs massive, pillar-like and semiplantigrade.
5. Incisors of great size and form tusks. Canines and premolars absent, molars lophodont.
6. Gregarious, live in herds of 10 to 100.
7. Feed on grass and bamboos.
8. Love dark and shady forests.

Examples. Only two species, *Elephas maximus* (the Indian elephant) and *Loxodonta africana* (African elephant).

Order 14. Perissodactyla

1. Includes hoofed mammals, the odd-toed ungulates like horse, tapir, rhinoceros, zebra, etc.
2. Feet unguligrade, hoof formed of uneven number (1 or 3) of toes. Axis of the limbs passes through the middle of the third digit.
3. Horns absent, size large, legs long.
4. Herbivorous mammals. Incisors present, molars and premolars alike and form a continuous series. Stomach simple.

Examples. *Tapirus* (malayan tapir), *Equus equus* (horse), *E.asinus* (wild ass), *E.zebra* (zebra), *Rhinoceros* (rhino).

Order 15. Artiodactyla

1. Includes even toed, hoofed ungulates like sheep, pig, hippopotamus, etc.
2. Herbivorous, usually of large size.
3. Digitigrade. The axis of the limbs passes between 3 and 4 digits. Number of toes even (2 or 4).

4. Check teeth bunodont or solenodont. Incisors and canines or upper jaw usually lacking. Stomach 4-chambered.
5. Many with antler or horns on head.
6. All but pigs ruminate or chew the cud.

Suborder *i*. Bunodontia

Includes pigs and its allies. Horns or antlers absent. Teeth 38 to 44, molars bunodont and provided with blunt cusps, canines enlarged to form curved tusks which is used for defence.

Examples. *Sus* (pig), *Hippopotamus amphibius* (hippopotamus) *Pecari tajacu,* peccaries or New World pigs.

Suborder *ii*. Pecora

It includes ruminants, the cud-chewing mammals like camel, deer, giraffe, antelope, cattle, etc. The skull bears horns which are outgrowths from the frontal bones. Teeth 32, canines small or absent, upper incisors absent but the gum hardened and the lower incisors bite against the hardened gum. Stomach four chambered (rumen, reticulum, omasum and abomasum).

Examples. *Camelus* (camel), *Axis axis* (spotted deer), *Moschus* (musk deer), *Giraffe* (giraffe), *Antilope* (black buck), *Ovis* (sheep), *Capra* (goat), *Bubalus bubalis* (buffalo), *Bos grunniens* (Yak), *Bison* (bison).

Order 16. Primates

1. Includes lemurs, monkeys, apes and man.
2. Limbs long, hand and feet enlarged, each with five digits with flattened nails. The innermost toe and thumb or both usually opposable, an adaptation for grasping and climbing.
3. Orbit usually directed forward and surrounded by bony ring. Eyes large, vision binocular.
4. Hairs cover the entire body except palm, sole and face.
5. Testes enclosed in scrotum. Two pectoral mammae.
6. Functional development of nervous system and intelligence remarkable.

Suborder *i*. Lemuroidea

Small, solitary, primitive, nocturnal, omnivorous, arboreal mammals with a long non-prehensile tail (sometimes lacking). Snout usually elongated. Eyes usually lateral and vision poor. Both pollex and hallux opposable, second toe with claw and others with flat nails. Brain case low, cerebellum not covered by cerebrum, orbit confluent with temporal fossa and nasals large. Dental formula $\frac{2.1.3.3}{2.1.3.3}$ = 36. Uterus bicornuate, placenta diffused and non-deciduate. Found in Madagascar, Africa and South-East Asia.

Examples. *Loris* (loris), *Lemur* (lemur), *Tupaia* (tree shrew).

Suborder *ii*. Tarsioidea

Slightly advanced, arboreal, nocturnal, solitary, insectivorous mammals with long thin tail tufted at the end. Eyes and ears are large and vision good. Snout shortened. Tarsus very long, fingers and toes ending in flattened discs, two toes with claws and three with nails. Orbits largely separated from temporal fossa. Dental formula $\frac{2.1.3.3}{1.1.3.3} = 34$. Upper incisors large, molars with numerous pointed cusps. Uterus bicornuate and placenta metadiscoidal and deciduate. Found in Indo-Malaya, Archipelago and Philippines.

Example. Single genus *Tarsius spectrum* (tarsier).

Suborder *iii*. Anthropoidea

Most advanced mammals including monkeys, apes and man. The snout short and eyes large and forwardly directed, with binocular vision. Quadripedal and plantigrade gait. Pollex and hallux opposable. All digits carry nails. Cranial cavity spacious, orbits rounded, look forward and separated from temporal fossa by a bony partition. Cerebrum covers cerebellum. Tail prehensile when present. Dental formula $\frac{2.1.2.3}{2.1.2.3} = 32$. Premolars and molars bunodont. Mammae two, thoracic. Testes in scrotum. Uterus simple, Placenta metadiscoidal. Have a world wide distribution.

Examples. *Hylobatis* (gibbon), *Satyrus* (orang-utan), *Gorilla* (gorilla), *Pan (Anthropopithecus) troglodytes* (chimpanzee), *Homo-sapiens* (man).

SOME COMMON MAMMALS

1. Ornithorhynchus

It is an egg-laying mammal commonly known as the **duck-billed platypus** (Fig. 24.1A). It is found in Australia and Tasmania. It is as large as a common cat. The body is stout and divided into head, trunk and short dorsoventrally flattened tail. The head is small with laterally directed eyes and a pair external ear openings. External ear or pinna absent. The snout is broad and elongated forming a flattened beak which is covered with leathery skin produced into a fold at the base of the snout. The body is covered with dense soft fur. Limbs are very short, each with five digits ending in strong claws. A web of skin is present between the digits (aquatic adaptation). Adult has no teeth, jaws are covered with horny plates. Ventrally a pair of slightly depressed spots is present on skin in the middle of the body. On these spots open the ducts of tubular mammary glands. The young animal merely licks the milk as it exudes through the pores (as nipples are absent). Cloaca present. Testes abdominal. Oviduct distinct, ureters and vagina absent. The duck-billed platypus is semi-aquatic and burrows into the bank of stream or pond. The burrows are 12–15 metres long. The entrance of burrow is under water but the inner end is high and dry. It feeds on aquatic animals. The female makes nest of roots and leaves during spring in the burrow and lays two eggs.

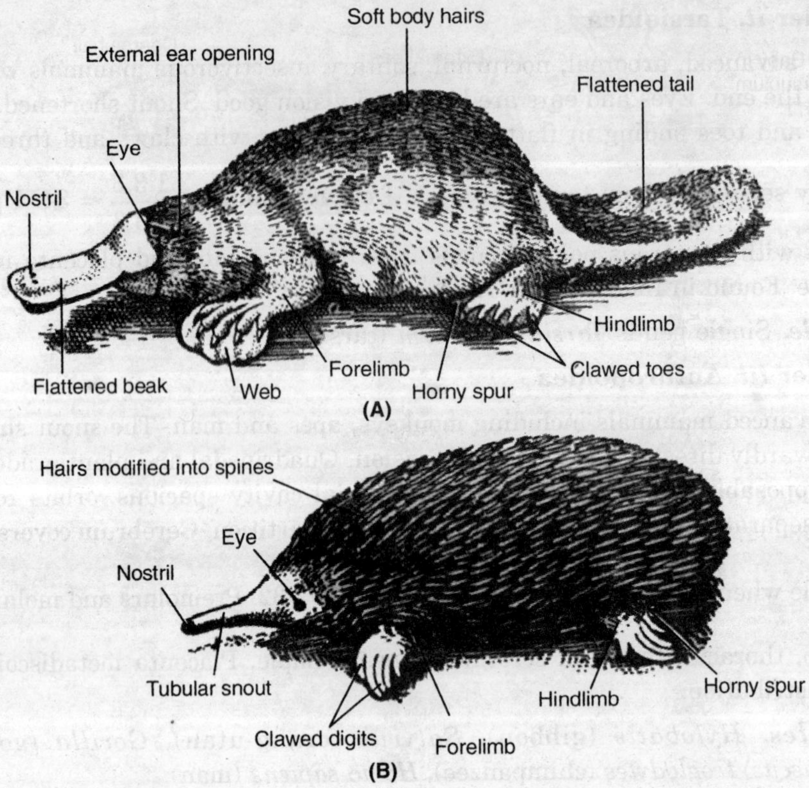

Fig. 24.1 *Egg-laying mammals.* **(A)** *Ornithorhynchus*; **(B)** *Tachyglossus*

2. Tachyglossus = Echidna

It is commonly known as the **spiny anteater** (Fig. 24.1B). It is a terrestrial, burrowing, nocturnal animal found in Australia, Tasmania and New Guinea. The spiny-anteater is as large as common cat. The body is covered with strong spines and coarse hairs are present at the base of spines. The head is small and produced into an elongated and tubular snout. Tail is very short. Eyes small, nostril at the tip of snout and pinna absent. Tongue is long, sticky and protrusible, teeth absent in adult. Feeds on ants. Limbs short, each have five clawed digits, digits not webbed. Sharp claws are used for digging into ant hills and excavating deep burrows. The second claw of each hind leg forms a long and curved **toilet claw** to clean spines. The area in which the ducts from milk glands open is depressed to form shallow pocket and a few long hairs project from the pocket possibly to guide the exuding milk into the mouth of the young whose beak is inserted into the pocket. A fold of skin in the vicinity of the mammary pockets forms a temporary pouch in which the egg is incubated and the young is retained for a time after hatching. Male echidna also has mammary glands secreting milk to feed young one. This condition is called **gynaecomastism.**

3. Macropus

Macropus is commonly known as **kangaroo** (Fig. 24.2A). The male is about 6 feet tall and weighs around 200 lbs and female is about 4 feet tall and weighs 65 kgs. It is a terre-

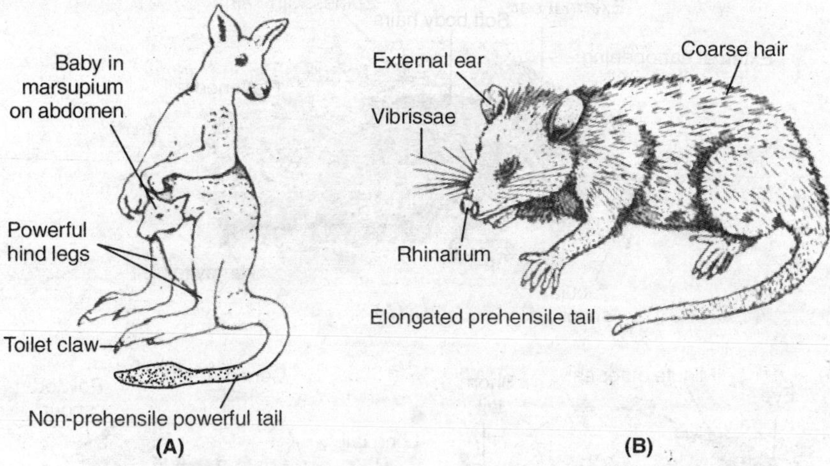

Fig. 24.2 *Marsupial mammals.* **(A)** *Macropus;* **(B)** *Didelphis*

strial, herbivorous and gregarious animal. Head is small with large ears and eyes. Hind legs and feet very long and powerful and used for leaping. Forelimbs small. Hallux absent, Second and third toes syndactylus (united), the fourth toe greatly enlarged and bears a powerful claw used for fighting. The tail is large, non-prehensile, usually thick at the base and tip hairy. The animal nearly sits on its tail during rest. A marsupium is present on the abdomen. The female gives birth to one offspring which is immature, naked and about 3 cm long at birth. The mother carries the baby for 4–6 months in the marsupium where it feeds on milk to complete development. Gestation period is of 35–45 days. It is hunted for flesh fur and sport.

4. Didelphis

Didelphis is commonly known as **opossum** (Fig. 24.2B). It occurs in America. It is a tree-dwelling nocturnal animal about 50 cm in size. The body is covered with coarse fur. The head is small with an elongated snout, large black eyes on a white face and large pinna. The tail is scaly long and prehensile. Pollex and hallux are opposable. The upper jaw has five pairs of incisors and lower has four pairs. If feeds on fruit, grain, carrion, birds, etc. (omnivorous). When disturbed it feins death which is called "playing possum". The gestation period is shortest (12 to 14 days). It gives birth to 7–13 young ones, about 1 cm in size. They grow attached to the nipples for about 50–60 days. The mother carries the young one on her back. It is hunted for flesh and fur.

5. Erinaceus

Erinaceus is commonly known as **hedgehog** (Fig. 24.3A). It is nocturnal, omnivorous animal living in holes and bushes. It hibernates during winter. When disturbed it rolls its body like a spiny rounded ball. The dorsal surface of the body is covered with sharp, backwardly directed spines while the ventral surface has soft fur. The head is small, conical and produced into a small snout bearing nostrils at the tip. Mouth small bearing 36 sharp pointed teeth. Eyes and pinnae small. Limbs short, with clawed digits. Tail very short and stumpy. It is found in India (Rajasthan), Africa and West Indies. In India it is commonly called **jhauchuha.**

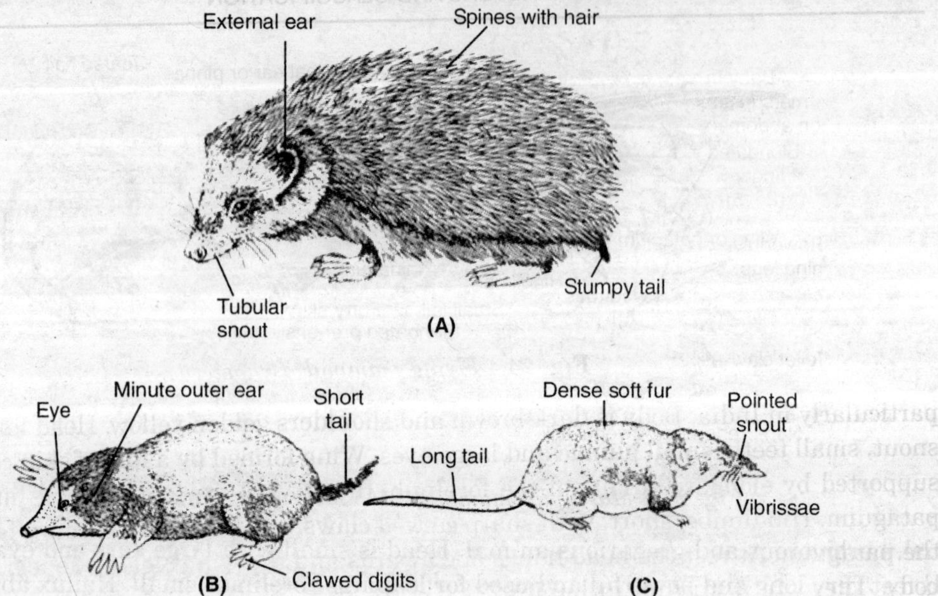

Fig. 24.3 *Insect eating mammals.* **(A)** *Erinaceus;* **(B)** *Talpa;* **(C)** *Sorex*

6. Talpa

Talpa is commonly known as **mole** and **chuchunder** (Fig. 24.3B) in Hindi. It is a nocturnal, burrowing animal living in tunnels. It feeds on worms, insects and seeds. Body is thick, cylindrical and covered with velvet-black fur. Head wedge-shaped bearing eyes which are very much reduced and covered with skin, pinnae absent. The elongated snout contains bristles or vibrissae at the tip and is highly sensitive. Tail is short and without hair. Forelimbs are enlarged, heavily clawed, flattened and turn permanently outwards, an adaptation for digging. Hindlimbs large and broad and bear 5 digits. It is found in India (Assam) and Western Himalayas.

7. Sorex

Sorex is commonly known as **shrew** (Fig. 24.3C). It is nocturnal, lives in burrows and eats insects, small invertebrates and some rodents. It has a world wide distribution. The pygmy shrew is the smallest mammal known. It is about 4–5 cm long and weighs about 3 grams. The body is slender and covered with short and soft steel-grey fur. The head is small, produced into a sharply pointed snout with vibrissae or whiskers and bears small, rudimentary eyes. Pinnae absent. Limbs short, with five clawed digits. Dental formula $\frac{3}{2}\frac{1}{0}\frac{3}{1}\frac{3}{3} = 32$. First pair of incisors long, lower canines absent and all teeth are red in colour. Tail long and covered with scales.

8. Pteropus or Cynopterus

These belong to the suborder Megachiroptera. They are commonly called **fruit bat** (Fig. 24.4) or **flying fox** and chamgader in Hindi. They are found in South Eastern Asia

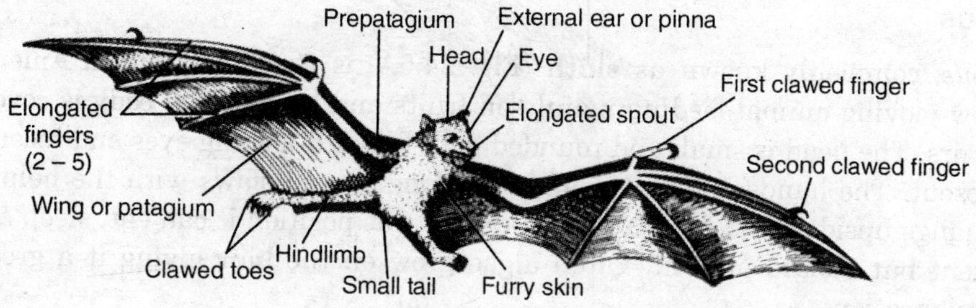

Fig. 24.4 *Flying mammal. Cynopterus*

particularly in India. Body is dark brown and shoulders golden yellow. Head fox-like with long snout, small teeth, small pinnae and large eyes. Wing formed by a fold of skin called pataguim supported by elongated digits of the forelimb, the claw of first and second fingers free from pataguim. Hindlimbs, short, with sharp curved claws, tail small and stumpy, both included in the pataguim. At rest the head hangs downwards and the wings folds like a cloak around the body. They are nocturnal in habit.

9. Desmodus

Desmodus is commonly known as **vampire bat.** It feeds on fresh blood of domestic animals and even human beings (sanguinivorous habit). The incisors are sharp like a razor with the help of which they make a slit on naked skin of the prey, blood starts oozing out and is lapped up. Because of this habit the lumen of oesophagus is small and solid food cannot pass through it. It transmits paralytic rabies. *Desmodus* belongs to the suborder Microchiroptera. It is small in size. The first digit of forelimb bears claw. Pinnae large, with a lobe called targus. Tail present. Molars have sharp cusps. It sleeps during the day with head hanging downwards.

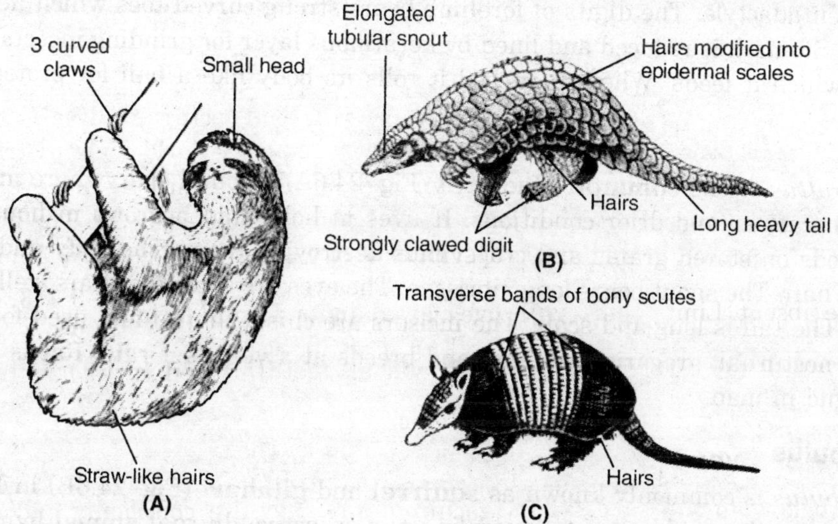

Fig. 24.5 *Toothless mammal.* **(A)** *Bradypus;* **(B)** *Manis;* **(C)** Dasypus

10. Bradypus

Bradypus, commonly known as **sloth** (Fig. 24.5A) is found in tropical America. It is arboreal, slow-moving animal feeding mainly on fruits and leaves. The body is covered with straw-like hairs. The head is small and rounded with forward-looking eyes and inconspicuous ears. Tail absent. The hands and feet bear 2 or 3 long curved claws with the help of which the animal hangs upside down from a branch and in this position it can eat, sleep and move. Incisors absent but molars present. Often algae grow on the hair giving it a green colour (protective colouration).

11. Dasypus

Dasypus, commonly known as **nine-banded armadillo** (Fig. 24.5C) is found in South and Central America. Body is about 75cm long, covered with well developed dermal skeleton of body scutes arranged in transverse bands. These are covered with bony plates or scales having hairs connected by flexible skin. Head large, conical, covered with bony scutes and contains a pair of eyes, prominent conical ears and nostrils present at the tip of the snout. Tail long and covered with bony scutes. Limbs hairy. Forelimb with strong claws for burrowing. When disturbed it rolls up into a ball for protection. Gestation period 18 weeks. A single fertilized egg produces 4–8 young ones of the same sex (a case of polyembryony).

12. Manis

Manis is commonly known as **scaly anteater or pangolin** (Fig. 24.5B). It is a nocturnal and burrowing mammal found in Africa and South Eastern Asia particularly in hilly areas. It ranges from 30–150 cm in size. Body is long covered with thick horny overlapping scales arranged in longitudinal rows. Coarse hairs are present between the scales. The ventral surface is also hairly. Head small, with pointed snout and bears small eyes and external ears. Tail is long tapering and prehensile. Tongue is long, sticky and extensile. Teeth absent. Limbs strong and pentadactyle. The digits of forelimb bears strong curved toes which help to open nests of ants. Stomach is reduced and lined by keratinous layer for grinding insects, ants, and termites on which it feeds. When disturbed it rolls its body like a ball for protection.

13. Rattus

Rattus rattus or the **common black rat** (Fig. 24.6A) is found everywhere in the world preferably in warmer and drier conditions. It lives in holes and burrows in houses and in fields and feeds on stored grains and crops, thus destroying them. The body and limbs are covered with hair. The snout bears long vibrissae. The eyes are sharp and ears well developed and conical. The tail is long and scaly. The incisors are chisel-shaped and used for gnawing. The rat is a nocturnal, gregarious animal and breeds at a very fast rate. Rat is a vector of bubonic plague in man.

14. Funambulus

Funambulus is commonly known as **squirrel** and **gilahari** (Fig. 24.6B) in Hindi. It is found throughout the world except, Australia. It is an active diurnal animal living on trees and feeds on fruits and seeds. The back of the body contains three alternate white and grey stripes. Head contains a snout with moustache, nostrils, large eyes and well developed external ears. Limbs well developed with clawed digits. Tail long and bushy. Incisors chisel-like and exposed, diastema present, canines absent. It destroys fruit crops.

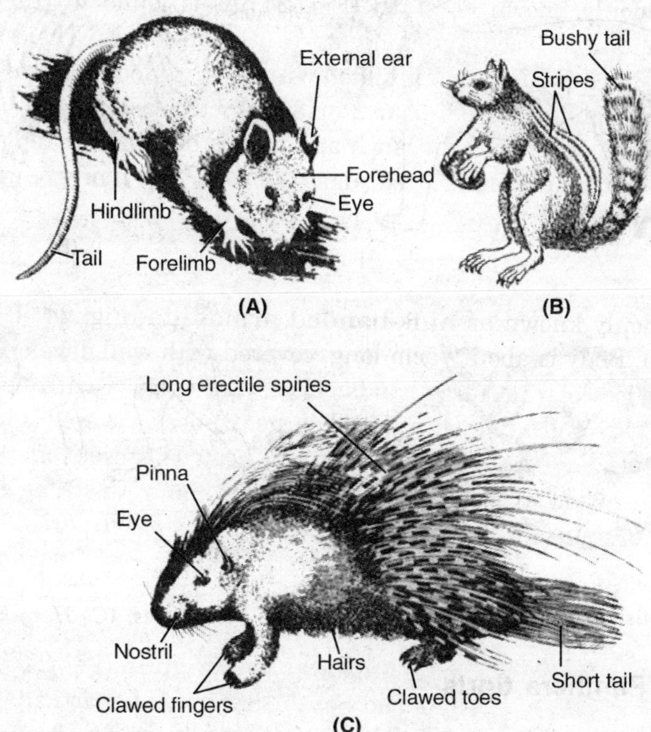

Fig. 24.6 *Gnawing mammals.* **(A)** *Rattus;* **(B)** *Funambulus;* **(C)** *Hystrix*

15. Hystrix

Hystrix, commonly known as **porcupine** and **sahi** (Fig. 24.6C) in Hindi is found in Africa and India. It inhabits forests, rocky hills, river banks and nallah. It is nocturnal, lives in burrows during the day and feeds on crops, vegetables and roots of plants. *Hystrix* is a large, short legged rodent with a short tail and body covered with coarse hairs. Its unique feature is the presence of long, sharp, erective, backwardly directed black and white spines or quills on the back. It frightens its enemies by erecting its spines which may penetrate the flesh of its enemies. It harms crops of potato, carrot, cabbage, etc. It produces grunting sound like a pig.

16. Panthera leo

Panthera leo or **lion** (Fig. 24.7A) or **sher** (in Hindi) is a huge carnivore found in Africa and South-West Asia. Now it is quite, rare. In India it is found in Gir forest of Gujarat where it is being conserved. The body is about 2 metres long and fawny yellow in colour. The male is larger than the female and can easily be distinguished by the **mane,** a ruff of long hairs around the neck and shoulders. Limbs well-developed, feet powerful, with thick pads below the digits which have sharp claws retractile into sheath. Lion preys on deer, buffalo, zebra, goat, etc. It is known for its great strength and roar. It lives in a group called **pride.**

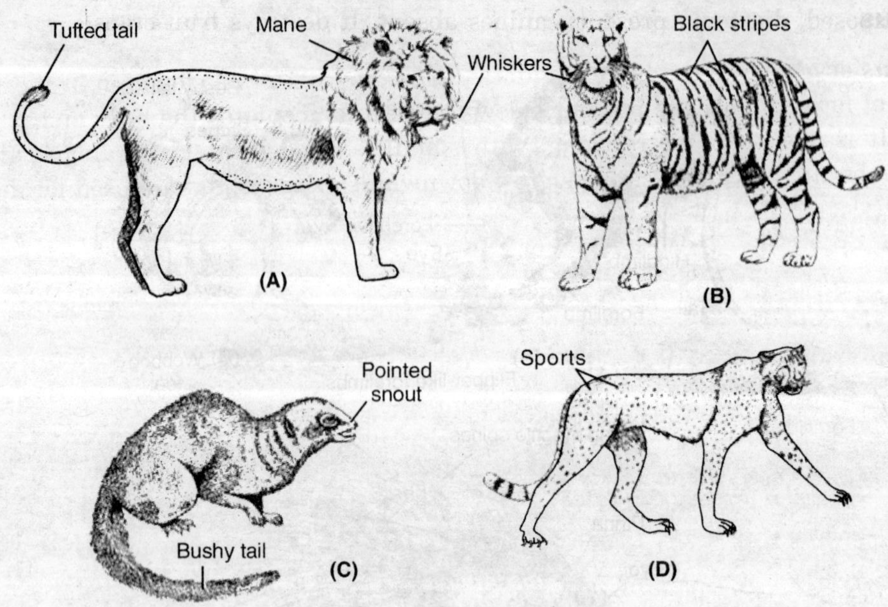

Fig. 24.7 *Carnivorous mammals.* **(A)** *Panthera leo;* **(B)** *Felis tigris;* **(C)** *Herpestes;* **(D)** *Acinonyx*

17. Felis tigris or Panthera tigris

Felis tigris or **tiger,** as it is commonly known, is a large carnivore living in forest in Asia (Fig. 24.7B). It is being protected under the "Project Tiger" in many sanctuaries and national parks like, Sariska, Ranthambore, Corbett, etc., in India. It is the national animal of our country. The skin is yellow with black stripes. It is a solitary animal. It is highly predaceous and can attack man also.

18. Acinonyx

Acinonyx jubatus, commonly known as **cheetah** or **hunting leopard** (Fig. 24.7D), lives in grasslands and semideserts of Africa and South-West Asia. In India it has become extinct. The animal is about 1.5 metres long, leanly built, with a small head having a distinct black stripe from the inner corner of each eye to the upper lip. The entire body has black spots. Feet padded with non-retractile claws. It is the fastest running land animal. It lives and hunts in groups. It preys on small birds and antelope. It can be tamed as pets.

19. Herpestes

Herpestes or **mongoose** or **neola** (in Hindi) is a small-size carnivore (Fig. 24.7C) found in Asia and Africa. It lives in burrow. It is nocturnal and feeds on small mammals, reptiles and birds. Body is covered with grey hairs. Limbs are short and bear five clawed digits adapted for-burrowing. Tail elongated and bushy. Pinnae small and rounded. It is known to kill snake.

20. Odobenus

Odobenus or *walrus* is the largest aquatic carnivore (Fig. 24.8A) found in Arctic ocean. A thick layer of fat or blubber is present below the skin. It insulates the body and protects it from cold. It is about 3.5 metres long, with spindle-shaped body. Hairs almost absent. External ears absent. The upper canines grow downwards as tusks and are used for digging

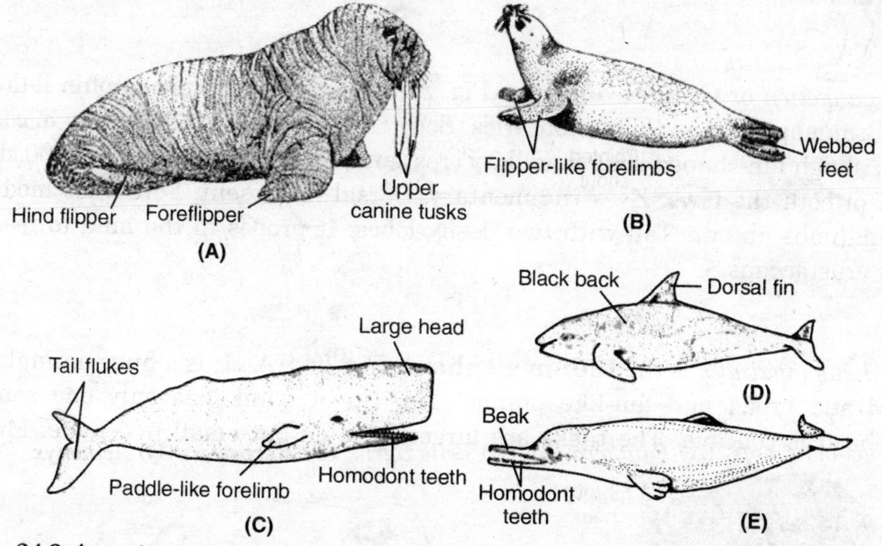

Fig. 24.8 *Aquatic mammals*–**(A)** *Odobenus*; **(B)** *Phoca;* **(C)** *Physeter;* **(D)** *Orcinus*; **(E)** *Platanista*

its prey (molluscs and crustaceans) from muddy bottom and for climbing ice from water. Upper lip is thick and fleshy and bears sensory bristles. The hind flippers are used for locomotion on land. It comes to land for breeding once in three years.

21. Phoca

Phoca or **common seal** (Fig. 24.8B) is found in Arctic and Antarctic regions. Like walrus it also has a blubber below the skin which insulates the body. It has a stream-lined body with limbs adapted for swimming. Forelimbs modified into flippers and hind feet are webbed. External ears absent. They are gregarious in habit and move about in groups in which one male (bull) is the lord and there are more than fifty females (cows). They are hunted for their skin and blubber.

22. Physcter

Physeler or **sperm whale** (Fig. 24.8C) is found in southern seas. It is a huge, heavy bodied whale more than 60 feet long. The female is nearly half in size than male. It is the largest toothed whale. It has no dorsal fin. The head is large and produced into a barrel-shaped *rostrum* about one-third the total length of body. It contains about one ton of **spermaceti oil**. The oil is not edible but is used as a lubricant in industries, in making candles and as a base in cosmetics. It makes the animal hydrostatic. The upper jaw has vestigeal teeth while the lower jaw is long and narrow and bears numerous conical teeth which fit into grooves of the upper jaw. It feeds on squids. The secretion of stomach forms a stone-like substance called **ambergris** which is used in perfumes.

23. Orcinus

Orcinus or **killer whale** (Fig. 24.8D) has a fish-like body. It is black in colour with white and yellow spots. The male is about 30 feet long with a triangular dorsal fin and the female is nearly half in size with a recurved dorsal fin. It has conical teeth in both the jaws. It is a fierce hunter, feeding on fish, birds, seal, propoise and other marine animals and even attacks larger whales.

24. Platanista

Platanista gangetica or **Ganges dolphin** (Fig. 24.8E) is a fresh water dolphin found in Ganges, Indus, Brahmaputra and their tributaries. Body 6–9 feet long and dark grey in colour. Head small with a high fan-shaped bony maxillary crest and a long beak with about 200 sharp homodont teeth in both the jaws. Eyes rudimentary. Dorsal fin absent. Forelimbs modified into paddles, hindlimbs absent. Tail with two fleshy lobes. It probes in the mud to feed on shell fishes and crustaceans.

25. Elephas

Elephas indicus maximus is the **Indian elephant** (Fig. 24.9 A). It is a huge animal with a rounded head and broad and fan-like pinnae. The tip of trunk has only one sensory projection. Forehead is concave. The tusks are larger in male but small in female. Molars

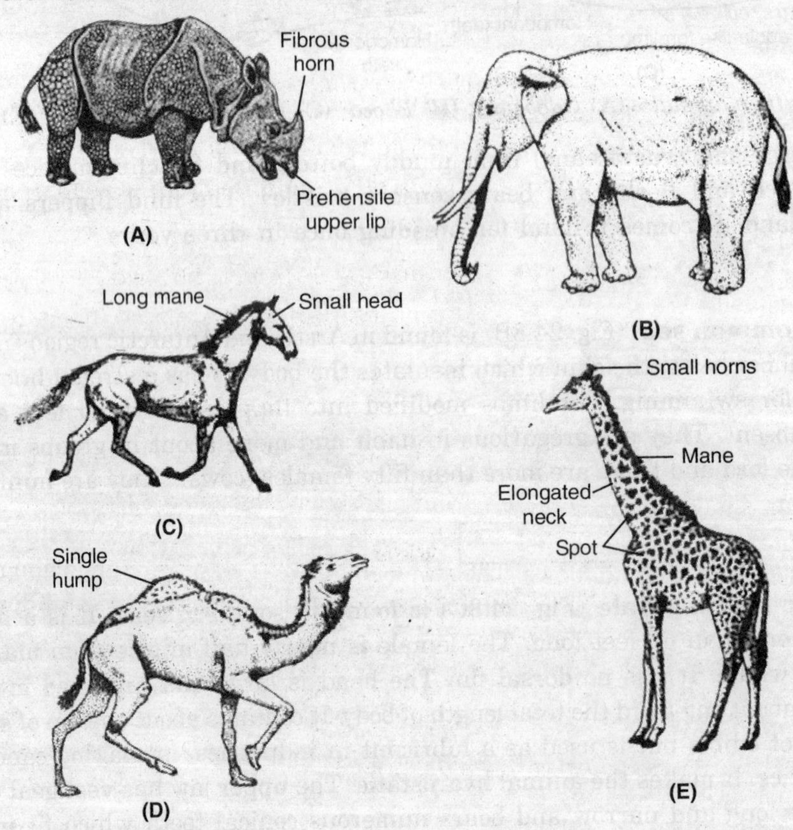

Fig. 24.9 *Some mammals.* **(A)** *Elephas. Hoofed mammals;* **(B)** *Rhinoceros;* **(C)** *Equus;* **(D)** *Camelus;* **(E)** *Giraffa*

have transverse parallel ridges. Characteristic feature is the long trunk or proboscis formed by nose and upper lip and with nostril at the tip. It serves as a flexible arm enabling the animal to lift water and food into the mouth. Legs pillar-like with club-like feet, with elastic pads behind toes which support the body weight. Tail is small. Eye sight poor but sense of smell and hearing well developed. The African elephant *Loxodonta* is bigger than the Indian elephant.

26. Rhinoceros

Rhinoceros, commonly called **gainda** (Fig. 24.9B), is found in Southern Asia and Africa. In India it is found in Bengal and Assam. The African rhinos live in grasslands and have two horns on the tip of the snout (*Rhinoceros* (= *Diceros*) *bicornis*). The Indian rhinos live in swampy jungles in the foot-hills of Himalayas and have single horn (*R.unicornis*). They are heavily built, thick-skinned animals having many folds. The head is large with a prehensile upper lip and fibrous horn (a skin derivative) at the tip of the snout. Eye sight is poor but sense of smell is very strong. Tail short, with a tuft of hair at the tip. They are solitary, swift and agile. *Rhinoceros* are valued for their horn which is an **aphrodisiac.** Therefore, the Indian species has reached endangered status.

27. Equus caballus

Equus caballus is the modern horse (Fig. 24.9C). It is a swift animal and has been domesticated for centuries. Limbs have only the third toe which bears a hoof of keratin having a large soft pad or heel. A horny callosity on the inner side of each limb represents the vestigeal hoof. Other toes atrophied to rudimentary splints. Head is small. There is a well developed mane on the long neck. Teeth adapted for grazing. It eats grass. Tail short, covered with hairs directed downwards.

28. Camelus dromedarius

Camelus or **camel** (Fig. 24.9D) is a desert animal mainly found in Arabian countries. It is called "ship of the desert". It is one of the tallest animal. The back of camel has one or two humps. The hump is fatty and serves as a source of reserve food. *Camelus dromedarius,* the Arabian camel, of North Africa and South West Asia has one hump and the bactrian camel, *Camleus bactrianus* of cold rocky deserts of Central Asia has two humps. Each foot contains two toes with leathery pads which support body weight, provide surface while walking on sand and also protects from hot sand. The snout is slender and the upper lip has a cleft. The limbs are elongated and well developed. The long eyelashes and muscular nostrils can be closed when sand blows. The stomach has four chambers and enough space for storing water. It can survive without water for two weeks. Camels are useful domesticated animal.

29. Giraffa camelopardalis

Giraffa or **giraffe** (Fig. 24.9E) is found in East Africa in grasslands having thick population of trees. It is the tallest living animal. Females are shorter (3.5–4.5 metres) than males (4 – 6 metres). The skin is yellow and has brownish spot. Head has 3–5 small horns covered with velvety skin. Lips are leathery. Tongue is prehensile and can extend upto 1.5 metres.

Neck is greatly elongated and has a mane of short erect hairs. There are only seven cervical vertebrae. Legs long, forelegs longer than the hindlegs. It escapes enemies by running very fast.

30. Loris

Loris (Fig. 24.10A) is a solitary, nocturnal and arboreal primate found in India and Sri Lanka. Body covered with thick, woolly, brownish fur. Head small and produced into a small snout. Eyes large, bulging and closely packed. Pinnae small and conical. Tail long and non-prehensile. Teeth thecodont and heterodont. Limbs long. Thumb and first toe opposable. It is often found hanging upside down.

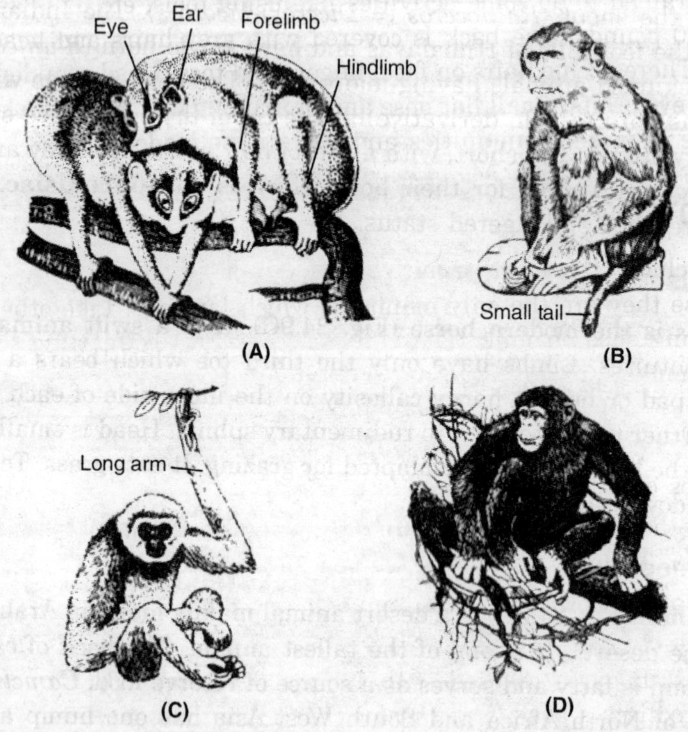

Fig. 24.10 *Primates*– **(A)** *Loris;* **(B)** *Macaca;* **(C)** *Hylobates;* **(D)** *Pan trogrodytes*

31. Macaca

Macaca is commonly called **Rhesus monkey** or **bandar** (Fig. 24.10B). It is found in North India, Southern China and Vietnam. It is of great value in biological studies. Rh blood factor was first discovered in monkeys. Body covered with brownish or golden hairs. Internal cheek pouched for storing food. Stomach simple. Tail hairy. It is diurnal, arboreal and social animal.

32. Hylobates

Hylobates is commonly called **gibbon** (Fig. 24.10C). It is found in South-East Asia and Indonesia. In India it is found in the rain-forests of Assam. Body is slender about 3 feet tall and thickly furred. Males are black while females are yellowish. Head is rounded and has no crest, eye orbits sunken and nose flat. Tail absent. Limbs are well developed. The long and muscular forelimbs are used to swing around trees with great speed. In standing position the arms touch the ground. The feet are used to hold objects. Gibbons live in trees and rarely descend on ground. They eat fruits. They are strongly territorial and make loud calls. It is peculiar among apes in having ischial callosities.

33. Pan troglodytes

Pan troglodytes is commonly known as **chimpanzee** (Fig. 24.10D). It is found in the forests of West Africa. It is arboreal but mostly lives on ground. Among the apes it is most intelligent and can be trained to do some activities like using tools, etc. It is about 4$^1/_2$ feet tall and weighs 150–170 pounds. The back is covered with grey hairs and head, neck and limbs with black hairs. There are no hairs on face, fingers and toes. Head rounded, with large prominent ears, sunken eye orbits, small flat nose and protruding lips. Arms are long and legs short. Chimpanzees live in loose communities and there is no leader.

SUBCLASS PROTOTHERIA

Subclass Prototheria includes the most primitive type of mammals. They are called egg-laying mammals because they are the only mammals which lay eggs. Like other mammals they also feed their young ones on milk. They are regarded as a connecting link between reptiles and higher mammals.

General organisation

The salient features of subclass Prototheria are:

Distribution: Prototheria shows **isolated distribution**. It is found only in the Australian region living in Australia, Tasmania and New Guinea.

Habits and habitat: Prototherians are semi-aquatic (*Ornithorhynchus*) or terrestrial (*Tachyglossus*), burrowing, nocturnal and insectivorous animals.

External features: Body is covered with hairs and spines. The snout is pointed (*Tachyglossus*) or produced into a beak (*Ornithorhynchus*). Pinna absent. Limbs short, digits clawed and webbed (in aquatic form) and used for burrowing.

Mammary glands modified sweat glands and without teats. Tail short. Males have a poisonous *tarsal spur* on each hind leg. Skin grandular, sweat and sebaceous glands present.

Body cavity: Body cavity is divided into two cavities, anterior thoracic and posterior abdominal, by a diaphragm.

Endoskeleton:

Skull: Skull has two occipital condyles, sutures between skull bones absent. Lacrimal and alisphenoid absent but pterygoid present, jugal reduced, nasal and premaxilla are drawn out into a rostrum. Orbits continuous with temporal fossae. Tympanum ring-like. Ear ossicles three. Each half of lower jaw (mandible) composed of a single dentary bone.

Vertebral column: Vertebrae without epiphyses. Cervical vertebrae seven, ribs single-headed, with only capitulum.

Girdles: Pectoral girdle built on a reptilian pattern, scapula without a spine, coracoid large, T-shaped interclavicle and precoracoid present, acromion process well developed. **Pelvic girdle** with perforated acetabulum, epipubic bone present, ischia and pubes fuse at a ventral symphysis. Marsupial bone (epipubic) present.

Digestive system: Tongue long and sticky, teeth absent in adults, replaced by horny beak. Anus absent, rectum opens into a cloaca.

Circulatory system: Heart four-chambered. Right auriculo-ventricular valve muscular and incomplete. Right valve tricuspid (not bicuspid). Chordae tendinae absent. Abdominal vein present. Only left systemic aorta present. RBC circular and non-nucleated. Temperature of the body low (25°C to 28°C) hence incompletely warm-blooded.

Nervous system: The brain is small and has no corpus callosum. Cerebrum is smooth. Optic lobes four. Cochlea of internal ear simple not spirally coiled.

Excretory system: As in other mammals kidneys metanephric. Ureters open into the cloaca on a urinogenital sinus.

Reproductive system: Testes abdominal, penis retractile and only passes out sperms. Oviducts open into cloaca independently. Vagina and uterus absent.

Development: Females oviparous, eggs cleidoic and megalecithal, cleavage meroblastic, placenta absent. Immature young ones feed on mother's milk in a temporary abdominal pouch developed during the breeding season.

Classification

Prototheria is a subclass of class Mammalia. It includes only one living order Monotremata. It is regarded as mammal because it has hairs and secrete milk. Order Monotremata is divided into two families.

Family 1. Ornithorhynchidae: It is represented by only one genus *Ornithorhynchus* commonly called duck-billed platypus. It is semi-aquatic, burrowing animal with clawed and webbed feet. Snout is produced into a beak like that of a duck. The body is covered with soft hairs.

Family 2. Tachyglossidae: It is also represented by single genus *Tachyglossus* (= *Echidna*) commonly known as spiny ant eater. It is terrestrial, burrowing animal with clawed feed. Snout is pointed. Body is covered with hairs and spines.

Affinities

Prototheria shows resemblance with reptiles, birds and mammals.

Affinities with Reptilia

1. Pinna absent.
2. Cloaca present.
3. Body temperature low and variable.
4. Skull with indistinct sutures, large pterygoid and ectopterygoid, dumb-bell shaped prevomers, ring-like tympanic bones. Tympanic bulla and alisphenoid absent. Zygomatic arch perforated by a temporal canal which resembles the post-temporal fossa of reptiles.

5. Ribs single-headed, cervical ribs present, vertebrae without epiphyses.
6. Pectoral girdle with well developed clavicles, large plate-like coracoid, epicoracoid and precoracoid and a T-shaped interclavicle.
7. Acetabulum perforated. Pubes and ischia form ventral symphyses.
8. Odontoid process long.
9. Auriculo-ventricular valve muscular and incomplete. Abdominal vein present.
10. Brain without corpus callosum. Cochlea of internal ear simple.
11. Ureters open into the urinogenital sinus.
12. Testes abdominal. Penis passing out only sperms.
13. Uterus and vagina absent. Oviducts open separately into the cloaca.
14. Females oviparous. Eggs cleidoic and megalecithal.
15. Cleavage meroblastic.

Affinities with Aves

1. Beak of *Ornithorhynchus* similar to that of bird.
2. Absence of teeth.
3. Feet of platypus webbed.
4. Skull without sutures.
5. Presence of tarsal spur and oil gland.

Affinities with Metatheria

1. Marsupial (epipubic) bone present.
2. Cloaca present.
3. Clavicle well developed.
4. Corpus callosum absent.
5. Allantoic placenta absent.

Affinities with Eutheria

1. Body covered with hairs.
2. Mammary glands, oil glands and sweat glands present.
3. Skull with two occipital condyles.
4. Each half of lower jaw made of a single dentary bone.
5. Jaw suspensorium craniostylic.
6. Palate present.
7. Ear contains three ear ossicles.
8. Seven cervical vertebrae in the neck.
9. Sternum segmented.
10. Diaphragm present.

11. Heart 4–chambered, only left aortic arch present.
12. RBC non-nucleated.
13. Optic lobes four (corpora quadrigemina).
14. Warm blooded.
15. Fertilization internal.

Conclusion

Prototherians are the only hairy animals which lay eggs. Presence of hair is a mammalian character and laying eggs is a reptilian character. The prototherians also feed their young ones on milk like higher mammals. Hence Prototheria is regarded as a connecting link between reptiles and higher mammals.

Prototherians possess primitive and degenerative features. They are not in the main line of mammalian evolution. It is concluded that Prototheria originated as a side line from mammal-like reptile ancestor and continued to survive in isolation retaining those primitive characters through which higher mammals have passed. It is not the ancestor of higher mammals.

INFRACLASS METATHERIA

Infraclass Metatheria includes pouched mammals called **marsupials**. The young ones are born in a very undeveloped state and are sheltered in a pouch called **marsupium** present in the abdomen of female. They are intermediate between monotremes and placental mammals.

General organisation

The salient features of infraclass Metatheria are:

Distribution: Metatheria were widely distributed till the Miocene period, almost 90 million years ago, but today they show discontinuous distribution as they are confined to Australia and America.

Habits and habitat: Metatherians are terrestrial, arboreal, burrowing, rarely aquatic, herbivorous or carnivorous animals.

External features: Body covered with hairs. Pinna well developed. Mammary glands modified sebaceous glands and have teats. Tail well developed, used for balancing, sometimes prehensile. Forelimbs short, hindlimbs long, digits clawed. Skin glandular. Female with a marsupium in the abdomen supported by marsupial or epipubic bones and enclosing teats.

Body cavity: Muscular diaphragm divides the body cavity into anterior thoracic and posterior abdominal cavities.

Endoskeleton: Skull dicondylic and with distinct sutures. Top of skull is flat. Paroccipital process elongated. Orbital and temporal fossae confluent as postorbital bar is absent. Zygomatic arch complete. Tympanic bulla absent. Ectopterygoid bone present. In some cases alisphenoid large and forms alisphenoid bulla. Jugal large and reaches up to glenoid articulation. Palate fenestrated. Mandible with a single dentary bone. Vertebrae with epiphyses. Seven Cervical vertebrae in the neck, atlas incomplete, thoracic vertebrae thirteen, lumbar seven and caudal vertebrae with a chevron bone. Thoracic ribs bicephalous. Hindlimb has a narrow rod-like parafibula bone on the outer side of fibula. Fourth toe of hindlimb largest and serves as the axis of the foot. Coracoid well developed in embryo but reduced in adult. Acromion, metacromion and scapular spine present. Clavicle well developed. Interclavicle absent. Pelvic girdle has epipubic or marsupial bones to support the marsupium.

Digestive system: Dentition heterodont, thecodont or monophyodont. Number of teeth always more than 44. Number of incisors always variable. Dental formula $\frac{5, 1, 3, 4}{4, 1, 3, 4} = 50.$ Anal and urinogenital apertures open separately into a shallow cloaca surrounded by a common sphincter. Stomach variable in shape and size.

Circulatory, respiratory and excretory systems like that of higher mammal.

Nervous system: Brain small and has less convolutions. Corpus callosum is absent. Anterior commissure is prominent. Olfactory lobes large and cerebral hemispheres small. Cerebellum small and exposed. Middle ear has 3 ossicles. Cochlea of internal ear spirally coiled.

Reproductive system: Testes lie in scrotrum in front of the penis which is bifurcated at the tip in some animals. In females there are two or three vaginae. When three in number one is used for the birth of young one and the lateral two are used for copulation. There are two uteri and clitoris may also be double.

Development: Fertilization internal, cleavage holoblastic and development internal. Gestation period short, 2 weeks in opossum and 5 weeks in kangaroo. Placenta develops for the first time in Metatheria, yolk sac placenta in all metatherians and allantoic placenta in *Perameles.* Young ones are immature at birth, about 2 cm long and 1 gm in weight. They are naked and blind and transferred to the marsupium. It feeds and sleeps inside the pouch for about 190 days.

Classification

Infraclass metatheria includes only one order **Marsupialia** represented by about 240 living genera. It is divided into two **suborders** on the basis of their dentition and structure of foot.

On the basis of dentition:

Suborder 1. Polyprotodontia: Each half of upper jaw has more than three incisors. It includes four families. **Didelphidae, Dasyuridae, Notoryctidae** and **Peramelidae.**

Suborder 2. Diprotodontia: Each half of upper jaw has less than three incisors. It includes four families. **Caenolestidae, Phalangeridae, Phascolomidae** and **Macropedidae.**

On the basis of structure of foot:

Suborder 1. Didactyla: Foot with all toes separate. Marsupium opens posteriorly. It includes families: **Didelphidae, Dasyuridae, Notoryctidac Cacnolestidae.**

Suborder 2. Syndactyla: Second and third toes of hind foot united by a common sheath. Pouch opens forwards. It includes families **Peramelidae, Phalangeridae, Phascolomidae Macropodidae.**

Family 1. Didephidae: *Didelphys* (America opossum), *Chironectes* (water opossum).

Family 2. Dasyuridae: *Dasyurus* (Australian cat), *Thylacinus* (Tasmanian wolf), *Myrmecobius* (banded ant eater).

Family 3. Notoryctidae: *Notoryctes* (marsupial moles).

Family 4. Peramelidae: *Perameles* (bandicoots of Australia).

Family 5. Caenolestidae: *Caenolestes, Rhyncolestes* (shrew-like marsupials).

Family 6. Phalangeridae: *Petaurus* (flying phalanger), *Phascolarctos* (Koala).

Family 7. Phascolomidae: *Phascolomys* (wombat).

Family 8. Macropodidae: *Macropus* (kangaroo), *Dendrolagus* (tree kangaroo), *Petrogale* (rock kangaroo).

Adaptive radiation in metatheria

Metatherians are found in Australia and America. In Australia, due to absence of competition with higher mammals they inhabit all kinds of habitat and have developed different kinds of adaptations. Development of different kinds of adaptations to live in different habitats by members of a single group of animals is called **adaptive radition**. They show following adaptations:

1. Arboreal: Arboreal metatherians live on trees. The most striking feature is the grasping great toe or **hallux** on hind foot. The *hallux* opposes the fourth digit to form an organ of grasping the tree. They have a prehensile tail. Common examples are *Didelphys* (opossum) and *Phascolarctus* (koala).

It is quite probable that all marsupials had passed through a tree inhabiting stage before their great radiation began. The marsupials are either all arboreal today or they give evidence in their anatomy of arboreal descent.

2. Volant or flying: Flying metatherians have a patagium, a fold of skin along the sides of the body between forelimbs and hindlimbs. A weak fold is present in front of the foreleg. The patagium helps them to glide from one tree to another. No true volant forms which can be compared with bats exist in the Australian marsupial fauna. A few genera showing volant adaptations are *Petaurus* and *Petauroides* (flying phalangers) and *Acrobates*.

In *Petauroides* the flying membrane extends from wrist to ankle. The tail is very bushy except the prehensile tip which is naked on the underside. The tail and long fur supplement the buoyancy of patagium. *Petaurus* has a much broader patagium. Tail is very large and bushy but there is no naked prehensile tip. *Acrobates* has a very narrow patagium extending from elbow to knee. The long fringing hairs of patagium and tail help in flight.

3. Cursorial: The fast running metatherians show cursorial adaptations. The best example is Tasmanian wolf *Thylacinus* which is dog-like in contour, development of limbs and digital reduction. The herbivorous bandicoots (*Perameles*) and kangaroo (*Macropus*) are also highly speedy forms. The bandicoot is quadrupedal whereas kangaroo is bipedal when running. The forelimbs of kangaroo are short and hindlimbs are long and used for leaping. It can make a leap upto 10 metres long and jump a fence about $2^1/_2$ metres high.

4. Fossorial: Fossorial metatherians are burrowing animals. They have short, powerful and furred limbs, each with five toes with strong claws for burrowing. The tail is short and covered with horny skin. The eyes are vestigeal. The marsupial mole *Notoryctes* and *Phascolomys* (wombat) are fossorial forms.

5. Aquatic: Aquatic metatherians are generally amphibious in habit and have webbed hind feet. The only known example is *Chironectes*, the water opossum of South America.

Parallel evolution in metatheria

Metatherians have evolved along several lines showing parallel evolution similar to higher mammals. When two groups of related animals undergo similar types of changes independently, it is called *parallel evolution*. Although Metatheria is a small group it has invaded almost all the niches like placental mammals showing parallel evolution. Some common examples are:

S. No.	Metatheria (Marsupial mammals)	Adaptation	Eutheria (Placental mammals)
1.	Kangaroo (*Macropus*)	Cursorial	Horse (*Equus*), Deer (*Antelope*)
2.	Flying phalanger (*Petaurus*)	Volant	Flying squirrel (*Glaucomys*)
3.	Wombat (*Phascolomys*)	Fossorial	Ground hog (*Mamota*)
4.	Marsupial mole (*Notoryctes*)	Fossorial	Mole (*Talpa*)
5.	Tasmanian wolf (*Thylacinus*)	Cursorial	Wolf (*Canis*)
6.	Marsupial bandicoot (*Perameles*)	Cursorial	Bandicoot (*Bandicota*)
7.	Banded ant eater (*Myrmecobius*)	Digging	Scaly ant eater (*Myrmecophaga*)

Affinities

Metatheria possesses primitive as well as advanced characters, so it shows relationship with Prototheria and Eutheria.

Affinities with Prototheria:

Similarities

1. Cloaca present.
2. Clavicles present, epipubic or marsupial bones present, tympanic bone ring-like, tympanic bulla absent.
3. Brain with large olfactory lobes and anterior commissure and without corpus callosum.
4. True allantoic placenta absent.

Differences

1. Permanent marsupial pouch present.
2. Mammary glands with teats.

3. External ears well developed.
4. Adults with teeth.
5. Vertebrae have epiphyses, ribs bicephalous, interclavicle absent, coracoid reduced.
6. Cochlea of internal ear spirally coiled.
7. Testes lie in scrotal sacs, penis with bifid tip.
8. Viviparous, young ones develop in uterus.

Affinities with Eutheria:

Similarities

1. External ear or pinna present.
2. Cochlea of internal ear spirally coiled.
3. Mammary glands modified sebaceous glands, with teats.
4. Dentition heterodont.
5. Four optic lobes (corpora quadrigemina) present in the brain.
6. Interclavicle absent, coracoid reduced and ribs bicephalous.
7. Testes present in scrotal sacs, penis present.
8. Uterus and vagina present.
9. Viviparous, eggs microlecithal, cleavage holoblastic, development internal, placenta present.

Differences

1. Presence of cloaca.
2. Presence of marsupial pouch.
3. Small flat cranium, tympanic bulla absent, epipubic bone present, jugal extending back, palate with perforations, incisors more.
4. Brain without corpus callosum.
5. Two vagina and uterus.
6. Scrotal sac in front of the penis, penis with bifid tip.
7. True allantoic placenta absent.
8. Gestation period short.

CONCLUSION

Metatheria is more advanced than the primitive egg-laying Prototheria and is close to the placental mammals but not close enough to belong to the same line of evolution. It is now believed that the Prototheria is a side line of evolution from mammal-like reptile and Metatheria and Eutheria have evolved independently from a common Pantotherian ancestor.

25

CHAPTER

CLASS MAMMALIA:
GENERAL ACCOUNT

ORIGIN OF MAMMALS

Mammals occupy the top position in the evolutionary tree as they are the final product of evolution. They originated in the Triassic period of Mesozoic era about 200 millions ago, but for over 100 million years they remained an unimportant group of animals. In the Palaeocene to Pleistocene period of the Cenozoic era the mammals became dominant, as such this period is known as the 'age of mammals'. Within this period of about 55 million years many ancient mammals have disappeared leaving no descendants.

First mammals

First mammals lived in the upper Triassic period almost 200 million years ago. Fossil remains do not reveal much about them but they possessed the following salient features:

1. They were tiny creatures not bigger than rats and mice.
2. They were nocturnal, thus avoided competition with large-sized diurnal reptiles.
3. They were burrowing or arboreal in habit and ate insects.
4. Body was covered with hair.
5. They were homeothermic and endothermic.
6. Brain was large and were intelligent.
7. They carried their young ones in pouches for further development after birth.

Probable ancestor

Mammal-like reptiles are believed to be the probable ancestors of mammals. *Cynognathus* (dog-jaw), a mammal-like reptile, belonging to suborder **Theriodontia,** order **Therapsida** and subclass **Synapsida** is regarded as the probable mammalian ancestor. It lived about 230 million years ago in the early triassic period. It was an advanced, wolf-sized, carnivorous animal showing both mammalian and reptilian characters.

Mammalian characters

1. Skull dicondylic, with enlarged temporal fossa and secondary palate.
2. Lower jaw with enlarged dentary.
3. Dentition heterodont, with incisors, canines, premolars and molars.
4. Limbs mammal-like, capable of fast speed.

Reptilian characters

1. Cranium small, parietal foramen present, quadrate and quadrato-jugal reduced.
2. Middle ear with single bone instead of three.
3. Lower jaw had many bones.

Theories of origin of mammals

Three theories have been proposed to explain the origin of mammals.

1. Amphibian ancestry: This theory was proposed by T.H.Huxley (1880) on the basis of presence of two occipital condyles in the skull of both Amphibia and Mammalia. This was, however, not acceptable because the condyles were derived from exoccipitals in Amphibia and básioccipitals in Mammalia.

2. Reptilian ancestry: Fossils of extinct reptiles and mammals provide enough evidences to show that mammals had a reptilian ancestry. As already mentioned *Cynognathus*, a theriodont therapsid, belonging to subclass Synapsida (mammal-like reptile) is the probable ancestor of mammals. This theory is further supported by the fact that the monotremes (primitive mammal) resemble living reptiles in many anatomical features.

3. Polyphyletic theory: Class Mammalia has been divided into two subclasses according to their reproductive habit **Prototheria** or primitive egg-laying mammals and **Theria** or viviparous mammals. Theria is further divided into two infraclasses: **Metatheria** or marsupials and **Eutheria** or higher placental mammals. It is believed that mammal-like reptiles (Theriodontia, suborder **Therapsida**) are the probable ancestors of mammals, but it is difficult to point out the exact mammalian ancestor among advanced theriodonts. Some theriodonts were more advanced than mammals in certain characters but comparatively primitive in others. Therefore, it was concluded that mammals had a polyphyletic origin, that is, different mammal-like reptiles gave rise to early mammals. It is generally assumed that present mammals originated from at least two mammal-like reptiles. Prototheria probably originated from the decodonts, while Metatheria and Eutheria originated independently from pantotherians by the end of Cretaceous period.

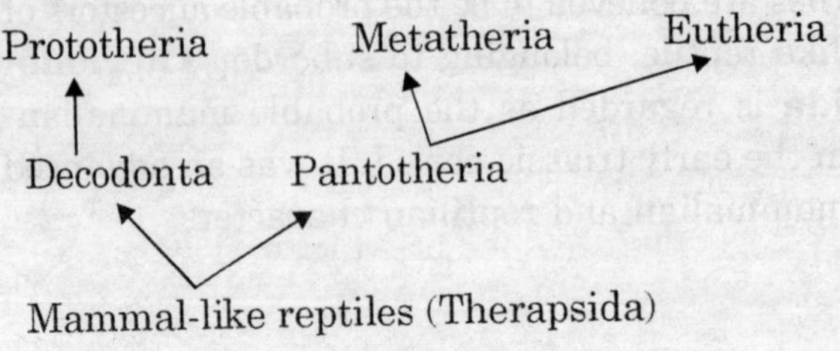

Polyphyletic origin of mammals

Evolutionary trends

When the mammal-like reptiles transformed into mammals, following changes occurred:

1. Hair developed on the body.
2. External ear (pinna) developed.
3. They became warm-blooded.
4. Mammary glands developed.
5. Diaphragm, a muscular partition between thoracic and abdominal cavities, developed.
6. Dentition became heterodont and thecodont.
7. Heart became completely four-chambered.
8. Females transformed from oviparous to viviparous condition.
9. Male developed scrotal sacs.
10. Skull became dicondylic.
11. Lower jaw bones reduced to a single large dentary bone.
12. Brain developed corpus callosum.

DENTITION IN MAMMALS

The arrangement of teeth in the upper and lower jaws is called **dentition.** Teeth are hard, usually pointed structures borne on the premaxillae and maxillae bones of the upper jaw and dentary of the lower jaw. The primary function of teeth is to hold the prey in the mouth cavity and to grind the food. Teeth are present in all mammals except spiny anteater (*Tachyglossus*), duck-bill platypus (*Ornithorhynchus*), great anteater (*Myrmecophaga*) and whalebone whale (*Balena*).

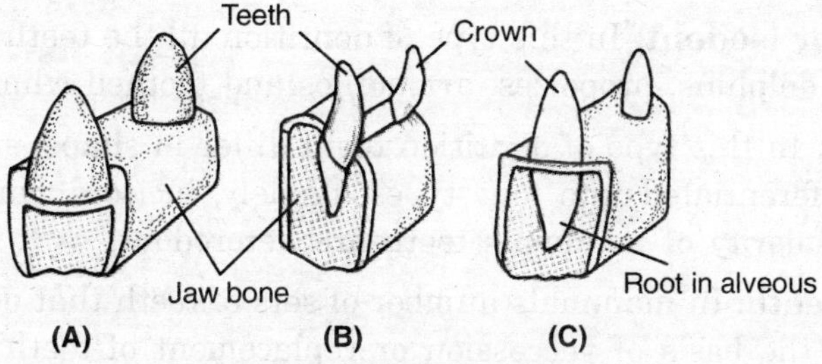

Fig. 25.1 *Mammalia*–Attachment of teeth. **(A)** Acrodont; **(B)** Pleurodont; **(C)** Thecondont

Types of dentition

Based on the manner of attachment of teeth to the jaw bones, shape of teeth and succession of teeth during life time, dentition may be of the following types:

Attachment of teeth: The manner of attachment of teeth to the jaw bones is quite variable among the vertebrates.

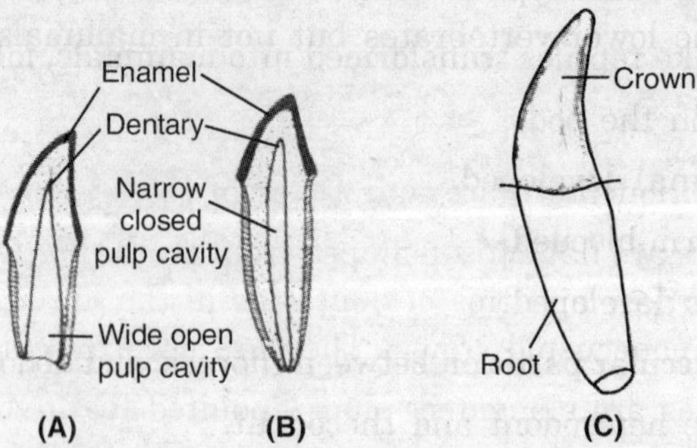

Fig. 25.2 *Mammalia*–**(A)** *Open rooted immature incisor*; **(B)** *Close rooted mature incisor*; **(C)** *Canine*

1. Acrodont: In this type of dentition (Fig. 25.1A) the teeth have no roots and they are fused with the free surface of the jaw bones. Such teeth usually break off easily but are replaced. Acrodont dentition occurs in lower-vertebrates like shark and frog.

2. Pleurodont: In this type of dentition (Fig. 25.1B) the teeth have no roots and are attached by their base and one side to the inner side of the jaw bone. Pleurodont teeth are found in urodeles and lizards.

Since both acrodont and pleurodont teeth are rootless, the nerves and blood vessels enter the pulp cavity along lateral side.

3. Thecodont: In this type of dentition (Fig. 25.1C) the teeth have roots which are embedded in individual sockets called **alveoli** or **theca,** in the jaw bones. Thecodont teeth are characteristic of mammals and among other vertebrates they are found in crocodiles and some fishes.

Shape of teeth: On the basis of shape of teeth dentition is of two types:

1. **Homodont or isodont:** In this type of dentition all the teeth are similar in shape, for example, dolphins, propoises, armadillos and toothed whales.
2. **Heterodont:** In this type of dentition teeth differ in shape, size and function. The teeth are differentiated into four types, namely, incisors, canines, premolars and molars. In majority of mammals teeth are heterodont.

Succession of teeth: In mammals number of sets of teeth that develop during the life time varies. Thus on the basis of succession or replacement of teeth dentition is of three types:

1. Monophyodont: In this type of dentition only one set of teeth develop during the life time. It is found in platypus, marsupials, moles, squirrels, toothless whales, sirenians, etc.

2. Diphyodont: In this type of dentition two sets of teeth develop during the life time. The first set of dentition is called **deciduous** or **lacteal** or **milk dentition.** The milk teeth usually come out soon after birth but in bats and guinea pigs they appear and are lost even before birth. In milk dentition molars are absent. The milk teeth are replaced by a set of **permanent teeth** which last throughout life. If lost they are not replaced.

3. Polyphyodont: In this type of dentition many sets of teeth develop during the life time. It is found in some lower vertebrates but not in mammals.

Kinds of teeth

In heterodont condition there are four types of teeth:

1. Incisors: Incisors are flat, chisel-shaped front teeth (Fig. 25.2A and B) present on the premaxillae in the upper jaw and tips of dentaries in the lower jaw. They are long single-rooted, having a sharp monocuspid crown. They are used for cutting and biting.

2. Canines: Canines are elongated, sharp, pointed teeth with monocuspid crown and single root (Fig. 25.2C). They are present just outside the incisors. There is single canine in each half of each jaw. Upper canines are borne by the maxillae. They are used for piercing and tearing the prey.

3. Cheek teeth (premolars and molars): The teeth outside the canines are called cheek teeth. There are two types of cheek teeth, the **premolars** followed by the **molars.** They have two or more roots. The crowns are broad and bear complex **cusps** or **tubercles.** Based on the number of cusps present the cheek teeth may be **monocuspid** (one cusp), **dicuspid** (two cusps) or **tricuspid** (three cusps). The premolars generally have two roots and two cusps and are preceded by milk teeth in early life. The molars have more than two roots and several cusps. They appear late in life and have no milk predecessors. The cheek teeth are collectively called **grinders** as they are used to grind the food.

Types of cheek teeth

Based on the shape and arrangement of cusps cheek teeth are of following types:

1. Triconodont: It was found in fossil mammals (Fig. 25.3A). There were three cone-shaped cusps arranged in a linear series.

2. Trituberculate: It was also found in fossil mammals (Fig. 25.3B). The three cone-shaped cusps were arranged in the form of a triangle.

3. Bunodont: It is found in man (Fig. 25.3C), monkey, pig, etc. The cusps are blunt or rounded and remain separated from each other. They are used for crushing food.

4. Secodont: In carnivorous mammals secodont (Fig. 25.3 D and E) condition occurs. The cusps have sharp cutting edges and are used for cutting and tearing prey (carnassial teeth).

5. Selenodont: It is found in herbivorous grazing mammals. The cusps are in the form of vertical crescentic ridges of hard enamel enclosing soft dentine. They help in grinding the food. Selenodont teeth with low crown and long root are called **brachyodont** (Fig. 25.3 F), as in ground sequirrel. In horse and cattle, the condition is called **hypsodont** (Fig. 25.3G). The teeth have high prism-shaped crown and low root.

6. Lophodont: It is found in elephants (Fig. 25.3 H and I). The teeth have fine folding of enamel and dentine. The crescentic enamel cusps are connected by masses of dentine in the form of transverse ridges called **lophos.** A single large lophodont molar (30 cm by 10 cm) is present in each half of each jaw. They are used to grind all kinds of plant materials.

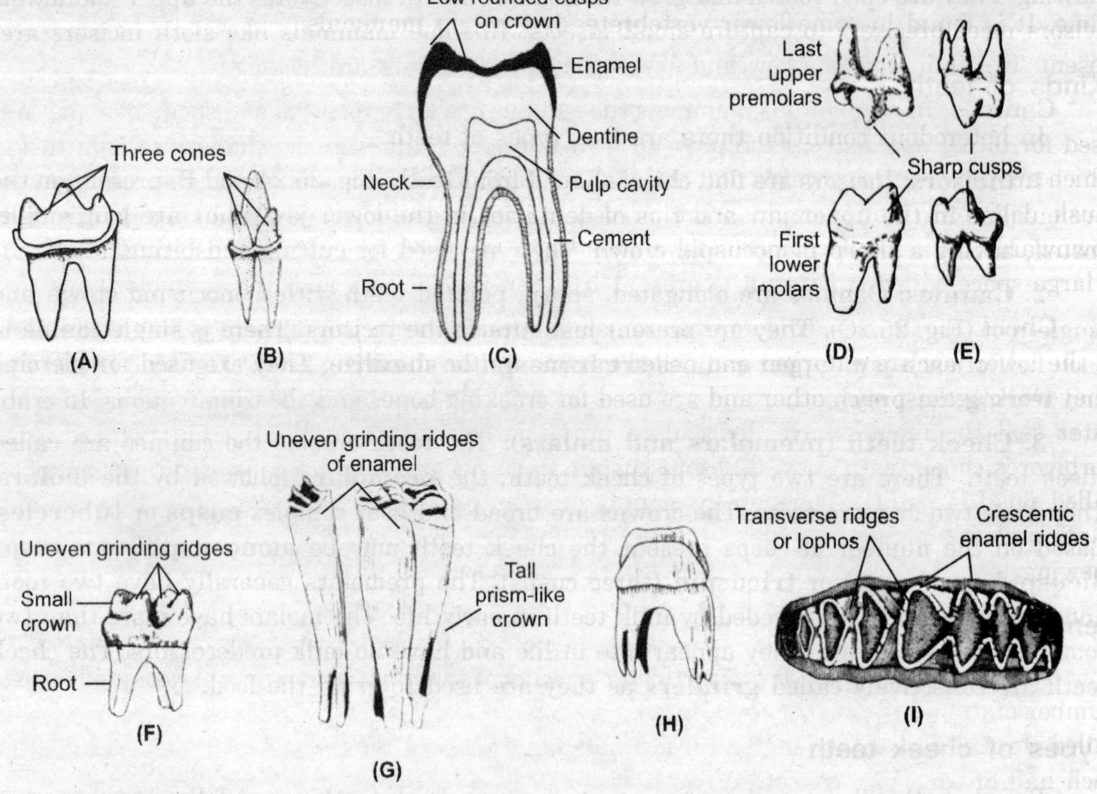

Fig. 25.3 *Mammalia–Types of check teeth:* **(A)** *Triconodont;* **(B)** *Trituberculate;* **(C)** *Bunodont molar (V.S.);* **(D)** *Secodont Carnassial teeth of cat;* **(E)** *Sam of dog;* **(F)** *Brachydont selenodont molar;* **(G)** *Hypsodont selenodont molar;* **(H)** *Lophodont molar;* **(I)** *Same, surface view*

Theories of evolution of cheek teeth

There are two theories to explain the evolution of cheek teeth:

1. According to the first theory the cheek teeth have evolved by the fusion of several simple cone-like teeth. This theory is not accepted due to lack of embryological evidence.
2. According to the second theory the cheek teeth were derived from simple conical reptilian teeth by the development of additional **cusps** or **tubercles.** The lateral cusps first arose in a linear series giving rise to the **triconodont condition.** In the next stage the cusps shifted and became arranged in a triangular pattern giving rise to **trituberculate** condition. Later bell-like extensions and additional cusps developed from these tubercles giving rise to varied mammalian cheek teeth. This theory is accepted because it has palaentological evidence.

Modifications of teeth

Incisors: In elephants incisors are modified into tusks. In lemurs incisors are comb-like and are used for cleaning fur. In rodents and lagomorphs incisors are chisel-like and used for gnawing. They are open rooted and grow throughout life. In insectivores the upper and lower incisors meet precisely to capture small insects. In some mammals like sloth incisors are absent. In ruminants like cow and sheep the upper incisors are lacking.

Canines: In carnivorous mammals the canines are large, pointed and spear-like and are used for biting and tearing the prey. In walrus upper canines are modified into long tusks which are directed downwards and are used for digging molluscs and walking on ice. In male musk deer only upper canines are present. They are elongated tusk-like and directed downwards. In some herbivorous mammals like rat, rabbit and ox, canines are absent and a large space called **diastema** is present between the incisors and first premolar.

Cheek teeth: In carnivorous mammals last premolar in the upper jaw and first molar in the lower jaw are enlarged and called **carnassial** or **shearing** teeth (Fig. 25.3 D and F). They work against each other and are used for cracking bones and shearing tendons. In crab eater seal the molars have denticulate processes which are used to strain plankton. In herbivores cheek teeth show different cusp pattern for grinding grass. In man last molar is called **wisdom tooth.** It generally errupts quite late in life or may be absent. Its eruption is often slow and painful. It is a vestigeal structure, because if present it is of no use in chewing.

Dental formula

The number of teeth for a particular species of mammal is remarkably constant. The number of different types of teeth in a particular species can be represented by an equation called **dental formula.** The dental formula expresses the number of each type of teeth in each half of each jaw. Since the two halves of each jaw are identical, the number of teeth in one half is recorded and is multiplied by 2. The teeth of upper jaw are placed as numerator and those of lower jaw as denominator. The letters i, c, pm and m represent incisors, canines, premolars and molars respectively.

$$\text{Dental formula} = \frac{\text{Number of teeth in one half of upper jaw}}{\text{Number of teeth in one half of lower jaw}} \times 2$$

A typical mammalian dentition includes 44 permanent teeth, for example, horse, pig and mole.

$$i\,\frac{3}{3},\ c\,\frac{1}{1},\ pm\,\frac{4}{4},\ m\,\frac{3}{3} = 22 \times 2 = 44.$$

$$\text{Simplified formula is } \frac{3.1.4.3.}{3.1.4.3} = 22 \times 2 = 44$$

If any tooth is absent it is indicated by a zero (0). Dental formula of some common mammals:

$$\text{Dog } \frac{3.1.4.2}{3.1.4.3} = 42 \qquad \text{Kangaroo } \frac{3.1.2.4}{1.0.2.4} = 34$$

$$\text{Man } \frac{2.1.2.3}{2.1.2.3} = 32 \qquad \text{Cow, sheep and goat } \frac{0.0.3.3}{3.1.3.3} = 32$$

Rabbit $\frac{2.0.3.3}{1.0.2.3} = 28$ Rat $\frac{1.0.0.3}{1.0.0.3} = 16$

Elephant $\frac{1.0.0.3}{0.0.0.3} = 14$ Cat $\frac{3.1.3.1}{3.1.2.1} = 30$

Structure of tooth

The structure of a tooth in all vertebrates is more or less the same. Tooth is embedded in a **socket** or **alveolus** in the jaw bone. A typical tooth consists of 3 parts: **crown, neck** and **root** (Fig. 25.4). The part of the tooth which is embedded inside the socket is called the root, the part above the root which is held in the thick fleshy gum is called the neck and the part that projects over the gum is called the crown. The root is fixed in the socket by

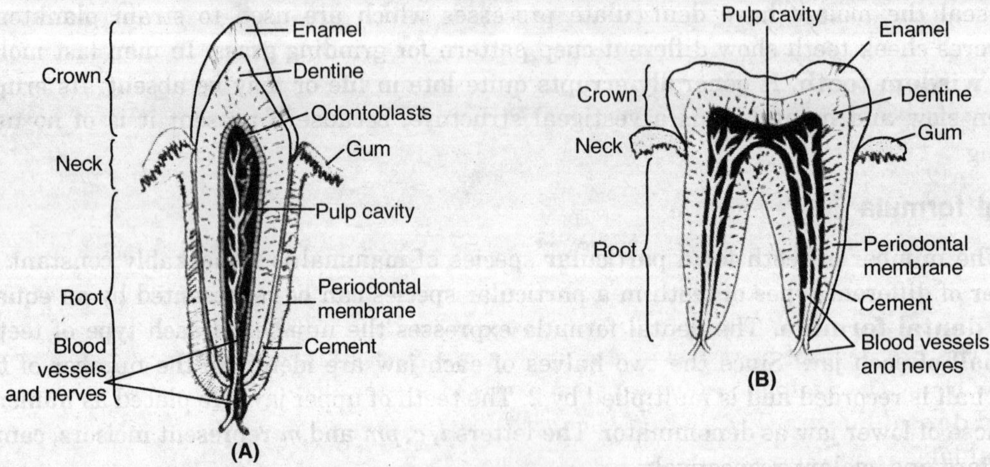

Fig. 25.4 *Mammalia*–Structure of tooth (V.S.) **(A)** *Incisor*; **(B)** *Molar*

a non-cellular layer of **cement** also known as **crusta petrosa.** It is very much like bone but has no Haversian canals. In between the cement and bone of alveolus there is a vascular **periodontal** or **alveolar membrane** of strong connective tissue fibres called **Sharpey's fibres.** The body of a tooth is made up of hard dermal bony material called **dentine** (ivory). Running through the dentine are hollow **dentinal tubules** or **canaliculi.** The crown of teeth is covered by a thin layer of shiny substance called **enamel** which is harder than dentine. It is the hardest substance in the body of vertebrates. It is secreted by the epidermal ameloblasts. A tooth encloses a space within, called **pulp cavity** which in life is filled with a soft, pulp consisting of blood capillaries and nerve fibres. Lining the pulp cavity there is a layer of mesenchyme cells, the **odontoblasts** which secrete dentine. The odontoblast cells send protoplasmic processes into the canaliculi of dentine.

In most of the mammals the pulp cavity becomes closed basally at a certain age, so nourishment stops and further growth does not take place. Such teeth are called **close-rooted.** But the incisors of rodents, lagomorphs and elephants grow throughout life because the pulp cavity remains open basally. Such teeth are called **open-rooted.**

Development of a tooth

In all vertebrates development of a tooth is similar to that of a placoid scale. A tooth develops in the gum covering the jaw bones. Both epidermis and dermis take part in the development of a tooth. The enamel is ectodermal in origin while remaining part of the tooth is mesodermal in origin.

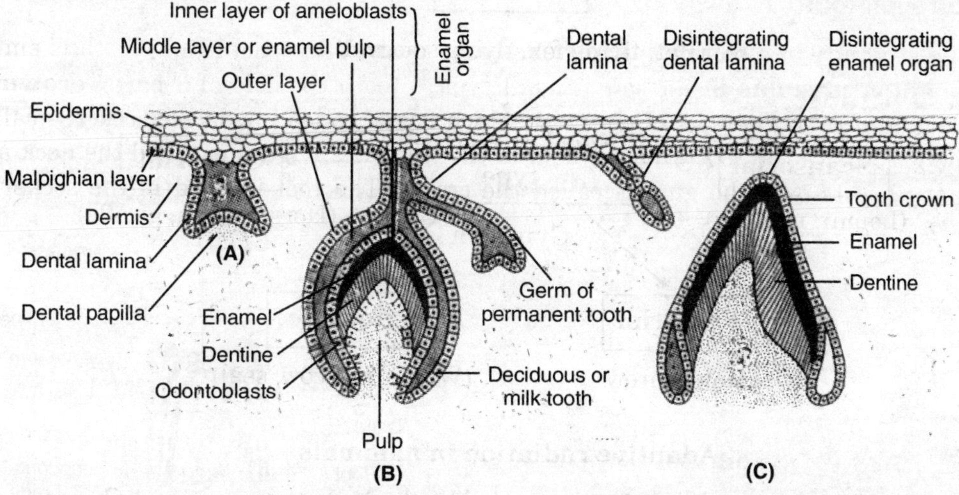

Fig. 25.5 *Mammalia*–Development of tooth

The first sign of appearance of a tooth is the thickening of ectoderm along the margin of jaw bone. The Malpighian layer of epidermis forms a solid vertical ridge-like invagination called **dental lamina** (Fig. 25.5) in the underlying dermis. The mesodermal cells beneath the dental lamina multiply rapidly and form a series of bud-like outgrowths called **tooth germs.** The tooth germ pushes down into the underlying mesenchyme, the cells of which press against it, as a result the tooth bud takes the form of an inverted cup. The inverted cup-like epithelial cap secretes enamel, so it is called **enamel organ.** The aggregation of mesenchymal cells beneath the enamel organ is called **dermal** or **dental papilla.** The outer columnar cells of dental papilla become differentiated into **odontoblasts** which secrete dentine. The inner epithelial layer of enamel organ becomes **ameloblasts** which secrete enamel over the dentine only in the crown part of the tooth. The central cavity forms the pulp cavity and the dental papilla is retained as pulp. The blood vessels and nerve fibres enter the pulp cavity through the basal aperture. Later eruption of tooth takes place by cutting the overlying epidermis. Cement appears around the root. When tooth is fully formed the odontoblasts become inactive.

ADAPTIVE RADIATION IN MAMMALS

The law of adaptive radiation or divergent evolution was given by Osborn which states "Each isolated region, if large and sufficiently varied in its topography, soil, climate and vegetation will give rise to a diversified fauna". There is a central or focal type from which animals diverge in different directions. The impelling causes of adaptive radiation are the need of food and safety which lead towards aquatic, amphibious, aerial, fossorial or arboreal habitat.

The earliest placental mammals arose during Cretaceous history as primitive insectivores and there was a wide range of adaptive radiation from this base through the Coenozoic era. With the disappearance of dinosaurs of the Cretaceous period, there was an "evolutionary explosion" of mammals to fill the ecological niches which were left by the dinosaurs. These mammals became adapted to many aspects of varied environment.

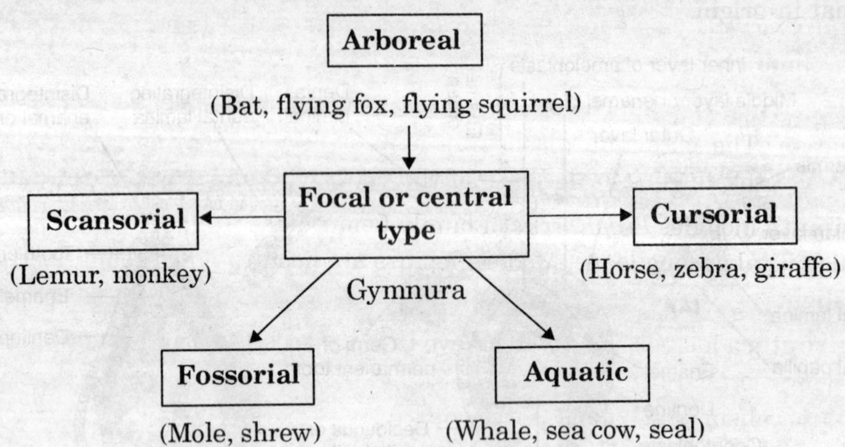

Adaptive radiation in nammals

1. Central or focal type: It is represented by the Malayan insectivore *Gymnura* which belongs to the hedgehog family Erinaceidae. Most of the insectivores have been small and they have survived with little change from Cretaceous times to the present. They had flat plantigrade feet. *Zalambdalestes* from Mangolia is regarded as ancestral insectivore. Recent insectivores include the modern elephant shrews of Africa and their relatives, the old world hedgehogs, shrews, moles and related forms and the golden moles and their relatives.

2. Cursorial types: The cursorial types include the Eutherians with amazing speed, running upon the toes. The **digitgrade** forms are dogs, swift fox and other carnivores. The ungulates (walk on the tips of their toes) usually have hoofs to protect their feet and take up the shock of running over hard ground. The ungulates are fast runners and include deer, the African and Mangolian antelope, horses (extinct and modern), *Rhinoceros*, zebra, ass, tapir, etc., Hares and rabbits (order Rodentia) are also fast running mammals.

All speedy animals show certain peculiar features:

(i) Body is moulded externally in such a way as to offer least resistance to the medium through which they pass.

(ii) The hindlimbs are efficient drivers and better evolved than forelimbs which serve to get food as they are nearer to mouth and the extended hands pull the body forward in running.

(iii) The primitive terrestrial foot is plantigrade (*i.e.* entire palm or sole rests on ground) but for cursorial adaptation the feet became elongated and there is reduction in the number of digits and hoof is formed.

(iv) With the lengthening of the limbs the length of neck also increases so that the animal may reach the ground for food and drink.

3. Arboreal forms: These are represented by the squirrels which have learned to launch themselves into the air and take long roaring leaps supported by folds of skin between the limbs and the sides of the body and thus have become semi-aerial. The different orders which have arboreal forms are:

(a) **Rodentia** which includes *Anomalurus* with six species found in Africa, *Pteromys, Sciuropterus* (flying squirrels) and *Eupetaurus* which is not perfectly arboreal but can climb rocks.

(b) **Insectivora** which includes *Galeopithecus* (flying lemur). It shows highest degree of aviation. The pataguim reaches its highest development as it extends from head between the fingers to the base of the claws, forming a web as in aquatic forms.

(c) **Primata** includes *Propithecus* of family Lemuroidae. It has powerful hindlimbs with which it takes long leaps. Modern lemurs are nocturnal and roam through the trees at night.

(d) **Chiroptera** includes bats which have become perfectly modified for flight by flapping of wings formed by stretched skin supported by bones of the forelimbs. Their rate of metabolism is very high to provide them energy for flight. They are nocturnal and have an **echo-location system** which helps them to fly during darkness of night.

4. Scansorial forms: Some mammals adopted the climbing habit as a means of safety and for feeding. During the age of reptiles arboreal life was one of the means that protected the mammalian race and rendered its subsequent evolution possible. The different orders which have scansorial forms are:

(a) **Edentata**—tree sloth, anteater.

(b) **Carnivora**—cat, jaguar, racoon, brown bear.

(c) **Rodentia**—tree porcupine, flying squirrel.

(d) **Insectivora**—pen-tailed shrew, flying lemur, *Gymnura*.

(e) **Chiroptera**—bat.

(f) **Primates**—baboon monkey.

The scansorial forms show certain special modifications.

(i) The chest and ribs support the body while climbing. In sloth the ribs are numerous to support their viscera in inverted position.

(ii) The shoulder girdle has strong clavicles and scapula to withstand the compression of powerful breast muscles. The hip bone (ilium) is broadened to support the viscera.

(iii) The proximal limb elongates in direct ratio to the animals climbing powers. In gibbon (*Hylobates*) the arms are so long that the hands touch the ground when the animal stands erect.

(iv) Feet generally prehensile, with opposable digits to prevent the animal from slipping.

(v) Tail is prehensile or non-prehensile. If non-prehensile spines and scales are present on the underside as in flying squirrel *Anomalurus*.

5. Fossorial forms: There are a number of mammals which show fossorial (burrowing) adaptations. Some important subterranean forms are:

(a) **Edentata—**armadillos (*Dasypus*).

(b) **Insectivora—**common mole (*Talpa*), golden mole (*Chrysochoris*), shrew mole (*Scalopus*), water shrew (*Crossopus*), hedgehog (*Erinaceus*).

(c) **Tubidentata—**aardvark (*Orycteropus*).

(d) **Rodentia—**ground squirrel (*Spermophilus*), prairie dog (*Cynomys*), wood chunck (*Aroctomys*), mole rat, bamboo rat.

(e) **Lagomorpha—**hare (*Lepus*).

(f) **Carnivora—**otter (*Lutra*), skunk (*Mydaus*), American badger.

A few ungulates dig for food—swine, elephant and mastodons.

The fossorial mammals show some special modifications which will be described later.

6. Aquatic forms: A number of mammals have become amphibious or secondarily aquatic. In transformation from terrestrial to aquatic forms they have undergone a number of modifications to suit aquatic mode of life. These specialized modifications will be described later. The different orders which have aquatic forms are:

(a) **Insectivora**—aquatic shrew.

(b) **Rodentia**—musk rat, water rat.

(c) **Carnivora**—otter, mink, seal.

(d) **Cetacea**—porpoise, whale.

(e) **Sirenia**—sea cow.

(f) **Artiodactyla**—hippopotamus.

AQUATIC ADAPTATIONS IN MAMMALS

Mammals are basically terrestrial animals but some have become secondarily aquatic, most probably due to competition on land for food and shelter. Aquatic mammals have originated from terrestrial ancestors. They are lung breathers and do not have gills. Their aquatic adaptations are secondary but they are very well adapted to lead aquatic life. There are two categories of aquatic mammals:

1. Amphibious mammals: They are not completely aquatic. They live on land but go into water for food and shelter. They show incomplete aquatic adaptations. A few examples are seal, otter walrus, beavers, *Ornithorhynchus,* hippopotamus, etc.

2. Completely aquatic mammals: There are two groups of mammals which are completely aquatic. They are Cetacea and Sirenia. They do not come on land.

Aquatic adaptations

Completely aquatic mammals show many secondary aquatic adaptations which can be described under three heads:

I. Modification of original structures

II. Loss of structures

III. Development of new structures.

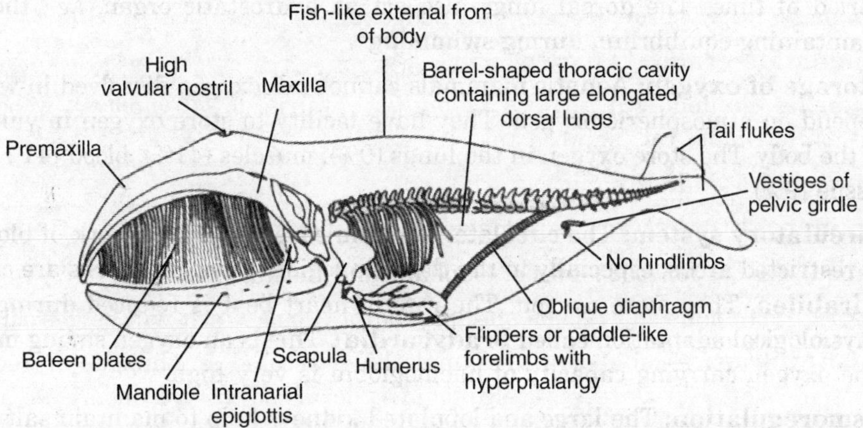

Fig. 25.6 *Mammalia*–Whale showing aquatic adaptations

I. Modification of original structures

1. **Shape of body:** The body of aquatic mammals is streamlined and boat-shaped like that of fish. This offers least resistance for movement in water.

2. **Body size and weight:** The body of aquatic mammals is usually large and heavy, but it does not create any problem for support in water due to buoyancy. Large size reduces skin friction and heat loss.

3. **Endoskeleton:** The bones are generally light and spongy and in Cetacea are filled with oil. In Sirenia the bones are heavy and solid, enabling the animal to stay under water for browsing on sea weeds. The cranium is shortened and produced into a pointed snout or rostrum which cuts the water efficiently. The neck is shortened and very much reduced. In whales the cervical vertebrae are fused. It helps the animal to pass through the water easily. The vertebrae are simple, the sacral region is not distinct due to reduction of hindlimbs. The ribs are arched dorsally to increase thoracic cavity.

4. **Paddles or flippers:** The forelimbs are modified into flattened and unjointed structures called paddles or flippers (Fig. 25.6). They have no indication of fingers. They can move as a whole from the shoulder joint. They act as balancers and provide stability during swimming. Hyperdactly (extra digits) and hyperphalangy (extra phalanges) serve to increase the surface area of flippers and thus help in swimming.

5. **Nostrils:** The nostrils are situated on top of the head so that the animal can breathe air easily without raising the head much out of water. They can be closed by a flap of skin (valve) to prevent entry of water.

6. **Diaphragm:** The diaphragm is oblique. This makes the thoracic cavity larger, dorsal and barrel-shaped, thus providing more space to lungs to expand.
7. **Lungs:** The lungs are large, highly elastic and extensible. Hence they can store large volumes of air. This helps the animal to remain under water for a considerable period of time. The dorsal lungs also act as hydrostatic organ, *i.e.*, they help in maintaining equilibrium during swimming.
8. **Storage of oxygen:** Aquatic mammals cannot use oxygen dissolved in water. They depend on atmospheric oxygen. They have facility to store oxygen in various parts of the body. The store oxygen in the lungs (9%), muscles (41%), blood (41%) and body tissue (9%).
9. **Circulatory system:** The circulatory system consists of a network of blood vessels in restricted areas, especially in the thoracic region. These networks are called **ritia mirabilea.** They store oxygen. The rate of heart beat is reduced during diving (a physiological adaptation called **bradycardia**). This is an oxygen saving mechanism. The oxygen carrying capacity of haemoglobin is very high.
10. **Osmoregulation:** The large and lobulated kidneys help to maintain salt balance of the body by excreting excess salt, thus the urine is hypertonic.
11. **Feeding young ones:** The ducts of mammary glands of aquatic mammals dilate to form reservoirs in which large amount of milk can be stored. The milk is forced down into the mouth of the baby by the contraction of efficient compressor muscles. This facilitates suckling of young ones under water.

II. Loss of structures

1. The hair covering of the body is absent except for a few bristles near the mouth. The skin surface becomes smooth and shiny.
2. Pinnae and hindlimbs absent. These external projections are a hindrance in movement in water.
3. All the skin glands (sweat and sebaceous) are absent as they are of no use in water.
4. Eyes without nictitating membrane and lacrimal glands.
5. Scrotal sacs absent, testes abdominal.

III. Development of new structures

1. **Flukes:** The tail develops large, lateral or horizontal expansions called tail flukes. Their up and down strokes propel the animal forward and also enable them to return rapidly to the surface for breathing.
2. **Median dorsal fin:** The median dorsal fin of Cetacea serves as a rudder during swimming.

   Both these structures are without skeletal support.
3. **Blubber:** A thick subcutaneous layer of fat called blubber is present below the skin. It acts as a heat retaining layer or heat insulator and a reservoir of food. It reduces the specific gravity of the animal and thus help in maintaining buoyancy.
4. **Baleen plates:** Whalebone whale is planktonophagous *i.e.*, it feeds on plankton.

Dentition is modified for filter feeding. Teeth are absent. There are numerous triangular horny plates called **baleen** (Fig. 25.6) or **whalebone** which hang down from the upper jaw in two transverse rows like the leaves of a book. During feeding large quantities of water is allowed to pass through these plates which serve as an effective sieve for filtering plankton.

5. **Melon:** It is a receptor organ which can detect pressure changes in water. It is situated in front of nostrils and consists of fat and muscle fibres.
6. **Foam:** From each middle ear cavity an inner pneumatic prolongation arises and both meet below the skull. It is filled with a fine emulsion or foam consisting of fat, mucus and gas. It improves hearing under water.
7. **Harderian glands:** The harderian glands produce a special fatty secretion which protects the eyes under water.

FLIGHT OR VOLANT ADAPTATIONS OF MAMMALS

There are some mammals which can fly in the air. The adaptations for flight are called volant adaptations.

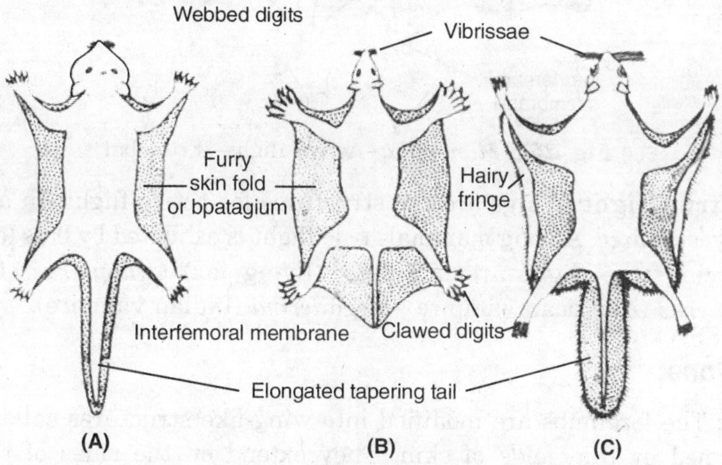

Fig. 25.7 *Mammalia*–Passive flying. **(A)** Flying phalanger; **(B)** Flying lemur; **(C)** Flying squirrel

Flight is of two types:

1. Passive flight or gliding.
2. Active flight or true flight.

Gliding or passive flight: Gliding is a smooth and gentle, horizontal or downward movement in the air. It is simply an aerial leap covering a distance of 10–20 metres. The animal glides from one tree to another with the help of its stretched patagium and limbs. Gliding does not require much energy. Energy is utilized only when gliding starts, no energy is utilized when the animal is in the air. Some common example of gliding mammals are:

Flying phalanger, *Petaurus* (order Marsupialia) (Fig. 25.7A), flying lemur, *Cynocephalus* (order Dermoptera) (Fig. 25.7B), wolly flying squirrel, *Eupetaurus* (Fig. 25.7C) and brown flying squirrel, *Petaurista* (order Rodentia).

Gliding adaptations

1. Body elongated, flattened and spindle-like.
2. Limbs long and equal.
3. Tail long and tapering.
4. Patagium is a fold of skin located on the side of the head, neck, limbs and tail. Sometimes it is supported by ribs. During gliding the patagium is expanded and at rest it is folded close to the body. It functions as a parachute.

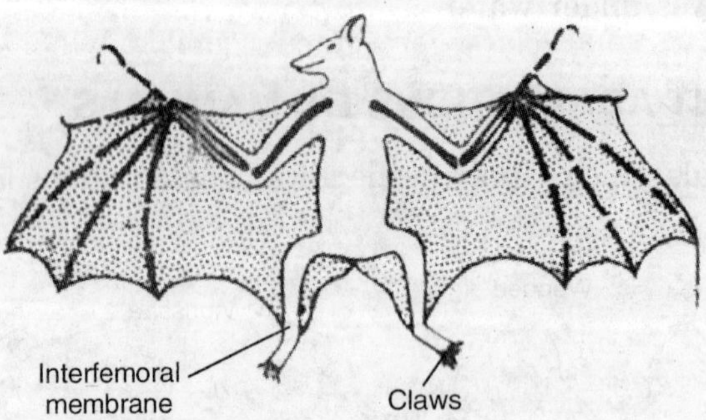

Fig. 25.8 *Mammalia*–Active flight—Fruit bat

Active or true flight: Wings are the structures for active flight. In active flight large amount of energy is utilized. Among mammals true flight is exhibited by bats (order Chiroptera). Some bats showing true flight are: *Pteropus* (flying fox) *Cynopterus* (fruit eating bat) (Fig. 25.8), *Vampyrus* (American vampire), *Megaderma* (Indian vampire).

Flight adaptations

1. **Wings:** The forelimbs are modified into wing-like structures called **patagia** which are formed by thin folds of skin. They extend on the sides of the body and are supported by skeletal elements of the forelimbs.
2. **Forelimbs:** The bones of forelimbs are greatly elongated. The digits are much elongated and support the wing. In Microchiroptera the first finger or pollex is short and free and bears a well developed curved claw. The remaining fingers are clawless. In Megachiroptera the second finger is also clawed. The forelimbs are also used for walking, climbing, holding food and killing prey.
3. **Hindlimbs:** A fold of membrane called **inter-femoral membrane** extends between the hindlimbs and also encloses the tail. The hindlimbs are rotated outwards so that the knee is directed backwards instead of forward. This helps in maximum spread of wing membranes. The digits are clawed. The claws help the bat to hang from trees.

4. **Antebrachial membrane:** An antebrachial membrane extends between the neck and humerus bone of upper arm.

   The pataguim, interfemoral and antebrachial membranes together form a continuous stretch of skin around the whole body.

5. **Endoskeleton:** Bones are light. The ribs are flattened and fused with the vertebrae. This gives large surface area for the attachment of flight muscles. The vertebrae of cervical region and also some of trunk region are devoid of neural spines. A spur of bone called **calcar** projects from the tarsus of each foot. It supports the interfemoral membrane.

6. **Flight muscles:** The flight muscles are well developed. They include pectoralis major and scapular muscles.

7. **Tail:** Tail may be small, large or scarcely visible. When large, it supports the interfemoral membrane which can act as a brake to flight, as a pouch to hold food and for the reception of a new born during flight.

8. **Dentition:** Teeth heterodont and diphyodont. Milk teeth of young ones are hooked. They help the young ones to grip the teats of mammary glands even while the mother is flying.

9. **Sensory system:** Bats are provided with an **echo-apparatus** or **radar system** or **ultrasonic sound system.** They fly with their mouth open and continuously produce ultrasonic sound waves. These waves strike on solid objects and are reflected back. These reflected waves and received by the ears. This helps the bats to locate the obstacles in their path. Bats are nocturnal animal. Due to this system they are able to fly easily and swiftly even in total darkness.

   The pinnae are long and provided with sensitive accessory lobes called **targus** which help to pick up echos. The nose is provided with leaf-like lobe called **nose leaf** which probably acts as an antenna and perceive air vibrations.

10. The wings are richly provided with blood vessels and nerve fibres. They probably act as tactile organs.

FOSSORIAL ADAPTATIONS IN MAMMALS

The animals which lead subterranean life are called **fossorial.** Fossorial adaptations are adaptations for **burrowing.** Fossorial mammals are of two types:

1. **Semifossorial:** These are partially fossorial. They do not live in burrows. They dig burrows only for food or for retreat.

   Mammals like swines, elephants and mastodon dig the ground for food. They dig with the help pointed snout (swine) or long tusks (elephant and mastodon). Some mammals like rabbit, hare, otter, badgers and skunk dig burrows for retreat. Their limbs are modified for digging.

2. **Completely fossorial:** These animals lead a completely subterranean life. Some examples are:

Prototheria—*Ornithorhynchus* and *Echidna*.

Metatheria—kangaroo, rat, marsupial mole, wombat and *Dasyurus*.

Eutheria—armadillo, aardvark, moles, shrews, hedgehog, ground squirrel, etc.

Fossorial adaptations

1. **Body shape:** The body is fusiform or spindle-shaped offering little resistance in coming and going out of burrows. The head tapers anteriorly forming a snout for digging. The shoulders are broad, strong and powerful and resist the great muscular strain while digging.
2. **Tail:** Tail is usually short (hedgehog) or vestigeal (moles), as a long tail is a hindrance in burrowing.
3. **Eyes:** The eyes are reduced or vestigeal as they are of no use in dark habitat.
4. **External ears:** The external ears tend to disappear as they might be an obstruction in burrowing.
5. **Digging mechanism:** The conical snout helps in digging. In mole a prenasal bone is found at the tip of the nose. Incisors and sometimes canines are directed forward **(procumbant)** for digging. Limbs are short and stout. The forelimbs are broad and provided with strong claws for digging.
6. **Hindlimbs:** Hindlimbs mainly push the body forward into the burrow and do not help in digging. The sacral vertebrae are fused together to provide solid support to hindlimbs.
7. **Hibernation:** Hibernation or winter sleep is a characteristic feature of fossorial animals. It is a means to tide over unfavourable conditions.

ECONOMIC IMPORTANCE OF MAMMALS

Mammals are very important economically as many of them are very beneficial to man. A few mammals are also injurious to man.

Beneficial mammals

Man, the most adaptable of mammals, has domesticated many other mammals that he needed. Some are used as beast of burden, some yield useful products and some are kept as pets.

1. **As food:** From very early times mammals have been hunted and consumed as food. Even today, many mammals serve as food as they provide various kinds of **meat** and **meat products.** The meat of animals like cow, buffalo, pig, goat, sheep, deer, etc. is eaten in all parts, etc., of the world. In some places meat of rabbit, rat, seal, whale, dog and bear is also eaten.

   Many domesticated animals like cow, buffalo, goat, sheep, camel, deer, etc. provide milk. Many dairy products like curd, cheese, butter, etc. are prepared from milk.

2. **Commercial products:** Many mammals provide products which are of great commercial value.

(a) **Hide:** The hide (skin) of many mammals yield leather and fur which are used for making clothes, shoes, purses, belts, bags, suitcases, harness, saddles, etc. Civilized races in northern countries have used animal pelts (raw skin) as clothing for many centuries. Eskimos in the Arctic use reindeer and bear skins for clothing, seal hide for footwear, and intestine and skin of various mammals for boats. The fur trade stimulated explorations and settlements. Much of the early explorations of North America was made by trappers in search of furs. Furs are still valued today and fur farming is a growing industry. Millions of fur-bearers like musk rat, skunk, opossum, otter, mink, weasels, beaver, fur seal, fox, lynx, bear, kangaroo, etc. are still destroyed for their fur of soft texture and pleasing colour. The rearing of black foxes and mink on fur farms has become successful. Pelts of rabbits, house cats and otters are dyed and altered to resemble the most expensive fur.

(b) **Hair and wool:** The wool of sheep is used to make variety of winter clothes. The hairs of goat, camel, etc. are also used to make wool. The hairs of some animals like pig, camel, horses, etc. are coarse and are used to make brushes.

(c) **Oil and fat:** The blubber of whalebone whale and seal yield edible oil. The Eskimos used the blubber as fuel oil. The oil of sperm whale is non-edible so it is used as a lubricant, in making candles and as a base in cosmetics.

(d) **Perfumes:** The **musk** obtained from the musk glands of musk deer is used in perfumes. A substance called **ambergris** obtained from the intestine of sperm whales is used as a base in perfumes.

(e) **Glue and gelatine:** The horns, hoofs and bones of cattle yield glue and gelatine.

(f) **Ivory:** Ivory is obtained from the tusks of elephant and walrus. It is used for making toys, decoration items and jewellery.

3. **Beast of burden:** Since the begining of civilization many mammals have been domesticated and used as beast of burden for transportation and carrying loads. Even today, inspite of so many automotive vehicles, many animals like cattle, horses, sheep, goat and mules are being used for work in plains and mountains, camel and llamas are used in deserts, and dogs and reindeers on ice in cold countries.

4. **In agriculture:** Many mammals like, buffalo, ox, yak, horses and camels are used for ploughing. The dung of cattle is used as manure. Bone meal prepared by crushing bones of some mammals like sheep, goat, cattle, etc is used as fertilizer. Many insectivorous mammals feed on harmful insects. Moles, shrews, and ground squirrel feed on insect larvae. Rodents are a serious pest of crop and they are preyed upon by mangoose.

5. **As pollinators:** Arboreal mammals like bat, squirrel and monkey help in bringing about pollination of flowers.

6. **As scavengers:** Hyaena, jackal and pigs are natural scavengers. They help in cleaning the environment.
7. **In recreation:** Mammals have a great recreational value. Many mammals like monkey, bear, rabbit, dog and cat are tamed and kept as pets. Zoos and national parks are places where a variety of animals are kept and protected. Many animals like lion, tiger, elephant, horse, monkey, hippopotamus, etc., are employed in circuses. Hunting is now more than a sport. It is a very enjoyable means of recreation. The moose, elk, deer, mountain sheep, bear, lion, tiger, rabbit, racoons, opossum, tree-squirrel, etc. are some important game animals. They are sought for their head, hide and meat. Many of the hunted animals are used for decoration purposes.
8. **Research:** There are many mammals which serve as study material in laboratory. Research and scientific studies are done on those animals. Some common examples are rat, guinea pig, cat, rabbit, pig, monkey, ape, etc.

Harmful mammals

1. **Damage crops:** There are many mammals which damage the crops of fields and gardens. Rabbits, wood chuncks and some mice forage in vegetable patches, gardens and fields and may gnaw the bark of trees. Squirrel damage grain and other crops and many fruits. Pocket gophers eat roots of plants. Bats damage guava groves and become a menace to keepers.
2. **Damage to property:** The house rats gnaw into buildings, consume all kinds of foodstuffs and destroy much property.
3. **Predators:** The carnivores destroy large number of farm animals. Wolves and hyaenas may attack sleeping persons and run away with small kids. The vampire (blood sucking) bats suck blood of cattle and even man.
4. **Diseases:** Some mammals harbour diseases that can be transferred to man. These diseases are bubonic plague, typhus, relapsing fever, tularemia, trichia, hydrophobia, etc. These disease are transferred by rodents, pigs, cats, dogs, etc.

INDEX